S0-CHI-760

Power Transistors: Device Design and Applications

OTHER IEEE PRESS BOOKS

VLSI: Technology Design, *Edited by O. G. Folberth and W. D. Grobman*
General and Industrial Management, *By H. Fayol: Revised by I. Gray*
A Century of Honors, *an IEEE centennial directory*
MOS Switched-Capacitor Filters: Analysis and Design, *Edited by G. S. Moschytz*
Distributed Computing: Concepts and Implementations, *Edited by P. L. McEntire, J. G. O'Reilly, and R. E. Larson*
Engineers and Electrons, *By J. D. Ryder and D. G. Fink*
Land-Mobile Communications Engineering, *Edited by D. Bodson, G. F. McClure, and S. R. McConoughey*
Frequency Stability: Fundamentals and Measurement, *Edited by V. F. Kroupa*
Electronic Displays, *Edited by H. I. Refioglu*
Spread-Spectrum Communications, *Edited by C. E. Cook, F. W. Ellersick, L. B. Milstein, and D. L. Schilling*
Color Television, *Edited by T. Rzeszewski*
Advanced Microprocessors, *Edited by A. Gupta and H. D. Toong*
Biological Effects of Electromagnetic Radiation, *Edited by J. M. Osepchuk*
Engineering Contributions to Biophysical Electrocardiography, *Edited by T. C. Pilkington and R. Plonsey*
The World of Large Scale Systems, *Edited by J. D. Palmer and R. Saeks*
Electronic Switching: Digital Central Systems of the World, *Edited by A. E. Joel, Jr.*
A Guide for Writing Better Technical Papers, *Edited by C. Harkins and D. L. Plung*
Low-Noise Microwave Transistors and Amplifiers, *Edited by H. Fukui*
Digital MOS Integrated Circuits, *Edited by M. I. Elmasray*
Geometric Theory of Diffraction, *Edited by R. C. Hansen*
Modern Active Filter Design, *Edited by R. Schaumann, M. A. Soderstrand, and K. B. Laker*
Adjustable Speed AC Drive Systems, *Edited by B. K. Bose*
Optical Fiber Technology, II, *Edited by C. K. Kao*
Protective Relaying for Power Systems, *Edited by S. H. Horowitz*
Analog MOS Integrated Circuits, *Edited by P. R. Gray, D. A. Hodges, and R. W. Brodersen*
Interference Analysis of Communication Systems, *Edited by P. Stavroulakis*
Integrated Injection Logic, *Edited by J. E. Smith*
Sensory Aids for the Hearing Impaired, *Edited by H. Levitt, J. M. Pickett, and R. A. Houde*
Data Conversion Integrated Circuits, *Edited by D. J. Dooley*
Semiconductor Injection Lasers, *Edited by J. K. Butler*
Satellite Communications, *Edited by H. L. Van Trees*
Frequency-Response Methods in Control Systems, *Edited by A. G. J. MacFarlane*
Programs for Digital Signal Processing, *Edited by the Digital Signal Processing Committee, IEEE*
Automatic Speech & Speaker Recognition, *Edited by N. R. Dixon and T. B. Martin*
Speech Analysis, *Edited by R. W. Schafer and J. D. Markel*
The Engineer in Transition to Management, *By I. Gray*
Multidimensional Systems: Theory & Applications, *Edited by N. K. Bose*
Analog Integrated Circuits, *Edited by A. B. Grebene*
Integrated-Circuit Operational Amplifiers, *Edited by R. G. Meyer*
Modern Spectrum Analysis, *Edited by D. G. Childers*
Digital Image Processing for Remote Sensing, *Edited by R. Bernstein*
Reflector Antennas, *Edited by A. W. Love*
Phase-Locked Loops & Their Application, *Edited by W. C. Lindsey and M. K. Simon*
Digital Signal Computers and Processors, *Edited by A. C. Salazar*
Systems Engineering: Methodology and Applications, *Edited by A. P. Sage*
Modern Crystal and Mechanical Filters, *Edited by D. F. Sheahan and R. A. Johnson*
Electrical Noise: Fundamentals and Sources, *Edited by M. S. Gupta*
Computer Methods in Image Analysis, *Edited by J. K. Aggarwal, R. O. Duda, and A. Rosenfeld*
Microprocessors: Fundamentals and Applications, *Edited by W. C. Lin*
Machine Recognition of Patterns, *Edited by A. K. Agrawala*
Turning Points in American Electrical History, *Edited by J. E. Brittain*
Charge-Coupled Devices: Technology and Applications, *Edited by R. Melen and D. Buss*
Spread Spectrum Techniques, *Edited by R. C. Dixon*
Electronic Switching: Central Office Systems of the World, *Edited by A. E. Joel, Jr.*
Electromagnetic Horn Antennas, *Edited by A. W. Love*
Waveform Quantization and Coding, *Edited by N. S. Jayant*
Communication Satellite Systems: An Overview of the Technology, *Edited by R. G. Gould and Y. F. Lum*
Literature Survey of Communication Satellite Systems and Technology, *Edited by J. H. W. Unger*

Power Transistors: Device Design and Applications

Edited by
B. Jayant Baliga
Manager, High Voltage Device Program
General Electric Company

Dan Y. Chen
Associate Professor of Electrical Engineering
Virginia Polytechnic Institute
and State University

A volume in the IEEE PRESS Selected Reprint Series, prepared under the sponsorship of the IEEE Aerospace and Electronic Systems Society

The Institute of Electrical and Electronics Engineers, Inc., New York

PRINTED IN THE UNITED STATES OF AMERICA

IEEE Order Number: PC01750

Library of Congress Cataloging in Publication Data
Main entry under title:

Power transistors.

(IEEE PRESS selected reprint series)
Bibliography: p.
Includes indexes.
1. Power transistors—Addresses, essays, lectures.
I. Baliga, B. Jayant, 1948- . II. Chen, Dan Y.
TK7871.92.P69 1984 621.31'7 84-19747
ISBN 0-87942-181-9

Contents

Introduction 1

Part I: Bipolar Power Transistors 19

Power Transistor Performance Tradeoffs, *P. L. Hower* (*IEEE Power Electronics Specialists Conference,* 1975) 21

Factors Limiting Current Gain in Power Transistors, *E. J. McGrath and D. H. Navon* (*IEEE Transactions on Electron Devices,* October 1977) 28

On the Proportioning of Chip Area for Multistage Darlington Power Transistors, *C. F. Wheatley, Jr., and W. G. Einthoven* (*IEEE Transactions on Electron Devices,* August 1976) 33

Stable Hot Spots and Second Breakdown in Power Transistors, *P. L. Hower, D. L. Blackburn, F. F. Oettinger, and S. Rubin* (*IEEE Power Electronics Specialists Conference,* 1976) 42

An Experimental Study of Reverse-Bias Second Breakdown, *D. L. Blackburn and D. W. Berning* (*IEEE International Electron Devices Meeting,* 1980) 55

Turn-Off Characteristics of Power Transistors Using Emitter-Open Turn-Off, *D. Y. Chen and B. Jackson* (*IEEE Transactions on Aerospace and Electronic Systems,* May 1981) 60

An Overview of Low-Lost Snubber Technology for Transistor Converters, *A. Ferraro* (*IEEE Power Electronics Specialists Conference,* 1982) 65

A 10-kW Series Resonant Converter Design, Transistor Characterization, and Base-Drive Optimization, *R. R. Robson and D. J. Hancock,* (*IEEE Power Electronics Specialists Conference,* 1982) 77

A Two-Quadrant Transistor Chopper for Electric Vehicle Drive, *R. L. Steigerwald* (*IEEE Transactions on Industry Applications,* July/August 1980) 89

Transistorized PWM Inverter-Induction Motor Drive System, *S. C. Peak and A. B. Plunkett* (*IEEE Industry Applications Society Meeting,* 1982) 96

Application of Power Transistors to Residential and Intermediate Rating Photovoltaic Array Power Conditioners, *R. L. Steigerwald, A. Ferraro, and F. G. Turnbull* (*IEEE Transactions on Industry Applications,* March/April 1983) 103

Pulse Battery Charger Employing 1000 Ampere Transistor Switches, *R. L. Steigerwald* (*IEEE Industry Applications Society Meeting,* 1977) 117

Part II: Gate Turnoff Thyristors/Latching Transistors 123

The Current Status of the Power Gate Turn-Off Switch (GTO), *M. Okamura, T. Nagano, and T. Ogawa* (*IEEE International Semiconductor Power Converter Conference,* 1977) 124

Investigations of Gate Turn-Off Structures, *H. W. Becke and R. P. Misra* (*IEEE International Electron Devices Meeting,* 1980) 135

Ultra High Voltage, High Current Gate Turn-Off Thyristors, *T. Yatsuo, T. Nagano, H. Fukui, M. Okamura, and S. Sakurada* (*International Power Electronics Conference,* 1983) 140

Gating Circuit Developed for High Power Thyristors, *N. Seki, Y. Tsuruta, and K. Ichikawa* (*IEEE Power Electronics Specialists Conference,* 1981) 150

Present Condition of High Power GTO Applications, *N. Seki, K. Ichikawa, Y. Tsuruta, and K. Matsuzaki* (*5th International PCI Conference on Power Conversion,* 1982) 161

Application Techniques for High Power Gate Turn-Off Thyristors, *R. L. Steigerwald* (*IEEE Industry Applications Society Meeting,* 1975) 174

Paralleling of Gate Turn-Off Thyristors, *H. Fukui, H. Amano, and H. Miya* (*IEEE Industry Applications Society Meeting,* 1982) 184

Application of Gate-Turn-Off Thyristors in 460 V 7.5–250 hp AC Motor Drives, *D. A. Paice and K. E. Mattern* (*IEEE Industry Applications Society Meeting,* 1982) 190

Part III: Power MOS Field Effect Transistors 197

Power MOSFETs—A Status Review, *B. R. Pelly* (*1983 International Power Electronics Conference,* 1983) 199

Modeling of the On-Resistance of LDMOS, VDMOS, and VMOS Power Transistors, *S. C. Sun and J. D. Plummer* (*IEEE Transactions on Electron Devices,* February 1980) 213

A Parametric Study of Power MOSFETs, *C. Hu* (*IEEE Power Electronics Specialists Conference,* 1979) 225

Second Breakdown of Vertical Power MOSFETs, *C. Hu and M.-H. Chi* (*IEEE Transactions on Electron Devices,* August 1982) 236

Understanding Power MOSFET Switching Performance, *S. Clemente, B. R. Pelly, and A. Isidori* (*Solid-State Electronics,* December 1981) 243

Improving the Reverse Recovery of Power MOSFET Integral Diodes by Electron Irradiation, *B. J. Baliga and J. P. Walden* (*IEEE Industry Applications Society Meeting,* 1981) 257

dv/dt Effects in MOSFET and Bipolar Junction Transistor Switches, *R. Severns* (*IEEE Power Electronics Specialists Conference,* 1981) 266
A 48V, 200A Chopper for Motor Speed Control, with Regenerative Braking Capability, Using Power HEXFETs, *S. Clemente and B. Pelly* (*IEEE Industry Applications Society Meeting,* 1980) 273
Solid State RF Generators for Induction Heating Applications, *W. E. Frank and C. F. Der* (*IEEE Industry Applications Society Meeting,* 1982) 285

Part IV: Emerging Transistors Technology 291
High-Voltage Junction-Gate Field-Effect Transistor with Recessed Gates, *B. J. Baliga* (*IEEE Transactions on Electron Devices,* October 1982) 292
Optimum Design of Triode-Like JFET's by Two-Dimensional Computer Simulation, *K. Yamaguchi and H. Kodera* (*IEEE Transactions on Electron Devices,* August 1977) 303
Characteristics of High-Power and High-Breakdown-Voltage Static Induction Transistor with the High Maximum Frequency of Oscillation, *M. Kotani, Y. Higaki, M. Kato, and Y. Yukimoto* (*IEEE Transactions on Electron Devices,* February 1982) 312
The Asymmetrical Field-Controlled Thyristor, *B. J. Baliga* (*IEEE Transactions on Electron Devices,* July 1980) 317
Electron Irradiation of Field-Controlled Thyristors, *B. J. Baliga* (*IEEE Transactions on Electron Devices,* May 1982) 324
High Gain Power Switching Using Field Controlled Thyristors, *B. J. Baliga* (*Solid-State Electronics,* May 1982) 331
Normally-Off Type High Speed SI-Thyristor, *Y. Nakamura, H. Tadano, S. Sugiyama, I. Igarashi, T. Ohmi, and J. Nishizawa* (*IEEE International Electron Devices Meeting,* 1982) 339
A New FET-Bipolar Combinational Power Semiconductor Switch, *D. Y. Chen, S. Chandrasekaran, and S. A. Chin* (*IEEE Transactions on Aerospace and Electronic Systems,* March 1984) 343
The Monolithic HV BIPMOS, *N. Zommer* (*IEEE International Electron Devices Meeting,* 1981) 350
The Insulated Gate Transistor (IGT)—A New Power Switching Device, *B. J. Baliga, M. Chang, P. Shafer, and M. W. Smith* (*IEEE Industry Applications Society Meeting,* 1983) 354
A 300 KHz Off-Line Switching Supply Using a Unique BI-MOS Switch Combination, *V. Farrow and B. Taylor* (*Proc. of the 2nd Annual International Power Conversion Conference,* 1980) 364

Appendix 375
High-Voltage Device Termination Techniques: A Comparative Review, *B. J. Baliga* (*Proceedings of the IEE,* October 1982) 376

Bibliography 383

Author Index 387

Subject Index 389

Editors' Biographies 393

Introduction

THE power bipolar transistor was developed over 25 years ago. During these years, a steady growth in its power handling capability has been encouraged by its use in a growing variety of power conditioning applications at increasing power levels. The 1982 power component market totaled nearly $2 billion of which power transistors represented $750 million. This market is expected to exceed $3 billion by the end of the 1980's.

Although the power thyristor is firmly entrenched for very high power low-frequency switching applications in which gate controlled turnoff is not required, the power bipolar transistor has been the workhorse for higher frequency circuits and, in particular, for circuits requiring gate controlled turnoff. Bipolar power transistors with 1500 V blocking capability and current ratings to up to 5 A have been developed for TV deflection circuits. At the higher current end, Darlington bipolar transistors capable of handling 200 A at blocking voltages of up to 500 V are also available today. These devices have been used for motor control and uninterruptible power supplies.

Due to the preeminence of the power bipolar transistor until the late 1970's, the term "power transistor" has been synonymous with "power bipolar transistors." However, in recent years several new device developments have begun to challenge the bipolar transistor. These new devices must now also be regarded as part of the transistor family. Consequently, in this book, we have also included the gate-turnoff thyristors (sometimes referred to as latching transistors), the power MOSFET's, and several new power transistors such as the static induction transistor and the insulated gate transistor. This book is intended as a reference which provides a comprehensive set of reprints dealing with this technology. The reprints include typical applications for these devices.

Here, the various types of transistors structures are introduced. The basic physics of device operation and its influence on the device output characteristics is also provided. This section serves to familiarize the reader with the terminology used in conjunction with each of the devices. This is followed by a part in which the devices are compared with each other in terms of power switching efficiency, gate drive requirements, and other device parameters. The third section deals with the ratings of available devices and provides a projection of how these ratings are expected to grow. The last section provides an overview of device applications and the suitability of each of the devices for these applications.

Power Transistor Overview

This review section provides an introduction to the different transistors discussed in the reprinted articles in this book. For each transistor, the device structure is first described with a discussion of its physics of operation and its output characteristics. Key features and limitations of each of the transistors are also highlighted. A brief discussion of the device fabrication technology is included to aid the reader in assessing the processing complexity associated with the development of each device. The design of devices will not be dealt with in this section. Instead, the design aspects of devices are discussed at the beginning of each of the reprinted sections with the aid of the references.

Power Bipolar Transistor

The basic structure of the power NPN bipolar transistor is illustrated in Fig. 1. Most power bipolar transistors are fabricated by starting with a highly conductive (n+) substrate and depositing an n-type drift layer on top using epitaxial growth. This drift layer is designed to support high voltages during forward blocking. The p-base region is then diffused into the drift layer followed by the diffusion of the n+ emitter regions. The emitter and base contacts are interdigitated as illustrated in Fig. 1 to provide high current handling capability as discussed below.

The power bipolar transistor is a current controlled device in which the base current (I_B) controls the output collector current (I_C). In the absence of the base current, junction $J1$ becomes reverse biased and provides the forward blocking capability. The maximum blocking voltage is determined primarily by the properties of the n-drift layer. To obtain higher voltages, the drift layer resistivity and thickness must both be increased in accordance with the plots provided in the Appendix (Fig. A). Unfortunately, this increases the series resistance during current conduction and reduces the current handling capability of bipolar transistors when their blocking voltage is increased. A good rule of thumb that can be used for device design is that the saturated on-resistance of the power bipolar transistor will increase as the square of the blocking voltage capability.

Current conduction in the power bipolar transistor is achieved by driving a base current (I_B) into the device. This base current causes the injection of electrons from the n+ emitter into the p-base region. These electrons must then diffuse through the base region and are collected at junction $J1$. The current gain of the bipolar transistor is, therefore, determined by the injection efficiency of the emitter and the base transport factor. At high current levels, power bipolar transistors typically operate at a current gain of 10–20. This results in the need to supply substantial base drive currents during steady-state current conduction. This is one of the major disadvantages of the bipolar transistor. However, at high current levels, the injected electron density can exceed the background doping level of the n-drift layer. The additional free carrier density lowers the drift region resistance and provides the bipolar transistor with its low on-state saturated resistance (R_{sat}) compared with the intrinsic drift region resistance (R_D). This results in efficient current conduction in these devices.

As illustrated in Fig. 1, it is necessary to interdigitate the n+ emitter regions with the p+ base contacts. One reason for this is that at high forward base drive currents, a significant lateral voltage drop occurs in the p-base region under the n+ emitter. This voltage drop debiases the middle of the emitter finger

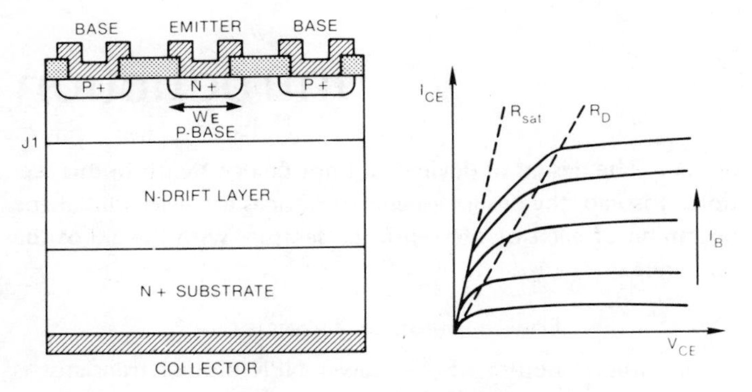

POWER BIPOLAR TRANSISTOR

Fig. 1. Power bipolar transistor structure and its current controlled electrical output characteristics.

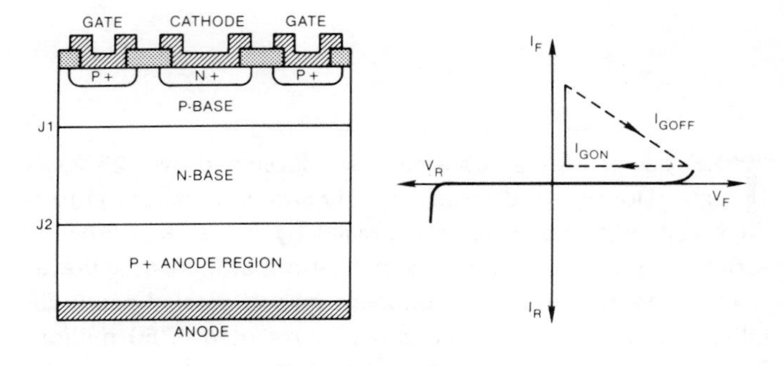

GATE TURN-OFF THYRISTOR

Fig. 2. Gate-turnoff thyristor structure and its electrical output characteristics.

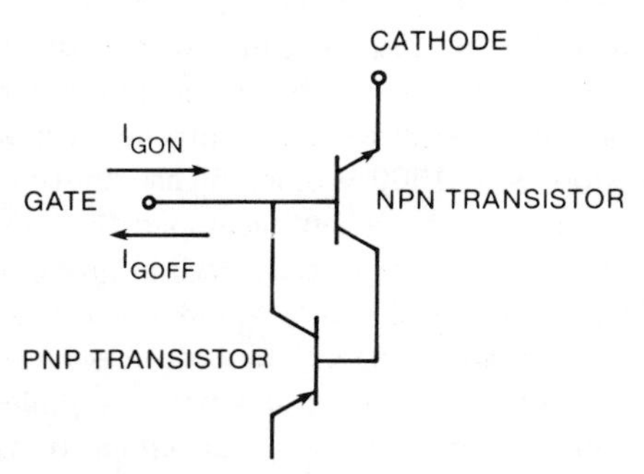

Fig. 3. Equivalent circuit of the gate-turnoff thyristor.

resulting in electron injection being confined to the edges of the emitter fingers closest to the p+ base contacts. Since the central portions of the emitter are inactive, the design and fabrication of these devices requires the formation of narrow emitter fingers interdigitated with p+ base contacts. Typical emitter finger widths (W_E) in the state-of-the-art power bipolar transistors range from 10 to 25 μ in size. The use of narrow emitter finger widths is also desirable for high-speed turnoff. During turnoff, a negative base drive current must be supplied to pull out the stored charge. Since this base current accesses the outer perimeter of the emitter fingers, emitter current can constrict to the center of the finger during turnoff causing destructive failure. Thus, narrow emitter fingers enhance bipolar transistor ruggedness.

In summary, the power bipolar transistor is a current controlled device with the attractive feature of a low on-resistance during current conduction. However, it requires relatively high base drive currents during its on-state and during high-speed turnoff. In addition, the on-resistance of bipolar transistors decreases with increasing temperature. This makes the paralleling of devices difficult and can cause current localization internal to the device leading to destructive failure. It is also worth noting that the rapid increase in the on-resistance of the bipolar transistor with blocking voltage has limited the device ratings to below 1000 V except for 1500 V low current bipolar transistors used in TV deflection circuits. These drawbacks in the bipolar transistor characteristics have encouraged the development of other devices with gate-turnoff capability.

Gate-Turnoff Thyristor (Latching Transistor)

The gate-turnoff thyristor (GTO) is a four-layer thyristor structure with a special gate design which provides forced gate-turnoff capability. A typical GTO structure is shown in Fig. 2. This structure consists of two (an NPN and a PNP) coupled transistors which are regeneratively connected as shown in Fig. 3. When a positive voltage is applied to the gate to drive a turn-on gate current (I_{GON}) into the p-base region, electrons are injected from the n+ cathode into the p-base region. These electrons diffuse through the p-base of the upper NPN transistor and are collected at junction $J1$. This electron current acts as a base current for the lower PNP transistor and causes injection of holes from the p+ anode region. If the sum of the current gains (α) of the NPN and PNP transistors exceeds unity, the above process is self-sustaining and the device latches into its forward conducting state. To achieve forced gate turnoff, a negative gate current (I_{GOFF}) must be applied. This gate current must be sufficiently large to lower the gain of the upper NPN transistor until the sum of the current gains (α) of the NPN and PNP transistor is less than unity in order to extinguish the regenerative forward current flow. As in the case of the power bipolar transistor, this requires the design and fabrication of interdigitated cathode and gate fingers with narrow cathode finger widths. In state-of-the-art GTO's, gate currents ranging from one half to one fifth of the anode current are essential to achieve forced gate turnoff. However, only a gate current pulse is necessary to achieve turnoff and no steady-state gate current is needed.

In summary, gate-turnoff thyristors are four layer latching devices with the capability of forcing current turnoff by using reverse gate drive. These devices have the lowest on-state voltage drops among high voltage devices. However, they require a very large gate drive pulse current to achieve forced gate turnoff. Further, since these devices latch up into the conducting state, they do not provide the capability for gate controlled current limiting. At present, the GTO is primarily used for high power applications in which devices with blocking voltage capability of over 1000 V are required. The inherent reverse blocking capability of the GTO also makes it an attractive device for alternating current (ac) traction applications.

Due to the higher voltage capability of GTO's, they are often fabricated by starting with high resistivity n-type silicon wafers and then diffusing from both sides of the wafer to form the anode, p-base and cathode regions. Due to the high quality of the bulk starting material, large area, high current GTO's with current ratings up to 1000 A have been developed. In these devices, the blocking voltage in both polarities is supported primarily in the n-base region by the reverse bias on junctions *J*1 and *J*2. Due to the existence of the PNP transistor in the lower portion of the GTO, the width of the n-base region must be made larger than the depletion layer width predicted in Fig. A by approximately one diffusion length for minority carriers (typically 25 μ). To aid in the design of the n-base region of GTO's, plots of the blocking voltage as a function of the n-base region doping concentration and width are provided in Fig. C of the Appendix. As an example, in Fig. C, a blocking voltage capability of 1500 V can be achieved using a n-base width of 200 μ and an n-base doping of about 7×10^{13} per cm^3. Despite this relatively wider n-base width compared with power bipolar transistors, the GTO exhibits superior current handling capability because in these devices the n-base region resistance is drastically reduced during current conduction by the very high concentration of injected carriers. This conductivity modulation makes the development of high voltage GTO's feasible because the current density in these devices decreases at a milder square root dependence upon the blocking voltage capability.

Power MOSFET

The power metal-oxide-semiconductor field-effect transistor (MOSFET) evolved from integrated circuit technology in the 1970's in response to the need to develop power transistors that can be controlled using much lower gate drive power levels compared to the existing power bipolar transistors. The structure of the power MOSFET is shown in Fig. 4. This device structure is fabricated by starting with a highly conductive n+ substrate and then growing an n-type drift layer whose design determines the forward blocking capability of the device. The design of this n-drift layer is similar to that for the bipolar transistor and can be obtained from Fig. A. The MOS gate structure is then formed in the n-drift layer by first growing a very thin (typically 1000 Å) gate oxide followed by the deposition of a refractory gate material—usually polysilicon. After patterning the gate material, the p-base region and the n+ source region are diffused into the n-drift layer using the gate material as a mask. This double diffusion from a common edge allows precise control over the channel length in the power MOSFET process. (The device shown in Fig. 4 has a DMOS structure. Another device structure called VMOS also exists but is rapidly falling out of favor among device manufacturers.)

The power MOSFET blocks current flow across junction *J*1 in the absence of gate bias. If a positive voltage is applied to the gate with respect to the source and p-base region (note that the p-base is electrically shorted to the n+ source by the source contact metal), then electrons are attracted to the surface of the p-base region. When the gate bias exceeds a threshold level, an n-type conductive channel forms at the surface of the p-base below the gate region and links the n+ source to the n-drift region. The device can now sustain source-to-drain current flow limited by the resistance of the n-drift region and the channel region. In high-voltage power MOSFET's, the on-resistance is primarily determined by the drift region resistance. This on-resistance increases as the 2.5 power of the blocking voltage capability because of the increasing drift region resistivity and thickness. This high on-resistance of high-voltage power MOSFET's has limited their development to devices capable of blocking only up to 600 V. At the low-voltage end (less than 100 V), the drift region resistance becomes small and the channel resistance makes a large contribution to the on-resistance. In these devices it is important to keep the channel length (determined by the difference between p-base and n+ source lateral diffusion) as short as possible. Typical channel lengths range from 0.5 to 2 μ. It is also important to obtain a large channel width by repeating the basic cell structure shown in Fig. 4. Devices with channel widths of several meters have been developed by integrating thousands of cells that are typically 25 μ wide. The high level of integration required to fabricate power MOSFET's has at present limited the current handling capability of these devices to about 10 A for 500 V devices. Higher current devices are available at lower voltage ratings.

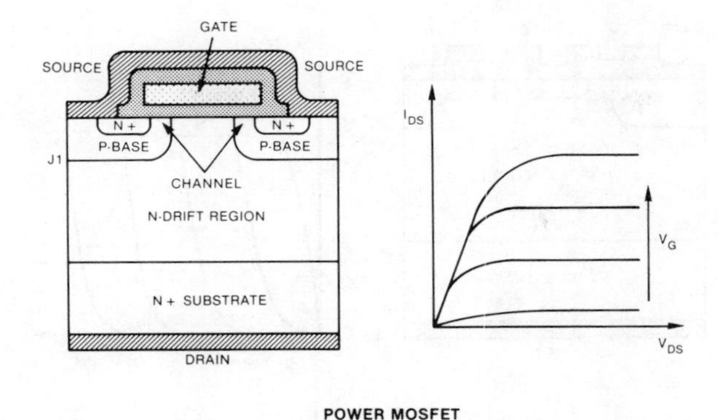

POWER MOSFET

Fig. 4. Power MOS gated field effect transistor (MOSFET) structure and its voltage controlled electrical output characteristics.

In summary, the power MOSFET is a voltage controlled device whose output current can be controlled using very low gate drive power levels. In fact these devices can be maintained in their on- or off-state with essentially no input gate power and require only relatively small gate pulse currents to charge and discharge their input gate capacitance during switching. In addition, the on-resistance of these devices increases with increasing temperature. This allows paralleling of power MOSFET's and prevents the formation of current localization inside the devices. These devices have also been found to have greater ruggedness than bipolar transistors. Further, since current conduction in these devices occurs without minority carrier injection, their inherent switching speed is very high. Devices have been developed capable of power switching at over 100 MHz. The major limitations, which have prevented the power MOSFET from rapidly displacing the bipolar transistor, are their higher on-resistance and the relatively higher cost of device fabrication.

Power JFET

The high-voltage power junction field-effect transistor (JFET) has also been called the static induction transistor (SIT). The basic structure of these devices is illustrated in Fig. 5. This device structure is fabricated by starting with a highly conductive n+ substrate and growing an n-type drift region on it. This drift region must support the sum of the drain and gate bias voltages. Its design is similar to that of the power MOSFET

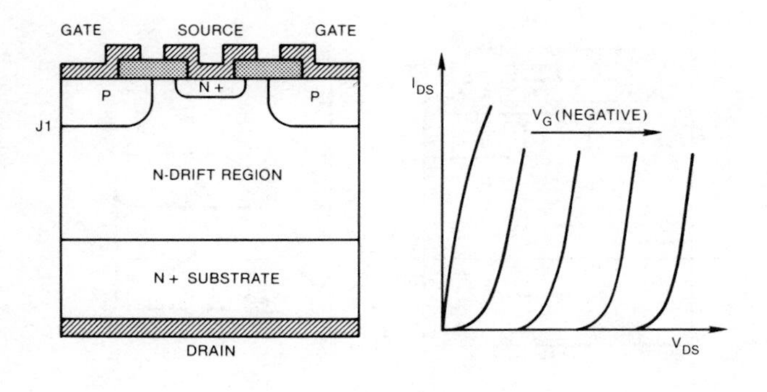

Fig. 5. Power junction gate field effect (JFET) transistor and its voltage controlled triodelike electrical output characteristics.

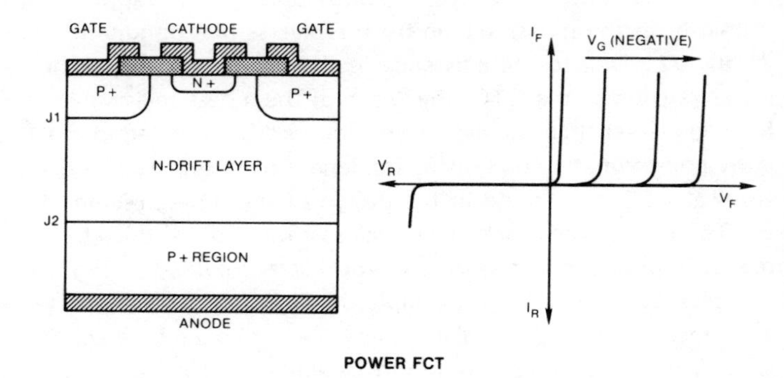

Fig. 6. Power field controlled thyristor (FCT) structure and its voltage controlled triodelike electrical output characteristics.

and can be derived from Fig. A. The n+ source region and p-type gate regions (which must surround the n+ source region completely) are then formed in the n-drift layer. In the absence of a gate bias voltage, the power JFET will conduct current between source and drain limited by the resistance of the n-drift region. In order to block current flow, a negative gate bias must be applied. This reverse biases the gate junction *J1* whose depletion layer then extends under the n+ source. When sufficient gate bias is applied, the depletion layers of the adjacent p gate regions intersect and establish a potential barrier for electron current flow between source and drain. This potential barrier then prevents drain current flow and provides the JFET with forward blocking capability. However, as the drain voltage is increased, the potential barrier setup by a given gate voltage becomes reduced and source-drain current flow can occur as shown in Fig. 5. Thus, for each gate bias voltage, drain current flow commences above a certain drain voltage. This results in the triodelike characteristics observed in these devices as illustrated in Fig. 5. An important design parameter for these devices is the blocking gain which is defined as the ratio of the drain voltage, above which drain current flow occurs, to the gate bias voltage. The blocking voltage capability of JFET's was initially limited by the low blocking gains (less than 5) of planar gate structures. Improved gate designs have allowed increasing the blocking gain to over 20 which has made the fabrication of 600 V devices feasible. However, the current rating of these devices is low due to the difficulty of fabricating these highly interdigitated devices with typical repeat spacings of 20 μ.

In summary, the power JFET is a voltage controlled device with triodelike characteristics. These devices conduct current in the absence of gate bias and require negative gate voltages to maintain them in the blocking state. The normally-on characteristics of these devices has severely limited their application. Their primary feature is a very high frequency capability which has been used for the construction of high-performance audio and radio frequency (RF) amplifiers.

Power FCT

The power field controlled thyristor (FCT) has also been called the static induction thyristor (SIT). The structure of this device is illustrated in Fig. 6. These devices are usually fabricated by starting with an n-type silicon wafer and then forming the p+ anode region on one side of the wafer, and the n+ cathode and p+ gate regions on the opposite surface. As in the case of the power JFET, the p+ gate regions must completely surround the n+ cathode region because the forward blocking capability of the FCT is achieved in the same manner as the JFET, i.e., an applied negative gate voltage is used to form a potential barrier under the n+ cathode to prevent anode current flow. However, the operation of the FCT in the on-state is substantially different from the JFET because of the presence of the p+ anode region. This p+ region injects a very high concentration of minority carriers into the n-drift layer during device current conduction. This severely reduces the n-drift layer resistance and allows the FCT to operate at high current densities similar to that of the GTO. The p+ anode junction *J2* in the FCT also provides reverse blocking capability in these devices. It should be noted that the n-drift layer of these devices must be designed in a manner similar to that described for GTO's using Fig. C (Appendix) because of the open base PNP transistor formed between the gate and the anode regions.

The first reported FCT devices had very low blocking gains (less than 5). Improvements in the gate structure have now led to devices with blocking gains of over 50. Their advantages are a high forward conduction current density similar to that of the GTO and the absence of a regenerative four-layer structure, which allows the FCT to operate at much higher (about 10 times) *dv/dt*'s than GTO's. These devices have also been shown to be capable of operating at higher temperatures and to exhibit superior radiation tolerance. The use of these devices in power switching applications has, however, been curtailed by their normally-on characteristics.

Power MOS-IGT

The power MOS-IGT represents a new class of power devices recently developed by combining bipolar and power MOSFET technologies. The basic structure of the insulated gate transistor (IGT) is shown in Fig. 7. This device can be fabricated by starting with a p+ substrate and then growing the n-type drift layer on it. The DMOS gate structure is then formed on this drift layer by using a process similar that described for power MOSFET's. Due to the presence of an open base PNP transistor in the structure, the n-drift layer must be designed in a manner similar to that used for the GTO using Fig. C (Appendix). Since this device contains two back-to-back high voltage junctions

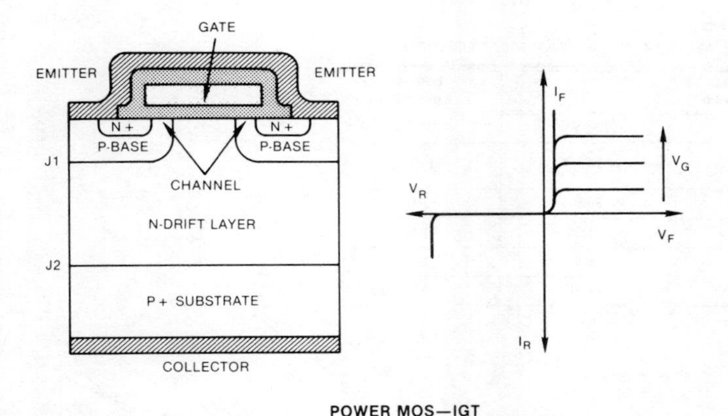

POWER MOS—IGT

Fig. 7. Power MOS gated insulated gate transistor (IGT) and its voltage controlled electrical output characteristics.

*J*1 and *J*2, it is capable of symmetrical forward and reverse blocking capability. The forward blocking is achieved at junction *J*1 when the gate electrode is shorted to the emitter contact. To turn on the devices, a positive gate voltage must be applied to form an n-type channel at the surface of the p-base region under the gate. This channel then links the n+ emitter to the n-drift layer as in the case of the power MOSFET. However in the IGT, the positive bias on the collector causes the p+ region to inject a high concentration of minority carriers into the n-drift layer which severely reduces its resistance. Thus, in the on-state, the IGT behaves like a forward biased p-i-n rectifier and is capable of operating at high current densities equivalent to that of the GTO. At the same time, the current in the IGT can be controlled using the applied gate bias voltage. This feature of the IGT not only allows gate turn-off capability but can be used to limit the collector current. Thus, the IGT exhibits the best feature of the power MOSFET, namely, voltage controlled operation with very high input impedance, as well as the best feature of bipolar devices, namely, high forward conduction current density. It has also been found that the on-state voltage drop of the IGT increases with increasing temperature. This allows paralleling of these devices and ensures good current distribution within each device.

One of the outcomes of bipolar operation in the IGT is a reduced switching speed compared with the power MOSFET. The first IGT devices were reported to have a typical forced gate turnoff time ranging from 10 to 50 μs. This feature would have limited IGT to very low frequency circuits. However, by using minority carrier lifetime control techniques, IGT's with gate turnoff times as low as 0.25 μs have recently been announced. This ability to tailor the IGT characteristics over a broad range of frequencies to optimize its characteristics is another important feature of the device. Further due to the strong modulation of the n-drift layer resistance, the IGT current density varies relatively slowly as the square root of the blocking voltage capability. This makes it an attractive device for high voltage applications. The combination of high input impedance and high conduction current density in the IGT makes it a superior power device in many circuits where bipolar transistors have been used in the past. Consequently, the IGT can be expected to displace the bipolar transistor in many of its applications. Power MOS-IGT's with current ratings of 10 A and 25 A at a blocking voltage capability of 500 V have already become available. A rapid growth in both the current and voltage ratings of these devices can be anticipated.

Device Comparison

With the development of many alternative devices to the power bipolar transistors, the circuit designer is faced with the task of making a judicious choice between these devices. As an aid to device selection, the transistors discussed in the previous section are compared here on the basis of several criteria. To begin with, normally-on devices have been found to be undesirable for power switching because of the need to ensure that a negative gate drive is available during circuit power-up. Since the power JFET and FCT exhibit this characteristic, their use is relegated to those applications in which some of their unique characteristics (such as the very high frequency response of the JFET or the high *dv/dt* and radiation tolerance of the FCT) are necessary. Since these requirements do not exist in most applications, the circuit designer is left with a choice between the remaining normally-off devices—namely, the bipolar transistor, the GTO, the power MOSFET, and the IGT. For very high power levels, such as ac traction, only GTO's with sufficient current and voltage ratings are available making the choice quite limited. However, at lower power levels, where gate controlled operation of the devices is highly desirable, the circuit designer must choose between the bipolar transistor, the power MOSFET, and the IGT. The relative merits of these three devices are discussed below.

One of the important criteria for selection a power device is its gate drive power requirements. Since the power bipolar transistor is a current controlled device with a typical current gain of 10 to 20, it requires relatively high gate drive power during steady-state current conduction as well as during turnoff. The gate drive circuitry for the power transistor is, therefore, complex and expensive. In contrast, both the power MOSFET and the power MOS-IGT are voltage controlled devices with very high input impedance. The gate drive power required to control these devices is, therefore, small. This eliminates complexity in the gate drive and often allows control of these devices directly from an integrated circuit since the gate circuit must merely provide enough current to charge and discharge the input capacitance of these devices. In this regard, the IGT is even superior to an equivalent power MOSFET because its input capacitance is an order of magnitude smaller for the same power rating. Further, since the technology for the fabrication of the power MOSFET and IGT is similar, the IGT offers a lower cost to the circuit designer because the chip area is an order of magnitude smaller than for the power MOSFET.

A graphic illustration which can aid in the selection between these devices is provided in Figs. 8–10. In these diagrams, the power dissipation in each of the three devices is compared as a function of the switching frequency. In performing these calculations, it is assumed that all the devices are operating at a current density of 100 A per cm^2 at a duty cycle of 50 percent. The turnoff speeds of the devices are provided in Table I. In computing the switching losses an inductive load has been assumed. For purposes of comparison, devices with three blocking voltage capabilities are considered. In order to mini-

TABLE I
TRANSISTOR PARAMETERS USED FOR CALCULATION OF POWER DISSIPATION

Device	Forward Voltage Drop At 100 A/cm²			Forced Gate-Turn-Off Time
	100 V	600 V	1200 V	
Bipolar Transistor	0.20 V	6.65 V	26.7 V	1.0 μs
MOSFET	0.55 V	20.0 V	80.0 V	0.1 μs
IGT (A)	1.10 V	1.45 V	1.70 V	1.5. μs
IGT (B)	1.75 V	2.45 V	2.85 V	1.0 μs
IGT (C)	2.50 V	3.80 V	4.80 V	0.25 μs

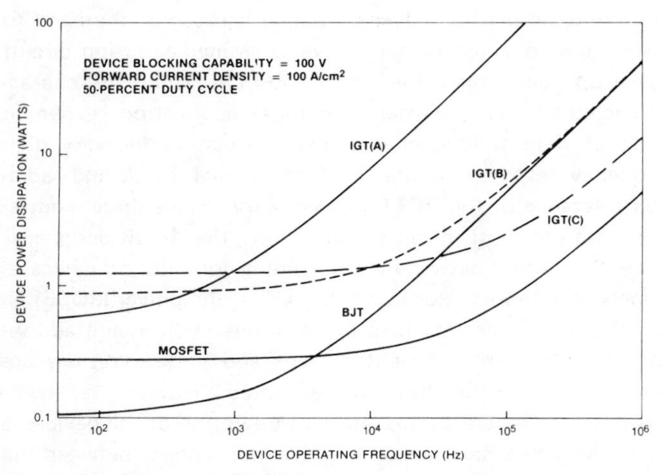

Fig. 8. Comparison of the power dissipation as a function of switching frequency for 100 V power bipolar transistor, power MOSFET and, three types of IGT's.

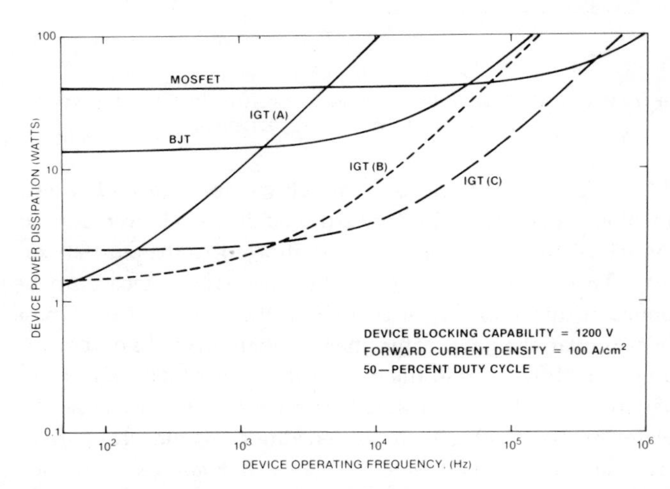

Fig. 10. Comparison of the power dissipation as a function of switching frequency for 1200 V power bipolar transistor, power MOSFET, and three types of IGT's.

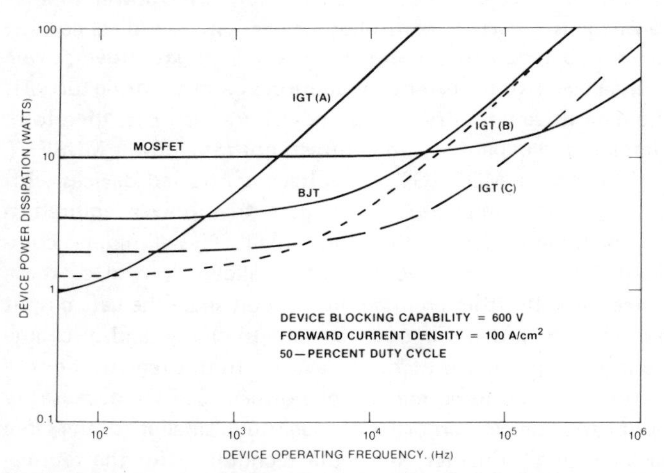

Fig. 9. Comparison of the power dissipation as a function of switching frequency for 600 V power bipolar transistor, power MOSFET, and three types of IGT's.

mize power dissipation to obtain optimum system efficiency, it is clear from these calculations that the power MOS-IGT is the best device for high voltage circuits up to frequencies as high as 100 kHz. The unique ability to tailor the switching speed of the IGT can be seen to play an important role in making these devices superior to the bipolar transistor and the power MOSFET over a broad range of frequencies. From these calculations it can also be seen that at frequencies above 100 kHz, the power MOSFET is the only device that can provide adequate switching speed. Further, it is apparent that for low-voltage power circuits in which the devices must block less than 100 V, the on-resistance of the power MOSFET drops to very small values. Since the power MOS-IGT has a rectifier-like forward conduction characteristics, it does not conduct current efficiently unless the forward drop exceeds 1 V. Consequently, in low-voltage power circuits, the power MOSFET becomes the best choice. This conclusion applies to direct current (dc) circuits. For circuits where reverse blocking capability is essential, the power MOS-IGT must be used irrespective of blocking voltage capability because it is the only device with the desired output characteristics.

As a further aid to the circuit designer in performing a comparison between all the transistors discussed in the book, Table II provides a summary of the features of each of the devices. This table is intended to provide a rapid comparison between the different devices to help the circuit designer in making the choice. After narrowing the choice among the devices, the reader should refer to the reprint sections for details regarding the design and performance for each of the devices.

DEVICE RATINGS

Power transistors are required for a very wide variety of power conditioning applications that range in power ratings from 100 W to over a megawatt as well as over a broad range of frequencies ranging from 60 Hz to over 1 MHz. In this section, the ratings of commercially available transistors are summarized in terms of their blocking voltage and current handling

TABLE II

COMPARISON OF OPERATING CHARACTERISTIC OF GATE-TURNOFF TRANSISTORS

Device Characteristic	Power Bipolar Transistor	Gate Turn-Off Thyristor	Power MOSFET	Power JFET/SIT	Power FCT/SITh	Power MOS-IGT
Normally On/Off	Off	Off	Off	On	On	Off
Reverse Blocking Capability (Volts)	<50 V	500-2500 V	0 V	0 V	500-2500 V	200-2500 V
Blocking Voltage Range (Volts)	50-500 V	500-2500 V	50-500 V	50-500 V	500-2500 V	200-2500 V
Forward Conduction Current Density[1] (A/cm^2)	40	200	10	4	200	200
Surge Current Handling Capability[2]	3x	10x	5x	5x	10x	5x
Maximum Switching Speed	50 kHz	10 kHz	20 MHz	200 MHz	20 kHz	50 kHz
Gate Drive Power	High	Medium	Low	Low	Medium	Very-Low
dV/dt Capability	Medium	Low	High	High	High	High
dI/dt Capability	Medium	Low	High	High	Medium	High
Maximum Operating Temperature	150°C	125°C	200°C	200°C	200°C	200°C
Radiation Tolerance	Poor	Very Poor	Moderate	Good	Good	Moderate

[1]The forward conduction current densities are compared for 500 V devices operating at a forward voltage drop of 2 V.
[2]The surge current handling capability is given here as a multiple of the forward conduction current density.

capability. A projection of the growth in the ratings of these devices is also provided.

The ratings of commercially available power transistors in 1983 are given in Fig. 11 in which the lines for each device define the boundary within which devices are available. As was mentioned earlier, for high voltage applications (above 1500 V) at high power levels (over 100 kW), only GTO's with adequate ratings are available today. Power bipolar transistors have the next highest power ratings. Devices that can handle power levels of up to about 50 kW have been developed using the Darlington configuration. The power MOSFET ratings have been growing rapidly but the highest power handling capability today is limited to about 5 kW. Although the power MOS-IGT was commercially introduced very recently in 1983, its power handling capability already exceeds that of the power MOSFET's. Devices are now available which can switch between 10 to 15 kW of power.

It can be anticipated that the power handling capability of all these devices will grow with improvements being made in device processing technology. However, the development of the new high input impedance power MOS devices will have a strong impact upon GTO's and bipolar transistors. It can be projected that power MOS-IGT's will be available with voltage blocking capabilities up to 2500 V which rival that of the GTO. Since these high voltage devices are expected to have a current handling capability ranging up to 10 A as indicated in Fig. 12, they will replace low current GTO because of their simpler gate drive requirements. The GTO device ratings can, therefore, be expected to be confined to over 100 A and 2500 V. A similar conclusion can be made regarding power bipolar transistors with current ratings of less than 100 A. These devices are not expected to compete with the power MOS-IGT due to their relatively high gate drive power requirements. However, they will continue to have a role at higher power levels that cannot be served by the power MOS-IGT or power

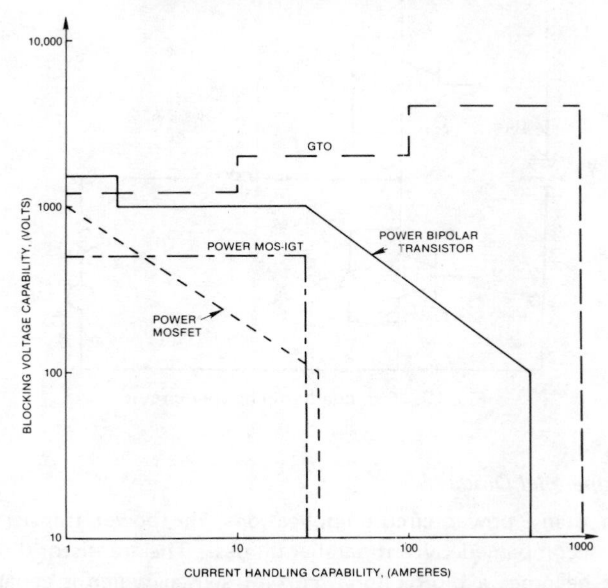

Fig. 11. Power ratings of commercially available power transistors.

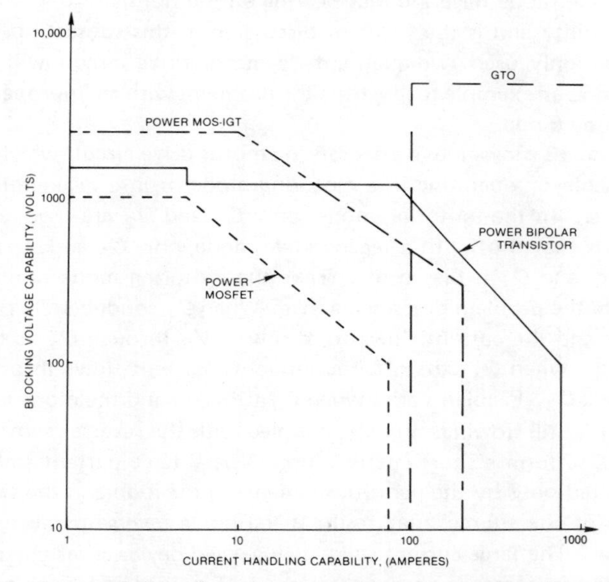

Fig. 12. Projected power ratings of several power transistors.

MOSFET. The ratings of the power MOSFET are also expected to grow to serve the low voltage and high frequency applications but this growth will occur at a more modest pace since it is restrained by the ability to fabricate larger area devices with adequate yields.

DEVICE APPLICATION CHARACTERISTICS

In previous sections, basic characteristics of the various devices have been described. To have efficient and reliable applications, however, it is important to investigate the devices from the user's point of view. In this section discussions will be focused on the following device application characteristics: a) antiparallel diode and its effect on the transistor devices, b) *dv/dt* effects and causes, c) snubbing circuit techniques, and d) device paralleling techniques.

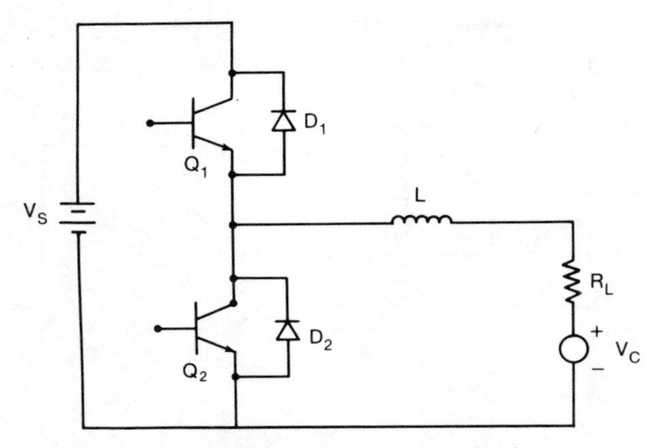

Fig. 13. Two-quadrant chopper circuit.

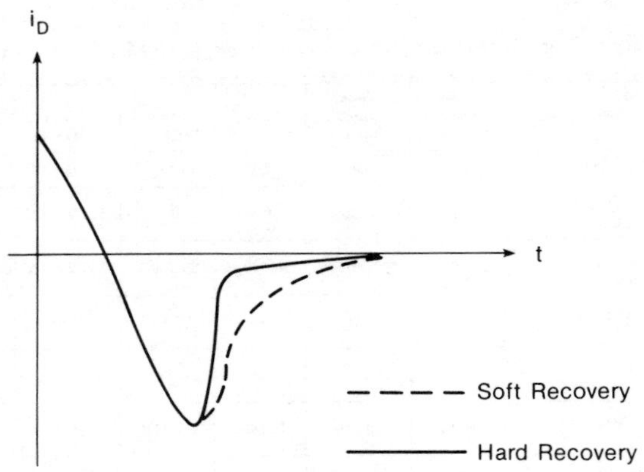

Fig. 14. Hard and soft reverse recovery of a diode.

Antiparallel Diode

In many power circuit applications, the power transistors are accompanied by antiparallel diodes. The transistor-diode pair becomes a bidirectional current switch which is capable of conducting reactive current and as such, the circuit is capable of regenerative operation. The characteristics of the antiparallel diode have a direct bearing on the transistor operating reliability and is the focus of discussion in this subsection. A commonly used two-quadrant dc motor drive circuit will be used as an example to illustrate the problems with an improperly chosen diode.

Fig. 13 shows a two-quadrant dc motor drive circuit which is capable of operating in a motoring mode or in a regenerative mode. In the motoring mode, only Q_1 and D_2 are used, and Q_2 is biased off. In a regenerative mode, only Q_2 and D_1 are used, and Q_1 is biased off. Take the motoring mode to illustrate the problem described above. When Q_1 conducts, D_2 cuts off, and the current flows from source V_S through Q_1 to the load. When Q_1 cuts off, the inductive current flows through diode D_2. Problems arise when Q_1 turns on and the diode current is still flowing. V_S, Q_1, coupled with the reverse recovery of D_2, form a short circuit loop. A very large current spike, limited only by the parasitic elements in the loop and the voltage of the source, may result if the diode reverse recovery is slow. The large current spike may exceed device surge current rating in some devices such as the BJT and the FET, or may cause latching in other devices such as the GTO and MOS-IGT. The problem described above also occurs when the circuit is operated in a regenerative mode, except that the devices involved in this case are Q_2 and D_1.

Even if the transistors survive the large surge current, the device power dissipation would be unacceptable especially for high frequency applications. Selection of a fast recovery diode is therefore crucial for a reliable and efficient circuit operation. Another important consideration in the choice of the diode is the reverse recovery "softness." Soft recovery, meaning small *di/dt* during the reverse recovery, minimizes the diode-turnoff voltage spike caused by inevitable parasitic inductance associated with diode lead. Fig. 14 shows the reverse recovery current waveform of a diode. It should be noted that the induced voltage spike applies a reverse bias to the base-emitter (B-E) junction of the transistor, which has breakdown voltage normally below 20 volts. Thus, it is a real possibility that transistor B-E junction may be broken down if the diode has hard turnoff characteristics.

The antiparallel diode is, in some cases, integrated in the device structure for economic reasons. In other cases, the diode is an inevitable parasitic in a device such as MOSFET structure where base-source metalization overlap is necessary. However, the diodes formed in such fashion often suffer from slow reverse recovery unless the device is selectively irradiated. In summary, an ideal antiparallel diode should have both fast and soft recovery characteristics. Some diode manufacturers are starting to respond to this request.

dv/dt False Triggering

dv/dt false triggering is a well-known phenomenon in a p-n-p-n structure such as SCR's and GTO's. The false triggering is caused by displacement current which functions as a gate current pulse. An auxiliary circuit for controlling the *dv/dt* (across device anode-to-cathode) must be used to ensure the off-state of the device. In three layer devices, *dv/dt* may cause problems for different reasons as will be described below.

In an FET circuit, *dv/dt* false triggering is caused by the "Miller effect." When the device is subjected to large *dv/dt*, gate voltage may be pulled up to exceed threshold gate voltage and the device temporarily turns on. The magnitude of *dv/dt*, device gate-to-drain capacitance and the design of gate drive circuit are all related to *dv/dt* problem. Theoretically speaking, if the gate drive impedance is zero, the capacitive current resulting from *dv/dt* can be sunk by the gate circuit. Consequently, gate voltage should not rise and the *dv/dt* problem can be avoided. In a practical application, reverse voltage gate drive is often used to alleviate the problem.

In a bipolar Darlington configuration, the possibility of *dv/dt* turn-on is enhanced by the high current gain and transistor reverse conduction before *dv/dt* is applied. The chopper circuit shown in Fig. 13 is used again for illustration. When the inductive current has already been established and Q_1 is off, the inductor current flows not only through D_2 but also through Q_2 in reverse direction if the base drive for Q_2 is kept high during this period as is commonly employed in circuit design. If Q_2 is a Darlington with base-emitter resistors as

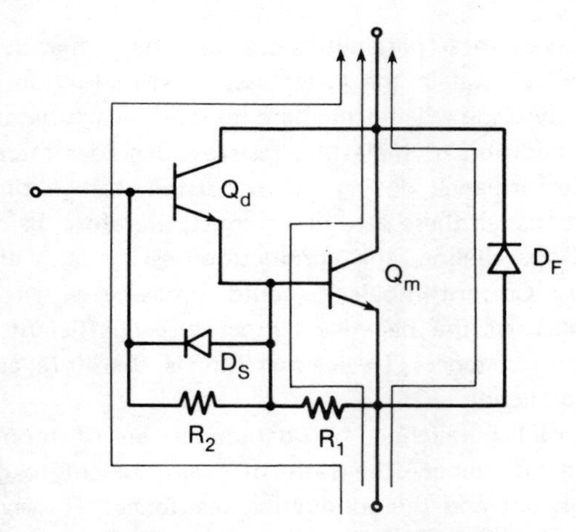

Fig. 15. Reverse current flow in a Darlington transistor.

shown in Fig. 15, the inductive current can also flow through R_1, R_2 and B-C junction of drive transistor Q_d and R_1 and B-C junction of Q_m. In any event, the stored charge in the device right before *dv/dt* is applied makes the device more susceptible to *dv/dt* false triggering. If reverse-bias is kept for Q_2 during the period when Q_1 is off, the situation would improve but this complicates the base drive timing because it is load dependent.

In the case of the MOS-IGT, there are two kinds of *dv/dt* that may cause problems. One is the *dv/dt* associated with the collector-to-emitter voltage, like the other devices described above. The other is the *dv/dt* associated with gate-to-emitter voltage. Both have significant effect on the device latching current level. The larger the *dv/dt*, the smaller the latching current. Proper control of *dv/dt* by circuit parameters is necessary to fully utilize the device capability.

Snubber Circuit

In a power transistor circuit, the use of switching-aid circuits are often necessary to relieve the power switch from overstress during switching. The switching circuit is often called a snubber circuit. There are three purposes of using snubber circuits. One of the purposes is to reduce the device switching power losses. The second purpose is to avoid device second breakdown. The third purpose is to control the device *dv/dt* to avoid latching in pnpn devices such as the GTO and the MOS-IGT, and to avoid temporary turn-on in BJT's and FET's as described in the previous section. There are numerous snubber circuit configurations reported. Rather than trying to cover the various configurations, three basic types of snubber circuit will be discussed. These are the dissipative snubber, the nondissipative snubber, and the active snubber.

Dissipative Snubber

In this type of snubber circuit, a capacitor is used to slow down device voltage rise during turnoff. The stored energy in the capacitor is then dumped in a resistor during device turn-on. Fig. 16 shows a polarized snubber. If the diode is not used, then the circuit is a nonpolarized snubber. Resistor R_S is used to limit transistor current spike at turn-on. The nonpolarized snubber is simpler but is less effective as compared with a polarized snubber. In both cases, however, the power

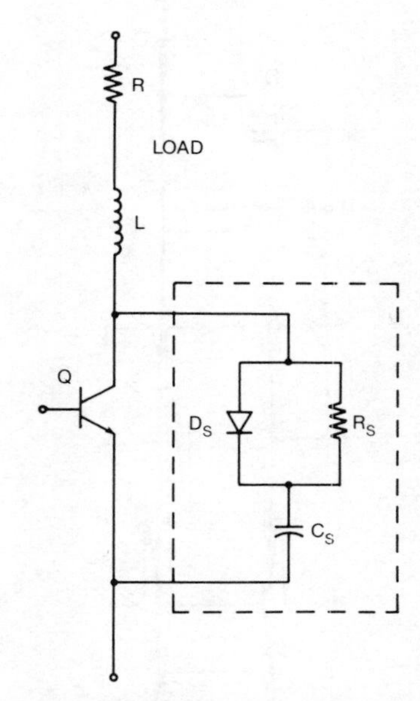

Fig. 16. Dissipative snubber circuits.

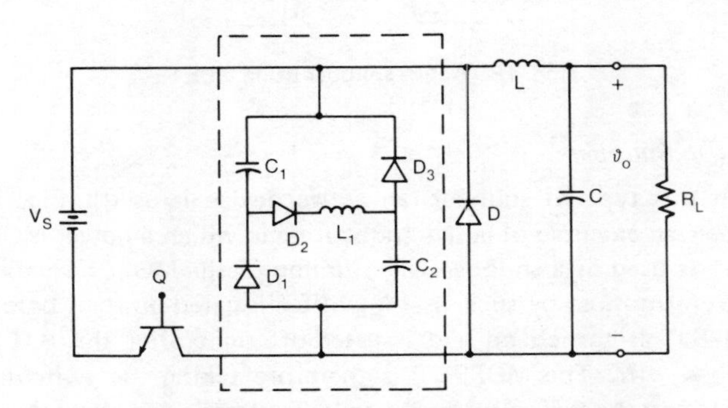

Fig. 17. Buck converter using a nondissipative snubber.

loss is $|\frac{1}{2}|\,CV_S^2 \cdot f$, where f is the operating frequency and V_s is the source voltage. The selection of capacitor size is a compromise between snubber power loss and the desired *dv/dt*. The larger the transistor current, the larger the required capacitance and so is the snubber power loss. As frequency and power level increases, snubber power loss increases rapidly. This type of circuit is normally used in low-power circuits or in a circuit in which efficiency is of secondary concern.

Nondissipative Snubber

A capacitor, inductor and diode are normally used in this type of snubber circuit. Fig. 17 shows a buck converter using this type of snubber. When the transistor Q conducts, C_1, C_2, D_2, and L_1 form a resonant loop with the source. Both C_1 and C_2 are charged until current of L_1 is zero. At that instant, D_2 cuts off and both C_1 and C_2 are charged up to nearly the source voltage. When the transistor turns off, both C_1 and C_2 discharge through D_1 and D_3, respectively, to the inductor L and the load. Transistor turnoff *dv/dt* is therefore controlled by C_1, C_2 as well as the transistor current at turnoff. By the choice of capacitor size, proper snubbering of the transistor is accomplished. It is noted that the energy stored in the capacitor is released to the load when the transistor turns off, unlike the case of the dissipative snubber in which energy stored in the capacitor is dumped in the resistor.

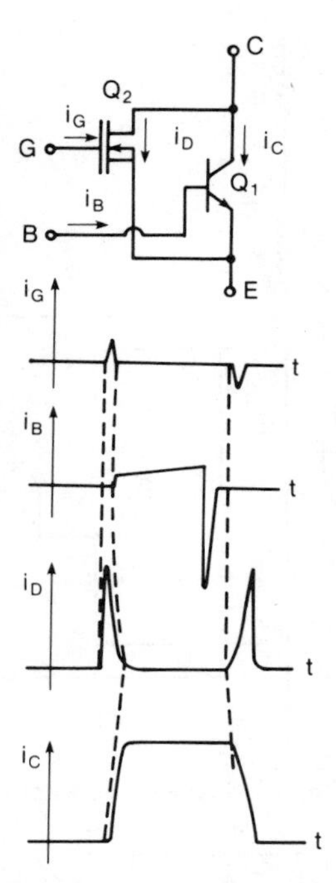

Fig. 18. Active snubber using a FET.

Active Snubber

In this type of snubber, an active device is used. Fig. 18 shows an example of active snubbering in which a power MOSFET is used as a snubber. The timing of the MOSFET gating waveform must be such that MOSFET is gated-on right before the BJT is turned on and is gated-off right after the BJT is turned off. The MOSFET is therefore taking the switching stress and the BJT is subjected to low voltage during switching. During the conduction phase, the BJT conducts most of the load current because of lower conduction resistance. Thus, the MOSFET conducts the load current only during the turn-on and turnoff switch period. The MOSFET fits this purpose very well because of its fast switching capability and freedom from second breakdown. Because the snubber MOSFET only conducts current pulses during switching, small chip area is required. The disadvantage of this snubber technique is that a precise gating waveform is required to achieve the purpose.

Device Paralleling

In some applications, device paralleling is necessary to handle the required current level. Generally speaking, to have proper current sharing among the paralleled devices, a certain degree of device parameter matching and certain circuit means must be employed. To have a successful device paralleling, proper current sharing must be maintained in both the conduction state and the switching period.

Conduction State

In BJT paralleling, three device parameters are mainly responsible for current imbalance during the conduction state. In the case of hard saturation operation, the parameters are base-emitter voltage-current characteristics and saturation resistance. In the case of quasi-saturation operation, such as the output transistor of a Darlington, the parameters of importance are base–emitter characteristics and current gain. The temperature coefficients of these parameters are such that run-away conditions could prevail in the latter case. This situation can be alleviated by using a ballasting base resistor in each base. However, the addition of ballasting resistors degrades the current sharing performance during the transistor storage time. A trade-off between these parameters must, therefore, be made.

In GTO paralleling, the conduction resistance is the key parameter. Current imbalance could run away with temperature because of the negative temperature coefficient of the conduction resistance. Device matching is, therefore, essential in GTO paralleling.

In MOSFET paralleling, two parameters are of importance, the transconductance (the ratio of drain current to gate-to-source voltage) and the conduction resistance. However, the temperature coefficients of both parameters are positive. Any current imbalance does not run away. Temperature difference among the paralleled devices force the current sharing.

For the MOS–IGT, the two sources for current imbalance are similar to those in the MOSFET, namely, the transconductance and conduction voltage drop. The temperature coefficient of transconductance is positive, but the temperature coefficient of conduction drop may be positive or negative, depending upon the current level. At the high current end of device operation, it is positive. Consequently, in MOS–IGT paralleling, current imbalance does not run away.

Switching Period

In all of the four devices mentioned above, current sharing is affected by both device parameters and circuit parasitics during the switching period. In the case of the BJT and the GTO, storage time is the key parameter that affects the current sharing during turnoff. To make things even worse, as the device that takes more current heats up, the storage time becomes even longer. This aggravates the situation. In both cases, a direct couple of the base terminals of the parallel devices improves the situation. The device with longer storage time automatically draws more reverse base current than the faster device when the faster device enters its fall time and, therefore, shortens the storage time of the slower device. This forces better current sharing. In the case of the MOSFET, both the gate source threshold voltage and the gate capacitance play roles in affecting the current imbalance. Slowing down the switching time by increasing the gate resistance reduces the duration of current imbalance and improves the situation. In the case of the MOS–IGT, the situation is similar to that for the MOSFET.

In all cases of device paralleling, current sharing during switching can be improved by insertion of a small inductance in series with the collector of the device. However, this causes larger collector voltage spikes at turnoff. The circuit layout also plays an important role in device paralleling. Circuit symmetry with respect to each parallel device should be maintained as much as possible.

Power Transistor Circuits and Power Applications

There are various power electronic circuit topologies reported in literature. However, the many varieties can be classified into

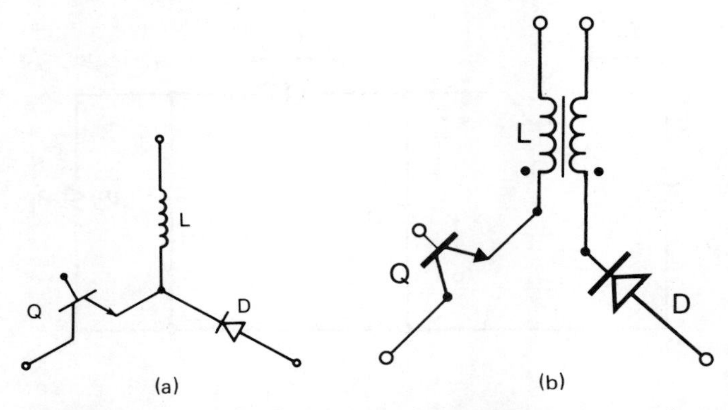

Fig. 19. Power circuit building block—point-connected transistor, diode and inductor combination.

several basic types. The choice of a circuit topology depends on application power level, circuit complexity, component stress level, ease of control strategy and cost. In this section, basic power circuit families will be reviewed. The circuit operation, the special features and the applications of each power circuit discussed will be summarized. Because of its gate-turn-off capability, transistor devices are normally used for applications with dc sources. The discussion in this section is therefore limited to dc applications, either dc to ac inversion or dc to dc conversion.

While the power circuit topology may appear to vary widely, there are commonalities. Understanding the commonalities should lead to a better appreciation of the similarities and differences among the circuits. Four common features of the power circuit are described below:

1) There are no dissipative element in the circuit except the nonideal characteristics of the components. The power semiconductor switches in the circuit are operated in switching mode.
2) The point-connected transistor-diode-inductor combination shown in Fig. 19 is the basic building block for many power circuit topologies. The transistor is used to control the duty cycle of the power flow. The inductor represents either a discrete current smoothing inductor or the leakage inductance of a transformer or the inductance of the load. And the diode is to provide a path for inductive current to flow when the transistor cuts off, and for that reason, is called a free-wheeling diode. The inductor can be a multiwinding inductor or a transformer with proper winding polarity as shown in the same figure.
3) Bidirectional switches are needed for a power circuit with regenerative capability. Such a switch is normally accomplished by having a transistor and an antiparallel diode as a pair. During the period when the power source and the load are connected through a transistor, or transistors, the power flow is from source to load. During the period when the source/load connection is made through a diode or diodes, the power flow is from load to source. It should be noted that the antiparallel diode is not the same as the free-wheeling diode mentioned in item 2). In some circuits, however, a diode serves both as the free-wheeling diode for a transistor and as the antiparallel diode for another transistor.
4) The dc and ac inversion process is inherent in the power circuit operation regardless of the application, (dc-to-ac or dc-to-dc). The power transformation is accomplished by the transformer or the inductor action and the transistor duty cycle control.

Circuit Topology

The commonly used power transistor circuits can be classified into three basic configurations: A. *Flyback Configuration*, B. *Parallel or Push-Pull Configuration*, and C. *Bridge Configuration.*

A. Flyback Configuration

Fig. 20 shows three basic dc-to-dc converter configurations. Fig. 20(a) is a voltage step-up converter or is often called a "boost converter." Fig. 20(b) is a voltage step-down converter or is often called a "buck converter" or "chopper." Fig. 20(c) is a voltage step up/step-down or buck/boost converter. Notice the point-connected transistor diode-inductor combination mentioned earlier. When the transistor conducts, the diode is off and energy is stored in the inductor. When the transistor cuts off, the diode conducts and part or all of the energy stored in the inductor is released to the output. The name "flyback converters" is descriptive of the inductive energy flyback action typically encountered in this type of converter operation.

In the case when only part of the inductor stored energy is released in the operating cycle, the converter is said to operate in a continuous conduction mode. Otherwise, the converter is said to operate in a discontinuous mode. Notice that the waveforms shown in Fig. 21 apply to all of the three configurations.

Figs. 22, 23, and 24 show three converter circuits which are variations of the three basic configurations shown in Fig. 20. Fig. 22 is an isolated version of a buck/boost converter. The operation of this circuit is identical to a buck/boost converter. The use of transformer provides not only electrical isolation but also additional design freedom. However, the transistor voltage spike at turnoff is greatly affected by the closeness of the transformer winding coupling. A large transformer leakage inductance may damage or greatly reduce the reliability of the power transistor. Fig. 23 shows a forward converter, which operates like a buck converter. Transformer windings N_1 and N_2 provide electrical isolation and voltage transformation, and winding N_3 is used to reset the magnetic core when the transistor cuts off. When transistor Q conducts, D_1 conducts and D_2 and D_3 cut off. When Q cuts off, D_2 is free wheeling with the inductive energy of L and D_3 is free wheeling with magnetizing energy of the transformer. Notice that there are two point-connected transistor-diode-inductor combinations: (Q, D_3, transformer N_1, N_3) and (D_1, D_2, L). D_1 is a diode but its conduction is controlled by Q and can be considered as a transistor. Because of the reset requirement of the transformer core, the maximum duty cycle of the forward converter is limited by $N_1/(N_1 + N_3)$. The main power transformation in this circuit is accomplished by transformer action through N_1 and N_2, unlike the case of the isolated buck/boost converter in which the power transformation is accomplished by flyback action. For this reason, a forward converter normally requires a smaller transformer than a comparable isolated buck/boost converter.

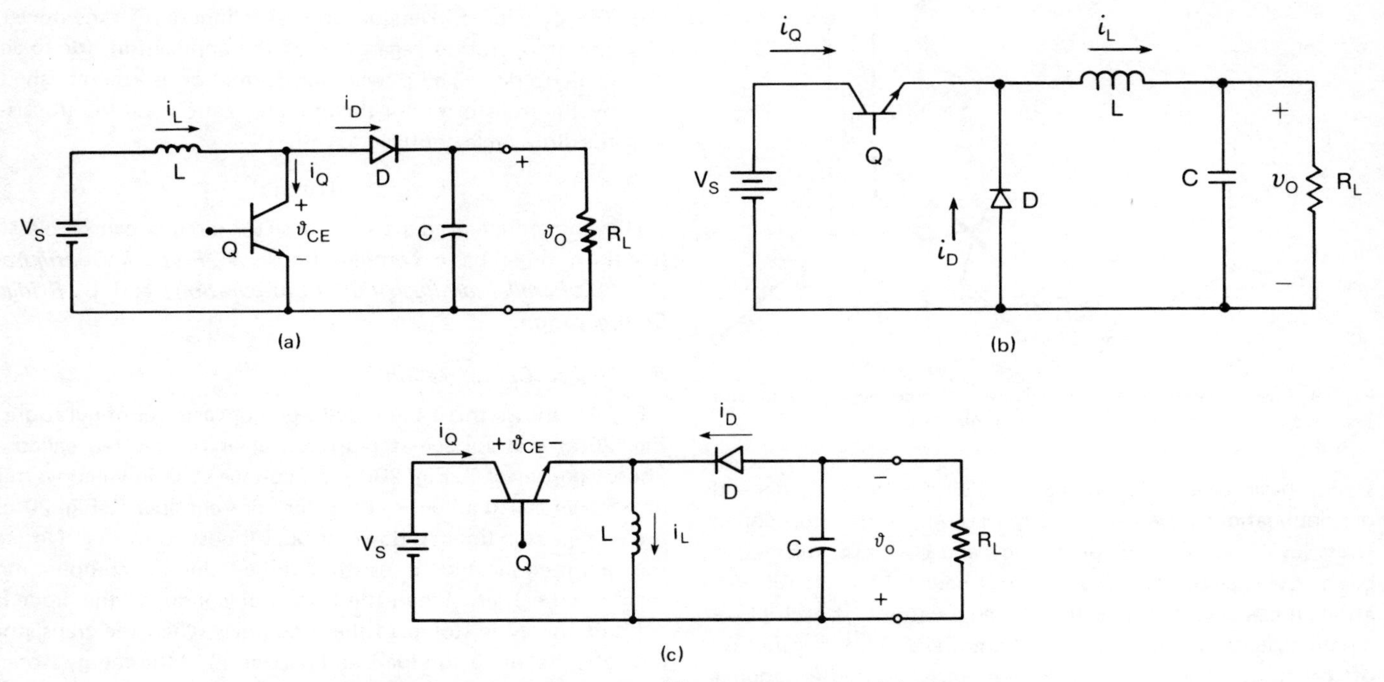

Fig. 20. Flyback configuration. (a) Boost converter. (b) Buck converter. (c) Buck/boost converter.

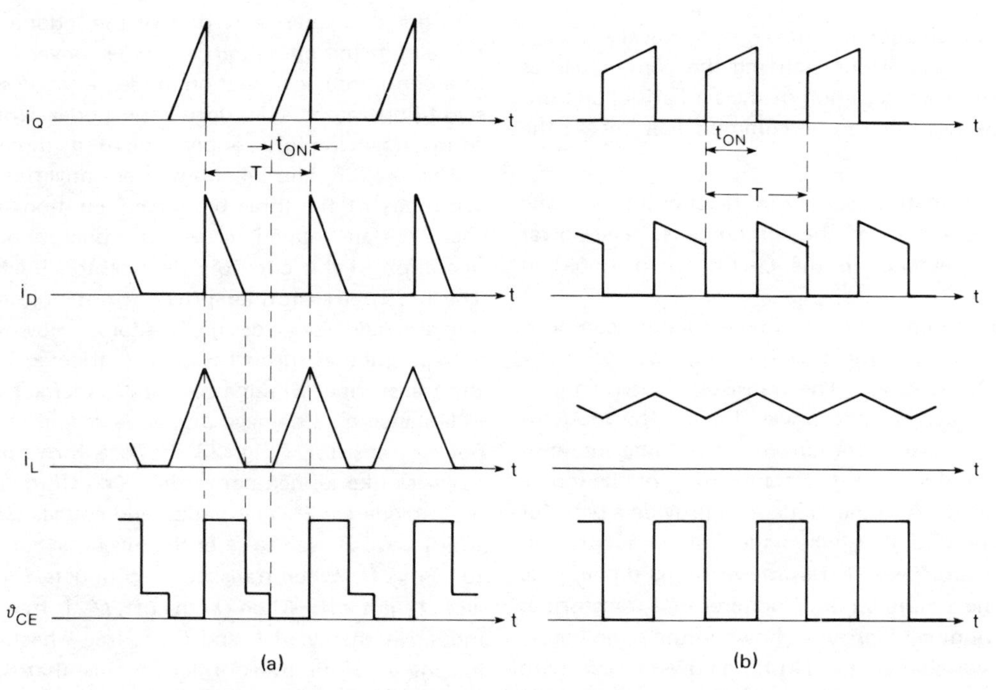

Fig. 21. Operating waveforms for flyback configuration. (a) Discontinuous mode. (b) Continuous mode.

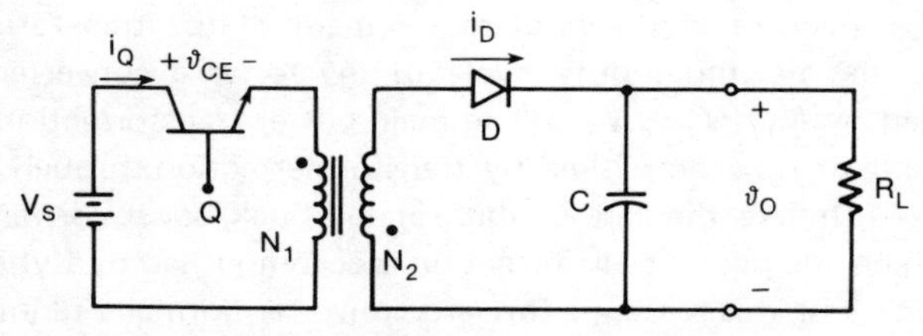

Fig. 22. Two-winding buck-boost converter.

Fig. 24 shows a Cuk converter, after the name of the inventor. The key difference between this converter and the rest of the family, from the operation point of view, is that a capacitor, rather than an inductor, is used for energy storage and transfer to accomplish power transformation. From this point of view, the Cuk converter is a capacitive energy flyback converter. In fact, the Cuk converter and the buck/boost converter are electrical duals of each other. The point-connected transistor-

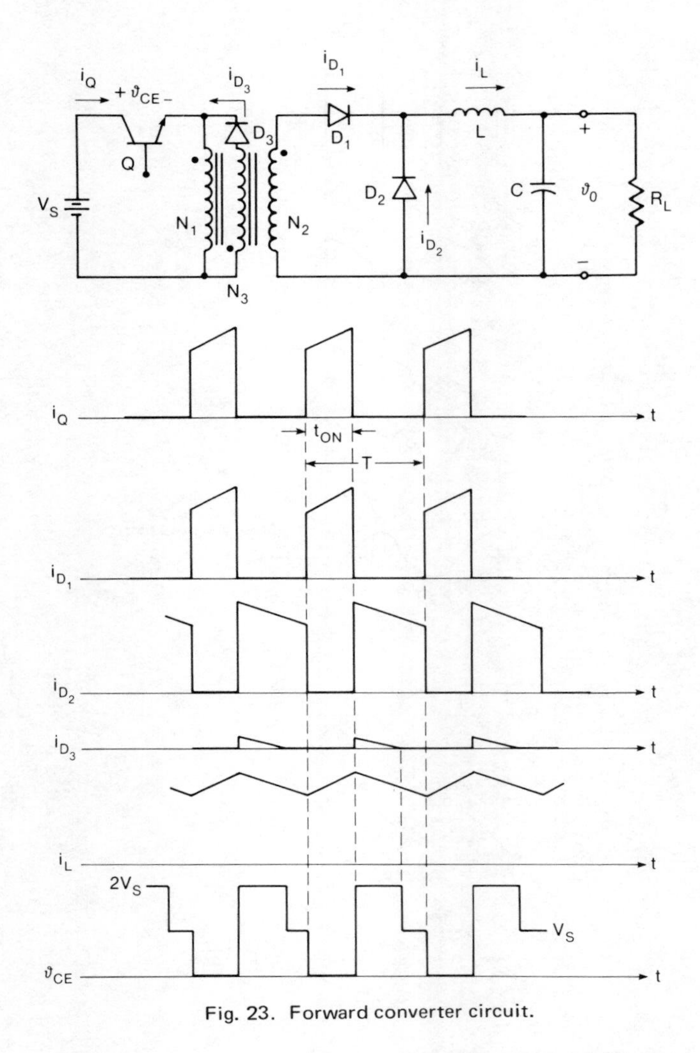

Fig. 23. Forward converter circuit.

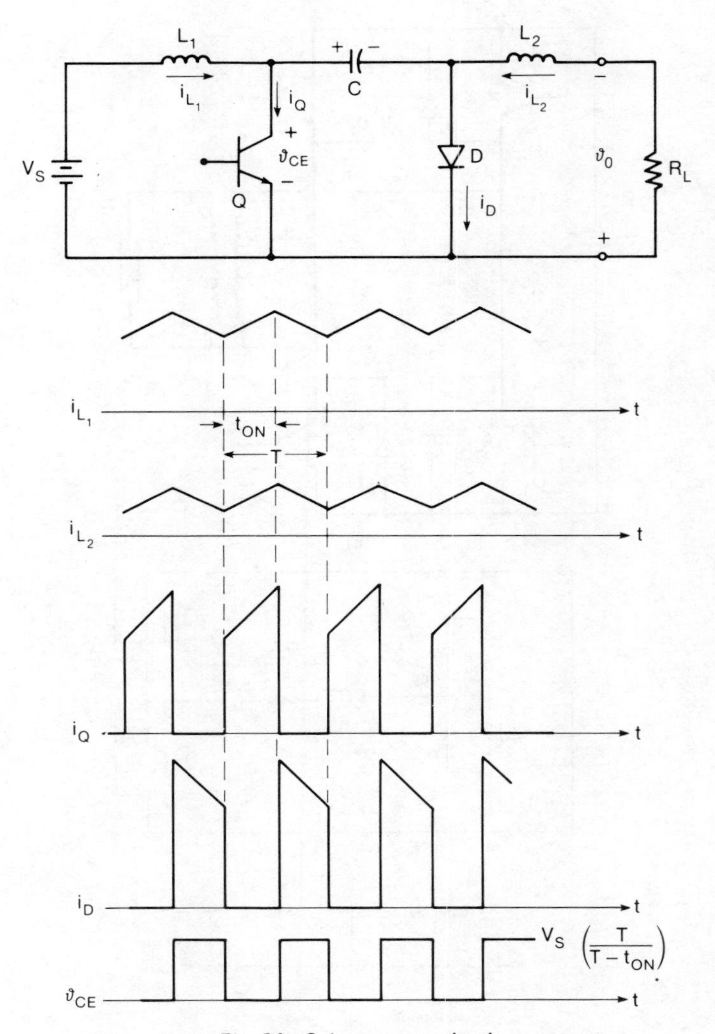

Fig. 24. Cuk converter circuit.

inductor-diode combination is replaced by the dual serially connected diode-capacitor-transistor combination. A unique feature of the Cuk converter is that both the input and the output current are nonpulsating. The input or the output current ripple can be made to approach zero if the inductors L_1 and L_2 are properly coupled.

B. Bridge Configuration

The basic bridge circuit configuration is shown in Fig. 25 in which four power transistors are used. Depending upon the timing of base drive waveforms, the circuit can be used for dc-to-ac inversion applications or dc-to-dc conversion applications. When the circuit is used for inversion applications, it is called a full bridge inverter, and when the circuit is used for conversion applications, it is called four-quadrant chopper. The name "four quadrant" refers to a dc motor drive application in which the chopper-motor system is capable of a motoring mode, a regenerative mode, a reverse motoring mode and a reverse regenerative mode of operation as depicted in Fig. 25(b). In other words, the chopper is capable of transforming power from source to motor and vice versa in either direction of motor rotation. Notice the four sets of transistor-diode-inductor combinations in this circuit. An antiparallel diode should not be confused with the free-wheeling diode. For example, in Fig. 25(a), Q_1, D_3 and transformer form the point-connected combination as mentioned earlier. D_3, therefore, serves as the free-wheeling diode for Q_1. In the meanwhile, D_3 also serves as the antiparallel diode for Q_3 for conducting reactive current. The combination of four sets of point-connected *Q-D-T* and the four sets of bidirectional current switches makes the circuit versatile.

Except for rare case of discontinuous mode of operation, the circuit output V_0 can be considered as a voltage source once the base drive waveforms are determined. Four basic output voltage waveforms are shown in Fig. 26(a), (b), and (c) in which (a) and (b) are for ac applications and (c) is for the dc motor drive application. In the figure, the current waveforms for different loads are sketched with indication of the conducting semiconductors in a complete cycle. Notice the current commutation sequence for each case. The sequence of current commutation makes a great deal of difference on the electrical stress of the various power devices. For example, in the cases of the *RL* load and the leading *RLC* load of (a) and (b), the transistors are always turned on at zero current and are turned off at high current and the diodes are just the opposite. Therefore, the transistors are subjected to larger turn-off stress than the diodes. The problems associated with large current spikes due to diode slow reverse recovery as described

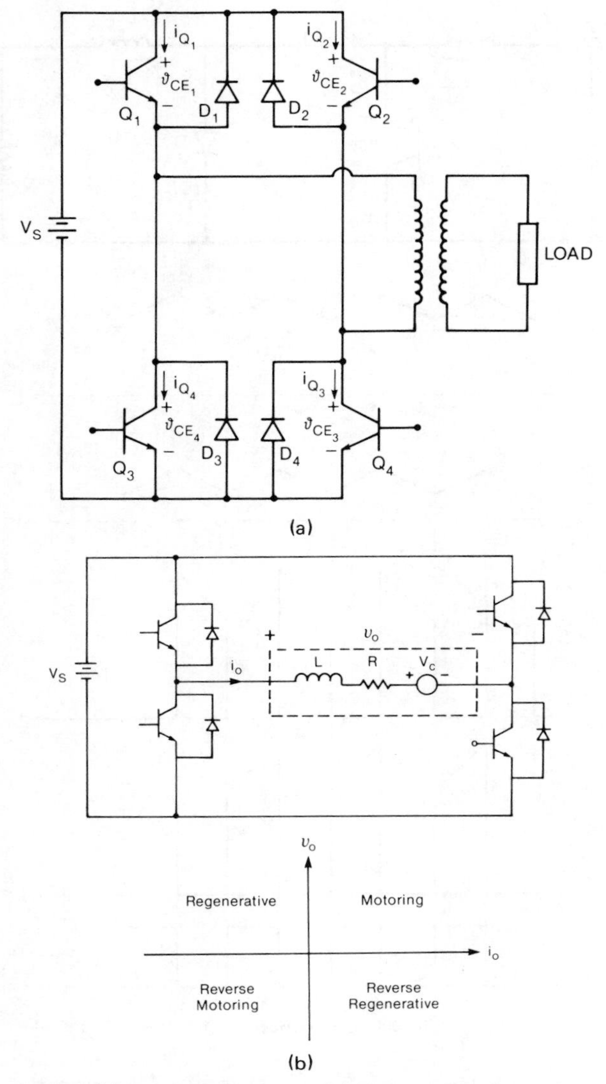

Fig. 25. (a) Full bridge inverter. (b) Four-quadrant chopper.

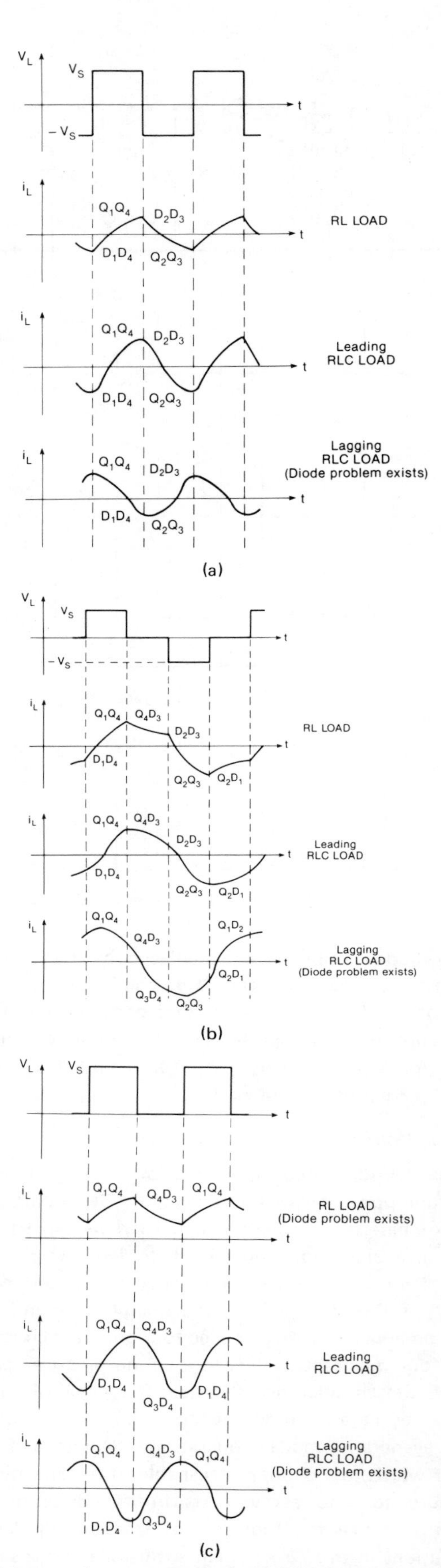

Fig. 26. Voltage and current waveforms of full bridge inverter and four quadrant chopper. (a) Square wave. (b) Quasi-square wave. (c) Unipolar square wave.

earlier, therefore, do not exist in these cases. For a lagging *RLC* load, the situation is just the opposite. The transistors are turned off at zero current and diodes are turned off at high current. Therefore, the reverse recovery characteristics of the diode is of importance in this case.

There are several disadvantages of using a bridge configuration. One of the disadvantages is that isolated base drive circuits are required. Proper care must also be taken in the base drive to ensure that the two transistors on the same totem pole are never turned on at the same time to avoid a catastrophic short circuit. Another disadvantage is that there is a tendency in the inverter application, for the transformer core flux to "walk" toward one-sided saturation which may cause catastrophic current rise in the device. The problem of flux walking is caused by an unbalanced volt-second across the transformer primary winding, due either to an asymmetrical base drive or to a variation in the device conduction drop and storage time. In a practical inverter circuit, this problem is solved by adding a dc blocking capacitor in series with the transformer primary winding or by electronics control to limit the extent of "walking."

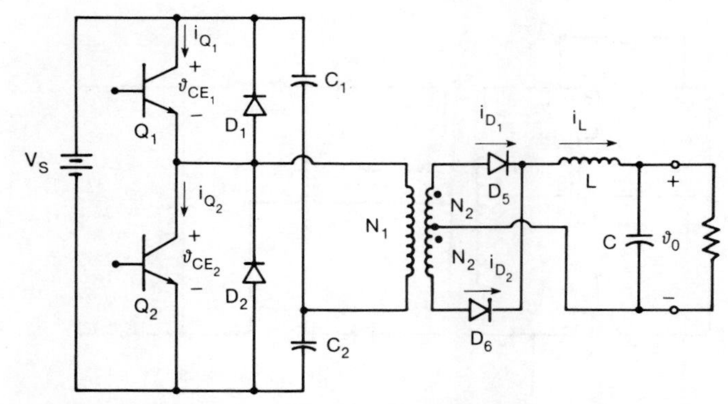

Fig. 27. Full bridge dc-to-dc converter.

Full Bridge dc-to-dc Converter: Fig. 27 shows a dc-to-dc converter in which a bridge inverter converts dc to ac and the transformer-rectifier-filter converts ac back to dc. The voltage transformation is achieved by the transformer turns ratio and the transistor duty cycle control. The operation and the features of the circuit are similar to the full bridge inverter described earlier and are not repeated here.

Three-Phase Bridge Inverter: Fig. 28 shows a dc to three phase ac bridge inverter that is commonly used for ac motor drive applications. Fig. 28(b) shows the base drive waveforms to achieve a three-phase output voltage. This drive scheme is normally referred to as "six-step" drive for motor control application. Notice that the transistors switch at the same frequency as the motor phase voltage. In some applications where lower harmonic distortion is required, the output voltage waveforms are pulse width modulated (PWM) as shown in Fig. 28(c). To have significant harmonics reduction, the transistor switching frequency is normally at least ten times the phase voltage frequency. Thus, from the circuit point of view, the PWM drive scheme impresses much larger stress on the semiconductor devices than the six-step scheme. From the motor point of view, however, the PWM scheme is preferred because of smoother rotation and less eddy current power losses. Generally speaking, for very low speed or servo-applications, the PWM scheme is preferred for the reasons of less motor cogging and faster dynamic response.

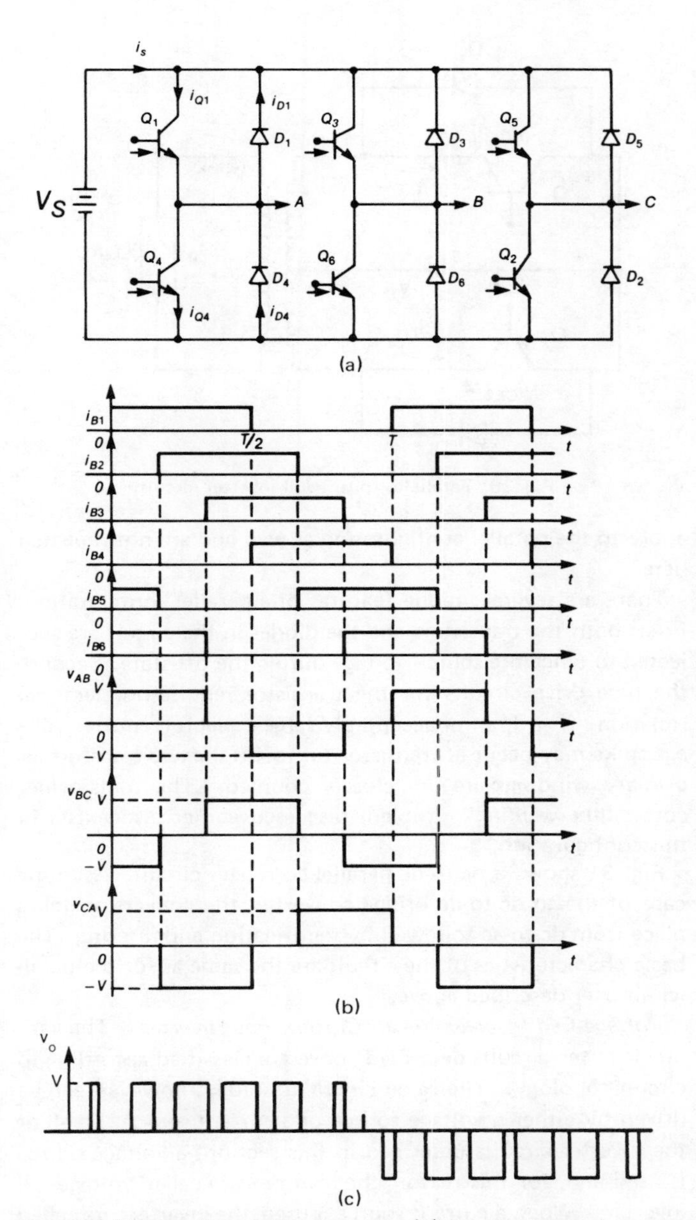

Fig. 28. (a) Three-phase bridge inverter. (b) Base drive waveforms for six-step output voltage. (c) Pulse-width modulated output voltage.

Half Bridge Inverter: Fig. 29 shows a half bridge inverter circuit in which two transistors and two capacitors are used to form a bridge. From the circuit performance point of view, the main difference between the half bridge and the full bridge inverter is that the problem associated with the transformer core "flux walking" in the full bridge inverter is alleviated in the half bridge inverter. The two capacitors tend to balance out the volt-second of the transformer primary. However, a half bridge circuit is functionally less versatile. The output voltage is limited a to square waveform at half the magnitude of the source voltage.

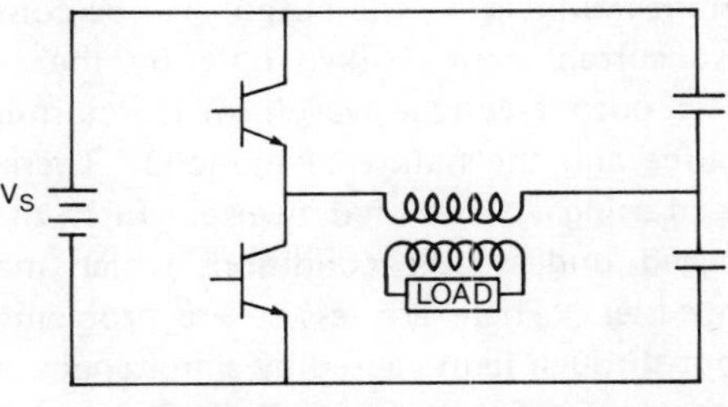

Fig. 29. Half bridge inverter circuit.

C. Parallel or Push–Pull Configuration

This type of power circuit uses two power transistors that are alternately turned on. Fig. 30 shows the basic inverter circuit diagram. Notice in the circuit the point-connected combination of (Q_1, D_2, T) and (Q_2, D_1, T). During the time when the transistor conducts, power flows from source to load and during the time when diode conducts, power flows from load to source. As in the case of the bridge configuration, the output voltage of the parallel configuration can be considered as a voltage source. The current waveforms and the current commutation sequence depend upon the nature of the load. The discussions, given in the section on the bridge configuration, regarding device electrical stress for different load conditions

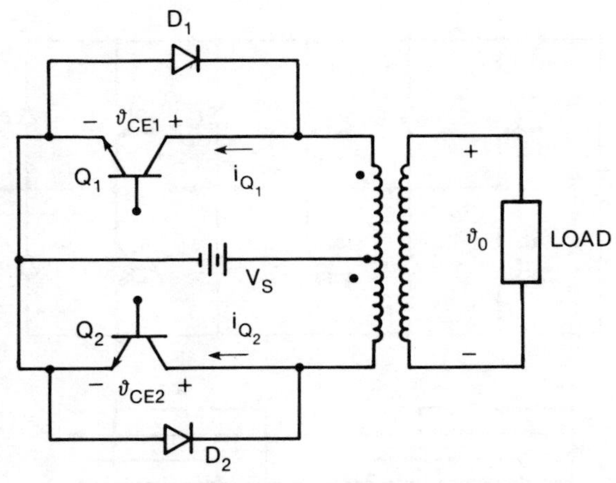

Fig. 30. Parallel or push-pull inverter circuit.

apply to the parallel configuration as well and are not repeated here.

There are several unique features of a parallel configuration. First, both the transistors and the diodes in the circuit are subjected to twice the source voltage during the off-state. Second, the base drive circuits for the transistor require no electrical isolation. Third, an unacceptably large collector-emitter voltage spike may occur at transistor turnoff if the two transformer primary windings are not closely coupled. The transformer core "flux walking" problem discussed earlier also exists in this configuration.

Fig. 31 shows a dc-to-dc parallel converter circuit. As in the case of the to dc-to-dc bridge converter, the conversion takes place from dc-to-ac followed by rectification and filtering. The basic characteristics of the circuit are the same as for the parallel inverter described above.

Voltage-Fed Inverter Versus Current-Fed Inverter: The various inverter circuits described above are classified according to circuit topology. The same circuit topology, however, can be driven by either a voltage source or a current source. In all of the inverter circuits described in this section, a voltage source is used and, for this reason, the inverters are called voltage-fed inverters. When a current source is used, the inverters are called current-fed inverters. Normally, a current source is obtained by putting a large inductance in series with a voltage source.

In a current-fed inverter, the output can be considered as a square-wave current source fixed only by the gating waveforms. The output voltage waveform is determined by the current source and the nature of the load. There are several advantages of using a current-fed inverter. First, the power circuit is rugged under fault conditions which makes circuit protection easier. There are less severe problems associated with a shoot-through fault caused by simultaneous conduction of the two transistors on the same "totem pole." Second, antiparallel diodes are not needed in a circuit driving a relatively constant load. However, the dc link inductor is bulky and expensive. The dynamic response of the inverter circuit also suffers because of the large inductor. The choice between voltage-fed and current-fed inverters depends a lot on the nature of the load. Generally speaking, if the load represents a low impedance or high power factor, the current-fed inverter is preferred. Otherwise, the voltage-fed inverter operates more satisfactorily.

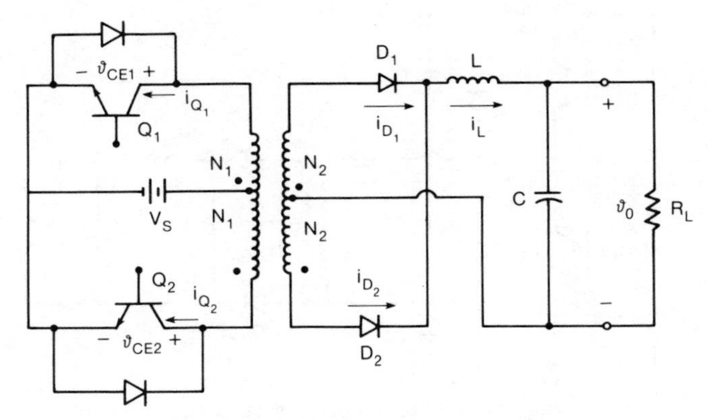

Fig. 31. Parallel dc-to-dc converter circuit.

Resonant Inverters and Converters: When an inverter circuit is used to drive a resonant load, the transistor switching power loss is minimal. As an example, for a voltage-fed inverter the output current waveform is nearly sinusoidal if the load circuit is in resonance with the fundamental voltage driving frequency. The transistors in the inverter circuit turn on and turn off at nearly zero current level and the switching stress is small. In the case of current-fed inverters, the transistors turn on and turn off at a fixed current level and at nearly zero voltage. In both cases, the switching stress in the devices is much reduced as compared with other load conditions.

Resonant inverters and converters refer to inverters or converters operating under resonant condition as described above. The load itself may not be a resonant circuit, but resonant operation can be achieved by placing reactive elements in series with or in parallel with the load.

Resonant operation has traditionally been used in SCR inverter circuits to provide load commutation. Recently, however, it has attracted much attention from transistor users especially for very high frequency applications. Besides the advantage of low power dissipation, electromagnetic noise generation is also much reduced. The disadvantage of the resonant inverters is that the operation is load dependent. This complicates the circuit design. It is expected that more literature will be available in this area in the future.

Power Circuit Applications

Table III summarizes the major features and the applications for the various power circuits described in this section. The choice of the power circuit for a particular application depends upon many considerations such as output power level, circuit complexity, drive requirements, component availability, electromagnetic compatibility (EMC) and cost, etc. The applications listed in Table III and the discussions to follow are for general reference only and should not be taken as rigid rules.

In the discussions below, three application areas will be focused on: A. *Switching Power Supplies*, B. *Motor Drives*, and C. *Consumer and Industrial Applications.* These three areas cover most of the power device applications.

A. Switching Power Supplies

Energy efficiency and small physical size are the two main attributes of a switching power supply. In this application, switching frequency is normally between 20 kHz and 200 kHz.

TABLE III
POWER CIRCUITS AND APPLICATIONS

Configuration	Power Circuit	Features	Applications
FLYBACK CONFIGURATION	DC-DC CONVERTER (Buck, Boost, Buck/Boost)	• Circuit is simple • Large input and/or output current ripple	• Low to medium power regulated power supply • Battery charger
	ISOLATED BUCK/BOOST CONVERTER	• Circuit is simple • Large input and output current ripple • Requires good winding coupling between the transformer primary and secondary windings	• DC motor drive application (Buck configuration) • Automotive power supplies
	CUK CONVERTER	• Circuit is simple • Very small input or output current ripple is possible • Requires a capacitor that conducts large AC current	• Low power regulated supply • Battery charger
	FORWARD CONVERTER	• Circuit is simple • Transformer is smaller than other Flyback types • It requires good coupling between transformer primary and reset windings	• Medium power DC supply • Battery Charger
PARALLEL CONFIGURATION	PARALLEL INVERTER	• Transistor rating must be larger than twice the source voltage • Need a center-tap transformer with very good coupling between the two primaries	• Electronic fluorescent lamp ballast
	PARALLEL CONVERTER	• Base drive needs no isolation • There is a possibility of transformer core walking • Proper care must be taken to avoid simultaneous conduction of the two transistors	• Isolated regulated power supply
BRIDGE CONFIGURATION	FULL BRIDGE CHOPPER (4 Quadrant Chopper)	• Transformer design is simple • Transistor is subjected to source voltage during off state	• Four-quadrant DC motor drive
	FULL BRIDGE DC-DC CONVERTER	• Requires isolated base drives	• High power DC supply
	SINGLE PHASE BRIDGE INVERTER	• Proper care must be taken to avoid simultaneous conduction to transistors on the same leg.	• Induction heating and welding • Switching Amplifier
	THREE PHASE BRIDGE INVERTER	• There is a possibility of transformer core flux walking due to unsymmetry of volt-sec.	• Induction motor synchronous motor and DC brushless motor drives
	HALF BRIDGE CONVERTER	• Same as the features listed under Full Bridge configuration except the last one.	• DC power supply
	HALF BRIDGE INVERTER	• Transformer core flux walking is avoided • It requires two capacitors that can handle large DC current	• Induction Heating and welding • Switching Amplifier

Operation at 500 kHz or higher is possible with MOSFET devices and in fact is the trend of the future.

All the dc-to-dc converters described in this section can be used for switching power supply applications. The flyback configuration is normally employed for low power applications typically below 500 W. The parallel configuration is normally suitable for low voltage applications because the transistors in the circuit must be rated as more than twice the input voltage. For applications at over 1 kW, the bridge configuration is typically the choice because power losses can be shared by four devices. Recently, converters operating in the resonant mode have received more and more attention for power supply applications. With this operation, the transistors are turned on and turned off at zero or low current levels. This reduces transistor switching losses and electromagnetic noise generation. Resonant converters operating above 500 kHz or higher should become more popular in the future.

B. Motor Drive Applications

In most motor drive applications, device switching frequency is below 5 kHz. Conduction power losses and device surge current capability are of major concern. The devices to be used are often sized according to a locked rotor conditon, or a motor acceleration or deacceleration condition. The device surge current requirement is normally much larger than the continuous current requirement. Devices with large surge-current to continuous-current ratio, such as the GTO and the MOS-IGT, normally have the advantages.

For high power applications, power Darlingtons and GTO's are used. Up to this writing, Darlington inverters of 50 kVA and GTO inverters of 300 kVA have been reported. For application below 200 V, MOSFET's may be considered because of their low conduction drop and drive simplicity. In the future, the MOS-IGT should find wide acceptance for motor drive applications with input voltage between 200 V and 1200 V.

DC Motor Drives: Chopper circuits are normally employed for dc motor drives. Depending upon the requirements, single quadrant or two-quadrant or four quadrant choppers may be used. The two quadrant chopper is suitable for transportation drives since it provides both motoring and regenerative capability. For servo-applications, a four-quadrant chopper may be used.

AC Motor Drives: The three phase bridge inverter is commonly used for ac motor drive applications, including induction motor drives, synchronous motor drives, and brushless dc motor (or electronically commutated motor) drives. To avoid stator saturation, a constant ratio of motor voltage to frequency must be maintained. The control of ac voltage is accomplished by the inverter internal switching control or by a chopper circuit preceding the inverter. In another control strategy, a current loop feedback is used to control the magnitude of motor current instead of the voltage. In any event, there are basically two drive schemes for ac motor speed control—a six-step drive and a pulse-width modulated drive. In a six-step drive, there are six pronounced steps in the phase current of the motor and the power transistors switch at the same frequency as the motor phase current. In a PWM drive, the transistors switch at a much higher frequency, normally ten times the motor current frequency, and the resultant motor current is much closer to a pure sinusoidal current. Because of a nearly sinusoidal waveform, the PWM drive has less motor cogging, especially at very low speed. Motor dynamic response in servoapplications also improves in PWM drives because of the much higher sampling rate. However, the device switching power losses are much larger in PWM drives. This should be taken into account in the choice of power transistors, diodes and the snubber. Recent progress made in both power semiconductor devices and microelectronics technology really makes ac drive technology realistically attractive.

C. Consumer and Industrial Applications

In recent years, power semiconductors have found applications in consumer electronics. The electronic fluorescent ballast and the induction cooking range are two examples. Both applications require dc-to-ac inversion but at different power levels. The ballast application normally requires less than 100 W and the induction range requires over 1 kW. Both the bridge inverter and parallel inverter can be used here. Efficiency and controllability are the main attributes of using the power electronics approach. To avoid acoustical noise, transistors are normally switched above the ultrasonic range. In all consumer applications, electromagnetic noise generated by the switching power circuit is of great concern. Noise filtering and product packaging techniques are normally needed to meet the government electromagnetic interference (EMI) regulations. The choice of circuit operating frequency is sometimes dictated by these EMI considerations.

In the industrial area, RF generators, induction heating or welding, and broadcasting transmitters are examples of power transistor applications. Bridge inverters are often used for such applications because of the high power levels involved. In the case of the RF generators, MOSFET and static induction transistors are the only choice because of their high frequency capability. A current-fed inverter operating in resonant mode at frequencies above 100 kHz is a common choice for induction heating or welding applications. In the applications described above, the purpose of using power transistors is either to replace vacuum tubes in high frequency applications or to replace SCR's in low frequency applications.

Further Reading

[1] B. J. Baliga, "Switching lots of watts at high speeds," *IEEE Spectrum*, pp. 42–47, Dec. 1981.

[2] B. R. Pelly, "Power semiconductor devices—A status review," in *Proc. IEEE Int. Semiconductor Power Converter Conf.*, May 1982, pp. 1–19.

[3] P. L. Hower, "A comparison of bipolar and field-effect transistors as power switches," in *Proc. IEEE Industry Appl. Soc. Meeting*, Oct. 1980, pp. 682–688.

[4] M. S. Adler and S. R. Westbrook, "Power semiconductor switching devices—A comparison based upon inductive switching," *IEEE Trans. Electron Devices*, vol. ED-29, pp. 947–952, June 1982.

[5] R. Blanchard, "Status and emerging directions of MOSPOWER technology," in *1983 PCI Proc.*, Apr. 1983, pp. 162–174.

Part I
Power Bipolar Transistors

The papers in this part provide a guide to power bipolar transistor design and provide examples of its applications. The power bipolar transistor has been under developed for a longer period than the other transistors discussed in this book. As a result, a large body of literature has accumulated on the design of these devices as well as their use in a large variety of applications. Due to limitations in space, only a few selected reprints are included here to serve as a guide to the reader in understanding the various trade-off that must be made in designing and applying these devices. To obtain a more detailed understanding of the device physics responsible for these trade-offs, the reader is referred to a book by S. K. Ghandhi entitled *Semiconductor Power Devices*.

The first reprinted paper in this section provides an overview of the trade-offs that must be made between the blocking voltage capability, the maximum forward current, the current gain, the storage time and the chip area. The impact of these parameters upon the forward biased safe-operating-area (SOA) are also discussed. Some of the important considerations defined in this paper are the rapid drop in operating current density with increasing blocking voltage capability, the need to increase device size to reduce storage time and the impact of using built-in emitter ballast resistance to improve the safe-operating-area at the expense of the forward voltage drop during current conduction. The design of the bipolar transistor also requires careful analysis of the emitter and base diffusion profiles in order to obtain a high current gain. Recent modeling and experimental work indicates that band gap narrowing and Auger recombination in the heavily doped emitter region play an important role in determining the emitter injection efficiency. Since the base transport factor in power devices is high, the emitter injection efficiency becomes the dominant factor controlling current gain. These effects are described in the second paper. An important conclusion made by the authors is that an optimum doping level exists for the emitter and it should not be arbitrarily doped at very high concentrations as was commonly done in the past.

As the current ratings of bipolar transistors has grown, it has become increasingly difficult to supply the proportionately higher base drive currents. To address this problem, the Darlington power bipolar transistor was introduced. In these devices, the chip area is partitioned into several segments consisting of the output transistor and the driver transistor. Most power Darlington bipolar transistors are made in the two stage configuration. The third paper discusses the design of these transistor stages. The partitioning of the chip into the driver and output stages is shown to provide the device designer with much greater flexibility in achieving a high output current at a higher current gain than is achievable in single stage bipolar transistors.

The next two papers in this section discuss forward and reverse bias second breakdown in power bipolar transistors. Since second breakdown causes destructive failure in these transistors due to localization of current flow, this is an important design consideration. The occurrence of forward biased second breakdown is an inherent property of all bipolar transistors and arises from the positive temperature coefficient of the collector current. In general, forward biased second breakdown can be prevented by emitter ballasting and making the emitter fingers narrower to decrease current crowding. The mechanism responsible for reverse bias second breakdown is quite different and as explained in the fifth paper is believed to be initiated by avalanche injection. This phenomena occurs when the reverse base drive causes the emitter current to constrict to the center of the emitter fingers. The resulting increase in current density causes the free electron concentration in the collector drift region to rise. When the free electron concentration is sufficiently large, the electric field peaks at the interface between the collector drift region and the substrate and leads to avalanche breakdown at this interface. This process leads to a negative resistance region in the collector characteristics and causes a collapse in the collector voltage. To avoid reverse bias second breakdown it is important to make the emitter fingers narrow to prevent current crowding. In addition, a significant improvement can be achieved by using a drift region whose doping concentration is increased near the substrate interface in order to prevent the electric field from rising at this interface.

As pointed out earlier, reverse bias second breakdown ruggedness of a transistor can be enhanced by modification of device design parameters. However, this is normally accomplished at the expense of device turnoff switching speed. The emitter-open turnoff scheme described in paper six can be used to significantly increase the device switching speed and the reverse bias breakdown ruggedness at the same time. Since the emitter is open during the turnoff process, the emitter current constriction problem is eliminated and reverse bias second breakdown can be prevented.

Snubber circuits are often used in a transistor converter circuit for several purposes. One purpose is to shift the device switching power loss to the snubber circuit itself. The second is to shape the turnoff load line to avoid second breakdown or unacceptable voltage overshoot and the third is to control device *dv/dt*. Paper seven presents a comprehensive overview of snubber circuit classification. Three categories-polarized and unpolarized, resonant and nonreasonant, dissipative and nondissipative are discussed.

The next four papers deal with various power transistor applications including dc and ac motor drive, photovoltaic application, and battery charger. Paper eight discusses a dc chopper for dc motors used in electric vehicles. The chopper is capable of operating in both the motoring mode and the regenerative mode. Paper nine describes a three-phase bridge inverter drive for an ac induction motor also used in electric cars. The inverter is pulse-width modulated to provide a controllable quasi-sinusoidal current to the induction motor. In both cases, the transistors switch one hundred amperes of load current in the low kilohertz range. High frequency switching using the gate turn-off capability of transistors makes the overall system performance superior to a SCR inverter drive. Paper ten describes transistor inverters for photovoltaic power conversion. Several power circuit configurations are described along with discussions of device electrical stress for each configuration. Paper eleven discusses a experimental battery charger employing water-cooled 1000-ampere transistor switches. The charger operates with variable duty cycle, dc to 1 kHz, to determine the best methods of charging large battery cell. Practical problems associated with switching large currents at high speeds with power transistors are also discussed.

In most of the transistor inverter and converter applications, the devices are forced to turn off at high levels of current. Recently, resonant converters have received attention. A distinct characteristic of a resonant converter is that the transistor is turned off at nearly zero current crossing. Because of this, the device turnoff stress is much reduced and circuit electromagnetic interference (EMI) performance improves. Characterization of transistors operating under such conditions is not available now as pointed out in the last paper.

POWER TRANSISTOR PERFORMANCE TRADEOFFS*

P. L. Hower

Westinghouse Electric Corporation

Pittsburgh, Pennsylvania

ABSTRACT

Recent advances in the understanding of bipolar device physics permit the derivation of a number of quantitative relationships which are useful for improving existing designs and also for assessing the feasibility of proposed devices. This paper describes three important tradeoffs which apply to the design and performance of high-voltage transistors in inverter circuits.

INTRODUCTION

When selecting a device for a circuit, it is conventional practice for the circuit engineer to rely on existing data sheets, some preliminary measurements, and in certain cases, additional information from the manufacturer. Ordinarily this procedure works well, particularly for the low-power portions of the circuit.

Problems arise when existing devices do not meet the needs of the circuit engineer. In this situation the conventional approach provides only limited information. In particular, it cannot answer basic questions regarding possible tradeoffs between various characteristics of existing devices or whether a new and larger device must be developed.

To provide these answers it is necessary to have available accurate, but also relatively simple, device models which permit the device designer to relate one performance characteristic to another in terms of "device variables." At present, adequate models exist for describing most of the important portions of the collector characteristic, and to a certain extent the switching performance can also be predicted. In addition, it is possible to predict the forward safe operating area, provided one has adequate knowledge of the effective emitter ballast resistance R_E and the thermal resistance $R_{\theta JC}$.

As a result of this knowledge, it is possible to define certain "tradeoffs" that occur among the various device characteristics. It is the purpose of this paper to describe three distinct tradeoffs which involve the transistor collector characteristic, the storage time, and finally the forward safe operating area or SOA. In describing these tradeoffs we purposely avoid any detailed discussion of device physics. The emphasis is mainly on outlining results that will be useful for the circuit engineer. Nevertheless, it is advisable that the circuit engineer gain some appreciation of those aspects of device physics which do limit circuit performance. For this reason, mention is occasionally made of relevant transistor phenomena which may not be totally familiar to some circuit engineers.

*This research was supported in part by Lewis Research Center, National Aeronautics and Space Administration, Cleveland, Ohio, under Contract NAS 3-18916.

$BV_{CEO}(sus)$, $h_{FE}(I_C,V_{CE})$ TRADEOFF

In most inverter applications the maximum load voltage is directly determined by the transistor sustaining voltage $BV_{CEO}(sus)$, and similarly the maximum load current is determined by the collector current I_C that can be controlled at a desired current gain h_{FE} and collector voltage V_{CE}. These terms can be displayed on the collector characteristic as shown in Fig. 1.

Curve 680489-A

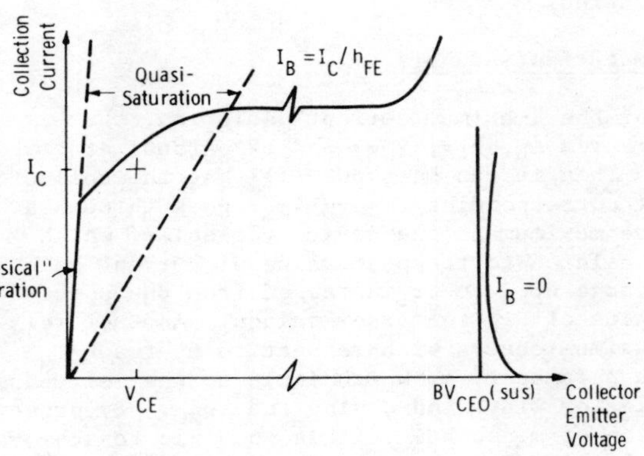

Fig. 1. Transistor collector characteristic, showing $I_B = 0$ and $I_B = I_C/h_{FE}$ curves which meet a given $BV_{CEO}(sus)$ and $h_{FE}(I_C,V_{CE})$ specification.

For high power applications, where device reliability and circuit efficiency are important, it is usually desirable to have $BV_{CEO}(sus)$, I_C, and h_{FE} large and V_{CE} small. In an earlier paper (1),

Reprinted from *IEEE Power Electron. Specialists Conf.*, 1975, pp. 217–223.

it was shown that such desires lead to conflicting requirements on the relevant device variables and the only alternative available is to increase device size while simultaneously making an optimum selection of the impurity profile.

The optimum design concept can be demonstrated by a "DC performance matrix" which is shown in Fig. 2. In this matrix each column corresponds

Characteristic / type of solution	h_{FE}	I_C	V_{CE}	$BV_{CEO}(sus)$	A_E
Minimum area*	S	S	S	S	min
Maximum blocking voltage	S	S	S	max	S
Minimum saturation voltage	S	S	min	S	S
Maximum current	S	max	S	S	S
Maximum gain	max	S	S	S	S

*Chip size will be approximately proportional to emitter area A_E.

Fig. 2. DC performance matrix.

to a particular characteristic and the symbol "S" indicates that the characteristic of that column is specified. For each row, four characteristics are specified and one is left unspecified. The "min" or "max" refers to the behavior of the unspecified characteristic as the device variables traverse their allowed ranges. As an example, we consider the solution of the fourth row in the next section.

Maximum current example

For the "maximum current" solution, the emitter area A_E, h_{FE}, V_{CE}, and $BV_{CEO}(sus)$ are given. Then it can be shown (1) that the collector current corresponding to $h_{FE}(V_{CE})$ goes through a relative maximum as the device variables, which can be related to the peak value of current gain h_{FEO}, cause h_{FEO} to be increased from the specified value of h_{FE} to larger values. Another relative maximum occurs with respect to a "reach-through parameter" m, which includes the influence of collector width and doping and h_{FEO}. By proper selection of m and h_{FEO}, it is possible to achieve an absolute maximum in I_C and at the same time to determine the device variables for this "optimum" design.

Suppose we examine how this procedure can be applied to a practical situation. For example, a particular application may require that the collector current be as large as possible but that a certain package size not be exceeded. This is equivalent to assuming a limit on the emitter area A_E. If the package size is a TO-3, for example, the limit on A_E is approximately 0.2 cm^2, assuming A_E is about one-half the die area. Let us further suppose that the circuit requires $BV_{CEO}(sus) \geq 400$ V, and $h_{FE} \geq 10$ at $V_{CE} = 5$ V. Then the optimization procedure shows that the maximum possible collector current is 13.5 A. The relevant plots are shown in Fig. 3 where it

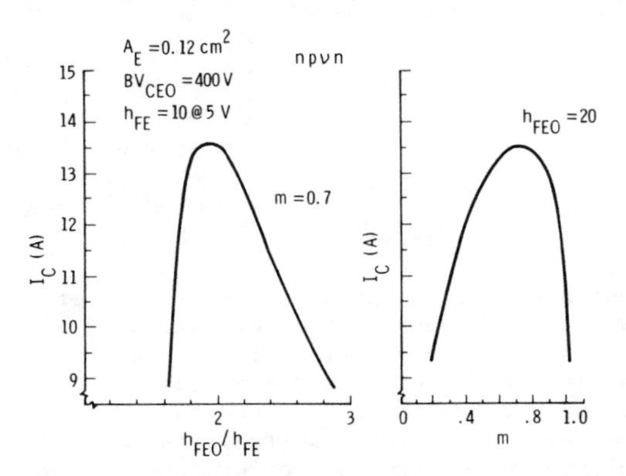

Fig. 3. Example of the maximum collector current solution.

can be seen that I_C reaches a maximum when the design is chosen in a way that makes h_{FEO} approximately two times h_{FE}. Our experience shows that this type of analysis can be quite valuable, particularly during the early stages of circuit design.

Maximum current density vs. sustaining voltage

Another question that frequently arises is— how large does a device have to be if it is to turn on a certain collector current at a particular h_{FE} and block a certain voltage in the "off" state. This question can be answered by carrying out the solution of the first row of the performance matrix. In this solution, it turns out that I_C and A_E always appear together as the ratio I_C/A_E (it is necessary to restrict A_E to the "effective emitter area" in this case. Current crowding effects are considered in the next section). This result means that it is possible to eliminate I_C and calculate a maximum "controllable" current density as a function of $BV_{CEO}(sus)$, h_{FE}, and V_{CE}. The calculation is carried out for the case $h_{FE} = 10$ in Fig. 4.

This figure also shows that the current density falls off quite steeply as BV_{CEO} increases. This means that for a given $I_C BV_{CEO}$ product, the device will become more expensive as BV_{CEO} increases, simply because larger values of A_E will be required to handle the same volt-ampere product.

An extension of these results can be made by noting that for most high-voltage transistors, the product $h_{FE}I_C$ tends to approach a constant when $h_{FE} < h_{FEO}/2$. In other words, the curves of Fig.4

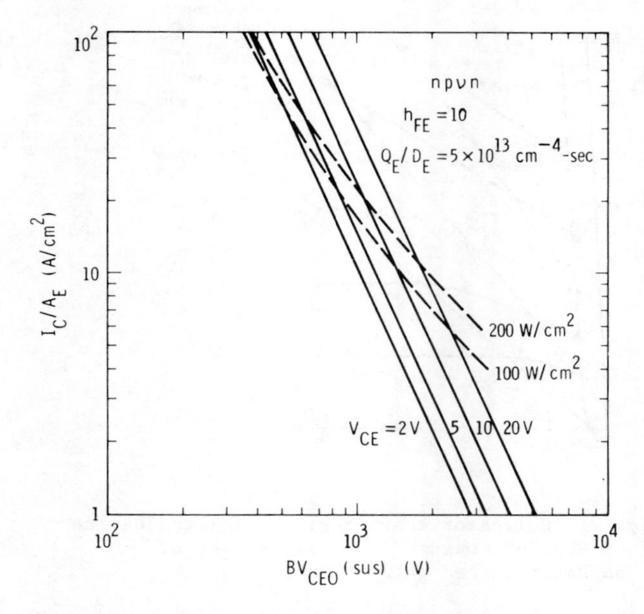

Fig. 4. Maximum controllable current density vs. collector-emitter sustaining voltage.

can be used for values of h_{FE} other than $h_{FE} = 10$. For example, if $BV_{CEO} = 800$ V and $V_{CE} = 5$ V, the maximum I_C/A_E is 25 A/cm^2 with $h_{FE} = 10$. If h_{FE} is reduced, say to $h_{FE} = 5$, then I_C/A_E will increase to 50 A/cm^2.

Current crowding effects

As noted previously, the term A_E refers to "effective emitter area," which will be very nearly equal to the metallurgical emitter area A_{EM} at small values of I_C. As the collector current increases beyond a certain value, A_E will decrease from A_{EM}. As is well known, this reduction occurs because of the transverse voltage drop which develops along the base-emitter junction. For small values of I_C this effect is unimportant, but as I_C increases beyond a certain critical value, the current will concentrate or "crowd" near the edge of the emitter because the local value of base-emitter voltage is greater in this region.

To account for emitter current crowding, we introduce the term L_E, which is related to the total emitter perimeter Z and A_{EM} by the equation

$$L_E = A_{EM}/Z \qquad (1)$$

(In an interdigitated transistor, L_E is approximately one-half the emitter stripe width.) The onset of current crowding occurs when I_C exceeds a "cross-over" current I_{C2} which is given by the approximation

$$I_{C2} = 4.1 \times 10^{-2} \frac{A_{EM}}{L_E^2} \qquad (2)$$

where I_{C2} is in amperes, A_{EM} in cm^2 and L_E in cm.

Equation (2) is the result of a solution of the current crowding problem which accounts for widening of the base into the collector region (2,3). It has been checked for a number of emitter geometries and has been found to accurately predict the onset of current crowding.

The fact that the emitter may be operating under conditions of current crowding does not invalidate Fig. 4; however, it would be more useful to have plots of I_C/A_{EM} rather than I_C/A_E because the first quantity can be related to the actual device size. This modification can be made by making use of the fact that for $I_C > I_{C2}$, A_E decreases as I_C^{-1}. The results of the analysis are shown in Fig. 5 for the case of $h_{FE} = 10$ and $V_{CE} = 5$ V.

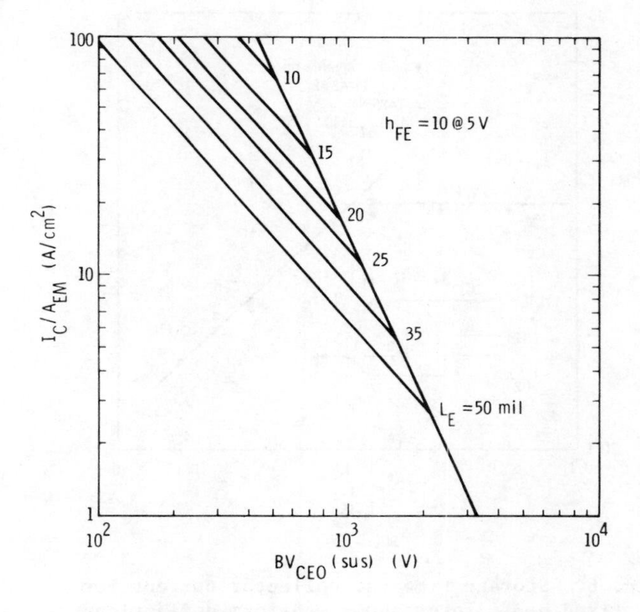

Fig. 5. Maximum controllable current density vs. collector-emitter sustaining voltage taking into account emitter current crowding. L_E is the metallurgical emitter stripe half-width.

An important conclusion here is that high-voltage devices do not require a large degree of interdigitation, which tends to offset the disadvantage of reduced current density I_C/A_E that occurs as BV_{CEO} increases. That is, from the standpoint of meeting the DC requirements, there is no point in making emitter fingers finer than is required by the particular $BV_{CEO}(sus)$.

Let us consider an example. Suppose the specifications are $BV_{CEO}(sus) \geq 700$ V and $h_{FE} \geq 10$ at 5 V. Then from Fig. 5, the maximum I_C/A_E is approximately 34 A/cm^2, and this value will be achieved provided L_E does not exceed 15 mil (0.0381 cm). If, for example, L_E is twice this value, then $A_E < A_{EM}$ and the maximum I_C/A_{EM} is reduced to 17 A/cm^2.

STORAGE TIME vs. OPTIMUM DC PERFORMANCE

The effects of storage time t_s become important when the transistor is driven into the quasi-saturation region. For example, this might occur when it is desired to minimize V_{CE}, and the corresponding conduction loss in the "on" state. Usually, it is desirable to keep t_s less than some specified value, although this characteristic does not usually receive the same degree of emphasis as $BV_{CEO}(sus)$ and $h_{FE}(I_C,V_{CE})$.

Storage time model

For a fixed value of forced gain, h_{FE} ($=I_C/I_{B1}$), the dependence of t_s on I_C typically shows the behavior of Fig. 6. Increasing I_C will

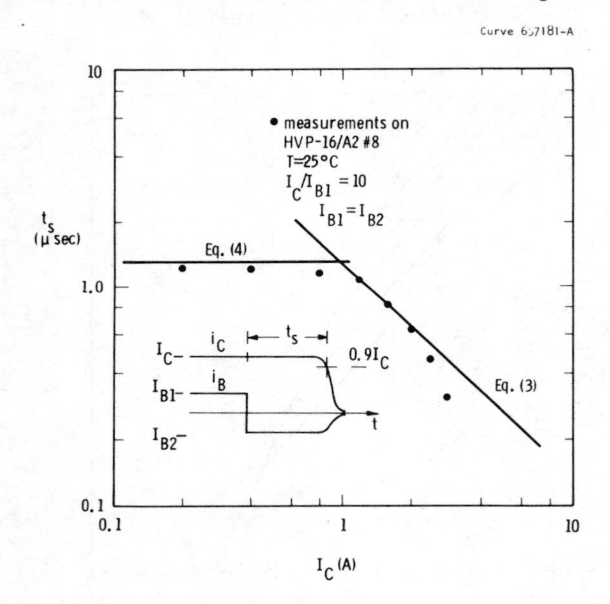

Fig. 6. Storage time vs. collector current for fixed h_{FE}. Inset shows waveform definitions.

shift the device operating point toward the active region and away from the "classical saturation" boundary. Another way of viewing this behavior is that less excess charge is stored in the collector region as I_C increases along a constant h_{FE} locus, which is indicated in Fig. 7.

The solid lines of Fig. 6 are theoretical calculations based on a simple charge control analysis (4). As the I_C, V_{CE} point in the "on"

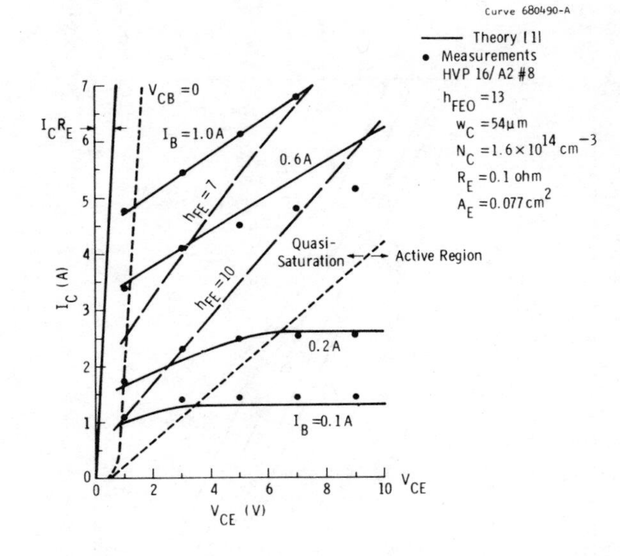

Fig. 7. Collector characteristic for the device of Fig. 6. Dashed lines are curves of constant forced gain.

state moves away from the classical saturation boundary, t_s will begin to decrease. The charge control analysis shows that t_s can be approximated by

$$t_s = \left(1 - \frac{h_{FE}}{h_{FEO}}\right) \frac{Q_E}{D_E} \frac{D_B}{2} \frac{qA_E}{I_C} \ln\left(1 + \frac{I_{B1}}{I_{B2}}\right) \quad (3)$$

in the quasi-saturation region. In this equation I_{B1} and I_{B2} are the forward and reverse base currents and Q_E/D_E is a device parameter that can be treated as a constant together with the base minority carrier diffusion coefficient D_B.

When the transistor is "overdriven," and the operating point approaches the classical saturation boundary, t_s tends to remain constant, as the data of Fig. 6 shows. In this region t_s can be approximated by

$$t_s = \frac{h_{FE} W_C^2}{8D_C} \ln\left(1 + \frac{I_{B1}}{I_{B2}}\right) \quad (4)$$

where W_C is the metallurgical width of the collector and D_C is the diffusion coefficient of electrons in the collector region.

Limitations due to t_s on the optimum DC performance

The important point to be noted in these results is that the same device variables which determine the DC performance are also involved in determining the storage time. This means that by adding a specification on t_s, the number of degrees of freedom on the device variables can be reduced and it may no longer be possible to achieve the same (optimum) DC performance that occurs when t_s is left unspecified.

To demonstrate the effects of adding a t_s specification, we consider first a minimum area solution corresponding to the first row of the DC performance matrix of Fig. 2. It is then shown how this solution must be modified as the t_s limit is changed. Suppose the specified characteristics are:

h_{FE}	I_C	V_{CE}	$BV_{CEO}(sus)$	A_E	
10	10A	3V	600 V	(to be minimized)	(5)

As noted in (1), the equations which describe $BV_{CEO}(sus)$ and $h_{FE}(I_C,V_{CE})$ allow the device variables to have "two degrees of freedom." This means that by reducing the number of variables to two, A_E can be represented as a surface, which, due to the conflicting requirements of h_{FE} and $BV_{CEO}(sus)$ goes through a minimum. An example of this type of solution for the specified values is shown in Fig. 8. The two device variables are the same ones used in conjunction with Fig. 3. The peak current gain is the first variable and it can be related by a constant to the total base doping per unit area. The second variable is the reachthrough factor m, which is also dimensionless.* From Fig. 8, it can be seen that there is a relatively broad minimum in the neighborhood of $h_{FEO} = 25$ and $m = 0.55$.

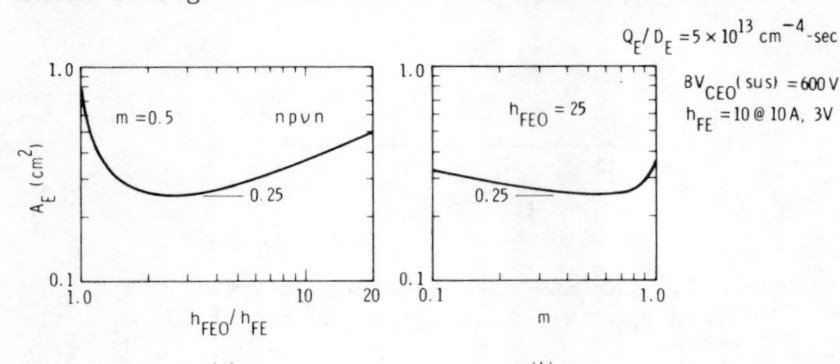

Fig. 8. Minimum area solution. (a) Emitter area vs. h_{FEO}/h_{FE}. (b) Emitter area vs. the reachthrough factor.

Because t_s is a function of the same device variables that determine the DC performance, we can plot constant t_s surfaces using the same coordinates h_{FEO} and m. Constant t_s curves are shown in Fig. 9. If t_s is to be ≤ 1.2 µsec, then A_E, m, and h_{FEO}/h_{FE} values must be chosen to be within the shaded areas corresponding to $t_s = 1.2$ µsec. From Fig. 9, it is clear that it is just barely possible to obtain a minimum area solution when $t_s = 1.2$ µsec. For smaller values of t_s, it is necessary to go to larger A_E than the minimum of 0.25 cm^2. Furthermore, the analysis shows that for $t_s < 0.76$ µsec, there will be no value of A_E that can simultaneously meet the DC requirements and the storage time specification.

*This device variable describes the shape of the collector field profile when $V_{CB} = BV_{CBO}$. For m = 1 the profile is a triangle, corresponding to the onset of reachthrough, while for $m \rightarrow 0$, the profile becomes a rectangle which means reachthrough occurs at very low voltages.

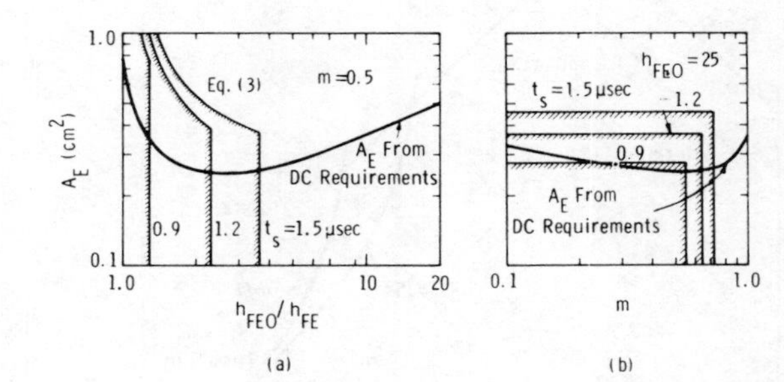

Fig. 9. (a),(b). Superposition of storage time specifications on the minimum area solution. To meet a t_s specification, the (A_E, device variable) point must lie within the shaded region corresponding to the particular value of t_s.

This is a valuable result and one that would be tedious and also expensive to obtain using cut-and-try design procedures. In addition, the tradeoff analysis of this section permits a determination of the price paid, in terms of increased A_E, as t_s is reduced. The analysis also shows that unless some care is taken in drawing up both the DC and storage time specifications, the transistor may be physically unrealizable.

FORWARD SOA, $V_{CE}(sat)$ TRADEOFF

Forward safe operating area (SOA) is important in inverter design because it poses a limit on the upper frequency of operation, where switching losses begin to dominate conduction losses. In addition, the SOA will limit the power handling capability at a fixed frequency, with the actual limit depending on the $i_C(t)$ and $v_{CE}(t)$ waveforms.

An example of the boundaries which determine forward SOA is shown in Fig. 10. The I_{Cmax} boundary is determined mainly by the current handling ability of the bonding wires and package. The constant power boundary is determined by the junction-to-case thermal resistance $R_{\theta JC}$ and the difference between the maximum junction temperature T_{Jmax} and case temperature T_C. An accurate and useful model for predicting $R_{\theta JC}$ has yet to be developed; however, it is known that the thermal resistance is largely determined by the temperature drop that occurs in the package and is only a weak function of the transistor die thickness.

Thermal instability

Beyond the constant power locus, the onset of thermal instability becomes important and

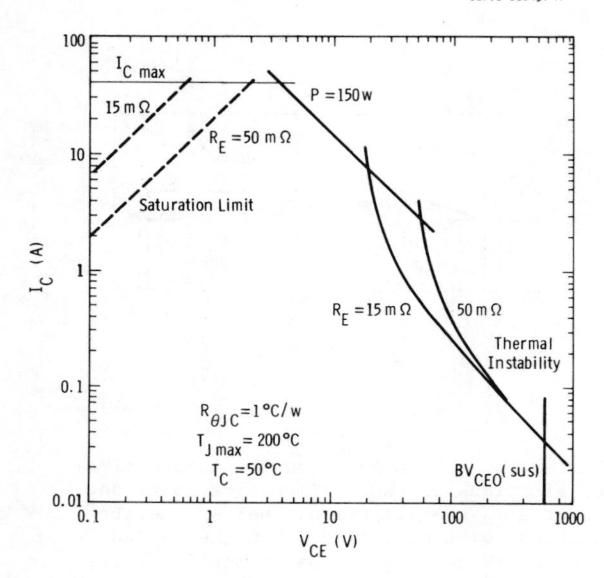

Fig. 10. Example of the boundaries (solid lines) that determine the forward safe operating area, showing the effect of increasing emitter ballast resistance R_E.

determines the safe operating area boundary.* Although this boundary can be predicted as a function of the device variables, the dominant terms are $R_{\theta JC}$ and R_E, the lumped value of series emitter resistance. The effect of increasing R_E is shown in Fig. 10, where the dashed line indicates the additional voltage drop caused by $I_C R_E$.

It is common practice for device designers to devote considerable effort to selecting and modifying mask geometries and impurity profiles so that the desired BV_{CEO}(sus) and $h_{FE}(I_C, V_{CE})$ values are obtained. However, the SOA that results from these design changes is generally accepted as an accomplished fact and not available for adjustment. This need not be the case. By "building-in" emitter ballast resistance and by accepting an increment $I_C R_E$ in the saturation voltage, it is possible to significantly improve the forward SOA. An example of the tradeoff that occurs is shown in Fig. 11.

*We are assuming here that second breakdown will occur immediately following thermal instability and the resultant current localization. In some cases it is known that stable transistor operation can occur following the onset of thermal instability (5). From the standpoint of device reliability, operation in such a mode is undesirable because local device temperatures can greatly exceed T_{Jmax}.

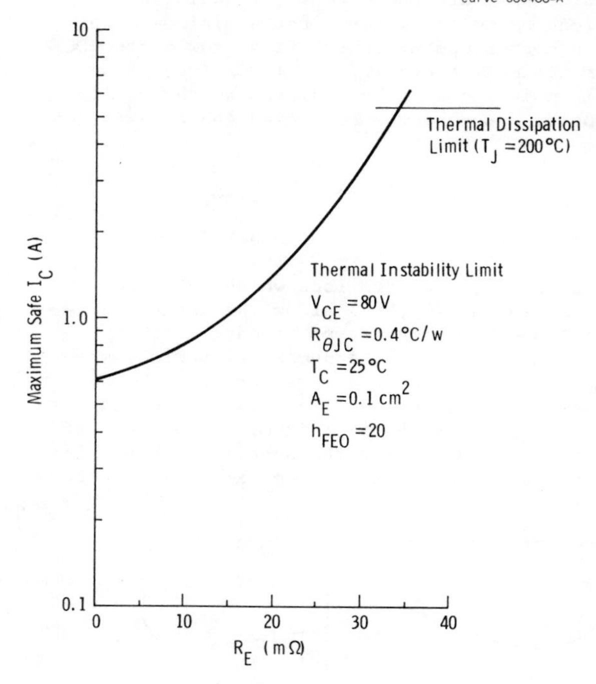

Fig. 11. Maximum safe collector current at fixed V_{CE} as a function of emitter ballast resistance.

Comparison of Forward SOA and the minimum area solution

There is an important link between the forward SOA, V_{CE}(sat) tradeoff and the minimum area solution results of Fig. 5. To demonstrate this, let us consider the case of fixed A_{EM} and an application that requires the volt-ampere product $I_C BV_{CEO}$(sus) to be maximized, with a wide range of values acceptable for BV_{CEO}(sus). Then it is tempting to reduce BV_{CEO}(sus) to as small a value as possible, because the I_C/A_{EM} curve has a steep slope ($I_C \propto BV_{CEO}^{-2.3}$) which will then make $I_C BV_{CEO}$ proportional to $BV_{CEO}^{-1.3}$.

There are several reasons to be concerned about this approach. First of all, to maintain $A_E \simeq A_{EM}$ at the larger values of I_C requires more interdigitation, that is, a smaller value of L_E. A decrease in L_E may necessitate refinements in device technology with consequent undesirable trends in device yields and costs.

If these problems can be surmounted, there is another which is related to the device physics. As BV_{CEO}(sus) is reduced and I_C increased, R_E must be reduced to keep $I_C R_E$ less than the V_{CE}(sat) requirement, which is typically in the 1 to 3 V range. This reduction in R_E will have a strong adverse effect on the thermal instability locus, an effect which is not generally realized.

Thus simultaneous improvements in $I_C BV_{CEO}$ and forward SOA cannot be achieved while maintaining A_{EM} fixed. The only alternatives are to parallel individual transistors (which can pose synchronization problems) or to increase device size, which is the more promising approach for the long term.

CONCLUSION

Although a truly comprehensive bipolar design theory is still not a reality, it is now possible to design devices "on paper" to meet the more important characteristics of high-voltage switching transistors. As demonstrated in this paper, relevant design tradeoffs can be expressed in terms that are meaningful to the circuit engineer. This knowledge can be valuable both in device selection and circuit design, and it is hoped that it will aid in narrowing some of the communication gaps that now exist between device and circuit engineers.

REFERENCES

(1) P. L. Hower, "Optimum design of power transistor switches," IEEE Trans. on Electron Devices, V. ED-20, pp. 426-435, April 1973.

(2) J. Olmstead, et al., "High-level current gain in bipolar transistors," RCA Rev., V. 32, pp. 222-246, 1971.

(3) P. L. Hower, unpublished work.

(4) P. L. Hower, unpublished work.

(5) F. F. Oettinger and S. Rubin, "The use of current gain as an indicator for the formation of hot spots due to current crowding in power transistors," Tenth Annual Proceedings, Reliability Physics 1972, Las Vegas, Nevada, pp. 12-18.

(6) P. L. Hower and P. K. Govil, "Comparison of one- and two-dimensional models of transistor thermal instability," IEEE Trans. on Electron Devices, V. ED-21, pp. 617-623, October 1974.

Factors Limiting Current Gain in Power Transistors

EDWARD JOHN MCGRATH, MEMBER, IEEE, AND DAVID H. NAVON, SENIOR MEMBER, IEEE

***Abstract*—The combined effect of sidewall injection, bandgap narrowing, and Shockley–Hall–Read and Auger recombination in determining emitter efficiency in n-p-n power transistor structures is demonstrated by utilizing a two-dimensional transistor model. The relative importance of each of these effects is calculated as a function of emitter junction depth, emitter surface doping, and injection level. It is shown that in a practical transistor design the reduction in emitter efficiency due to the increased injection of holes into the emitter, resulting from bandgap narrowing caused by heavy doping, is not dominated by the emitter sidewall. Auger recombination is seen to be especially important when bandgap narrowing is present. Enhanced Auger-type recombination is due both to increased minority carrier injection in the emitter as well as current crowding effects. The predictions of the model are compared with results of the measurement of current gain versus current level characteristics on existing devices.**

I. Introduction

THE mechanisms which determine emitter efficiency and therefore the current gain in bipolar transistors have been the subject of a great deal of discussion in the literature. DeMan, Van Overstraeten, and Mertens [1], [2], [3] published a series of papers in which they showed that in devices with a shallow emitter and heavy doping concentrations, bandgap narrowing due to heavy doping could account for a decrease in emitter efficiency with increased emitter doping. Separately, Mock [4] using an alternative formulation also attributed this effect to bandgap narrowing. Mock's formulation yielded quantitatively different results at heavy doping concentrations. The more recent work of Heasell [5] agrees with the results of Mock. New measurements of bandgap narrowing in silicon transistor emitters by Slotboom and De Graff [6] roughly confirm the Mock computations over a limited range of doping levels.

Recent experiments [7], [8] show that recombination in heavily doped silicon can be dominated by the Auger process and Sheng [9] postulated that this would have an important effect on the current gain in deep emitter devices where bandgap narrowing becomes less important. Adler *et al.* [10], using the formulation of Van Overstraeten *et al.*, have investigated the relative effects of Shockley–Hall–Read (SHR) recombination, Auger recombination, and bandgap narrowing. The study is limited to one-dimensional effects. Also this model requires using SHR lifetime parameters in the emitter region which are ap-

Manuscript received March 31, 1977; revised June 23, 1977. This work was supported in part by the National Science Foundation under Grant ENG 75-09334, and in part by the IBM Corporation, Poughkeepsie, NY.

The authors are with the Department of Electrical and Computer Engineering, University of Massachusetts, Amherst, MA 01002.

Reprinted from *IEEE Trans. Electron Devices*, vol. ED-24, pp. 1255–1259, Oct. 1977.

proximately one percent of the lifetime parameters in the base and collector regions in order to correctly predict current gain versus current characteristics.

Mock [4] has suggested that the two-dimensional effects associated with the emitter sidewall have a dominating effect in heavily doped devices both for carrier diffusion and SHR recombination. It should also be noted that the relative importance of SHR and Auger recombination and bandgap narrowing are highly dependent on the values of SHR lifetimes chosen and the bandgap narrowing model used.

This paper reports the results of a two-dimensional transistor model analysis extending the one-dimensional treatment of Adler *et al.* and based on the bandgap narrowing formulation of Mock. No down adjustment of emitter lifetime is made. Instead, the normal variation of carrier lifetime with doping and injection level is used. The relative effects of bandgap narrowing and SHR and Auger recombination are calculated for various device geometries and surface doping concentrations. Finally, the theoretically predicted current gain versus current characteristics are compared with experimentally measured data.

II. Theory

A. Bandgap Narrowing

In a typical bipolar transistor, bandgap narrowing occurs near the heavily doped surface of the emitter region. This narrowing results in a quasi-field tending to pull both electrons and holes into the emitter. The increased flow of holes into the emitter degrades the emitter efficiency and therefore the current gain. The variation of the bandgap narrowing parameters with doping level were derived from Mock [4].

B. Recombination

Recombination in indirect gap semiconductors such as silicon can occur either through SHR and Auger processes. The relative effects of SHR and Auger recombination are expected to vary both with injection level and doping concentration. Since the Auger rate is quadratic with electron concentration, it is expected to be large at heavy doping concentrations and high injection levels. The effective lifetime for SHR recombination increases with injection level while the effective Auger lifetime decreases. It is therefore expected that Auger recombination will become increasingly dominant with increasing injection level.

The SHR lifetime parameters assumed were $\tau_{p0} = 0.35$ μs and $\tau_{n0} = 1.1$ μs.[1] The Auger coefficient used was 2×10^{-31} cm^6 s^{-1} [7].

C. Numerical Solution Method

The technique used to solve the problem is an extension of that of Alwin *et al.* described elsewhere [12]. Briefly, the method used is to simultaneously solve the governing

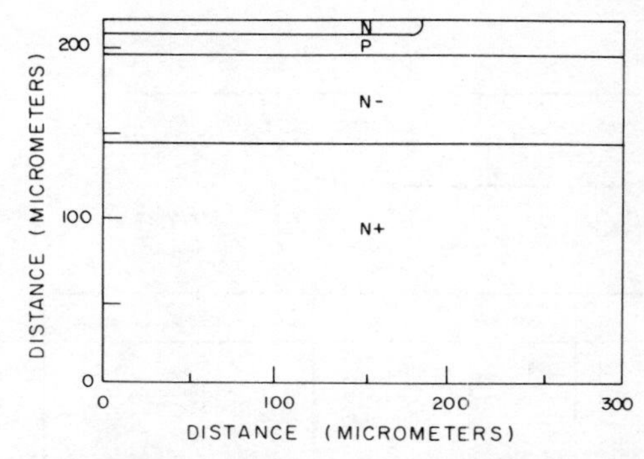

Fig. 1. Cross section of the portion of the transistor structure analyzed.

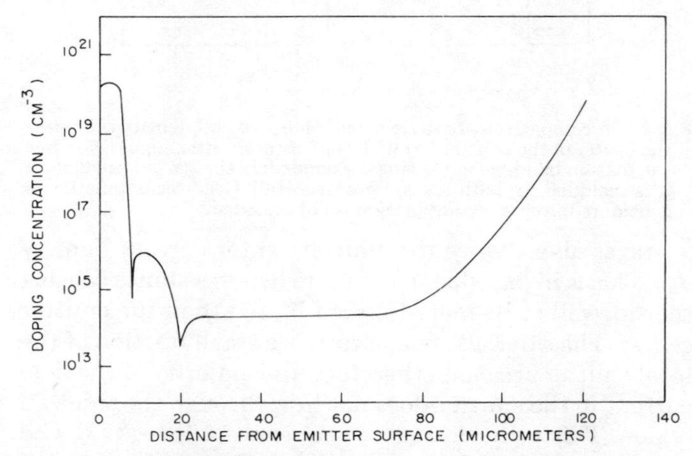

Fig. 2. Distribution of impurities in the bulk of the transistor structure of Fig. 1, from emitter to collector, along the center line of the emitter.

charge carrier flow differential equations in discretized form using a nonuniform grid constructed on a two-dimensional power transistor structure. The grid used contains 28 longitudinal lines by 12 transverse lines. The discretized equations are linearized using Newton's method and the resulting equations are solved in matrix form utilizing the SLOR technique. The dependence of the carrier mobilities on the total impurity ion density, as well as the electrical field, is included in the model [11]. A typical doping profile and device geometry are shown in Figs. 1 and 2.

III. Numerical Results and Analysis

A. Two-Dimensional Model of Emitter Injection Efficiency

In a modern bipolar transistor a major factor in determining current gain is emitter injection efficiency. A measure of this efficiency is the ratio of electron to hole current density at the emitter–base junction. Fig. 3 shows this calculated factor at the center of the emitter (A) and at the sidewall (B), both with and without bandgap narrowing. It is seen that the reduction in emitter efficiency with the introduction of bandgap narrowing is very nearly the same at the sidewall as at the emitter center. Examination of the current densities at the emitter center and sidewall show that the sidewall hole current density J_{hx}

[1] Much of the device data used here was supplied by the IBM Corporation, Poughkeepsie, NY.

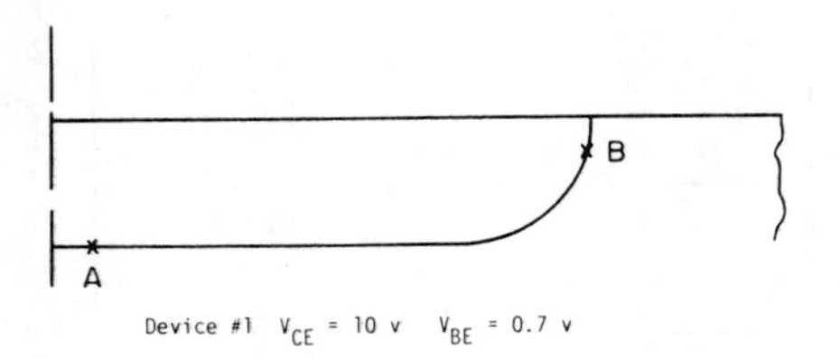

Device #1 $V_{CE} = 10$ v $V_{BE} = 0.7$ v

	With Bandgap Narrowing	Without Bandgap Narrowing
at A	$\frac{J_{ey}}{J_{hy}} = \frac{6.71}{0.051} = 131$	$\frac{J_{ey}}{J_{hy}} = \frac{9.04}{0.032} = 286$
at B	$\frac{J_{ex}}{J_{hx}} = \frac{18.7}{0.237} = 78.9$	$\frac{J_{ex}}{J_{hx}} = \frac{19.2}{0.106} = 181$

Auger recombination not included.

Fig. 3. A comparison of the electron to hole current density ratio near the center of the emitter (A) with that at the emitter edge (B). In one calculation bandgap narrowing is excluded; in the second calculation it is included. In both cases, Shockley–Hall–Read recombination is considered; Auger recombination is not included.

averages about twice the emitter center current density J_{hy}. This is in part due to the finite base resistance causing the sidewall to be more forward biased than the emitter center. The sidewall area, in fact, is a small fraction of the total emitter area and, therefore, the majority of the hole current in the emitter does not flow through the sidewall region. This is in contrast to the results of Mock [4] who sees a much more increased importance of the sidewall when bandgap narrowing is introduced. This difference is probably due to the fact that the emitter–base junction in Mock's device was heavily doped on both sides (see [4, fig. 6]). Also, Mock later showed [4] that the charge neutrality equilibrium assumption used in the calculation of the bandgap narrowing parameters overestimates the bandgap narrowing for heavily doped compensated regions.

It is therefore expected that for power transistor structures, of which the device shown in Figs. 1 and 2 is typical, the importance of the sidewall region in determining emitter efficiency is approximately in proportion to its area.

B. Shockley–Hall–Read and Auger Recombination

Fig. 4 shows the relative effects of SHR and Auger recombination as a function of injection level without including bandgap narrowing effects. As expected, the Auger component becomes increasingly important as the injection level is increased. The Auger component increases from about 1 percent of the SHR component at low injection to about 50 percent at high level injection. Even more dramatic, however, is the case when bandgap narrowing is introduced. In Fig. 5 it is seen that, while at low levels of injection, the Auger component is still about 1 percent of the SHR component; at high level injection, the Auger component is approximately two times the SHR component.

The increased importance of Auger recombination with bandgap narrowing is due to two effects. First, increased

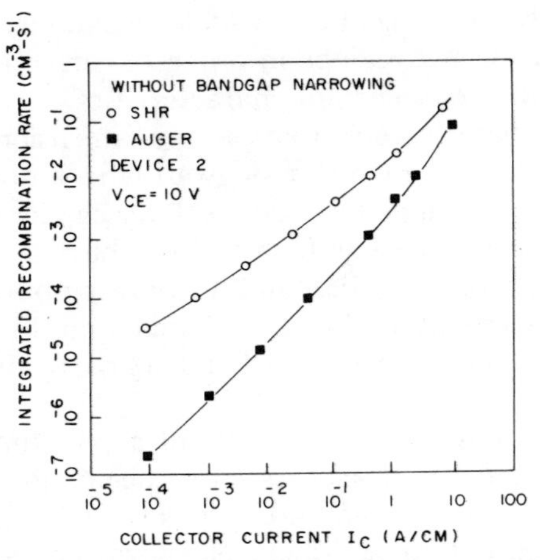

Fig. 4. The relative effects of Shockley–Hall–Read and Auger recombination as a function of a collector current level. The effect of bandgap narrowing is not included.

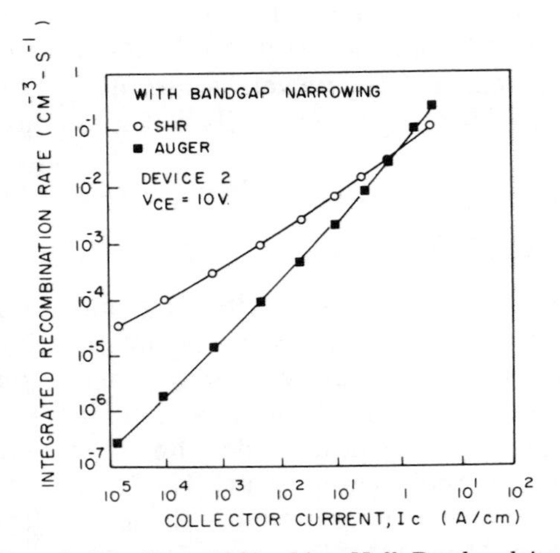

Fig. 5. The relative effects of Shockley–Hall–Read and Auger recombination as a function of collector current with the effect of bandgap narrowing included. Comparison with Fig. 4 illustrates the enhancement of Auger recombination with bandgap narrowing.

injection of holes into the emitter contributes to the Auger rate since the Auger rate is expected to be large in the emitter, where the doping is greatest; this is because the Auger rate is proportional to n^2p. The second effect is a secondary effect. The increase in transverse base current flow due to the introduction of bandgap narrowing causes an increase in current crowding under the emitter corner. At high levels of injection the electron density can get very high in this region and, therefore, the Auger rate becomes very large. The results of the calculation show that for a device with a current crowding factor of four,[2] the "crowding" in the Auger rate is about ten. For example, a device with a wide emitter was modeled which exhibited a current crowding factor of about 70. This crowding effect was magnified in the Auger rate which exhibited a crowding factor of about 2000. The effect of Auger recombination on the current gain can be seen in Fig. 6. The small difference between the top two curves shows the

[2] The crowding factor is defined as the ratio of the current density at the emitter edge to the emitter center. The "crowding" in the Auger rate refers to the ratio of the Auger recombination rate at the emitter edge compared to the emitter center.

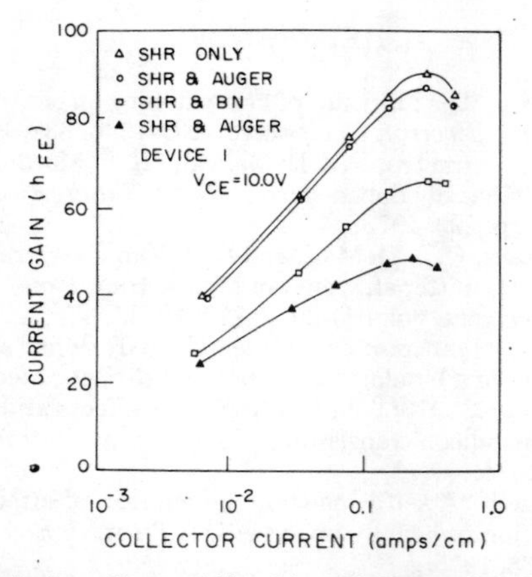

Fig. 6. The relative effects of Shockley–Hall–Read and Auger recombination as well as bandgap narrowing on transistor current gain.

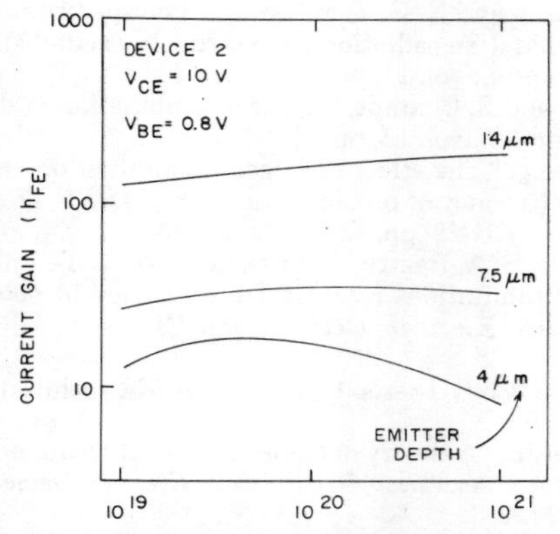

Fig. 7. Transistor current gain versus emitter surface impurity concentration for various emitter depths.

nearly negligible effect of Auger recombination in the absence of bandgap narrowing. The bottom two curves, in contrast, show the very large enhancement in the Auger recombination effect when bandgap narrowing is introduced, especially at high current density when current crowding is a factor. Adler *et al.* [10] reported much less importance of Auger recombination. The difference can be attributed to two factors. Their use of a one-dimensional model ignores the current crowding effect. In addition, their use of very low lifetimes in the emitter region causes the injected holes to recombine via the SHR mechanism before they reach the heavily doped region where Auger recombination is expected to be large.

C. Current Gain Versus Emitter Surface Doping

The traditional method for increasing the emitter efficiency and current gain is to dope the emitter region more heavily. However, it is expected that since bandgap narrowing and Auger recombination reduce the emitter efficiency at high doping concentrations, there would be some doping concentration which would yield an optimum current gain. In Fig. 7, the current gain is plotted as a function of emitter surface doping concentration for three emitter depths, beginning with the geometry of Fig. 1 and leaving the base doping unchanged. It is seen that for the 4-μm depth emitter device, there is a distinct maximum

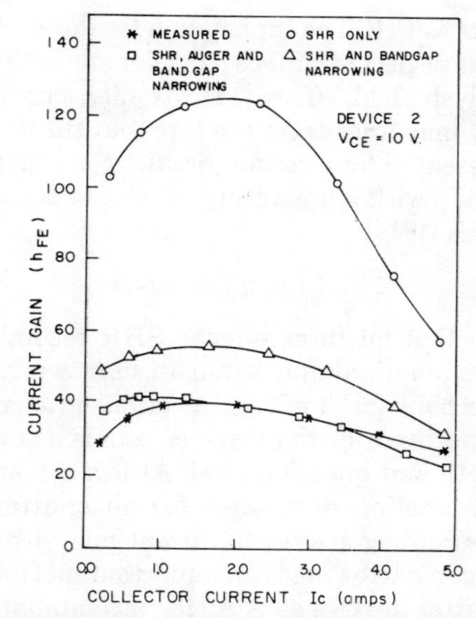

Fig. 8. Comparison of calculated with measured transistor current gain versus collector current.

occurring at about 5×10^{19} cm^{-3}, a very broad maximum at about 1.5×10^{19} cm^{-3} for the 7.5-μm device, and no maximum at all under 10^{21} cm^{-3} for the 14-μm device. These results are consistent with the expectation that bandgap narrowing and Auger recombination effects become less important as the emitter depth is increased; the minority holes recombine via SHR recombination before they reach the heavily doped Auger-dominant bandgap-narrowed region.

The one-dimensional treatment of Adler *et al.* [10] reports very similar results except that correspondence occurs at emitter depths of about 25 percent of those reported here. The difference can be explained by their postulation of a very small lifetime in the emitter region.

D. Measured and Calculated Current Gain Characteristics

Current gain characteristics were measured on a lot of 50 devices for which device geometry and doping profiles are known and are shown in Figs. 1 and 2. Device characteristics were generated for this specific design for three different sets of assumptions: 1) only SHR recombination, 2) SHR and bandgap narrowing, and 3) SHR and Auger recombination and bandgap narrowing. Fig. 8 shows the results of the analysis. It is clear that both bandgap narrowing and Auger recombination must be considered for accurate current gain prediction. The increasing gain with increasing injection level at low currents can be explained by SHR effects dominating while the falloff at high current is due to both Auger recombination and base stretching [12]. A discrepancy is seen between the measured and predicted characteristics at both high and low currents. Smaller SHR lifetime parameters would tend to reduce the gain at low currents where the gain is controlled by SHR recombination but it would also reduce the overall gain causing a lack of agreement with the gain at moderate and high currents as well. The explanation for this discrepancy may lie in the neglect of surface recombination [13] and the Franz–Keldysh field effect [14]. At high currents, the discrepancy may arise from the fact that the isothermal analysis presented here cannot predict the expected rise in current gain with temperature which can occur at high current levels [11].

IV. Conclusions

It is seen that all three effects, SHR recombination, Auger recombination, and bandgap narrowing, are important mechanisms limiting injection efficiency. The relative importance of the three effects is a function of emitter depth and injection level. At low current levels, SHR recombination dominates for all emitter depths considered, resulting in a rise in current gain with injection level. Bandgap narrowing is an important mechanism for shallow emitter devices as is Auger recombination. The Auger recombination rate is shown to be of the same order of magnitude and at certain times larger than SHR recombination rate for moderate and high injection levels. Auger recombination becomes especially important when bandgap narrowing is considered. The enhancement in the Auger rate is due to both increased minority carrier injection into the emitter and current crowding effects under the emitter corner. It is also seen that the importance of the sidewall region in determining emitter efficiency is minimal, except perhaps in the case of heavily doped base regions.

Acknowledgment

Many thanks to Dr. V. C. Alwin, L. J. Turgeon, and D. Spaderna for valuable discussions and suggestions.

References

[1] H. J. DeMan, "The influence of heavy doping on emitter efficiency," *IEEE Trans. Electron Devices*, vol. ED-18, p. 833, 1971.

[2] R. J. Van Overstraeten, H. J. DeMan, and R. P. Mertens, "Transport equations in heavily doped silicon," *IEEE Trans. Electron Devices*, vol. ED-20, p. 290, 1973.

[3] R. P. Mertens, H. J. DeMan, and R. J. Van Overstraeten, "Calculations of the emitter efficiency of bipolar transistors," *IEEE Trans. Electron Devices*, vol. ED-20, p. 772, 1973.

[4] M. S. Mock, "Transport equations in heavily doped silicon and the current gain of a bipolar transistor," *Solid-State Electron.*, vol. 16, pp. 1251–1259, 1973; "On heavy doping effects and the injection efficiency of silicon transistors," *Solid-State Electron.*, vol. 19, pp. 819–824, 1974.

[5] E. L. Heasell, "A self consistent calculation of effective intrinsic concentration in heavily doped silicon," *Int. J. Electron.*, vol. 38, pp. 127–135, 1975.

[6] J. W. Slotboom and H. C. De Graff, "Measurements of bandgap narrowing in Si bipolar transistors," *Solid-State Electron.*, vol. 19, pp. 857–862, 1976.

[7] N. G. Nilsson and K. G. Svantesson, "The spectrum and decay of the recombination radiation from strongly excited silicon," *Solid State Commun.*, vol. 11, pp. 155–159, 1972.

[8] J. D. Beck and R. Conradt, "Auger recombination in silicon," *Solid State Commun.*, vol. 13, pp. 93–97, 1973.

[9] W. W. Sheng, "The effect of Auger recombination on the emitter injection efficiency of bipolar transistors," *IEEE Trans. Electron Devices*, vol. ED-22, pp. 25–27, Jan. 1975.

[10] M. S. Adler, B. A. Beatty, S. Krishna, V. A. K. Temple, and M. L. Torreno, "Limitations on injection efficiency in power devices," *IEEE Trans. Electron Devices*, vol. ED-23, pp. 858–863, Aug. 1976.

[11] V. C. Alwin, Ph.D. thesis, University of Massachusetts, Amherst, June 1976.

[12] C. T. Kirk, Jr., "A theory of transistor cutoff frequency (f_T) falloff at high current densities," *IEEE Trans. Electron Devices*, vol. ED-9, pp. 164–174, 1962.

[13] N. G. Chamberlain and D. T. Roulston, "Determination of minority-carrier lifetimes of bipolar transistors from low current h_{FE} fall-off," *IEEE Trans. Electron Devices*, vol. ED-23, pp. 1346–1347, 1976.

[14] E. S. Rittner, "An improved theory of the silicon p-n junction solar cell," IEDM, Washington, DC, *Tech. Digest*, pp. 69–70, 1976.

On the Proportioning of Chip Area for Multistage Darlington Power Transistors

C. FRANK WHEATLEY, JR., SENIOR MEMBER, IEEE, AND WILLEM G. EINTHOVEN, MEMBER, IEEE

***Abstract*—A Model has been proposed and solved in which all Darlington circuits may be represented to a first order approximation by five constants, one of which may be normalized. Experimental verification has been provided offering excellent agreement with theory. Several orders of magnitude improvement in current-handling ability have been shown to exist for multistage Darlington circuits over conventional discrete transistors. The allocation of chip area for each stage is extensively discussed as a design aid.**

I. Summary

THE DARLINGTON connection [1] of two or more transistors has been an often used method of obtaining very high current gain [2]–[4]. Recently, monolithic Darlington power transistors offering this higher composite gain at costs comparable to discrete power transistors have become commercially available [5], [6]. A greater degree of chip effectiveness may be realized by operating the individual stages at high current densities and increasing the number of stages, thereby increasing the collector current capability an order of magnitude while retaining a moderately high current gain. An alternative is a much smaller chip for a given collector current.

This paper examines the implications of multistage Darlington power transistors. The analysis is made from the circuit viewpoint and assumes that the chip is composed of many isolated and very small discrete power transistors, each identical in characteristics. If each of these transistors is connected in parallel, they form the model of a discrete transistor having a beta versus collector-current characteristic I_C which may be defined by three parameters: maximum beta ($\beta_{\max}$), beta knee (β_{knee}), and merit current (I_{merit}). The many small transistors may also be connected in Darlington circuits comprising two, three, or more stages and analyzed. In addition, the geometry ratio of the stages may be varied, i.e., the output stage may be made smaller than, equal to, or larger than the input stage. The beta versus I_C characteristic may be varied as well. Results of experiments with both discrete models and monolithic circuits are presented and shown to be in good agreement with theoretical results. Conclusions are drawn and a working example is discussed.

II. Derivation

The current gain of a power transistor is plotted as a function of its collector current in Fig. 1 for two different collector voltages. It is well known that a good fit may be approximated by three straight line segments [7]:

$$\beta = \beta_{\max} \tag{1}$$

$$\beta = I_{\text{merit}}/I_C \tag{2}$$

$$\beta = (1/\beta_{\text{knee}})(I_{\text{merit}}/I_C)^2 \tag{3}$$

which represent the zero, negative-one [8], [9], and negative-two slopes, respectively.

$\beta_{\max}$ is evident from Fig. 1; I_{merit} is that collector current where the negative-one sloped line crosses the unity beta value, and β_{knee} is that beta value where the negative-one and negative-two sloped lines intersect. For this discussion, best fit should be attempted in the range of beta varying between 0.5 and 20.

The transistor is assumed to be designed and distributed such that any fractional part behaves as that fraction of the whole. As an example, the curve of Fig. 1 was obtained

Manuscript received November 5, 1975; revised March 1, 1976. This paper was presented at the IEEE Power Specialists Conference, Los Angeles, CA, June 1975.

C. F. Wheatley, Jr., is with the RCA David Sarnoff Research Center, Princeton, NJ.

W. G. Einthoven is with the RCA Solid State Division, Somerville, NJ.

Reprinted from *IEEE Trans. Electron Devices*, vol. ED-23, pp. 870–878, Aug. 1976.

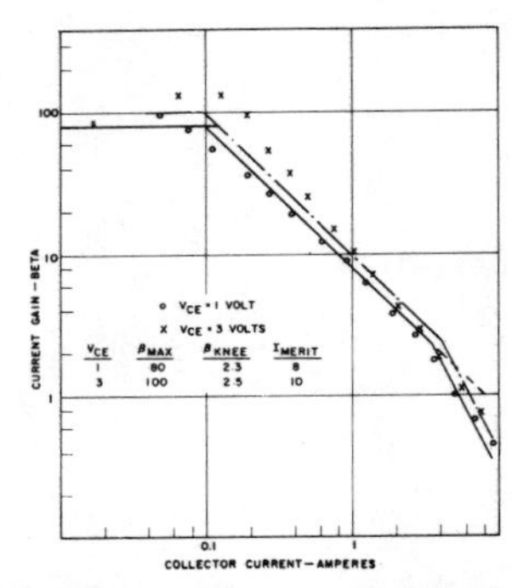

Fig. 1. Current gain of a power transistor versus collector current for two collector voltages and piece-wise linear approximation.

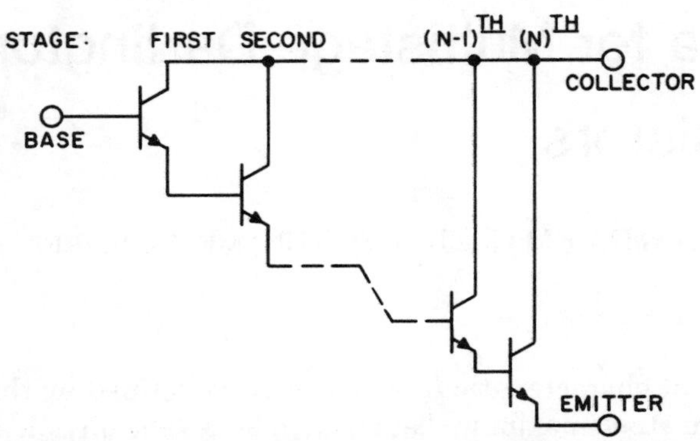

Fig. 2. General multistage Darlington composite structure.

by paralleling seventeen matched discrete transistors, each having a β_{max} of 80, a β_{knee} of 2.3, and an I_{merit} of 0.47 A (one seventeenth of the 8-A combined value).

If the power chip is divided into mini-transistors and reconnected in the general multistage Darlington composite structure of Fig. 2, each stage may be defined and all currents calculated using only the three defining parameters of the chip (β_{max}, β_{knee}, and I_{merit}), the number of stages N, and the proportioning of chip area for each stage. In order to simplify treatment of the infinite possibilities of this proportioning, a restriction has been made in the study that limits the area allocation to a geometric progression; i.e., 1:1:1:1, 1:2:4:8, etc. This limitation permits classification of the area allocation with a single parameter, R_{geom}. That is, the progression takes the general form:

$$(R_{geom})^0:(R_{geom})^1:\cdots:(R_{geom})^{(N-2)}:(R_{geom})^{(N-1)}. \quad (4)$$

Then, the total chip area must equal the sum of the stages. Therefore, each stage area is related to the total chip by a fraction: $m_1, m_2, \cdots m_{N-1}, m_N$ where

$$m_1 = \frac{(R_{geom})^0}{\sum_{i=0}^{i=N-1} (R_{geom})^i} \quad (5)$$

$$m_2 = \frac{(R_{geom})^1}{\sum_{i=0}^{i=N-1} (R_{geom})^i} \quad (6)$$

$$m_N = \frac{(R_{geom})^{N-1}}{\sum_{i=0}^{i=N-1} (R_{geom})^i}. \quad (7)$$

Each stage of the composite transistor may now be defined by β_{max}, β_{knee}, and mI_{merit}, where

$$m = f(N, R_{geom}). \quad (8)$$

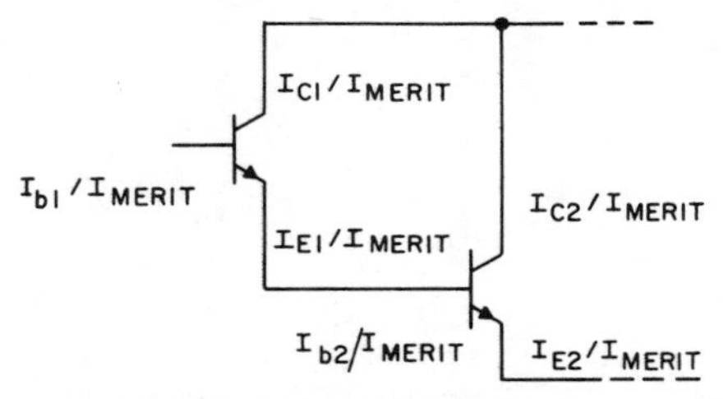

Fig. 3. Arbitrary base current, normalized to I_{merit}, applied to input of composite structure.

An arbitrary base current, normalized to I_{merit} is applied to the input, Fig. 3. Knowing that

$$I_C = \beta I_b \quad (9)$$

and modifying equations (1), (2), and (3) such that I_{merit} is multiplied by m, and combining, base current$_1$ is

$$I_{b1}/I_{merit} \quad (10)$$

collector current$_1$ is the least value of

$$\frac{I_{C1}}{I_{merit}} = (\beta_{max})\left(\frac{I_{b1}}{I_{merit}}\right); \quad (11)$$

$$\frac{I_{C1}}{I_{merit}} = \left[(m_1)\left(\frac{I_{b1}}{I_{merit}}\right)\right]^{1/2} \quad (12)$$

$$\frac{I_{C1}}{I_{merit}} = [(m_1)^2(1/\beta_{knee})(I_{b1}/I_{merit})]^{1/3} \quad (13)$$

and emitter current$_1$ is

$$(I_{E1}/I_{merit}) = (I_{b1}/I_{merit}) + (I_{C1}/I_{merit}). \quad (14)$$

But

$$(I_{E1}/I_{merit}) = (I_{b2}/I_{merit}). \quad (15)$$

Therefore, the currents of the second stage may be found in like manner using subscript 2 in place of 1 for both I_b and m in (11), (12), (13), and (14).

The total collector current of the Darlington is

$$(I_C/I_{merit}) = \frac{(I_{EN} - I_{b1})}{I_{merit}} \quad (16)$$

which produced a total circuit gain of

$$\beta = \left(\frac{I_{EN}}{I_{b1}}\right) - 1. \quad (17)$$

This process is repeated over an appropriate range of I_{b1}/I_{merit}, such that plots of current gain versus I_C may be established for any values of β_{max}, β_{knee}, R_{geom}, N, and I_{merit}.

III. Experimental Results

Before any computer printouts were evaluated, the model was tested. This was accomplished by two differing approaches.

A. Discrete Transistors—Two Stages

Approximately seventy-five small high-voltage transistors were obtained from a common wafer. These were matched for collector current at three points: $\beta = 1$, $\beta = 10$,

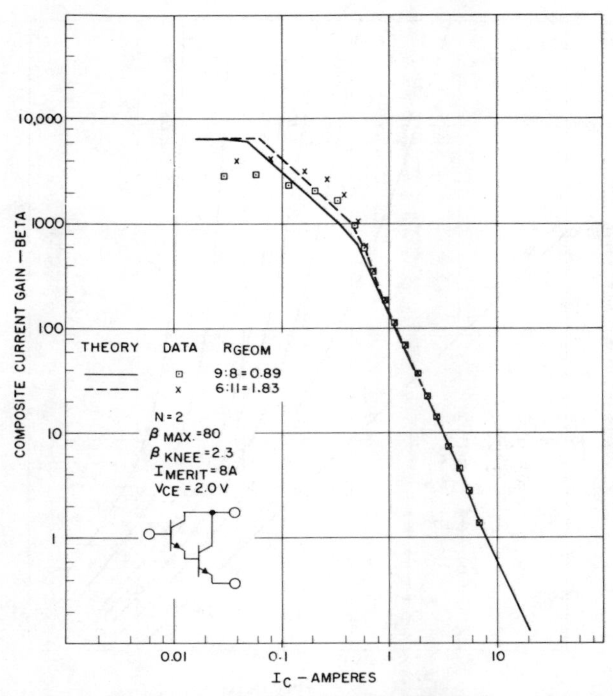

Fig. 4. Composite two-stage Darlington current gain as a function of collector current; R_{geom} = 0.89, and 1.83; discrete transistors.

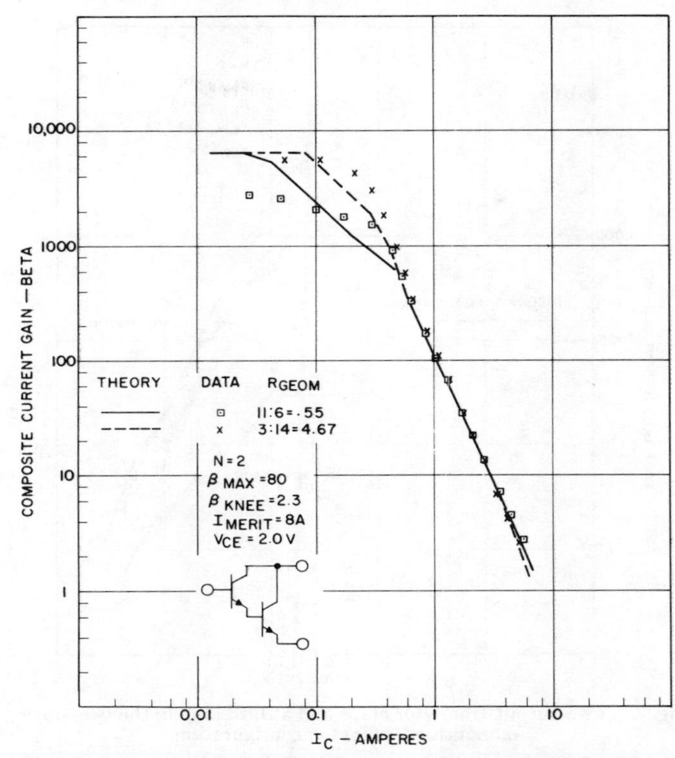

Fig. 5. Composite two-stage Darlington current gain as a function of collector current; R_{geom} = 0.55, and 4.67; discrete transistors.

and β_{max}. Seventeen transistors were obtained with 10-percent match for the first two parameters and a 30-percent match for β_{max}. A plot of beta versus I_C has been presented in Fig. 1 for these parallel-connected transistors. Darlington connections were made ($N = 2$) with driver to output device counts of 9:8, 6:11, 11:6, and 3:14 (R_{geom} = 0.89, 1.83, 0.55, and 4.67, respectively). Using the values of β_{max}, β_{knee}, and I_{merit} for 1 V, as obtained from Fig. 1, the theoretical and empirical curves presented in Figs. 4 and 5 were constructed. The disparity of the curves at the high current-gain end is probably a result of the output stage being operated at 2 V where the high beta curve of Fig. 1 departs from the negative-one slope. The added voltage also increases I_{merit} for the output fraction and results in an increased current handling capability of approximately 5 percent.

The failure to reach the predicted maximum beta is the result of an observed low-current beta fall off which departs from the zero slope used in the calculation. It is interesting to note that the highest current data point of Fig. 4 requires the individual transistors to operate at a current density equivalent to a beta of 0.6.

B. Discrete-Transistor—Four Stage

Fifteen of the seventeen transistors were connected into a four-stage Darlington circuit, $N = 4$, in a 1:2:4:8 and 8:4:2:1 sequence (R_{geom} = 2 and 0.5, respectively). The measurements were made at 4 V; data and calculated values are presented in Fig. 6. Note that I_{merit} has changed since only fifteen transistors were used.

The reason for the slight departure from theory is similar to that for the two-stage results. The high current performance for R_{geom} = 2 requires that all transistors operate at $\beta = 0.5$, approximately. In the case of R_{geom} =

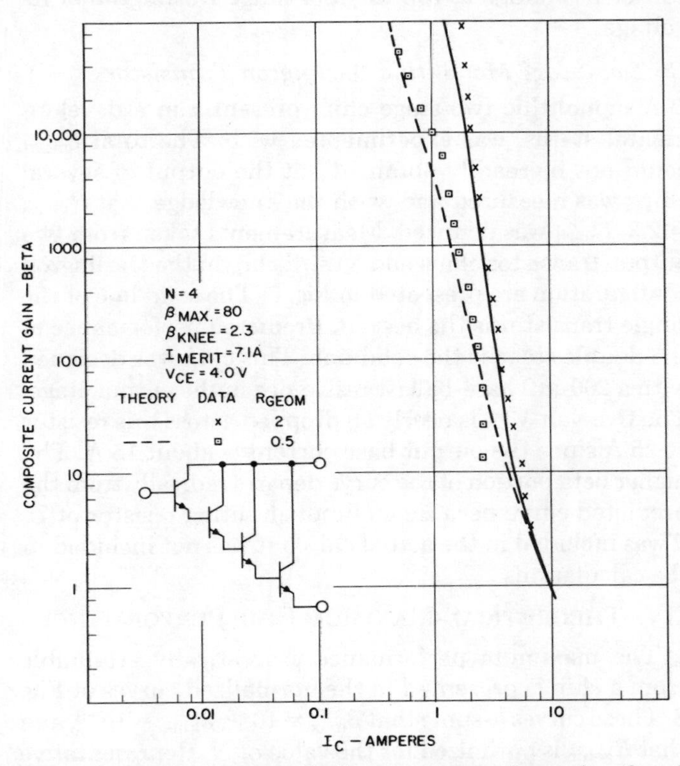

Fig. 6. Composite four-stage Darlington current gain as a function of collector current; R_{geom} = 0.5 and 2; discrete transistors.

0.5, it should be noted that all current must flow through a single discrete emitter. The measured high current curve departs from the theoretical curve when $\beta_N = 0.25$. (β_N is the beta of the last, or fourth, stage.) It is felt that ohmic parasitic emitter resistance and wiring resistances dropped

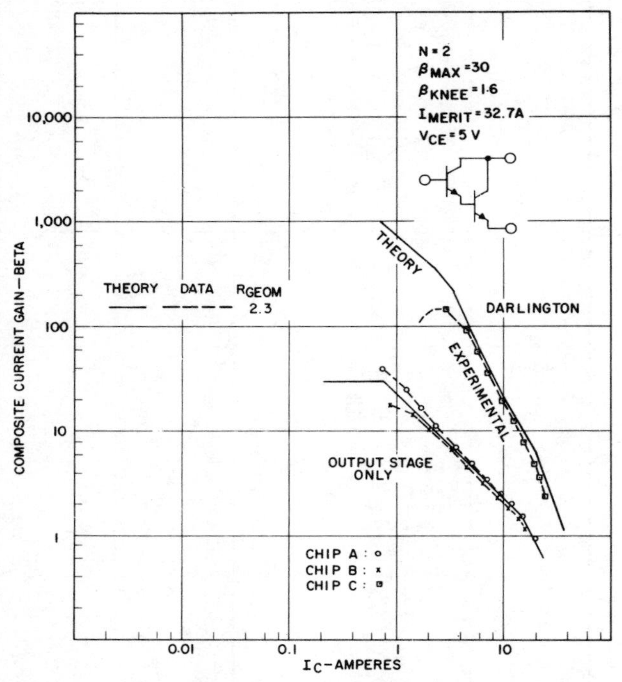

Fig. 7. Two output-transistor chips and a third chip in the two-stage monolithic Darlington configuration.

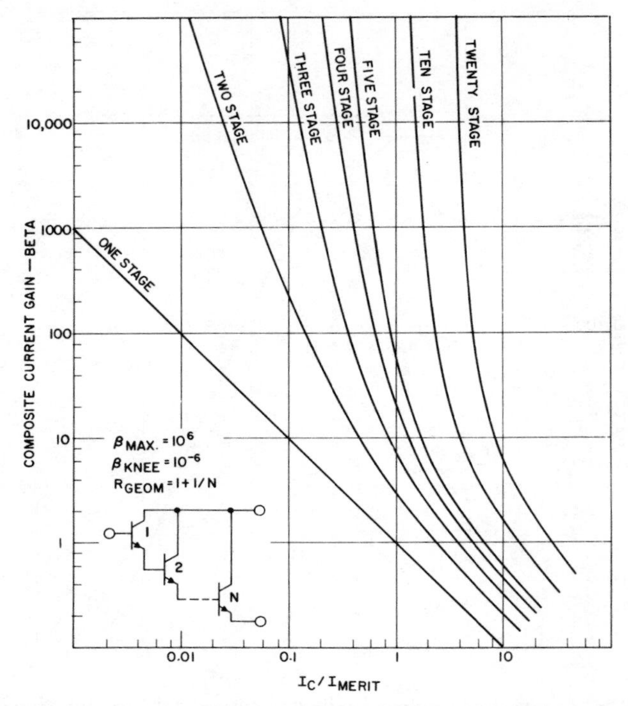

Fig. 8. Beta versus collector current for Darlingtons with different numbers of stages but same total area.

sufficient voltage to rob the four-stage Darlington of its voltage.

C. *Two-Stage Monolithic Darlington Transistor*

A monolithic two-stage chip, presently in a developmental status, was experimented with. The total I_{merit} could not be readily obtained but the output of several chips was measured and, with the knowledge that $R_{\text{geom}} = 2.3$, I_{merit} was deduced. Measurements taken from two output-transistor chips and a third chip in the Darlington configuration are presented in Fig. 7. The solid line of the single transistor is the best fit. Predicted performance of the double stage is the solid line. This chip was designed with a 200-mΩ base-ballast resistance in the output stage. The five-volt V_{CE} is nearly all dropped across this resistor at 25 A since the output base current is about 15 A. The higher beta portion of the curve departs radically from the predicted curve because an input shunting resistor of 75 Ω was included in the actual chip but was not included in the calculations.

IV. Theoretical-Maximum Chip Performance

The maximum performance theoretically attainable from a chip is presented in the normalized curves of Fig. 8. These curves assume that $\beta_{\max} = 10^{+6}$, $\beta_{\text{knee}} = 10^{-6}$, and that R_{geom} is optimized for the value of N. Representative values of N from one to twenty are presented. The optimum value of R_{geom} is approximately[1]

$$R_{\text{geom}_{\text{opt}}} \approx 1 + 1/N, \qquad \text{for } N > 1. \tag{18}$$

[1] Maximizing R_{geom} as a function of I_C/I_{merit}, where the composite current gain is fixed, results in a solution depending upon N and I_C/I_{merit}. Equation (18) was chosen as a best fit over the range of composite current gain varying between 10 and 1000 when N varies between 2 and 20.

The implications of Fig. 8 are impressive. By way of illustration, assume a high-voltage transistor is to be designed with a current gain of twenty at 5 A of collector current. Referring to the one-stage curve of Fig. 8, a current gain of twenty requires that the chip operate at an I_C/I_{merit} value of 0.05. The requirement of 5 A demands an I_{merit} of 5/0.05 or 100 A. Currently available high-voltage transistors featuring an I_{merit} of 100 A employ a chip approximately 0.3 cm^2 (220 × 220 mils).

Option 1—By dividing the above chip to form a two-stage Darlington, a gain of 1500 is available at 5 A ($I_C/I_{\text{merit}} = 0.05$).

Option 2—The same chip could be operated at higher current levels as a two-stage Darlington, yielding the same current gain requirement of twenty at an I_C/I_{merit} level of 0.32, or a composite collector current of 32 A.

Option 3—A two-stage Darlington could be used with an I_{merit} of sixteen A and operated at the I_C/I_{merit} current level of 0.32, thereby meeting the original requirement for a current gain of twenty at 5 A. This lesser value of I_{merit} would require a much smaller chip, in fact, 0.05 cm (90 × 90 mils).

Option 4—The beta of twenty at 5 A could be met with an I_{merit} of 16 A for a two-stage Darlington, as noted above. Rather than reduce the chip size, however, the chip could be kept the same size and the blocking voltage could be increased by a factor of two or three [10] by increasing the collector thickness and resistivity.

Option 5—A four-stage Darlington could be operated at an I_C/I_{merit} level of 0.32, permitting a reduction of chip size from 220 × 220 mils to 90 × 90 mils and simultaneously increasing the 5-A current gain to 5000.

Option 6—Should a ten-stage circuit be chosen, the

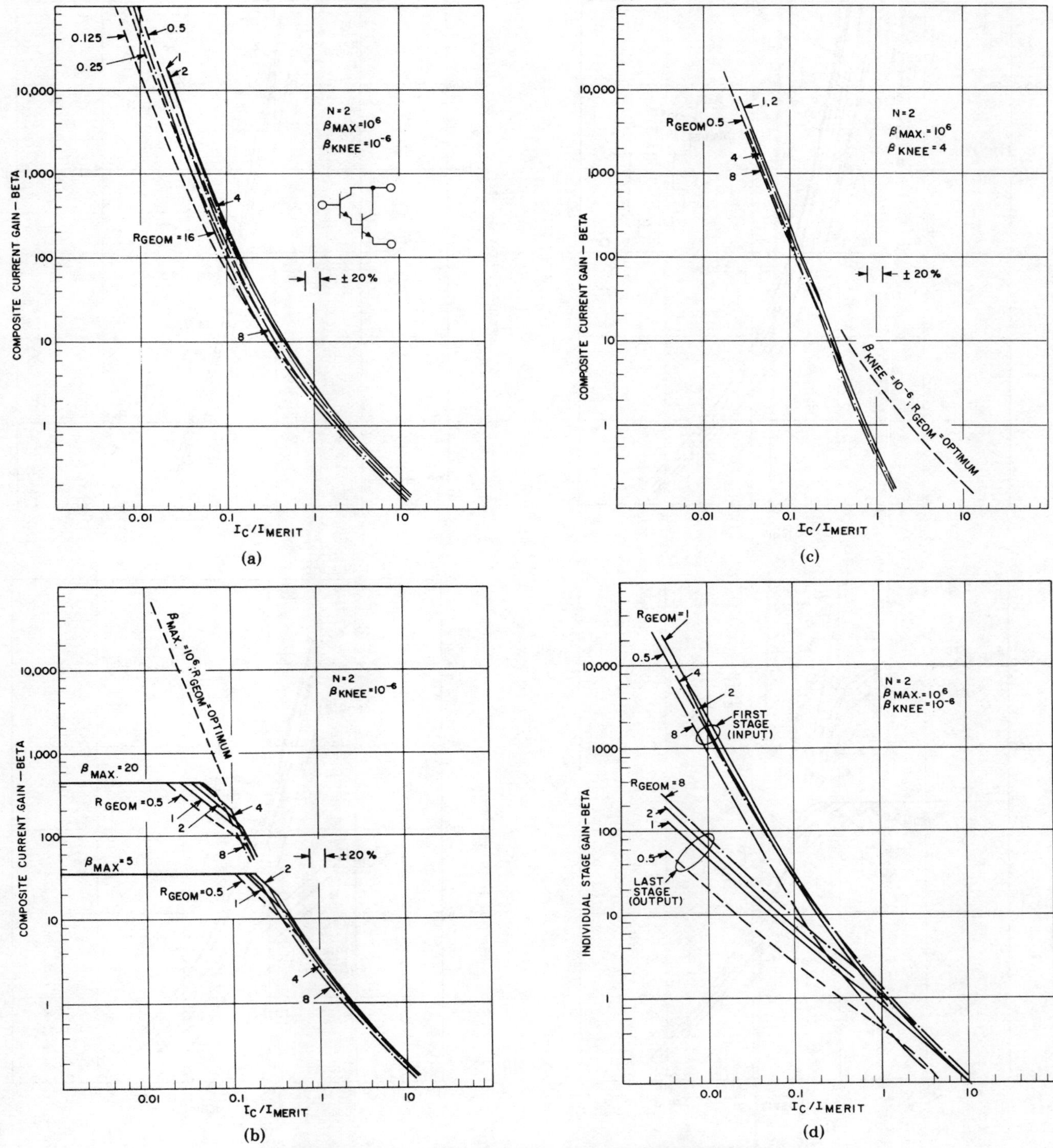

Fig. 9. Beta versus collector current for a two-stage Darlington. (a) Ideal as a function of R_{geom}. (b) Influence of β_{max}. (c) Influence of β_{knee}. (d) Beta of first and last stage.

aspects of operation at an I_C/I_{merit} level of 1.7 may be considered. Here, the advantages of option 5 may be chosen along with a blocking voltage improvement by a factor of two or three.

Option 7—A combination of trade-offs of the above options could be chosen, offering increased gain, increased voltage, increased current, and reduced chip area.

The discussion, to this point, neglects the effects of high current densities [11], [12]. Figs. 9 through 11 present a detailed study of the circuit behavior for $N = 2$, 3, and 4, respectively, as R_{geom} is varied. The specific influence of β_{max} and β_{knee} is reviewed as well as the current densities of the stages.

V. Two-Stage Darlington Design Curves

Fig. 9 presents the circuit behavior of two-stage Darlington transistors. Fig. 9(a) presents a normalized display of current gain versus collector current as a function of R_{geom}. β_{max} and β_{knee} are chosen as 10^6 and 10^{-6} to guar-

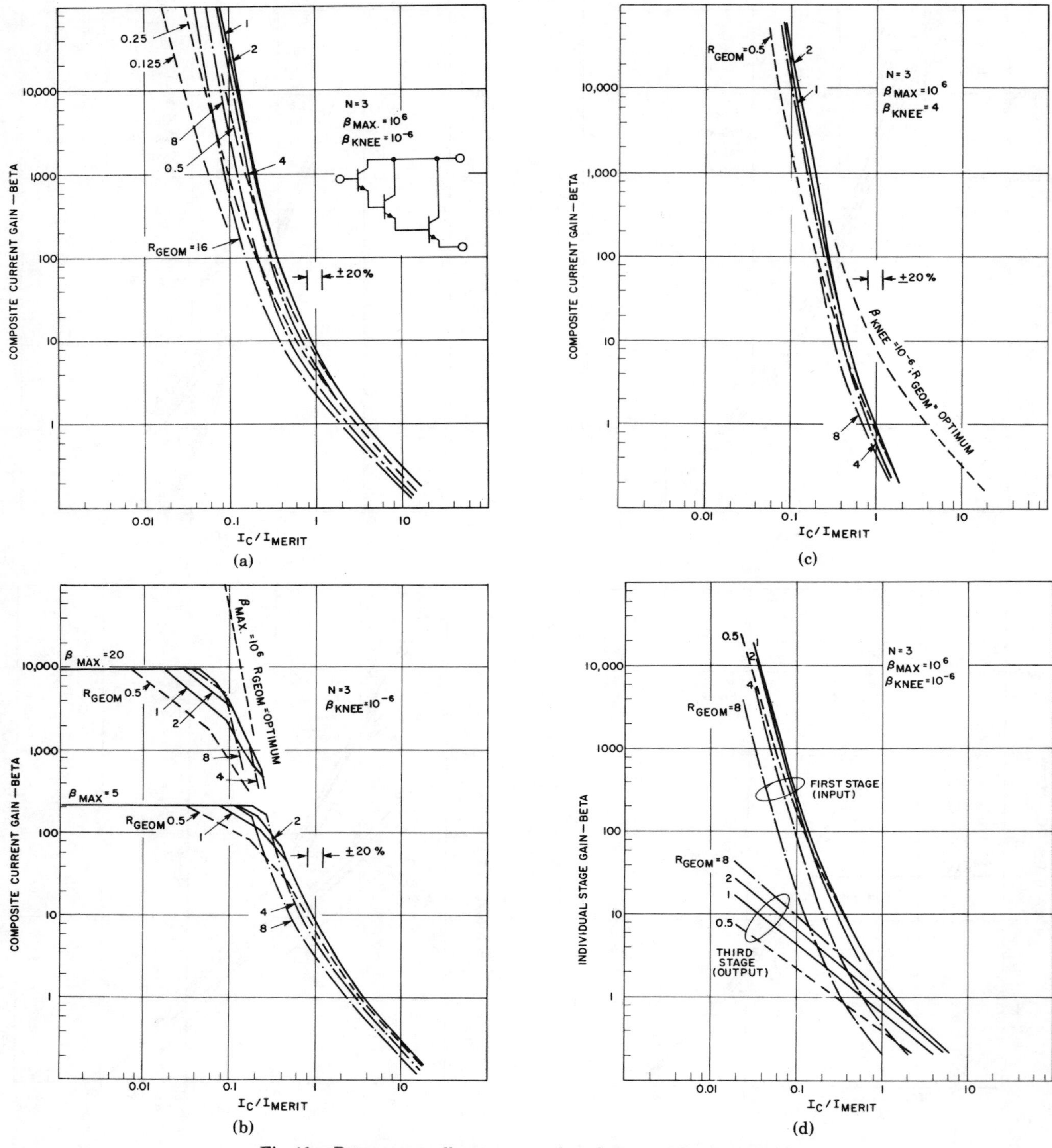

Fig. 10. Beta versus collector current in a three-stage Darlington. (a) Ideal as a function of R_{geom}. (b) Influence of β_{max}. (c) Influence of β_{knee}. (d) Beta of first and last stage.

antee that they will not influence the curves. The curves were calculated two decades above and two decades below the range of most anticipated interest. Values of desired R_{geom} are expected to lie in the range from 1 to 8 [10]. It should be noted that the collector current is relatively insensitive to the choice of R_{geom} in Fig. 9(a), varying only ±15 percent over the range of maximum concern.

Fig. 9(b) presents the influence of β_{max} upon chip performance. Again, the characteristics are not particularly sensitive to R_{geom}, for the most part. It should be noted that the peak Darlington gain is higher than β_{max} [2]. For convenience, the dashed curve at the upper left serves as a reference to the maximum theoretical chip performance for a two-stage circuit as presented in Fig. 8.

As high current densities build, the negative-two slope becomes a factor. This influence is presented in Fig. 9(c). The value of 4 was chosen for β_{knee} because it was felt that it was fairly representative of typical designs. (This value

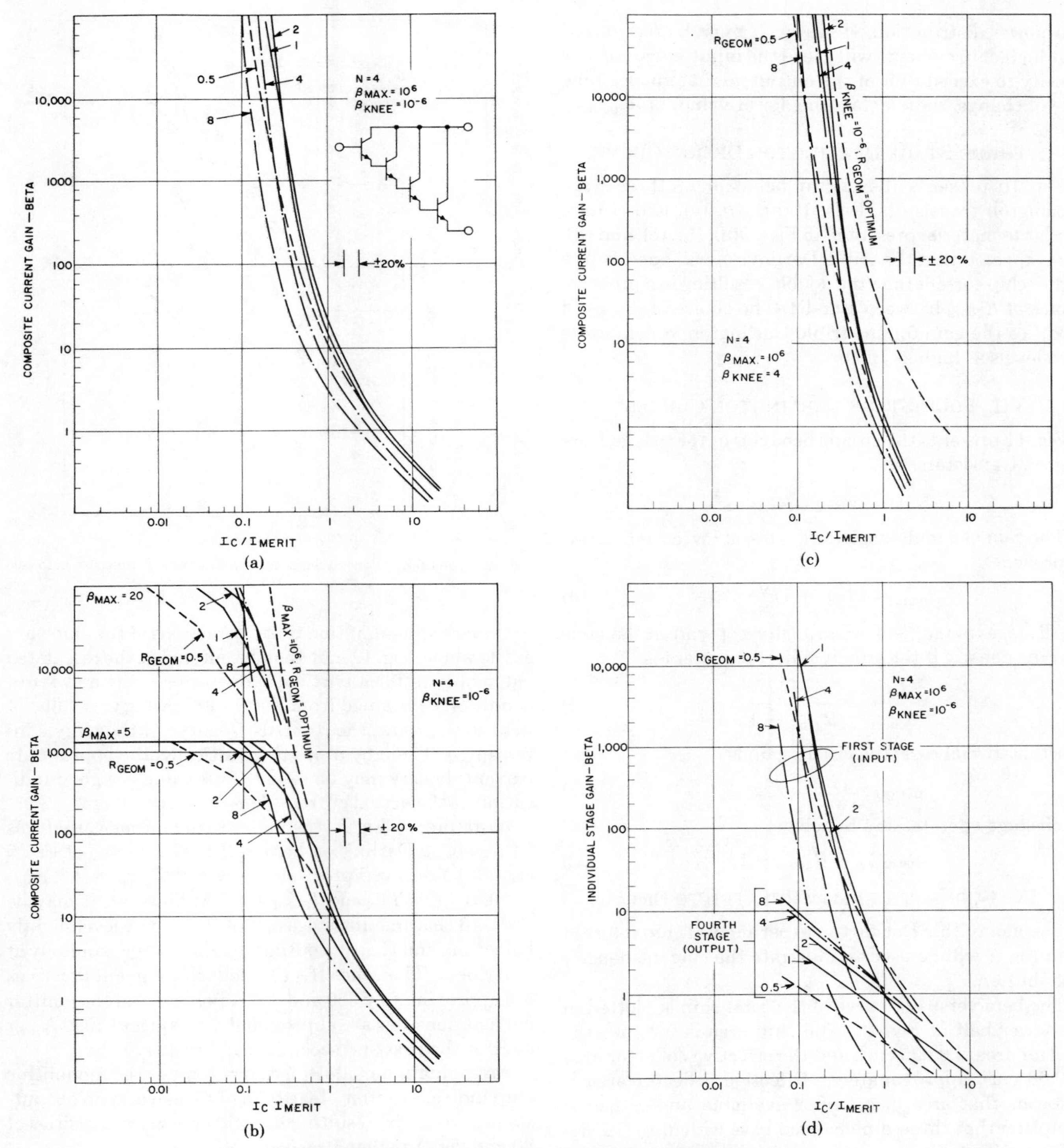

Fig. 11. Beta as a function of collector current for a four-stage Darlington. (a) Ideal as a function of R_{geom}. (b) Influence of β_{max}. (c) Influence of β_{knee}. (d) Beta of first and last stage.

of β_{knee} is usually caused by metallization resistance, which results in nonuniform current injection.) Departure from the curves of Fig. 9(a) starts with a beta approximately equal to 50. Again, the dashed curve (lower right) serves as a reference to the maximum performance of Fig. 8.

The guide to current-density management is offered in Fig. 9(d). Assuming the conditions of Fig. 9(a) ($\beta_{max} = 10^6$, $\beta_{knee} = 10^{-6}$), the operating beta of both stages is presented as a function of I_C/I_{merit}. Since the current density of the stage varies in an inverse relation to the stage beta, the current-density distribution may be deduced. It should be noted that moderate values of I_C/I_{merit} operate the two stages at similar current densities for large values of R_{geom}. Hence, a slight reduction in chip current capability resulting from a large value of R_{geom} may pay off in the overall scheme with respect to stored charge, turn-off-time, and power distribution. However, it must be recognized that high chip currents will cause the input-stage current density to exceed that of the output, and at an alarming rate of change, thereby favoring lower values of R_{geom}.

VI. Three-Stage Darlington Design Curves

Fig. 10 presents the circuit behavior of three-stage Darlington transistors. Figs. 10(a), (b), (c), and (d) are similar to material presented in Figs. 9(a), (b), (c), and (d). It is expected that the triple Darlington will operate at a higher chip current than the double, resulting in a probable choice of R_{geom} between 1 and 4. The choice of $\beta_{knee} = 4$ modifies the gain for the triple Darlington at composite beta levels as high as 1000.

VII. Four-Stage Darlington Curves

Fig. 11 presents the circuit behavior of four-stage Darlington transistors.

VIII. Special Cases

The gain of a multistage Darlington at low current levels approaches:

$$\beta_{peak} = (\beta_{max} + 1)^N - 1. \tag{19}$$

All stages of multistage Darlingtons operate at the same current density if the circuit collector current is

$$\frac{I_{CEQ}}{I_{merit}} = \frac{1}{R_{geom} - 1}. \tag{20}$$

For this special case, the circuit gain is

$$\beta_{darlEQ} = (R_{geom})^N - 1. \tag{21}$$

Each stage operates at a beta level

$$\beta_{stageEQ} = R_{geom} - 1. \tag{22}$$

IX. Current Density Related to Beta

The monolithic Darlington experimental chip referred to in Fig. 7 will be used to illustrate the current-density distribution.

The beta versus I_C curve of the total chip is plotted in the lower half of Fig. 12. The chip area is 0.2 cm^2, the emitter area is 0.025 cm^2, and the effective collector area is 0.08 cm^2, approximately. (Effective collector area is taken as that area of collector available under the assumption that three-dimensional base widening is a significant effect. This assumption results in an area somewhat between that of the chip and that of the emitter.)

The electrode currents may be plotted vertically versus collector current by using the beta versus I_C curve. (The 45° line of collector current versus collector current is included only as a reference.) When operating at high currents, the emitter current starts to differ from the collector current, while the base current rapidly approaches and exceeds the collector current.

By combining the electrode currents with the device areas, the emitter, effective collector, and chip-current densities may be plotted as a function of I_C.

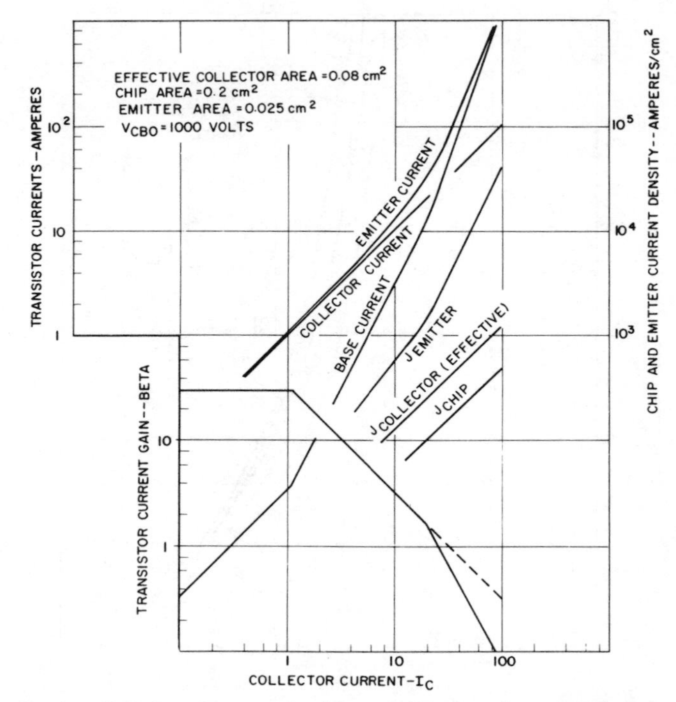

Fig. 12. Relation of current densities as a function of transistor beta (see text).

Under the assumption that all fractions of the chip vary as the whole, Fig. 12 may then be entered at the calculated value of beta for a particular operating stage and set of conditions (obtained from Fig. 9). Following this value of beta to the right, the β versus I_C curve of the chip is intercepted. Then, by projecting vertically, the appropriate current density may be found. This value is a good indication of the actual current density of that stage.

Example: In Fig. 7, the highest current measured was 24 A, with a Darlington beta of 2.4. The curves of Fig. 9 cannot be resolved precisely for $N = 2$, $R_{geom} = 2.3$, $\beta_{max} = 30$, $\beta_{knee} = 1.6$, and $I_{merit} = 33$ A; however, it may be deduced that the first stage operates at a beta level slightly below one and that the output operates approximately at unity beta. Then the effective collector current densities are approximately 350 and 300 A per cm^2 and the emitter current densities are approximately 3000 and 2000 A per cm^2 for the first and second stages, respectively.

An evaluation made at 5 A, $\beta = 100$, for the monolithic chip indicates a first-stage beta of fifteen and an output-stage beta of six, resulting in emitter current densities of 80 and 250 A per cm^2, respectively.

X. Other Considerations

The base–emitter voltage of a Darlington circuit is the sum of the V_{BE}'s of the stages. The V_{BE} of a stage at very low beta (high emitter current density) may be prohibitive. Although this material is not presented, it must be considered.

The circuit saturation voltage for silicon is approximately equal to $(N - 1)$ V if all ohmic effects are well controlled. This situation makes a high value of N less desirable then a lower value.

The output stage has more voltage than the $(N - 1)$ stage, etc. As a result, I_{merit} is high for those stages. In addition, the output stage is operated with less need for base widening, and, hence less stored charge and less "turn off tail." The output stage is generally operated at higher current densities, however, partially offsetting these advantages.

When working with high-current transistors, ohmic drops of the metallization must always be considered. When using Darlington power transistors at these high current densities, metallization and interconnect resistances may seldom be neglected.

Shunting resistors are normally connected from base to emitter of each stage to enhance the turn off time and to bypass high-temperature junction leakage currents. These resistors are often unavoidable as they are integral with the processing and appear as parasitic components. As a consequence, Darlington circuit gain in excess of several thousand is difficult to attain, particularly if a usable power device is to result.

Bond pads, isolation between components, shunting resistors, a commutation diode (collector to emitter), and chip edge treatment must be considered as second order effects upon I_{merit}.

Constructional limitations may exclude larger values of R_{geom} for large values of N. As an example, if $N = 10$, and $R_{geom} = 2$, the input stage would have to be 1/1000th the chip area, discounting isolation.

XI. Recommendations

The maximum recommended value of R_{geom} is one for which the worst case collector current equals that of equation 20. This choice of R_{geom} assures that the current density of any stage is always less than that of a succeeding stage. The minimum recommended value of R_{geom} is the value determined by equation (18). An additional physical restriction of R_{geom} is fabrication. Generally, R_{geom} should be sufficiently small so that

$$1 + R_{geom} + R_{geom}^2 + \cdots R_{geom}^{(N-2)} + R_{geom}^{(N-1)} \ll 1000. \qquad (23)$$

As larger values of N are employed, the impact of β_{knee} becomes more significant.

As larger values of N are employed, high values of β_{max} are no longer needed for gain. In fact, they are detrimental to some degree in that they degrade sustaining voltage breakdown, speed, and temperature performance.

XII. Conclusions

A model has been proposed and solved in which all Darlington circuits may be represented to a first order approximation by five constants, one of which may be normalized. Curves for Darlington circuits cascading two, three, and four stages have been presented. Experimental verification has been provided offering excellent agreement with theory. Several orders of magnitude improvement in current handling ability have been shown to exist for multistage Darlington circuits over conventional discrete transistors. Recommendations have been offered for allocation of chip area relative to each stage.

References

[1] S. Darlington, U.S. Patent #2,663,806, Semiconductor signal translating device, Dec. 22, 1953.

[2] P. Cutler, *Semiconductor Circuit Analysis*. New York: McGraw Hill, 1964, pp. 356–360.

[3] J. Watson, *Semiconductor Circuit Design*. Princeton, NJ: Van Nostrand, 1966, pp. 93–96.

[4] John D. Lenk, *Handbook of Modern Solid State Amplifiers*. Englewood Cliffs, NJ: Prentice Hall, 1974, pp. 234–244.

[5] RCA Solid State 1976 Data Book Series, Power Transistors and Power Circuits, information on Monolithic Darling Power Transistors, RCA Corp., 1976.

[6] *The Semiconductor Data Library*, Motorola Inc., 1973, 4th ed., pp. 5–76.

[7] J. Olmstead et al., "High level current gain in bipolar transistors," *RCA Review*, vol. 32, p. 221, June 1971.

[8] W. M. Webster, "On the variation of junction-transistor current-amplification factor with emitter current," *Proc. IRE* 42, p. 914, 1954.

[9] L. E. Clark, "High current density beta diminution," *IEEE Trans. Electron Devices*, vol. 17, pp. 661–666, Sept. 1970.

[10] P. L. Hower, "Optimum design of power transistor switches," *IEEE Trans. Electron Devices*, vol. 20, pp. 426–435, Apr. 1973.

[11] C. F. Wheatley, "Observation of the influence of base widening upon beta and f_T," *RCA Review*, vol. 32, pp. 247–250, June 1971.

[12] H. C. Poon, H. K. Gummel, and D. L. Scharfetter, "High injection in epitaxial transistors," *IEEE Trans. Electron Devices*, vol. 16, pp. 455–457, 1969.

STABLE HOT SPOTS AND SECOND BREAKDOWN IN POWER TRANSISTORS

Philip L. Hower
Westinghouse Research Laboratories
Pittsburgh, PA 15235
and
David L. Blackburn, Frank F. Oettinger, and Sherwin Rubin
Institute for Applied Technology
National Bureau of Standards
Washington, DC 20234

ABSTRACT

The mechanism of hot spot formation in transistors is examined from both experimental and theoretical viewpoints. It is shown that after the device becomes thermally unstable the device may restabilize in a hot spot mode of operation. The I_C, V_{CE} thermal instability locus can accurately be predicted assuming the current density is uniform prior to hot spot formation. A new model is proposed which explains why the device may restabilize in a hot spot mode and why devices exhibit "thermal hysteresis". It is also shown using thermal mapping techniques that emitter current crowding exists in the stable hot spot mode. Finally, the experiments support the idea that second breakdown occurs when the current density within the hot spot reaches a critical value.

INTRODUCTION

The existence of hot spots in transistors has been known since the early days of power transistor development [1]. More recent investigations [2] have shown that the hot spot can be stable and that its area is typically a few percent of the metallurgical emitter area [3]. In the stable hot spot mode, current becomes localized to a small area of the device but base control of collector current is still maintained, even under conditions of DC operation. Perhaps the most remarkable feature of this behavior is that it is brought about by conditions that would normally lead to thermal runaway and catastrophic device failure.

Previous investigations of hot spot behavior concentrated on experimental observations and no mechanism has been proposed which satisfactorily explains the stable nature of the hot spot. Such a model is proposed in this paper, and it is shown that emitter current-crowding and base-widening play an important role in determining hot spot stability.

Experimental confirmation of emitter current-crowding in the stable hot spot is demonstrated by means of infrared thermal mapping techniques. Additional data show that the observed hot spot size and temperature are in good agreement with theoretical predictions. In addition, the observed onset of second breakdown is consistent with the hypothesis that a critical current density triggers a low voltage mode of operation.

The first part of the paper shows how the concept of the stability factor, S, can be used to describe various regions of stable and unstable operation. In addition, hot spot formation is related to the forward safe operating area (SOA) of the transistor. This is followed by a description of the experimental techniques used to investigate the thermal and electrical properties of hot spots. The final part of the paper deals with a comparison of theory and experiment for two representative transistor types used in power switching applications. A list of symbols is also provided.

DESCRIPTION OF THE MODEL

Uniform current

It is worthwhile to review what happens in a transistor as V_{CE} is increased with I_C held fixed, where the path being considered is labeled as A, B in Fig. 1. The forward biased safe operating area limits are indicated where the limit defining the onset of thermal instability has replaced the more commonly used second breakdown limit.

In most transistors, the current is uniformly distributed over the metallurgical emitter area, A_{EM}, for operation along the path of constant current from point A to point B. The temperature of the active portion of the transistor, that is the so-called junction temperature T_J, increases linearly with V_{CE} in this region (Power = $I_C \cdot V_{CE}$) and

$$T_J - T_C = \frac{I_C \cdot V_{CE}}{K \cdot A_{EM}} \quad (1)$$

where T_C is the case temperature (held constant) and $R_{\theta JC} = (K \cdot A_{EM})^{-1}$ where $R_{\theta JC}$ is the junction-to-case thermal resistance and K a constant of proportionality. A plot of $T_J - T_C$ vs. V_{CE} is shown in Fig. 2 for two different values of A_{EM}.

Stability factor

At some critical voltage, V_{CE1}, the tran-

Reprinted from *IEEE Power Electron. Specialists Conf.*, 1976, pp. 234–246.

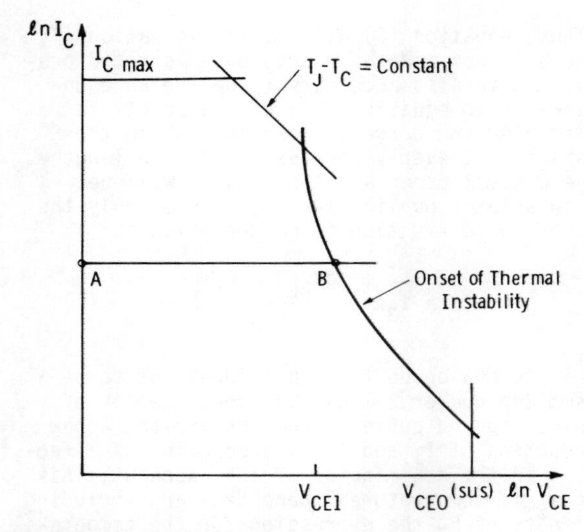

FIG. 1 - Forward SOA boundaries.

sistor becomes thermally unstable. This instability arises because of the positive temperature coefficient of collector current and it is an inherent property of all transistors.

For the calculations to follow, use is made of the stability factor, S [4,5,6], which accounts for the thermal feedback which exists between the device and its heat sink:

$$S = R_{\theta JC}V_{CE} \cdot \frac{\partial I_C}{\partial T_J} = \frac{T_J - T_C}{I_C} \cdot \frac{\partial I_C}{\partial T_J} \quad (2)$$

For $S < 1$, the system is stable, for $S \geq 1$ it is unstable. Using the Ebers-Moll equation (assuming $e^{qV_{BEJ}/kT_J} >> 1$),

$$I_C = I_1 \exp\left(\frac{qV_{BEJ} - E_g}{kT_J}\right), \quad (3)$$

where

$$I_1 = \frac{qn_{io}^2 A_{EM}}{Q_B/D_B}, \quad (4)$$

one can derive from equation (2) an expression for the junction temperature, T_s, corresponding to $S = 1$ [6]. If it is assumed that I_1 is temperature independent and that the junction voltage is related to the terminal voltage by $V_{BEJ} = V_{BE} - I_C R_E$*, then:

$$T_s = \frac{I_C R_E + T_C \cdot \frac{k}{q} \cdot \ell n\,(I_1/I_C)}{\frac{k}{q}\,[\ell n(I_1/I_C) - 1]}. \quad (5)$$

*The equations of Hower and Govil [6] are used in this paper with the exception that base resistance is ignored because its effect on S can be neglected for most transistors.

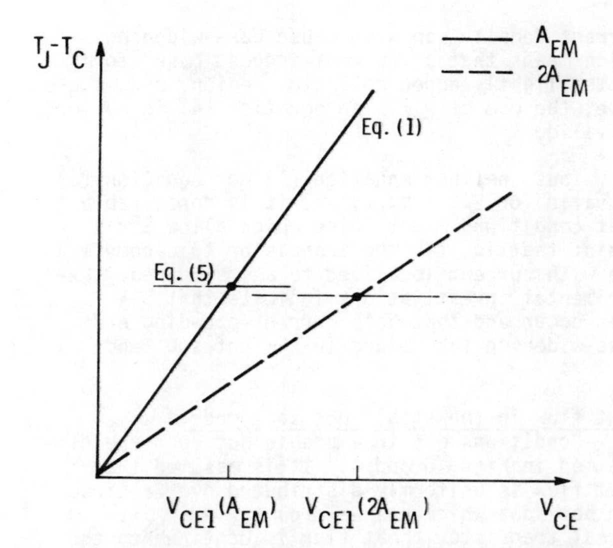

FIG. 2 - Effect of doubling metallurgical emitter area on the critical voltage V_{CE1}.

The constant temperature lines of Fig. 2 are obtained from equation (5) and help to determine the onset of thermal instability. On this plot, the critical voltage V_{CE1} is determined by the intersection of the constant temperature, $S = 1$, line with the linearly increasing temperature line as determined by equation (1). Fig. 2 also shows how changes in A_{EM} affect V_{CE1}.

An interesting observation here is that most of the variation in V_{CE1} with A_{EM} is due to the change in $R_{\theta JC}$. Although I_1 is proportional to A_{EM}, the fact that $I_1 >> I_C$ (for a typical TO-3 power switching transistor, $I_1 \cong 10^7$ A and lies in the range of 10^6 to 10^8 A for nearly all devices) means that the $\ell n(I_1/I_C)$ term in equation (5) undergoes only a small fractional increase in response to a doubling of A_{EM}. This behavior with A_{EM} will be referred to later when comparing the stable hot spot mode of operation with the uniform mode.

<u>Stable hot spot mode</u>

As V_{CE} exceeds V_{CE1}, the transistor will become thermally unstable and current will begin to localize to the highest temperature region. This localization has several consequences. One is that heat flux is confined to an area A_{HS} that can be a small fraction of A_{EM}. As a consequence, $R_{\theta JC}$ is no longer a function of A_{EM} but $R_{\theta JC} \propto A_{HS}^{-1/2}$ and equation (1) must be revised accordingly. Also, as the active area of the transistor decreases, the current density increases. Depending on the amount of interdigitation, the current can become non-uniformly distributed across the emitter stripe. That is, transverse voltage gradients in the base cause the emitter current to "crowd" to the edges of the emitter. The onset of current crowding means that the effective emitter area, A_E, is now dependent upon the current density and will be less than the area A_{HS} defined by the hot spot boundary. Increases in current density can also cause base-widening, which means that a "current-induced base" forms in the lightly-doped collector region, and, therefore, the use of Q_B/D_B in equation (4) is no longer valid.

Thus, neither equation (1) nor equation (5) is valid for $V_{CE} > V_{CE1}$, and it is conceivable that conditions might arise which allow S < 1 again; that is, for the transistor to become stable with current localized to a small area. Experimental investigations indicate that S < 1 does occur and that both current-crowding and base-widening take place in the hot spot mode.

<u>Heat flow in the stable hot spot mode</u>

Conditions within a stable hot spot are diagrammed in Figs. 3 and 4. It is assumed that heat flux is uniformly distributed over a circular hot spot which has an area $A_{HS} = \pi r_{HS}^2$. In a real transistor, heat flux is confined to the individual emitter fingers, and as a result there will be some "ripple" in the temperature profile, as noted in Fig. 3.

The assumption of uniform heat flux within the hot spot permits the use of the analysis of Carslaw and Jaeger [7] for the temperature difference $T_p - T_C$, which can be written as

$$T_p - T_C = \frac{I_C \cdot V_{CE}}{K_{HS}\sqrt{\pi A_{HS}}} . \qquad (6)$$

That is, the peak junction temperature of the hot spot, T_p, is proportional to the applied power and inversely proportional to the hot spot radius, r_{HS}. The quantity K_{HS} is the thermal conductivity of the region surrounding the hot spot.

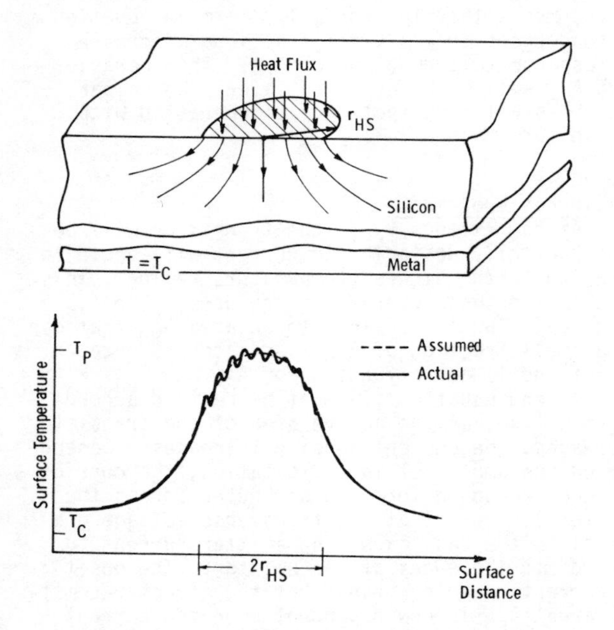

FIG. 3 - Diagram of heat flux and surface temperature for a hot spot.

Thus, equation (6) will replace equation (1) for the hot spot mode of operation. As noted previously, it is also necessary to derive an equation similar to equation (3) which accounts for base-widening and current-crowding. While the analysis is not overly complex, it is too lengthy for the present paper and the details will be given in a later publication [8]. Here, only the result of the derivation is needed which is

$$I_C = I_3 \exp\left(\frac{qV_{BEJ} - E_g}{4kT_J}\right) \qquad (7)$$

where I_3 is analogous to I_1 but turns out to be a much smaller number. Thus, the consequences of base-widening and current-crowding are the apparent reduction of I_1 and the introduction of a factor of 4 in the denominator of the exponent. Assuming I_3 is temperature independent and including the effects of R_E the expression for the temperature at which S = 1 becomes:

$$T_s = \frac{I_C R_E + T_C \frac{4k}{q} \ln(I_3/I_C)}{\frac{4k}{q}[\ln(I_3/I_C) - 1]} . \qquad (8)$$

Note that equation (8) differs from equation (5) only in that I_3 replaces I_1 and 4k has replaced k. For purposes of calculating transistor behaviour, it is assumed that $T_J \cong T_p$ within the hot spot.

The term I_3 is given by

$$I_3 = \frac{qA_{HS}D_C}{K_A L_E}\left(\frac{2K_C n_{io} D_C \; Q_E/D_E}{W_C}\right)^{1/2} . \qquad (9)$$

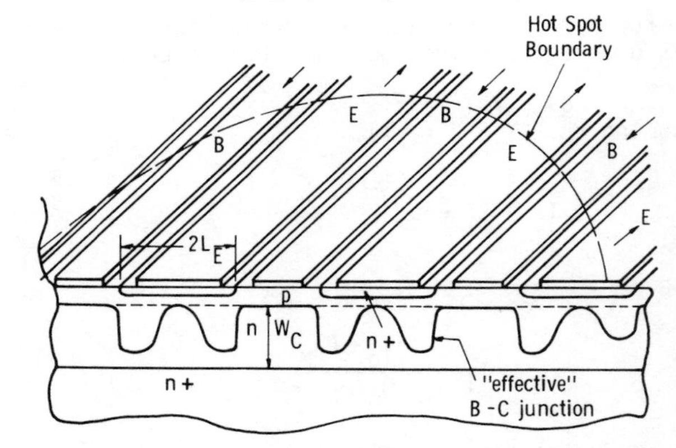

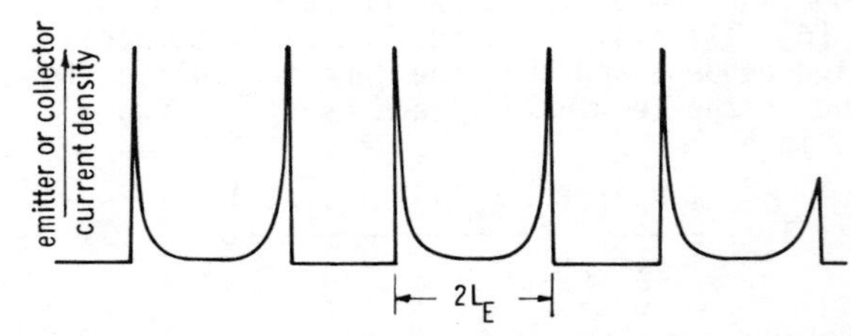

FIG. 4 - Cross-section through the hot spot showing the current-induced-base and current density profiles. Direction of current flow within the metallized emitter and base fingers is indicated by the arrows.

In equation (9), K_A is a geometrical ratio defined by:

$$K_A = \frac{A_{HS}}{A_E} \cong 1 + \frac{L_s}{2L_E} \qquad (10)$$

and is introduced to account for the fact that A_{HS} will be larger than the diffused emitter area within the hot spot.* Also, K_C is a dimensionless numerical term which is a result of the current crowding analysis [8]. For most npn transistors $K_C \cong 12$.

Several of the parameters which comprise I_3 are temperature dependent and these variations should be considered when evaluating equation (9). The terms n_{io} and D_C will vary approximately as $T^{3/2}$. (In addition, D_C is slightly dependent upon injection level.) The term Q_E/D_E, which accounts for minority carrier recombination in the emitter, is proportional to $\exp(-\Delta E/kT_J)$, where ΔE is usually in the range of 0.02 to 0.1 eV. The value of Q_E/D_E is usually in the range of $2\text{-}6 \cdot 10^{13}$ $cm^{-4} \cdot s$ at $T_J = 298$ K.

Increment in emitter ballast resistance

One more deviation from the uniform case is taken into account and that is the increase in R_E that occurs with current localization. The situation is diagrammed in Fig. 5. The term R_E is

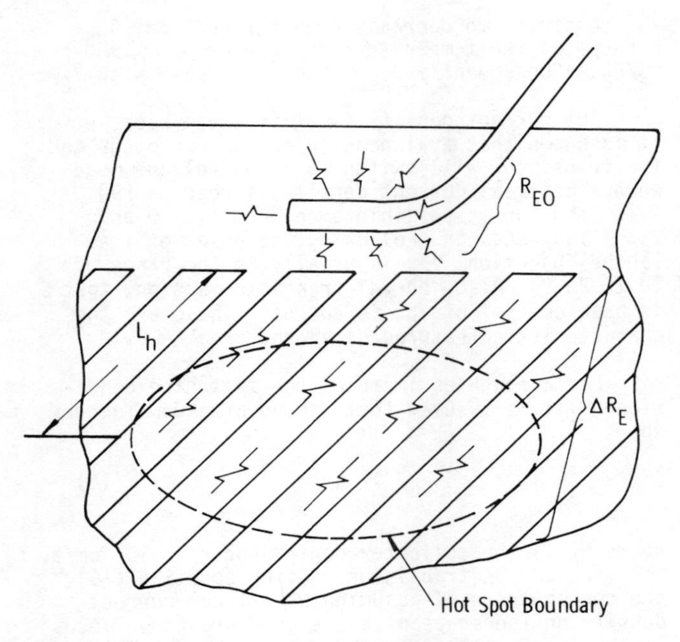

FIG. 5 - Diagram indicating the location of the two components of the emitter ballast resistance R_E (= $R_{EO} + \Delta R_E$).

*The approximation is valid when there is a large number of contact fingers within A_{HS}. When there are only two or three fingers making up A_{HS}, more detailed information is necessary to determine K_A.

taken to be the sum of two components, R_{EO} and ΔR_E. The first component, R_{EO}, is due to bonding wire resistance and the spreading resistance through the emitter metallization in the neighborhood of the wire bond. The value of R_{EO} is usually in the range of 20 to 40 mΩ for a single bond. For calculations of the thermal instability locus, R_{EO} can be determined directly from measurements to be mentioned. The quantity ΔR_E accounts for the fact that there will be just a few fingers which carry current from the emitter bus or pad to the hot spot. Assuming the hot spot is located a distance approximately equal to L_h from the emitter bus, ΔR_E can be approximated by

$$\Delta R_E = \frac{R_{sm}\, L_h\, K_A \sqrt{\pi}}{2\sqrt{A_{HS}}} \qquad (11)$$

where R_{sm} is the sheet resistance of the emitter metallization. Typically $1/3 \lesssim \Delta R_E/R_{EO} \lesssim 1$, depending on the particular emitter geometry, hot spot area, and location.

Behavior in the $T_J - T_C$, A_{HS} plane

Consider now what happens to the right of point B of Fig. 1. As the conducting area decreases, I_C is confined to a region with an emitter-base perimeter that is also decreasing. Eventually, a point is reached where emitter current-crowding begins to occur. This particular A_{HS} can be approximated by [8]

$$A_{HS}\Bigg|_{\substack{\text{for the onset}\\ \text{of current}\\ \text{crowding}}} = A_{HS2} = \frac{K_A L_E^2}{K_C}\,\frac{I_C}{qD_C^2\, Q_E/D_E} \qquad (12)$$

For $A_{HS} < A_{HS2}$, two things occur. The current density begins to increase rather dramatically along the edge of the emitter, and the temperature T_s also increases. The behavior of T_s is in marked contrast to the uniform situation, where T_s, from equation (5), is insensitive to A_{EM}. The reason for this change in behavior of T_s is that $I_3 << I_1$. As A_{HS} continues to decrease, I_3 also decreases according to equation (9) and a point may be reached where $\ln(I_3/I_C) = 1$. By equation (8) this means that the temperature required for stable operation is infinite, which is, of course, only a mathematical limit. Prior to this point the transistor will switch to a low voltage due to the onset of second breakdown or destructive alloying of the metal and silicon will take place.

Thermal hysteresis

Permissible paths of operation in the T_J-T_C, A_{HS} plane are diagrammed in Fig. 6. This diagram offers an explanation of the experimental I_B, V_{CE} plot given in Fig. 4 of Oettinger and Rubin [2], which shows "thermal hysteresis". Along path A,B, $V_{CE} < V_{CE1}$, current density is uniform, and the device is stable. As V_{CE} exceeds V_{CE1}, the transistor becomes unstable with A_{HS} decreasing to $A_{HS}(V_{CE1})$, where stability is achieved at point C. For small changes in V_{CE} about V_{CE1}, A_{HS} will ad-

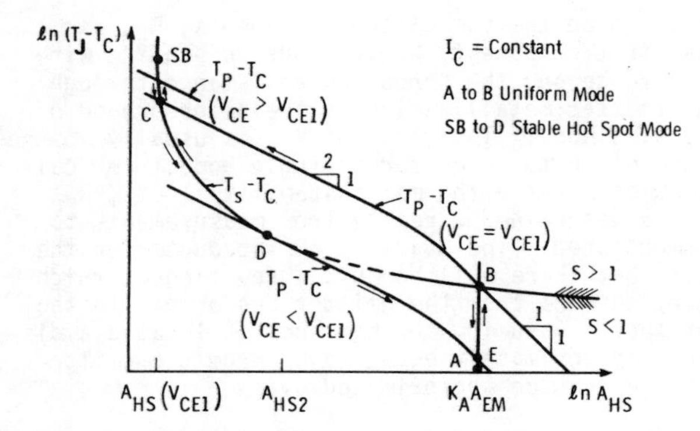

FIG. 6 - Thermal hysteresis diagram in the T_J- T_C,A_{HS} plane. Arrows indicate permissible movement of the operating point.

just itself so that stability is maintained.

The point labeled SB indicates that second breakdown will occur for a V_{CE} increase which reaches this point on the $T_s = T_C$ curve. This is discussed further below. As V_{CE} is decreased from V_{CEI}, a point is reached at D where no $T_s = T_p$ solution is possible and the device will fall out of the stable hot spot mode and switch to point E.

Onset of second breakdown

Consider now what happens as V_{CE} continues to increase beyond V_{CEI} from point C. That is, the device is in the stable hot spot mode and V_{CE} is increased. From Fig. 6, it can be seen that the stable operating point will move along the $T_s(A_{HS})$ curve with temperature increasing greatly, but there is only a slight reduction in A_{HS}. Also, the average and peak current densities, $\overline{J}$ and J_{peak}, will increase.* From the current crowding analysis [8],

$$\overline{J} = \frac{I_C{}^2\, K_A}{A_{HS}\, I_{C2}} \tag{13}$$

and

$$J_{peak} = \overline{J}\left(3 + \frac{I_{C2}}{I_C}\right) \tag{14}$$

where I_{C2} is the "cross-over current" given by

$$I_{C2} = \frac{K_C\, A_{HS}}{K_A L_E{}^2}\, qD_C{}^2\, Q_E/D_E \tag{15}$$

For $I_C \gtrsim I_{C2}$, the current crowding solutions (13) and (14) are valid.

Typical behaviors of the quantities of interest are diagrammed in Fig. 7. As V_{CE} increases,

*It is assumed that emitter and collector current densities are approximately equal.

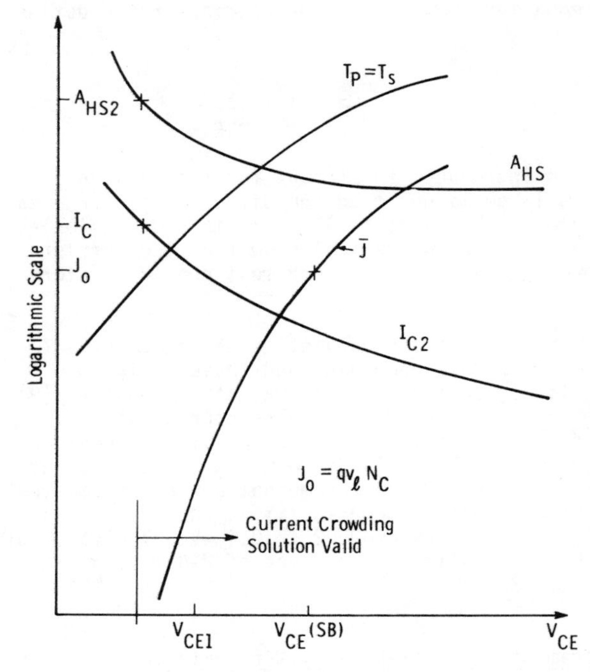

FIG. 7 - V_{CE} dependence of hot spot area, peak temperature, average current density, and cross-over current in the stable hot spot mode.

I_{C2} continues to decrease even for constant A_{HS} because of the temperature dependence of D_C and Q_E/D_E. Consequently J_{peak} and $\overline{J}$ increase with V_{CE}.

The current density is of interest because it is known that avalanche injection can occur and the transistor will switch to a low voltage mode when a critical current density is reached [9]. This behavior is possible even when $I_B > 0$ and $V_{CE} < V_{CEO}(sus)$.** Following the onset of avalanche injection, V_{CE} is usually in the range of 10 to 20 V. Also, normal transistor action, that is base control of I_C, is absent. These are the characteristic features of second breakdown.

For a rough estimate of the switching condition*** it is assumed that second breakdown occurs when

$$\overline{J} = J_0 = qv_\ell N_C \tag{16}$$

where N_C is the collector doping and $v_\ell = 10^7$ cm/s, assuming an npn transistor. Using equation (16), the consequences of assuming a critical current density on the shape of the second breakdown locus in the I_C, V_{CE} plane can be demonstrated. This

**For example see Figs. 17 and 18 of reference [9].

***A more accurate determination of the critical current density is desirable, but the analysis is complicated by a number of factors including the two- (and almost three-) dimensional nature of current flow plus the lack of an accurate model for avalanche injection in the $I_B > 0$ region.

shape is indicated in Fig. 8, which shows the HS locus, where the hot spot begins to form, and also the SB locus corresponding to the onset of second breakdown. The shaded region is where stable hot spots can exist. As shown in Fig. 8, the loci cross, and for small currents the HS locus lies above the SB locus.

This crossing of the loci can have important consequences when investigating the experimental behavior of transistors. For example, if the path A,B of Fig. 1 is chosen so that $I_C < I_{C4}$, upon reaching the HS locus, a hot spot will begin to form; however, it will not reach a stable state and at some A_{HS} and T_J, second breakdown will occur.

MEASUREMENTS

Several techniques were used to determine the physical, electrical, and thermal properties of the two types of power transistors studied. Measurements were made to characterize the devices in the region where current density is substantially uniform and also in the stable hot spot mode. Table 1 lists various parameters needed to solve the two stability criteria given by equations (5) and (8). Also listed are other measurements made to test the predictions of these equations and to determine the temperature distribution within and around the stable hot spot.

Two different types of commercially available transistors were studied. The devices are labeled A and B and a summary of their characteristics is given in Table 2. Both types had triple-diffused profiles, were mounted in TO-3 packages, and had conventional comb-like interdigitated emitter-base contact patterns. The devices differed mainly in the value of $V_{CEO}(sus)$, W_C, L_E, A_{EM}, and die-size.

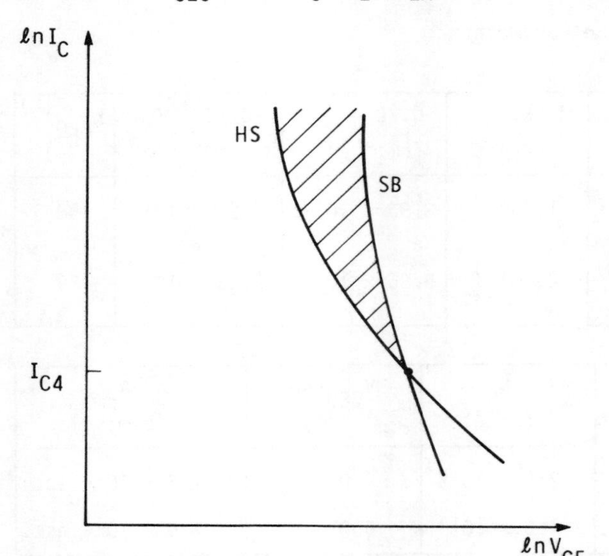

FIG. 8 - Loci corresponding to the onset of thermal instability (HS) and second breakdown (SB). The shaded area indicates where stable hot spots can form.

Measurements required for prediction of the HS locus

The device parameters required to predict the onset of thermal instability are the emitter resistance R_E, the uniform current coefficient I_1, and the junction-to-case thermal resistance $R_{\theta JC}$. The value of R_E was measured using the floating collector method discussed by Giacoletto [10]. This measurement can be made using a conventional transistor curve tracer. The measurement of the current I_1 can also be made using a curve tracer operating in the I_C vs. V_{BE} display mode with base current applied. (For sufficient resolution of V_{BE}, the capability of an offset voltage and magnification of the V_{BE} axis is desirable.) I_1 is then determined using equation (3).

The final device parameter necessary for the prediction of the HS locus is $R_{\theta JC}$, which is measured using the standard procedure [11]. The value of $R_{\theta JC}$ is measured at the I_C for which the initiation of instability is to be determined, but with V_{CE} 20 to 60 V below the critical value V_{CE1}.

Measurement of the HS locus

Either the common-emitter current gain, h_{FE}, or the emitter-base voltage V_{BE} can be used to determine the values of I_C and V_{CE} for which the initial instability occurs. These two quantities are both sensitive to changes in junction temperature and changes in current density.

The method for monitoring h_{FE} for determining the onset of device instability has been discussed in detail by Oettinger and Rubin [2]. In this technique, I_C is maintained constant while V_{CE} is increased and the base current $I_B(= I_C/h_{FE})$ is monitored. When the instability occurs, a sudden increase in I_B is usually observed.

One problem with the use of base current for detection is that there are competing influences on h_{FE} when current localization occurs. Increasing current density causes h_{FE} to decrease and increasing temperature causes h_{FE} to increase. These effects can sometimes lead to uncertainties in determining the V_{CE} for which instability occurs. This is particularly true for the instabilities which occur at large values of I_C. In this case the $T_s - T_C$ curve of Fig. 6 is shifted to the right and the stability factor never deviates far into the region where $S > 1$ during the transition from B to C. For this situation there is no sudden "jump" into a stable hot spot, but only a gradual constriction of the current as V_{CE} is increased. For these more gradually forming or "amorphous" hot spots, the I_B vs. V_{CE} plot is rounded (as noted in Fig. 4 of reference [2]), and the gradual increase in I_B may lag the actual initiation of the instability by as much as 10 to 20 V.

The other parameter for detecting instability (V_{BE}) also has conflicting dependencies upon current density and temperature, but because the temperature dependence is dominant, V_{BE} is always observed to decrease upon formation of the hot spot. The value of V_{BE} can be measured by applying a

Table 1. Summary of Measurements Made

Parameter or Property	Symbol	Method	Purpose
Emitter Resistance	R_E	Curve Tracer-Floating Collector [10]	Used in the stability equations (5) and (8)
Critical Current	I_1	Curve Tracer-I_E Versus V_{BE}	Used in equation (5) and to determine Q_B/D_B
Thermal Resistance	$R_{\theta JC}$	EIA Recommended Standard RS-313-B [11]	Used in equation (1)
Conditions of Initial Instability	V_{CE1}, I_C	h_{FE} or V_{BE} as a function of V_{CE} at constant I_C	To test the predictions of equation (5)
Hot Spot Area	A_{HS}	Cooling Curve Technique [3]	Used for comparison with calculated A_{HS}
Peak Junction Temperature	T_p	Infrared Microradiometer	Used to test predictions of the device conditions at restabilization.
Surface Temperature	$T_J(x,y)$	Automated Infrared Microradiometer	To infer qualitatively the current distribution and the severity of current crowding in the hot-spot. To determine the hot spot thermal conductivity, K_{HS}.
Activation Energy	ΔE	Curve Tracer-h_{FE} Versus T_J	To determine the temperature dependence of Q_E/D_E in expression for I_3.
Collector Width	W_C	Spreading Resistance	Used in expression for I_3.
Emitter Finger Half Width	L_E	Metallurgical Microscope	Used in expression for I_{C2}, equation (15) and I_3 equation (9).
Second Breakdown Locus	I_C, V_{CE} (SB)	Sudden Drop in V_{CE}	To determine locus of $(I_C, V_{CE})_{SB}$ for comparison with $J = J_o$ criteria.

Table 2. Measured Device Parameters

Device	Chip Size	A_{EM} (cm^2)	L_E (cm)	L_S (cm)	K_A (--)	Q_E/D_E (2) (cm^{-4} s)	Q_B/D_B (2) (cm^{-4} s)	h_{FEO} (2) (--)
A	6.35 × 6.35 mm (250 × 250 mil)	0.176	1.1×10^{-2}	0.95×10^{-2}	1.43	2.83×10^{13}	4.92×10^{11}	58
B	4.06 × 4.06 mm (160 × 160 mil)	0.0615	2×10^{-2}	1.75×10^{-2}	2.30 (1)	5.86×10^{13}	2.02×10^{12}	29

Device	ΔE (eV)	L_h (cm)	K_{HS} (W/°C cm)	W_C (μm)	N_C (cm^{-3})	V_{CEO}(sus) (V)	R_{sm} (ohm/sq)
A	0.0575	3.0×10^{-2}	0.78	27	2.5×10^{14}	270 est.	1.5×10^{-2} est.
B	0.024	6.4×10^{-2}	0.78	60	1.5×10^{14}	650	1.5×10^{-2} est.

Notes:

(1) This value is obtained from the temperature contours using $K_A = A_{HS}/A_{EM}$. The approximation of equation (10) is not valid because only one base finger and two emitter fingers are contained within the observed hot spot, see Fig. 12.

(2) T_C = 25°C, 300 μs pulsed measurement.

steady state bias, as with the h_{FE} technique, or by using a pulsed technique whereby a pulse of approximately 1 s duration is applied to the transistor and V_{BE} is monitored either 1 ms or so before conclusion of the pulse or 10 μs after the pulse (with a low level measurement current present) [12]. The circuitry and measurement procedure used for the latter method are identical to those of the standard $R_{\theta JC}$ technique.

Although the low level measuring current method is more sensitive, both pulsed V_{BE} techniques indicate the $V_{CE}(I_C)$ for which instability occurs, with sufficient accuracy, even for gradually forming hot spots. Fig. 9 shows plots of V_{BE} and I_B vs. V_{CB} for device B. The condition for which the instability occurs is clearly evident by the knee in the V_{BE} curve at V_{CB} = 60 V.

Measurements required for stable hot spot calculations

The device parameters required for solution of equation (8) are the emitter efficiency term, Q_E/D_E, and its associated activation energy, ΔE; the collector width and impurity concentration, W_C and N_C; and the emitter finger width $2L_E$ and spacing L_s. In addition, it is desirable to know the hot spot location, L_h, and the sheet resistance of the contact metallization, R_{sm}, in order to calculate ΔR_E from equation (11). (Accurate values for L_h and R_{sm} are not usually necessary because ΔR_E is normally less than R_{EO} and acts only as a correction term.)

With the determination of these quantities, the current I_3 can be calculated as a function of hot spot area, A_{HS}, using equation (9). Assuming the hot spot thermal conductivity is known,* the value of A_{HS} which gives stable operation is then found by simultaneously solving equations (6) and

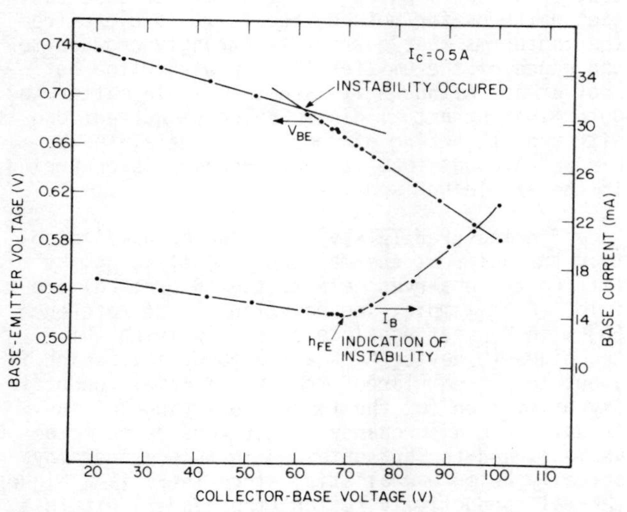

FIG. 9 - Comparison of the use of V_{BE} and I_B for the detection of the initiation of thermal instability. Note that V_{BE} detects the actual onset of hot spot formation while I_B lags by about 5 to 10 V.

(8). This solution also determines the temperature for which $T_s = T_p = T_J$. The calculated A_{HS} and T_J can then be compared with measured values as is shown later in the paper.

The technique for experimentally determining A_{HS} is described by Blackburn and Oettinger [3]. As in the case of the V_{BE} measurements, this technique also uses the $R_{\theta JC}$ measurement circuit. The method is based upon the fact that for about the first 250 μs of cooling from steady state, power transistors appear to cool one-dimensionally. The value for A_{HS} can then be found from the slope, m_T, of the cooling response curve when the measured $T_J - T_C$ is plotted as a function of the square root of time. This value of A_{HS} is the effective hot spot area for which a uniform current density would exist in the hot spot. (Because a relatively small sampling current is used, the measurement current is not crowded.)

An example of the cooling response for device A is shown in Fig. 10 for I_C = 0.3 A and V_{CE} = 60 V. The hot spot area of 6.6 × 10^{-3} cm^2 is about 3.8 percent of the metallurgical emitter area. The area is found from the equation

$$A_{HS} = \frac{2 I_C V_{CE}}{m_T \sqrt{\pi\rho K_{si} C_p}} . \quad (17)$$

For the calculation of A_{HS}, K_{si} = 0.7 W/cm·°C (at 200°C) and C_p = 0.71 J/gm·°C, and ρ = 2.33 g/cm³. The term Q_E/D_E is related to the peak current gain h_{FEO} by

$$h_{FEO} = \frac{Q_E/D_E}{Q_B/D_B} \quad (18)$$

The denominator of equation (18) is obtained from equation (4) using the measured current I_1 and the metallurgical emitter area A_{EM}, which in this case is determined from photomicrographs. (The emitter

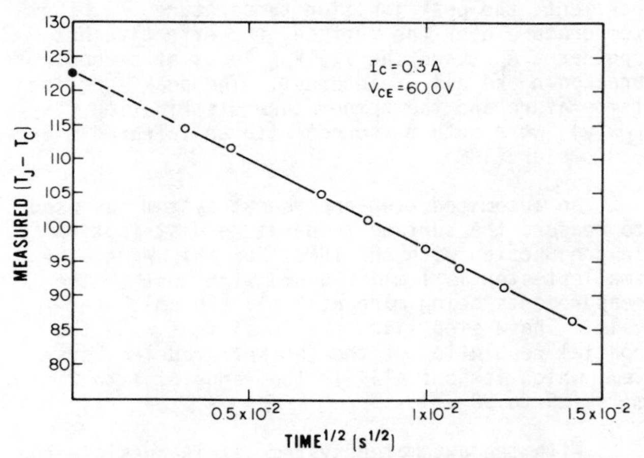

FIG. 10 - The electrically measured cooling response plotted vs. the square root of time for device A.

*As shown in the next section, K_{HS} can be determined from the measured $T_J(x,y)$.

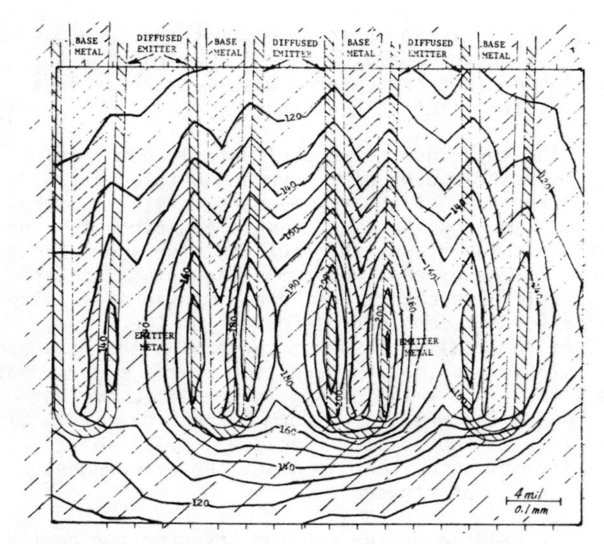

FIG. 11 - Temperature contour map (°C) for device A (V_{CE} = 60 V, I_C = 0.3 A). Contours are obtained from a linear interpolation of IRM data. Step size is 2 mil (50 μm).

finger width $2L_E$ and spacing L_s were also determined in this manner.)

The activation energy ΔE is determined by measuring h_{FEO} as a function of temperature and making a least squares fit to the data assuming $h_{FEO} \propto \exp(-\Delta E/kT_J)$.

Finally, the width of the collector region, W_C, and the collector impurity concentration, N_C, are determined from spreading resistance measurements.

<u>Behavior in the stable hot spot mode</u>

For the purposes of comparing theory and experiment, the peak junction temperature, T_p, the temperature over the surface, the effective hot spot area A_{HS}, and the I_C, V_{CE} locus of second-breakdown are all of interest. The peak junction temperature and the temperature distribution, $T_J(x,y)$, were both measured* with an infrared microradiometer (IRM).

An automated step and repeat system was used to measure the surface temperature distribution in conjunction with the IRM. For this work, the smallest step is 1 mil (25 μm) with most of the measurements being made at 2 mil (50 μm) intervals. These step sizes are consistent with the spatial resolution of the IRM infrared lens system, which is nominally in the range of 1 to 2 mils (25 to 50 μm).

With the automated system, it is possible to make acceptably accurate temperature measurements

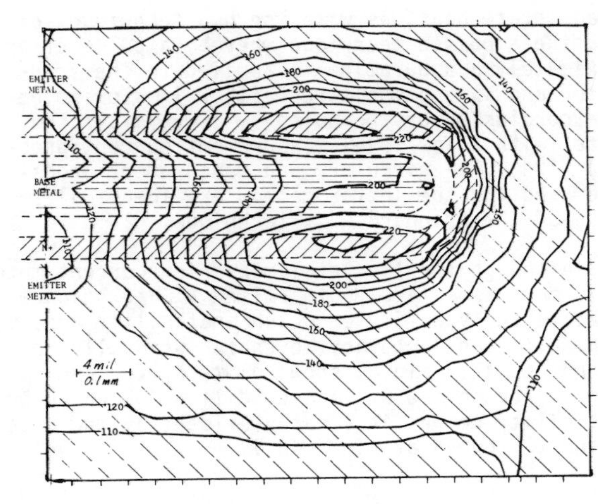

FIG. 12 - Temperature contour map for device B. V_{CE} = 125 V, I_C = 0.3 A. Other conditions are the same as for Fig. 11.

without coating the device to equalize the surface emissivity. The procedure used here requires a calibrating run at an elevated case temperature to obtain the correction for emissivity variations. The repeatability of the system permits a correction to be made for emissivity at each point. In principle, one can scan manually, but there are practical difficulties in repeating each point with sufficient precision for each emissivity measurement and the subsequent temperature correction. This precision is needed when one is scanning over regions of rapidly changing emissivity which will always exist on a power transistor.

In addition to measuring T_p, the infrared scanning system gives direct evidence that current crowding is occurring in the stable hot spot. This can be seen in Figs. 11 and 12 which show isotherms superimposed upon the emitter-base contact patterns for both devices. It is clear from the isotherms that current is strongly crowded to the edges of the emitter fingers within the hot spot area. Although it is not a simple matter to determine the actual distribution of current density from this kind of temperature data, there can be no doubt that current crowding is occurring in the stable hot spot.

The measured $T_J(x,y)$ can also be used to obtain the hot spot thermal conductivity, K_{HS}, by fitting the observed data to the theoretical profile. For example, see equation (2) of reference [6] with R_{TH} set equal to zero. Applying this technique to device A, K_{HS} = 0.78 W/cm°C, which is about 10 percent larger than the thermal conductivity of silicon for the temperature range of interest. The discrepancy is not considered to be serious, and the K_{HS} obtained in this manner may actually be more realistic, since there is a higher thermal conductivity region (the header) within a radius of the hot spot.

Finally, measurements were made of the operating conditions which led to the onset of second breakdown. Following conventional practice, the

*Heat transfer from the chip surface is negligible and it is reasonable to assume that the measured surface temperature is equal to the junction temperature.

measuring circuit diverted power from the device under test when a sudden decrease in V_{CE} was detected. Although it was intended that this test be performed nondestructively, most devices deteriorated with successive second-breakdown measurements.

COMPARISON BETWEEN MEASUREMENTS AND THEORY

Initiation of thermal instability

The initiation of thermal instability is relatively easy to determine, both experimentally and analytically. As an example, the thermal instability or HS locus for a device of type B is shown in Fig. 13, where it can be seen that good agreement between theory and experiment is demonstrated.

As a further indication of the accuracy and usefulness of the model, results are listed in Table 3 for four different transistors. These are two pairs of outwardly identical devices; however, they differ greatly in the measured points of instability, V_{CEI}, for each type. As indicated in the table, the reason for this difference is the relatively large difference between the thermal resistances. The devices with the larger thermal resistance are suspected of having voids in the die-attach, which leads to the larger values of $R_{\theta JC}$ and the corresponding reduction of V_{CEI}.

Stable hot spot mode

A quantitative check of the stable hot spot model is not as straightforward as checking the HS locus; however, a number of comparisons can be made. Fig. 14 shows the predicted and measured behavior in the hot spot mode for a device of type A, where it can be seen that the measured temperature and hot spot area agree fairly well with the predicted values.

The theoretical curves of Fig. 14 are obtained from a simultaneous solution of equations (6), (8), and (9), using the measured device data of Table 2. One necessary parameter, which cannot be directly determined, is the electron diffusion

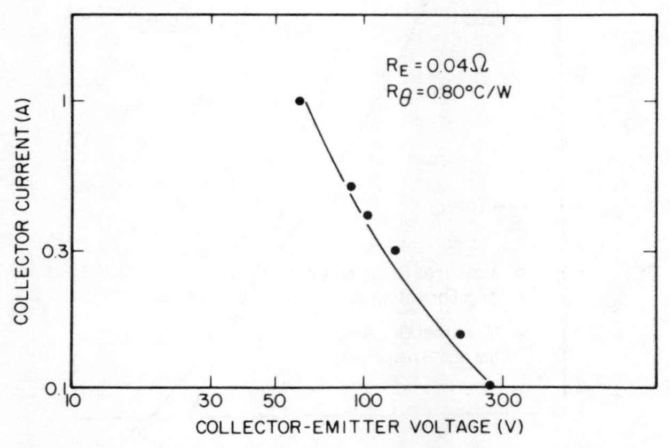

FIG. 13 - A comparison of the measured and predicted locus corresponding to the onset of thermal instability for device B.

Table 3. Variation of the Instability Point Within a Given Device Type

Device Type	Device No.	I_C (A)	$R_{\theta JC}$ (°C/W)	V_{CEI}(calc.) (V)	V_{CEI}(meas.) (V)
A	1	1.0	0.6	48	49
A	2	1.0	1.0	29	31
B	3	0.5	0.8	88	90
B	4	0.5	1.2	60	60

coefficient in the collector, D_C. In the hot spot mode, D_C will be reduced due to increases in lattice temperature, high electric fields, and large carrier concentrations.

As noted previously, the temperature dependence is taken into account by using the measured temperature dependence of electron mobility in silicon. The room temperature or T_C value of D_C is then adjusted to give the best fit with the $T_P - T_C$ and A_{HS} vs. V_{CE} data. For device A, $D_C(T_C)$ obtained in this manner turns out to be about one-half the low-field value, which is estimated to be 28 cm^2/s and corresponds to an average carrier concentration of 3×10^{16} cm^{-3}. A reduction of D_C by roughly a factor of two seems reasonable since the average field in the collector is approximately 2.2×10^4 V/cm, which is well beyond the constant mobility range.

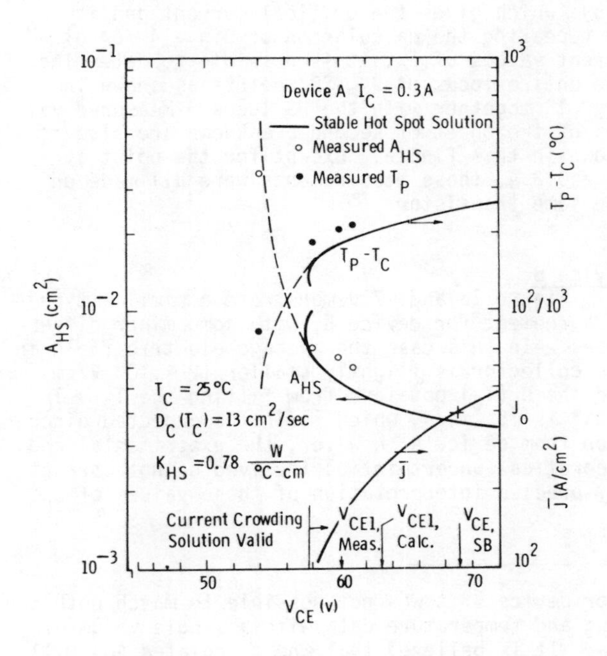

FIG. 14 - Predicted and measured behavior in the stable hot spot mode for device A.

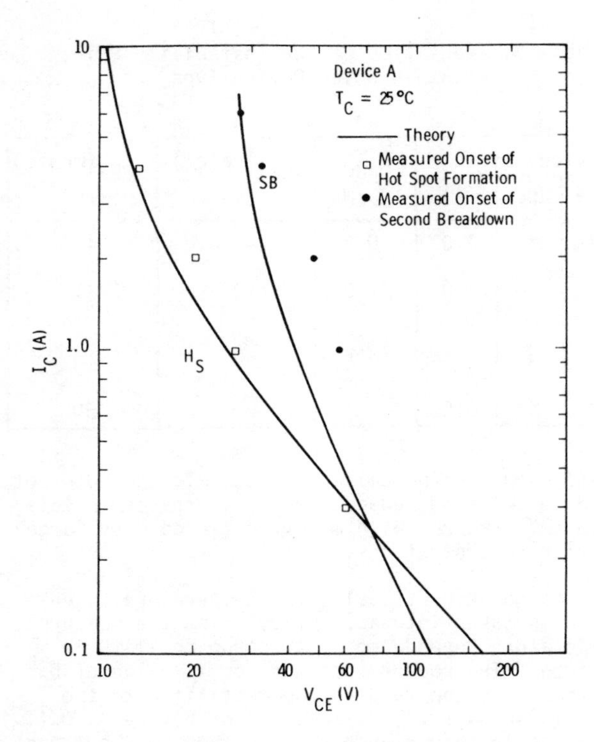

FIG. 15 - Predicted and measured HS and SB loci for device A.

HS and SB loci - device A

Also shown in Fig. 14 is a plot of the average current density $\bar{J}$ vs. V_{CE}, which is obtained from equations (13) and (15) using the calculated A_{HS} and T_s values. The V_{CE} corresponding to second breakdown is then determined from equation (16), which gives the critical current density J_o. By repeating the calculation of Fig. 14 for different values of I_C, it is possible to determine the entire locus of V_{CE}(SB) points as shown in Fig. 15 together with the HS locus. Measured values of the onset of second breakdown are also shown in this figure. Except for the point at I_C = 0.3 A, these measurements were all made on the same transistor.

Device B

Figure 16 and 17 demonstrate a similar type of agreement for device B, with some minor differences. In this case the average electric field in the collector is slightly smaller (2 × 10^4 V/cm) and the $D_C(T_C)$ obtained from fitting the T_p - T_C data* is 15 cm^2/s, which is in the expected direction from device A; however, the experimental and theoretical uncertainties involved do not warrant any precise interpretation of these values of D_C.

*For device B it was not possible to match both the A_{HS} and temperature data with a single value of D_C. It is believed that the calculated A_{HS} will be too large due to the strong departure from the uniform heat flux assumption within the hot spot (see Fig. 12).

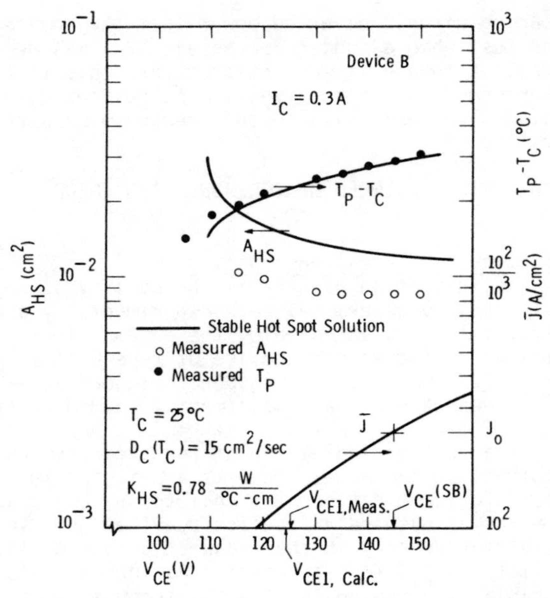

FIG. 16 - Predicted and measured behavior in the stable hot spot mode for device B.

Figure 17 shows the HS and SB loci for device B. In this case the predicted SB locus shows better agreement with the measurements. It was also observed that no stable hot spot formed for I_C = 0.1 A. That is, the transistor immediately went into second breakdown. This behavior is in agreement with the theory which says that the loci will eventually cross, and stable hot spots will not form at low currents.

Discussion

The measured SB points do not agree as well with the theory as do the HS points. This is not surprising because the stable hot spot model which yields the SB points is more complex and contains

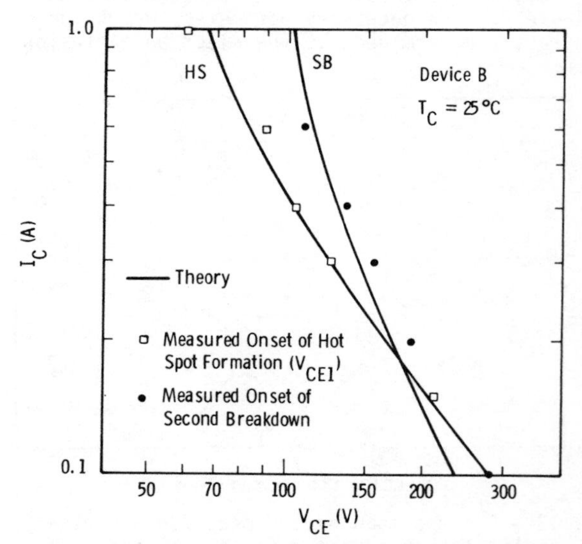

FIG. 17 - Predicted and measured HS and SB loci for device B.

more simplifying assumptions than does the uniform mode calculation which yields the HS points. For example, the heat flux may not be sufficiently uniform over A_{HS} as is assumed in equation (6). The use of a low-field model to calculate current crowding can also be questioned. In addition, experimental verification of the $\exp(-\Delta E/kT_J)$ behavior of Q_E/D_E has not been obtained in the temperature range 250 to 300°C.

Nevertheless, the calculated A_{HS} and $T_J - T_C$ curves do show reasonably good agreement with measurements and the predicted SB locus shows the proper behavior with V_{CE}. Thus, while the stable hot spot model may require some refinements in order to give a more exact description of some of the hot spot variables, it does appear to give the proper explanation of the reasons for hot spot stability, and, in addition, it also provides a satisfactory picture of the physical behavior within a stable hot spot.

CONCLUSION

Two models have been described in this paper. The first model gives an accurate prediction of the locus for which the device becomes thermally unstable. The second model shows why the hot spot stabilizes and gives an approximate picture of the physical behavior within the hot spot.

It is worth emphasizing the relative simplicity of the measurements required, especially when determining the conditions for which the initial instability occurs. The quantities needed for this calculation, R_E, I_1, and $R_{\theta JC}$ are all easily determined for any transistor. Experimental determination of the HS locus is also straightforward and makes use of the same type of equipment needed for the $R_{\theta JC}$ measurement.

Of the measurements performed, only the step and repeat infrared measurements might be considered to be out of the ordinary; however, the IRM system was used mainly to check the validity of the models and is not normally needed to calculate the HS and SB loci.

The measurements described in this paper were obtained for relatively long pulse times (1 s or longer), and one may question how the loci shift with shorter pulse times. It is reasonable to expect that the HS locus will shift to higher power levels in a manner indicated on most transistor data sheets for the SOA "second breakdown" boundary. Movement of the SB locus is less certain, but it is likely to remain fixed. This means that the current I_{C4} of Fig. 8 will increase with decreasing pulse time. It also means that second breakdown detection may be a valid technique for determining the HS locus for short pulse times but not necessarily for 1 s pulses.

This paper has concentrated on improving the understanding of transistor behavior when it is stressed beyond the useful range of operation under conditions of forward base drive. The results have been extended to the point where it is possible to define two important boundaries on the I_C, V_{CE} plane. While additional information is usually required to determine which boundary is applicable for "safe operation", the HS locus is believed to represent the best choice* for most applications. Operation within the HS boundary assures that current is never localized and that the junction temperature can be well approximated using a fixed $R_{\theta JC}$ value. Standard pulse train analyses are also applicable within this region. Finally, for long pulse times, points on the HS locus can usually be measured without taking the transistor into second breakdown, thereby avoiding the possibility of device degradation.

LIST OF SYMBOLS AND DEFINITIONS

Symbol	Definition
A_E	Effective Emitter Area (area of hot spot to which current is crowded - $A_E < A_{HS}$) (cm^2)
A_{EM}	Metallurgical Emitter Area (cm^2)
A_{HS}	Hot Spot Area (cm^2)
D_B	Base Minority Carrier Diffusion Coefficient (cm^2/s)
C_p	Specific Heat of Silicon (J/g°C)
D_C	Electron Diffusion Coefficient in Collector (cm^2/s)
D_E	Minority Carrier Diffusion Coefficient (cm^2/s)
ΔE	Activation Energy - Describes Temperature Dependence of Q_E/D_E (eV)
E_g	Forbidden Energy Gap of Silicon (1.1 eV)
I_C	Collector Current (A)
I_{C2}	Crossover Current for Current Crowding (A)
I_1	Defined by Equation 4 (A)
I_3	Defined by Equation 9 (A)
$\bar{J}$	Average Current Density in Hot Spot (A/cm^2)
J_{peak}	Peak Current Density in Hot Spot (A/cm^2)
k	Boltzmann's Constant (1.38×10^{-23} J/K)
K	Constant of Proportionality Relating $R_{\theta JC}$ and A_{EM} ($W/cm^2 \cdot °C$)
K_A	Geometrical Ratio Defined by Equation 10
K_C	Dimensionless Constant ($\sim$12 for most npn transistors)
K_{HS}	Thermal Conductivity of Region Surrounding Hot Spot (W/cm·°C)
K_{si}	Thermal Conductivity of Silicon (W/cm·°C)
L_E	Emitter Finger Half Width (cm)
L_h	Distance of Hot Spot from Emitter Bus (cm)

* A similar argument is used by Gates and Ballard [13] who describe a transistor rating procedure which in essence limits device operation to within the HS locus

L_s Emitter Finger Spacing (cm)

m_T Slope of Transistor Cooling, Plotted as $T_J - T_C$ vs. $t^{\frac{1}{2}}$ (°C/$s^{\frac{1}{2}}$)

n_{io} $n_i \cdot \exp\frac{E_g}{2kT_J}$ where n_i is the intrinsic carrier density ($n_{io} \sim 2.46 \cdot 10^{19}$ cm^{-3} at 298 K)

Q_B Total Impurity Concentration in the Metallurgical Base per Unit Area (cm^{-2})

Q_E Effective Total Doping Concentration in the Emitter - Always appears as the ratio Q_E/D (cm^{-2})

R_E Emitter Resistance (Ω)

ΔR_E Incremental Increase in Emitter Resistance in Hot Spot Mode (Ω)

$R_{\theta JC}$ Thermal Resistance (°C/W)

S Stability Factor

T_C Case Temperature (K)*

T_J Junction Temperature (°C)*

T_p Peak Junction Temperature Determined by Equation 6 (K)*

T_s Junction Temperature at which S = 1 (K)*

V_{BE} Base-Emitter Terminal Voltage (V)

V_{BEJ} Base-Emitter Junction Voltage (V)

V_{CE} Collector Emitter Voltage (V)

V_{CEI} Collector Emitter Voltage at which Instability Occurs (V)

W_C Collector Width (cm)

ρ Mass Density of Silicon (g/cm^3)

REFERENCES

[1] R. M. Scarlett, W. Shockley, and R. H. Haitz, "Thermal Instabilities and Hot Spots in Junction Transistors," *Physics of Failure in Electronics,* v. 1, Eds. M. E. Goldberg and J. Vaccaro, Spartan Books, Baltimore, Md., 1963, pp. 194-203.

W. Schroen and R. M. Scarlett, "Second Breakdown in Simplified Transistor Structures and Diodes," *IEEE Trans. on Elec. Dev.* vol. ED-13, pp. 619-622, July 1966.

*The temperatures needed in all equations are in the units of K. When data are presented and discussed, though, the temperature is given in units of °C.

[2] Frank F. Oettinger and Sherwin Rubin, "The Use of Current Gain as an Indicator for the Formation of Hot Spots Due to Current Crowding in Power Transistors," *IEEE 10th Annual Proceedings, Reliability Physics,* pp. 12-18, April 1972.

[3] David L. Blackburn and Frank F. Oettinger, "Transient Thermal Response Measurements of Power Transistors," *PESC 74 Record, 1974 IEEE Power Electronics Specialists Conference,* pp. 140-148, June 1974.

[4] R. M. Scarlett and W. Shockley, "Secondary Breakdown and Hot Spots in Power Transistors," *IEEE Int. Conv. Rec.* vol. 11, part 3, pp. 3-13, 1963.

[5] F. Bergmann and D. Gerstner, "Thermisch bedingte Stromeinschnürung bei Hockfrequenz-Leistungstransistoren (Ein Beitrag zum Problem des 'Second Breakdown')," *Arch. Elek. Übertragung,* vol. 17, pp. 467-475, 1963.

[6] P. L. Hower and P. K. Govil, "Comparison of One- and Two-Dimensional Models of Transistor Thermal Instability," *IEEE Trans. on Elec. Dev.* vol. ED-21, pp. 617-623, October 1974.

[7] H. S. Carslaw and J. C. Jaeger, *Conduction of Heat in Solids,* 2nd Ed., London, Oxford Press, 1950, p. 216.

[8] P. L. Hower, to be published.

[9] P. L. Hower and V. G. K. Reddi, "Avalanche Injection and Second Breakdown in Transistors," *IEEE Trans. on Elec. Dev.* vol. ED-17, pp. 320-355, April 1970.

[10] L. J. Giacoletto, "Measurement of Emitter and Collector Series Resistance," *IEEE Trans. on Elec. Dev.* vol. ED-19, p. 692-693, May 1972; ibid., p. 1224, November 1972.

[11] "Thermal Resistance Measurements of Conduction Cooled Power Transistors," EIA Recommended Standard RS-313-B (revision of RS-313-A), October 1975 (Electronics Industries Association, 2001 Eye Street, N. W. Washington, D. C. 20006).

[12] F. F. Oettinger, D. L. Blackburn, and S. Rubin, "Thermal Characterization of Power Transistors," *IEEE Trans. on Elec. Dev.* vol. ED-23 (to be published, August 1976).

[13] T. W. Gates and M. F. Ballard, "Safe Operating Area for Power Transistors," *Mullard Technical Communications* vol. 13, pp. 2-25, April 1974.

AN EXPERIMENTAL STUDY OF REVERSE-BIAS SECOND BREAKDOWN*

D. L. Blackburn and D. W. Berning

Electron Devices Division
National Bureau of Standards
Washington, DC 20234

ABSTRACT

Experimental results showing the influence of reverse-base current, case temperature, collector inductance, and peak collector current on the reverse-bias second breakdown (RBSB) behavior of high-voltage n^+-p-n^--n^+ power transistors are presented. The results are in qualitative agreement with the theory that avalanche injection initiates RBSB. The inductance and peak collector current results are in conflict with the theory that RBSB is initiated at a critical temperature. It is concluded that for these devices for the condition studied, RBSB is not initiated at a critical temperature. It is shown that the theory of current focusing, in conjunction with the theory of avalanche injection, does not accurately predict the RBSB conditions during device sustaining. It is proposed that other mechanisms in addition to current focusing contribute to the nonuniformity of current during transistor turnoff.

INTRODUCTION

The reverse-bias (turn-off) characteristics of high voltage, fast switching power transistors have become more important as these devices are used in increasing numbers as high speed switches in power conditioning applications. Research to improve the device and circuit operating characteristics [1-3] and to better model and understand their operation [4,5] is continuing. Perhaps the most severe impediment to this research is the continued lack of documented experimental results on which to base device design changes and with which to compare and improve the prediction of models. The scarcity of experimental data results primarily because of the difficulty in obtaining repeatable, accurate measurements in the presence of the large voltage and current transients that occur during transistor turnoff and because of the destructive nature of reverse-bias second breakdown (RBSB). The phenomenon of RBSB determines the reverse-bias limits of safe operation, but the physical mechanisms of RBSB are only partially understood.

The purpose of this paper is to present experimental results of the RBSB behavior of high voltage, fast switching power transistors obtained using the nondestructive reverse-bias safe operating area test circuit developed at the National Bureau of Standards (NBS) [6]. Earlier results have previously been reported [7,8]. This paper reports results of measurements of the RBSB behavior of n^+-p-n^--n^+ power transistors as a function of the reverse-base current, case temperature, collector load inductance, and peak collector current. It is anticipated that the new results will be valuable to device and circuit designers and modelers for testing and improving their designs and models. In addition to helping to develop a better understanding of the switching and RBSB characteristics of power transistors, the objective of the NBS work is to develop a basis for the improved characterization of these devices.

THEORETICAL BACKGROUND

It is generally believed that the mechanism of avalanche injection is the dominant initiating mechanism of RBSB in n^+-p-n^--n^+ high-voltage power transistors [4,10]. This theory assumes that the collector current density becomes large enough locally that the charge density in some region of the collector is comprised primarily of the free, current-carrying charges. The net effect of this is that the peak electric field in the collector occurs at the collector-substrate (n^--n^+) interface rather than at the base-collector (p-n^-) junction. If the field is peaked at the collector-substrate interface and is simultaneously large enough for significant carrier multiplication by impact ionization to occur, avalanche injection will be initiated [9]. This forces the device to operate in a negative resistance region (current density increases as voltage decreases) which is inherently unstable [11]. The instability manifests itself as RBSB. The electric field peaks at the collector-substrate interface whenever

$$j_C \geq q \cdot v_1 \cdot N_C , \quad (1)$$

where

j_C = collector current density (A/cm^2),
q = electronic charge (1.6 x 10^{-19} C),
v_1 = scattering-limited drift velocity ($\sim 10^7$ cm/s), and
N_C = collector dopant density (cm^{-3}).

The large current density required for avalanche injection to occur is thought to be achieved as a

*This work was done as part of the Semiconductor Technology Program at the National Bureau of Standards and was supported by Lewis Research Center, National Aeronautics & Space Administration, Cleveland, Ohio under NASA Order No. C-32818-D.

Reprinted from *IEEE Int. Electron Devices Meet.*, 1980, pp. 297-301.

result of focusing of the emitter current to the centers of the emitter fingers [4,10]. When the transistor is turned off from a saturated operating condition, the charges that have been stored in the base and collector regions are extracted via the base terminal. Because the active base beneath the emitter acts as a resistive component from and through which most of the stored charge is extracted, a voltage gradient is created along the width of the base-emitter junction. The gradient is such that the center of the emitter is more strongly forward biased than the edges and thus injects a larger current density. Expressions which have been developed for calculating the current density as a function of position along the emitter width [4,10] predict that the current density at the center of the finger, $j_e(0)$, increases as emitter current, I_E, or reverse-base current, I_{BR}, is increased.

EXPERIMENTAL RESULTS

All of the results to be discussed were generated for the transistor being turned off with an inductive load at its collector terminal. Unless otherwise noted, the magnitude of the reverse-base current, I_{BR}, was held constant during turnoff. The test circuit and conditions were the same as have been described previously [6-8].

Measurements were typically made for both clamped and unclamped conditions. The clamped condition is one for which a circuit external to the device does not allow the collector voltage to rise above a specified value, and the unclamped condition is one for which the voltage is allowed to rise with no clamping external to the transistor. Because the only significant difference observed in the RBSB behavior between clamped and unclamped conditions is in the magnitude of the collector voltage at which RBSB occurs, V_{SB}, and not in the functional dependence of this voltage on other parameters, only unclamped results are presented.

The occurrence of RBSB manifests itself as a sudden collapse of the collector voltage of the transistor from the maximum voltage achieved to about 200 V. The collapse occurs in less than 10 ns, and in this work the voltage is reduced to zero within an additional 40 ns by the protection circuit [6]. Because of these short transition times, RBSB is observed on most of the oscilloscope traces presented in this work as an abrupt halt of the collector voltage waveform with an apparent discontinuous transition to zero voltage.

The results presented in this paper represent a sampling of results of numerous measurements on numerous devices. Each of the oscilloscope tracings presented shows the results of several measurements made on a single device. Typically, between 10 and 100 similar measurements have been performed on each of these devices. The repeatability of these measurements is within the resolution of the oscilloscope presentation. The large number of RBSB measurements is made possible by the protection circuit that is used [6]. For the device types discussed in this paper, hundreds of RBSB measurements can be made on a single device with no apparent degradation of the device's electrical parameters.

Temperature

The temperature of the transistor may have a strong influence on its RBSB behavior. The collector current and voltage waveforms for a device for the case temperature T_C equal to 25°, 50°, 75°, and 100°C are shown in figure 1. It has been observed that in almost every instance, for all operating conditions, as T_C is increased, the voltage at which RBSB occurs, V_{SB}, is increased. This behavior is consistent with the avalanche injection theory of RBSB. Because the ionization coefficients for electron and holes decrease as the temperature is increased, the critical electric field required for significant impact ionization to occur is increased. This requires that the collector voltage at which the critical field is reached is also increased, i.e., V_{SB} is increased.

Reverse Base Current - I_{BR}

The magnitude of I_{BR} has a strong but varied effect on the RBSB behavior of high-voltage transistors. The measured V_{SB} is shown as a function of I_{BR} for three devices in figure 2. The variation of V_{SB} with I_{BR} can be partially explained with the aid of collector voltage and current waveforms for these same devices, which are shown in figure 3. For some values of I_{BR}, the collector voltage reaches the reverse-bias sustaining voltage, $V_{CEX(SUS)}$, prior to V_{SB}. The sustaining condition is evident in figure 3 as the voltage reaching a plateau and the current decaying at a relatively slow rate and approximately linearly with time. Because of the time scales of figure 3, it is not obvious for devices A and B at I_{BR} = 0.2 A and device C at I_{BR} = 1 A that the voltage reaches $V_{CEX(SUS)}$. By expanding the time scale for these waveforms, as done for figure 4 for device C, it can be seen that the devices do reach the sustaining condition for a brief time prior to RBSB.

Sustaining Conditions - If the device voltage reaches $V_{CEX(SUS)}$ prior to RBSB, as I_{BR} is increased, V_{SB} may either increase or decrease. For devices such as device A which remain in the sustaining condition for only a short time prior to RBSB and for which $V_{CEX(SUS)}$ is constant in time (does not change as I_C decays), V_{SB} will usually increase as I_{BR} is increased for sustaining conditions. Most often, though, $V_{CEX(SUS)}$ is not constant with time, but tends to increase as I_C decays during the sustaining condition. Devices of this type tend to remain in the sustaining condition for a rather long time prior to RBSB for the lowest values of I_{BR}. For these devices, examples of which are devices B and C, as I_{BR} is increased for sustaining operation, V_{SB} usually decreases. The V_{SB} of device C experiences both types of behavior, first decreasing and then increasing as I_{BR} is increased. The reasons for the differences in the sustaining behavior of these devices are not known. Devices A and B are identical devices made by the same manufacturer with the same

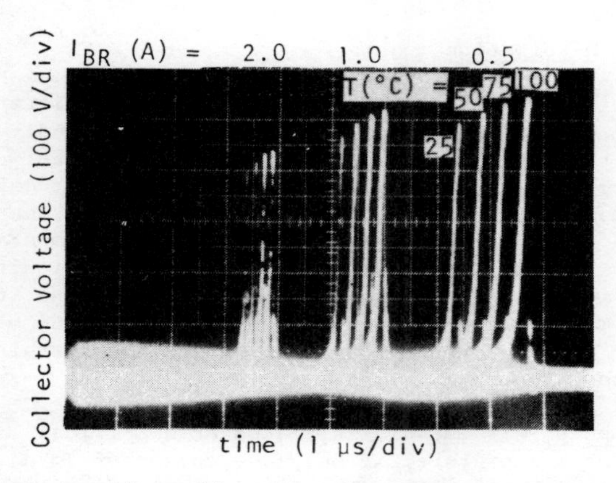

Figure 1. The collector voltage waveform for a device for three values of reverse-base current and the case temperature at 25°, 50°, 75°, and 100°C. For each I_{BR}, increasing T is from left to right. RBSB occurs at the peak of each voltage waveform for which the waveform abruptly halts.

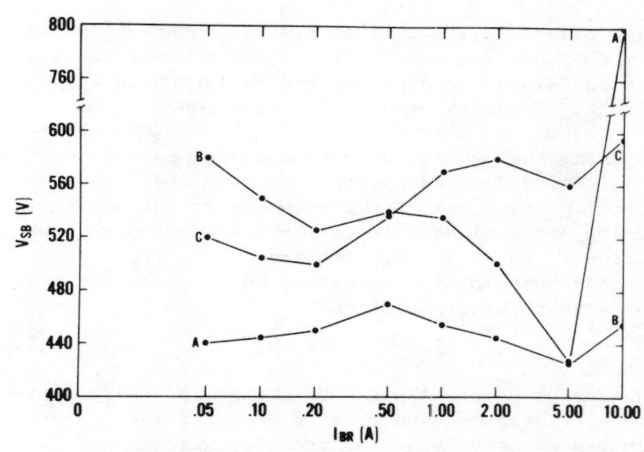

Figure 2. The measured voltage at which RBSB occurred, V_{SB}, for a collector current of 10 A for various values of reverse-base current, I_{BR}, for three devices, A, B, and C.

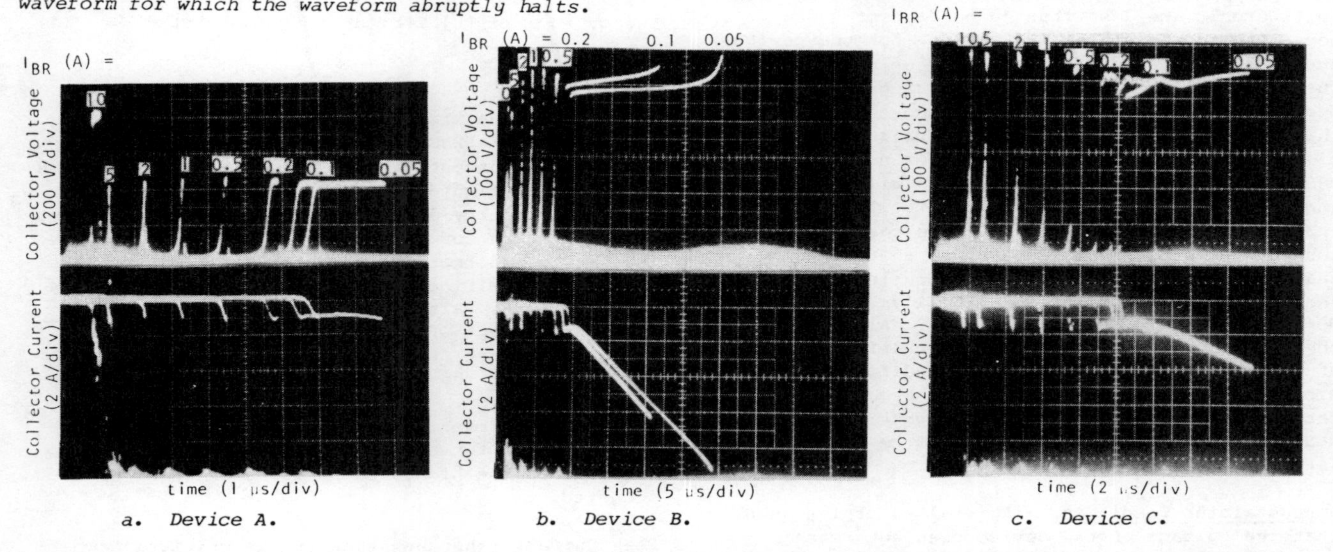

a. Device A. b. Device B. c. Device C.

Figure 3. The collector voltage and current waveform for the same devices as in figure 2. RBSB occurs when the voltage waveform appears to halt and discontinuously goes to zero.

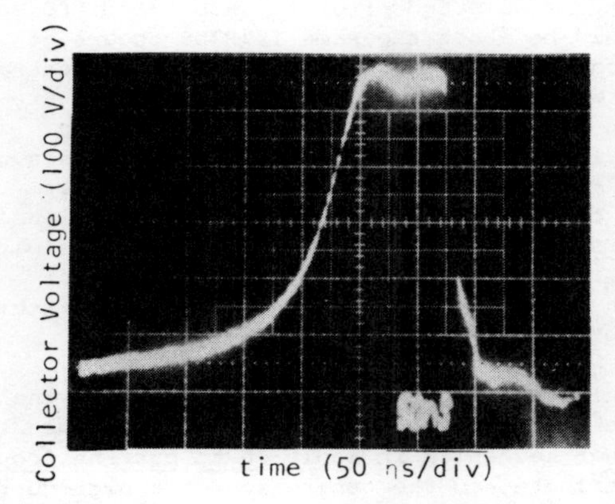

Figure 4. The collector voltage waveform at an expanded time scale for device C, I_{BR} = 1 A. The voltage collapses to about 200 V after RBSB. The voltage waveform after RBSB is determined both by the device and the protection circuit.

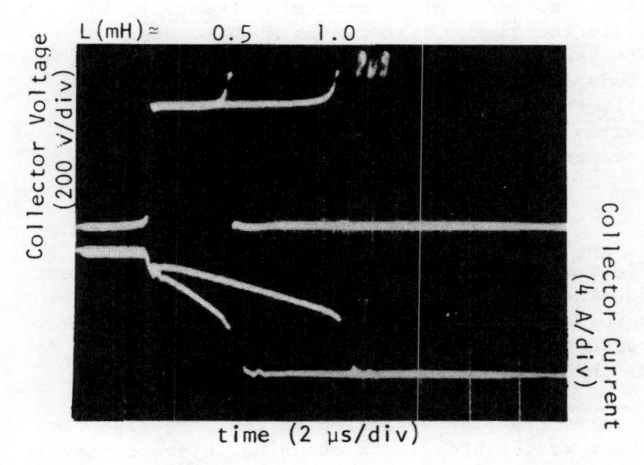

Figure 5. The collector voltage and current waveform for I_{CM} = 10 A, I_{BR} = 0.1 A, and L = 0.5 mH and 1.0 mH.

date code. Device C is of similar construction to A and B, but was made by a different manufacturer. For sustaining conditions, the variation in V_{SB} with I_{BR} is not a predictable function.

By using the expression developed for calculating $j_e(0)$ [4,10], it is possible to test the quantitative predictions of the theories of current focusing and avalanche injection for sustaining conditions. For small I_{BR} compared to I_C ($I_E \simeq I_C$) and for sustaining conditions, eq (1), the criteria for RBSB to occur, becomes

$$j_e(0) = q \cdot V_1 \cdot N_C \quad . \tag{2}$$

The device physical parameters required for computing $j_e(0)$ can be determined from spreading resistance measurements and emitter dimension measurements. Typically, for the smallest values of I_{BR} and for sustaining operation, it has been found that the computed values of $j_e(0)$ for the observed second-breakdown conditions, such as shown in figure 3, are less than 10 percent of those necessary to satisfy eq (2). Also, for the devices and conditions studied, the theory of current focusing predicts that during sustaining, as I_C is decreasing, $j_e(0)$ is also decreasing.* Thus, the theories of current focusing and avalanche injection predict that if the device does not experience RBSB before sustaining begins, RBSB will not occur and the device will safely turn off. The results in figure 3 show this not to be the case. One reason for these discrepancies may be that other mechanisms in addition to current focusing can cause the current to constrict to a locally high density. More will be said about this later. Also, because the device may be dissipating a significant amount of energy during the sustaining condition, the possibility was investigated that a thermal mechanism rather than avalanche injection might initiate RBSB. These measurements will also be discussed later.

Nonsustaining Conditions - For values of I_{BR} such that the voltage of the device does not reach $V_{CEX(SUS)}$, the behavior of V_{SB} with varying I_{BR} is as predicted by the theory of current focusing and avalanche injection. That is, as I_{BR} is increased, for a given emitter current, $j_e(0)$ should increase. This causes RBSB to occur at a decreased voltage. Eventually, though, as the magnitude of I_{BR} approaches that of the maximum collector current, I_{CM}, the magnitude of the maximum emitter current, I_{EM}, approaches zero because:

$$I_{EM} = I_{CM} - I_{BR} \quad . \tag{3}$$

As $I_{EM} \rightarrow 0$, $j_e(0) \rightarrow 0$. Because the emitter effectively is turned off and there is no current focusing, the breakdown voltage increases and can approach V_{CBO}, the open emitter, collector-base breakdown voltage. This occurs for $I_{BR} = 10$ A for all three devices of figures 2 and 3.

*A correction for variation in the effective base sheet resistance due to changing injection levels was made in performing these calculations.

Energy Dissipation

There is ample evidence that for devices of different structures and for different operating conditions than those studied here, RBSB may be initiated by a thermal mechanism [12]. When this mechanism dominates, RBSB should be initiated when the device locally reaches a critical junction temperature. Because the energy dissipated may be significant, measurements have been made to determine if a thermal mechanism is responsible for the RBSB behavior for high-voltage devices during the sustaining condition.

Inductance - If RBSB is initiated at a critical junction temperature, if only the collector load inductance L is varied, the total energy dissipated by the transistor prior to RBSB should not change. This energy is given by:

$$\text{Energy} = \int_{t_o}^{t_{sb}} I_C(t)\, V(t)\, dt = \int_{I_{CM}}^{I_{sb}} L\, I_C(t)\, dI \tag{4}$$

where t_o is the time that turnoff begins and t_{sb} is the time RBSB occurs. The collector voltage and current waveform for L = 0.5 and 1.0 mH are shown in figure 5. From these waveforms, the calculated energy dissipated prior to RBSB is about twice as great for L = 1.0 mH as for L = 0.5 mH. This indicates that the temperature at which RBSB occurs is significantly greater at L = 1.0 mH than at L = 0.5 mH. Although this is not in agreement with the concept of RBSB being initiated at a critical temperature, the results are in qualititative agreement with the prediction of the theory of avalanche injection. That is, L should have almost no effect on the magnitude of V_{SB}. The slight increase in V_{SB} for L = 1 mH in figure 5 is probably due to the higher temperature at which RBSB occurred.

Peak Current - Another test of the critical temperature theory is the RBSB behavior as I_{CM} is varied, with all other parmeters held constant. The energy dissipated prior to RBSB for different I_{CM} should be about the same if RBSB occurs at a critical temperature. The collector current and voltage waveforms for several values of I_{CM} are shown in figure 6. As I_{CM} is increased, the energy dissipated prior to RBSB also increases until I_{CM} = 15 A for which the device no longer reaches the sustaining condition, but experiences RBSB first. Neither the varying energy dissipation with I_{CM} nor the sudden change in behavior at 15 A can be explained by the critical temperature theory of RBSB.

The results of figure 6 can be explained by the theory of avalanche injection if it is assumed that some other mechanism <u>in addition to current focusing</u> contributes to the generation of a high-current density. The discontinuity in behavior can occur as I_{CM} is increased to $\sim$15 A if the value of $j_e(0)$ at the instant the device reaches the sus-

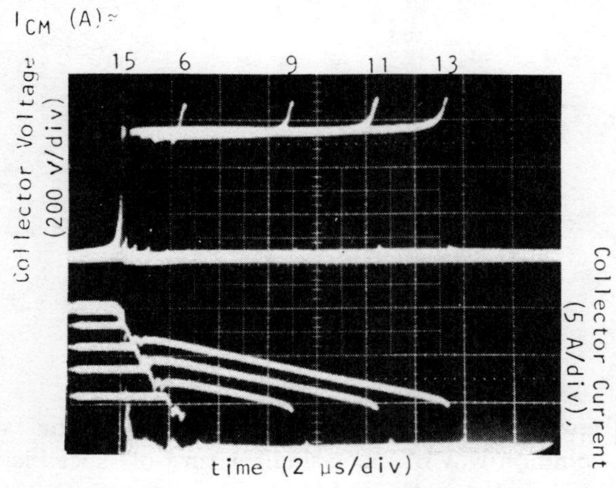

Figure 6. The collector current and voltage waveform for various values of I_{CM} and I_{BR} = 0.1 A, all other parameters held constant. When I_{CM} = 15 A, the device no longer sustains prior to RBSB.

taining condition satisfies eq (1) with the equality sign. If I_{CM} is increased further, the inequality in eq (1) is satisfied and RBSB occurs prior to sustaining. Below I_{CM} = 15 A, when the device reaches the sustaining condition $j_e(0)$ does not satisfy eq (1). As the sustaining condition continues, other mechanisms in addition to current focusing may cause the current to begin to localize within the transistor. Perhaps a phenomenon similar to the thermal instability [13] that is known to occur for forward-bias operation causes the current to begin to localize, or perhaps some regions of the transistor do not "turn off" as readily as others because of their location with respect to the base or emitter leads or because of the device geometry. If such mechanisms do occur, then the effective active area of the transistor may begin to decrease and the current density increase even though the total emitter current is decreasing during sustaining. This can cause $j_e(0)$ to increase and eventually become large enough to satisfy eq (1).

CONCLUSIONS

Measurements showing the influence of reverse-base current, case temperature, collector inductance, and peak collector current on the reverse-bias second-breakdown (RBSB) behavior of n^+-p-n^--n^+ high-voltage power transistors have been presented. The inductance and peak collector current results are in conflict with the concept that RBSB may be initiated at a critical temperature. It is concluded, for these devices and for the operating conditions studied, that the theory that RBSB is initiated at a critical temperature is incorrect. All of the results are in qualitative agreement with the theory of avalanche injection as the initiating mechanism of RBSB. The quantitative predictions of the theory of current focusing during sustaining are shown not to be accurate in predicting the RBSB behavior. It is speculated that other mechanisms contribute to the nonuniformity of the current during this type of operation.

REFERENCES

(1) W.R. Skanadore, "A New Bipolar High Frequency Power Switching Technology Eliminates Load-Line Shaping," Proc. Powercon 7, 7th National Solid State Power Conversion Conf, pp. D2.1-d2.14, Mar. 1980.

(2) B. Jackson and D. Chen, "Effects of Emitter Open Switching on the Turn-Off Characteristics of High-Voltage Transistors," PESC '80 Record, 1980 IEEE Power Electronics Specialists Conf., pp. 147-154, June 1980.

(3) K. Owyang and P. Shafer, "A New Power Transistor Structure for Improved Switching Performances," 1978 IEDM Technical Digest, 1978 IEEE Int. Electron Devices Meeting, pp. 667-670, Dec. 1978.

(4) B.A. Beatty, S. Krisna, and M.A. Adler, "Second Breakdown in Power Transistors Due to Avalanche Injection," IEEE Trans. Electron Devices ED-23, pp. 851-857, Aug. 1976.

(5) P.L. Hower, "Collector Charge Dynamics and Second Breakdown Energy of Power Transistors," PESC '74 Record, 1974 IEEE Power Electronics Specialists Conf., pp. 144-153, June 1974.

(6) D.W. Berning, "Semiconductor Measurement Technology: A Reverse-Bias Safe Operating Area Transistor Tester," NBS Special Publication 400-54, April 1979.

(7) D.L. Blackburn and D.W. Berning, "Some Effects of Base Current on Transistor Switching and Reverse-Bias Second Breakdown," 1978 IEDM Tech. Digest, 1978 IEEE Int. Electron Devices Meeting, pp. 671-675, Dec. 1978.

(8) D.L. Blackburn and D.W. Berning, "Reverse-Bias Second Breakdown in Power Transistors," Electrical Overstress/Electrostatic Discharge Symp. Proc., pp. 116-121, Sept. 1979.

(9) P.L. Hower and V.G.K. Reddi, "Avalanche Injection and Second Breakdown in Transistors," IEEE Trans. Electron Devices ED-17, pp. 320-335, Apr. 1970.

(10) S. Krishna and P.L. Hower, "Second Breakdown of Transistors During Inductive Turn-Off," Proc. IEEE 61, pp. 393-394, Mar. 1973.

(11) M.W. Muller and H. Guckel, "Negative Resistance and Filamentary Currents in Avalanching Silicon p^+-i-n^+ Junctions," IEEE Trans. Electron Devices, ED-15, pp. 560-568, Aug. 1968.

(12) H.A. Schafft, "Second Breakdown - A Comprehensive Review," Proc. IEEE 55, pp. 1272-1288, Aug. 1967.

(13) P.L. Hower, D.L. Blackburn, F.F. Oettinger, and S. Rubin, "Stable Hot Spots and Second Breakdown in Power Transistors," PESC '76 Record, Proc. 1976 Power Electronics Specialists Conf., pp. 234-246, June 1976.

Turn-Off Characteristics of Power Transistors Using Emitter-Open Turn-Off

DAN Y. CHEN
Virginia Polytechnic Institute and State University

BARRY JACKSON
General Electric

As compared with conventional reverse-biased turn-off, emitter-open turn-off provides superior transistor turn-off characteristics. Not only are the storage time and the fall time of the power transistors much reduced, but also the device reverse-biased second breakdown phenomenon, commonly associated with turn-off of inductive load, is eliminated. Furthermore the storage time tolerance due to device variation and temperature change is minimized.

I. Introduction

The turn-off characteristics of a power transistor are very important consideration in the design of power electronic circuits. They affect both the circuit energy efficiency and overall system reliability. The transistor should be turned off rapidly and held off firmly during the entire off period. The turn-off characteristics of a transistor depend not only on the intrinsic characteristics of the transistors, but also very much on the base-drive strategy used. Conventionally, a power transistor is turned off by applying a reverse-biased voltage to the base-emitter junction of the transistor. In so doing, a reverse base current is withdrawn from the transistor during turn-off. The most common way of increasing the turn-off speed is by drawing a large reverse base current from the transistor. However a danger arises, especially in the case of inductive load, because the transistor may enter into reverse bias second breakdown which may lead to the destruction of the transistor and possibly the rest of the circuit as well. Recently a new transistor circuit turn-off technique, referred to as emitter-open switching, was reported, which promises to improve both the device second breakdown capability and the turn-off speed at the same time [1, 2].

The purpose of the present paper is to investigate the turn-off characteristics of high voltage power transistors under the condition of emitter turn-off and compare the results with conventional reverse-biased turn-off. Effort was focused upon the comparison of the transistor turn-off storage time, fall time, and the second breakdown ruggedness.

Manuscript received October 24, 1980.

Authors' addresses: D.Y. Chen, Department of Electrical Engineering, Virginia Polytechnic Institute and State University, Blacksburg, VA 24061; B. Jackson, Teletype Corporation, Skokies, IL.

II. Emitter-Open Turn-off of a Power Transistor

Fig. 1(A) shows the circuit arrangement and the base current waveform associated with conventional reverse-biased turn-off, in which the reverse base current I_{BR} is always smaller than the collector current to be turned off. Fig. 1(B) shows the circuit arrangement of the emitter-open turn-off, where Q_1 is the main power transistor and Q_2 is a low voltage high speed transistor, rated for the full emitter current of Q_1. To turn on Q_1, Q_2 must be turned on at the same time. The turn-off of Q_1 is accomplished by open circuit of the emitter terminal, i.e., by turning off Q_2. During turn-off when Q_2 is open, the reverse base current I_{BR} of Q_1 immediately assumes a value equal to the collector current and flows through the diode strings D_1, D_2, and D_3 to ground. The turn-off gain is therefore equal to unity by conventional definition.

It is important to note that Q_1 is a high voltage transistor and Q_2 is a low voltage one. In the present investigation, the transistor collector-base breakdown voltage rating BV_{CBO} ranges from 600 V to 1500 V for Q_1 and is only about 50 V for Q_2. Because of much different voltage ratings, the device structure

Reprinted from *IEEE Trans. Aerosp. Electron. Syst.*, vol. AES-17, pp. 386–391, May 1981.

TABLE I
Devices Types Investigated

Device Type	Manufacturer	Catalog Number	BV_{CBO} (v)	$I_{C,max}$ (Continuous) (A)
A	Toshiba	2SC1892	1500	2.5
B	Toshiba	2SC1172B	1500	7.0
C	Amperex	BU208A	1500	5.0
D	AEG-Telefunken	BU207	1300	5.0
E	AEG-Telefunken	BU208	1500	5.0
F	AEG-Telefunken	BU209	1700	4.0
G	Motorola	BU108	1500	5.0
H	General Electric	D56W2	1400	5.0
I	Texas Instruments	TIP54	600	5.0
J	General Electric	D44TE	600	4.0

for Q_1 and Q_2 are quite different and so are the turn-off speed. For the BV_{CBO} ratings of Q_1 and Q_2 mentioned above, it is not unusual to have a ten-to-one difference in turn-off speed in favor of low voltage device Q_2. Q_2 can be turned off rapidly by reverse biasing of the base-emitter junction or even by just grounding the base terminal.

The main advantages of using emitter-open turn-off, as is shown in the experimental data in the next section, are the much improved turn-off speed and the enhanced reverse-bias second breakdown capability of Q_1. On the surface it may seem that the total transistor losses increase in the emitter-open turn-off circuit, because Q_2 contributes to additional losses. However, because Q_2 is a low voltage device, both the switching losses and the conduction loss of Q_2 are insignificant as compared with the turn-off loss of Q_1. This is especially true in the high voltage and high frequency system in which the turn-off loss of Q_1 is predominant. Therefore in the emitter-open turn-off, the gain in reducing the turn-off loss of Q_1 much offsets the loss due to Q_2.

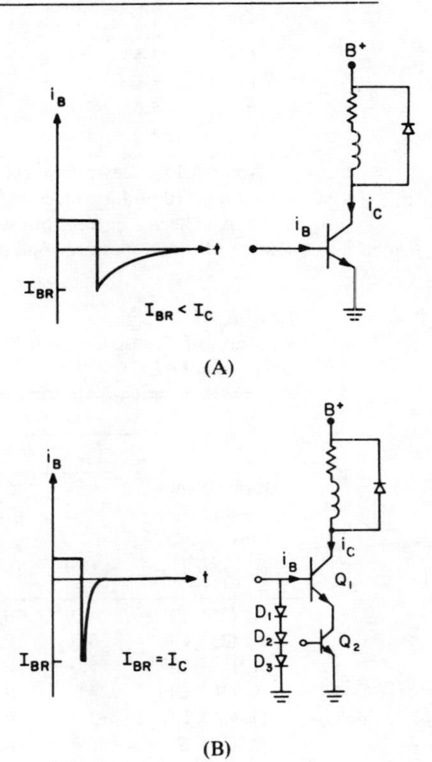

Fig. 1. (A) Reverse biased turn-off. (B) Emitter-open turn-off.

III. Comparisons of Experimental Results Between Reverse-Biased Turn-Off and Emitter-Open Turn-Off

For the purpose of comparison, two testing circuits were constructed, one for the reverse-biased turn-off test and the other for the emitter-open turn-off test. The efforts were focused on the measurement of transistor storage time, collector current fall time, and the device second breakdown capability. Second breakdown may not be destructive to the device provided that the duration of the breakdown is kept short. To accomplish such a feature, a crowbar shut down circuit is implemented in both of the testing circuits. The function of the crowbar circuit is to remove the collector current from the transistor under test as quickly as possible after the initiation of the second breakdown and therefore to prevent any permanent damage to the device [2, 3].

Ten different transistor types were chosen in the present investigation. Table I lists the 10 types with manufacturer's catalog number and voltage and current ratings. All the devices are n-p-n silicon triple-diffused high voltage switching power transistors.

A. Transistor Storage Time and Collector Current Fall Time

The transistor storage time depends not only on the transistor intrinsic characteristics and the degree of saturation at the instant of turn-off, but also on the magnitude of reverse base drive. For the same device under the same degree of saturation, the larger the reverse base current, the shorter the storage time. Table II summarizes the results obtained from both reverse-biased turn-off and emitter-open turn-off. All the devices except I and J were tested at 2.5 A collector current with a turn-on gain of 10 and collector-emitter voltage of 600 V. Devices I and J are of lower

TABLE II
Representative Data Showing Storage Time (t_s) and Fall Time (t_f) as a Function of Collector Current and Reverse Base Current for Both the Reverse-Biased and the Emitter-Open Turn-Off

	Reverse-Biased Turn-Off				Emitter-Open Turn-Off	
	I_{BR} = 0.5 A Turn-Off β = 5		I_{BR} = 1.0 A Turn-Off β = 2.5		I_{BR} = 2.5 A Turn-Off β = 1	
Test Device	t_s (μs)	t_f (μs)	t_s (μs)	t_f (μs)	t_s (μs)	t_f (μs)
A	4.0	0.3	2.4	0.225	1.2	0.125
B	12.0	0.3	6.7	0.25	2.7	0.15
C	9.0	0.35	6.0	0.28	2.2	0.2
D	12.0	0.4	6.2	0.3	2.8	0.15
E	6.0	0.25	3.4	0.2	1.7	0.15
F	2.5	0.175	1.7	0.15	0.8	0.15
G	2.5	0.14	1.6	0.1	0.75	0.09
H	6.0	0.35	3.3	0.175	1.7	0.15
I	5.0	0.30	3.0	0.18	1.4	0.10
J	4.0	0.30	2.8	0.18	1.2	0.10

Note: All the devices except I and J were tested at 2.5 A collector current with a forced β of 10 and a collector-emitter voltage of 600 V. Devices I and J were tested also at 2.5 A collector current but with a forced β of 15 and a collector-emitter voltage of 450 V. All the tests were conducted at room temperature.

TABLE III
Variation of Transistor Turn-Off Time for Both the Reverse-Biased and the Emitter-Open Turn-Off

	Reverse-Biased Turn-Off				Emitter-Open Turn-Off	
	I_{BR} = 0.5 A		I_{BR} = 1.0 A		I_{BR} = 2.5 A	
Device Types and Case Temp (° C)	t_s (μs)	t_f (μs)	t_s (μs)	t_f (μs)	t_s (μs)	t_f (μs)
A_1 (25° C)	4.0	0.3	2.4	0.225	1.2	0.125
A_2 (25° C)	6.0	0.4	3.4	0.18	1.7	0.15
A_1 (70° C)	5.0	0.5	3.0	0.17	1.6	0.15
A_2 (70° C)	8.0	0.45	4.7	0.27	2.0	0.18
D_1 (25° C)	12.0	0.4	6.2	0.3	2.8	0.15
D_2 (25° C)	8.0	0.35	4.8	0.28	2.3	0.19
D_1 (75° C)	17.0	0.55	8.0	0.4	3.2	0.17
D_2 (75° C)	11.0	0.4	6.0	0.3	2.7	0.1
H_1 (25° C)	6.0	0.35	3.3	0.175	1.7	0.15
H_1 (-20° C)	4.0	0.25	2.4	0.22	1.3	0.13

voltage rating and were tested at 2.5 A collector current and turn-on gain of 15 and a collector-emitter voltage of 450 V. All the results shown in this table were obtained at room temperature. The reverse-biased turn-off tests were conducted for the two cases when the turn-off gain is 5 and 2.5. In other words, the reverse base current is 0.5 A in one case and is 1.0 A in the other. Several remarks can be concluded from the results shown in the table. 1) Using the emitter-open turn-off, the storage time t_s is reduced by a factor of 2-4.5 as compared with the case when the turn-off gain is 5. 2) Collector current fall time t_f is reduced by a factor of 1.5-2.6 as compared with reverse-biased turn-off. 3) In the emitter-open turn-off, the reduction of storage time and fall time is more pronounced for slower devices such as types B and D. The results are not surprising because a large reverse base current, equal to collector current, is withdrawn from the base during the emitter-open turn-off, and the excessive carriers in the base region are swept out rapidly primarily by the large reverse base current rather than by recombination.

B. Temperature Effect on Storage Time and Fall Time

Both the storage time and the fall time of a transistor vary significantly with temperature and device. However, the extent of variation depends on the turn-off strategy used. Table III summarizes the test results of the variation of transistor turn-off time for both the reverse-biased and the emitter-open turn-off. In the table a transistor is designated with a device type and a subscript number which is used to distinguish different devices within the same device type. For example, transistors A_1 and A_2 are different devices but both are type A devices. As can be seen from the table, the range of tolerance of transistor storage time and fall time, due to either temperature change or device variation, is minimized in the emitter-open turn-off. The effect is especially pronounced for a slow device such as transistor D_1. The results shown are plausible because it is the reverse base current, not the recombination, that plays the dominant role in sweeping out the excessive carriers in the emitter-open turn-off.

C. Reverse-Biased Second Breakdown Ruggedness

In this paper the device second breakdown ruggedness is characterized by V_{CEP}, the peak voltage blocking capability of the collect-emitter terminal during turn-off. For a given device, the value of V_{CEP} depends on the condition of I_{BR} and collector current I_C. The tests of device second breakdown ruggedness are focused on the display of collector-emitter voltage waveforms during turn-off and the measurement of V_{CEP} from the displayed waveforms. Figs. 2 to 5 show the collector-emitter waveforms for different types of devices for both the reverse-biased and the emitter-open turn-off. As can be seen from Fig. 3(A), during the turn-off period of type A device, V_{CE} rises from the conduction voltage drop V_{CESAT} to about 680 V, at which second breakdown initiates and V_{CE} drops rapidly to second breakdown level of about 100 V and stays there for about 160 ns until the collector current is rapidly removed by the testing circuit. During the second breakdown period, large power dissipation occurs in the device due to the coexistence of high current and high voltage. But 160 ns is short enough that any permanent damage to the device is avoided. For the same device under emitter-open turn-off, however, second breakdown phenomenon has never been observed. As can be seen from Fig. 3(B), V_{CE} rises from V_{CESAT} all the way up to BV_{CBO} without losing voltage blocking capability. Figs. 3 and 4 show similar results for device types D and F.

Table IV summarizes the results obtained from second breakdown tests for both turn-off conditions. Because of the nondestructive nature of the tests, the same transistor can be tested under different conditions. As can be seen from the table, for device B, the value of V_{CEP} depends on I_C and I_{BR}. For the same I_C, the value of V_{CEP} could increase or decrease with I_{BR} in the case of reverse-biased turn-off, which is consistent with the results reported in [3]. However, in all the tests conducted for the emitter-open turn-off, the second breakdown phenomenon has never been observed.

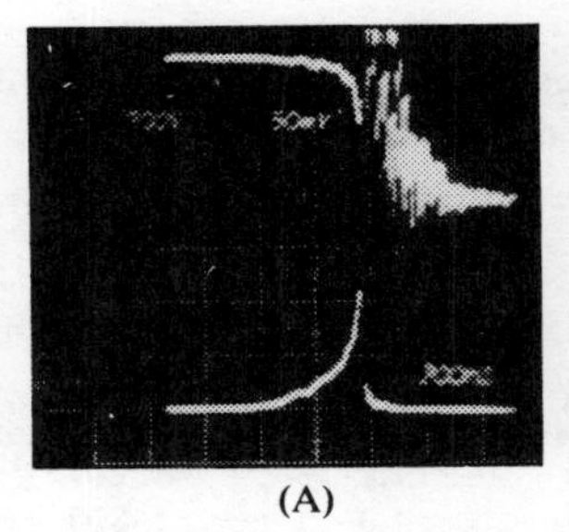

(A)

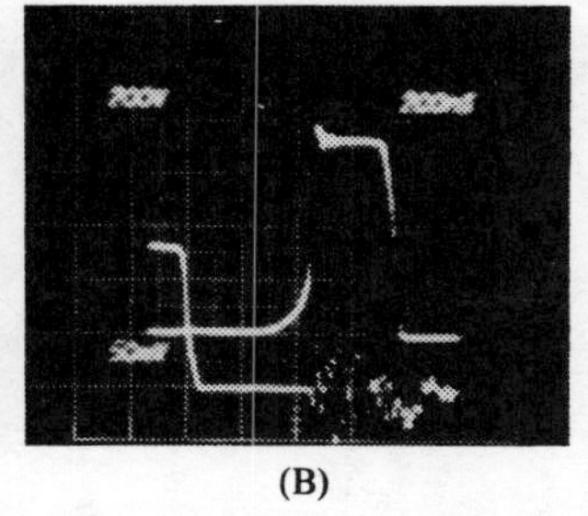

(B)

Fig. 2. Waveforms associated with second breakdown tests for type A device. (A) Reverse bias turn-off. Parameters are I_C = 2 A/cm; V_{CE} = 200 V/cm; time = 400 ns/cm. (B) Emitter-open turn-off. Parameters are V_{CE} = 400 V/cm, I_E = 1A/cm, time = 1 μs/cm.

Fig. 3. Waveforms associated with second breakdown tests for type D device. (A) Reverse bias turn-off. Parameters are I_C = 1 A/cm, V_{CE} = 400 V/cm, time = 200 ns/cm. (B) Emitter-open turn-off. Parameters are V_{CE} = 400 V/cm, I_E = 1 A/cm, time = 1 μs/cm.

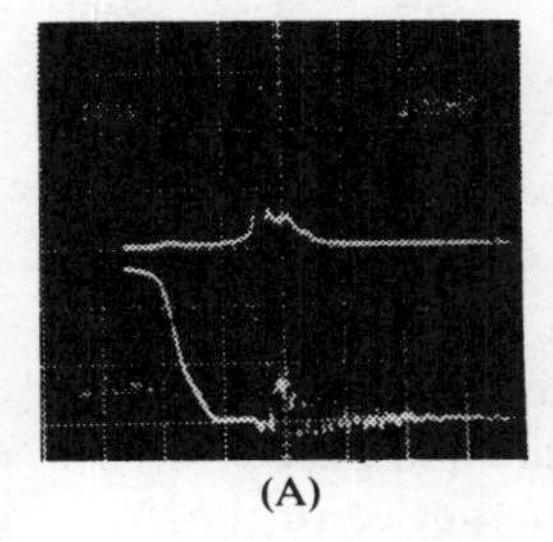

(A)

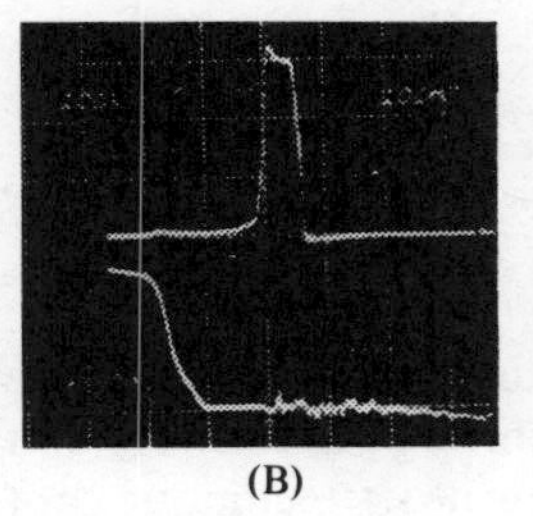

(B)

Fig. 4. Waveforms associated with second breakdown tests for type F device. (A) Reverse bias turn-off. Parameters are I_C = 2A/cm, V_{CE} = 400 V/cm, time = 400 ns/cm. (B) Emitter-open turn-off. Parameters are V_{CE} = 400 V/cm, I_E = 1 A/cm, time = 400 μs/cm.

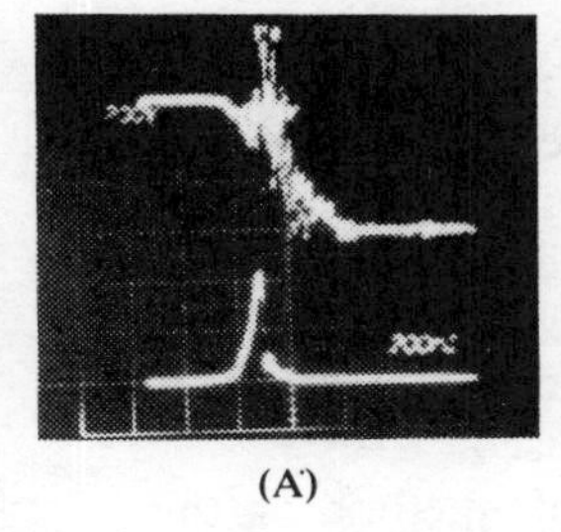

(A)

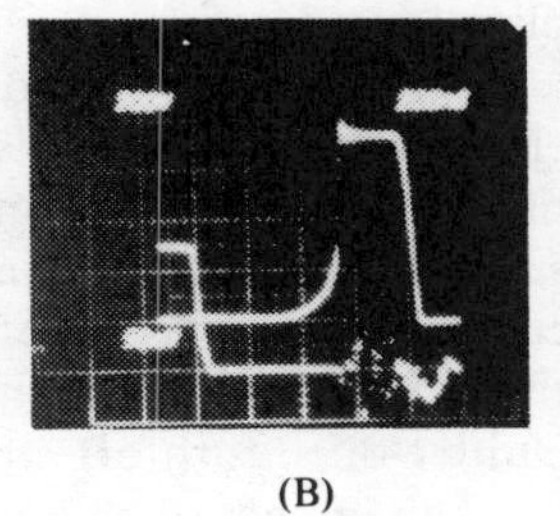

(B)

The results obtained are somewhat surprising but are explainable. According to theory, the second breakdown phenomenon in the reverse-biased turn-off of inductive load is attributed to the emitter current constriction during the turn-off process [4, 5]. The constriction of the current is caused by the potential drop laterally along the base-emitter junction as a result of the base region lateral resistance and the reverse-base current. In the case of the emitter-open turn-off, the emitter terminal is open during turn-off, and the collector current is diverted out of base terminal. Therefore current constriction phenomenon does not occur and the second breakdown is avoided.

V. Conclusion

While the power transistor manufacturers constantly try to fabricate faster and more rugged devices, the design parameters are normally such that the device ruggedness is achieved at the expense of

TABLE IV
Device Second Breakdown Ruggedness for Both the Reverse-Biased and the Emitter-Open Turn-Off

		Reverse Biased Turn-Off ($I_{BR} < I_C$)		Emitter-Open Turn-Off ($I_{BR} = I_C$)	
Device Type	I_C (A)	I_{BR} (A)	V_{CEP} (V)	$I_{BR} = I_C$ (A)	$V_{CEP} = BV_{CBO}$ (V)
A	2.5	1.1	680	2.5	1500
B	2.5	1.0	800	2.5	1500
B	2.0	1.0	910	2.0	1500
B	2.5	0.8	760	2.5	1500
B	2.5	0.3	840	2.5	1500
C	2.5	1.3	900	2.5	1500
D	2.5	1.6	870	2.5	1500
E	2.5	0.9	820	2.5	1500
E	2.5	0.6	720	2.5	1500
E	2.0	0.9	900	2.0	1500
F	2.5	0.7	860	2.5	1500
G	2.5	1.5	1100	2.5	1500
H	2.5	1.0	1000	2.5	1400
I	2.5	1.0	390	2.5	600
J	2.5	1.0	400	2.5	600

turn-off speed or vice versa. The test results reported in this paper show that, by using the emitter-open turn-off technique, not only can the device turn-off speed be significantly increased but also the reverse bias second breakdown phenomena can be eliminated. The very same device can then be utilized to its full potential for higher frequency and higher voltage applications.

In addition to the advantages mentioned above, emitter-open turn-off also provides the designers practical means of minimizing the variations of device storage time, fall time, and second breakdown capability. The agony of matching transistor storage time in some applications may become unnecessary if emitter-open turn-off strategy is used.

The impact of this new turn-off method on the design of both the device and the circuit remains to be evaluated. From the circuit designer's point of view, the emitter-open turn-off method provides a practical means of significantly improving both the turn-off speed and the second breakdown ruggedness of the device at the same time. From the device designer's point of view, the rule of tradeoff in designing a high voltage power transistor may need some modifications since the device turn-off characteristics are significantly improved by the use of the emitter-open turn-off method. The turn-off characteristics may be traded for other transistors parameters such as current gain, leakage current, and saturation voltage drop.

References

[1] J.D. Walden
U.S. Patent 3 956 684
DC-AC inverter Having Improved Switching Efficiency, Overload and Thermal Protection.

[2] Jackson, B., and Chen, D. (1980)
Effects of emitter-open switching on the turn-off characteristics of high voltage power transistors.
Presented at the Power Electronics Specialist Conferences, June 1980.

[3] Jahns, T.M. (1974)
Investigation of reverse-bias second breakdown in power transistors.
M.S. Thesis, Massachusetts Institute of Technology, Cambridge, May 1974.

[4] Scarlett, R.M., and Schockley, W. (1963)
Secondary breakdown and hot spots in power transistors.
In *IEEE International Convention Record*, Part 3, 1963, *11*, 3-11.

[5] Hower, P.L., and Reddi, V.G.K. (1970)
Avalanche injection and second breakdown in transistors.
IEEE Transactions on Electronic Devices, 1970, *ED-17*, 320-335.

AN OVERVIEW OF LOW-LOSS SNUBBER TECHNOLOGY FOR TRANSISTOR CONVERTERS

Angelo Ferraro

Corporate Research and Development
General Electric Company
Schenectady, N.Y.

ABSTRACT

A study of the state-of-the-art for low-loss snubber technology has been completed, and the results are summarized. The review begins with the purpose and philosophy in the use of snubbers, and ambiguities of their benefits are discussed. Finally, a system of classification for snubbers is proposed as a means of defining snubber operation more exactly.

I. INTRODUCTION

The reliable application of power semiconductors is very often determined by the attention given the techniques for stress relief. The methods available to the circuit designer are numerous and involve not only the manner of control, but very often include additional circuitry which is recognized by several terms including snubber, stress relief, load line shaping, voltage clamps, and stress reduction networks. These networks are employed for both turn-on and turn-off stress relief. The purpose of this paper is to review and assess the state of snubber technology for transistor converters.

Although the use of snubbers in the stress relief of power semiconductors is common to both thyristors and transistors, the philosophy of operation is quite different. The turn-off snubber for the thyristor is primarily intended to limit or control the rate of rise of the voltage applied at the end of a conduction cycle. The requirement to limit the rate of rise of voltage, the *dv/dt,* is based on the thyristor's ability to successfully commutate the conduction current in a prescribed length of time. Although the design and operation of snubbers for thyristors is not essential to this study, the methods are well documented and can be useful in the analysis of present circuits of interest [1] [2] [3]. The purpose, design, and operation of snubbers for transistors is significantly different and less definitive in absolute performance requirements. To further obscure the purpose of snubbers and their parameters for design, several of the newer devices such as power JFET, MOSFETs, or the reemerging gate turn-off thyristor (GTO) possess unique snubber requirements. The discussion to follow will attempt to reveal the existing ambiguities and, where possible, establish a basis for evaluation of low-loss snubber technology. A system of classification will also be proposed as a point-of-reference for future developments. The review, however, will set out with a discussion of snubber philosophy.

II. SNUBBER PHILOSOPHY

The intention of utilizing stress-relieving networks in power conversion equipment as mentioned previously is multifarious. The snubber function as applied to modern power devices would include:

- switching loss transfer
- overvoltage suppression
- rate of rise control of voltage and current
- noise and electromagnetic interference abatement
- secondary breakdown avoidance.

All of these functions will be individually discussed, although each one is dependent on the others.

The networks that might be considered the conventional snubbers are given in Table I and are presented not only by when they afford the power semiconductor protection (i.e., turn-on, turn-off) but also by the polarized derivative forms. The transistor Q in each instance is the device being protected by the passive snubber components R_s, L_s, C_s. The load that the transistor is controlling is not shown but can be connected to either collector or emitter terminal shown. The operation of the snubber is facilitated by Fig. 1, which shows typical voltage and current waveforms for a transistor without protection; Fig. 2 illustrates the device with both turn-on and turn-off snubbers in place. When the transistor turns on, the capacitor C_s discharges through R_{sc} and is shown as a step in the transistor current. The load current, however, must be developed through the turn-on snubber L_s, resulting in the gradual increase of collector current. Upon turn-off, the load current is shunted by C_s, allowing the transistor to turn off with reduced stress. The turn-on snubber, however, must now be discharged, and this can be seen as the collector-to-emitter voltage transient above the supply voltage. The experienced snubber designer will immediately recognize techniques to moderate the circuit performance with additional reactive elements shown as dotted components in Fig. 2. A more detailed analysis, however, must be delayed at this point to more closely examine the designer's impetus to formulate low-loss snubber approaches.

The operation of conventional dissipative snubbers at low frequencies (utility frequencies) or low power levels (converter power < 250 W) can generally absorb this power loss without serious consequence. The situation clearly changes, however, as the converter operating points increase. As the system frequency is increased substantially above utility frequencies, the individual turn-on and turn-off snubbers must be discharged of their stored energy for every transistor cycle. It should be recalled that the energy stored in an inductor is determined by $1/2\ L\ I^2$, and the stored energy of a capacitor is $1/2\ CV^2$. This energy is not dependent on the

Reprinted from *IEEE Power Electron. Specialists Conf.*, 1982, pp. 466–477.

Table I

CONVENTIONAL DISSIPATIVE SNUBBERS

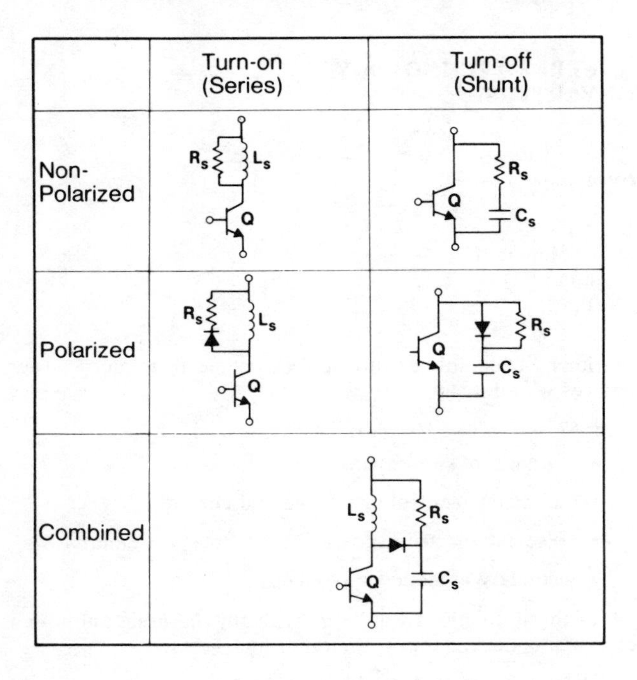

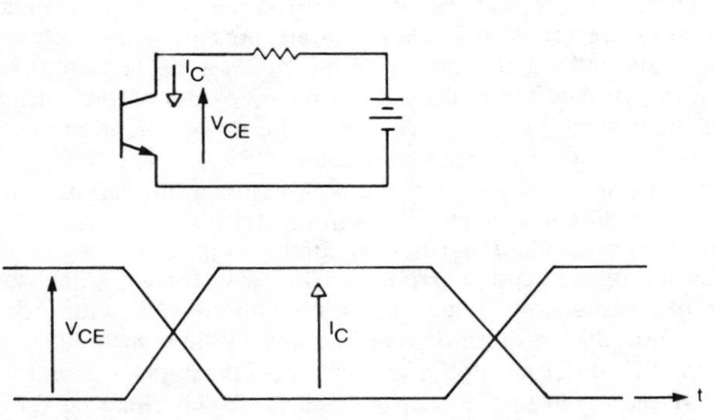

Fig. 1. Transistor Without Snubber Protection

value of the discharge resistor chosen. The power loss can then be determined by the frequency of the snubber discharge times the trapped energy; for example,

$$P_s = f_s\ (1/2\ CV^2) \tag{1}$$

is the power lost by the turn-off snubber of Fig. 2. The graph of Fig. 3 illustrates typical inefficiency contributed by such a snubber as a function of frequency and demonstrates why the conventional snubber loss cannot be tolerated at the frequencies the designer is using today. In addition, the increase in power bandwidth product of today's transistors is being reflected not only in speed but also in the voltage and current levels that can be handled [4]. This encroachment by transistors into the former thyristor domain has allowed new power control strategies and directed the development of new stress-relieving techniques [5]. With this basis for stress relief established, a more detailed discussion of snubber philosophy can follow.

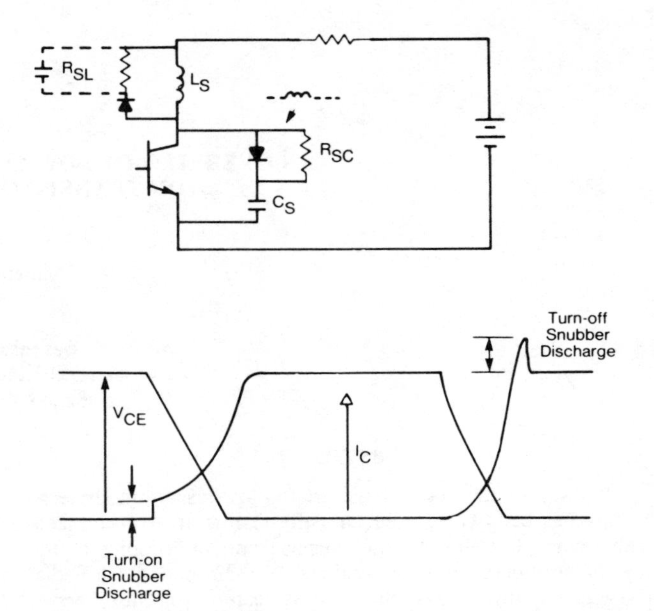

Fig. 2. Transistor with Turn-On, Turn-Off Protection

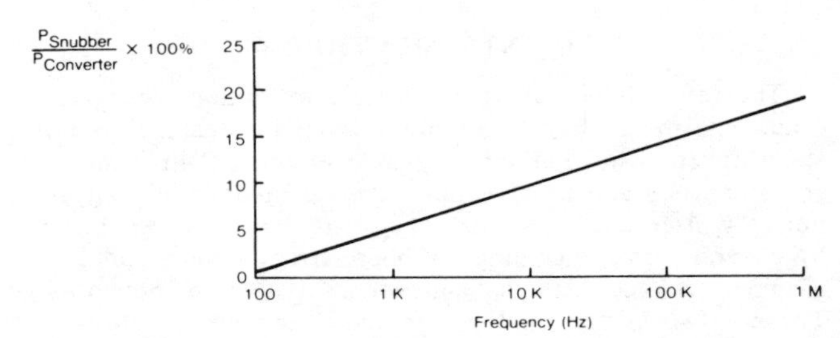

Fig. 3. Typical Snubber Loss as a Function of Frequency

Switching Loss Transfer

The power loss of the circuit in Fig. 1 is given by Fig. 4a, the entire amount which must be dissipated by the transistor. However, the proper selection of the snubber components will alter the load line of the transistor, as shown in Fig. 5, and result in the power dissipation curves for the transistor with snubber protection, Fig. 4b, and the snubber discharge dissipation, Fig. 4c. The sum of the two dissipations of Figs. 4b and 4c is the switching loss of the snubber protected system. A thorough analysis by McMurray [7] provides the results shown in Fig. 6 and illustrates the optimization that can be employed by the designer. It should be noted that not only is the switch loss transferred from the transistor to the discharge resistors, but the total power loss can be lowered when compared to the unprotected case. Calkin and Hamilton [8] demonstrated this in their work, which reduced the transistor switching loss by an order of magnitude while the transferred snubber loss resulted in half the total switching loss. Although the designer today may require more substantial improvements, it is clearly indicated that these optimization techniques should not be overlooked as a possible solution to low-loss snubbers.

Overvoltage Suppression

The result of a transistor switching an inductive load such as shown in Fig. 7a, can be the voltage surge of Fig. 7b, the

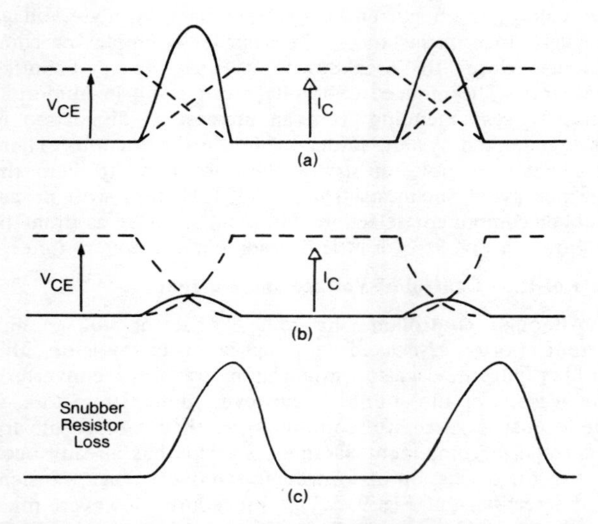

Fig. 4. Switching Loss with and Without Snubber Protection; (a) unprotected, (b) protected, (c) snubber resistor loss

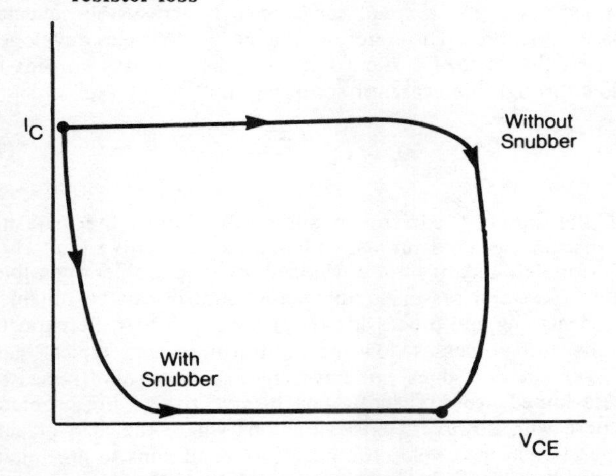

Fig. 5. Turn-Off Switching Load Lines for Transistor with and Without Snubber Protection

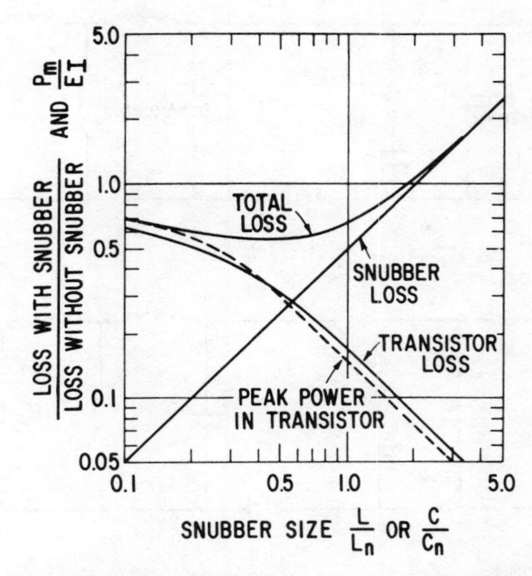

Fig. 6. Relative Switching Losses and Peak Transistor Power Dissipation as a Function of Snubber Size (from [7])

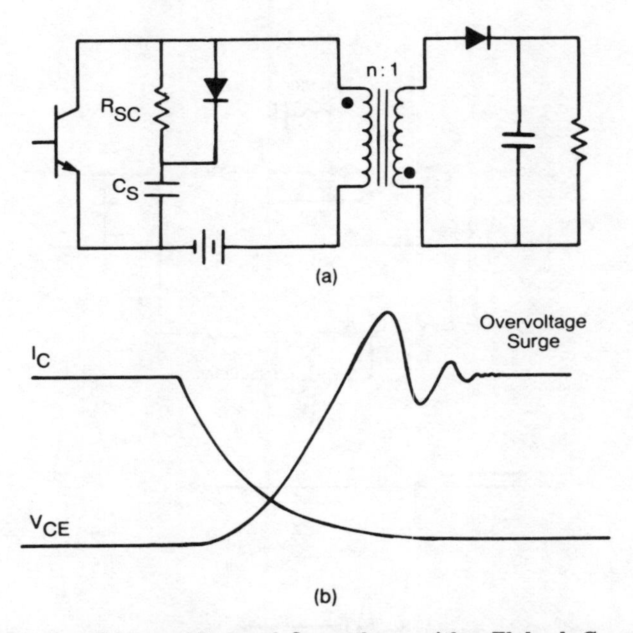

Fig. 7. Effects of Induced Overvoltage with a Flyback Converter; (a) Circuit, (b) Overvoltage at Turn-Off

exact nature of which is determined by the specific components in place. If there is no snubber, for instance, and the transistor is sufficiently fast, the voltage transient can easily exceed the transistor voltage breakdown. An analysis by Harada et al. [9] is given for a similar circuit with the use of a conventional nonpolarized snubber. The analysis begins with a high-frequency model of the circuit and, by the use of root locus methods, the procedure for an optimized solution is given. In this instance, the objective is to maintain specific control on the overvoltage surge while allowing an optimized selection of snubber components to control the snubber loss. Although the procedure is presented for the case of the nonpolarized RC snubber, its application can be extended to several other situations by judicious modification of the models. The case of the polarized snubber can be accommodated by considering the resistor with an asymmetric conductance resulting in models accounting for the diode forward conductance and for the resistor with the diode reversed biased. This would alter the circuit of Fig. 7a to the high-frequency models of Figs. 8a and 8b, where they correspond to the conduction and nonconduction mode of the snubber diode, respectively.

This occurrence of overvoltages can also be a result of differences in device recovery time, which are either inherent to the operation of the device or a result of the circuit's layout constraints. The converter of Fig. 9a is shown with the parasitic inductances of the circuit wiring and the transformer leakage inductance as seen by the secondary. The waveforms of Figs. 9b and 9c display the voltage induced across the rectifiers without and with snubber protection, respectively. This is a very real situation to the high-frequency designer of

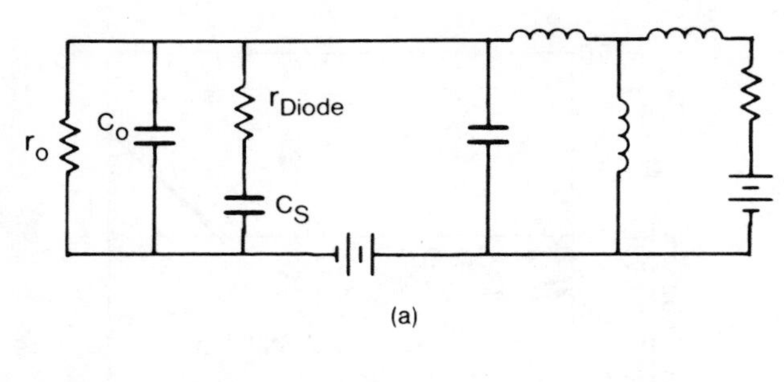

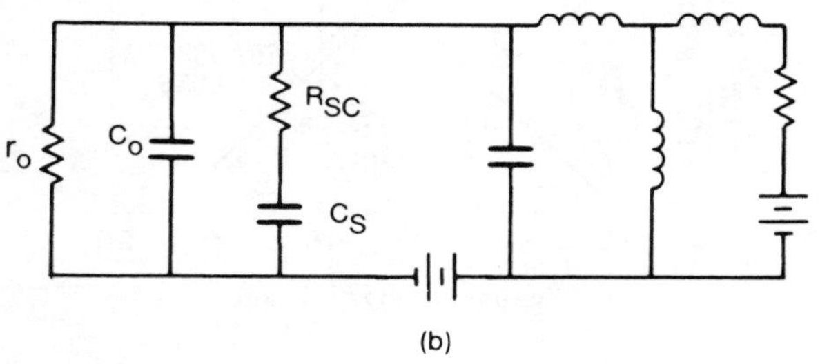

Fig. 8. High-Frequency Model of Circuit of Fig. 7 a) snubber diode conduction b) snubber diode non-conduction

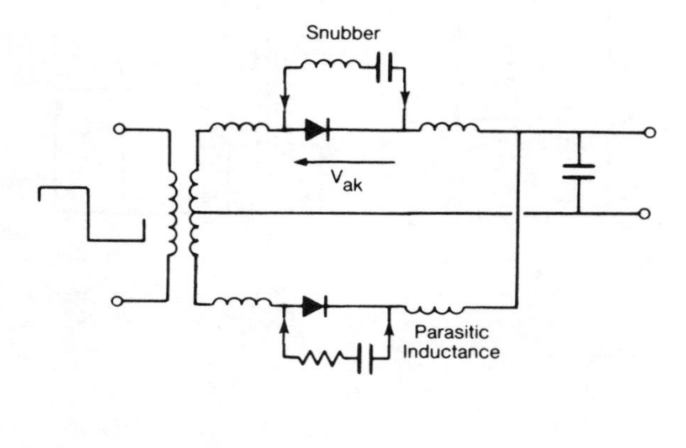

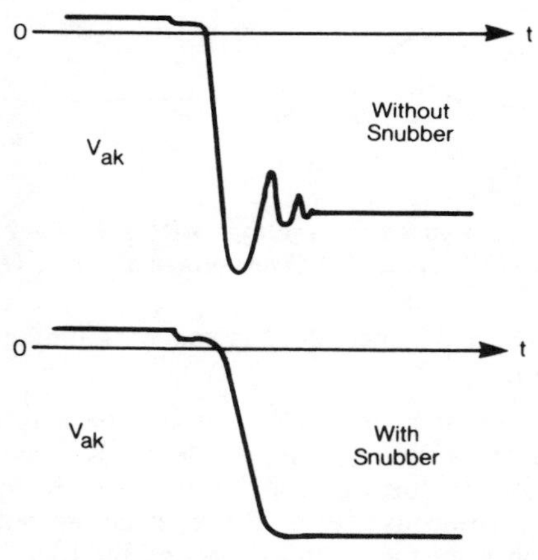

Fig. 9. Rectifier Stress Relief

low voltage, high current converters or very high voltage supplies. In both instances, the designer is employing components close to breakdown voltages (e.g., Schottky rectifiers). The induced overvoltage can result in numerous consequences extending from an increase in dissipation to noise-generated system faults to device destruction. There have been several industry application notes to help the designer avoid such catastrophy [10] [11] and, with proper snubber component selection, the results can be as dramatic as shown in Fig. 9c with little degradation in recovery time.

Rate of Rise Control of Voltage and Current

Although controlling the rate of rise of voltage and current is often associated with application to thyristors, this snubber purpose is also employed in transistor converters. The reason for the snubbers, however, unlike thyristors, is not to ensure successful commutation; they compensate for and equalize component parameters. This has already been seen in the discussion of controlling the overvoltage transient of the rectifiers of Fig. 9a. This procedure, however, must now be a consideration in the application of power MOSFETs to high-frequency converters. Figure 10 illustrates a parasitic low gain bipolar transistor that exists in power MOSFETs biased such that a small capacitance is across the collector base junction. The rate of change of voltage developed across the MOSFET (i.e., V_{DS}) will cause a base current to flow through the capacitor according to

$$i_p = C_p \frac{d\, V_{DS}}{dt} \tag{2}$$

If the induced current is sufficiently large, the parasitic transistor can be saturated, initiating a destructive fault [12]. Fortunately this has not developed into the problem possible, due to several reasons. Substantial control can be obtained by designing and processing the device such that the parasitic transistor possesses low gain. Furthermore, typical gate driver circuitry does not have the capability of driving the gate impedance sufficiently fast to give rise to the problem. There will, however, exist situations where the load or gate conditions can develop the necessary conditions to precipitate a dv/dt failure. Snubber use would be indicated; however, the application would likely be very high in frequency, and close attention to snubber loss would be essential.

The constant drive to higher frequency power systems also allows the use of distributed component snubbers. McMurray [7] refers to these as parasitic snubbers in the example of a flyback converter where the transformer leakage inductance and wiring inductance serve to limit the di/dt. It is also realized that the snubbers so obtained are seldom optimal unless control, if possible, is exercised over the causative parameters.

Noise and Electromagnetic Interference Abatement

Effective and cost-effective control of equipment-generated noise and electromagnetic interference (EMI) can be realized by the utilization of snubbers. In this application, the high-frequency model can be derived to include coupling mechanisms to the various structural elements of the circuitry, requiring either calculation or measurement of such parameters. (These parasitics typically include component ca-

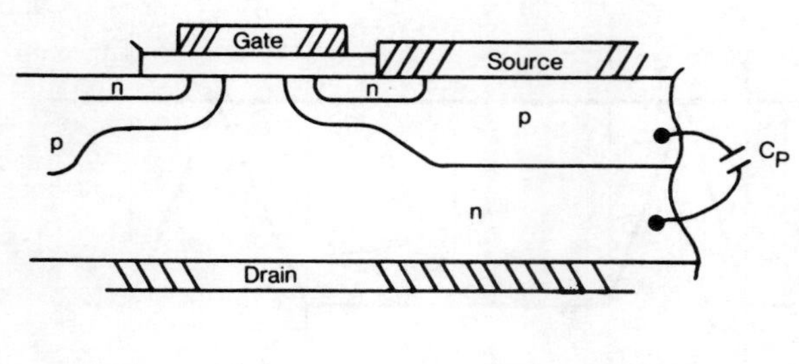

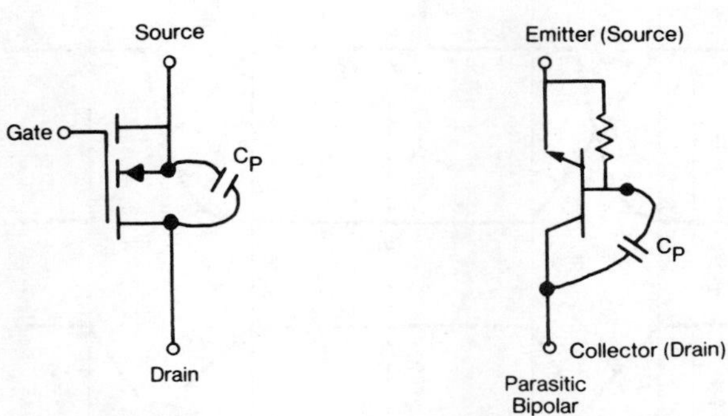

Fig. 10. Simplified MOSFET Section Illustrating Parasitic Bipolar Transistor

pacitance to ground, wiring inductance, and parasitic transformers due to physical nuances of a particular system.) In the converter of Fig. 11a, Harada and Ninomiya [13][14][15] have extensively evaluated a number of these mechanisms to produce high-frequency models, such as Fig. 11b, which allow voltage and current surge analysis resultant from the switching semiconductors. It is shown that eliminating the coupling mechanism (Cg approaches zero) also eliminates the noise; that is, the noise-conducting circuits are broken. The physical constraints of a system, however, do not always allow the luxury of such a solution and snubbers offer an alternative. The reduction of the magnitude and frequency content of the voltage and current surges reduces, in effect, the importance of the coupling mechanism, C_g, thereby reducing the system-generated noise.

Secondary Breakdown Avoidance

The most significant application of snubbers in transistor converters is in the avoidance of secondary breakdown. This phenomenon, although it may be generated by several distinct mechanisms, generally exhibits the common result of irreversible deterioration or destruction of the semiconductor. From the circuit designer's viewpoint, secondary breakdown can be classified by two biasing conditions of the device: forward or reverse bias base-to-emitter junction. The capabilities of a device can be characterized by a set of figures referred to as forward and reverse bias safe operating areas (FBSOA and RBSOA) in the plane of collector current and collector-to-emitter voltage such as in Figs. 12a and 12b.

The load line that results in the switching of the inductive load, Fig. 13a, can be mapped onto the FBSOA and RBSOA curves to reveal the operating stress, Figs. 13b and 13c. Clearly the turn-on interval shows the load line well within the FBSOA; however, the turn-off interval causes the load line to cross the RBSOA, indicating a circuit modification is necessary to ensure reliable operation. The circuit

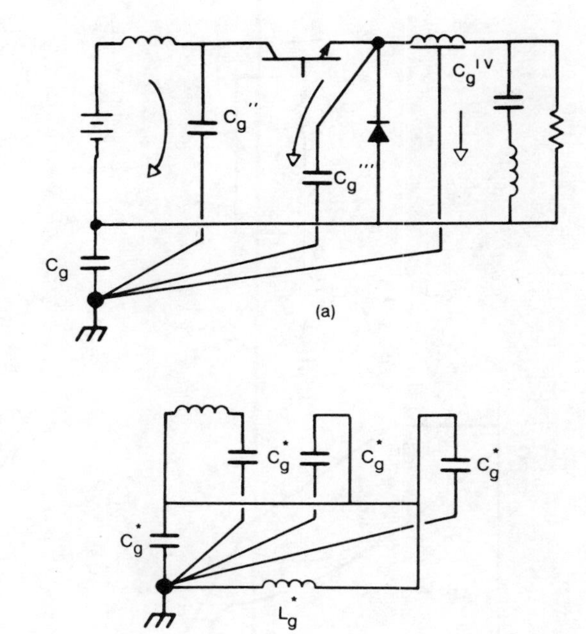

Fig. 11. (a) Converter Showing Parasitic Components Causing Ground Currents and Noise, (b) High-Frequency Model

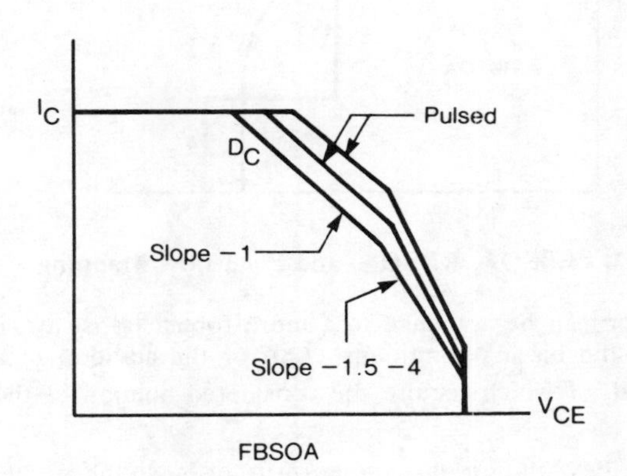

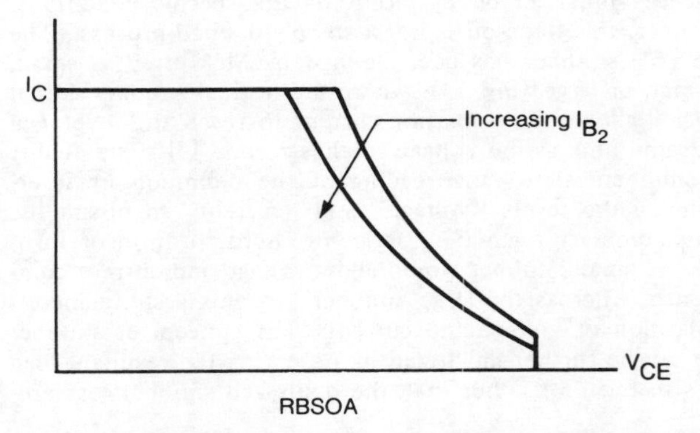

Fig. 12. Typical Forward and Reverse Bias Safe Operating Area for Transistors

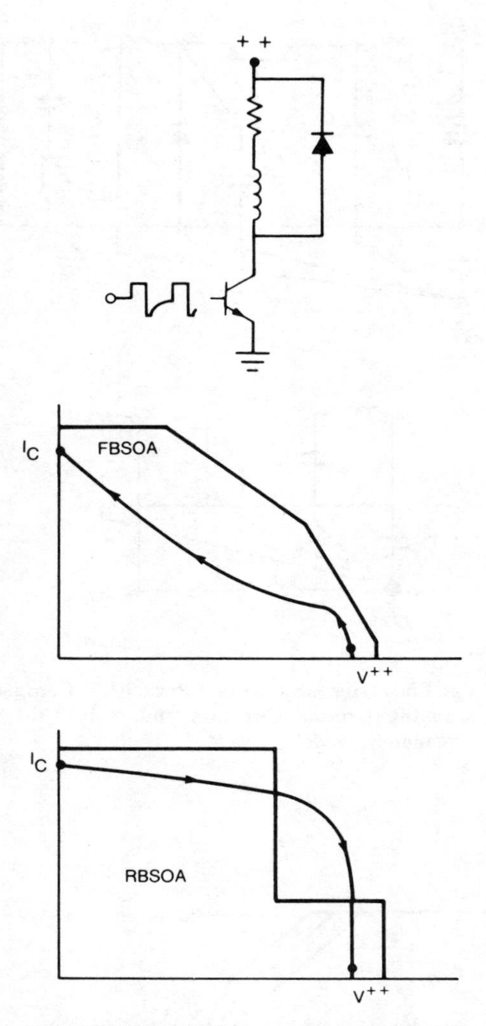

Fig. 13. FBSOA, RBSOA, and Load Line Mapping

modification can be a change to a more robust transistor, a change in the biasing conditions (I_{B_2}), or the addition of a snubber, all of which require the considered opinion of the designer.

When the solution takes the form of a snubber, the power loss, the stress relief, and numerous other pragmatic concerns must all be balanced for the specific conditions; however, this trade-off is not a sharply defined process. The size of a snubber has been defined by McMurray as small, normal, or large (Fig. 14), where the normal snubber is that which "allows the transistor current to reach the level I at the same time as the voltage reaches zero..."[7]. The duality of snubbers allows the reading of the definition by interchanging the terms "voltage" and "current" to obtain the complementary form (i.e., series or shunt, turn-on or turn-off). A small snubber would allow voltage and current coincidence, whereas the large snubber prevents a simultaneous application of voltage and current. The concept of snubber size can be further illustrated by observing the resultant load lines in Fig. 15.* Other than the optimized snubber loss pre-

* The exact shape of the load line would require detailed information on both the particular load and transistor.

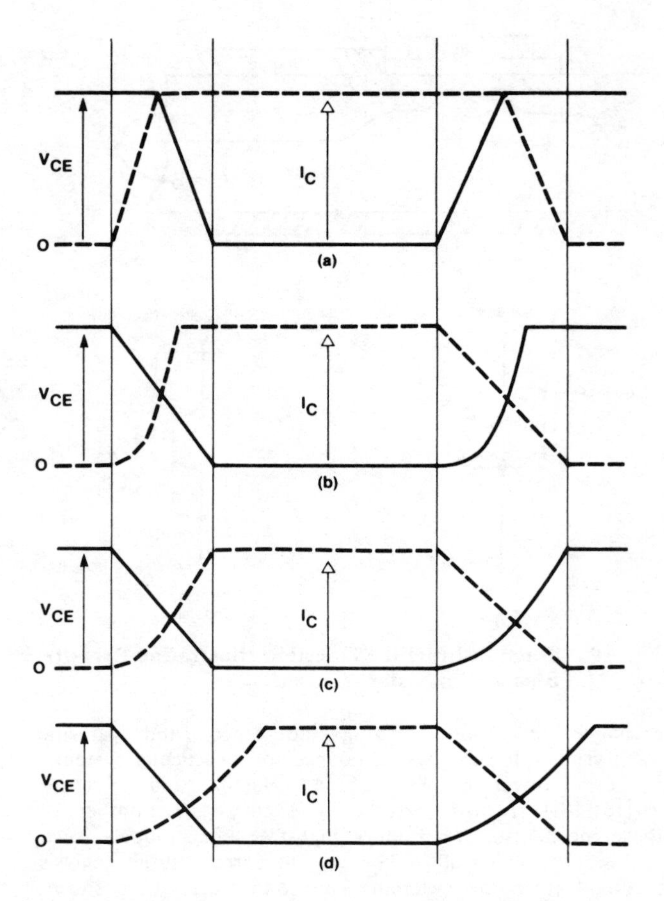

Fig. 14. Snubber Size Defined by McMurray. For an inductive load: (a) no snubber, (b) small snubbers, (c) normal snubbers, (d) large snubbers.

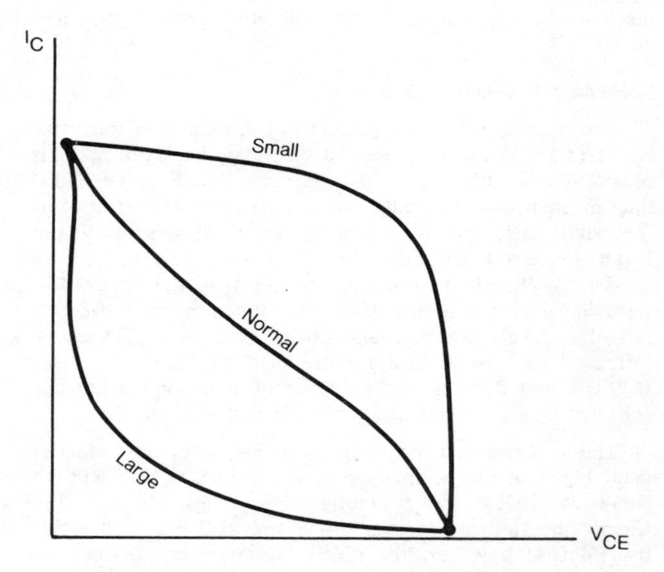

Fig. 15. Effect of Snubber Size on Load Line

viously discussed, generally the snubber power loss, weight, volume, parasitics, cost, and stress-relief all increase with size. It can therefore be asked if any conditions exist that benefit from the use of a larger snubber than that required to minimize the converter dissipation or maintain the load line within the safe operating area. If there is any significant advantage in eliminating the dissipation of the switching interval, there does not seem to be a consensus of opinion; furthermore, there does not appear to be any definitive evidence supporting either approach. Clearly, if the snubber is very large, the switching loci of the transistor will follow the coordinate axis with virtually no switching stress to possibly produce an enhanced reliability. Calkin and Hamilton reported such performance; however, the specific topology also allowed the improvement of efficiency previously discussed [8].

In contrast there are several sources that report secondary breakdown to be a precipitous mode of failure, and no degradation occurs if the initiation mechanisms for secondary breakdown are not formed [16][17][18]. It may be hypothesized, therefore, for similar devices operating at the same junction temperature, that reliability is neither enhanced nor degraded with snubber size if onset of secondary breakdown is not approached. To establish this assumption true would greatly reduce the ambiguities in determining the requirements for optimum snubbers. It should be realized, however, that the effects of lateral instabilities caused by device imperfections could result in premature initiation of second breakdown in circuits with no or very small snubbers, and the designer must address this possibility. The essential reality is that significant ambiguities exist in the trade-offs for a snubber utilized for secondary breakdown avoidance.

Minimal Snubber Technique

In addition to the philosophy of snubbers, an emerging philosophy of minimal or no snubber techniques also exists. At present a number of methods have been proposed in an effort to eliminate the necessity of snubbers; among these are emitter-open switching [19][20], drive control stress relief [21][22][23], and hybrid switches [24][25][26][27].

The technique of emitter-open switching is shown in Fig. 16, where transistor *Q1* is a high-voltage transistor in series with transistor *Q2,* a fast low-voltage device. Although both transistors have comparable current capability, they are each required to perform specific tasks in the switching function as follows. To turn on the structure, both *Q1* and *Q2* must be forward-biased, and the fact that *Q2* is a low-voltage device allows it optimization for speed and excellent saturation characteristics; therefore *Q2* does not contribute significantly to the loss in this series topology. The turn-off is accomplished by turning *Q2* off, effectively open circuiting the emitter of *Q1;* hence the name. The stored charge in the base of *Q1* allows the load current to continue where the reverse base current equals the collector current. The charge is finally depleted and transistor *Q1* can fully support its appraised voltage. The essential facet of the technique is that common mechanisms initiating secondary breakdown (i.e., emitter current crowding) are avoided while decreasing the transistor storage and fall times.

The work reported by Rischmueller [22] in Fig. 17 illustrates one example of drive control. The forward bias, I_{B_1}, method is conventional; however, when transistor *Q1* is re-

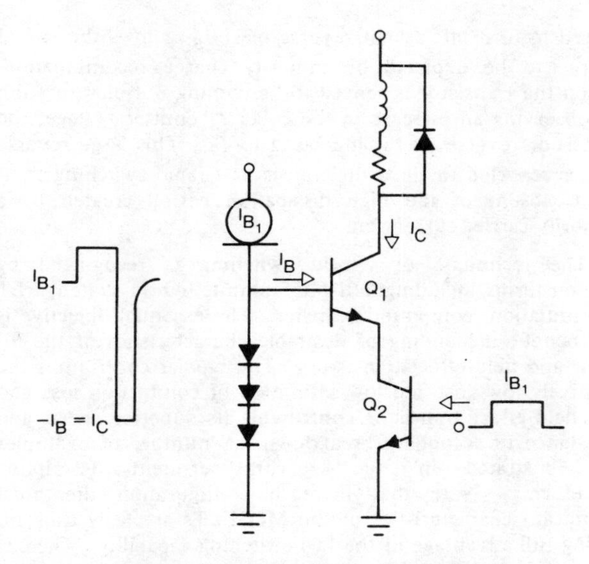

Fig. 16. Emitter Open Switching

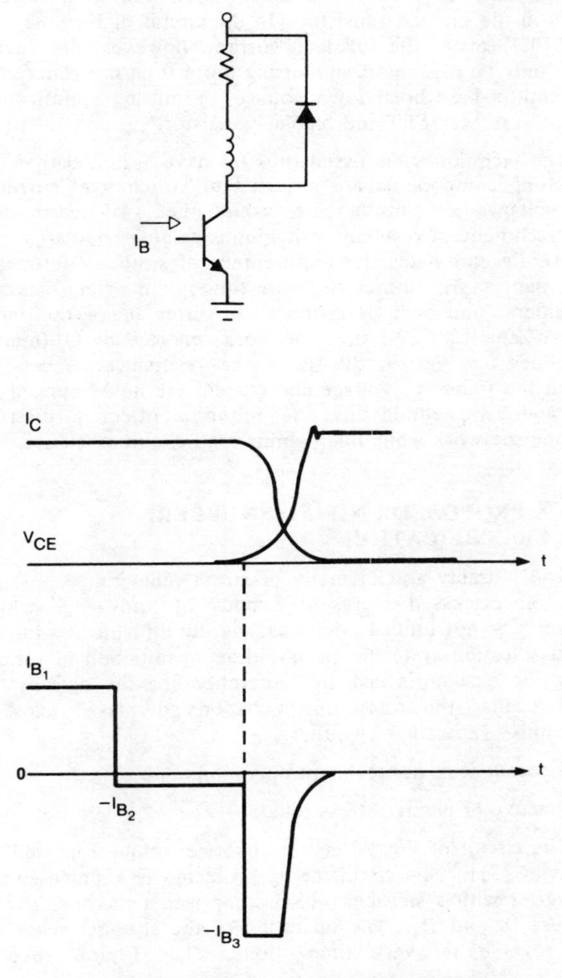

Fig. 17. Drive Control Stress Reduction

quired to turn off, a light reverse bias, I_{B_2}, allows the stored charge to be depleted by minority charge recombination. When the transistor is sensed to be coming out of saturation by observing an increase in the collector emitter voltage, the transistor reverse bias is increased to I_{B_3}. This large reverse bias is reported to allow the transistor a rapid switching transition absent of the high dissipation period created by a minority-carrier current tail.

The technique of hybrid switching is recognized by several terms including: BI-MOS shunt, hybrid switch, FET commutation, compound switches. The essential objective is the beneficial blending of desirable characteristics of the bipolar and field-effect transistors. The bipolar contributes the relatively low cost and low saturation or conduction loss and the field-effect transistor contributes its superior speed and resistance to secondary breakdown. A number of examples are illustrated in Fig. 18 with pertinent descriptive waveforms. Note that in each configuration the most significant characteristics of the MOSFETs are fully utilized, taking full advantage of the fast switching capability. The circuits of Figs. 18a and b permit large drain current only during the switching intervals, allowing the use of small area devices. The resultant higher resistance is not a problem and, furthermore, can be optimized in the Darlington configuration to prevent hard saturation and long storage times in the bipolar transistor. In the circuit of Fig. 18c, the MOSFET carries the full load current; however, the device need only be capable of supporting more than the zener plus the emitter-base breakdown voltage, permitting optimization of both the MOSFET and bipolar transistor.

The techniques discussed thus far have assumed that the switching components are required to switch load currents and voltages in a simultaneous fashion (i.e., switched mode). The technique of resonant switching is gaining popularity and inherently eliminates the requirement of snubber networks. The names are numerous with tuned converters, class E amplifiers, and switch resonant converter being the most recognizable [28][29][30]. The work reported by Gutmann, simplified in Fig. 19, illustrates the resonance concept in which the transistor voltage and current are never applied simultaneously, eliminating the requirement of a discrete snubber network while maintaining high circuit efficiency.

III. A PROPOSED UNIFIED SNUBBER CLASSIFICATION

The plurality and diversity of stress relieving techniques thus far discussed represent a body of knowledge which presently is not unified and consequently difficult to address. A classification system is proposed as an initiation in simplifying the designer's task of comprehending the options. It appears that the numerous variations discussed allow a descriptive separation by either

- resonant or nonresonant operation
- active or passive stress relief.

The circuit of Fig. 20 is a snubber technique reported by Weaving [31]. The circuit being protected is a single-ended converter with a snubber which comprises a resistor, R and diodes, D_s and D_1. The operation of the snubber relies on the reverse recovery time diode, D_s. Typical reverse

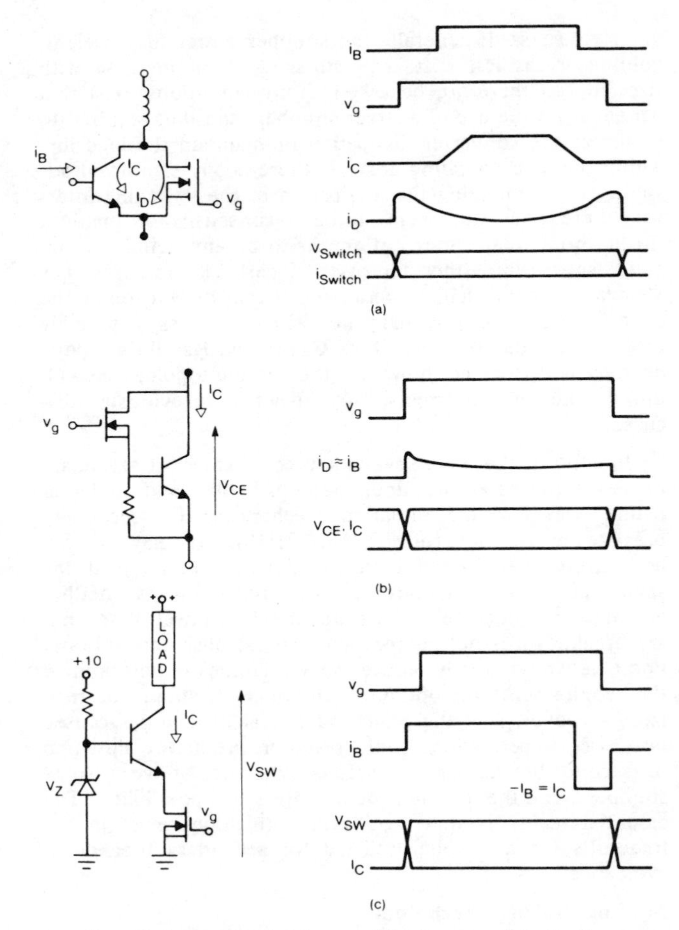

Fig. 18. Hybrid Switching; (a) Shunt (b) Darlington, (c) Compound

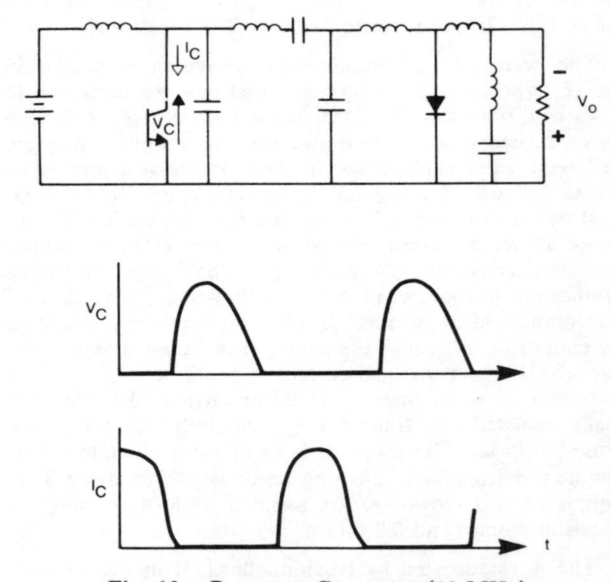

Fig. 19. Resonant Converter (10 MHz)

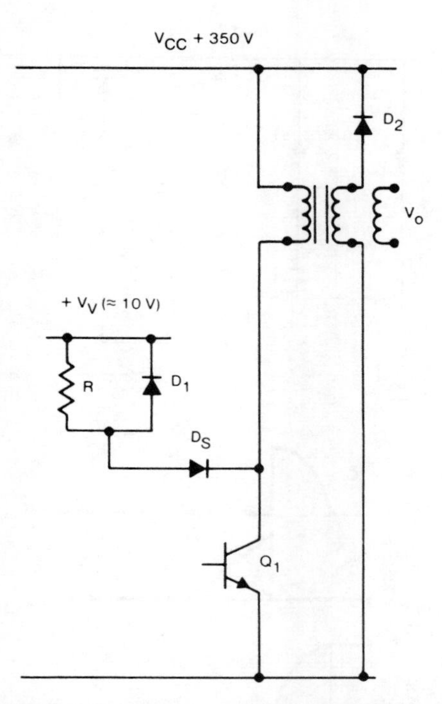

Fig. 20. Single Ended Converter Power Stage with Snubber Diode Circuit Reported by Weaving

recovery characteristics for a *p-n* junction given in Fig. 21 are characterized by

$$Q(t_s) = -I_R \tau_R \tag{3}$$

where $Q(t_s)$ is the excess stored charge, I_R is the reverse current as shown and τ_R is an effective reverse lifetime which can be experimentally determined. Furthermore manipulation and utilization of charge control principle yields the storage charge time

$$t_s = \tau_F \left[\ln \left(1 + \frac{I_F}{I_R} \right) - \ln \left(1 + \frac{\tau_R}{\tau_E} \right) \right] \tag{4}$$

where I_F is the forward current and τ_F is the effective forward lifetime. The fall time is given by the approximation

$$t_f \approx \frac{2{\cdot}3\ (\tau_R + R\ C_t^*)}{1 + \tau_R/\tau_F} \tag{5}$$

where R and C_t^* are the resistor and capacitor in Fig. 21. Both results are utilized in various forms by the circuit designer with appropriate simplifications [17]. The essential point is that the recovery of the diode, D_S, of Fig. 20 can be controlled to some extent by external circuit parameters such as forward and reverse current.

The operation of the present snubber takes full advantage of this phenomenon, usually considered an inconvenience to the circuit designer. The full load current is diverted to diode D_S during turn-off of the protected transistor *Q1*, allowing a relatively stress-free turn-off. The application of such a scheme, however, is not without its challenges, as acknowledged by Weaving. The reverse recovery characteristic is generally regarded as an effect to be minimized and avoided. Consequently, the application of the circuit requires

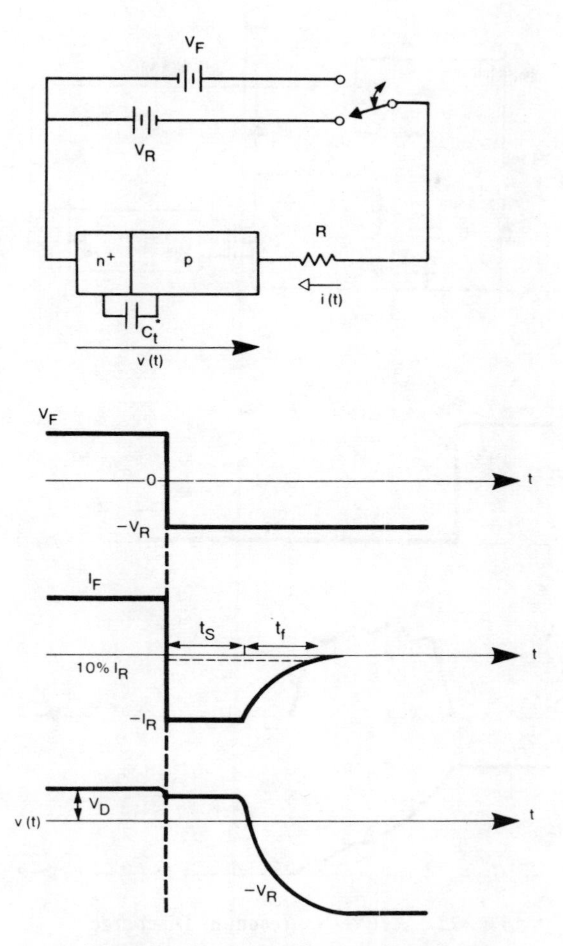

Fig. 21. Recovery of a *p-n* Junction

close coordination between the device and circuit designers to control the device parameters with a different than usual perspective. The results reported for the technique indicate a reduction in dissipation to 4 W, where the conventional snubber required 17 W.

The classification of the technique just discussed is termed an active, nonresonant, low-loss snubber. The nonresonant nature of the circuit operation is evident, and the primary dependence on the dynamic switching of a *p-n* junction defines it as an active network. The extent of the stress-relief can also be included as to small, normal, or large (as previously discussed) to complete the description of the network.

A similar application of the proposed classification system is possible with the various techniques reviewed, including the methods which eliminated the use of any discrete networks (i.e., emitter-open switching, device-control stress relief, etc.) and can, in general, be classified as active snubbers.

Further resolution and description of the system can be shown by separating the charging and discharging operations. The low-loss snubber of Fig. 22 performs the transistor turn-off stress relief in a passive conventional manner via the diode D_1 and series capacitor C_S. However, the discharge process of the capacitor is performed by the active nonresonant network which comprises the inductor L_S, transistor

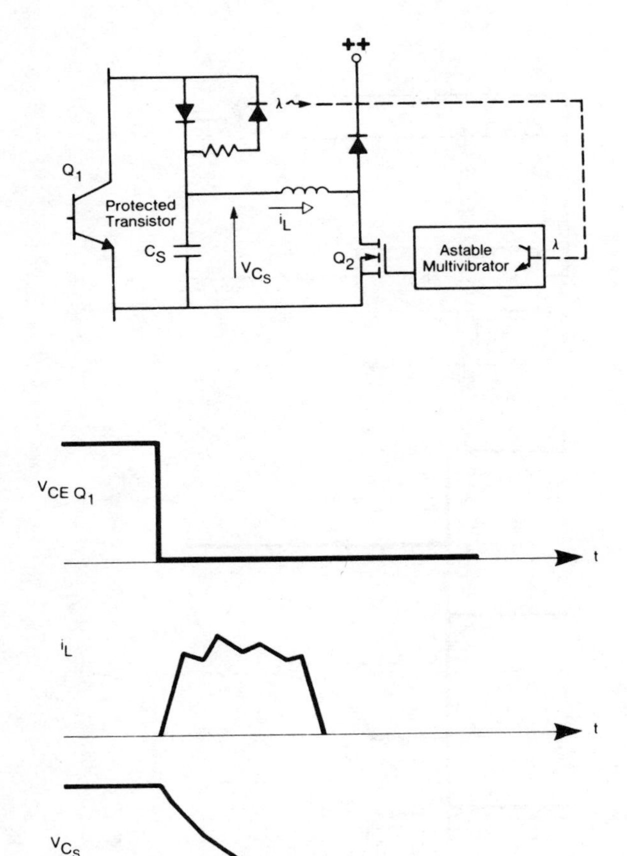

Fig. 22. Active Nonresonant Discharge

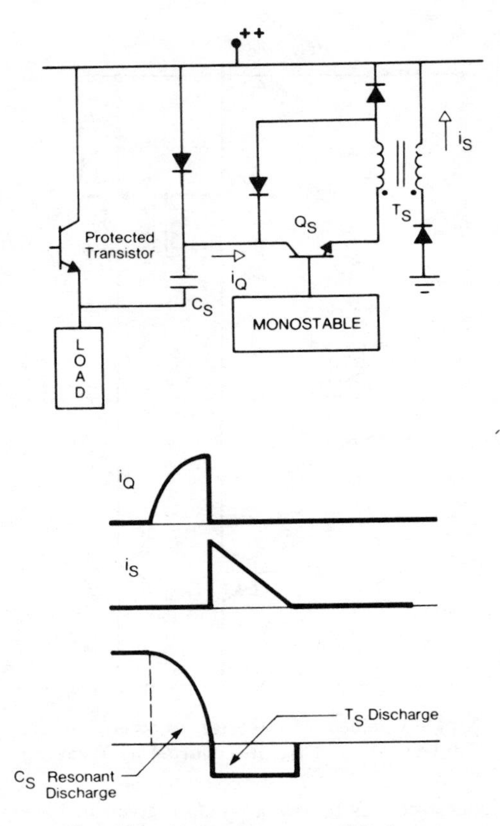

Fig. 23. Active Resonant Discharge

Q2 and the diode D_2. The transistor can be controlled by a number of circuits, but is shown with a very high-frequency constant duty-cycle oscillator and is illustrated with descriptive waveforms. The full classification of the present snubber, to be analyzed and reported in a future study by the author, is an example of a large passive snubber with low-loss active nonresonant discharge. The use of such a classification system would hopefully limit ambiguities; for example, an active resonant discharge has been described by Schroeder [32] and is shown in Fig. 23 with descriptive waveforms.

Passive resonant discharge has also been extensively reported [33][34][35] and two examples are illustrated in the simplified schematics of Figs. 24a and 24b for the buck and forward converters, respectively.

There are obviously a myriad of possibilities, and Table II will hopefully allow a basis for organization and encouragement for future work. Although the system allows immediate application to any simple snubber, as previously discussed, a hierarchy must be assigned for compound or nested snubbers. Examples of nested snubbers can be seen in Fig. 25 with the approximate waveforms for an inductive load. The local snubber is very close electrically to the transistor to mitigate wiring inductances, whereas the major snubber actually performs most of the load line shaping. However, due to physical constraints on the components, parasitic lead inductance is an unavoidable condition. The use of nested snubbers would likely be seen in large transistor converters [7] where the parasitics are controlled by the physical size of components even with optimum layout techniques. The use of such nested arrangements attempts to blend the advantages of the techniques. The efficiency analysis trade-off is given by McMurray [7] for the balance between conventional shunt and clamp snubbers similar to the circuit of Fig. 25a. The application of the classification can be made with confidence by addressing the various networks in the order of their localization (i.e., the most local are first to be named.) This will allow nested snubbers of any order to be systematically identified.

CONCLUSION

The philosophy and intent of low-loss snubber technology as applied to transistor converter methods has been reviewed. It is hoped and intended to serve more as a starting basis for the germination of new ideas than as a compilation of all methodologies reported. In respect to the purpose of the snubber, the ambiguities of secondary breakdown avoidance have been discussed. To date, the field has addressed only the optimization of efficiency. It is hoped that future work will deal with the unanswered questions of the designer regarding the trade-offs of snubber size, load line shaping, and reliability enhancement.

To the end of organizing and evaluating the state of the technology, a classification system has been proposed. The system allows the full description of snubber with respect to size, polarization, active or passive, and resonant or non-

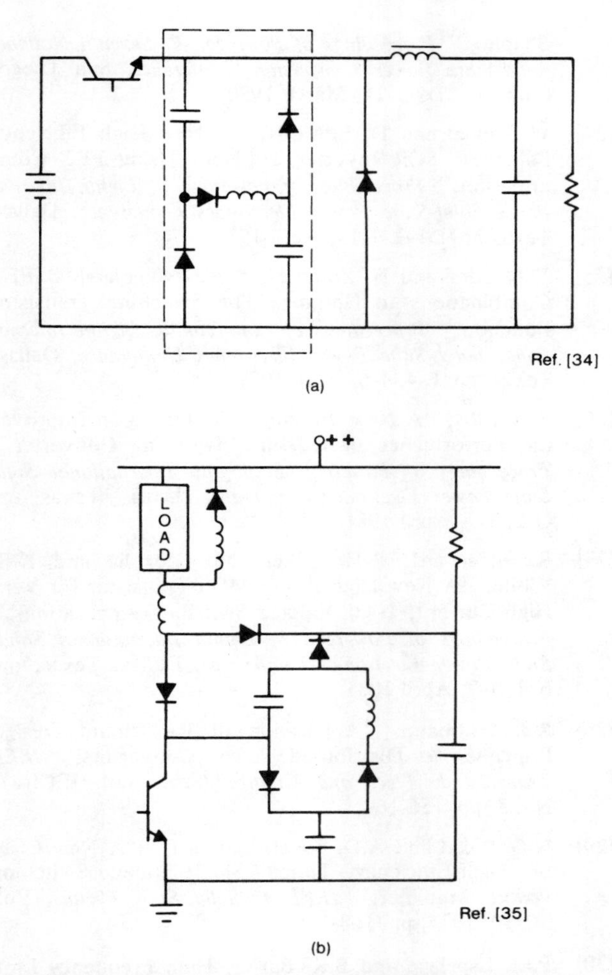

Fig. 24. Examples of Passive Resonant Discharge for (a) Buck, (b) Forward Converters

Table II

TURN-ON (SERIES), TURN-OFF (SHUNT) SNUBBER CLASSIFICATION

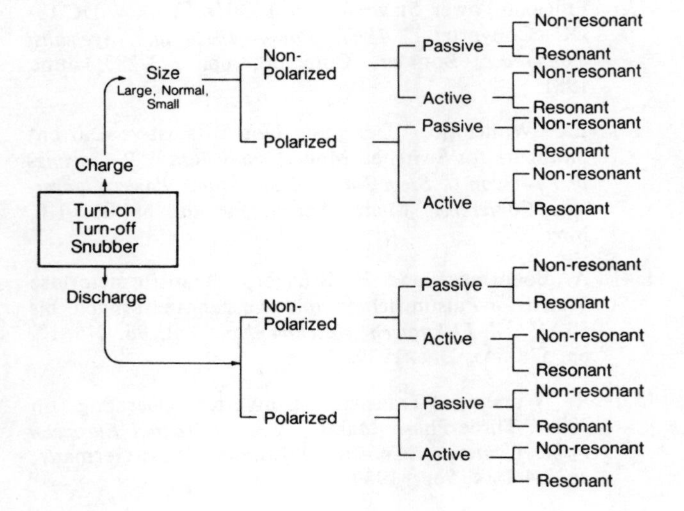

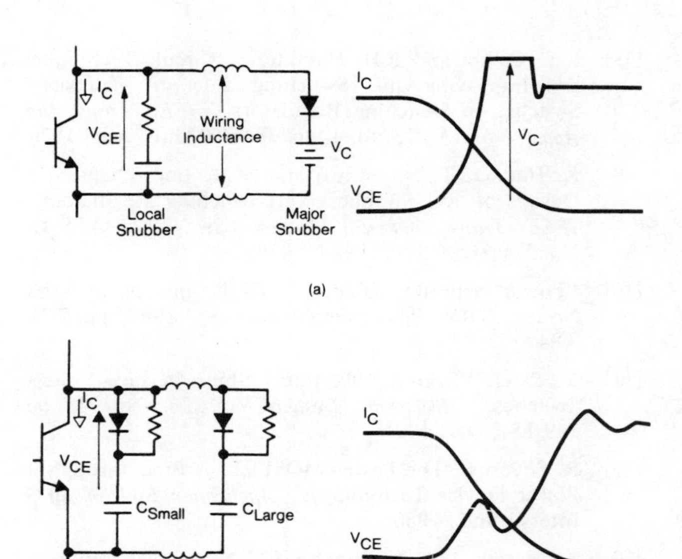

Fig. 25. Nested Snubbers;
(a) Nested Small Conventional Snubber with a Large Clamp

(b) Nested Conventional Polarized Snubber

resonant charging and discharging operation. It is hoped that the descriptions will allow meaningful conveyance of information but also not be so restrictive as to prevent its use or modification. The example of nested snubbers was shown as an example of evolution by the introduction of a hierarchy.

REFERENCES CITED

[1] W. McMurray, "Optimum Snubbers for Power Semiconductors," *IEEE Trans. Ind. Appli.*, Vol. IA-8, No. 5, pp. 593-600, Sept./Oct. 1972.

[2] D.R. Grafham and F.B. Golden, "S.C.R. Manual, Sixth Edition," General Electric Company, 1979, pp. 83, 139-142, 151-158, 481-488.

[3] J.B. Rice, "Design of Snubber Circuits for Thyristor Converters," *IEEE IGA Conference Record,* 1969, pp. 483.

[4] B.J. Baliga, "Switching Lots of Watts at High Speed," *IEEE Spectrum,* Vol. 18, No. 12, pp. 42-48, Dec. 1981.

[5] R.L. Steigerwald, A. Ferraro, and F.G. Turnbull, "Application of Power Transistors to Residential and Intermediate Rating Photovoltaic Array Power Conditioners," *1982 International Semiconductor Power Converter Conference,* Orlando, Florida.

[6] T.E. Anderson and J.P. Walden, "Inverter Having Forced Turn-Off," U.S. Patent No. 3, 953, 780.

[7] W. McMurray, "Selection of Snubbers and Clamps to Optimize the Design of Transistor Switching Converters," *IEEE Trans. Ind. Appli.*, Vol. IA-16, No. 4, pp. 513-523, July/Aug. 1980.

[8] E.T. Calkin and B.H. Hamilton, "Circuit Techniques for Improving the Switching Loci of Transistor Switches in Switching Regulators," *IEEE Trans. Ind. Appli.*, Vol. IA-12, No. 4, pp. 364-369, July/Aug. 1976.

[9] K. Harada, T. Ninomiya, and M. Kohno, "Optimum Design of RC Snubbers for Switching Regulators," *IEEE Trans. Aerosp. Electron. Syst.*, Vol. AES-15, No. 2, pp. 209-218, March 1979.

[10] "Power Schottky Diodes," *TRW Application Notes*, No. 15, TRW Power Semiconductor, 1980, pp. 7.77-7.84.

[11] R. Patel, "Schottky Rectifiers Shine in Low-Voltage Switches," *Electronic Design*, Vol. 29, No. 25, pp. 149-154, Dec. 1981.

[12] R. Severns "The Power MOSFET, A Breakthrough in Power Device Technology," *Application Bulletin A033*, Intersil, Inc., 1980.

[13] K. Harada and T. Ninomiya, "Noise Generation in DC-to-DC Converters," *The U.S.-Japan Cooperative Science Seminar on Analysis and Design in Power Electronics*, Tokyo, Japan, pp. 99-108, Nov. 1981.

[14] T. Ninomiya and K. Harada, "Common-Mode Noise Generation in a DC-to-DC Converter," *IEEE Trans. Aerosp. Electron. Syst.*, Vol. ASE-16, No. 2, pp. 130-137, March 1980.

[15] K. Harada and T. Ninomiya, "Noise Generation of a Switching Regulator," *IEEE Trans. Aerosp. Electron. Syst.*, Vol. AES-14, No. 1, pp. 178-184, Jan. 1978.

[16] W.B. Smith, D.H. Pontius, and P.P. Budenstein, "Second Breakdown and Damage in Junction Devices," *IEEE Trans. Electron Devices*, Vol. ED-20, No. 8, pp. 731-744, Aug. 1973.

[17] S.K. Ghandhi, *Semiconductor Power Devices: Physics of Operation and Fabrication Technology*, John Wiley & Sons, New York, N.Y., 1977, pp. 15-29, 172-183.

[18] S.K. Ghandhi, *Semiconductor Power Devices: Physics of Operation and Fabrication Technology*, John Wiley & Sons, New York, N.Y., 1977, pp. 100-105.

[19] H.A. Schafft, "Second Breakdown — A Comprehensive Review," *IEEE Proceedings*, Vol. 55, No. 8, pp. 1272-1285, Aug. 1967.

[20] B. Jackson and D. Chen, "Effects of Emitter-Open Switching on the Turn-Off Characteristics of High Voltage Power Transistor," *Power Electronics Specialist Conference*, June 1980.

[21] A. Pshaenich, "The Effects of Base Drive Conditions On RBSOA," *Proceedings 2nd International Powerconversion Conference*, Munich, West Germany, pp. 4A.4-1 - 19, Sept. 1980.

[22] K. Rischmueller, "An Improved Base Drive Method Eliminates Switching Aid Networks in Transistorized High Power Converter Circuits," *Proceedings of Powercon 8, Eighth International Solid-State Power Electronics Conference*, Dallas, Texas, pp. G1-1, 1-9, April 1981.

[23] W.R. Skanadore, "A New Bipolar High Frequency Power Switching Technology Eliminates Load-Line Shaping," *Proceedings of Powercon 7, Seventh National Solid-State Power Conversion Conference*, San Diego, Calif., pp. D2-1-14, March 1980.

[24] W. Rippel and D. Edwards, "A New High Efficiency Polyphase SCR Inverter Topology Using FET Commutation," *Proceedings of Powercon 8, Eighth International Solid-State Power Electronics Conference*, Dallas, Texas pp. D1-3, 1-15, April 1981.

[25] J. Meador and N. Zommer, "Using Bipolar-MOSFET Combinations to Optimize The Switching Transistor Function," *Proceedings of Powercon 8, Eighth International Solid State Power Electronics Conference*, Dallas, Texas, pp. F-4, 1-6, April 1981.

[26] P. Bardos, "A New Switching Configuration Improves the Performance of Off-Line Switching Converter," *Proceedings of Powercon 8, Eighth International Solid State Power Electronics Conference*, Dallas, Texas, pp. G-2, 1-7, April 1981.

[27] R. Blanchard, R.H. Baker, M. Glozolja and K.E. White, "A New High-Power MOS Transistor for Very High-Current, High Voltage Switching Applications," *Proceedings of Powercon 8, Eighth International Solid-State Power Electronics Conference*, Dallas, Texas, pp. N-1, 1-7, April 1981.

[28] R.J. Gutmann, "Application of RF Circuit Design Principles to Distributed Power Converters," *IEEE Trans. Ind. Elec. and Control Instr.*, Vol. IECI-27, No. 3, pp. 156-164.

[29] N.O. Sokal and A.D. Sokal, "Class E — A New Class of High-Efficiency Tuned Single-Ended Switching Power Amplifier," *IEEE J. Solid State Cicuits*, Vol. SC-10, 1975, pp. 168-176.

[30] P.M. Espelage and B.K. Bose, "High Frequency Link Power Conversion," *IEEE Trans. Ind. Appl.*, Vol. IA-13, No. 5, pp. 387-394, Sept./Oct. 1977.

[31] R.J. Weaving, "Snubber Technique for H.V. Switched Mode Power Supplies," *Electronic Engineering*, pp. 39, 41, and 45, Oct. 1978.

[32] R.E. Schroeder, "Analysis and Design of a Highly Efficient Power Stage for an 18 kHz, 2.5 kW DC-to-DC Converter," *IEEE Power Electronics Specialist Conference*, Boulder, Colorado, pp. 273-283, June 1981.

[33] E.C. Whitcomb, "Designing Non-Dissipative Current Snubbers for Switched Mode Converters," *Proceedings of Powercon 6, Sixth National Solid-State Power Conversion Conference*, Miami Beach, Florida, pp. B1, 1-6, May 1979.

[34] A. Boehringer and F. Brugger, "Transformatorlose Transistor-Pulsumrichter mit Ausgangsleistungen bis 50 kVA," *Elektrotech Maschinenbau*, Vol. 96, No. 12, pp. 538-545, Dec. 1979.

[35] R. Würslin, "Transistor Converter Operating on 380 V Three-Phase Mains," *Second Annual European Power Conversion Conference*, Munich, West Germany, pp. 2.4-1-14, Sept. 1980.

A 10-kW SERIES RESONANT CONVERTER DESIGN, TRANSISTOR CHARACTERIZATION, AND BASE-DRIVE OPTIMIZATION

R.R. Robson and D.J. Hancock

Hughes Research Laboratories
3011 Malibu Canyon Road
Malibu, CA 90265

This work was sponsored by
NASA-Lewis Research Center
21000 Brookpark Rd.
Cleveland, Ohio 44135
Contract NAS3-22471

ABSTRACT

The Westinghouse D60T and development D7ST transistors are characterized for use as switches in resonant circuit applications. A base drive circuit to provide the optimal base drive to these transistors under resonant circuit conditions is developed and then used in the design, fabrication and testing of a breadboard, space-borne-type 10-kW series resonant converter.

INTRODUCTION

To meet the goals of multi-hundred-kilowatt space power systems planned for the middle and late 1980s, advanced power processing technology is required to convert the power available from solar arrays or other space-borne power sources to the various voltage and/or current levels required by the spacecraft loads. This technology can be built on the series resonant conversion technology that has been brought to a high state of development with the use of thyristors as the basic switching element (1,2). However, new technology is needed to increase the power level and improve the performance. The use of transistors in place of the thyristors promises to meet both of these goals. Although present thyristors offer higher power capabilities than transistors, their losses are high, their operating frequency is limited, and additional protection and commutation circuitry is required.

Hughes Research Laboratories has been involved in a contractual program to develop transistorized series resonant conversion technology at the 10-kW power level. As a first step in this development, the switching characteristics of the Westinghouse D60T and developmental D7ST transistors required measurement for sinusoidal collector current conditions. Switching characteristics are normally measured with a resistive load where the collector current has a square-pulse type of waveform. In a resonant converter, however, the current is switched on by the transistor, and thereafter varies as a sinusoid.

It is the unusual condition of half-sinusoid collector current with the transistor in saturation that makes conventionally measured switching parameters unsuitable for dynamic analysis of series resonant circuits. Manufacturers have not, as yet, begun to characterize their power switches for use in this type of application. Therefore, it was first necessary to obtain the proper transistor parameters before proceeding to the design and development of the converter.

The 10-kW power level is the next logical step in the development of series resonant converter technology, which is presently at the 2.5-kW level for space-borne systems. The goals of the program were to develop a single-stage 10-kW converter employing transistors as the switching elements, having an input-voltage range of 230 to 270 Vdc, an output-voltage range of 200 to 500 Vdc, and an output current-limit range of 0 to 20 A.

TRANSISTOR CHARACTERIZATION

Five Westinghouse D60T and five Westinghouse D7ST transistors were tested and characterized for use as switches in a Series Resonant Converter. The tests were performed using a test circuit that imposed sinusoidal collector current on the transistor under test while allowing the base-drive characteristics to be varied. The base drive consisted of regenerative (or proportional) feedback of the collector current, a leading edge pulse, and a trailing edge pulse. Table 1 lists the peak collector currents and resonant frequencies for which the transistors were tested. The D60Ts were not tested at 140 A and 40 kHz, and the D7STs were not tested at 350 A and 40 kHz because the saturation voltages were prohibitively high under these conditions (greater than 6 V).

The base-drive parameters that were varied during this testing were the amplitude of the leading-edge pulse, the duration of the leading-edge pulse, the regenerative-feedback turns ratio, and the amplitude of the trailing-edge (negative I_B) pulse. The transistors were tested in both

Reprinted from *IEEE Power Electron. Specialists Conf.*, 1982, pp. 33-44.

Table 1. Base Drive Parameters for Minimum Total Device Dissipation For the D60T Transistor and the D7ST Transistor

RESONANCE FREQUENCY, f_r, kHz	D60T							D7ST						
	PEAK I_C, AMPS	SATURATED			UNSATURATED			PEAK I_C, AMPS	SATURATED			UNSATURATED		
		LEADING EDGE PULSE		TURNS RATIO	LEADING EDGE PULSE		TURNS RATIO		LEADING EDGE PULSE		TURNS RATIO	LEADING EDGE PULSE		TURNS RATIO
		AMPS	µS		AMPS	µS			AMPS	µS		AMPS	µS	
10	40	2.5	32.0	8:1	7.0	6.0	6:1	100	7	22.0	10:1	15	40.0	10:1
10	60	3.0	31.0	8:1	2.0	3.0	5:1	150	12	30.0	10:1	14	40.0	6:1
10	100	4.0	30.0	5:1	6.0	44.0	6:1	250	20	32.0	8:1	28	42.0	8:1
10	140	6.0	30.0	4:1	4.0	44.0	3:1	350	16	28.0	4:1	26	30.0	5:1
20	40	4.0	12.5	8:1	13.0	19.0	8:1	100	15	9.0	10:1	18	10.0	10:1
20	60	7.0	13.5	7:1	12.0	19.0	6:1	150	22	10.5	10:1	18	13.0	10:1
20	100	11.5	14.0	7:1	5.0	20.0	4:1	250	28	12.0	7:1	28	15.0	7:1
20	140	14.5	15.0	5:1	10.0	20.0	3:1	350	30	13.0	4:1	30	15.0	4:1
40	40	10.0	4.0	8:1	8.5	8.0	8:1	100	24	4.0	10:1	28	4.0	10:1
40	60	11.5	5.0	8:1	12.0	8.0	8:1	150	30	5.0	10:1	30	6.0	10:1
40	100	15.0	5.5	5:1	15.0	5.5	4:1	250	30	5.5	4:1	30	4.5	3:1

a saturated and unsaturated (Baker clamp) condition. One of the D7ST transistors (number 13) developed a collector-to-emitter short approximately 10 sec after being installed in the test setup; therefore, no data on it is available.

Figure 1 shows some typical base drive waveforms and defines the base drive parameters. The following sections discuss each parameter that was measured and give a summary of the data collected. Further data is available in Reference 3.

Base-Drive to Minimize Total Device Dissipation

The base-drive required to minimize total device dissipation was determined by monitoring the total device dissipation with an electronic power-measuring circuit and varying the various base-drive parameters to obtain the lowest reading. This electronic circuit calculated the average of $(V_{BE} \times I_B) + (V_{CE} \times I_c)$ and was calibrated against a balance-type calorimeter.

Figure 2 is a typical set of curves showing that the minimum power dissipation is nearly independent of the regenerative feedback ratio (forced β) at the lower current levels and shows a much more pronounced minimum at the higher current levels.

Table 1 lists the base-drive parameters for the D60T and D7ST that resulted in minimum total device dissipation. The data to follow for delay time, rise time, storage time, fall time, V_{CE} (SAT), and total device dissipation were taken under these base-drive conditions, and a negative reverse base bias of 7 V for the D60Ts, and 8 V for the D7STs. These negative reverse base bias voltages were chosen since they minimized storage time.

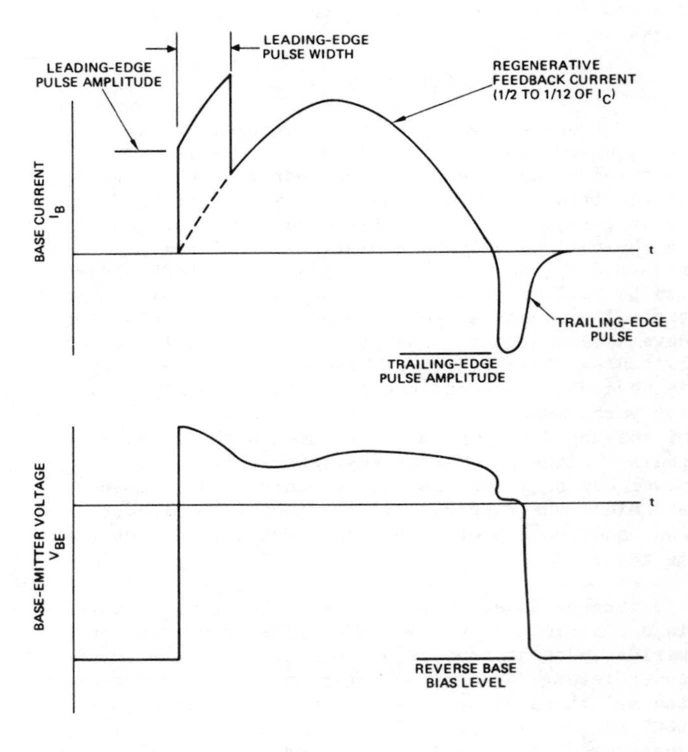

Figure 1. Typical base drive waveforms generated by the test circuit.

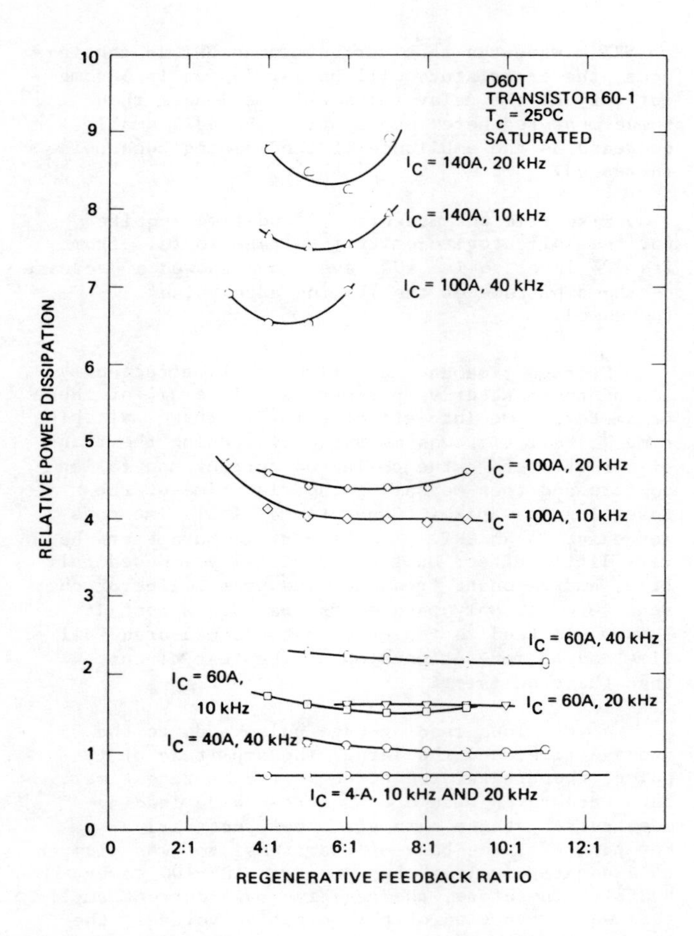

Figure 2. Variation of relative power dissipation with regenerative feedback ratio.

Total Device Dissipation

Total device dissipation was measured using the electronic power-measuring circuit which has an accuracy of approximately ±10%. The data presented represent the power that would be dissipated in the transistor for a full-wave-rectified sinusoidal collector current. The maximum power that a transistor would dissipate in a series resonant inverter (SRI) is 50% of the values shown. Figures 3 and 4 show how the power dissipation in a typical D60T and D7ST, respectively, varies with peak collector current and resonant frequency. These figures indicate that the power dissipation for the unsaturated condition with the D60T is considerably higher than for the saturated condition, while the power dissipation for the D7ST is almost the same for either condition. This indicates that the D7ST does not have time to get into hard saturation during the collector-current pulse. In general, the total device dissipation increases with increasing peak-collector current and increasing resonant frequency (Figures 3 and 4).

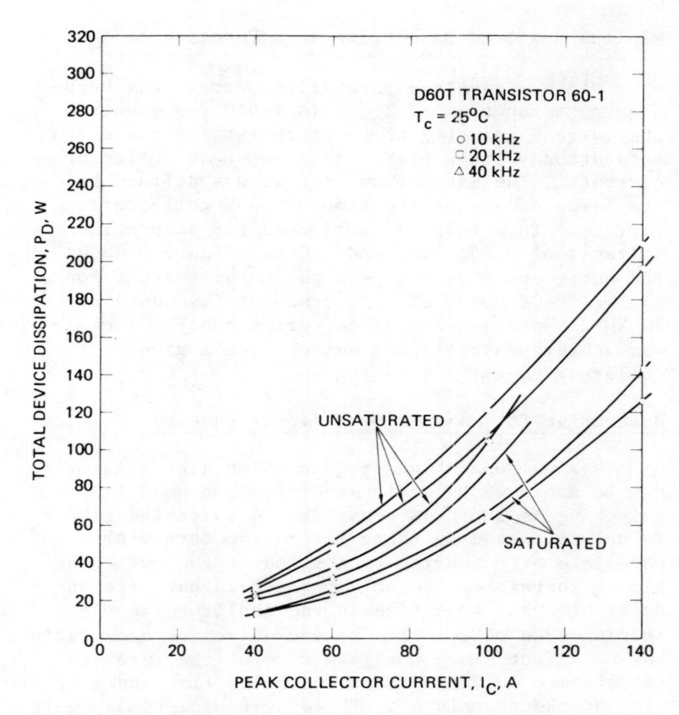

Figure 3. Variation of total device dissipation with peak collector current and resonant frequency.

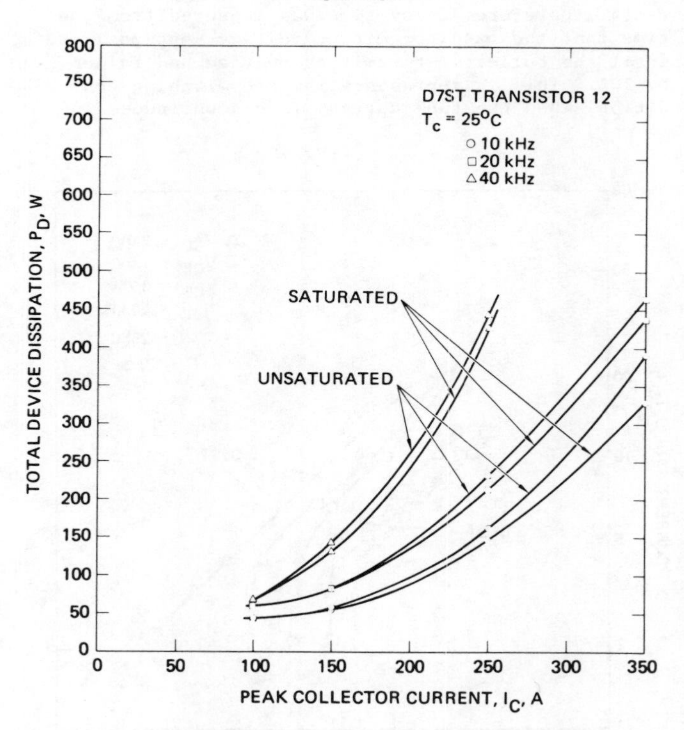

Figure 4. Variation of total device dissipation with peak collector current and resonant frequency.

Minimum Required Base-Drive as a Function of I_C

Minimum required base-drive current was measured as a function of I_C; this ratio represents the current gain (β) of the transistor for a given saturation voltage over a range of peak collector currents. The saturation voltage was defined as the value of V_{CE} at the time of peak collector current. This test was performed for saturation voltages of 0.75, 1.0, and 2.0 V. Figure 5 shows the curves of β versus peak collector current for a typical D60T and D7ST at a resonant frequency of 20 kHz. Both sets of these curves exhibit a maximum achievable collector current for a given collector-to-emitter voltage.

Base-Drive for Maximum Operating Frequency

Maximum operating frequency for a transistor can be achieved by making the time consumed by switching as short as possible. A switching time is only affected by those base-drive parameters that come into affect prior to the end of the switching time. Therefore, the only parameter that affects delay time and rise time is the amplitude of the leading-edge pulse. The regenerative feedback ratio has no effect since the rise time of the feedback transformer is longer than the delay time and rise time of the transistor. These switching times could not be measured in the conventional manner (based on the collector current waveform) since switching was taking place at a time of zero collector current. Therefore, delay time was measured from the time that the base-to-emitter voltage went positive until the collector-to-emitter voltage had fallen by 10%. This is representative of switching conditions when the tank current is discontinuous in an SRI. When the tank current in an SRI is continuous, the transistors will be turning on into some current and the delay times will be longer than those measured here. Delay time showed a small decrease as the amplitude of the leading edge pulse increased.

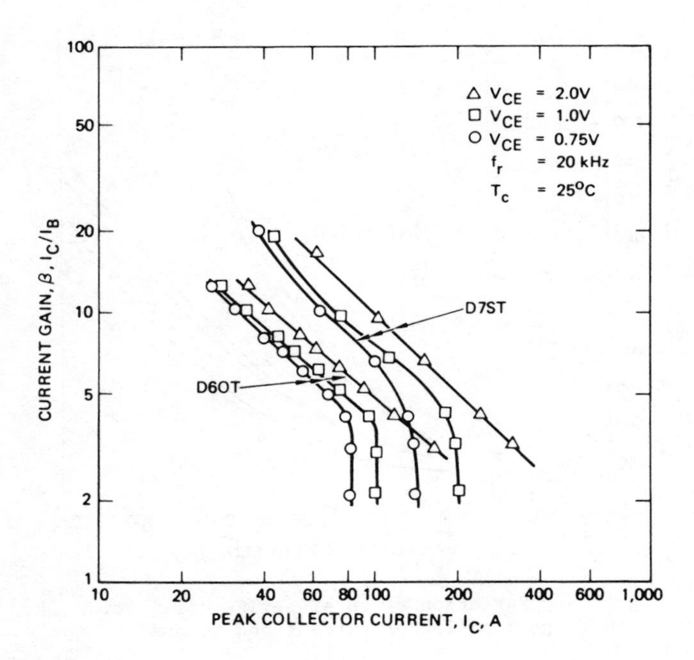

Figure 5. Current gain versus peak collector current for typical transistors resonant at a frequency of 20 kHz.

Rise time was measured as the time required for the collector-to-emitter voltage to fall from its 90% level to its 10% level, and showed a decrease as the amplitude of the leading edge pulse increased.

Storage time and fall time can be effected by all of the base-drive parameters, since all of these parameters come into effect prior to these switching times. Fall time was measured by turning the transistor off before the collector current had fallen to zero and then measuring the rise time of the resulting V_{CE} spike. Conventional fall time does not exist in an SRI. The base-drive parameters had very little effect on the artificially induced fall time, and resonant frequency and peak collector current were the only parameters that did affect it. This would tend to indicate that the measured fall time may be more a function of the test circuit than the transistor.

A very long leading-edge pulse adds to the storage time, and the larger the amplitude of the pulse, the greater the effect. The storage time data were taken with a 5 A, 5-μsec-wide leading-edge pulse; these parameters were not varied. For turn off, the base of the transistor was clamped to a negative voltage by use of an IRF-100 power MOSFET. Therefore, the negative base-current amplitude was a function of this negative voltage, the base resistance of the transistor under test, and the IRF-100 drain-to-source resistance. The only parameter that could be varied was the negative voltage. A negative 7 V gave the minimum storage time for the D60T transistor, while a negative 8 V was required for the D7ST. In general, the storage time increased with increasing collector current, increasing frequency, and decreasing feedback turns-ratio. Storage time was defined as the time between the 10% points on the negative base-current pulse.

Switching Times Under Minimum Total Device Dissipation Conditions

Delay time, rise time, storage time, and fall time were measured under the base-drive conditions that resulted in minimum total device dissipation. These base-drive conditions are listed in Table 1. Typical switching times for the D60T and D7ST are listed in Table 2 for various peak collector currents and resonant frequencies. From a comparison of these data with data for the value of base-drive required to maximize operating frequency, it was determined that the transistors can be operated under the base-drive conditions that produce minimum total device dissipation without compromising the maximum operating frequency.

Saturation Voltage, V_{CE} (SAT)

The variation of the average V_{CE} (SAT) with peak collector current and resonant frequency is shown

Table 2. Measured Switching Times for the D60T and D7ST Transistors

RESONANT FREQUENCY, f_r, kHz	D60T									D7ST								
	PEAK I_C, AMPS	SATURATED				UNSATURATED				PEAK I_C, AMPS	SATURATED				UNSATURATED			
		t_d ns	t_r ns	t_s μs	t_f ns	t_d ns	t_r ns	t_s μs	t_f ns		t_d ns	t_r ns	t_s μs	t_f ns	t_d ns	t_r ns	t_s μs	t_f ns
10	40	65	80	1.3	350	60	70	0.4	210	100	180	100	4.0	600	150	80	1.5	270
20	40	65	95	1.6	360	60	65	0.7	270	100	140	70	4.3	400	130	70	2.6	320
40	40	60	65	1.8	270	55	70	0.8	210	100	125	60	4.1	220	125	60	3.2	220
10	60	65	90	1.4	360	75	105	0.5	200	150	140	90	4.3	530	150	95	1.8	290
20	60	65	90	1.8	360	65	80	0.7	200	150	130	70	4.3	300	140	80	3.0	250
40	60	55	75	1.8	230	60	75	1.0	160	150	125	60	4.5	280	125	65	4.2	300
10	100	70	95	2.1	370	65	95	0.7	200	250	135	75	4.6	310	125	70	2.4	260
20	100	60	75	1.9	250	65	95	0.9	160	250	130	70	5.0	330	130	65	3.5	390
40	100	55	70	2.2	200	60	70	1.3	180	250	120	65	6.3	–	125	65	5.2	–
10	140	65	90	2.4	290	75	105	0.7	200	350	140	90	6.4	–	130	80	2.8	280
20	140	60	75	2.5	180	60	90	1.2	200	350	125	75	6.8	–	120	75	4.5	–

in Figures 6 and 7 for the D60Ts and D7STs, respectively. The saturation voltage was defined as the collector-to-emitter voltage at the time of peak collector current. The saturation voltages were measured for the transistors under the base-drive conditions that minimized total device dissipation. Saturation voltages measured under these conditions are the optimum tradeoff between base-drive power and collector-emitter dissipation. The saturation voltage increased as either the peak collector current or resonant frequency was increased (Figures 6 and 7).

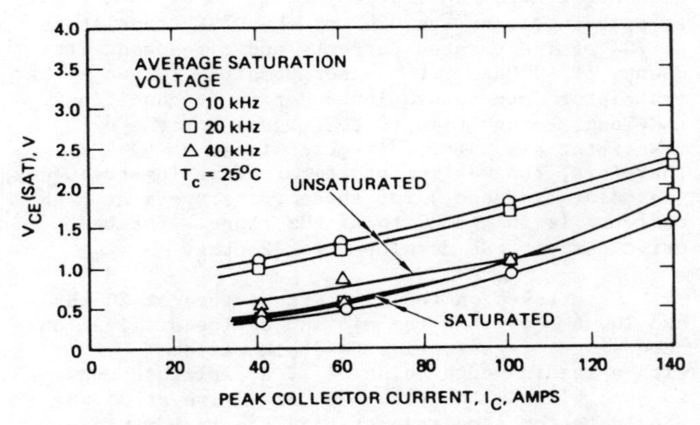

Figure 6. Variation of saturation voltage with peak collector current and resonant frequency for the D60Ts (average of five transistors).

Output Capacitance, C_{OB}

The construction of high-power transistors necessarily requires the use of a large-area junction to handle the high currents. Large-area junctions result in a very undesirable characteristic (i.e., large junction capacitances). Published data for the D60T give a small-signal, 1-MHz measurement of C_{OB}, which does not accurately depict the nonlinear character of C_{OB} on a large-signal basis, such as will be encountered in high-power inverter applications. The large signal C_{OB} measured here averaged 2140 pF for the D60Ts, and 3696 pF for the D7STs.

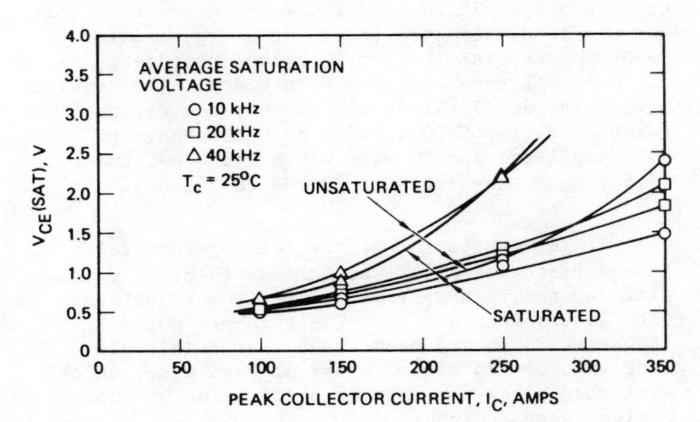

Figure 7. Variation of saturation voltage with peak collector current and resonant frequency for the D7STs (average of three transistors).

Input Capacitance, C_{IB}

The maximum operating frequency of a switching transistor is a direct function of the time required to charge (delay time) and discharge (storage time) the base-to-emitter capacitance (C_{IB}). To properly design the base-drive circuit to adequately handle the peak-base-drive currents required, a worst-case value for C_{IB} is required. The input capacitances of the transistors tested averaged 76,100 pF for the D60Ts and 118,500 pF for the D7STs.

BASE-DRIVE DEVELOPMENT

A base-drive strategy was developed for both the D60T and D7ST transistors when they are used as switches on SRIs. The strategy is the same for both types of transistors and is described in the following section.

Strategy

Regenerative (or proportional) feedback of the collector current is used since it minimizes the required power from the rest of the base-drive circuitry. At the same time, the transistor is maintained at a constant β (except during the leading-edge pulse), which saves on base-drive power. The base current is allowed to go to zero as the collector current goes to zero, which minimizes the storage time. The base-drive parameters from the transistor characterization that resulted in minimum total device dissipation are employed, and the base-emitter junction is kept reverse-biased during the transistor off-time, which eliminates dV/dt turn-on caused by C_{OB}.

Requirements

A study of the data obtained in the transistor characterization indicates that it is not practical to operate either the D60T or the D7ST transistors at 70% of their rated currents and a resonant frequency of 40 kHz. It was not possible to get either transistor into saturation under these conditions. The long storage time (6 to 7 μsec) for the D7ST transistor also limits its usefulness at 40 kHz. Therefore, the maximum practical operating frequency (resonant frequency) for these transistors at high currents is in the 20 to 25 kHz range. The base-drive circuit was developed for 25 kHz.

Data taken on the D60T transistors at 20 kHz and 100 A show that the minimum device dissipation occurs at a regenerative feedback ratio of 7:1, with a leading-edge pulse of 12 A amplitude and 15 μsec width. The required base-drive at 25 kHz should be the same regenerative feedback ratio of 7:1, and a leading-edge pulse of 12 A amplitude that is 12 μsec wide (reduced from 15 μsec by the ratio of 20 kHz to 25 kHz). Data taken on the D7ST transistors at 20 kHz and 250 A show that the minimum device dissipation occurs at a regenerative feedback ratio of 10:1, with a leading-edge pulse of 30 A amplitude and 12.5 μsec width. The required base-drive at 25 kHz should be the same regenerative feedback ratio of 10:1, with a leading-edge pulse of 30 A amplitude and 10 μsec width (reduced from 12.5 μsec by the ratio of 20 kHz to 25 kHz).

It is important that the leading-edge pulse have a rise time of 1 μsec or less in order to minimize the turn-on time and get the transistor into saturation quickly. The turn-off pulse (negative I_B pulse) needs to be large in amplitude in order to minimize the storage time, which will then allow for maximum utilization of the series resonant tank.

Circuit Development

Conventional transformer-coupled base-drive circuits will not provide a 30-A pulse with a rise time of 1 μsec or less because of the leakage inductance of the transformer, base-emitter inductance of the transistor, and stray inductance of the wiring. Direct switching of the 30 A current pulse into the base was considered, but rejected because of the poor efficiency (less than 50%) for that type of base-drive. It was decided to use a transformer-coupled base-drive circuit and overcome the inductance problem by "brute forcing" it with voltage.

The circuit of Figure 8 was developed to provide a 30 A pulse with a rise time of 1 μsec or less and a 7:1 (10:1 for the D7ST) regenerative feedback ratio. Referring to Figure 8, Q_5-X forms a constant-current source that charges capacitor C_6-X to the voltage potential of V_2. When Q_6-X is turned on to apply a leading-edge pulse to QX, the high voltage charge (V_2 = 75 V) on C_6-X is applied to the primary of T_4-X. This high voltage overcomes the effect of the leakage inductance of T_4-X, the stray wiring inductance, and the base-emitter inductance of QX, allowing the base current to rise to 30 A in approximately 1 μsec. After the charge on C_6-X has decayed to potential V_1 (12 V), the remainder of the leading-edge pulse is supplied from V_1 through CR_5-X. The width of the leading edge pulse is controlled by the on-time of Q_6-X.

The regenerative feedback is supplied by transformer T_3-X. A separate transformer is used for the regenerative feedback so that the leakage inductance of T_4-X (supplying the leading-edge pulse) can be minimized. Transistors Q_7-and Q_8-X are used to isolate the transformers so that the base of QX can be held at a negative bias during the time that it is turned-off. Q_9-X supplies the turn-off pulse to the base of QX, and holds it at the negative bias level of -7 V. T_6-X is used to provide an isolated turn-off pulse to Q_9-X during the on-time of QX.

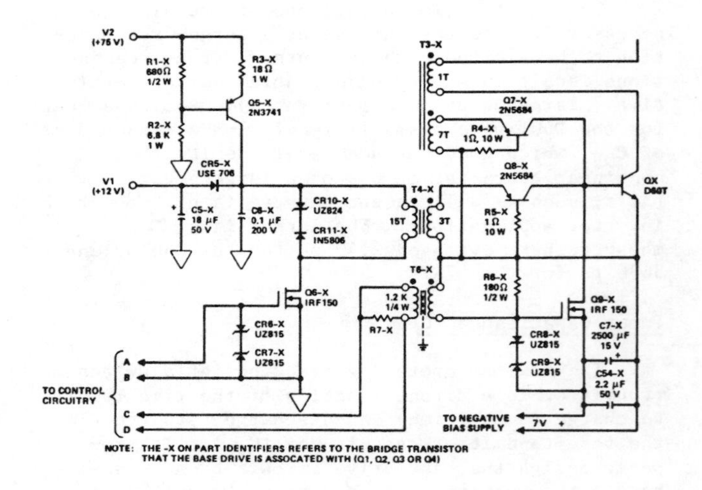

Figure 8. 10-kW base drive circuitry.

Figure 9 shows the resonant tank current, collector current, and base current for a D60T transistor using this base-drive circuit.

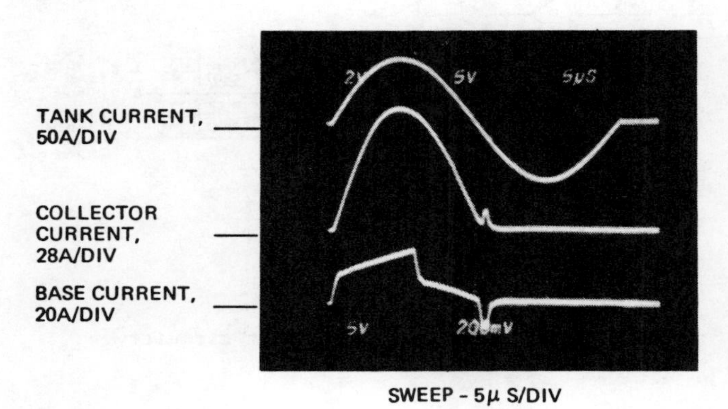

Figure 9. D60T transistor waveforms during test of the base-drive circuit.

CONVERTER DESIGN

The 10-kW series resonant converter was designed to meet the following specifications:

- Resonant frequency — 25 kHz
- Main bus voltage — 230 to 270 Vdc
- Output voltage — variable from 200 to 500 Vdc.
- Output current limit — adjustable from 0 to 20 A
- Output power — 10-kW minimum at 500 Vdc output
- Output voltage regulation — ±2% for input voltage variations of ±10% and/or load variations between 10% and full load current
- Output ripple — less than 1% rms
- Remote voltage sensing — terminals provided on the rear of the chassis
- Isolation — all output terminals are isolated from the chassis, and either the positive or negative terminal may be floated ±100 Vdc from chassis ground.
- Mode indication — a local and remote indication is supplied to indicate when the inverter is in the current-limited mode
- Meters — output voltage and current meters on the front panel
- AC power control — a front panel switch for control power and front panel breaker for the main power
- Cooling — forced air
- Mounting — standard 19-in. EIA rack.

The decision was made to use the D60T transistor as the basic switch for this design because of its shorter storage time and because of the very limited availability of the D7ST transistors. The use of the D60T transistors at the 10-kW level requires a full-bridge circuit which uses twice as many switches as a half-bridge (4 versus 2), but results in a peak tank current that is one-half that for a half-bridge. The lower peak tank current gives rise to a more efficient design because of lower I^2R losses and lower transistor saturation voltages.

Bridge and Tank Circuitry

A schematic of the bridge and tank circuitry is shown in Figure 10. Transistors Q1 through Q4 are the four switches of the full-bridge, and T3-1 through T3-4 provide the regenerative feedback base drive for these transistors. SR1 and SR2 are saturable reactors that limit the di/dt that Q1 through Q4 see, allowing these transistors to saturate quickly, and thereby reduce power losses. SR1 and SR2 saturate in approximately 500 nsec, after which they are effectively out of the circuit. Diodes CR1 through CR4 provide the paths for returning excess energy in the tank circuit to the source. Diodes CR54-1 through CR54-4 were added to suppress voltage spikes caused by SR1, SR2, and stray wiring inductance. They are mounted as close to Q1 through Q4 as possible, as are C3 and C4, which provide a low impedance AC clamp for CR54-1 through CR54-4 to work into.

The series resonant tank is composed of C1, L1, T1, and T2. C1 is made from four polypropylene capacitors in parallel and has a total capacitance of 0.875 μF. Capacitor C1 resonates with the inductance of L1 and the primary leakage inductance of T1 at a resonate frequency of 25 kHz. L1 was fabricated by winding 18 turns of 165/30 (16,500 circular mils) Litz wire on an Indiana General-type 8200 ferrite core with a 12.9 cm^2 cross-sectional area. Ferrite was used for both L1 and T1 because of its lower losses. The design of these components could be further improved by using Ceramic Magnetics Inc., type MN60L ferrite instead of the

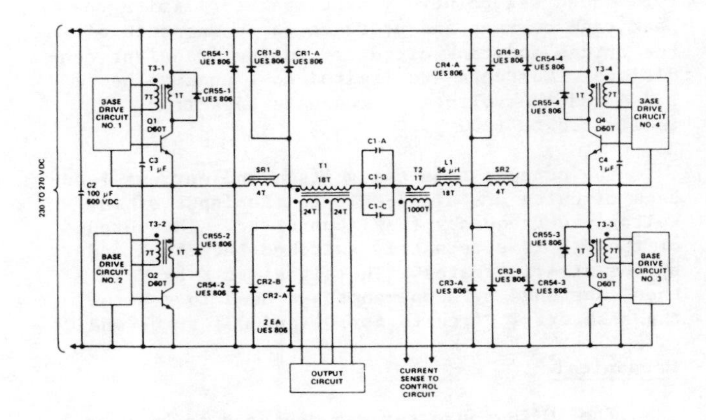

Figure 10. 10-kW bridge and tank circuitry.

Indiana General, type 8200. T1 is used to remove energy from the tank circuit and supply it to the load. The primary of T1 was wound with 18 turns of 165/30 Litz wire and each of the two secondaries was wound with 24 turns of two 150/36 Litz wires in parallel (a total of 7500 circular mils). The core used was the same as for L1. T2 is a current transformer that provides a tank-current feedback-signal to the control circuit.

CONTROL AND OUTPUT CIRCUITRY

The control and output circuitry is shown in Figure 11. The AC current supplied by T1 is rectified by two full-bridge circuits and filtered by C8 to provide the output voltage. The output voltage is sensed as a feedback-signal by R10 and R11 and isolated from the floating output by the isolation amplifier. This feedback signal is then compared against the output voltage reference signal (V_{REF}) and the difference integrated by integrator, AR1. Comparator AR2 senses when the output voltage is more than 25 V higher than the referenced level and immediately phases the inverter off until the output voltage drops down to the referenced level. This keeps the output voltage under control during transient conditions.

The average output current is related to the average current in the series resonant tank circuit by the turns ratio of T1. Therefore, the average tank current can be sensed and used to control the output current while at the same time protecting the bridge and tank circuit components. The tank current is sensed by current transformer T2, and converted to a voltage by R12. This voltage is then compared against the output current reference-signal (I_{REF}) and the difference integrated by AR3. The I_{REF} signal is summed with the voltage feedback signal and an offset reference to linearly phase the output current back to 6 A as the output voltage falls to zero. The static output operating envelope produced in this manner is shown in Figure 12. Comparator AR5 senses when the peak tank current exceeds 120 A and immediately resets the integrator (AR3) to zero, which in turn phases the inverter off. The integrator can immediately start to integrate back up and phase the inverter back on to the referenced set point. This comparator limits the peak tank current and protects the components of the bridge and tank circuitry during transient conditions. Currents are limited to a peak value of 120 A and the voltage on resonate capacitor, C1, is limited to 1200 V.

The outputs from the voltage and current feedback circuits are diode-OR'd to the input of the voltage-to-frequency (V/F) converter. The output of the V/F is alternately switched between X and Y by the steering gates. The signal at X or Y is then shortened by a one-shot and used to control the base drive circuits for Q1 and Q3, or Q2 and Q4.

Mechanical

The 10-kW converter was designed to be used in a laboratory type environment with forced air as the cooling medium. The unit, shown in Figure 13, is contained in a rack-mountable chassis that is 48.25 cm by 31.1 cm x 46 cm (19 in. x 12.25 in. x 18 in.). It weights 52 kg (115 lb) and can be mounted at a convenient height for operation of its controls and reading of the meters.

Figure 14 is an interior view of the converter showing the locations of the D60T and commutating diode heat sinks, output rectifiers, control circuitry, base-drive, series resonant capacitor, output transformer, and series resonant inductor. The input capacitor, output capacitor, housekeeping supplies and fans are mounted in the bottom of the chassis and cannot be seen in this photograph.

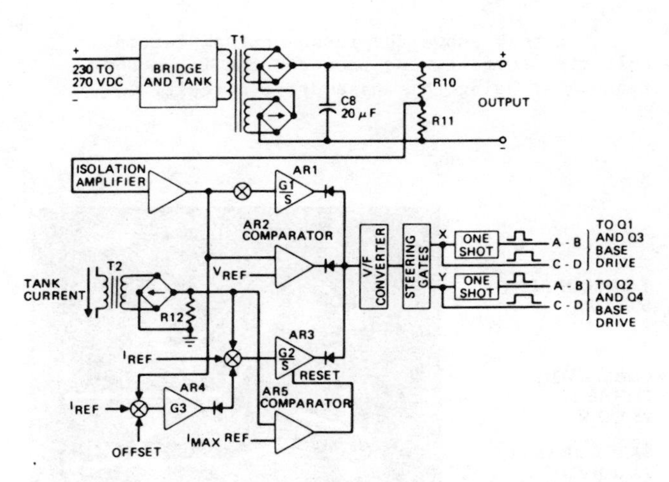

Figure 11. Control and output circuitry.

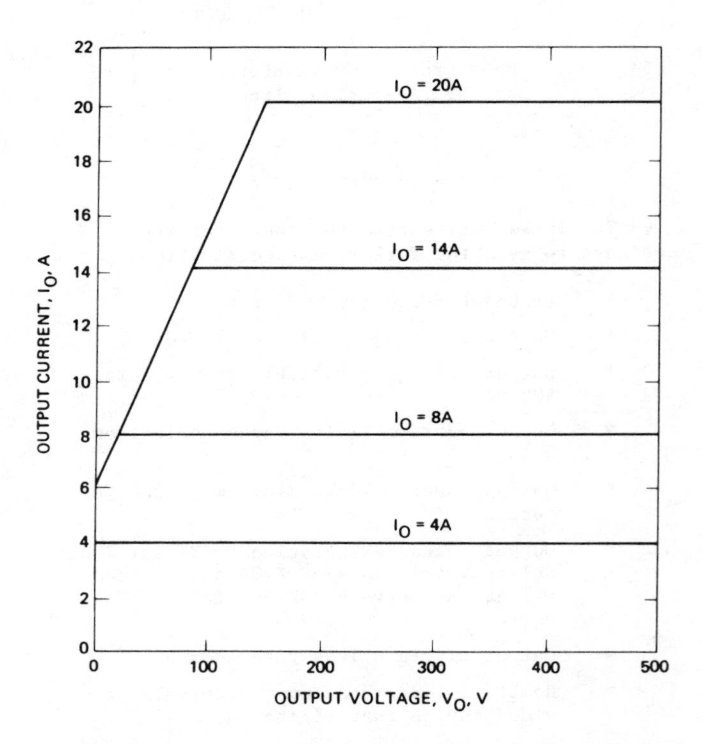

Figure 12. 10-kW SRI output operating envelope.

MC14014

Figure 13. 10-kW series resonant inverter.

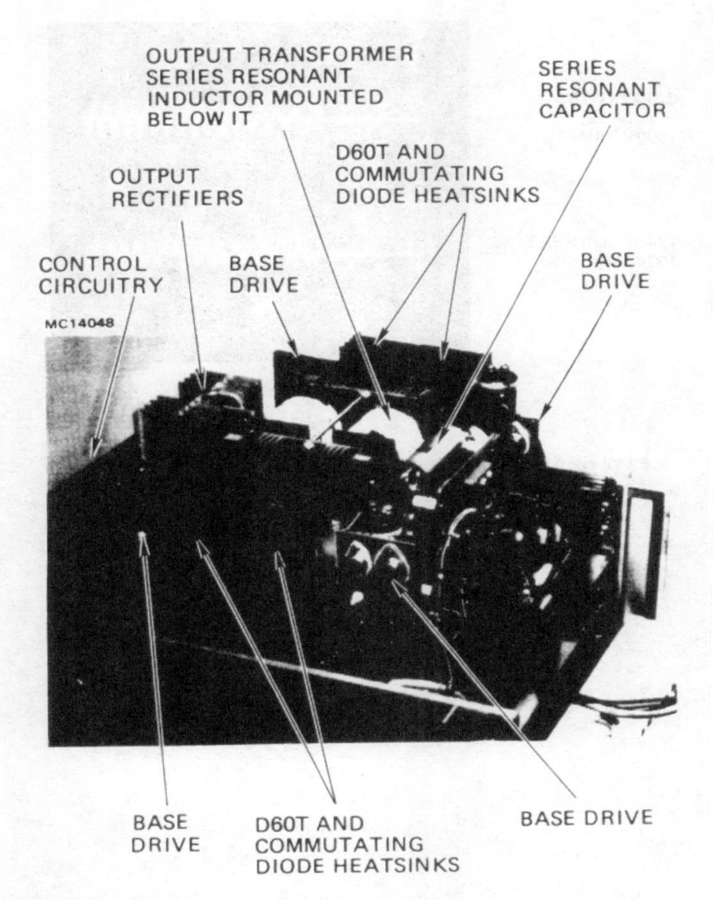

Figure 14. Interior view of the 10-kW series resonant inverter.

TESTING

The 10-kW converter that was designed and fabricated was also tested under a variety of conditions to determine its operational characteristics. The parameters tested for were stability, steady-state waveforms, output ripple, regulation, transient waveforms, and efficiency. The test results for each of these parameters are discussed in the following paragraphs.

The voltage and the current control loops both have integrators in their forward loops to provide very high DC gain, necessary for good regulation. In addition to the integrator, the voltage control loop has a lead-lag network for loop compensation. The bandwidth of this loop increases and the stability decreases as either the output voltage increases or the load resistance decreases. The worst case gain margin is 10 dB, and the worse case phase margin is 65 deg.

The current control loop does not have any compensation in addition to the integrator. The bandwidth and the stability of this loop are fairly constant for the conditions tested with a worst case gain margin of 15 dB and a worst case phase margin of 90 deg.

The steady-state resonant tank waveforms (500 V output with a 25 Ω load) are shown in Figure 15. From these it can be seen that for a 10-kW load the peak tank current is approaching 100 A and the peak capacitor voltage is approximately 700 V.

The tank current and resonant capacitor voltage response to a load transient of 500 V at 20 A to short circuit is shown in Figure 16. This was the response before the peak tank current limiting circuitry was added to the control circuit, and shows the tank current peaking at 190 A and the capacitor voltage peaking at 1700 V. These peak levels were unacceptably high and therefore the circuit was modified. Figure 17 shows the response of the tank current and capacitor voltage to the same transient condition after the peak tank current limiting circuitry was added. The tank current is now limited to 120 A and the capacitor voltage to 1000 V, which are acceptable levels.

The tank current and output voltage response to a load transient of short circuit to open circuit (350 V setpoint) is shown in Figure 18. The tank current response is acceptable, but the output voltage overshoots the 350 V setpoint by 200 V, which is unacceptable. The output voltage limiting circuitry at this time was a fixed level (≈550 V), optically coupled circuit that phased the V/F converter back when the output voltage exceeded 550 V. This circuitry was eliminated and a circuit whose trip level is a function of the programmed setpoint was added. The response of the output voltage to the same transient condition after this change in the control circuitry was incorporated is shown in Figure 19. The output voltage now overshoots the 350 V setpoint by only 50 V.

The efficiency of the inverter was measured under static conditions while operating into a resistive load. The efficiency versus output power

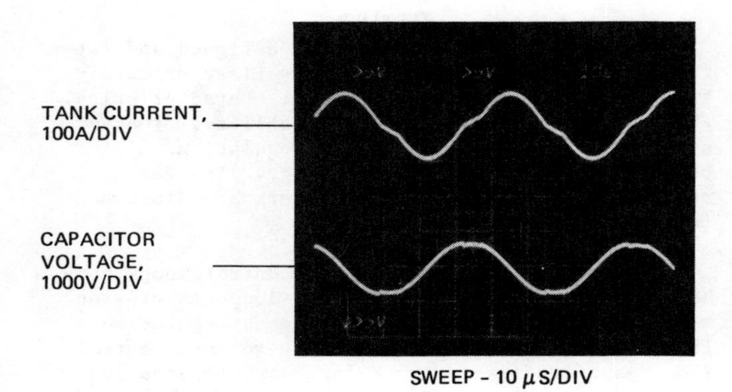

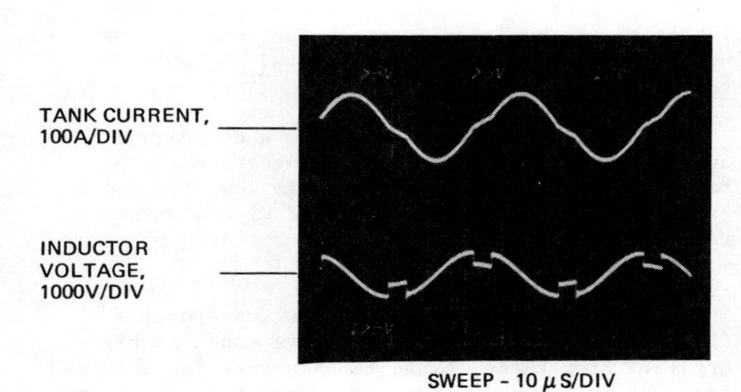

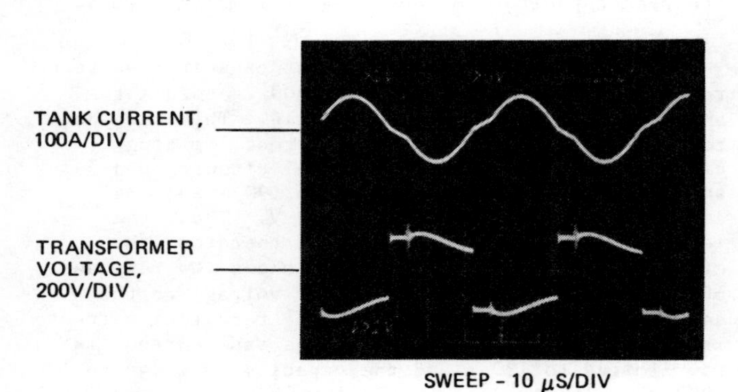

Figure 15. Steady-state resonant tank waveforms for a 500-V output with a 25-Ω load.

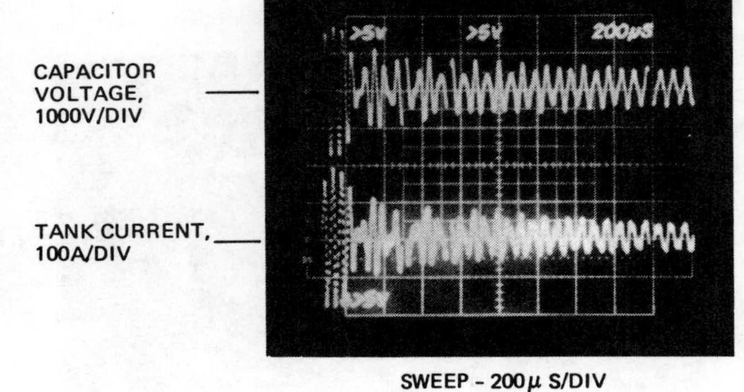

Figure 16. Tank current and resonant capacitor voltage response to a load transient of 500 V at 20 A to short circuit without the peak tank current limiting circuitry.

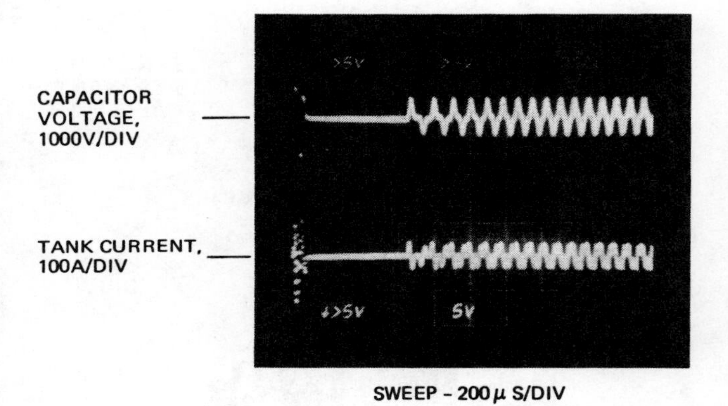

Figure 17. Tank current and resonant capacitor voltage response to a load transient of 500 V at 20 A to short circuit with the peak tank current limiting circuitry.

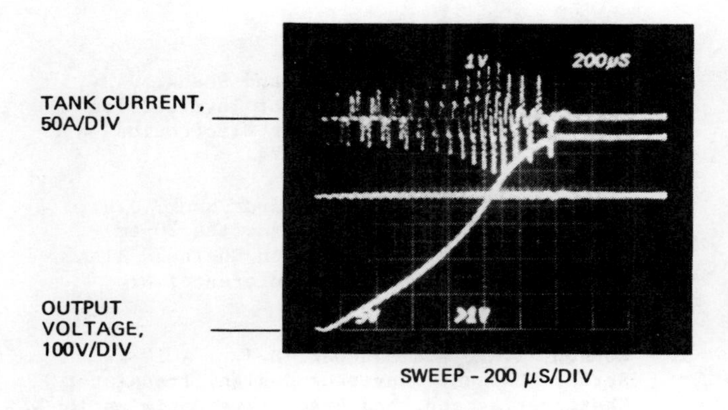

Figure 18. The tank current and output voltage response to a load transient of short circuit to open circuit (350 V setpoint) with original output voltage limiting.

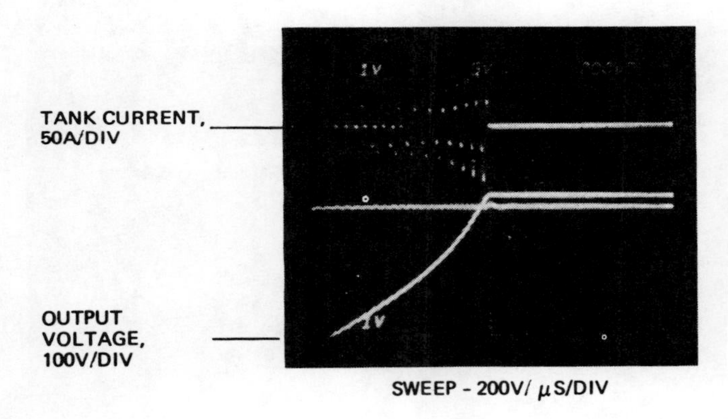

Figure 19. The tank current and output voltage response to a load transient of short circuit to open circuit (350 V setpoint) with improved output voltage limiting.

for loads of 25 Ω and 50 Ω is shown in Figure 20. This figure shows that the efficiency is almost constant for output powers above the 2-kW level. These curves represent the efficiency of the main DC input-power bus to the output of the converter.

The output ripple of the converter is given in Table 3. The open circuit, 327 V condition produced the largest peak-to-peak ripple observed. The peak-to-peak ripple could be reduced by adding more output capacitance, but this would also decrease the bandwidth of the frequency response. Line and load regulation measurements showed that the regulation is better than or equal to 1 V.

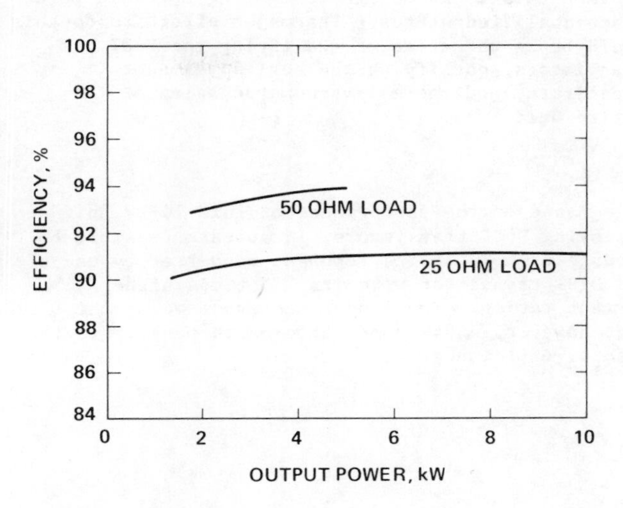

Figure 20. Efficiency versus output power for the 10-kW inverter.

Table 3. Output Ripple of the 10-kW Converter

OUTPUT VOLTAGE, V_O, V	OUTPUT LOAD R_L, Ω	OUTPUT RIPPLE, PEAK-TO-PEAK	
		VOLTS	% OF V_O
180	25	4.0	2.2
250	25	4.5	1.8
500	25	4.5	0.9
327	∞	34.0	10.4
500	∞	25.0	5.0

CONCLUSIONS

The suitability of the developed base drive circuit and of the D60T transistor to function in a series resonant converter was demonstrated by the successful development and testing of a 10-kW, full-bridge series resonant converter. The unit achieved an electrical efficiency of 91% at its full power level of 500 V and 20 A, and an efficiency of 93.7% at a half-power load of 500 V and 10 A.

There are no inherent problems that would prevent this 10-kW design from being upgraded to a space-qualified status. The major effort to do this would be in the areas of qualifying the D60T transistors, qualifying the series resonant capacitors, and thermal-vacuum packaging of the entire unit.

Based on the performance of this 10-kW unit employing D60T transistors, it appears feasible to produce a 25-kW series resonant converter by use of the D7ST transistor with its 2.5 times higher current rating. The longer storage time of the D7ST may, however, cause some performance penalty in the area of efficiency.

REFERENCES

1. Biess, J.J., Inouge, L.Y., and Shank, J.H., "High Voltage Series Resonant Inverter Ion Engine Screen Supply," Power Electronics Specialist's Conference, 1974.

2. Biess, J.J., Inouge, L.Y., and Shank, J.H., "Thyristor Power Processor for the 30-cm Mercury Electronic Propulsion Engine," AIAA 11th Electric Propulsion Conference, New Orleans, LA, 1975.

3. Robson, R.R., and Hancock, D.J., "A 10-kW Series Resonant Converter Design, Transistor Characterization, and Base-drive Optimization," NASA Contractor Report CR-16554, 1981.

A Two-Quadrant Transistor Chopper for an Electric Vehicle Drive

ROBERT L. STEIGERWALD, MEMBER, IEEE

Abstract—**A two-quadrant dc–dc transistor converter capable of delivering 400 A motoring current and of generating 200 A braking current is described. The chopper operates from a 108-V dc source (54 lead-acid cells) and supplies the armature current of a separately excited dc machine in an electric vehicle application (3000-lb commuter-type vehicle). The chopper employs high-current transistors specifically developed for the application and power diodes packaged together in power module form. Snubber networks which reduce both turn-on and turn-off device stresses are employed. The interaction of the snubber networks for the motoring and braking transistors is described and design considerations presented. It was found that for these snubbers a minimum on-time and a minimum off-time for the transistors must be maintained to ensure that the transistors' dynamic load lines never enter into the region of forward bias or reversed bias second breakdown. A technique is described which instantaneously detects a transistor failure and initiates the appropriate action in order to prevent machine overcurrent and overtorque. Factors are discussed which are crucial to ensure proper transitions from motoring to braking and to inhibit device power dissipation due to parasitic currents. The selection of a variable-frequency/variable-pulsewidth switching strategy and protection and control techniques unique to high-current transistor choppers are discussed.**

INTRODUCTION

IN a separately excited dc motor drive, machine armature current and, hence, torque can be controlled by field current alone once the machine has reached or exceeded base speed. In this case the machine counter electromotive force (CEMF) is approximately equal to the dc source voltage, and the dc source is applied directly across the machine armature. Below base speed the armature current must be controlled and limited by means other than field control. These techniques include contactor switched resistors (sometimes used in combination with stepped battery voltage levels under contactor control) and armature chopper control. The switched resistor scheme tends to be lossy unless a machine of sufficiently low base speed is employed. Such a machine, however, becomes larger and more expensive as the base speed decreases, which is undesirable in an electric vehicle application. Alternately, the switched resistor scheme can be used in conjunction with a shifting transmission which effectively raises the base speed of the machine to more manageable levels. However, the inconvenience and/or additional weight may be unacceptable. In addition, smooth regenerative braking is difficult to obtain with this scheme.

In the case of an armature chopper the armature current is controlled and limited in an efficient manner, and the base speed of the machine can be selected at reasonably high values (although the armature chopper rating decreases with lower base motor speeds). A highly efficient smooth lightweight drive system results from this scheme. In addition, by employing a two-quadrant chopper, smooth regenerative braking is easily obtained.

A two-quadrant transistor armature chopper which performs this function is described. By employing transistors, circuit simplification results, and a high chopper frequency can be used compared with more conventional thyristor choppers. By employing a sufficiently high chopping frequency, additional inductance in series with the relatively low inductance armature winding is avoided. The power circuit characteristics as well as snubber circuit selection are discussed. Control and protection schemes unique to this transistor chopper are given, and experimental data are presented to illustrate chopper operation.

DESCRIPTION OF CONTROL SYSTEM

The control system in which the armature chopper is employed is shown in Fig. 1. The approach is based on a scheme which is discussed in [1]. The microcomputer control and additional details of this system are discussed in [2]. The approximate torque-speed envelope for this drive system is shown in Fig. 2. At motor speeds below base speed, the armature chopper supplies an adjustable armature current up to 400 A positive for motoring and up to 200 A negative for generating. The base speed is defined here as that speed at which the counter EMF is approximately equal to the battery voltage (i.e., armature current can be controlled by field control only). The base speed in generating is normally higher than the base speed in motoring because motoring current reduces battery terminal voltage, while braking (charging) current increases the battery voltage. The maximum motor armature current rating is selected such that a vehicle acceleration from 0–30 mi/h can be accomplished in 9 s. The maximum generating current is limited mainly by the maximum battery charging current which can be accepted repeatedly without inhibiting battery life. The corner points (i.e., motor base speed) were selected as the best compromise between the armature chopper rating and the motor size, and commutator design. During the armature control mode the field current is maintained at rated value by a low-current (≅5 A) transistor chopper. In the field control mode the armature chopper is completely bypassed by a contactor, and the armature current is controlled by the field chopper. The overall system is controlled by a microcomputer

Paper SPCC 79-22, approved by the Static Power Converter Committee of the IEEE Industry Applications Society for presentation at the 1979 Industry Applications Society Annual Meeting, Cleveland, OH, September 30–October 4. Manuscript released for publication March 17, 1980. This work was supported by the U.S. Department of Energy, Washington, DC, under Contract DE-AC03-76CS51294 for Phase II of the Near-Term Electric Vehicle Program.

The author is with the Corporate Research and Development Center, General Electric Company, Schenectady, NY 12301.

Reprinted from *IEEE Trans. Ind. Appl.*, vol. IA-16, pp. 535–541, July/Aug. 1980.

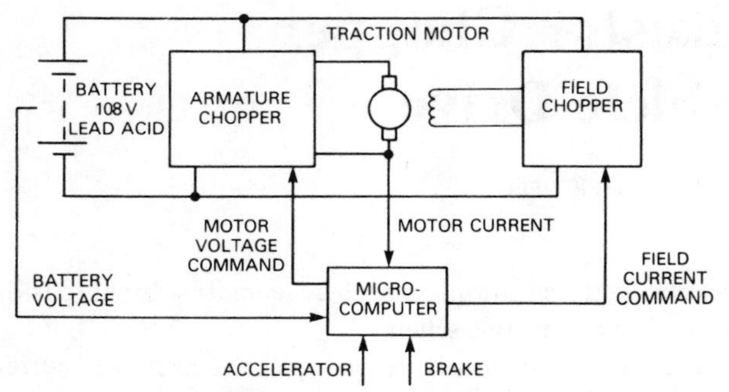

Fig. 1. Electrical drive system block diagram.

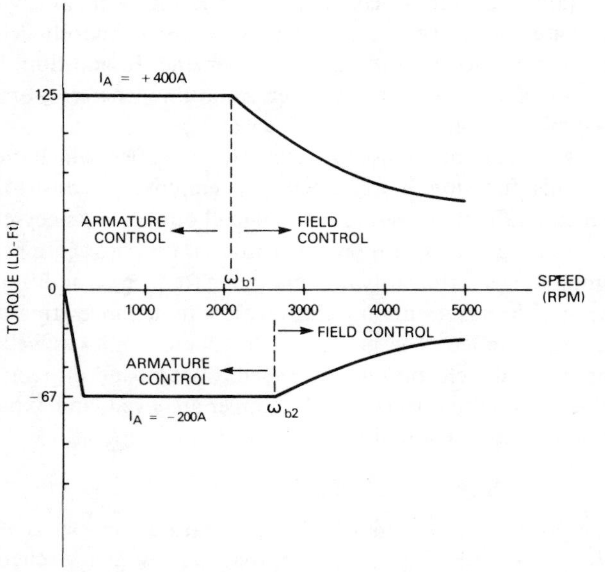

Fig. 2. Torque-speed envelope of drive system.

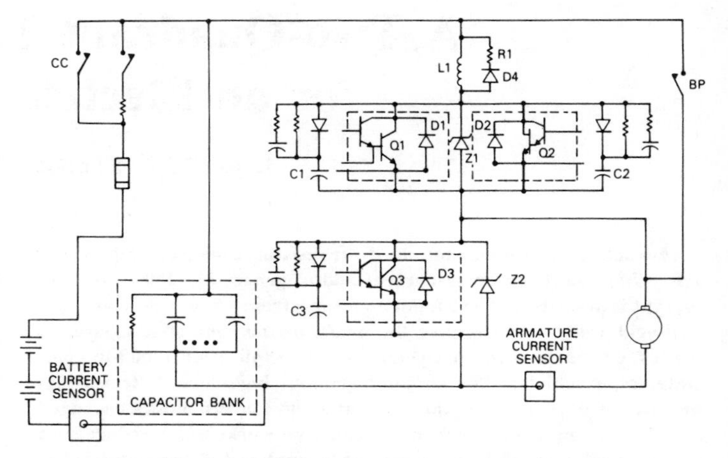

Fig. 3. Armature chopper power circuit.

[2]. Note that the armature control loop is closed through the microcomputer software. This fact creates some interesting and unique interface requirements (such as armature chopper fast-acting current limiting) which will be discussed below.

ARMATURE CHOPPER POWER CIRCUIT

The armature chopper power circuit is shown in Fig. 3. Three power modules (shown within the dotted line) are employed in the circuit. Each power module contains a set of two Darlington power transistors in parallel (only one is shown for simplicity) and a feedback high-speed power diode. Two parallel modules are used to obtain the two parallel transistor sets (Q1 and Q2) needed for the higher motoring current rating (400 A), while only a single module transistor set (Q3) is needed to meet the generating current rating (200 A).

Transistors Q1 and Q2 along with diode D3 comprise the motoring (voltage stepdown) chopper, while transistor Q3 and parallel diodes D1 and D2 comprise the generating (voltage stepup) chopper. A 1200-μF capacitor bank provides a path for the high-frequency current required by the chopper and limits voltage transients each time the transistors turn off, which otherwise would occur due to the energy stored in the battery cable inductance ($\cong$10 μH). A main contactor CC connects the armature circuit to the motor and serves as an emergency disconnect in the event of a fault. A small resistor and relay in parallel with the CC serve to charge the capacitor bank before the CC is allowed to close.

A magnetically sensitive current sensor measures armature current which is used for system control and for a fast-acting current limit for armature chopper protection. A battery current sensor provides information for fuel gauging, battery charging, and battery fault current which shuts the drive down (i.e., opens the CC).

A turn-off polarized snubber consisting of C1 and C2 in parallel and their series diodes provide turn-off stress reduction for the motoring power transistors, while C3 and its series diode provide turn-off stress reduction for transistor Q3. In parallel with each of these series snubber diodes is a resistor which serves to discharge the snubber capacitors when its appropriate transistor turns on. It was found that an additional small *RC* snubber across each series diode was needed to limit voltage transients each time the diode snapped off at the end of a capacitor charging interval (i.e., after a transistor has turned off). A detailed description of turn-off stress reduction networks can be found in [3]-[8].

A 2-μH inductor, L1, is placed in series with the chopper transistors in order to limit transistor turn-on stress. The inductor serves to limit the magnitude of the reverse recovery current of the feedback diodes as well as the peak snubber capacitor charging current. For example, when Q3 turns on, the reverse recovery current of D1 and D2, as well as the charging current of C1 and C2, is limited by L1. When the diodes have recovered and the snubber capacitors have charged, the current trapped in L1 now circulates and dies out in its parallel connected resistor diode combination.

Zener diode voltage clamps are employed across each transistor to limit the voltage magnitude applied to the transistor after turn-off. The voltage spike is due to parasitic inductances, as will be discussed later. Other investigators have noted the importance of the circuit layout in reducing voltage transients [9] and have employed Zener clamps to limit their magnitude [10].

ARMATURE CHOPPER MICROPROCESSOR INTERFACE

The interface between the microprocessor and the armature chopper is illustrated in Fig. 4. The contactor commands, motor/brake command, and armature chopper malfunction are

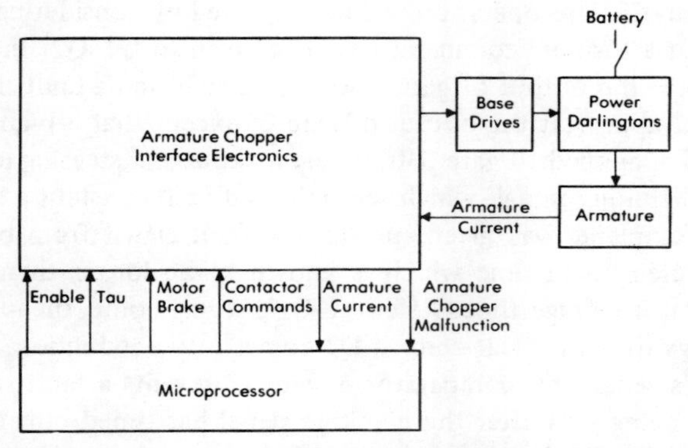

Fig. 4. Armature chopper/mircoprocessor interface.

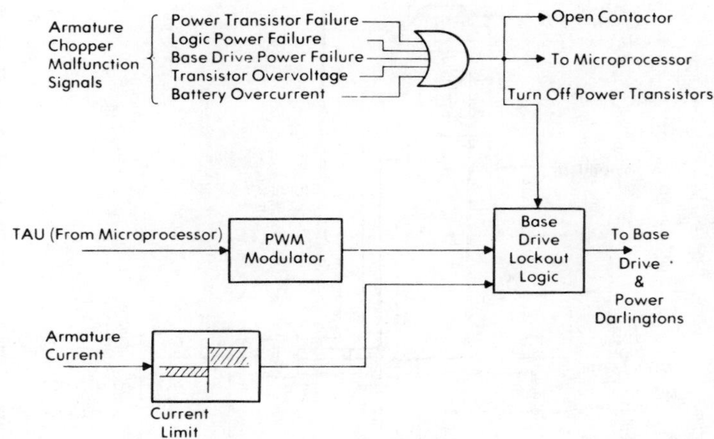

Fig. 5. Interface electronics.

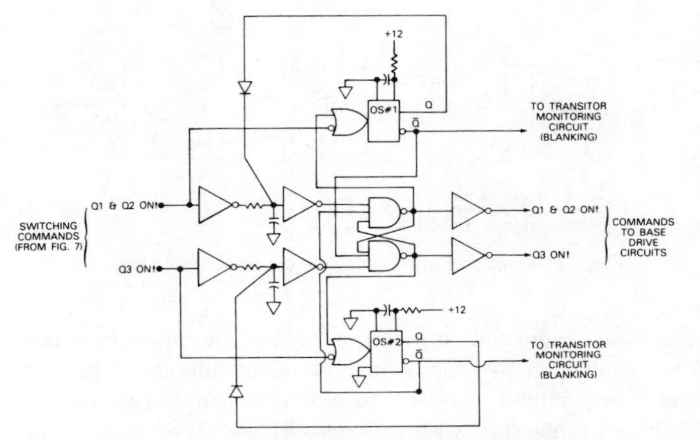

Fig. 6. Base drive lockout and steering logic.

digital signals, while the chopper duty cycle command, τ, and armature current are analog signals. It is the function of the interface electronics to convert these commands into modulated base drive signals, provide a local fast-acting current limit, and provide an analog armature current signal to the microprocessor. In addition, the interface electronics must sense an armature chopper malfunction and notify the microprocessor of such an event.

A more detailed description of the interface electronics is shown in Fig. 5. The pulsewidth modulator (PWM) is of the variable-voltage/variable-frequency type. The modulator frequency as a function of duty cycle, τ, is given by

$$f = \frac{E_d}{2L_a \Delta I_a} (\tau - \tau^2).$$

The frequency is a parabolic function of duty cycle (as shown in Fig. 9). This type of modulator results in constant peak-to-peak ripple current for all motor speeds and armature currents and has the characteristic that the chopping frequency is never higher than necessary for a given condition, and thus average snubber losses are minimized [11]. For the parameters of this system, a 2000-Hz peak frequency was employed which resulted in a peak-to-peak ripple current of approximately 40 A. The motor armature inductance is approximately 250 μH.

A local fast-acting current limit is also provided by the interface electronics. If the current exceeds approximately 440 A motoring or 240 A braking, the power transistors are turned off until the current falls approximately 80 A at which time normal switching is again allowed to occur. The current limit will be discussed in more detail later.

Several parameters are sensed to determine if there is an armature chopper malfunction as seen in Fig. 5. A power transistor failure detection circuit senses the transistor voltage after every transistor turn-off to ensure that the transistor has not failed at turn-off. As will be discussed in more detail below, a transistor failure or any other malfunction causes the contactor CC to be opened, the power transistors to be turned off, and the microprocessor to be notified. Other signals which are sensed include the logic power supply voltage levels, base drive power supply, transistor overvoltage, and battery overcurrent. Any abnormal values of these parameters will cause a malfunction signal to be generated.

Fig. 6 illustrates the circuit which interfaces the PWM to the power transistor base drive circuit. The operation of this circuit is critical. The flip-flop (cross-coupled NAND gate) in the center of Fig. 6 ensures that Q3 and Q1-Q2 never conduct simultaneously, which would short the dc source. However, due to the transistor's storage times, the gating on of a transistor must be delayed until after the command to turn off the opposite transistor in order to allow for the off-going transistor's storage time. This function is accomplished by a simple time delay that is implemented with the two one-shot multivibrators (OS #1 and OS #2), which are cross-coupled to the NAND gate flip-flop. In addition, the one-shots are used for two other functions. First, a constant transistor off-time is accomplished by coupling the one-shot outputs through signal diodes back to the inverted input to the NAND gates. This has the effect of holding a power transistor off for the duration of the one-shot timing interval each time a transistor is turned off. The assurance of a constant off-time is important to allow the current which circulates in L1, D4, and R1 (Fig. 3) to die out before a transistor is again turned on. This is necessary since, if D4 is not cleared of stored charge, the turn-on snubbering effect of L1 will be ineffective which, in turn, causes high transistor turn-on stress. The other function of the one-shots is to provide a blanking signal to the transistor monitoring circuit as discussed below.

Fig. 7 illustrates the armature current switching logic and transistor monitoring circuit, the function of which will now be described. NAND gates 1 and 2 pass the modulator on/off

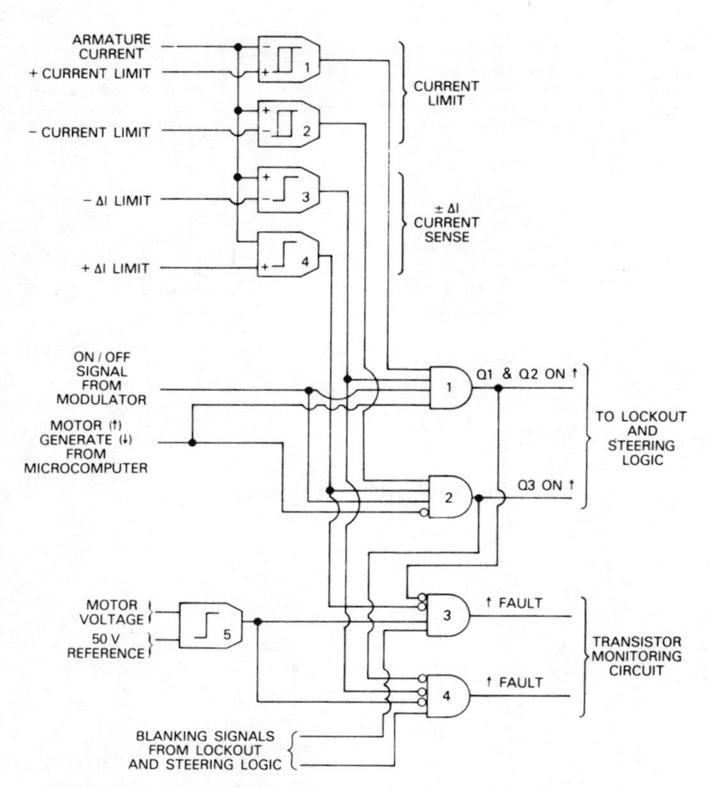

Fig. 7. Transistor monitoring circuit and current logic.

transistor command as determined by the motor/generate command from the microcomputer. Other inputs to gates 1 and 2 may inhibit transistor turn-on commands. Two current limiting comparators with hysteresis (1 and 2) serve as a local fast-acting current limit. The instantaneous armature current is compared against the peak current limit reference values. In the motoring case, if the peak armature current (which is also the peak transistor current) exceeds the current limit value of 420 A, the output of comparator 1 turns off Q1 and Q2 through the action of gate 1, independent of the command from the microprocessor. Q1 and Q2 are held off by comparator 1 until the armature current (and hence the transistor current) falls approximately 80 A, as set by the comparator hysteresis. After that point has been reached, the normal modulator signal is again allowed to pass through gate 1 until the current limit value is again exceeded. The current limit scheme for the generating chopper works in a similar manner, except that the limit value is set at –240 A. Note that with this scheme the transistor collector currents are never allowed to exceed a value beyond which they are rated and can successfully turn off. This limiting scheme is fundamentally different from other schemes [12] which measure the collector-emitter voltage to determine when a transistor is coming out of saturation and turns the transistor off when this condition occurs. With that scheme, however, the transistor peak current limit will be a function of transistor gain, and transistors with higher gain will be required to turn off higher currents. Since the reverse bias second breakdown capability of a power transistor does not normally increase with gain, that scheme may require transistor selection or derating to be practical.

The transistor monitoring circuit senses a transistor failure at turn-off. This operation will be explained by considering the instant a turn-off command has been given to Q1-Q2. In this instance the output of gate 1 would go high, and a fault signal would appear at the output of gate 2, except that a blanking signal is applied to gate 3 from the lockout and steering logic. This blanking signal, which was initiated at the instant a turn-off command was given, prevents a fault signal from being generated for a time which is known to be longer than the transistor storage time. After Q1-Q2's storage time, the motor voltage drops to near zero as D3 comes into conduction. This fact is sensed by comparator 5, which prevents a fault signal from being sent after the blanking signal has timed out. Note that if Q1-Q2 failed to turn off, the motor voltage would not have decreased, and comparator 5 would have allowed a fault signal to be sent and the drive to be shut down as soon as the blanking signal timed out. Thus a transitor failure is detected even if the current is well below the current limited value. It is noted that if the current is low enough to be discontinuous, the motor voltage will be at the counter EMF value after the current has gone to zero (rather than at zero due to diode D3 conducting). This would cause the transistor monitoring circuit to indicate a fault erroneously. To prevent this, the $\pm\Delta I$ current sense comparators prevent a fault signal from being sent if the current has fallen below a sufficiently low value (i.e., if the current is approaching the discontinuous condition). The transistor monitoring circuit for the generating transistor works in an identical manner except that the output of comparator 5 is inverted. The $\pm\Delta I$ current sensing logic serves one other function. If the current is in a module feedback diode, the $\pm\Delta I$ current sense signal prevents the transistor parallel with the feedback diode from being gated on and storing charge and possibly carrying current in the reverse direction. Such an operation can cause transient currents in the transistor when the opposite transistor is again gated. This charge appears as an abnormally long reverse recovery time feedback diode to the transistor being switched on.

ARMATURE CHOPPER PERFORMANCE

The armature chopper is shown in Fig. 8. The hardware on top of the casting comprises the armature chopper. The field chopper/battery charger, logic power supply, logic cards, accessory battery charger, filter capacitors, and contactors are mounted below the casting. In Fig. 9 the chopping frequency and computed losses are shown as a function of the chopper duty cycle. The highest frequency is needed at a duty cycle of 0.5 in order to maintain the constant 40-A peak-to-peak ripple current since at this point the highest fundamental ripple voltage is delivered to the motor. Since the switching losses are proportional to frequency, the highest loss also occurs at this operating point. The chopper efficiency varies between 92 percent at a duty cycle of 0.5 and 98 percent at a duty cycle of 1.0.

In Fig. 10 the transistor collector-emitter voltage and base drive waveform are shown. Note the voltage spike which occurs at transistor turn-off. This spike is due to parasitic inductance caused by the loop (Fig. 3) defined by the capacitor bank, R1, D4, Q1-Q2, D3, and back to the capacitor bank. The parasitic inductance in this case is approximately 0.1 μH. It was

Fig. 8. Armature chopper hardware.

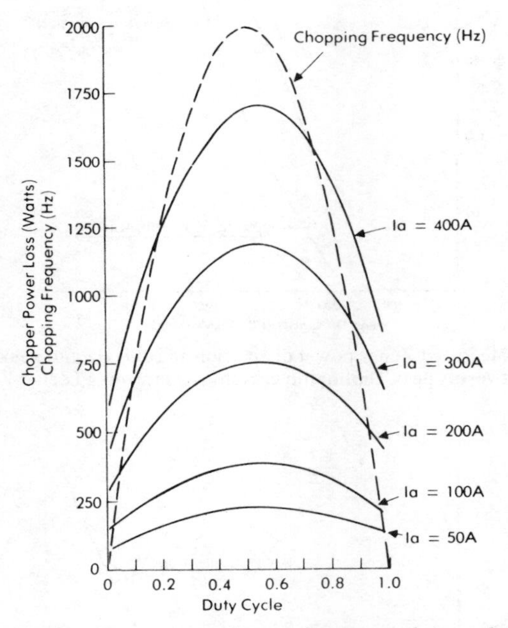

Fig. 9. Frequency and power loss versus duty cycle.

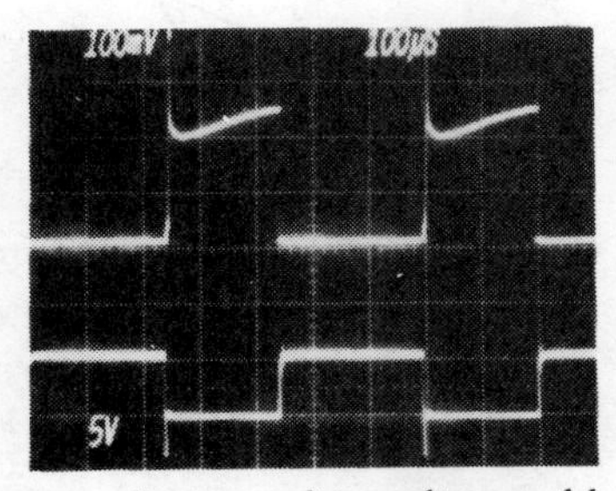

Fig. 10. Darlington collector–emitter voltage and base current versus time. Top: 50 V/cm (zero at center line). Bottom: 2 A/cm (0–1 cm from bottom). Time: 100 μs/cm.

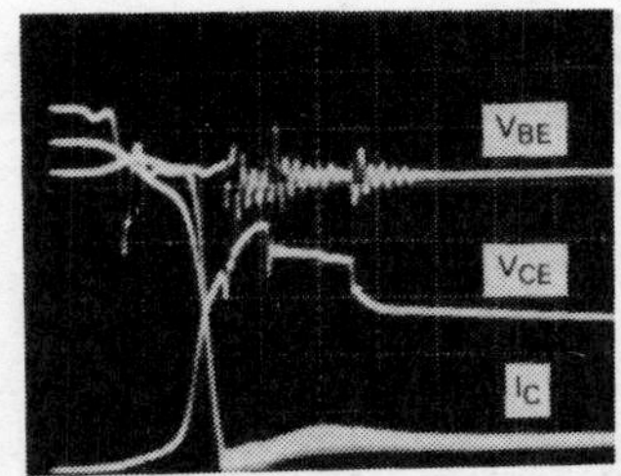

Fig. 11. Darlington waveforms at turn-off base–emitter voltage: 2 V/cm (0–2 cm from top). Collector–emitter voltage: 50 V/cm (zero at bottom line). Darlington collector currents: 40 A/cm (zero at bottom line). Time: 2 μs/cm.

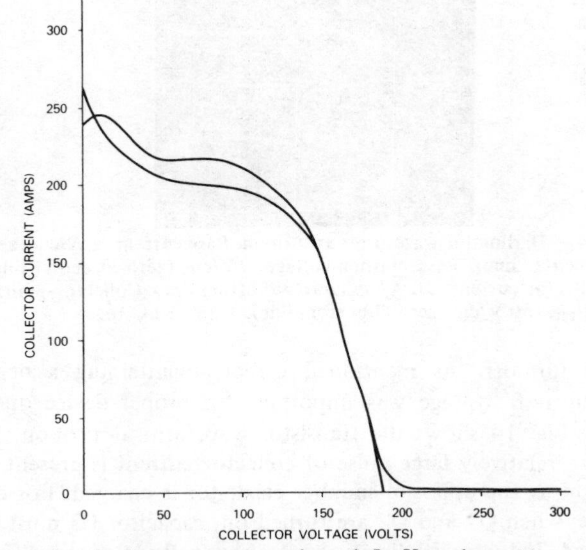

Fig. 12. Darlington transistor through I_C-V_{CE} plane at turn-off.

found that due to the fast transistor switching speed (fall time) of less than 1 μs acting in conjunction with the parasitic inductance, the Zener diode clamps (Z1 and Z2) were needed to limit the transistor voltage both at turn-off and turn-on due to the overcharging of the snubber capacitors. The base circuit employed is similar to that described in [13] and provides a 1.5-A turn-off pulse and maintains a –4-V reverse base bias during the transistor off-time.

In Fig. 11 the transistor turn-off waveforms are shown. The sharing of the two parallel transistors (Q1 and Q2) collector currents is shown. The fall time of less than 1 s is seen as well as the clamping action of Zener diodes which manifests itself as a voltage "backporch" of approximately 180 V immediately after turn-off. The base–emitter voltage is also shown to maintain a negative bias after turn-off. It was found that, for these devices, a negative base voltage was necessary to ensure reliable device turn-off. Fig. 12 shows the transition through the $I_C - V_{CE}$ plane at turn-off for the parallel transistors. From this curve it is seen that the turn-off dissipation is shared approximately equally. The snubber capacitors (C1–C3) were chosen as 1 μF and were judged to be a reasonable compromise between the snubber losses and transistor turn-off dissipation. Note that the snubber capacitors C1–C3 also act as snubbers for diodes D1–D3 and thus limit voltage transients when these diodes turn off.

Fig. 13 shows the detailed base voltage and current at tran-

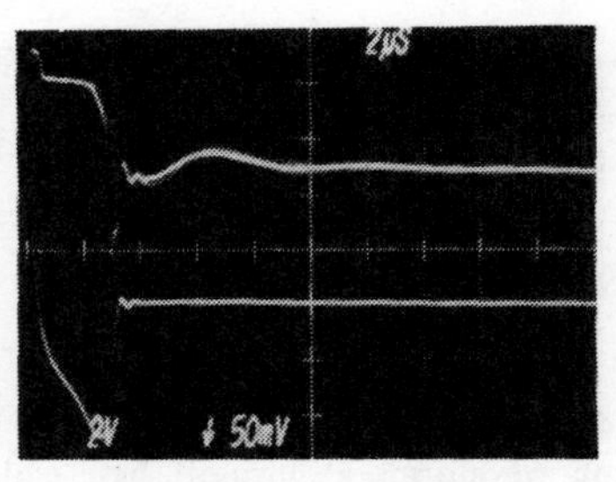

Fig. 13. Darlington base waveforms at turn-off. Top: Base–emitter voltage: 2 V/cm (0–1 cm from top). Bottom: Base current: 1 A/cm (0–3 cm from bottom).

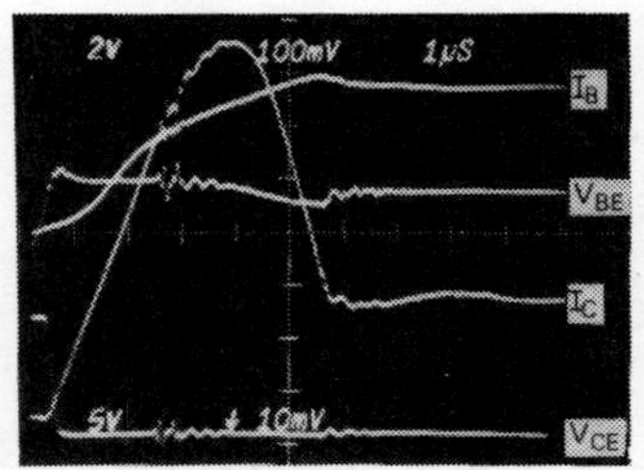

Fig. 14. Darlington waveforms at turn-on. Base current: 1 A/cm (zero at center line). Base–emitter voltage: 1V/cm (zero at center line). Collector current: 20 A/cm (zero at bottom line). Collector–emitter voltage: 50 V/cm (zero at bottom line). Time: 1 µs/cm.

sistor turn-off. As mentioned earlier, maintaining a negative turn-off voltage was important for proper device operation. Fig. 14 shows the transistor waveform at turn-on. As seen, a relatively large pulse of collector current is present at turn-on as the opposite snubber capacitor is charged. For example, when Q1 and Q2 are turned on, capacitor C3 must be charged. Inductor L1 limits the peak magnitude of this charging pulse to

$$I_{CP} = E_d/\sqrt{L1/C3}.$$

This pulse must be added to the motor current at that instant to determine the total transistor current. It may be important that the transistor not try to turn off at the peak of this current pulse since the reverse biased second breakdown rating of the device may be exceeded. Consequently, a minimum turn-on duration circuit can be used. For the equipment described here, it was found that the transistor storage time was sufficiently long to ensure that once the transistor is turned on it cannot turn off again until after the peak of the turn-on current pulse. As seen in Fig. 14, the turn-on dissipation is negligible since the presence of inductor L1 allows the collector voltage to fall to zero before the current has risen substantially.

Fig. 15 shows the Zener diode currents into the two parallel diodes, which comprise Z1, and their relationship to the collector–emitter voltage at transistor turn-off. As seen, the Zener limits the voltage to approximately 200 V. The measured power dissipated in the Zener diodes under the worst operating condition (2000 Hz) is illustrated in Figs. 16 and 17. As seen, the dissipated power increases substantially with current and battery voltage. Under the usual operating conditions, the dissipation of the Zener clamps is small. Only under hard accelerations (>200 A) and high battery voltage, due to high generation into a substantially charged battery, are the Zeners

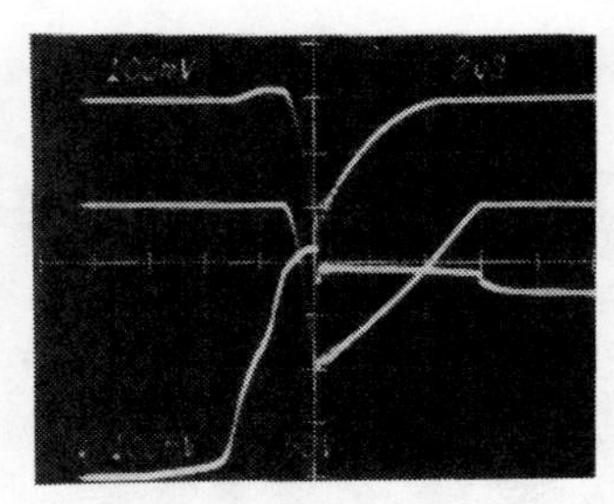

Fig. 15. Zener clamp currents and collector voltage at Darlington turnoff. Top: Zener current: 40 A/cm (0–1 cm from top). Middle: Zener current: 40 A/cm (0–1 cm above center line). Bottom: Collector–emitter voltage: 50 V/cm (zero at bottom line). Time: 2 µs/cm.

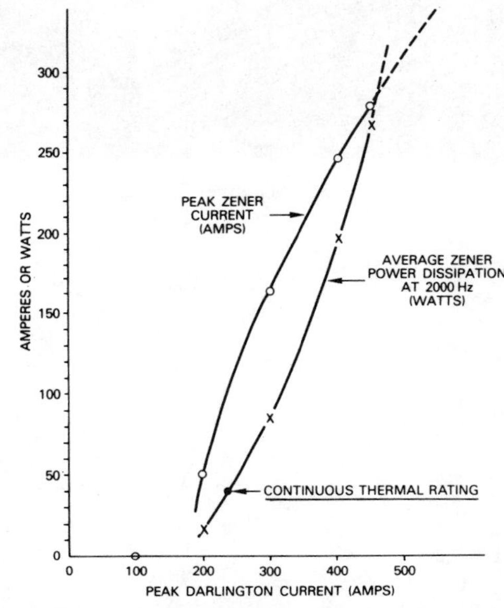

Fig. 16. Measured Zener power dissipation at 2000 Hz and peak Zener current versus peak Darlington current for motoring (Zener Z1).

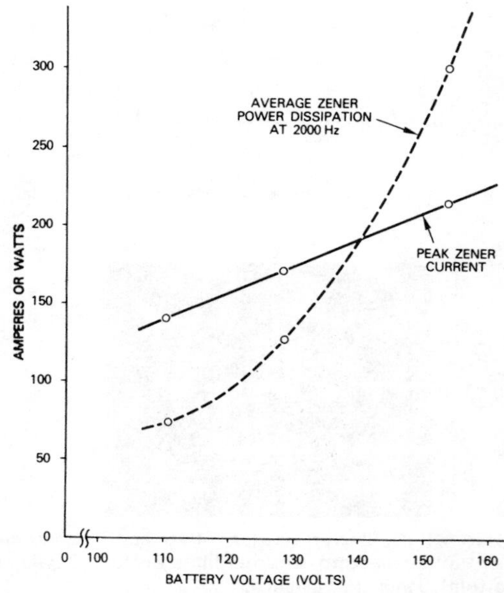

Fig. 17. Measured Zener power dissipation at 2000 Hz and peak Zener current versus battery voltage during regeneration for a 270-A peak transistor current (Zener Z2).

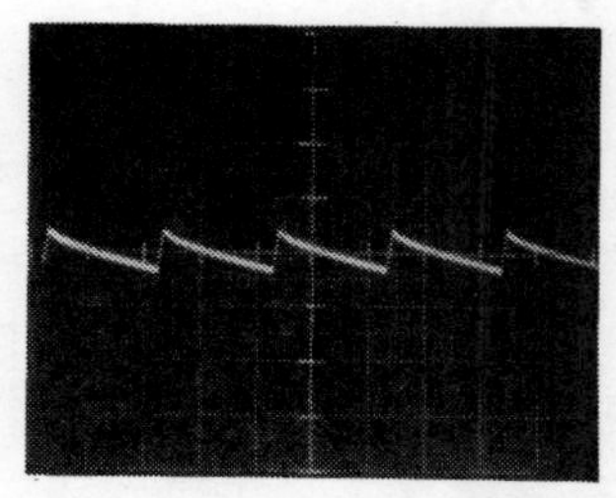

Fig. 18. Current delivered to stalled rotor. Armature current: 50 A/cm (zero at bottom line). Time: 1 μs/cm.

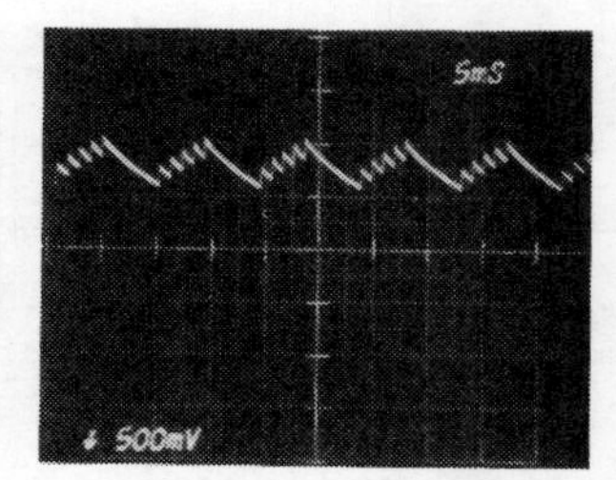

Fig. 19. Motor current during local current limit (stalled rotor). Armature current: 100 A/cm (0–2 cm from bottom). Time: 5 μs/cm.

called upon to dissipate substantial power. Since these are transient conditions, the thermal design and overall drive efficiency are not substantially affected by employing the Zener clamps.

Figs. 18 and 19 show the armature current delivered by the chopper under stalled rotor conditions for the normal and for the local current limited cases. In Fig. 19 it is seen that when the 400-A point is reached, the transistors are turned off (independent of the microprocessor command) until the current drops approximately 80 A, at which point normal chopping is resumed.

CONCLUSION

A two-quadrant transistor chopper with a peak current rating of 400 A motoring and 200 A generating and operating from a 108-V dc source has been described. A peak switching frequency of 2000 Hz obviated the need for a choke in series with the motor armature. For a two-quadrant chopper there is an interaction between the turn-off and turn-on stress reducing networks. The turn-off network, while reducing transistor dissipation at turn-off, causes a transient current pulse in the opposite transistor when it is switched on, thus increasing its turn-on duty. The turn-on snubber, while reducing turn-on losses, causes the transistor voltage to rise above its normal level at turn-off. Proper snubber selection must take these effects into account. A minimum transistor on-time and a minimum transistor off-time are needed to ensure proper snubber operation and minimize transistor stress.

A technique has been described which senses a power transistor failure within approximately 50 μs independent of the current load level. Other interface circuitry between the microcomputer control and the chopper has been described. A unique current control logic scheme, which ensures that the transistor is gated only when it can carry forward current, ensures that a nonconducting transistor does not incur switching losses due to inadvertent charge storage. A local fast-acting current limit which overrides the microcomputer command ensures that the chopper current does not exceed a value which the transistor can successfully turn off.

The chopper described here is being used in a commuter-type electric vehicle developed for the Department of Energy and serves to illustrate the practicality of transistor drives in the 40-kW range.

REFERENCES

[1] B.K. Bose and R.L. Steigerwald, "A DC motor control system for electric vehicle drive," *IEEE Trans. Ind. Appl.*, vol. IA-14, pp. 565–572, Nov./Dec. 1978.

[2] B.K. Bose and H. Sutherland, "A microcomputer based real time feedback controller for an electric vehicle drive system," *IEEE Ind. Appl. Soc. Conf. Rec.*, 1979.

[3] E.T. Calkin and B.H. Hamilton, "A conceptually new approach for regulated DC to DC converters employing transistor switches and pulse width control," *IEEE Trans. Ind. Appl.*, vol. IA-12, No. 4, pp. 368–377, July/Aug. 1976.

[4] E.T. Calkin and B.H. Hamilton, "Circuit techniques for improving the switching loci of transistor switches in switching regulators," *IEEE Ind. Appl. Soc. Conf. Rec.*, pp. 477–484, 1972.

[5] R.L. Steigerwald, "Application techniques for high power gate turn-off thyristors," *IEEE Ind. Appl. Soc. Conf. Rec.*, pp. 165–174, 1975.

[6] T.M. Undeland, "Switching stress reduction in power transistor converters," *IEEE Ind. Appl. Soc. Conf. Rec.*, pp. 383–392, 1976.

[7] PH. Lataire, G. Maggetto, and M. Hendrickx, "Variable speed drives with low and medium power DC motors and transistor choppers," in *Int. Conf. Electrical Machines*, pp. 131-1–31-10, 1976.

[8] R.W. Stokes, "High voltage transistor inverters for AC traction drives," in *Int. Semiconductor Power Converter Conf. Rec.*, pp. 270–294, Mar. 1977.

[9] J.D. van Wyk and J.J. Schoeman, "The application of transistorized switches to DC and AC machines for the control of battery vehicles up to 30 kW," in *Proc. Second IFAC Symp. Control in Power Electronics and Electrical Drives*, pp. 841–852, Oct. 1977.

[10] A.R. Daniels, V.S. Gott, and K.W. Howe, "A transistor controller for a battery driven vehicle," in *Int. Conf. Power Electronics—Power Semiconductors and Their Applications*, pp. 57–62, Dec. 1974.

[11] W. McMurray, "Stepless solid-state controls for battery-powered DC electric vehicles," in *IFAC Symp. Control in Power Electronics and Electrical Drives*, pp. 421–435, 1974.

[12] H. Knoll, "High-current transistor choppers, "in *Pro. Second IFAC Symp. Control in Power Electronics and Electrical Drives*, pp. 307–315, Oct. 1977.

[13] R.L. Steigerwald, "Pulse battery charger employing 1000 A transistor switches," in *IEEE Ind. Appl. Soc. Conf. Rec.*, pp. 1127–1132, 1977.

TRANSISTORIZED PWM INVERTER-INDUCTION MOTOR DRIVE SYSTEM

Steven C. Peak
General Electric Company
Corporate Research and Development
Schenectady, NY 12345

Allan B. Plunkett
General Electric Company
Corporate Research and Development
Schenectady, NY 12345

ABSTRACT

This paper describes the development of a transistorized PWM inverter-induction motor traction drive system. A vehicle performance analysis was performed to establish the vehicle tractive effort-speed requirements. These requirements were then converted into a set of inverter and motor specifications. The inverter was a transistorized three-phase bridge using General Electric power Darlington transistors. The description of the design and development of this inverter is the principal object of this paper.

The high-speed induction motor is a design which is optimized for use with an inverter power source. The primary feedback control is a torque angle control with voltage and torque outer loop controls. A current-controlled PWM technique is used to control the motor voltage. The drive has a constant torque output with PWM operation to base motor speed and a constant horsepower output with square wave operation to maximum speed. The drive system was dynamometer tested and the results are presented.

INTRODUCTION

Presently, the dc motor and dc controller combination is the dominant electric vehicle drive system configuration, with only a few vehicles using an ac system. However, recent studies comparing various electric vehicle propulsion system approaches have concluded that the most promising drive system for near term electric vehicle use is the ac induction motor with a pulse width modulated (PWM) transistor inverter based controller [1]. The impetus behind the ac drive system is the cost, maintenance, size, reliability, and efficiency advantages of the ac induction motor. The size, cost, and complexity of the controller for this motor represent the technical challenge to the potential and desirable advantages of the ac induction motor. Much of the complexity of the controller exists in the signal level controls, where advances in microelectronics technology will play a significant role in reducing cost and parts count. The evolution and downward price trend in high-power transistors will allow the inverter to be economically feasible and reliable.

The purpose of this development project was to design, fabricate, test, evaluate and cost analyze an engineering model ac motor controller for a variable speed traction ac polyphase induction motor. The feasibility and performance of a transistorized ac inverter based drive system was demonstrated. This work was supported in part by the U.S. Department of Energy under contract DEN3-59 with the NASA Lewis Research Center.

As a general requirement, the drive system is designed for use in urban electric vehicles according to the SAE J227a-Schedule D driving cycle. This determines the drive system thermal rating. In addition, a practical electric vehicle must satisfy performance requirements in excess of the J227a-D duty cycle in order to merge with traffic, start on a steep grade, pull out of a pothole, climb hills, and pass. The additional specifications define the maximum (peak) desired vehicle performance. All of these specifications are primarily vehicle performance requirements and must be analyzed to establish the vehicle tractive effort-speed requirements. These requirements were then converted into a set of motor and inverter specifications.

The overall system configuration, consisting of the inverter, induction motor, and control electronics, is shown in Figure 1. The control electronics include an improved method of PWM control [2]. This method adaptively modulates the inverter switching times so as to produce the maximum possible torque in the load motor for the minimum ac peak current. There is an instantaneous feedback current control with independent current control on each inverter output phase. The improvements in this current-controlled PWM method are to minimize peak transistor current and minimize motor loss with better sinusoidal waveforms.

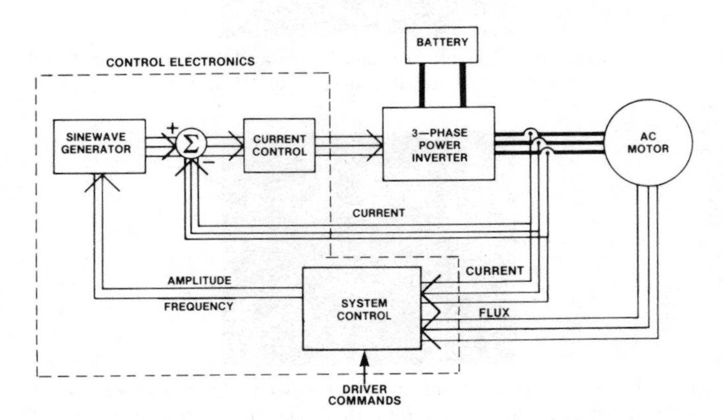

Figure 1. Simplified System Block Diagram

The system control method involves a transition of operation of the inverter from a current source to a voltage source, as the control transitions from the PWM mode to the square wave mode of operation. The system control must therefore be suitable for use in either mode of operation. A motor torque angle control is used to supply the primary system stabilization [2]. The system control generates the appropriate frequency and current amplitude signals by using the feedback signals of motor current and flux, and the driver commands [3]. The resulting system has operating characteristics similar to both voltage inverters and current inverters. As a hybrid of these two types, the system has the good low-speed performance of a current inverter and the good high-speed performance of a voltage inverter.

In dc machines, sensing of armature and field currents will yield a fairly accurate measure of the flux level in the machine and, hence, the developed torque. However, the flux and torque of an induction motor are not as easily measured, particularly when the voltage supplied to the terminals of the machine is not sinusoidal. The method used involves the sensing of the motor air gap flux and stator current. The air gap flux can be inferred from the motor terminal voltage but will suffer some inaccuracy for very low-speed operation due to the stator resistance voltage drop; or can be measured directly by air gap flux sensing coils wound around a motor stator tooth [4].

Figure 2 shows the three-phase inverter power circuit, which is a three-phase bridge connection with feedback diodes and uses six power modules. Each power module contains a combination of power Darlington transistors and antiparallel fast recovery diodes capable of switching 600 amps at up to 300 volts peak. There are four power Darlington transistors in parallel, each rated at 200 amps and 450 volts. There are two diodes in parallel, each rated at 250 amps and 600 volts. Figure 3 shows a photograph of the completed inverter, which is an engineering model for development tests.

A dc filter capacitor bank is connected across the dc input to the inverter and serves to filter the dc input voltage and provide a low impedance path for the high-frequency currents generated by the inverter during PWM switching. Each power module is driven by a separate base drive circuit with its own isolated power supply.

Reprinted from *IEEE Ind. Appl. Soc. Meet.*, 1982, pp. 892-898.

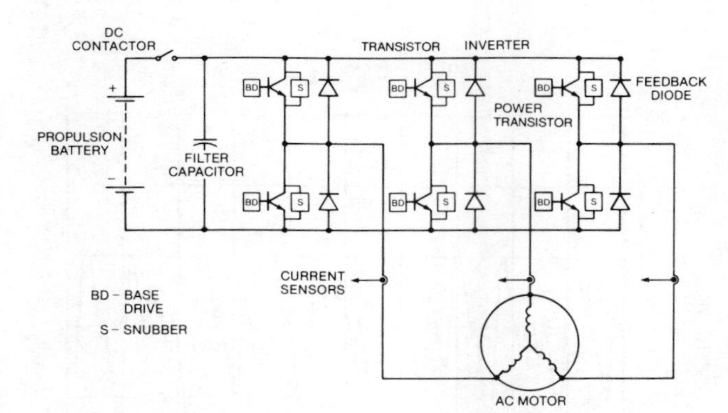

Figure 2. Inverter Power Circuit

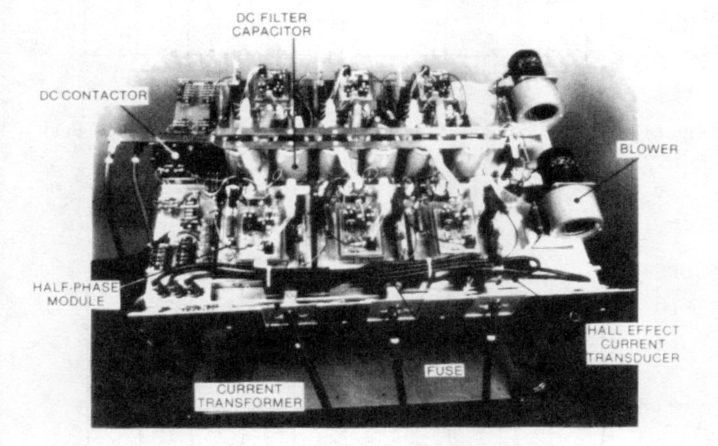

Figure 3. Engineering Model Inverter

A snubber is placed across each power module to maintain the instantaneous current and voltage switching locus within the ratings of the power Darlington transistors. The inverter is modularized into six identical half-phase modules built up on heatsink material, which is forced air cooled (Figure 4).

The PWM approach for motor voltage control is chosen because of the fixed dc battery bus, small dc filter, inherent regeneration capability, and dual use of the power switches for inversion and voltage control. The choice of transistors over thyristors was made because transistors obviate the need for commutation circuits and have the switching speed required by the current-controlled PWM method.

The design of the inverter and motor must take into account the presence of the extra time harmonics in the inverter output voltage which tend to cause increased motor heating and peak transistor current [5,6].

Figure 4. Half-Phase Module

HARMONIC CURRENT ANALYSIS

The control system used in this drive system combines the ability to control both ac frequency and ac voltage by pulse width modulation control of the inverter transistors. The pulse width modulation is accomplished by switching the transistors at a high frequency and then varying the switching duty cycle at the desired motor frequency.

The system used (Figure 1) is to feed back the motor current, compare with a sine wave (or other) reference wave, and switch the inverter to contain the error within a fixed band set by the current comparator hysteresis (current control). This method causes the inverter switching to be adaptive to load variations such as motor back emf and inductance. In addition, the inverter switching losses are proportional to frequency and must be controlled; thus the chopping frequency should, on the average, be limited. The rate of change of current and thus the chopping frequency is automatically controlled by the comparator hysteresis, the motor stator plus rotor leakage inductance, and the dc link voltage by the equation

$$\frac{di}{dt} \approx \frac{E_{dc} - E_m \sin\omega t}{L_{eq}}$$

where

E_{dc} = dc link voltage

L_{eq} = equivalent motor leakage inductance

$E_m \sin\omega t$ = motor back emf voltage

which, however, applies only approximately since the motor leakage inductance is a function of frequency and the voltage is actually the dc link voltage minus the motor back emf.

With the three-phase inverter shown in Figure 2 and three-wire wye motor, there is no neutral connection. Thus, the operation of the inverter is constrained because the sum of the three line currents must be zero at all instants,

$$i_a + i_b + i_c \equiv 0$$

The current in any one phase is completely determined by controlling the currents in the other two phases, rendering one phase of a current controller redundant. The switches of one phase should not, however, be removed and the circuit simplified because the conventional three-phase configuration is more efficient under square wave operation at high speed.

The effect of the high-frequency inverter chopping is to cause extra ripple in the motor ac current. This ripple current causes extra motor losses and increases the required inverter transistor current rating. A method of calculating these extra losses and peak currents has been developed for the more commonly used sine wave—triangle wave intersection voltage-controlled PWM and is described in References [5 and 6]. For a large number of chops per motor frequency cycle, the extra motor losses due to current-controlled PWM operation should be similar to that of sine-triangle voltage PWM, so that the method described in the two papers has been used to predict the increased motor losses and inverter peak current.

The motor used with this drive is a specially designed high-efficiency induction motor. The motor parameters are shown in Table I. The required peak load is a torque of 42.7 lb-ft. Applying this load to the inverter at the maximum speed, for which full PWM with no lost chops is possible, results in the operating conditions of Table II. The inverter design is based on the peak and average currents of Table II. The required chopping frequency is 1740 Hz; however, the motor always sees two phases chopping simultaneously which results in a motor torque ripple at 3480 Hz. Figure 5 shows the transistor current required for the peak load condition of Table II. Transistor conduction is limited to 180 degrees with the opposite transistor in the same phase conducting during the second 180 degree period. The notches in the transistor current represent the opposite diode conduction periods.

Figure 6 shows the motor torque ripple for two cases. Figure 6A shows the motor torque ripple at the peak load condition

Table I

MOTOR PARAMETERS

Parameter	
$R_1 = 0.00298\ \Omega$	STATOR
$L_1 = 12.69 \times 10^{-6}$ hy	
$R_2 = 0.00145\ \Omega$	ROTOR
$L_2 = 10.15 \times 10^{-6}$ hy	
$L_3 = 240 \times 10^{-6}$ hy	
$F_{BASE} = 180$ Hz	
$V_{BASE} = 45$ volts rms/phase	

Table II

PEAK LOAD OPERATING CONDITION
(Sine-Triangle PWM Used in Calculation)

DC Voltage	108 volts
AC Voltage	35.96 volts rms L-N
AC Current	321 amps rms
AC Frequency	145 Hz
Slip	1.7 Hz
Torque	42.7 lb-ft
Speed	4299 rpm
Peak Current	593 amps
Average Transistor Current	116 amps
Average Diode Current	30 amps
Average dc Link Current	262 amps
Chopping Frequency	1740 Hz/phase
Chopping Frequency	3480 Hz (Motor Effect)

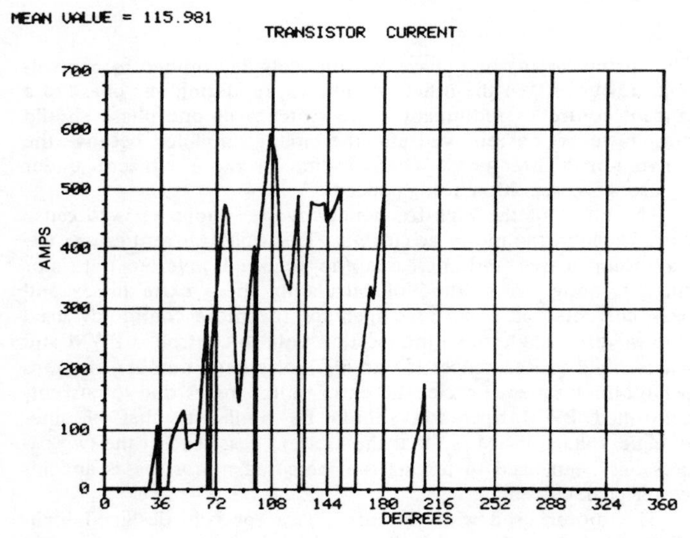

Figure 5. Calculated Transistor Current Waveform for the Peak Load Condition

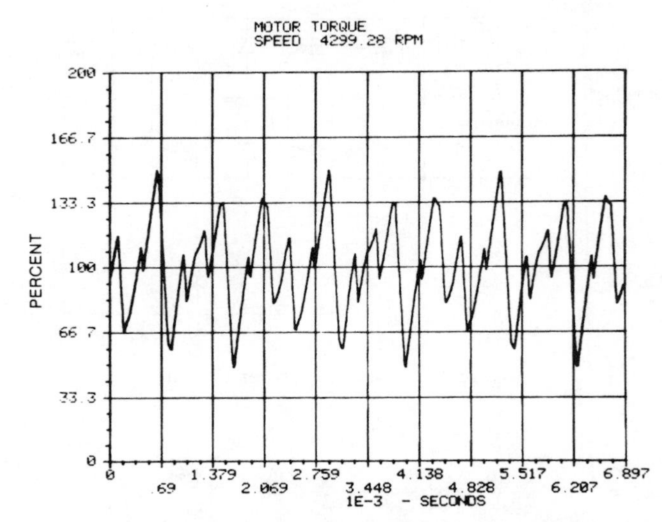

Figure 6A. Calculated Motor Torque Waveform for the Peak Load Condition

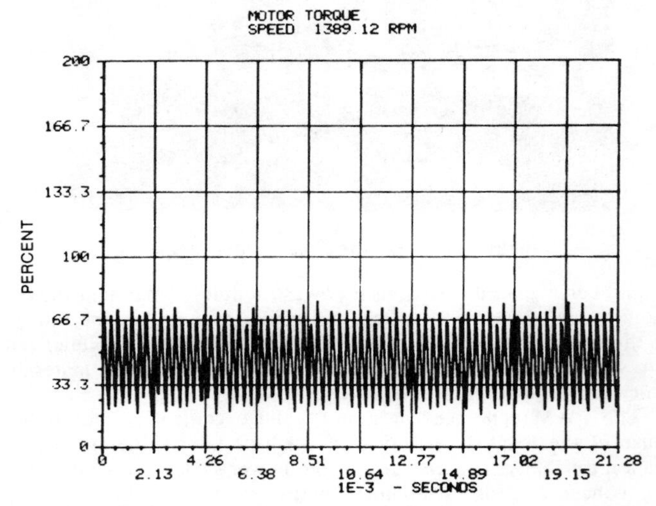

Figure 6B. Calculated Motor Torque Waveform for the Test Condition

of Table II. For the sine-triangle voltage PWM method, the torque ripple shows a considerable non-smooth ripple effect. However, at a lower speed of 47 Hz, the torque ripple as illustrated in Figure 6B shows a more uniform ripple character with the ripple frequency being twice the per phase chopping frequency.

A series of tests were performed on the completed drive system in order to test the predictions of the computer analysis. Figure 7 shows the calculated motor current for the selected test condition at a speed of 47 Hz. Figure 8 shows the actual measured motor current for the same operating condition. The current-controlled PWM system causes equal ripple current over the ac motor frequency cycle, while the sine-triangle voltage PWM causes a cyclic variation in motor current ripple. The operating conditions were matched by adjusting the calculated PWM frequency so that the measured and calculated fundamental currents match. Another interesting effect of current-controlled PWM can be observed by comparing the frequency spectra of motor current illustrated in Figures 9 and 10. Note that the chopping frequency in Figure 10 shows a broadened principal chopping frequency spectral line due to the expected cyclic variation in chopping frequency resulting from the motor back emf. Figure 9 shows spectral lines of 3.4 kHz for the sine-triangle voltage PWM chopping.

A comparison of the measured and calculated drive system performance at the test condition is shown in Table III. Very good correlation is observed except for a difference in predicted and measured chopping frequency, which is probably due to the approximation of using sine-triangle voltage PWM to predict current-controlled PWM performance.

The predicted currents of Table II are used for the inverter design with the expectation that the worst case sine-triangle voltage PWM with a few chops per cycle (145 Hz) will require somewhat more peak current than the actual current-controlled PWM.

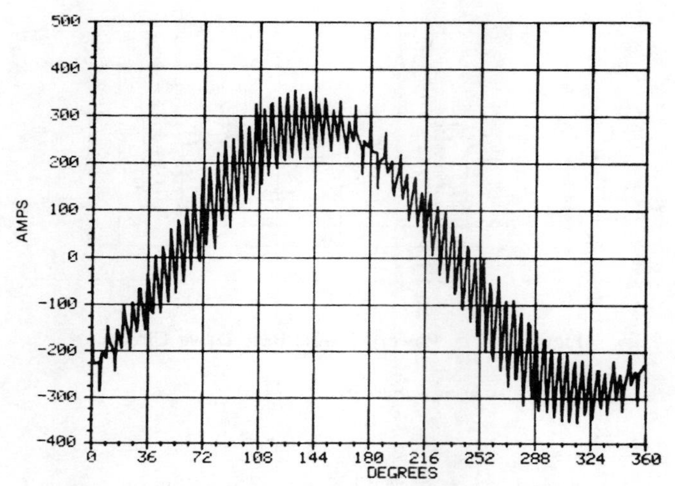

Figure 7. Calculated Motor AC Current Waveform for Test Condition

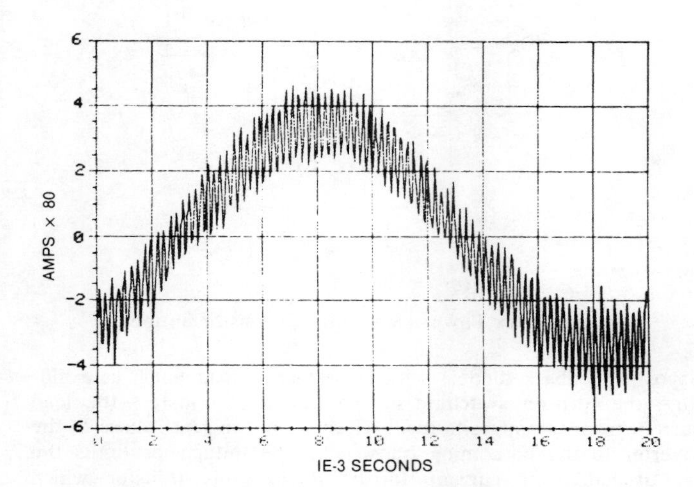

Figure 8. Measured Motor AC Current Waveform for Test Condition

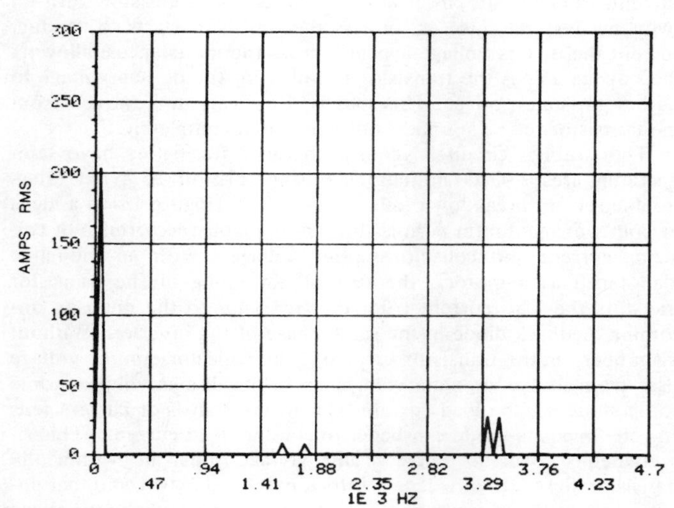

Figure 9. Frequency Spectrum of Calculated Motor AC Current Waveform for Test Condition

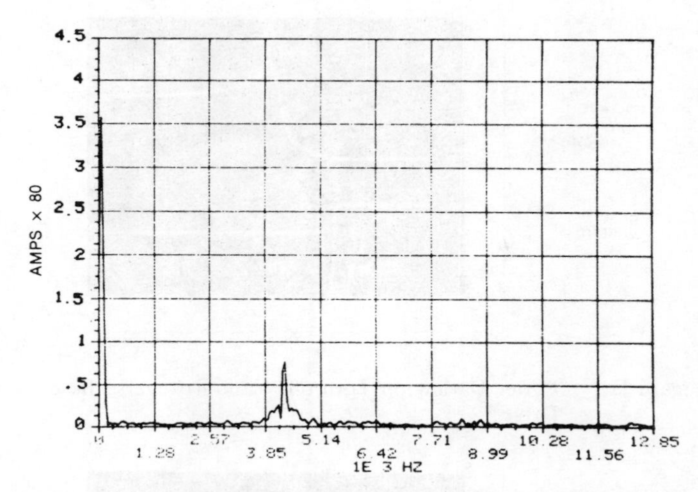

Figure 10. Frequency Spectrum of a Measured Motor AC Current Waveform for Test Condition

Table III

COMPARISON OF MEASURED AND CALCULATED RESULTS FOR TEST CONDITION

	Calculated	Measured
DC Voltage	108 volts	107.5 volts
AC Voltage	12.31 volts rms L-N	–
AC Current	203 amps rms	202 amps rms
AC Frequency	47.24 Hz	47.24 Hz
Slip	0.7 Hz	–
Torque	20.25 lb-ft	20.32 lb-ft
Speed	1389 rpm	1438 rpm
Peak Current	356 amps	366 amps
Chopping Ripple Frequency	3.4 kHz	4.3 kHz (avg)
Chopping Ratio	36	–
Current Band	150 amps	160 amps
Motor Harmonic Loss	68 watts	–
Motor Total I^2R Loss	495 watts	–
PWM Method	sine-triangle	current control

INVERTER DESIGN

Base Drive Design

Both bases, B1 and B2, of the power Darlington transistor are driven. B1 is the base of the npn driver transistor and B2 is the base of the npn output transistor. Figure 11 shows the collector to emitter voltage and the base 1 current of a single Darlington transistor switching off about 100 amps, with a 10 amp reverse base 1 drive and without any reverse base 2 drive. Note the initial slow rise of collector voltage as the device turns off and the relatively long storage time (from the base 1 current reversal to the steep rise of collector voltage). Figure 12 shows the result of applying a reverse base 2 current of about 1.5 amps to the same device, base 1 drive and switching conditions. The reverse base 2 drive is achieved by adding an external diode and resistor between base 2 and base 1. The base drive current in Figure 12 shows the sum of the two base currents. Note the initial slow rise of collector voltage is shortened and the storage time is reduced, thus reducing the transistor switching loss.

The power module base drive circuit (Figure 13) supplies a total forward (positive) base current into the power module of 4 amps peak and a total reverse (negative) current of 5 amps peak. This is based on the gain characteristic of the Darlington transistor, the base drive power supply requirements and the desired turn-off switching time. The diode and resistor between B1 and B2 are shown in Figure 13. There are four Darlington transistors

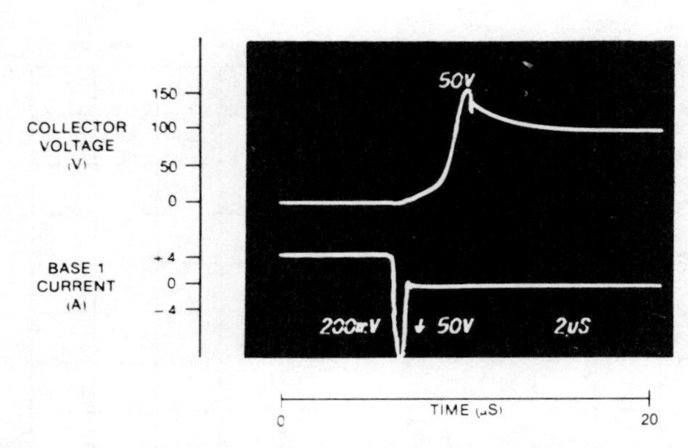

Figure 11. Power Darlington Turn-off Without Reverse Base 2 Drive

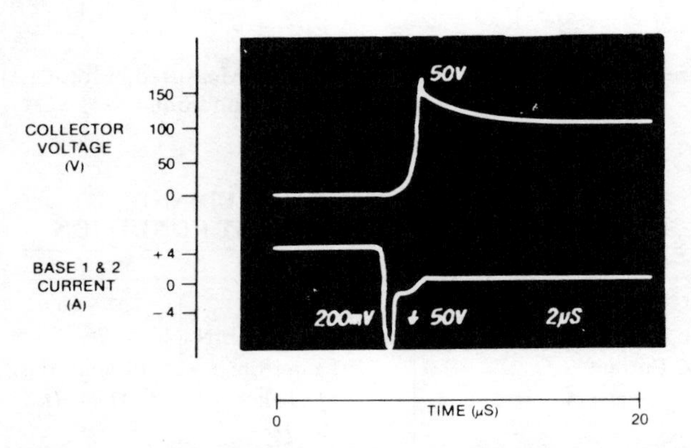

Figure 12. Power Darlington Turn-off With Reverse Base 2 Drive

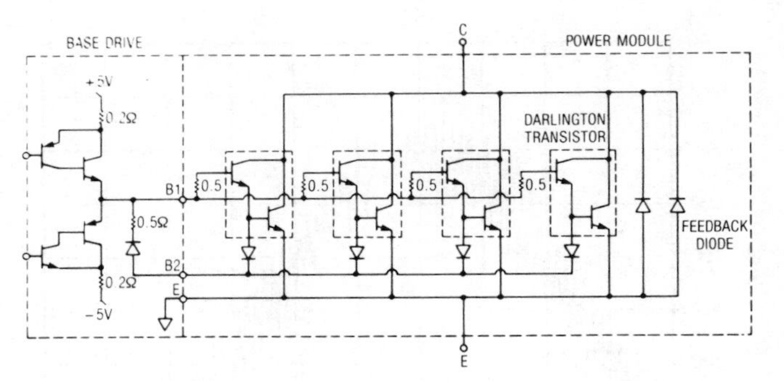

Figure 13. Power Module Base Drive Circuit

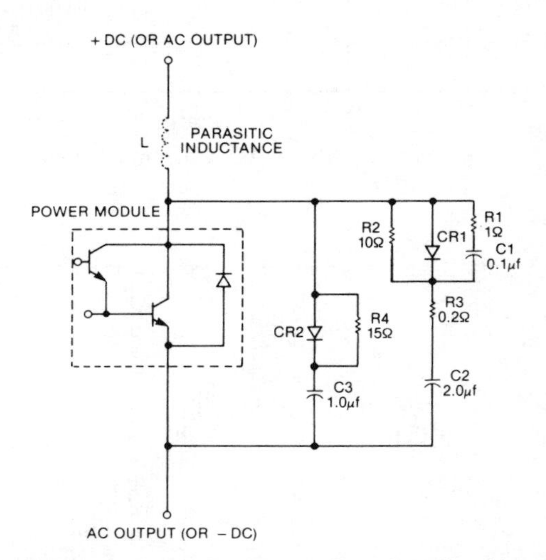

Figure 14. Power Darlington Transistor Snubber

in parallel in each power module. A 0.5-ohm resistor is in series with base 1 for each Darlington transistor in the power module, to assist in the sharing of the base current. A blocking diode is in series with each base 2, which allows a reverse base 2 drive but isolates base 2 during a Darlington transistor fault and prevents the consequential failure of the other transistors. Each power module base drive circuit has its own isolated base drive power supply.

Snubber Design

The snubbering for the power Darlington transistors (Figure 14) is composed of two circuits: a resistor-capacitor snubber circuit in parallel with the power module and a parasitic, but intentionally placed, snubber inductance in series with the power module. The snubber circuit is effective during the turn-off of the Darlington transistors and the snubber inductance is effective during the turn-on of the Darlington transistors, as described below. The two snubbers actually operate on the parallel combination of the four power Darlington transistors, i.e., the power module. The four Darlington transistors are matched so that each one is operated safely when the set is protected by the snubbers. The dc filter capacitor bank provides a low impedance path (much lower than the power cables and battery) for the high-frequency currents generated during switching.

The small parasitic inductances (approximately 0.25 μH) in series with the power modules (Figure 14) are obtained by the routing of the bus bars between the input capacitor bank and the half-phase modules, and between the upper and lower half-phase modules in each phase (Figure 3). These inductances are required during turn-on to limit the power module transistor current to a safe value. There are several currents which flow through the transistors during turn-on, with an inductive load: the load current, the snubber capacitor discharge current, the opposite snubber capacitor charging current and the recovery current of the opposite feedback diode. With an inductive load (such as a motor), the turn-on switching of the transistor transfers the load current from the opposite feedback diode in the same phase of the inverter to the on-coming transistor. The inductance limits the rate-of-change of current (di/dt) during this transfer which prevents an excessively large recovery current in the opposite feedback diode. The inductance also interacts with the capacitance of the opposite snubber to limit the peak magnitude of its charging current. Lastly, the inductance reduces the transistor turn-on switching loss, because as soon as the transistor starts to conduct current the dc bus voltage appears across the inductances, allowing the voltage across the transistor to fall from the dc bus voltage to its low on-state voltage. This loss is, however, only moved from the transistor to the snubber, not eliminated completely.

The snubber circuit is required because transistors have safe-operating-area (SOA) limitations during turn-off to avoid catastrophic second breakdown failure. The SOA (Figure 15) is a locus of points of maximum permissible simultaneous occurrence of collector current and collector-emitter voltage. With an inductive load (such as a motor), the turn-off switching of the transistor transfers the load current from the transistor to the opposite on-coming feedback diode in the same phase of the inverter. Without a snubber, as the transistor turns off, the collector-emitter voltage rises from its low on-state voltage up to the dc bus voltage, while still conducting the load current. Then, the transistor current falls and the opposite diode can begin to conduct load current. This simultaneous locus of current and voltage must be within the transistor SOA (Figure 15). Unfortunately, the two snubber inductances in the inverter phase now generate an overshoot voltage above the dc bus voltage, which appears across the transistor during the transistor current fall time (V_{pk} = 2L di/dt). This overshoot can be as high as 150 volts above the maximum 140-volt dc bus (Figure 15).

The action of the snubber circuit is to alter this locus of current and voltage to a safe condition with the SOA (Figure 15) and pro-

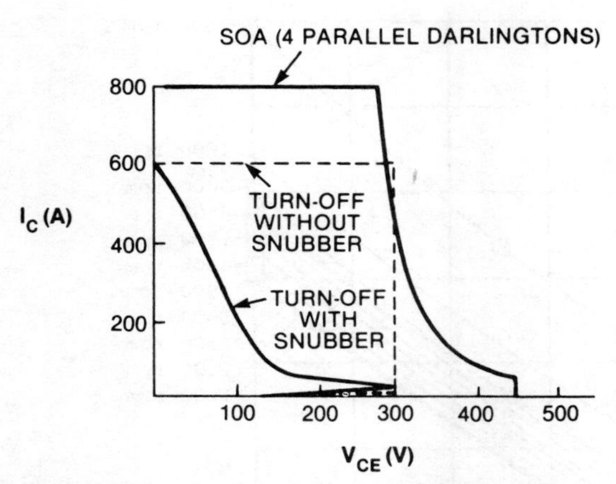

Figure 15. Power Darlington Transistor Turn-off Safe-Operating-Area (SOA) and Switching Locus

vide margin for paralleling. With a snubber, as the transistor turns off and the collector-emitter voltage begins to rise, the snubber capacitance begins to charge. The charging current is current diverted from the transistor and eventually all the load current is in the snubber, charging the snubber capacitance to the dc bus voltage. At this point the on-coming opposite feedback diode can begin conduction and the current transfers from the snubber to the diode based on the interaction of the snubber inductances and the snubber capacitors. The choice of snubber capacitance limits the peak overshoot voltage [7]. Therefore, the snubber displaces the transistor current and voltage so they are within the SOA (Figure 15). The snubber circuit also reduces the transistor turn-off switching loss, because as the collector-emitter voltage rises the collector current is falling instead of remaining constant. The loss is, however, only moved from the transistor to the snubber, not eliminated completely.

The snubbers, as can be seen for the above discussion, depend a great deal on the parasitic inductances of the layout of the inverter power circuit and of the snubber circuit itself. The final snubber design and component values were selected after experimentation, using the actual inverter package layout. A two-stage snubber is used (Figure 14). The first stage is the C3 capacitor and the second is the R3-C2 resistor-capacitor. The stages are polarized with a diode (CR1,CR2) so that large resistors (R2,R4) are in series with the capacitors to reduce the peak currents when they discharge during turn-on. The second stage diode is also snubbered (R1,C1)

Figure 16 is the collector-emitter voltage across a power module at the peak load condition (switching 600 amps). The first-stage snubber is tightly coupled to the power module and, in the first few microseconds, controls the reapplied voltage. This initial control of the rate of reapplied voltage (dv/dt) prevents the high gain Darlington transistors from turning on again. The second stage, less tightly coupled due to the component physical sizes, then comes into action to continue control of the reapplied dv/dt and peak voltage. The transient at 4 microseconds (Figure 16) is the CR2 diode recovery and the transient at 5 microseconds is the CR1 diode recovery. The CR1 diode is itself snubbered so that its snap off recovery does not turn on the high gain Darlington transistor.

Darlington Transistor Paralleling

The reverse breakdown of the Darlington transistors chosen for power module fabrication was at least 450 volts at 0.5 mA. The transistors were matched using the collector-emitter voltage V_{CE} for a base current of 2 amps and an emitter current of 140 amps, with the transistor mounted on a hot plate held at 125 °C. This higher temperature, rather than room temperature, was used for matching because of the gain falloff with temperature and desire to match at the more critical junction temperature. Values of V_{CE} at 140 amps varied from 0.88 volt to 1.20 volts, but

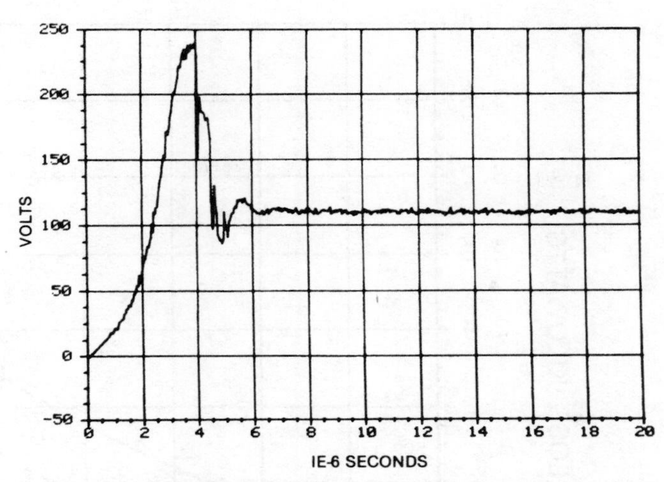

Figure 16. Power Module Collector-Emitter Voltage at Turn-off of Peak Load Condition

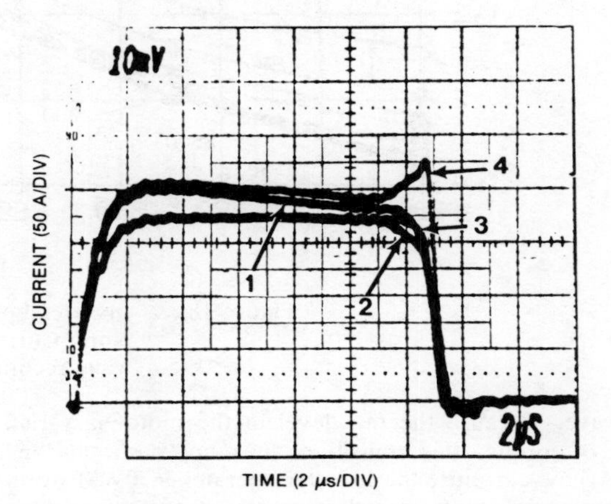

Figure 17. Turn-off of Four Matched Darlington Transistors

transistors could be chosen to match within 0.04 volt. In addition, values of h_{FE} and $V_{BE(SAT)}$ (base 2 to emitter) were matched in order to obtain similar switching speeds for the four transistors in parallel.

The maximum current through one power module of the three-phase inverter circuit is 600 amps. While individually mounted transistors could be operated at well over 200 amps and three transistors could, in principle, supply 600 amps, it was desirable to parallel four transistors to provide some safety margin for current sharing among the transistors. Furthermore, with four transistors there was better thermal dissipation to limit the junction temperature under the most demanding conditions of power delivery.

Figure 17 shows the turn off of four matched transistors each carrying between 150 and 200 amps. The steady-state current sharing among them is within 30 amps. However, during turn off, three of the transistors turn-off ahead of the fourth, causing the fourth to carry a transiently higher current. This illustrates the need for careful matching of transistors to maintain each transistor within its own SOA.

INVERTER LOSSES

The inverter losses versus speed and current are shown in Figure 18. The speed is the drive system output speed after the 2.923 to 1 reduction of the motor speed by an integral gear box. Depending on the current and speed, the inverter may be operating in the PWM or square wave mode. Below base speed (1833 rpm, 180 Hz), the inverter operates in PWM regardless of the current but, as the speed increases, the inverter transitions to

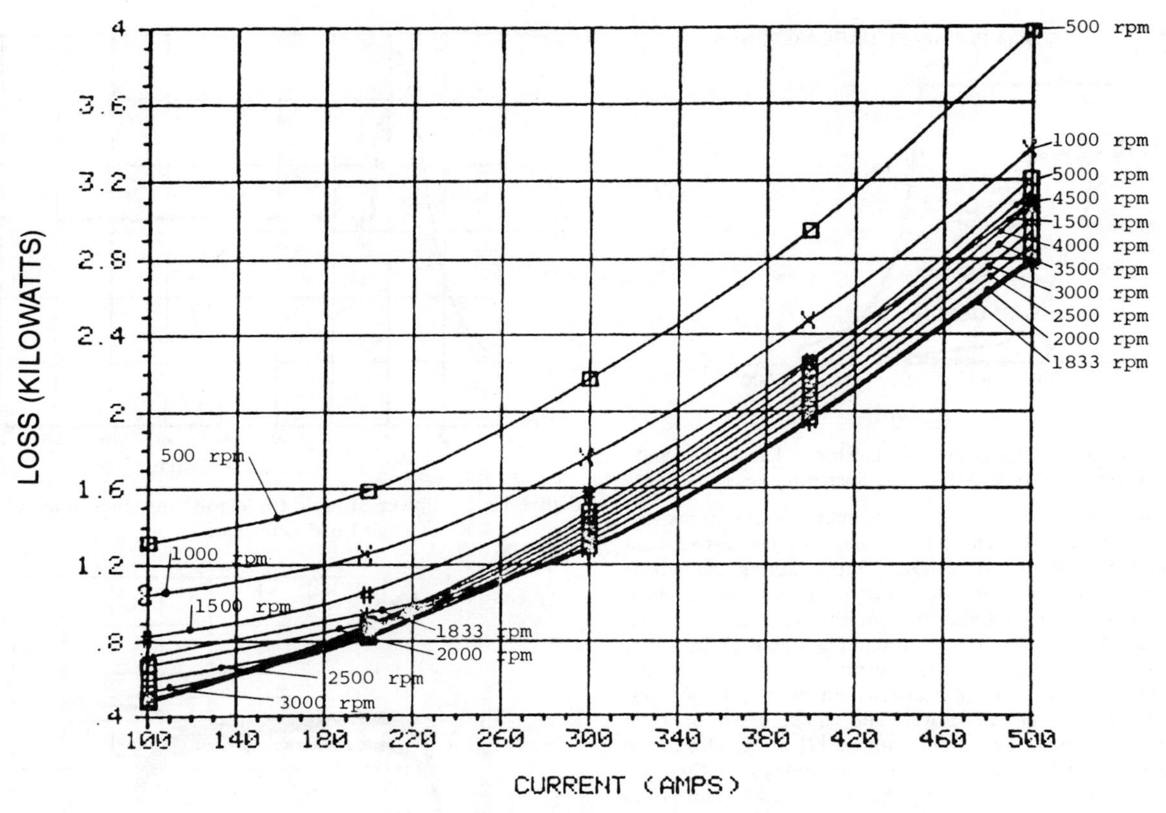

Figure 18. Inverter Losses as a Function of Speed and Current. Note: Current is the average of the three-phase full wave rectified motor current.

square wave. Because the flux level in the motor is varied as a function of torque, this transition does not occur at the same speed. At low currents, the inverter operates in PWM up to relatively high speeds, but at high currents, it transitions to square wave at base speed.

The inverter loss in PWM decreases with speed (for a constant current) as the inverter drops the number chops in the current-controlled PWM operation, then the loss in square wave increases with speed as the frequency increases. The inverter loss increases with current (for any constant speed). The inverter loss has several components: constant loss (blowers), loss proportional to current (conduction), loss proportional to current squared (conduction, parasitic resistance), loss proportional to current and frequency (switching), and loss proportional to current squared and frequency (snubber).

CONCLUSIONS

The feasibility of a three-phase ac transistorized inverter for an electric vehicle ac induction motor propulsion system application was demonstrated. This type of system has potential cost and maintenance advantages over a dc chopper-based system. The ac motor has simple yet rugged construction without commutators and brushes, requires no routine maintenance, is amenable to mass production, is capable of high speeds which reduces its size, and can be totally enclosed allowing advanced packaging and cooling techniques. The engineering model drive system weight was 100 pounds for the motor and 130 pounds for the inverter. The inverter utilizes the advances in high-power Darlington transistors to obviate the need for costly power level commutation circuits that would be needed for a thyristor approach. The inverter is consequently allowed to operate at higher switching frequencies to improve the current waveform delivered to the motor and thereby increase the motor efficiency. Four power Darlington transistors were successfully paralleled in the power module. Good agreement between the calculated sine-triangle voltage PWM parameters and the measured current-controlled PWM parameters was observed. The measured inverter losses were mapped over the full current and frequency range.

ACKNOWLEDGMENTS

The authors with to acknowledge the contributions of Dr. G.B. Kliman for the high-speed high-efficiency motor, A.J. Yerman for the power module, and A.J. Yerman and J.C. Driscoll for the power Darlington paralleling.

REFERENCES

[1] A.B. Plunkett and G.B. Kliman, "Electric Vehicle AC Drive Development," SAE Paper 800061.

[2] A.B. Plunkett, "A Current-Controlled PWM Transistor Inverter Drive," Conference Record 1979 IEEE Industry Applications Society Annual Meeting, October 1979.

[3] A.B. Plunkett, "Direct Flux and Torque Regulation in a PWM Inverter-Induction Motor Drive," IEEE Transactions on Industry Applications, Vol. IA-13, No. 2, March/April 1977.

[4] T.A. Lipo, D.W. Novotny, A.B. Plunkett, and V.R. Stefanovic, "Dynamics and Control of AC Drives," Course notes, University of Wisconsin Extension, November 3-5, 1976.

[5] G.B. Kliman and A.B. Plunkett, "Development of a Modulation Strategy for a PWM Inverter Drive," IEEE Transactions on Industry Applications, Vol. IA-15, No. 1, January/February 1979.

[6] G.A. Kaufman and A.B. Plunkett, "Steady-State Performance of a Voltage Source Inverter Synchronous Machine Drive System," Conference Record 1981 IEEE Industry Applications Society Annual Meeting, October 1981.

[7] W. McMurray, "Optimum Snubbers for Power Semiconductors," Conference Record 1971 IEEE Industry Applications Society Annual Meeting, October 1971.

Application of Power Transistors to Residential and Intermediate Rating Photovoltaic Array Power Conditioners

ROBERT L. STEIGERWALD, MEMBER, IEEE, ANGELO FERRARO, MEMBER, IEEE, AND
FRED G. TURNBULL, SENIOR MEMBER, IEEE

Abstract—An investigation into power conditioners that interface with photovoltaic arrays and utilities has been recently completed. The ratings for this investigation include residential systems (5–30 kW) that interface with a 240-V single-phase utility connection and intermediate systems (30–200 kW) that interface with a 480-V three-phase utility connection. Both systems mandated that an isolation transformer be provided between the array and the utility interface. A trade-off study was performed for many transistor and thyristor circuits and configurations. The weighting criteria included full- and part-load efficiency, size, weight, reliability, ease of control, injected harmonics, reactive power requirements, and parts cost. As the result of this study, a 10-kW high-frequency PWM transistor inverter feeding a high-frequency isolation transformer with a sinusoidally shaped current wave was selected. The output of the transformer is rectified with a diode bridge rectifier. Four thyristors, used as 60-Hz switches, reverse the polarity of the rectified current on every other half-cycle of the utility voltage. This reversal is accomplished slightly before the natural zero crossing of the voltage, thereby providing commutation for the thyristors. The results in the intermediate rating study targeted on a 100-kW design using power transistors in a full-bridge circuit connected to a 60-Hz three-phase transformer. Three bridge circuits are provided to interface with the polyphase utility. The switching strategy for the power transistors is based on a sine wave shape of current with a controlled deadband. The transistors operate over a range in frequency much higher than the utility frequency so that a simple inductor–capacitor filter provides a low harmonic content in the output voltage. The two systems are described in greater detail, and computer-generated waveforms of the input, output, and switching device current and voltage are provided together with computed system performance parameters. A free convection-cooled conceptual design for the residential system and a forced convection-cooled design for the intermediate system are presented.

Paper IPCSD 82-37, approved by the Static Power Converter Committee of the IEEE Industry Applications Society for presentation at the 1982 International Semiconductor Power Converter Conference, Orlando, FL, May 24–27. Manuscript released for publication September 28, 1982. This work was supported by the U.S. Department of Energy under Contract DE-AC02-80ET29310.

The authors are with the Corporate Research and Development Department, General Electric Company, P. O. Box 43, Schenectady, NY 12301.

INTRODUCTION

PAST STUDIES of power converters that interface between solar arrays and utility systems have concentrated on modification of existing power conversion equipment. Two types of systems have been proposed, the current-fed line-commutated thyristor inverter and the voltage-fed forced-commutated inverter. These two systems have their antecedents in phase-controlled rectifiers for dc motor drives and uninterruptible power supplies. The thyristor has been the preferred switching device for reasons of voltage, current, and power rating availability coupled with low cost. On the negative side, forced commutation in thyristor inverters requires a large quantity of power-rated auxiliary components (thyristors, capacitors, inductors, resistors, and controls). These additional components added to the system cost and adversely affected the system efficiency.

Recent advances in power transistor current, voltage, and power ratings, together with advances in fast-recovery rectifier ratings, have allowed these devices to be considered favorably in photovoltaic power-conditioning systems. Additionally, rapid advances in dc capacitor technology, spurred on by the switching power supply market, have greatly increased the

Reprinted from *IEEE Trans. Ind. Appl.*, vol. IA-19, pp. 254–267, Mar./Apr. 1983.

TABLE I
COMPARISON OF ALTERNATE SCHEMES FOR RESIDENTIAL POWER CONVERTERS

	Remarks	Relative Power Circuit Cost
Baseline (line-commutated inverter) with 60 Hz transformer	large dc choke, harmonic filters needed, poor power factor, 60 Hz transformer	1
Chopper-fed line-commutated inverter with current "waveshaping"	transistor chopper feeds inverter with half sine waves, no harmonic filters, good power factor, 60 Hz transformer	0.76
Full-bridge PWM	good power factor, no harmonic filters, 60 Hz transformer	0.99
High-frequency link (rectifier and reversing SCR's)	good power factor, no harmonic filters, high-frequency transformer, high efficiency	0.74
High-frequency link (cycloconverter)	good power factor, no harmonic filters, severe thyristor stresses, poor utilization of high-frequency transformer	0.86

capability of these components. Also, the advent of amorphous metals for power transformers holds promise for loss reduction and cost reduction. Finally, microprocessor based control systems allow for optimum control of photovoltaic/utility power conditioners. A recently completed conceptual design of a 10-kW and a 100-kW power conditioner is described in this paper. Both systems rely on power transistor modules, high-frequency switching technology, advanced passive components, and microprocessor based control systems. These advanced systems compare favorably in size, weight, efficiency, and cost projections to existing converter systems.

SELECTION OF CONVERTER APPROACH FOR RESIDENTIAL POWER RATING

Several approaches were considered for the residential 10-kW power converter [1]. The most promising approaches are listed in Table I along with their estimated relative costs. A line-commutated inverter with harmonic filters to achieve a total harmonic current distortion of less than five percent was selected as the baseline approach. In general, by "shaping" the line current using PWM action, large dc chokes and harmonic filters are eliminated, and a cost advantage is gained. For the high-frequency link system, better transformer utilization and more efficient device switching results by first rectifying the high-frequency ac and converting it to 60 Hz using a phase-controlled rectifier (compared to using a cycloconverter for the high-frequency ac to low-frequency ac conversion [2]).

Even though the high-frequency link system uses more power semiconductors, a net size, weight, and parts cost savings results compared to the other systems due to the elimination of the 60-Hz transformer. Both ferrite material and amorphous metal material were considered for the high-frequency transformer core. The amorphous metal material appears advantageous for this application due to its lower loss, the capability to build large cores, and its potentially low cost. A detailed description of the high-frequency link system is given in the next section.

High-Frequency Link Converter for Residential Rating

The selected system for which the 10-kW conceptual design was performed is shown in Fig. 1. Ideal waveforms are shown in Fig. 2. A full-bridge transistor inverter operating from 10 to 20 kHz inverts the dc voltage of the solar array and transforms this voltage via high-frequency transformer $T1$. In addition, as will be discussed below, by controlling the transistor switching properly, the magnitude of $T1$'s output current is controlled in PWM fashion, and saturation of the transformer $T1$ also is prevented. The high-frequency voltage at $T1$'s secondary is rectified by fast-recovery rectifiers $D1$-$D4$. The transistor PWM action is controlled so that half sine waves of current are forced in high-frequency inductor L. The four phase-controlled thyristors $T1$-$T4$ act as reversing switches to convert the half sine waves of current in inductor L to full sine waves which are delivered to the utility in phase with the line voltage. The current in the thyristors is allowed to go to zero near the end of each half-cycle so that the thyristors are allowed to commutate off naturally.

Although the power is converted three times using this scheme, two of the conversions are very efficient (rectification and phase-controlled inversion), and the transistor inverter will also be very efficient due to the lossless snubbers ($C1$-$C4$) which can be employed due to proper transistor control, as will be discussed. It is emphasized that the current waveshape in high-frequency inductor L is controlled by the inverter transistors so that a relatively pure sine wave of current is delivered to the ac line, thus eliminating the need for harmonic filters. Only an EMI filter (not shown) will be needed at the output of the inverter.

Power Circuit Control for Residential Inverter

Fig. 3 shows a block diagram of the power circuit control. The power circuit is shown along the top of the diagram with appropriate sensors and the important interfaces to the logic and control circuits. With the aid of the ideal waveforms of Fig. 2, the operation of the power and control circuit will now be explained.

The logic power supply and a synchronizing signal for the current reference are obtained from a 60-Hz transformer connected to the 60-Hz power line. A phase-locked loop (PLL) outputs a pure sine wave in phase with the line voltage and serves as the line current command signal. This current command signal has its magnitude controlled by the maximum power tracker that will be discussed below. The phase-lock loop sine wave also controls a peak flux reference wave ψ^*.[1] The instantaneous line current (which is equal to the inductor L current) is sensed by an inexpensive current transformer and fed back into the PWM modulator block. The transformer flux is determined by integrating the voltage obtained from a low-voltage sense winding on the high-frequency transformer and is fed into the modulator block. The instantaneous transistor current I_Q is read from an inexpensive current shunt when opposite corners of the transistor bridge are in conduction—this signal is fed to the modulator block and is used for instantaneous protection of the power transistors in the event of a fault or

[1] Denotes control reference.

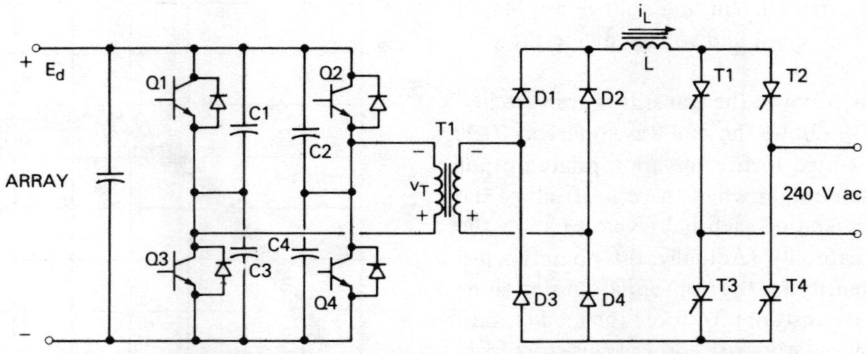

Fig. 1. Utility interactive power converter with high-frequency transformer.

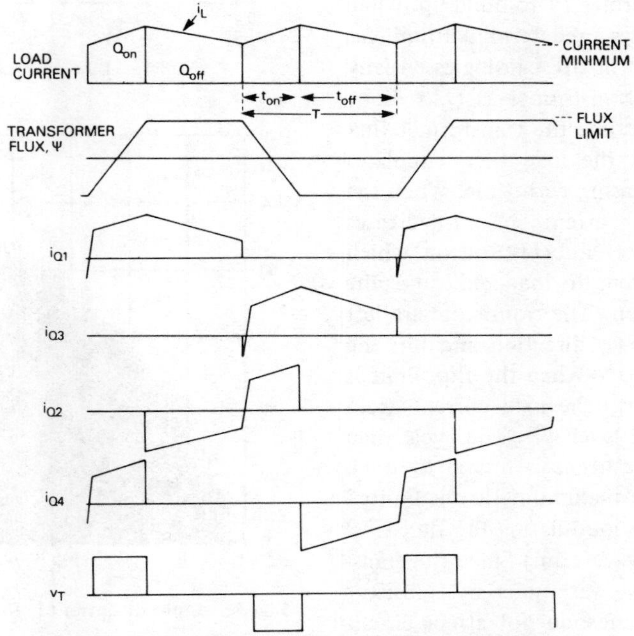

Fig. 2. Ideal high-frequency inverter waveforms.

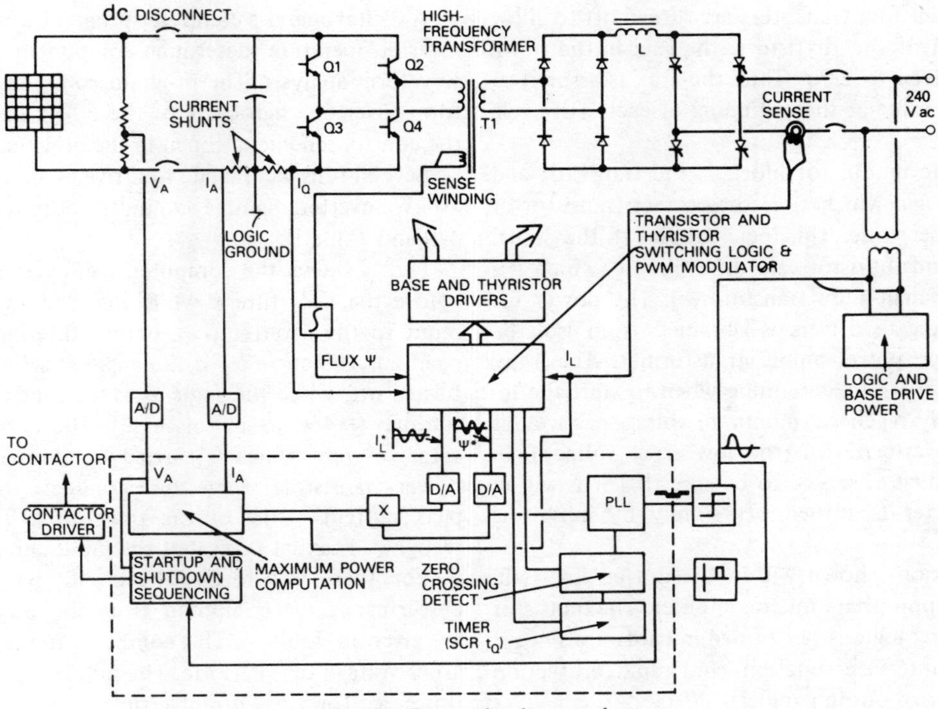

Fig. 3. Power circuit control.

circuit malfunction. The array current and voltage are sensed using an inexpensive current shunt and voltage divider, respectively.

The circuit operates as follows: the transistors are switched so that the load current I_L follows the sine wave reference $I_L{}^*$. The PLL zero crossing is used to fire the appropriate output thyristors, and a timer determines when to turn off all of the power transistors near the end of each half-cycle to allow the thyristors to commutate naturally. (Actually, this point is determined in software by identifying the appropriate point along the wave to turn off all transistors.) To force the load current to follow the reference wave, opposite corner transistors (e.g., $Q1$ and $Q4$) are first turned on, causing the current to build up in L and causing the flux in transformer $T1$ to build up. When the flux in transformer $T1$ reaches a predetermined limit (as determined by ψ^*) transistor $Q4$ turns off, causing essentially zero voltage to be applied to the transformer as $Q2$'s diode comes into conduction (during this time the transformer flux remains constant); the current in the load then circulates through L and the rectifiers, decreasing with time. When the load current (being sensed by the current sensor) decreases to the reference level I^*, $Q1$ turns off and $Q3$ turns on, which again causes the current to build up in the load and causes the transformer flux to reverse direction. The transistors are left on until the transformer flux reverses direction and hits the negative limit as determined by $-\psi^*$. When the flux limit is hit, $Q2$ turns off and stays off until the load current again falls and hits the current command level $I_L{}^*$. The cycle then repeats. In this manner, the load current is forced to track the reference current $I_L{}^*$, and the maximum flux is limited to a known value. (The reason for modulating the flux level is to attain a better load current waveform.) Since the transformer flux is monitored directly, variations in transistor conduction drop, and switching times do not affect circuit operation. Also, the maximum flux level is controlled, and transformer saturation is avoided. In addition, near the end of each cycle, when all four transistors are turned off to allow natural commutation of the thyristors, the flux in the transformer is rapidly driven to zero. Thus the flux is assured of starting at a known value at the beginning of each half-cycle of the 60-Hz line.

Also shown in the modulator block is the transistor and thyristor switching logic which fires the correct transistor or thyristor at the proper time. This logic couples to the actual base driver circuits and thyristor gate drive circuit (which is a single pulse from a small-pulse transformer). The power for the base drivers and gate drivers is obtained from isolated windings on the logic power-supply transformer. The array voltage is sensed in order to determine when to start the inverter (when the array reaches a minimum voltage) as well as when to shut the inverter down (too low array voltage). In addition, the line voltage is sensed to be sure that it is within tolerance. The inverter is turned off simply by inhibiting transistor conduction.

Most of the functions shown within the dotted lines will be performed by an appropriate microcomputer. The transistor and thyristor switching logic is performed in hardware since it contains fast-acting interface functions and rapid protection functions that are carried out in parallel.

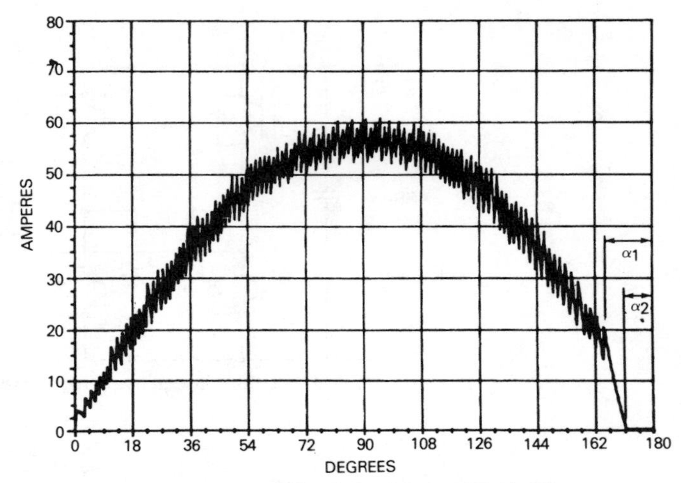

Fig. 4. Line current (conditions of Table II).

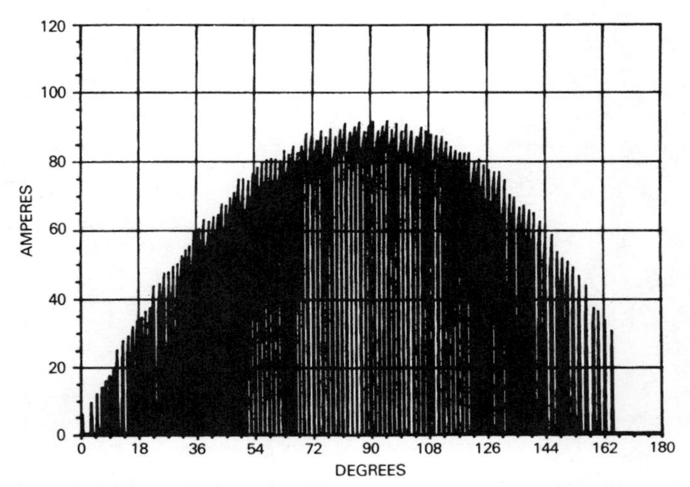

Fig. 5. Input dc current before filtering (conditions of Table II).

Results of Computer Simulation for Residential Inverter

A digital analysis computer program was developed to study inverter operation, determine component ratings, and perform waveform analysis. The program computes the real time waveforms over one half-cycle of the 60-Hz line wave according to the control scheme outlined in the previous section.

Several runs were made to arrive at an optimum design for a 10-kW inverter. Sample computer outputs are shown in Figs. 4-9 and Table II.

Fig. 4 shows the computed half-cycle of line current (i.e., before the EMI filter), while Fig. 5 shows the current at the input to the inverter (i.e., before filtering by the dc link capacitor). The angle α_1 is the angle at which all transistors are turned off, while the angle α_2 is the time allowed for thyristor turnoff ($\cong$460 μs in this case). The current in Fig. 5 is the actual current which is carried by alternate pairs of diagonal power transistors when they conduct (thus determining the peak current rating of the transistor). This current will be recognized as that typical of the input current to a chopper.

For these waveforms, the circuit parameters and circuit performance (as computed from the waveforms themselves) are given in Table II. This computer run is for a nominal input array voltage of 312-V dc. The inductor L (Fig. 1) is 1 mH for this case. The rms current rating of the input dc filter capaci-

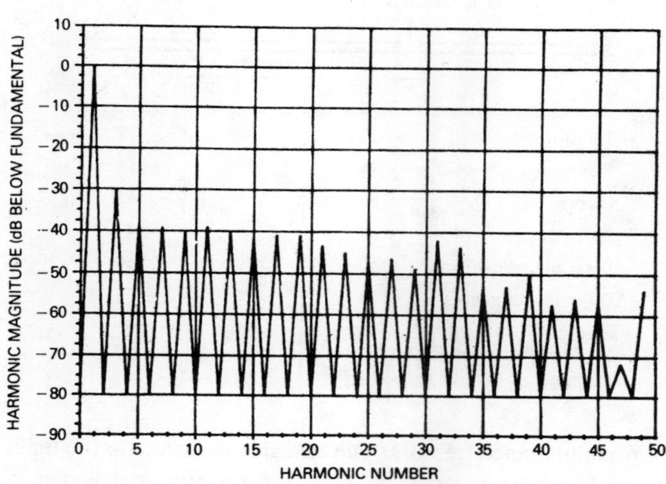

Fig. 6. Spectrum of inverter output current (log scale).

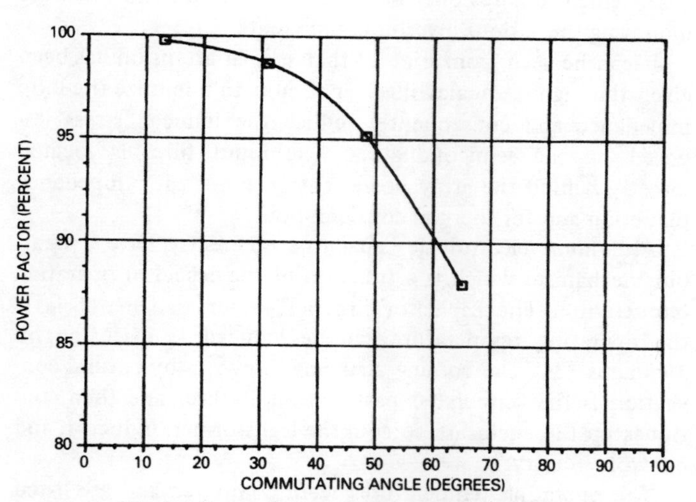

Fig. 7. Power factor versus commutating angle.

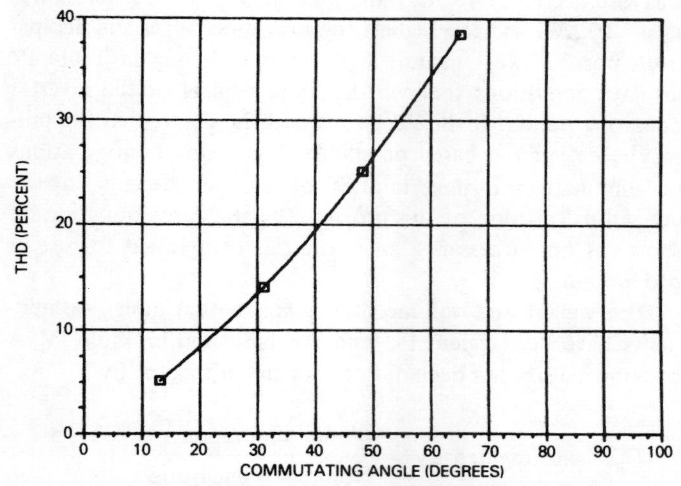

Fig. 8. THD (current) versus commutating angle.

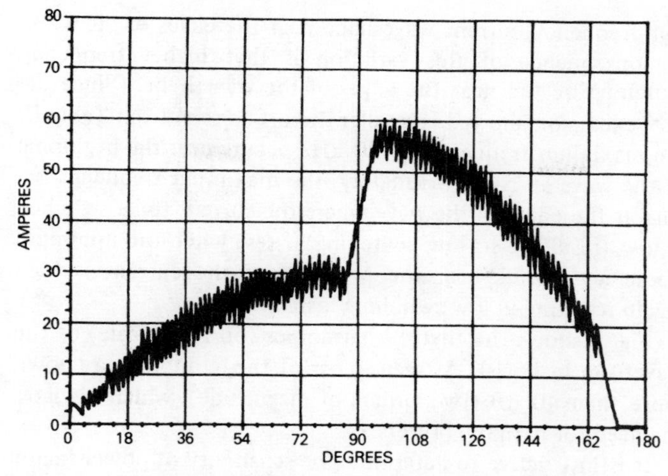

Fig. 9. Line current response to step increase in current demand.

TABLE II
DATA FOR NOMINAL ARRAY VOLTAGE

POWER DELIVERED=	10019.5
DC LINK VOLTS=	312
REFLECTED DC VOLTS =	471
PK LINE VOLTS =	339
RMS LINE VOLTS =	239.745
XL=	.377
L=	1.00000E-03
RMS CAPACITOR AMPS=	37.1057
DC LINK CURRENT=	32.1137
RMS OUTPUT CURRENT=	41.1937
FUND MAG OUTPUT CURRENT=	58.1775
IN PHASE CURRENT=	58.1762
DISPLACEMENT FACTOR=	.999962
POWER FACTOR=	.998619
DISTORTION FACTOR=	.998641
THD(CURRENT)=	.0521809
TRANSISTOR CONDUCTION LOSSES(ONE DEVICE)=	42.0229
TRANSISTOR SWITCHING LOSSES(ONE DEVICE)=	30.0865
NUMBER OF SWITCHES=	220
MAX FREQ=	20440
POINT AT WHICH MAX FREQ OCCURS (DEGREES)=	10.0391
EQUIVALENT MIN FREQ=	10224.2
MAX ON TIME(DEGREES)=	1.05674
POINT AT WHICH MAX ON-TIME OCCURS=	70.274
INDUCTOR VOLT-AMPS=	669.072
INDUCTOR POUNDS=	14.2079
TRANSFORMER VOLT-AMPS=	19402.2
TRANSFORMER POUNDS=	11.0485

tor is determined numerically by computing the rms value of the current wave of Fig. 5 (including the high-frequency component of current) minus its dc value (which is in the array). From Table II it is observed that the power factor (including the effect of harmonics) is near unity, and the total harmonic current distortion is 5.2 percent. The distortion assumes all of the ripple current shown in Fig. 5 is passed to the ac line. In practice, a small capacitor placed line-to-line will shunt a substantial portion of the higher frequency ripple current resulting in a line current distortion of less than five percent.

The transistor switching losses are determined by computing the energy dissipated during each individual switching of the inverter and totaling over a complete cycle to obtain the total switching losses. An equation for the switching losses was developed from actual switching loss measurements using the type of transistor to be used in the inverter. The minimum switching frequency is seen to be approximately 10 kHz (which determines the transformer size) and occurs near the center point of the wave at approximately 70°. (The maximum ON time determines the minimum frequency.) For these runs the flux wave reference was varied sinusoidally to improve the high-frequency current waveshape near the edges of the wave. A consequence of this variation is that higher frequency switching occurs near the edges of the waveshape. (Since the flux excursion allowed is smaller there.) As seen from Table II, the maximum frequency of 20 kHz occurs near the beginning of the wave at 10°. Fortunately, the maximum frequency occurs at the edges of the wave where the current to be switched is low (resulting in low switching losses) while the minimum frequency occurs near the peaks when the current is high (again resulting in low switching losses).

Fig. 6 shows the first 49 harmonics (on a log scale) of the waveform in Fig. 4. As seen, most of the harmonics are down more than 40 dB (two orders of magnitude), which obviates the need for harmonic filters.

It is instructive to determine the sensitivity of power factor and harmonic current distortion to the angle α_1 at which all transistor conduction is interrupted. Figs. 7 and 8 show such a calculation. As seen, the power factor is relatively insensitive to commutating angle. For example, operating at the large commutation angle of 50° (2.3 ms) results in a high power factor of nearly 95 percent. However, Fig. 8 shows that the current distortion is a much more sensitive function of commutating angle. Thus there is incentive to operate at as low a commutating angle as possible consistent with reliable thyristor commutation.

It is of interest to note that the inverter itself can respond very rapidly to step changes in current command. Fig. 9 shows the case where a step increase in current command occurs near the 90° point in the conduction cycle. As seen, the response is very rapid with no overshoot. It should be noted in Fig. 9 that during the rise in current the inverter switches several times. This is due to the fact that the maximum inverter on time (i.e., conduction of diagonal transistors) is limited in order to avoid saturation of the output transformer. In this case, the inverter is essentially running full square wave during the transition in order to apply maximum voltage to inductor L and the load in order to drive the current up to the commanded value.

Table III shows a breakdown of inverter losses and the resulting efficiency for three values of load.

TABLE III
POWER LOSS AND EFFICIENCY ESTIMATE FOR 10-kW INVERTER OPERATING AT THREE OUTPUT POWERS (P_0)

	Power Loss (W)		
	P_o = 10 kW	P_o = 8 kW	P_o = 5 kW
Transformer	50	48	35
Transistors	296	237	154
Diodes	85	68	42
Thyristors	80	64	40
Logic and base drive	50	50	50
Other (busbars, etc.)	30	20	8
Total Watts Loss	591	487	329
Efficiency (%)	94.1	93.9	93.4

Residential Inverter Package Design

The operation of this new inverter imposes design constraints beyond the normal power line frequency thyristor-based system. A package design guideline was established to provide

- mechanical protection to both the equipment and the user,
- optimum circuit layout for reliable operation and ease of maintenance and installation,
- a sound thermal management system,
- an economical solution in light of the residential cost sensitive application.

In an effort to meet the cost objectives it was felt that the thermal management should be an integral part of the structural elements. Furthermore, the need for natural convection cooling was also a constraint that was imposed to ensure a maximum maintenance-free life. The two side surfaces are the main heat sinks for the inverter and serve as main structural elements. These heat sinks are mechanically secured to the sheet metal base, which requires only nominal machine work to facilitate mounting the various inverter components.

It can be seen from Fig. 10 that special attention has been given the high-frequency inverter circuits to minimize the loop impedance and consequently reduce the induced stress imposed on the semiconductors. The input filter is located directly behind the array power entrance both for impedance reduction and for thermal considerations.

Aluminum electrolytic capacitors inherently have a wear-out mechanism which is a function of the capacitor operating temperature. The capacitor lifetime is increased manifold if the operating temperature can be reduced as discussed by McManus [3]. The cooling air being drawn in by natural convection is first circulated past the input filter and then continues up the enclosure to cool the transformer, inductor, and control circuitry.

The operating temperatures were computed and are listed in Table IV for both the worst case and for the probable point of operation. The ambient temperature for the control room was assumed to be 45° C, the output power of the inverter was set at 10 kW, and the highest thermal impedances and dissipations possible were assumed. The second listing in Table IV displays conditions that will be more typical of the inverter operating points. It should be realized, however, that the prototype design was based on standard heat sink configurations to eliminate the uncertainties of custom designs in the conceptual portion of this project. The standard configuration shown is not necessarily an optimized component for use in production.

The weight and volume of the conceptual design components have been calculated and are tabulated in Table IV. A packing density has been defined for the prototype by

$$D_p \equiv \text{packing density} = \frac{\text{volume of components}}{\text{volume of enclosure}} \times 100 \text{ percent}$$

and is approximately 42 percent. The volume of the enclosure

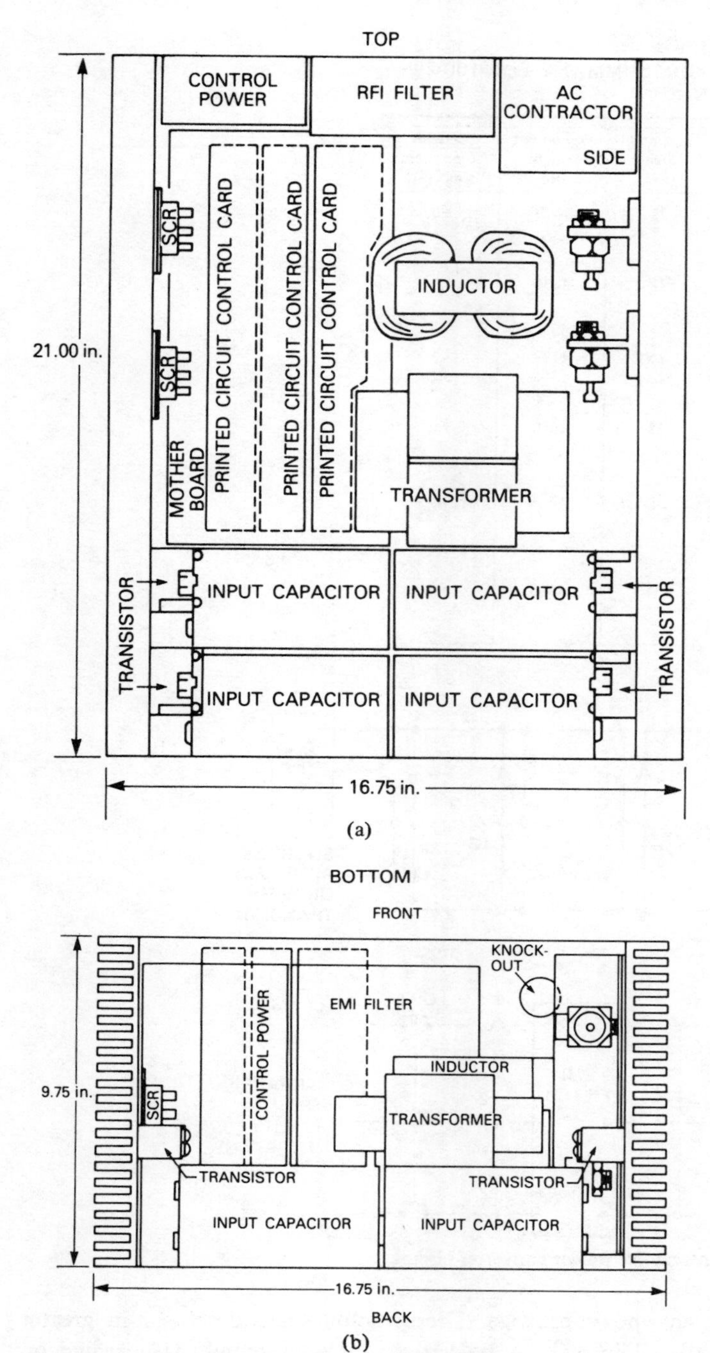

Fig. 10. (a) Internal front view. (b) Internal bottom view.

TABLE IV
10-kW INVERTER PACKAGE PERFORMANCE

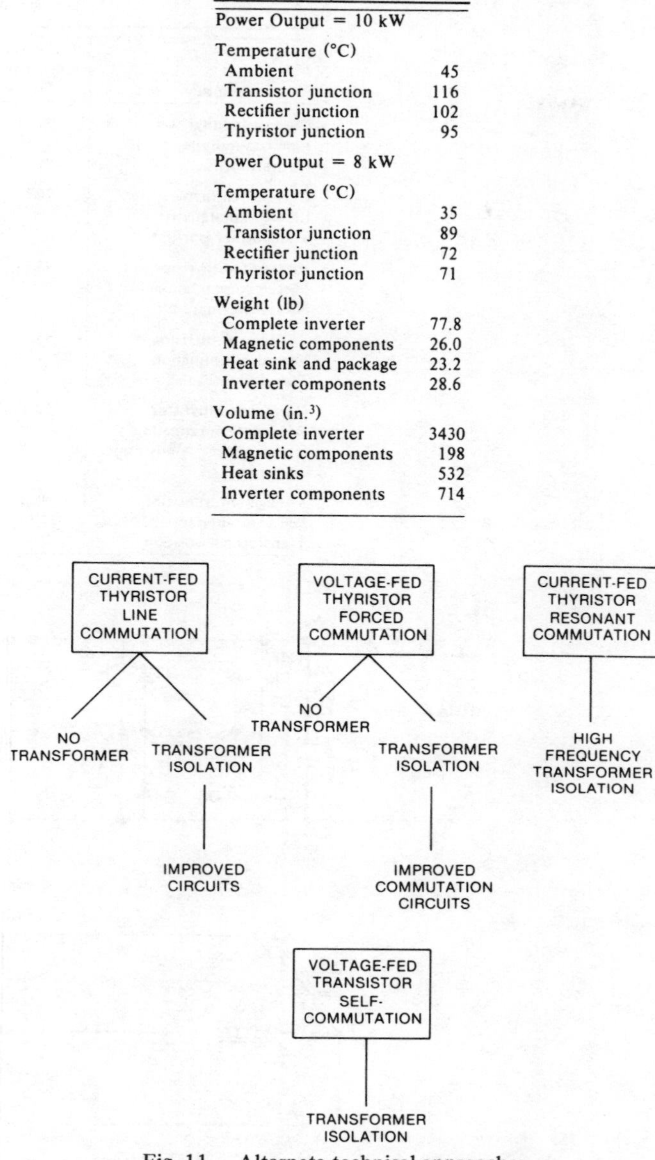

Power Output = 10 kW	
Temperature (°C)	
Ambient	45
Transistor junction	116
Rectifier junction	102
Thyristor junction	95
Power Output = 8 kW	
Temperature (°C)	
Ambient	35
Transistor junction	89
Rectifier junction	72
Thyristor junction	71
Weight (lb)	
Complete inverter	77.8
Magnetic components	26.0
Heat sink and package	23.2
Inverter components	28.6
Volume (in.^{3})	
Complete inverter	3430
Magnetic components	198
Heat sinks	532
Inverter components	714

Fig. 11. Alternate technical approaches.

is 3430 in^3. The size of a transformer rated at 10 kVA and 60 Hz would be approximately two-thirds the size of this entire prototype design. The weight of the prototype has been estimated at approximately 78 lb, which is virtually the same weight as a 10-kVA 60-Hz transformer. It has been estimated that the control circuitry will require two of the three printed circuit cards shown. The extra card has been included for any monitoring and data gathering interfaces that may be desired.

SELECTION OF CONVERTER APPROACH FOR INTERMEDIATE POWER RATING

A number of power converter circuit approaches were reviewed with respect to efficiency, size, weight, and per-unit parts cost based on the requirements discussed in the previous section [4]. A breakdown of the approaches is summarized in Fig. 11. It can be seen that both current-fed and voltage-fed systems were included as were nontransformer isolated, 60-Hz transformer isolated, and high-frequency transformer isolated systems. Both transistors and thyristors were chosen as the controllable semiconductor devices. Several modifications to the current-fed line-commutated thyristor system were proposed to improve its technical performance. Several modifications for thyristor-based forced-commutation circuits were proposed that overcome some of the technical limitations of conventional forced-commutating circuits. Table V summarizes the major findings for a 100-kW nominal design. The current-fed line-commutated thyristor system without an isolation transformer is the lightest and most efficient. When an isolation transformer is required, the voltage-fed PWM transistor system is the lowest cost and provides the potential for further cost reductions due to continuing reductions in semiconductor cost and continuing use of control circuit integration.

TABLE V
ALTERNATE SYSTEM COMPARISON SUMMARY FOR 100-kW RATING

	Full-Load Efficiency (%)	Equipment Size (ft^3)	Equipment Weight (lb)	Per-Unit Component Cost ($)
Current-fed thyristor Line commutation No transformer	98.1	22	1020	1.0
Current-fed thyristor Line commutation Transformer isolation	96.1	38	1495	1.3
Voltage-fed thyristor Forced commutation No transformer	95.9	26	1100	1.5
Voltage-fed thyristor Forced commutation Transformer isolation	93.7	28	1050	2.0
Current-fed thyristor Resonant commutation High-frequency transformer isolation	90.4	26	1574	1.6
Voltage-fed transistor Self commutation Transformer isolation	95.3	34	1300	1.1

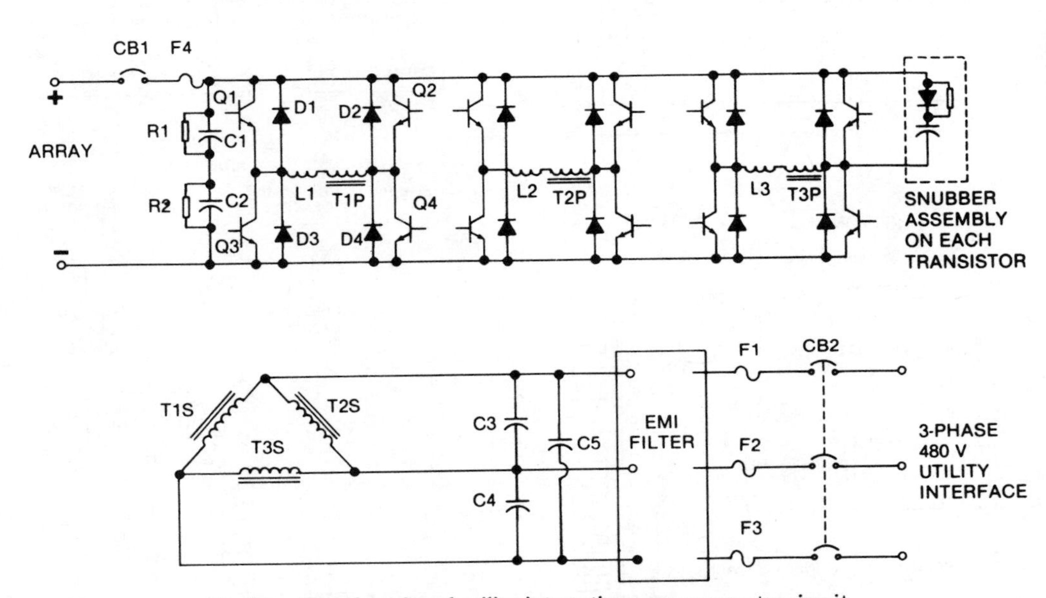

Fig. 12. Transistor based utility interactive power converter circuit.

Pulsewidth Modulated Converter for Intermediate Rating

Based on a trade-off study of alternate power conditioning systems summarized in the previous section, the PWM transistor-based inverter with transformer isolation was selected as the preferred system for the intermediate power rating. The power circuit diagram for a nominal 100-kW rating system is shown in Fig. 12. The dc current from the array flows through a dc contactor and fuse for safety and fault protection. An electrolytic capacitor on the dc bus provides both the necessary low impedance for the fast-rising currents required by the power transistors and provides a supply of reactive volt-amperes for the transformer magnetizing reactance. Each phase of the polyphase output is supplied by a full-bridge circuit of power transistors ($Q1$-$Q4$) and their associated feedback diodes ($D1$-$D4$). The power transistors are composed of several transistor chips connected in parallel with an associated feedback diode chip. The entire semiconductor assembly is packaged in an epoxy package. These modules are described in greater detail in [5]. A polarized snubber assembly is provided on each of the transistor modules for dynamic load line shaping to reduce the switching stress from the power transistor. The other two phases are supplied by a separate bridge circuit, resulting in a total of 12 modules. The three-phase transformer is connected in delta on the utility side. A single section filter ($L1$-$C3$) removes the transistor modulation harmonic currents, preventing them from flowing into the utility grid. The reactor $L1$ also limits the transistor switching frequency. A polyphase EMI filter bypasses the very high frequency components of the switching waveforms, preventing them from flowing into the utility. A line fuse and three-pole disconnect are provided for overcurrent protection and safety.

Power Circuit Control for Intermediate Rating

The power transistors in each bridge circuit will be controlled in a PWM strategy that produces a sinusoidal current wave-

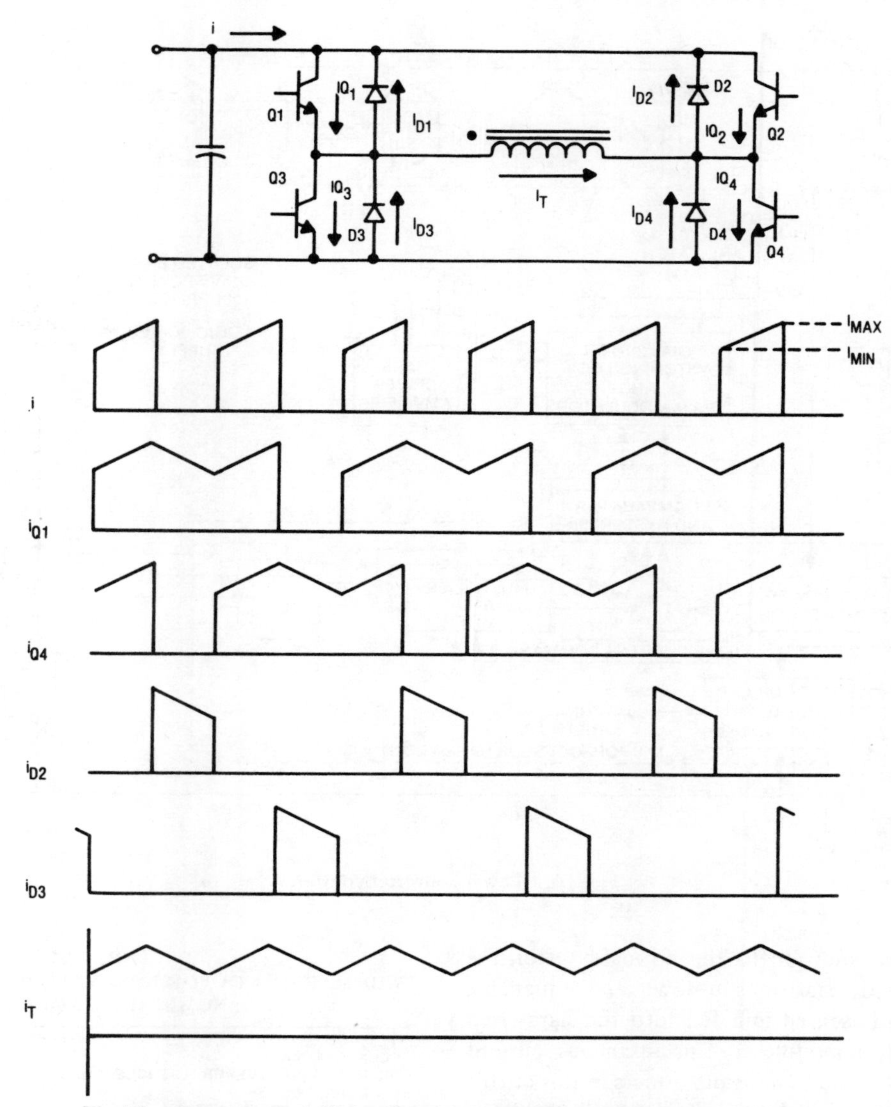

Fig. 13. Transistor and diode current waveforms for positive output current.

form in the transformer primary. Fig. 13 shows the operation of the transistors for a positive output current waveform of 60 Hz. Only a small portion of the 60-Hz waveform is represented as a dc positive reference level. The switching strategy is to sense the transformer current (i_T); when the current reaches an upper current reference (I_{max}*), the appropriate power transistor is turned off. The current coasts through one transistor, the reactor, the transformer, and one feedback diode. When the current has fallen to a lower current reference (I_{min}*), the appropriate two transistors are turned on again, causing the current to increase. The two conducting transistors ($Q1$ and $Q4$ in Fig. 13) are alternatively turned off. This reduces the switching duty to one-half and balances the conduction and switching losses of the devices. The alternate turn-off of transistors $Q1$ and $Q4$ is illustrated in Fig. 13 for a 50-percent duty cycle. During a negative current output half-cycle, transistors $Q2$ and $Q3$ are the active devices controlling the current. In this manner, all four transistors and diodes have the same dissipation at the end of one cycle of the utility voltage. The dc capacitor sums the currents (i in Fig. 13) for all three phases. The balanced nature of a polyphase system reduces the rms current demand from the dc electrolytic capacitor.

Fig. 14 provides a block diagram of the microprocessor-based control system. The overall strategy is to operate in a power-tracking mode that operates the array at its maximum power output point independent of changes in insolation and array temperature. The reference waveform for the transistor current is derived from the utility voltage with a phase-locked loop. This ensures that the ac current is in phase with the utility voltage, resulting in unity power factor operation. The magnitude of the current command is controlled by the maximum power tracker. The array power is measured as the product of array voltage and array current. This power signal controls the magnitude of the current command through a multiplier. This signal is the current reference which is summed with the actual transformer current. The error and a fixed deadband limit the transistor switching frequency. Both the array and the utility voltages are sensed, and if they are outside a specified window, the inverter is stopped and the dc and ac contactors are opened. All of these calculations are carried out in a microprocessor with the analog input signals converted

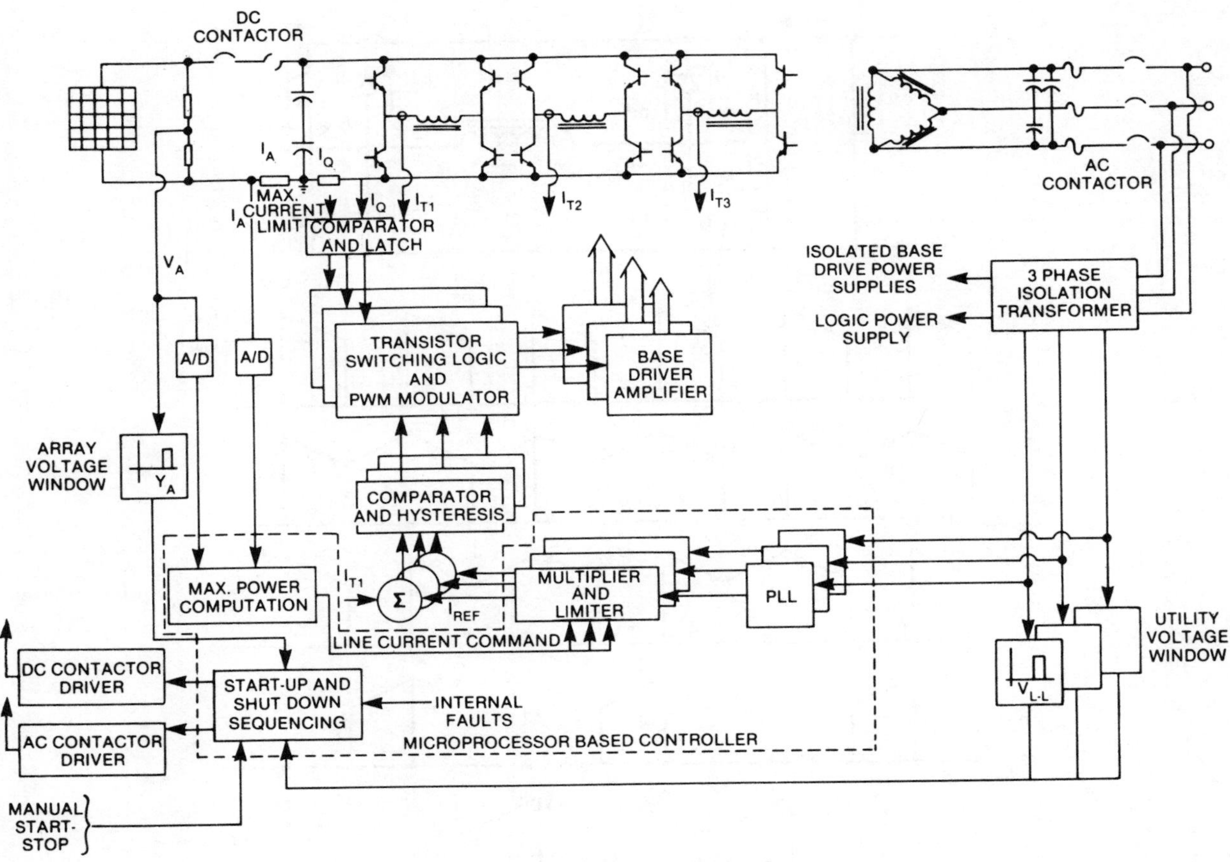

Fig. 14. Power converter control.

with A/D converters. In addition to the normal control, the microprocessor controls the startup, shutdown, and sequencing circuits. A dc current is sensed and fed into the hardwired transistor switching logic module for instantaneous current limit of the transistor current. The transformers in this circuit are designed to support the highest utility voltage at 60 Hz; therefore, instantaneous flux sensing to prevent transformer saturation was not incorporated.

Results of Computer Simulation for Intermediate Rating

A digital computer analysis program was developed to study inverter operation, determine component ratings, and perform harmonic analysis on waveforms. The control strategy for the digital computer program was the same as the proposed system, that is, pulsewidth modulation between a fixed current deadband. An example of a computer-generated summary is shown in Table VI. This particular summary is for nominal utility voltage (480 V), nominal array voltage (250 V), and nominal power (82 kW). The transistor, diode, and snubber losses are tabulated together with the maximum and minimum switching frequency. The maximum frequency occurs at 135° where the peak current is reduced from its maximum value occurring close to 90°. The rms capacitor current is calculated based on operation of all three inverter phases. The series inductor and the snubber capacitor were determined from many such computer runs. The transformer and ac reactor volt-ampere ratings and weights were generated by the computer using look-up tables for these components. Since the data given in Table VI are not from the worst case, the actual weights of these components are slightly greater than tabulated in the figure.

TABLE VI
SUMMARY DATA FOR NOMINAL ARRAY VOLTAGE AND NOMINAL AC VOLTAGE

PEAK REF AMPS =	80
DC WATTS DELIVERED (SINGLE PHASE) =	27413.7
DC LINK VOLTS=	250
REFLECTED DC VOLTS =	950
PK LINE VOLTS =	679
RMS LINE VOLTS =	480.198
COASTING INDUCTANCE (H) =	2.00000E-03
SNUBBER CAP (F) =	1.50000E-07
INPUT RIPPLE AMPS (ONE PHASE)=	129.039
RMS OUTPUT CURRENT=	56.7382
TRANSISTOR CONDUCTION LOSSES(ONE DEVICE)=	69.341
TRANSISTOR SWITCH-OFF LOSSES(ONE DEVICE)=	18.5306
TRANSISTOR SWITCH-ON LOSSES(ONE DEVICE)=	41.4674
SNUBBER RES LOSSES (PER DEVICE) =	29.16
DIODE LOSSES =	17.9307
NUMBER OF SWITCHES=	72
MAX FREQ=	5840.06
POINT AT WHICH MAX FREQ OCCURS (DEGREES)=	135.088
MIN FREQ=	2358.46
POINT AT WHICH MIN FREQ OCCURS (DEGREES) =	9.68689
INDUCTOR VOLT-AMPS=	2735.62
INDUCTOR POUNDS=	40.8522
TRANSFORMER VOLT-AMPS=	27241.4
TRANSFORMER POUNDS=	254.451

Figs. 15–17 show additional output from the computer program, in these cases, instantaneous current waveforms as a function of time for a positive half-cycle of the utility. Fig. 15

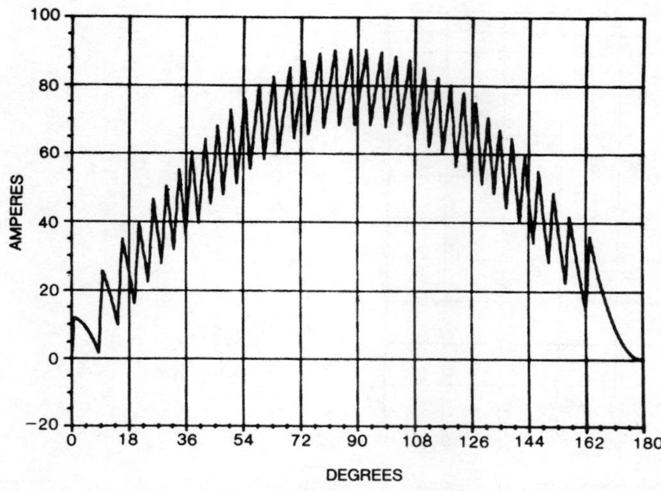

Fig. 15. Transformer secondary current for nominal array voltage and nominal ac voltage.

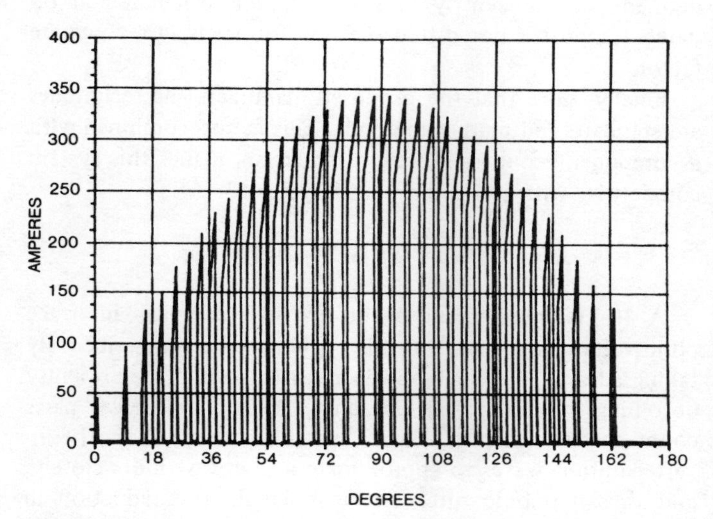

Fig. 16. Transistor current for nominal array voltage and nominal ac voltage.

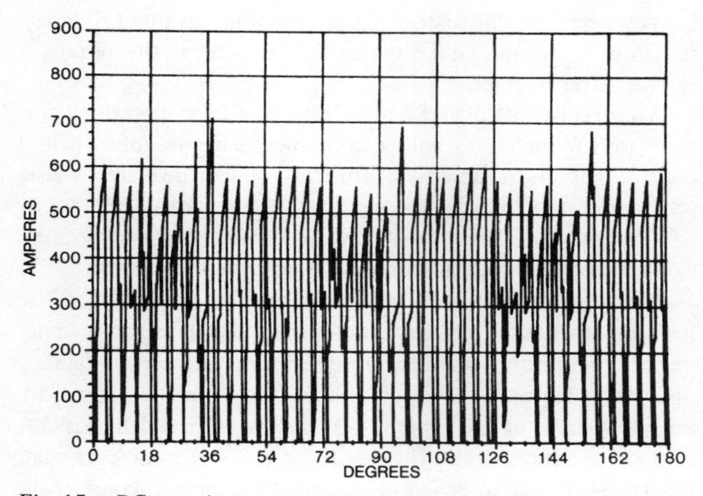

Fig. 17. DC capacitor current for nominal array voltage and nominal ac voltage.

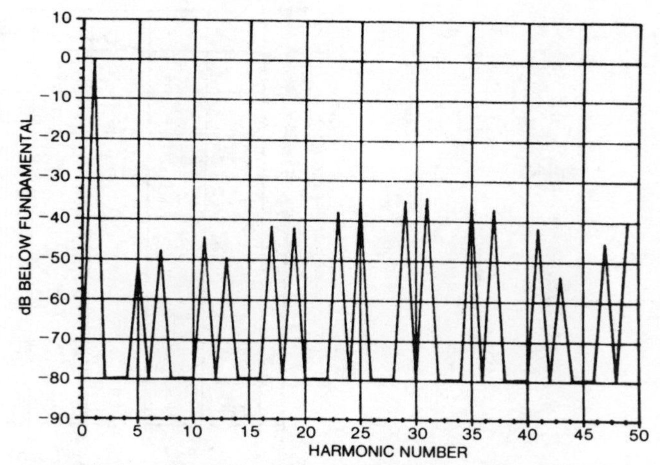

Fig. 18. Harmonic current versus harmonic number for nominal array voltage and nominal ac voltage.

shows the instantaneous transformer secondary phase current, in this case 80-A peak ± 10-A deadband. Fig. 16 shows the instantaneous transistor current for one half-cycle of the output frequency. The peak current is approximately 340 A for nominal conditions. The worst case current is 430-A peak. The portion of the current that is not shown is carried by the feedback diodes. The computer-generated dissipation data factored in the alternate turnoff concept are described earlier in this section. The conduction loss, switch-on, and switch-off calculations were based on test data obtained with similar transistor modules operating at these current and voltage levels. Fig. 17 shows the dc capacitor current for the three-phase inverter. The mean value of this current is supplied by the array. The rms ripple current is computed and is used to select the proper number of electrolytic capacitors and calculate their losses. The transistor switch times are not synchronized; therefore, the harmonic cancellation is not perfect. However, very little low-frequency harmonics associated with the 60-Hz utility are present in this waveform.

Fig. 18 shows a third output from the computer analysis program, a harmonic current spectrum. This spectrum is based on the waveform shown in Fig. 15 with the third and other tripled harmonics eliminated by the delta transformer connection. In this graph, –40 dB is a ratio of one percent of the harmonic to the fundamental. The total rms harmonic distortion for this waveform is calculated at 2.8 percent. These calculations do not include the effect of the filter capacitor ($C3$) in bypassing the harmonic currents. Therefore, the harmonic distortion will be less than the calculated value of 2.8 percent.

The power conditioner efficiency can be calculated as a function of output power with the aid of the tabulated data from Table VI and similar tables based on reduced power output. Transformer and reactor core and copper losses can be calculated based on conventional transformer laminations and construction techniques. The dc capacitor losses can be calculated from the equivalent series resistance (ESR) and rms current from Table VI. Additional losses, both fixed and variable, can be calculated. Examples are control power, fan

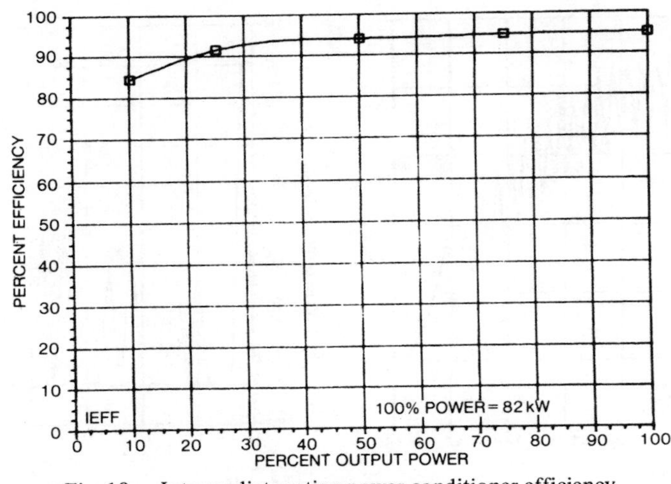

Fig. 19. Intermediate rating power conditioner efficiency.

power, I^2R loss in cables, bus bars, contactors, etc. The efficiency is plotted in Fig. 19 as a function of load. The efficiency is 85 percent at ten percent load and 95.3 percent at 100 percent load or 82.2 kW. At maximum conditions, the output power is 113 kW. This high value of light-load and full-load efficiency is due in part to the self-turnoff capability of power transistors, eliminating the need for full-current rated commutation circuits necessary for thyristor-based conversion systems.

Intermediate Inverter Package Design

The higher frequency of operation of this transistor-based converter, however, requires that more attention be given the component placement and package. The proper operation of power circuitry at high frequency requires a tight low-impedance loop to minimize the component stress during the switching transients. This requirement which ensures proper circuit operation can be seen to interfere with the desire to have a product which requires minimal effort and cost to construct, install, and maintain. The conceptual design resolves many of these conflicts with a modular approach shown simplified in Fig. 20(a). The inverter module shown includes the power devices, snubbers, base driver amplifiers, and power distribution bus required for one phase and is structurally supported by the thermal management system.

The maintenance of the system is optimized by a design that allows the inspection and replacement of most components without the removal or disturbance of any other component. This is especially true for the cooling fans and filter which can be replaced without entering the enclosure. It is believed that this feature will minimize the downtime in the maintenance schedule of the inverter.

The illustration of the interior of the complete inverter [Fig. 20(b)] indicates the major component placement. The three-phase power transformer is thermally isolated in the base of the cabinet to reduce the heat load on the remainder of the components in the upper portion of the enclosure. The specifications for the inverter weight, volume, and operating temperatures have been calculated and are given in Table VII. The defined packing density of this system is calculated at 65 percent, and it is noted that the transformer is a predominate factor.

Finally, note that the proposed manufacturing techniques are standard and mature methods. This factor, combined with a component count and weight reduction, makes this system competitive with comparably sized systems.

SUMMARY AND CONCLUSION

A 10-kW inverter and a 100-kW inverter which interface photovoltaic arrays with a utility grid have been described. By taking advantage of high-power transistors which are presently becoming available, high switching speeds and lower parts count can be achieved. Higher switching speeds allow the inverter output waves to approximate a sine wave more closely, and thus harmonic filters are eliminated. The reduction in filtering requirements and the absence of commutating circuits allow the transistor PWM inverter to enjoy a size, weight, and cost advantage in comparison with conventional thyristor inverters. In addition, for the 10-kW case, the switching frequency could be raised to a significantly high value (10 kHz) to allow a substantial reduction in the size of the required isolation transformer.

A current-controlled PWM scheme has been described for the 100-kW inverter, and a combined current control/flux control has been described for the 10-kW inverter. These schemes ensure that the transistor will switch known amounts of current and that the 10-kW high-frequency transformer cannot saturate.

Simulation results for both circuits were presented to illustrate inverter operation, determine device stresses, and perform waveform analysis. Furthermore, the power factor, distortion factor, and displacement factor are all close to unity by virtue of the control strategy and pulsewidth modulation of the transistors. The results of the study indicate that the transistor based schemes can meet the performance goals of utility interactive inverters.

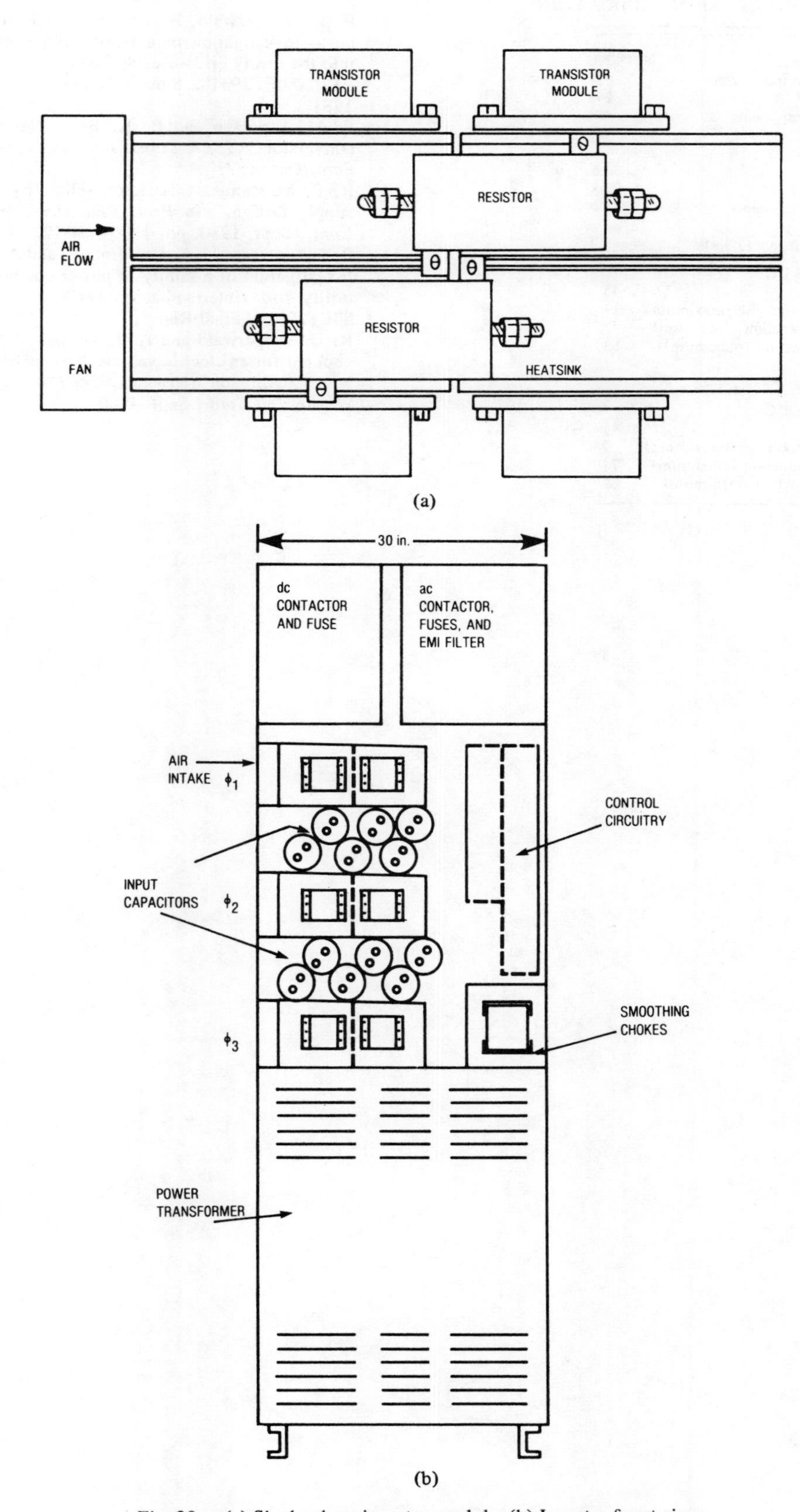

Fig. 20. (a) Single-phase inverter module. (b) Inverter front view.

TABLE VII
INTERMEDIATE INVERTER SPECIFICATIONS

Weight (lb)	
Total system	1370
Three-phase transformer	618
Cabinet	298
Inverter components	454
Volume (ft^3)	
Enclosure	51.6
Transformer	20.8
Inverter components	11.4
Power Output = 113.7 kW	
Temperature (°C)	
Ambient	45
Internal cabinet rise (maximum)	14
Transistor junction (maximum)	92
Rectifier junction (maximum)	83
Power Output = 82.2 kW	
Temperature (°C)	
Ambient	45
Internal cabinet rise (maximum)	9
Transistor junction (maximum)	74
Rectifier junction (maximum)	69

REFERENCES

[1] R. L. Steigerwald, A. Ferraro, and R. E. Tompkins, "Final report—Investigation of a family of power conditioners integrated into the utility grid—Residential power level," DOE Contract DE-AC02-80ET29310, Sandia National Lab., Rep. SAND81-7031, 1981.

[2] P. M. Espelage and B. K. Bose, "High-frequency link power conversion," *IEEE Trans. Ind. Appl.*, vol. IA-13, pp. 387–394, Sept./Oct. 1977.

[3] R. P. McManus, "Capacitor reliability in switched-mode power supply design," in *Proc. 2nd Annu. Int. Powerconversion '80 Conf.*, Sept. 1980, pp. 1.1-1–1.1-12.

[4] R. L. Steigerwald, F. G. Turnbull, and A. Ferraro, "Final report—Investigation of a family of power conditioners integrated into the utility grid—Intermediate power level," DOE contract DE-AC02-80ET29310, Final Rep.

[5] R. L. Steigerwald and A. J. Yerman, "A two-quadrant transistor chopper for an electric vehicle drive utilizing monolithic integrated power darlington transistors," in *Proc. 2nd Annu. Int. Powerconversion '80 Conf.*, Sept. 1980.

PULSE BATTERY CHARGER EMPLOYING 1000 AMPERE TRANSISTOR SWITCHES

R. L. Steigerwald
General Electric Company
Schenectady, New York

Abstract

A battery charger employing water cooled 1000 Ampere transistor switches has been developed for empirically studying methods for rapidly charging batteries.

Rapid charging of batteries is of considerable interest for electric vehicle applications in order to extend the useful vehicle range in a given day. To empirically establish the best method of charging (or quick-charging) new batteries, a new type of charger was developed. The charger is very flexible and allows a wide range of adjustment for use in gathering experimental data pertaining to new batteries in both charging and discharging modes.

The charger provides control inputs suitable for accepting a variety of limit signals and control variables, such as those derived from temperature, time, current, voltage, and gas flow or pressure. The charger supplies charge and discharge pulses independently variable from 0 to 1000 Amps. The charger operates with variable duty cycle, DC to 1000 Hz to empirically determine the best methods of charging large cells in the one to two volt range. Two water cooled 1000 Ampere transistor switches were implemented using commercially available devices for use in the charger. The charger circuitry and the implementation of the transistor switches is described. Practical problems associated with switching high currents rapidly with transistors are discussed.

Introduction

While in the immediate future the lead-acid battery is the only practical alternative for vehicle applications other systems appear practical and are being investigated. These systems include the Nickel Zinc and Nickel Iron systems.

Fundamental to the performance of a battery is the manner in which it is charged and discharged. For practical road vehicle applications it is of particular interest to accomplish a complete charging of the battery in as short a time as possible. The problem of rapid charging may be viewed as consisting of two parts:

- Limitation on charge acceptance for useful energy storage imposed by the properties of the battery at the time of charging
- Deleterious side effects on the battery by the energy not usefully stored (heat, gassing, etc.)

Even in the relatively well known field of lead-acid batteries it would be difficult to establish an optimum charging cycle without an experimental program designed around the specific application of concern. Several methods of rapidly charging batteries have been proposed and are summarized in reference [1] for the case of lead-acid batteries. Pulse charging is one method which has been proposed for rapidly charging batteries by periodically interrupting the charging current (or providing a discharge pulse) during the charging process [2], [3].

A charger has been developed which supplies charge and discharge pulses independently variable from 0 to 1000 Amps. The charger operates with variable duty cycle, DC to 1000 Hz to empirically determine the best methods of charging large cells in the one to two volt range. To obtain the desired charger flexibility a unique circuit was developed which employs two water cooled 1000 Ampere transistor switches (four commercially available devices in parallel per switch).

The charger can also be programmed as a conventional charger and in addition the discharge profile can be controlled by providing suitable control signals. Thus, dynamic loads can be simulated to study their effects. This paper describes the charger circuitry and the implementation of the transistor switches. Practical problems associated with switching high currents rapidly with transistors are discussed.

Principles Of Charger Operation

Simplified Power Circuit

The principles of operation of the power circuit will first be explained using the simplified schematic shown in Figure 1. A description of the detailed power circuit will then follow. As shown in Figure 1, the charger consists of a charging circuit and a discharging circuit controlled by appropriate logic.

Consider the charging circuit first. With transistor Q1 turned on, the charging current set by the adjustable charging current supply is circulating in the local path provided by Q1 and L_c (i.e., $i_{Q1} = I_c$). Because Q1 is on, D1 is back biased by the cell voltage, and no charging current is supplied to the cell. (For this particular discussion the semiconductor voltage drops are neglected.) When Q1 is turned off, constant current I_c is forced into the cell under test as diode D1 becomes forward biased. Current is thus supplied to the cell until Q1 is once again turned on, shunting the charging current away from the cell and back biasing D1. In the above manner charging pulses of current of varying width and magnitude are supplied to the cell. Ideal waveforms showing the conducting devices are also shown in Figure 1. Note that the current magnitude is adjusted by adjusting the D-C current power supply independently, while the pulse width is adjustable by varying the off time of Q1.

Now consider the discharge circuit alone. With Q2 off, the discharge current circulates in the local path provided by D2 and L_d. When Q2 is turned on, D2 becomes back biased by the cell under test, and constant current I_d is transferred from D2 to Q2 and the cell. Thus with Q2 on, constant current I_d is deliv-

Reprinted from *IEEE Ind. Appl. Soc. Meet.*, 1977, vol. 2, pp. 1127–1132.

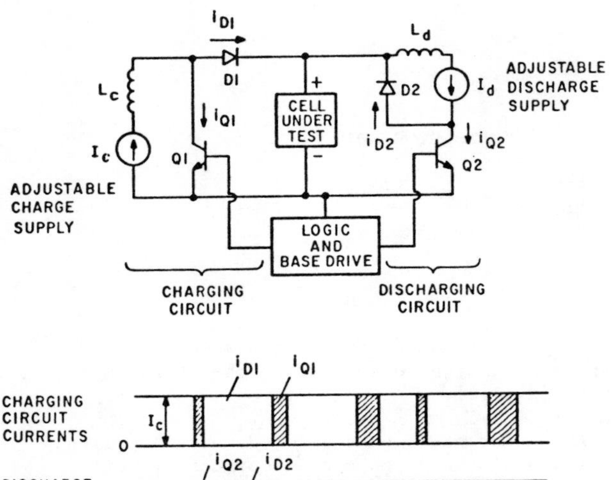

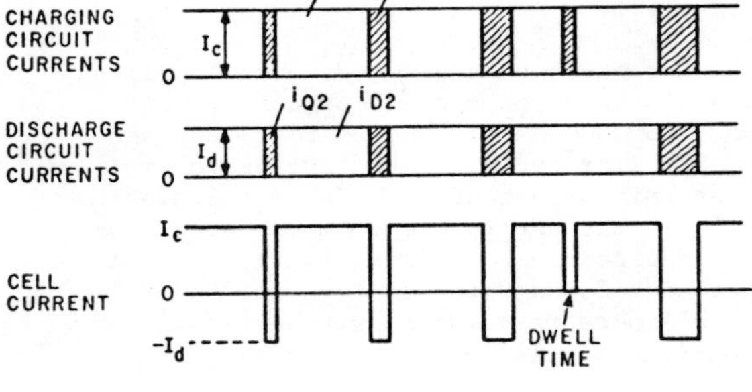

Figure 1 Simplified Circuit Diagram (top) and Ideal Waveforms

ered to the cell in the discharge direction (i.e., current I_d is out of the positive terminal of the cell). When Q2 is turned off again, the supply current, I_d, is diverted from the cell back to the local circulating path as D2 once again becomes conducting. In the above manner discharge pulses of controlled magnitude and width are applied to the cell.

The ideal waveforms for the discharge circuit currents are also shown in Figure 1. As seen, the steady current, I_d, from the adjustable discharge current supply is alternately switched between transistor Q2 and hence the cell (shown as the shaded portion of the waveform) and diode D2 (shown as the unshaded portion of the waveform).

When the charging and discharging circuits are operated together, the desired cell current waveforms result as shown in the bottom of Figure 1 for arbitrary pulse widths and amplitudes. Transistors Q1 and Q2 are turned on and off simultaneously unless a dwell time, providing zero current in the cell, is desired. If such a dwell time is desired it is accomplished by turning Q1 on and Q2 off.

In the above waveforms the magnitude of the charging and discharging currents are shown as constant; however, this condition is not necessary. The current supplies may be adjusted with time, if desired, as well as the pulse widths. This charger may also be used as a conventional charger. For example, Q1 and Q2 may be left off, resulting in the charging current continuously being supplied to the cell. Alternately, Q1 and Q2 may be left on resulting in a continuous adjustable discharge.

Detailed Power Circuit

The detailed circuit schematic is shown in Figure 2. Commercially available power supplies rated at 1000 Amperes, 8 volts are used for the external high current supplies. The current can be controlled either by a potentiometer or by an analog signal. An inductance is added in series with each supply to provide filtering of the ripple current as well as to absorb transient voltages during transistor switching and thus maintain a constant current.

The transistor switches, Q_1 and Q_2, are rated at 1000 amps D-C. The 1000 amp transistor switch consists of four Power Tech Inc. PT9501 transistors in parallel driven by a fifth PT9501 in a Darlington configuration. The transconductances (base to emitter voltage drops at a given collector current) of the four parallel transistors are matched to within 50 mV at 300 A collector current. Figure 3 shows the water cooled transistor switch.

The voltage drop across the transistor switch is in the range of 1.3 volts at 1000 amps. To be sure that this drop is not sufficient to keep D1 forward biased (in Figure 1), two diodes in series are used for D1 (D1A and D1B), as shown in Figure 2. For example if the cell voltage is 1.6 volts, then the conducting drop across Q1 would have to be approximately $1.6 + 2(0.7) = 3$ volts, assuming D1A and D1B need approximately 0.7 volt to become conductive. Since the drop across Q1 is 1.3 volts, charging current is effectively shunted from the cell.

For similar reasons, D2 consists of two diodes, D2A and D2B. For example, assuming a cell voltage of 1.6 volts, the drop across Q2 must only be less than $1.6 + 2(0.7) = 3$ volts in order for the discharge current, I_d, to be effectively shunted through the cell when Q2 is turned on.

D1A and D2A are fast recovery diodes which limit excessive reverse current transients which otherwise would occur in the diodes each time Q1 or Q2 was turned on.

Two diodes D3A and D3B are added to ensure that the discharge current power supply is always delivering power rather than absorbing power. The high current power supplies have uncontrolled rectifiers at the output and are thus unable to accept power. Referring back to Figure 1, it is seen that, assuming ideal components, the discharge supply actually must absorb power when Q2 is on. In Figure 2, diodes D3A and D3B are added to ensure that the discharge supply is always delivering power; that is, the drop across the two diodes D3 plus the conducting drop across Q2 will always be greater than the voltage due to the cell.

Diode string D4 is added as a protective feature to ensure that the voltage is limited in the event that the charger is operated without a cell; that is, current built up in L_c can circulate in diode string D4 in the event that no cell is attached and Q1 is off. Diodes D4 need only be rated to withstand the current on a transient basis. The presence of substantial current in diode string D4 is sensed and used to shut down the charger so that a cell must be present for the charger to operate. Diode D5 is connected to limit substantial reverse voltage from being applied to the cell.

The power semiconductors are water cooled, because this allows a compact design in order to minimize parasitic inductances. For example, the area of the loop defined by Q1-D1A-D1B and the cell (or diode string D4) must be as low as possible to prevent excessive voltages on Q1 when it turns off and rejects the current being delivered by the high current charging supply. To achieve the required low inductance, water cooling was used so that the circuit could be packaged in as small an area as possible. A loop inductance of ap-

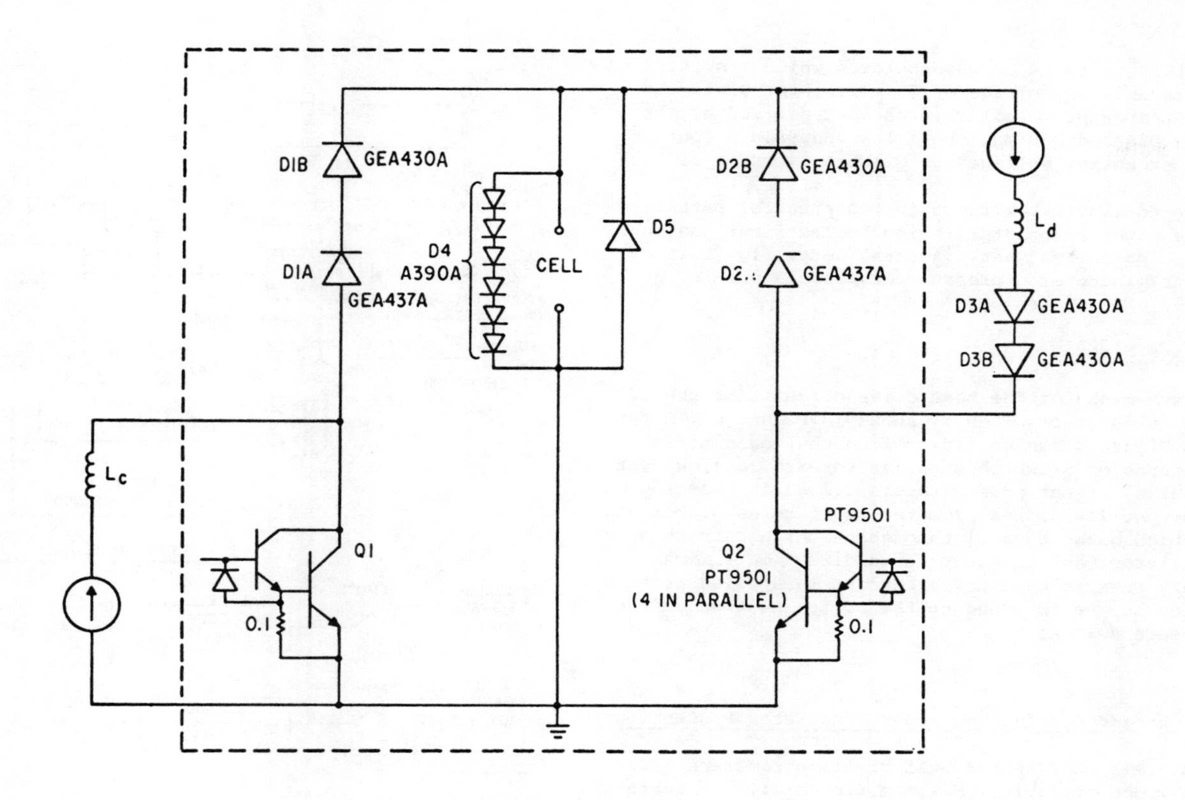

Figure 2 Detailed Power Circuit

Figure 3 Water Cooled 1000 Ampere Transistor Switch

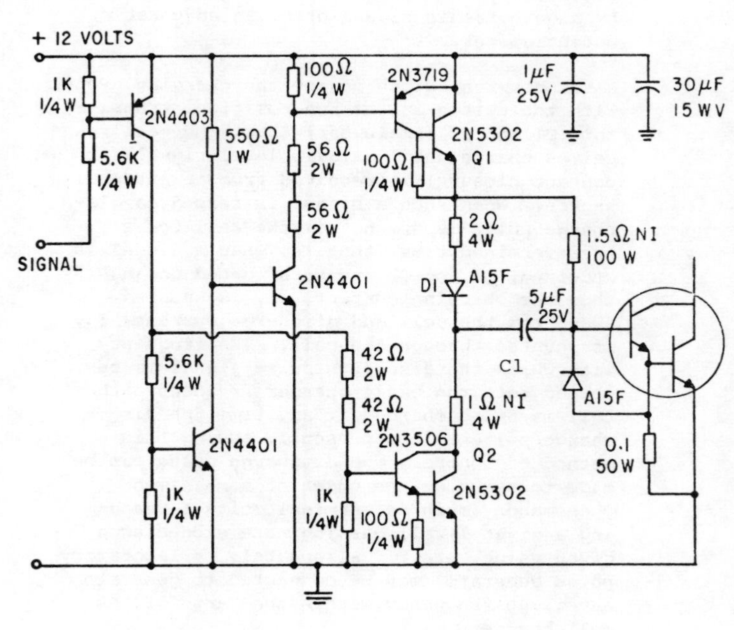

Figure 4 Base Drive Circuit

proximately .2 to .4 μH was achieved which resulted in a peak transistor voltage of 35 volts when Q1 turned off 1000 Amperes. Similarly the loop defined by the cell-Q2- D2A and D2B must be a low inductance loop to prevent excessive voltages on Q2 when it turns off.

The cooling water flows in two parallel paths with the power transistors being cooled first and the diodes being cooled last. A total water flow rate of 3 gallons/minute at a pressure drop of 25 psi is required.

Base Drive Circuit

A schematic of the base drive circuit for the 1000 Amperes transistor switch is shown in Figure 4. Referring to Figure 4 a high signal from the logic effectively turns on Q1 which supplies forward base current to the high current power transistors while charging C1. A subsequent low signal from the logic turns Q1 off and Q2 on which discharges C1 through the high current power transistor base in the reverse direction. Thus, a temporary reverse base drive current is applied at turn-off which serves to speed up the switching time of the 1000 Ampere switch.

Cell Charger Control and Power Transistor Protection

The cell charger has most of its parameters independently controllable. Maximum flexibility is achieved by allowing the charging/discharging parameters to be controlled by electronic signals or the cell charger may be operated without external electronic signals by adjustment of appropriate potentiometers. In particular, the cell charger is controlled as follows (Refer to Figure 5 for the following discussion).

- Both the charging and discharging current magnitudes are controlled from 0 to 1000 amperes by an electronic signal or by an adjustable potentiometer.

- A standby/run switch begins the charging cycle. With the switch in the run position, transistors Q1 and Q2 remain off; thus, the cell receives charge until a logic level signal (or contact closure) is received from an external source. When such a signal is sensed, a discharge pulse is drawn from the cell for a predetermined time; that is, when a signal is received, Q1 and Q2 are both turned on and thus the charging current, I_c, is shunted away from the cell and discharge current, I_d, is shunted through the cell. The frequency with which the discharge pulse signal is received sets the basic charger frequency (the cell is being charged at any time the discharge pulse is not present). Using this method of control, the discharge pulse can be made to occur at the onset of some given phenomenon (such as terminal voltage reaching a given level, gassing rate exceeding a given value, etc.). Alternately, a laboratory pulse generator may be connected to generate any given frequency within the range of the cell charger.

- The duration of the discharge pulse is adjustable using either a potentiometer or an analog signal. The duration of the discharge pulse is adjustable from a fraction of a millisecond up to several hundred milliseconds. An override switch is available if it is desired to continuously discharge the cell. This switch effectively locks Q2 and Q1 on.

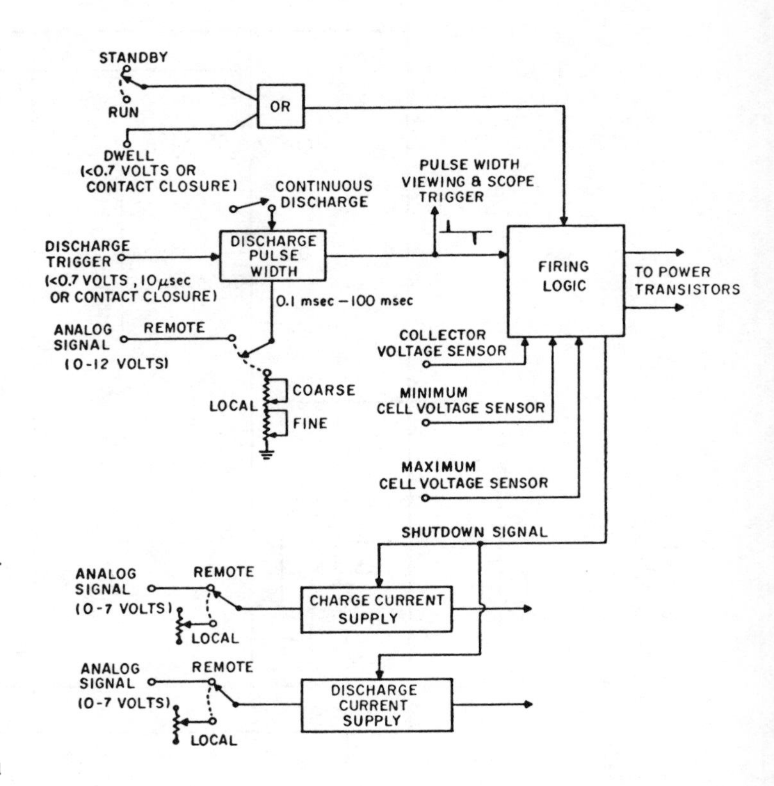

Figure 5 Charger Control and Power Transistor Protection

- When the standby/run switch is thrown to standby, currents into or out of the cell are terminated. This period corresponds to the dwell period in which Q1 is on and Q2 is off. When the switch is returned to run, charging will be resumed. Alternately, the dwell period may be initiated by a logic signal. The dwell period will last as long as the logic signal is present and may be given at any time.

There are three events which will shut down the charger:

1. Collector Voltage Sensor -- If the collector voltage on either transistor exceeds approximately 45 volts, both transistors are immediately turned on (effectively acting as a crowbar) and the current in the high current supplies reduced to zero.

2. Minimum Cell Voltage Sensor -- If the cell voltage falls below a preset minimum, both transistors are turned off and the high current reduced to zero.

3. Maximum Cell Voltage Sensor -- If the voltage as sensed at the positive battery terminal exceeds approximately 5 volts for 150 μseconds, (the terminal voltage is integrated), the transistors are turned off and the high current is reduced to zero. Thus, with no cell attached the charger will shut down since approximately a 6 volt drop will be sensed due to the current which will then be in diode string D4. Note that since the terminal voltage is integrated before shutdown occurs, a 30 volt terminal voltage transient lasting for 10 μsec will not shut the charger down. This mode of operation allows a temporary voltage transient of reasonable magnitude to exist such that the charging current can be forced into the cell rapidly even though there is some small but finite lead and/or cell inductance.

Operating Results

Figure 6 shows the pulse charger with its two high current supplies. Figures 7 and 8 show charge and discharge current pulses delivered by the charger for two different cases. These figures illustrate the transient characteristics of the charger as viewed from the load. In Figure 7 the ripple current (at 180 Hz) due to the commercial high current supplies as well as the transient response of these supplies as their load impedance charges during charger switching is illustrated. Figure 8 shows a narrow discharge pulse being delivered by the charger during the charging process. In general discharge pulses of controlled peak currents up to 1000 amps could be delivered for durations ranging from 100 μsec to infinity (continuous discharge).

Figure 9 shows the transistor switch turning off 1000 amps. A storage time of approximately 7 μsec and a fall time of approximately 10 μsec resulted when turning off 1000 amperes. Note in Figure 9 that a transient collector voltage of 33 volts above the normal voltage occurred at turn-off due to stray circuit inductance. Since the voltage rating of the transistor is 60 volts, caution must be exercised in insuring proper circuit design and layout to minimize parasitic inductances as discussed previously. As mentioned earlier, a stray inductance of only .2 to .4 μH in this case resulted in a peak overshoot of 33 volts - more than half the voltage rating of the device. This fact points out that additional device voltage capability goes hand in hand with faster fall times for high current low voltage type applications.

It was found that devices with matched V_{BE} (base to emitter voltage drop) at specified collector currents resulted in adequate matching of collector current fall times. For example, with V_{BE} matched within 2%, the fall time was matched within 15%. It is important that the switching characteristics of parallel transistors be matched in order to avoid a transient overload on a single device if, for example, its storage time is longer than the other parallel devices. In this case a single device can end up carrying the total load current at turn-off.

Figure 6 1000 Ampere Pulse Charger

With the transistor switches carrying 1000 amperes DC with a base drive of 6 amperes a total power dissipation of 1,350 watts resulted in a maximum case temperature spread between the coldest and warmest device of 3.9°C with the hottest device at 38.5°C. When switching 1000 amperes at 1000 Hz rate at 70% on time, the maximum temperature spread between the coldest and warmest device was 9.5°C after the whole system reached thermal equilibrium. At the 1000 Hz rate the hottest device had a case temperature of 46° resulting in an average junction temperature of approximately 100°C. Ambient heat sink temperature for these cases (water flowing) was 18.6°C .

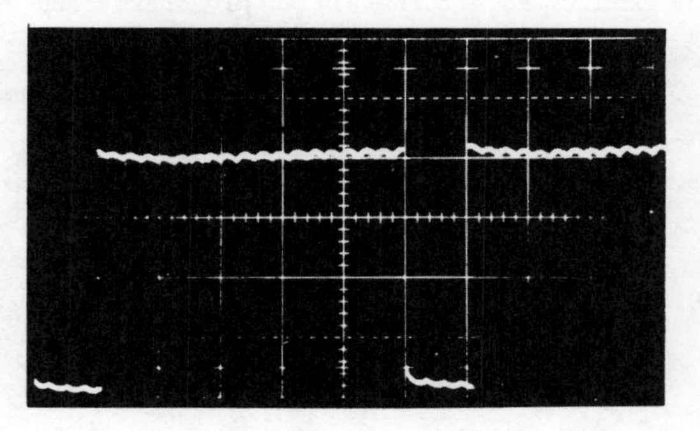

Figure 7 Typical Charger Output Current
200 A/cm (zero at center line)
20 msec/cm

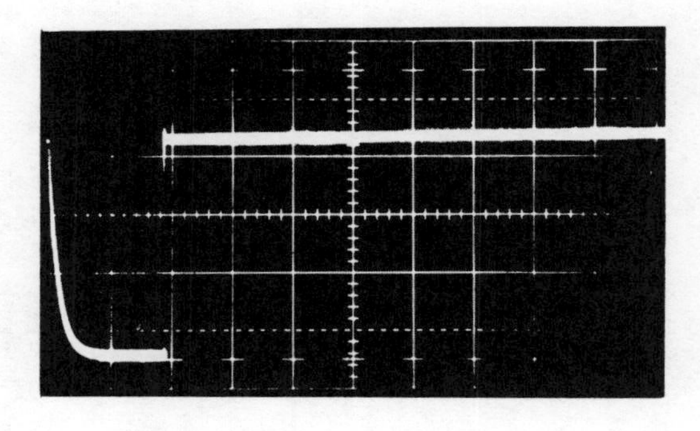

Figure 8 Charger Output Current
200 A/cm (zero at center line)
0.5 msec/cm

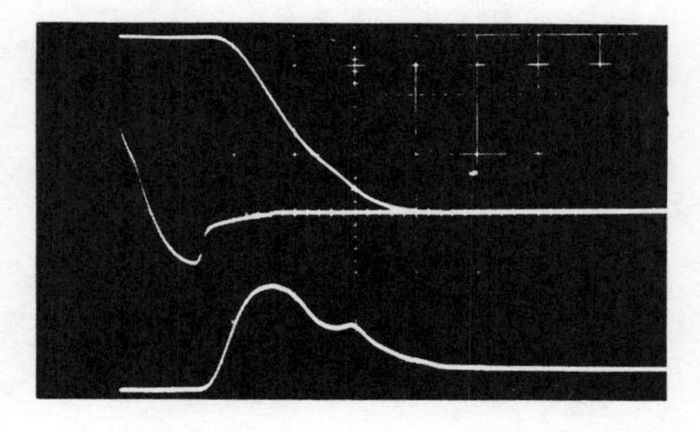

Figure 9 Transistor Turn-off Waveform,
Top: Collector Current - 333A/cm
Middle: Base Current to Darlington - 4 A/cm
Bottom: Collector Voltage - 20 volts/cm
Time: 5 μsec/cm

Conclusions

A pulse charger with a wide range of adjustments of charge current, discharge current, and pulse timing has been designed, built, and tested. The charger has the following characteristics:

1. Capability of deliverying a charging (positive) current from 0 to 1000 amps.
2. Capability of extracting a discharge (negative) current from 0 to 1000 amps.
3. Ability to alternately supply charge and discharge current pulses from DC to 1000 Hz rate.
4. Ability to continuously vary the discharge pulse width from 0.1 msec to infinity.
5. Ability to provide zero cell current (zero charge and discharge current) controlled by an electronic signal.
6. Ability to select the current magnitudes and pulse width either using potentiometers mounted on a control panel or by supplying electronic signals to the equipment.
7. Ability to operate either in a continuous charge mode or a continuous discharge mode.

To implement this charger, a water cooled high current transistor switch was developed which used parallel connected commercially available power transistors. The transistor switch is capable of carrying 1000 amps DC as well as switching 1000 amps at a 1000 Hz rate. All of the solid state power components are water cooled in order to obtain the compact, low parasitic inductance circuit configuration necessary to switch the high currents at the 1000 Hz rate.

Acknowledgment

This work was supported by the Lewis Research Center, National Aeronautics and Space Administration under Contract NAS3-19750.

References

[1.] S. Gross, "Rapid Charging of Lead Acid Batteries," in Conf. Rec. 1973 8th Annu. Meet. IEEE Ind. Appl. Soc., pp. 905-912.

[2.] P. G. Boyd, "Battery Charging -- A New Method Suggested," Electrical Times, Aug. 14, 1958.

[3.] J. A. Mas, "The Charging Process," Proceedings of the Second International Electric Vehicle Symposium, Atlantic City, N.J., Nov. 1971, Electric Vehicle Council, New York, N.Y., 1972, p. 228.

Part II
Gate Turnoff Thyristors/Latching Transistors

Papers dealing with the design and fabrication of gate turnoff thyristors (GTO's) are provided in this part. The first three papers provide guidelines for designing devices and the rest of the papers provide applications information. The design of the GTO is reviewed in the first paper which also provides a comparison between the GTO and the thyristor from the applications standpoint. One of the most important elements in designing these devices is the emitter width. During forced gate turnoff, the current tends to localize towards portions of the emitter located remote from the base contacts. In order to prevent current filamentation it is important not only to make the emitter finger narrow but to also maintain a uniform emitter size as has been achieved in the involute gate structure. The turnoff gate current also reverse biases the emitter-base junction. Since a higher turnoff gate current is required to increase the maximum controllable anode current, it is important to obtain high emitter-base breakdown voltages, for example, by using mesa etching and junction passivation.

One of the most important developments that has allowed increasing the power ratings of GTO's for direct current circuits is to use anode shorting. This should preferably be done by breaking up the p^+ anode region and using a Schottky contact to the n-base region which is aligned to the center of the n^+ emitter as discussed in the second paper. Since anode shorting destroys the reverse blocking capability, it is also customary to use a heavier doping in the n-base near the anode which allows making the total n-base width smaller. This design can significantly reduce the forward voltage drop of the GTO. This paper also describes an emitter ballasting technique which counteracts the formation of current filaments during forced gate turnoff. These design techniques have been successfully applied to the development of high power GTO's with breakdown voltages of up to 4500 V and controllable anode currents of 2000 A. The design considerations and characteristics of these devices are discussed in the third paper.

The next three papers deal with application techniques for high power GTO's including gate drive circuit design, device paralleling techniques and circuit techniques to alleviate the problems associated with high *dv/dt*'s and *di/dt*'s.

One of the difficulties of using high power GTO's is the gate drive circuit design. Because of a low turnoff current gain, typically in the range of 3-5, the gate drive circuit must provide a very large gate reverse current pulse at high *di/dt* rates driven by a relatively low voltage gate circuit power supply. Paper four describes several types of practical gate drive circuit configurations to achieve a reliable and fast turnoff of high power GTO's.

Device paralleling of bipolar devices is always troublesome. Circuit means must be employed in addition to a certain degree of device matching to achieve proper current sharing among the devices. Paper six describes a "gate coupling" technique for paralleling GTO devices. This technique requires direct connection of gate terminals of the parallel GTO's. The autobalancing nature of the technique cancels the mismatch of device switching time and is very effective in balancing current sharing during switching. However, this technique is not very effective in balancing the on-state current and thus requires matching on-state voltage of the devices.

Since high power GTO devices are relatively new to the user, many secondary yet practically important characteristics are not well known. In paper six, many such characteristics are illustrated and recommendations are made for device applications. The characteristics discussed includes the effect of gate pulse on device latching current, the effect of snubbers on maximum interruptible current, difference between direct current and pulse gating and the effect of gate voltage *dv/dt*, etc.

Although high power GTO devices are relatively new, many practical applications have been reported. The last two papers describes application examples including motor drives, uninterruptible power supply systems, midfrequency (500 Hz–2 kHz) inverters, and variable frequency inverters. Experimental data collected include inverter efficiency, inverter physical size and weight, and acoustical noise level, etc. This information should be helpful to would-be users.

THE CURRENT STATUS OF THE POWER GATE TURN-OFF SWITCH (GTO)

M. Okamura
T. Nagano
T. Ogawa
Hitachi Research Laboratory, Hitachi Ltd.,
Hitachi, Ibaraki, 319-12, Japan

Abstract

Design considerations, characteristics, and performances of high power GTO's are reviewed.

A better understanding of the gated turn-off operation was obtained by using the two-dimensional model which treats the effects related to current squeezing in turn-off processes as the essential phenomenon. Methods of increasing the capability of the GTO are discussed.

Recent studies have led to the successful developments of high power GTO's which can turn off 100 - 200 amperes in several microseconds.

The switching behavior of the GTO is strongly affected by the circuit conditions. The turn-off operations of the newly developed high power GTO are studied in conjunction with various circuit conditions.

The merits of the GTO in power electronics applications are reviewed.

Introduction

A gate turn-off switch (GTO) is a thyristor which can be turned off as well as on by the application of a gate drive. The GTO has a decided advantage over the conventional thyristor in direct current circuits owing to its gated turn-off ability, since the conventional thyristor needs a forced commutation circuitry which is expensive and bulky. Moreover, as the operation of the GTO in its gated switching mode is inherently faster than that of the conventional thyristor, the GTO is preferred in circuits operating at higher frequencies.

In the fields of high frequency applications, the GTO competes with the power transistor. The former has an advantage over the latter in the ability to provide high voltage and high current capabilities. Thus the development of high power GTO's are expected.

The feasibility of the gated turn-off operation of a p-n-p-n device was studied soon after the development of the thyristor, and an expression for the maximum turn-off gain, $G = \alpha_p/\alpha_p+\alpha_n-1$, was derived from a one-dimensional analysis.[1~3] Attempts were made to develop GTO's of high turn-off gain through various one-dimensional approaches,[1,3,4] but only low power devices were produced.

It was then recognized by many studies that the gated turn-off process was essentially a two-dimensional phenomenon and that the maximum current to be turned off was limited by the local concentration of current in the turn-off process which led to thermal failure, and by the breakdown of the n-emitter junction which resulted from the lateral voltage drop. Thus, theoretical and experimental works were done on the basis of a two-dimensional models,[5~9] and showed rather low value for the maximum turn-off gain. Fortunately, however, the power gain can be kept high enough even if the current gain is reduced moderately in the turn-off operation by the GTO. The essential problem then becomes reliable turn-off performance rather than high turn-off current gain. Thus, new design criteria have been sought. Recently, GTO's of the 10 amperes class have been produced commercially. Though higher power GTO's are not yet available generally, some papers have reported good performances of GTO's of the 100 - 200 amperes class.[8,10~12]

In this paper, many of the considerations that must be included in a practical design of high power GTO's are surveyed, and then the characteristics and the performances of high power GTO's are reviewed. Comparisons among the GTO, the conventional thyristor, and the power transistor are also made and some applications of high power GTO's are shown briefly.

Device Model

Several authors, using the two-transistor model, have derived the maximum turn-off gain of a GTO as

$$G = \frac{\alpha_p}{\alpha_n + \alpha_p - 1}, \qquad (1)$$

where α_p and α_n are the current gains of the n-p-n and p-n-p transistor parts of the GTO, respectively.[1~3] This equation has been derived from the condition for the critical state between the regenerative and the non-regenerative states; that is, the hole current supplied to the p-base is just equal to the sum of the hole current extracted to the gate and that needed for the recombination with the electrons in the p-base. This model makes no consideration of the lateral effect in the device.

When the base current is extracted from the gate, the electron injection from the emitter stops in the area adjacent to the gate, and the on-region is squeezed into the area remote from the gate. As the current is restricted only by the external load resistance, high current density filaments are formed at the centers of the emitter areas, and the localized heating may cause device failure. Another significant phenomenon is the breakdown of the n-emitter junction. The lateral current flow through the p-base layer results in a potential drop underneath the n-emitter, which causes the emitter junction to be reverse-biased at the edge adjacent to the gate. Eventually, the potential drop may reach the breakdown voltage of the junction. If the breakdown occurs at a gate current smaller than that required, the device fails to turn off.

A simple two-dimensional model has been introduced by Wolley.[5] He has derived the storage time which is required to squeeze the on-region until a one-dimensional model is applicable, as

$$t_s = (g-1)\,\frac{w_p^2}{2D_n}\,\ln\left[\frac{(SL_n/w_p^2)+(2L_n^2/w_p^2)-g+1}{(4L_n^2/w_p^2)-g+1}\right], \qquad (2)$$

where $g = I_A/I_G$, D_n is the diffusion coefficient of electrons, w_p is the width of the p-base layer, S is the width of the n-emitter, and L_n is the effective diffusion length of electrons in the p-base region. The maximum turn-off gain, G, is also derived as a

Reprinted from *IEEE Int. Semiconductor Power Converter Conf.*, 1977, pp. 39–49.

g value corresponding to infinite t_s, as follows:

$$G = 1 + 4L_n^2/w_p^2 \ . \quad (3)$$

An expression for the maximum anode current which can be turned off, I_{Aoff}, is written as

$$I_{Aoff} = 4GV_{GR}/R_B \ , \quad (4)$$

where V_{GR} is the reverse breakdown voltage of the n-emitter junction. R_B is the lateral resistance of the p-base layer under the emitter, and is given by

$$R_B = \rho S/w_pT \ , \quad (5)$$

where ρ is the average resistivity of the p-base and T is the length of the n-emitter.

A charge controlled model was used by Kao and Brewster[8], and gave more simple expressions for t_s and G, namely

$$t_s = \tau_n \cdot \ln(1/1-g\alpha_n) \ , \quad (6)$$

and

$$G = 1/\alpha_n \ , \quad (7)$$

where τ_n is the effective lifetime of electrons in the p-base. They suggested also the presence of a drift field in the base region during the turn-off period to explain the effect of the circuit condition on the turn-off performances.

Recently a CAD model of a GTO has been developed on the basis of the charge control equations.[7] This model gives a time transient response of charges, currents, and voltages.

All the previous models assumed that the α's of the device are constant. However, the values of α's depend on the current and the temperature as well as the junction structures. In the turn-off process, the current varies every moment and, consequently, the α's also vary with time. Therefore, the previous models do not provide sufficient knowledge about the dynamic turn-off process of the GTO.

To obtain an insight into the dynamic behaviour of the GTO, a computer simulation of the switching operation is now being developed in our laboratory. It is carried out by solving the basic equations numerically for given device parameters and circuit conditions to give the transient responses of the carriers and the potential profiles in the switching process as well as the voltage-current characteristics of the device. In our method, the complex relationships among the above parameters are included automatically, which could not be done in the previous studies. This simulation should provide sufficient insight into the switching mechanism and become a powerful aid in design of the GTO.

Design Consideration for a High Power GTO

In the early stage of the development of the GTO, Eq. (1) was regarded as the only design criterion. It indicates that the value of α_p should be close to unity and that of α_n just high enough to satisfy the on-state requirement, $\alpha_p + \alpha_n \geq 1$.

Then, in recognition of the two-dimensional effect, Eq. (4) has acquired more importance in the design of a higher power GTO. The requirements to satisfy Eq. (4) do not necessarily agree with the conditions to make α_p large. In this case, α_p should be as large as the gate drive condition allows and α_n a little larger than $1-\alpha_p$ so that small deviations of the α's from the chosen values have less effect on the turn-off gain.

Though these conditions are only qualitative guide, we must be satisfied with them at present.

Cathode Emitter Pattern

According to Eqs. (4) and (5), a long, narrow emitter is desirable to reduce the lateral resistance. This requirement can be realized by the emitter areas of strips surrounded by the p-base region. The widths of the strips should be the same everywhere to achieve uniform turn-off. Practically the width, S, is limited by the difficulty of electrode fabrication and is usually at least several hundred microns.

Equation (2) indicates that the storage time would be influenced by the emitter width. However, it has been shown experimentally that both the storage time and the fall time do not depend on the emitter width of less than 625 microns.[8]

P-base Structure

When the p-base and the n-emitter are formed by a double diffusion technique, the mesa structure will be preferred in order to obtain a high breakdown voltage of the emitter junction. Though the planar emitter junction usually gives low breakdown voltage, some improvement can be made by modifying the impurity concentration profile in the p-base layer.

The epitaxial growth technique will give a better p-base for the GTO, because the epitaxial base has a uniform profile, which results in a higher emitter breakdown voltage than that of a diffused base for the same average base resistivity.

Better performance can be achieved by using a double p-base which consists of a p^+ and p^- layer.[9] The p^+ layer adjacent to the center junction reduces the lateral resistance and the p^- layer adjacent to the emitter junction gives a high breakdown voltage. It has been reported that the GTO with this p-base structure shows a gate-cathode breakdown voltage as high as 80 volts.

Reduction of α_n

In design of a GTO, another major problem is to control α_n. The value of α_n is considered to be the product of the injection efficiency, γ_p, and the transport factor, β.

The transport factor in the n-base depends upon the hole lifetime, τ_p, and the base width, w_n, as follows:

$$\beta = \mathrm{sech}(w_n/\sqrt{D_p \cdot \tau_p}) \ . \quad (8)$$

A large base width gives a small transport factor, but at the same time, it gives rise to a high on-voltage and a slow switching speed. Hence, base widening is not a good way to reduce α_n.

Therefore, it becomes necessary to reduce the lifetime. The lifetime reduction is also necessary for fast switching operation. Gold diffusion is usually used to reduce the turn-off time of conventional thyristors. However, gold diffusion is very structure-sensitive and often causes localization of the on-region because of the non-uniform distribution of gold. For a GTO, the non-uniform distribution of lifetime will result in not only the non-uniform current distribution in the on-state but also the localized crowding of the current at the turn-off period, which causes failure of the device. Therefore, higher uniformity of lifetime is required for a GTO.

Recently high energy electron irradiation has been used for the lifetime control of silicon devices.[13~15] Precise and uniform control of the lifetime required for high turn-off gain might be achieved by the ease of controling dosage of irradiation.

The emitter efficiency is expressed as

$$\gamma_p = 1/(1-\sigma_B \cdot w_n/\sigma_E \cdot L_E) \ , \tag{9}$$

where σ_B and σ_E are the average conductivity of the n-base and the p-emitter, respectively, and L_E is the diffusion length in the p-emitter. This equation indicates that γ_p can be reduced by a decrease in $\sigma_E \cdot L_E$ or by an increase in $\sigma_B \cdot w_n$. The first case can be realized with a thin and/or lightly doped p-emitter. Recently, a low efficiency p-emitter has been reported to be provided by the use of the vapor phase deposition of boron-doped polycrystalline silicon,[16] and is expected to be available for use in a GTO.

The second method of γ_p reduction is to increase $\sigma_B \cdot w_n$. As a wide base is undesirable as discussed previously, an increase in σ_B is preferable. However, the increasing in σ_B is limited by the required blocking voltage. For a high voltage device, the thin p-emitter may result in a larger on-voltage since sufficient conductivity modulation in the n-base cannot be expected.

A p-n-i-p-n structure is an effective solution for providing a high blocking voltage and a low on-voltage to a GTO.[4] In practice, the high resistivity i-layer is slightly doped with donors and the thin n-layer adjacent to the p-emitter layer is doped rather heavily. This structure has a thinner total n-base width comparing with the p-n-p-n structure for the same forward blocking voltage. The additional advantage of reduced turn-off time will result from the reduction of the carrier storage.

The other method to reduce the effective emitter efficiency is the emitter junction shorting.[1] The emitter efficiency can be controlled by the configuration of shorting areas and the sheet resistance of the n-base layer. Figure 1 shows the influence of the shorting areas on the turn-off operation. It can be seen that the tail part of the anode current decreases as the shorting area increases. The effect of the emitter shorting will be remarkable if it is used in the p-n-i-p-n structure because the resistivity of the n-base layer is lower than that of the p-n-p-n structure.

For lower frequency applications, a GTO whose α_n is reduced only by the p-emitter shorting (without use of gold diffusion) can provide satisfactory switching performance. Such a device has advantages of lower on-voltage and less temperature dependence of gated

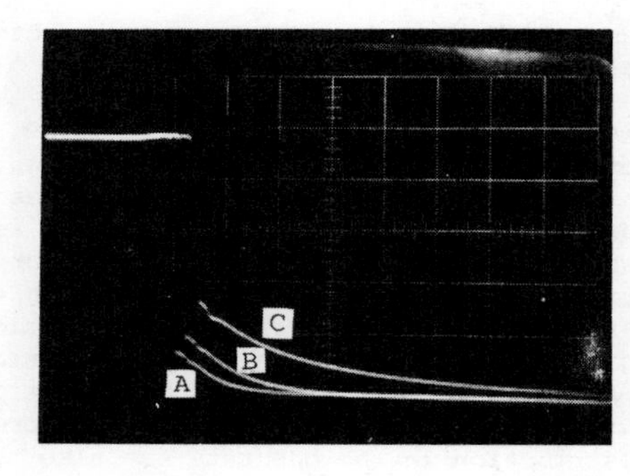

Fig. 1 Anode current responses at turn-off.

Hor. 2 μsec/div
Vert. 2 amps/div

A : largest shorting area
B : middle shorting area
C : smallest shorting area

turn-off characteristics than gold-doped devices.

A GTO with the shorted p-emitter can also be regarded as a composite device of a thyristor and an n-p-n transistor in parallel connection. The transistor section can be driven in saturation by the base current of the thyristor section during the on-state, but it will turn off by its non-regenerative nature when the thyristor section goes into the off-state. Becke and Neilson[9] have suggested that a non-regenerative section should be provided at the region remote from the gate contact so as to aid the extinction of the final current filament.

It should be noted that the reverse voltage blocking capability is lost when a thin or shorted p-emitter is used. However, the reverse blocking capability is not necessarily required for most of the applications of high power GTO's.

Cathode and Gate Electrodes

The cathode and the gate electrodes are important in device design of high power GTO's.

In this paper, thin metal layers which contact directly with the n-emitter or the p-base of the silicon wafer are referred to as the emitter, or, base electrodes and metal pieces which are connected to the cathode or the gate terminal are referred to as the cathode, or, gate electrodes.

The n-emitter of a high power GTO is usually composed of long, narrow strips which are surrounded by the p-base region. This configuration makes it difficult to provide them with electrodes. The n-emitter electrode should be in complete contact with the cathode electrode because the main current flows through it. At the same time, the p-base electrode should be placed so as to surround the emitter electrodes and to uniformly extract a large gate current from the base layer under the emitter strips. Non-uniform distribution of the gate current would cause the crowding of the anode current into the areas where less gate current is supplied and, conseqently, result in the failure of the device.

A non-uniform gate current would bring crowding also at the turn-on process. In this case, the area with high gate resistance might not turn on, and a localized on-region could make a hot spot if the anode current has a large di/dt value.

Although it is desirable for the p-base electrode to make a complete and direct contact with the gate electrode, it is usually impossible because of the presence of the cathode electrode. Therefore, the p-base electrode should be made as thick as possible to reduce the resistance.

The junction structure must also be taken into consideration in the design of the electrodes. When the n-emitter junction has the mesa structure, a flat cathode electrode can easily make contact with the emitter electrode at all points. However, the mesa emitter presents difficulties in fabrication and passivation. The etching of the p-base is very critical because variations in the etching depth result in non-uniform base resistivity and, consequently, in non-uniform gate current. For the planar junction, fabrication and passivation are easier, but achieving contact with the cathode electrode requires a complex electrode structure.

Structures and Characteristics of High Power GTO's

Developments of high power GTO's have been reported by Kao and Brewster[8,10] and by Wolley et al.[11,12] The authors also have developed GTO's which can turn off more than 100 amperes. In this section, the structures and the electrical characteristics of these high power GTO's are reviewed.

The electrical characteristics of the high power GTO's are listed in Table 1.

Figure 2 shows the emitter pattern of the device of Wolley et al.[11,12] It is termed an "involute emitter" and has been designed to make the distances between the emitter strips equal everywhere in order to obtain a good active area utilization. The diameter of the wafer is 33 mm. The amplifying gate is used to supply a large turn-on trigger current which is due to the large perimeters of the n-emitters. A by-pass diode is mounted on the p-base electrode to shunt the turn-off gate current from the amplifying gate. In order to reduce the turn-off time and to obtain high turn-off gain, gold is heavily doped. The n-emitter junction has the mesa structure. The p-base electrode is formed at the bottom of the involute grooves and it makes contact with the gate electrode at the center of the wafer. A pressure contact package is used to decrease the thermal resistance.

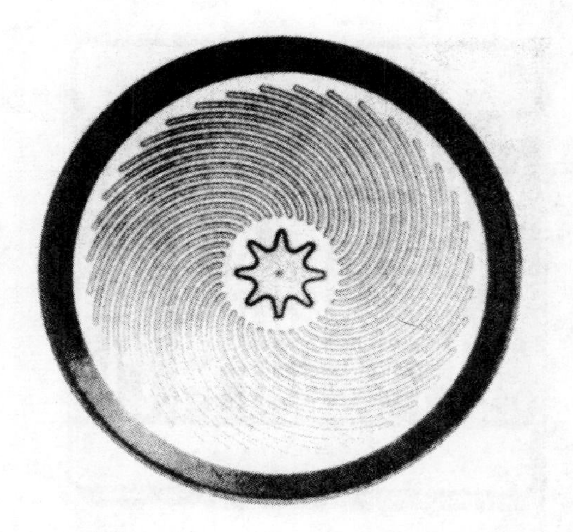

Fig. 2 Involute emitter pattern.[11,12]

Kao and Brewster reported two types of high power GTO's; one which could turn off 100 amperes[10] and the other, 200 amperes.[8] Figure 3 shows the n-emitter pattern of the 100-ampere device (diameter, 23 mm). The emitter junction has the planar structure. The device is packaged in a stud mount compression bonded encapsulation. For the larger devices, it was reported only that the devices were capable of blocking 1,000 volts and turn-off more than 200 amperes in 2 microseconds at 125°C, and the forward voltage drops were between 1.6 and 1.9 volts.

Table 1 Characteristics of high power GTO's.

Item (Unit)	ref.12	ref.10	ref.8	The Authors
Current Interrupt Capability (A)	>200	70-110	>200	>100
Forward Blocking Voltage (V)	600 -900	800 -900	1000	400
Reverse Blocking Voltage (V)	1200 -1400	900 -1000	1000	
Cathode Junction Breakdown Voltage (V)		20		18
On-state Voltage (V)	3.7-8.0 @200A	1.3-1.7 @100A	1.6-1.9 @200A	1.65 @100A
Minimum Triggering Gate Current (mA)		10-60		300
Turn-on time (μs)	<2	<2.5		<2
Storage Time (μs)		<2.0		4.5
Fall Time (μs)	0.6-0.9	1.0		0.6
Turn-off Gain	4.8-6.2	2-3		5
Static dv/dt (V/μs)	25->100	>400		
di/dt (A/μs)	>200	>150		

Figure 4 shows the geometry of the n-emitter of the authors' device. The pellet is a square whose length is 15.5 mm. The n-emitter consists of 12 individual strips which were diffused by the planar technique. The breakdown voltage of the emitter junction was higher than 15 volts. The center junction was passivated with glass to get high reliability. The emitter shorting was adopted for the p-emitter in order to reduce the tail of the anode current at turn-off. Figure 5 shows a schematic cross-section of the device and indicates the configuration of the n- and p-emitters as well as of the cathode and the anode electrodes. The bottom surface of the cathode electrode was fabricated so as to have ridges which were of the same configuration as the n-emitter strips and, where the contact with the n-emitter electrode was made by soldering. This device was mounted in a conventional stud type package.

Fig. 3 Radial emitter pattern.[10]

Fig. 4 Emitter pattern of the GTO developed by the authors.

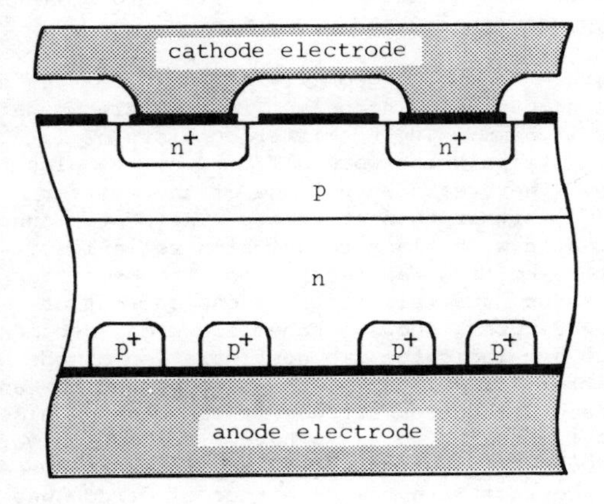

Fig. 5 Configuration of n-, p-emitters and electrodes.

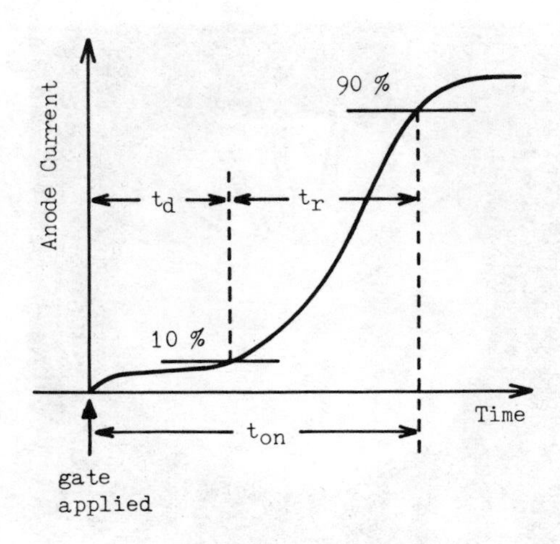

Fig. 6 Delay, rise, and turn-on times during gated turn-on.

Switching Performance

Gate Turn-on

The anode current of the GTO does not respond immediately to the applied gate signal. The turn-on response of the anode current is characterized by a turn-on time, t_{on}, which consists of a delay time, t_d, and a rise time, t_r, as shown in Fig. 6.

For high frequency applications, a short turn-on time, especially a short rise time is required to reduce the switching power loss. The power loss dissipated during the delay time is negligible because of the low anode current.

The GTO is expected to have a faster turn-on time than the conventional thyristor because of its narrower emitter width. In fact, the turn-on times of the GTO's in Table 1 are less than a few microseconds. Figure 7 presents the typical waveforms of the anode current and the anode voltage at turn-on of the authors' device, which has a turn-on time of about 2 microseconds.

Moreover, relatively high di/dt rating can be expected for the GTO compared with the conventional thyristor, because the initial on-region of the GTO occupies a considerable part of the n-emitter area.

The delay time decreases with increasing gate current, while the rise time does not vary so far as the gate current is much smaller than the anode current in the on-state. However, for larger gate drive, the rise time also decreases as shown in Fig. 8.

The turn-on time depends also upon the rising rate of the gate current; the faster the rising rate, the shorter the turn-on time.[10]

Gate Turn-off

To turn GTO's off, a negative voltage is applied to the gate, but the anode current does not respond immediately just as with the gate turn-on. Figure 9 shows a schematic anode current waveform, which is characterized by three time periods: the storage time, t_s, the fall time, t_f, and the tail time, t_t.

In the storage period, all three junctions in the GTO are still forward biased. As the gate current removes the excess carriers in the p-base region, the n-emitter recovers, followed by the coming out of saturation of the center junction. This causes a

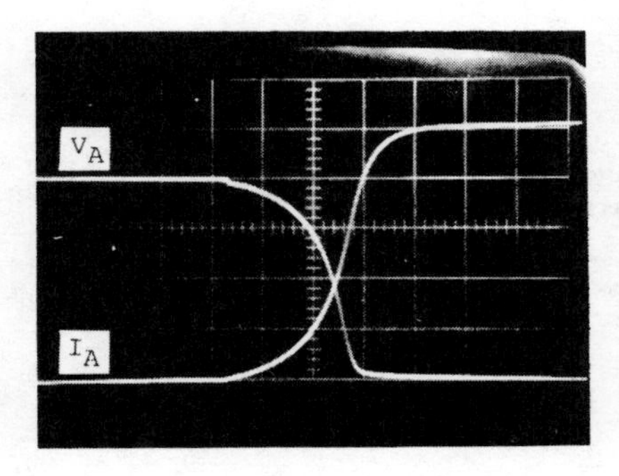

Fig. 7 Anode current and anode voltage responses at turn-on for I_G = 2 amperes.

Hor. 1 μsec/div
Vert. 50 volts/div - V_A
20 amps/div - I_A

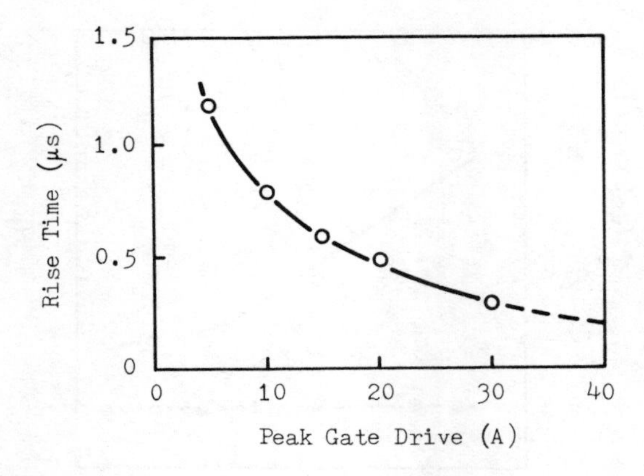

Fig. 8 The effect of the peak gate drive on the rise time.[10]

gate pulse width - 3 μsec
gate pulse risetime - 100-250 nsec
anode circuit risetime - 200 amps/μsec

sharp decrease of the anode current to a small value. After the center junction is recovered, the anode current tails. In this tail period, the p-n-p transistor section of the GTO is in its active region, and the residual charge carriers in the n-base act as a base drive for the transistor.

Figure 10 shows a simplified testing circuit used for the measurements of the gate turn-off characteristics in our laboratory. The gate turn-off process is strongly affected by the circuit conditions. In Fig. 10, a snubber circuit composed of a diode, a capacitor, and a resistor was connected parallel to the GTO. The purposes of the snubber circuit are[17] (1) to limit the power dissipation at turn-off by suppressing the anode voltage rise during the fall period, and (2) to limit the rising rate of the anode voltage during the tail period to prevent the re-ignition of the GTO. A simple RC snubber which is commonly used in thyristor circuits cannot sufficiently suppress the anode voltage rise at turn-off. Thus, the polarized snubber is necessary to obtain a full capability of current interruption of the GTO.

Figure 11 shows the typical waveforms of the anode current, I_A, the anode voltage, V_A, and the gate current, I_G, of the GTO developed by the authors. As the gate voltage is applied, the gate current starts to rise and then reaches a peak. During this period, t_s, of about 4 microseconds, the anode current is substatially unchanged. Then the anode current falls abruptly in approximately 0.5 microseconds. Finally the anode current tails for about 2 microseconds.

During the fall period, the anode voltage shows a spike of about 100 volts. The spike results from the high falling rate of the anode current acting with the inductance of the snubber circuit and the turn-on characteristics of the snubber diode. It is very important to reduce this voltage spike, because it causes a high switching power dissipation in a localized region of the GTO. Thus, a low inductance circuit and a fast turn-on diode should be used for the snubber. After the fall of the anode current, the anode voltage rises at a rate determined by the snubber capacitance and the load impedance to the source voltage.

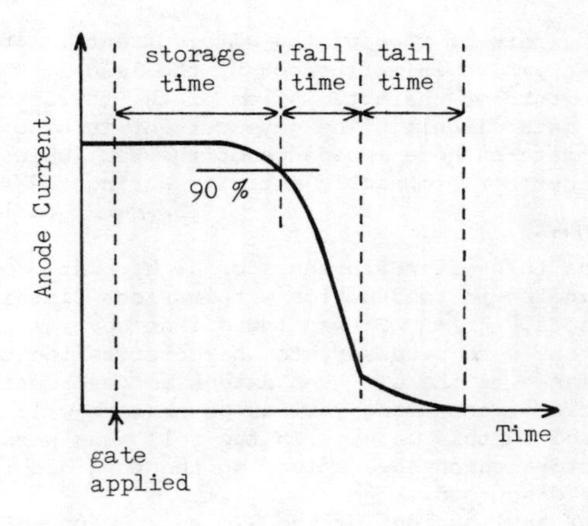

Fig. 9 Storage, fall, and tail times during gated turn-off.

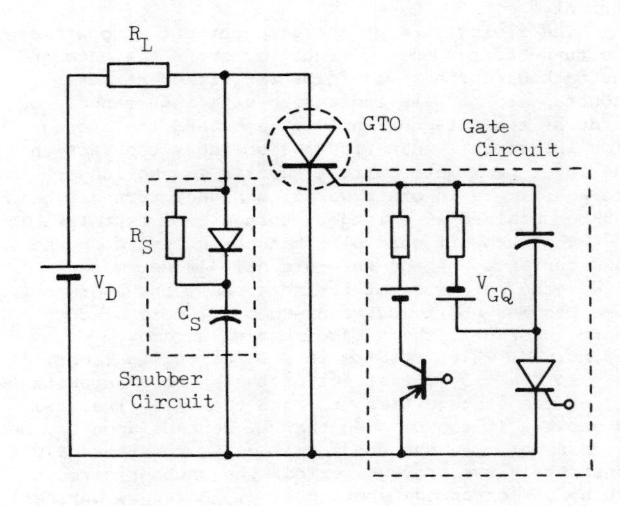

Fig. 10 Simplified GTO test circuit.

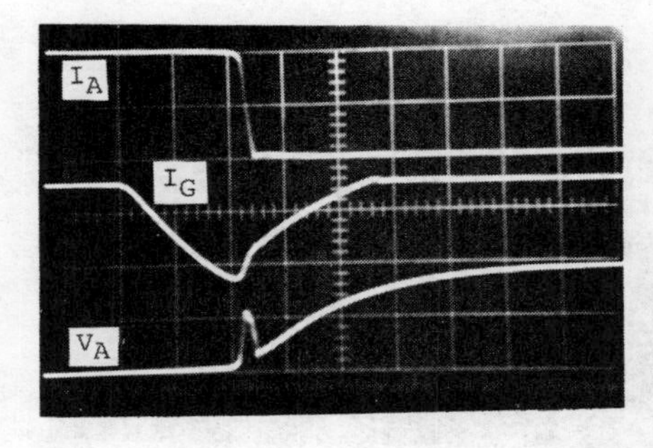

Fig. 11 Anode current, gate current, and anode voltage waveforms at turn-off.

Hor. 2 μsec/div
Vert. 50 amps/div - I_A
10 amps/div - I_G
100 volts/div - V_A

Circuit conditions:
V_D = 200 V, R_L = 2 Ω, R_S = 20 Ω,
V_{GQ} = 15 V, C_S = 2 μF.

The rate of rise of the gate current is about 5 A/μsec, which is influenced by the lead inductance and the turn-on characteristics of the thyristor used in the gate circuit. The gate current is also affected by the gate-cathode impedance of the GTO, which varies from about zero to nearly infinity during the turn-off period.

The turn-off performance of high power GTO's was investigated in conjunction with various circuit conditions.[8~11,17] It was found that several parameters were necessary to characterize the turn-off performance of the GTO. An extensive description of the correlations among various parameters will not be presented in this paper. In the following paragraphs, the factors which have effect on the turn-off response will be discussed.

For applications of the GTO's, the turn-off time should be kept small to minimize the switching power loss. It was reported that, for a given turn-off gain, the storage time, t_s, increased with increasing anode voltage but decreased with increasing anode current.[8]

The rising rate of the gate current also affects the turn-off response. Figure 12 shows the gate and the cathode current waveforms for different gate inductances.[10] With increasing gate inductance, the di/dt of the gate current decreases and the storage time increases. This figure also shows the fact that the lower peak gate current results in the longer storage time. In other words, a higher turn-off gain can be obtained if a longer storage time is tolerable.

The turn-off gain also has an influence on the tail current. As the turn-off gain is decreased (i.e., the gate current is increased), the storage time becomes short and, consequently, more of the charge stored in the n-base remains unremoved or unrecombined, which results in a large tail current.[11]

In Fig. 13, the effect of the snubber capacitance C_S, on the storage time, t_s, and the fall time, t_f, is shown. It can be seen that t_f depends upon C_S, but t_s does not. As the anode voltage is substantially constant in the storage period, the snubber circuit has no effect on the storage time. However, during the fall period, the anode current is shunted to the snubber capacitance through the snubber diode. The shunted current increases with increasing C_S, but it approaches a constant for larger C_S values because it is also affected by the switching speed of the snubber diode. Consequently, t_f decreases with increasing C_S and finally reaches a constant.

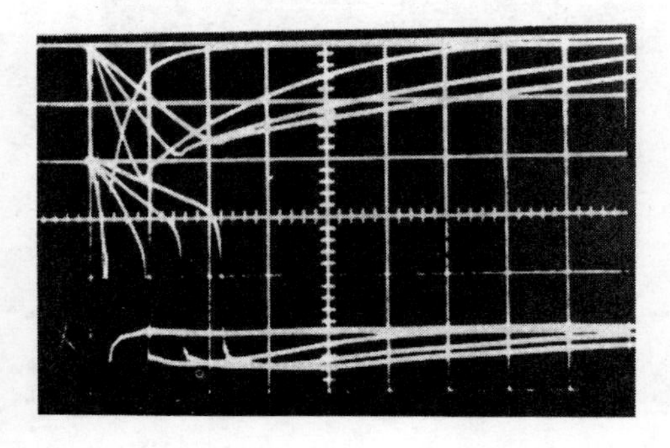

Fig. 12 Gate and cathode current waveforms at turn-off for series gate inductances of 2, 6, 12, and 20 μH.[10]

upper traces - I_G
lower traces - I_K

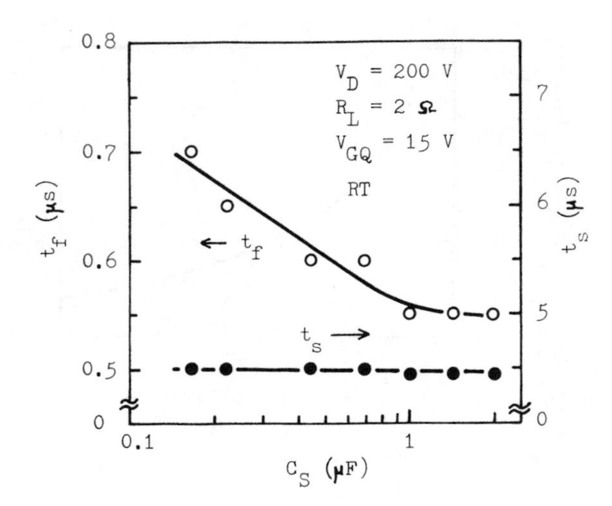

Fig. 13 The effect of the snubber capacitance (C_S) on the storage time (t_s), and the fall time (t_f).

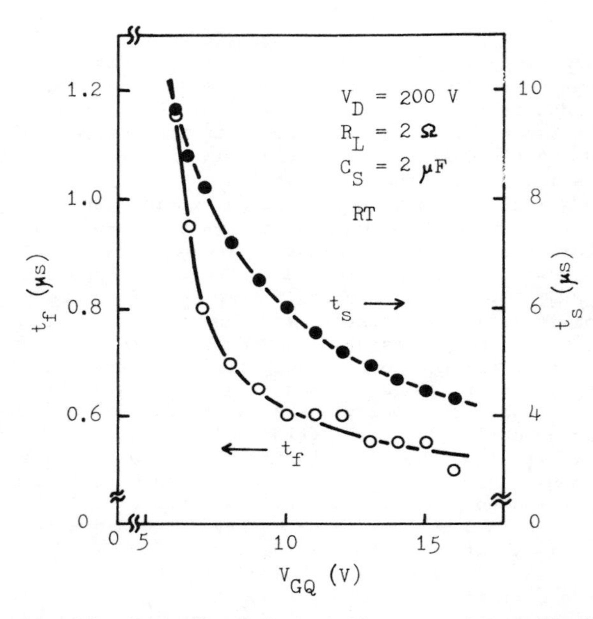

Fig. 14 The effect of the off gate voltage (V_{GQ}) on the storage time (t_s), and the fall time (t_f).

Figure 14 shows the effect of the off gate voltage, V_{GQ}, on t_s and t_f. A high V_{GQ} makes t_s and t_f short. But V_{GQ} should not exceed the breakdown voltage of the n-emitter junction.

Switching Loss

The switching power loss is a primary factor determining the maximum operating frequency of a GTO.

In order to estimate the maximum operating frequency of the GTO developed by the authors, a trial calculation was carried out for a simple DC-DC chopper circuit. The chopper was assumed to have a rating of 10 kVA, a source voltage of 200 volts, a load resistance of 2 ohms, and to be operated at 50% duty cycle. A V_{GQ} of 15 volts and C_S of 0.5 or 1.0 microfarads were used.

The energy loss per cycle at turn-on, E_{on}, and at turn-off, E_{off}, were measured from the waveforms of the anode current and the anode voltage. From Fig. 7, E_{on} was approximately estimated to be 6 mJ. E_{off} depends strongly on the off gate voltage, V_{GQ}, and the snubber capacitance, C_S. In Fig. 15, the measured E_{off} is plotted against V_{GQ} and C_S. The energy loss dissipated during the fall period, E_{fall}, is also shown there. Figure 15 shows that E_{off} decreases with increasing values of both V_{GQ} and C_S. Thus, for this GTO, the minimum values of E_{off} was approximately estimated to be 5 mJ.

In the estimation of the power loss of the GTO chopper, not only the power loss of the GTO but also that of the snubber circuit must be considered. Figure 16 shows the variation of the power losses with the operating frequency, f. In this figure, the solid lines show the power losses of the GTO at turn-on (P_{on}), at turn-off (P_{off}), at on-state ($P_{on\text{-}state}$), and the sum of these losses (P_{total}), whereas the broken lines show the snubber losses (P_S) for each C_S. These losses were calculated as follows:

$$P_{on} = E_{on} \cdot f ,$$

$$P_{off} = E_{off} \cdot f , \quad (10)$$

and $$P_S = C_S \cdot V_D^2 \cdot f/2 .$$

$P_{on\text{-}state}$ was estimated from the on-state voltage shown in Table 1. As this GTO could dissipate the power of 200 watts, this chopper would operate at 10 kHz with an efficiency of 96% for C_S of 1 microfarads.

Failure Modes

Two failure modes have been observed for the GTO.

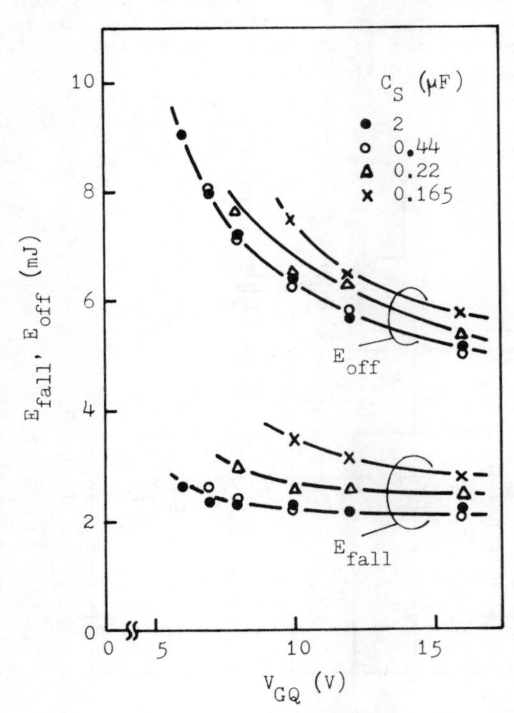

Fig. 15 Energy loss at turn-off as a function of off gate voltage (V_D = 200 V, R_L = 2 Ω).

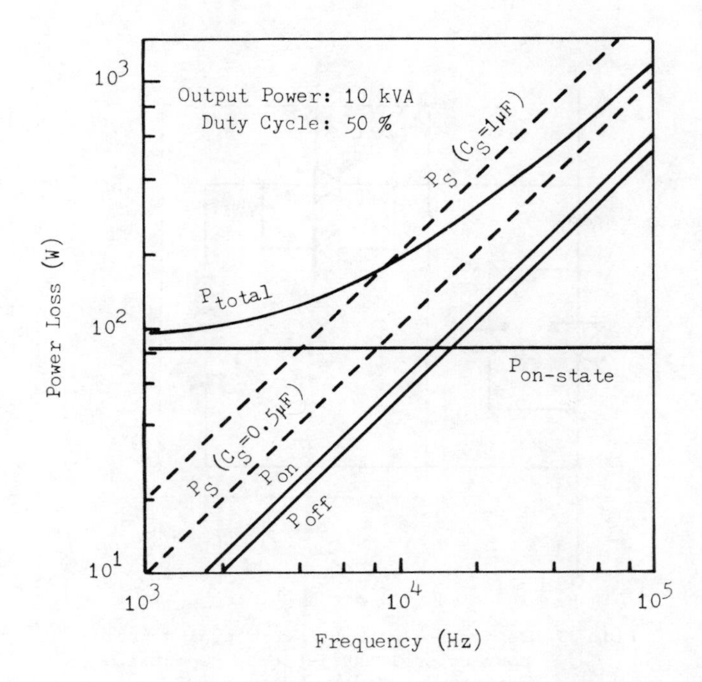

Fig. 16 Operating frequency vs. power loss.

One happens when the anode current exceeds its maximum interruptable current, while the other is caused by the dv/dt effect during the tail period.

The former is due to the switching loss dissipated at the turn-off period. Even though a snubber circuit is used, the anode voltage spike cannot be suppressed to zero. The more the anode current flows, the higher the spike becomes, and the more the switching power is generated. Unfortunately, this power is usually dissipated in a tiny spot because of current squeezing, and leads to the thermal failure.

The latter failure mode is due to a re-ignition, which is caused by the dv/dt acting with the tail current. If the on-region introduced by this re-ignition is wide enough, no failure occurs. However, the distribution of the residual charge carriers in the tail period is usually localized. As a result, the re-ignited anode current flows in a filament and causes a spot failure.

Application to Power Converter Equipment

Comparison of the GTO with Other Power Switches

GTO vs. high speed thyristor. In order to turn off the conventional thyristor in direct current circuits, a commutating circuit is used. Figure 17 shows a fundamental commutation circuit, which consists of a semiconductor switching device, a commutating capacitor, and an inductor.

The commutating capacitor and the auxiliary thyristor used in this circuit have to carry a current comparable to the main current and to block nearly twice as high as the source voltage. The turn-off capacitor and the auxiliary switch shown in Fig. 10 need only to carry a fraction of the main current and to block 20 - 30 volts. Moreover, the commutating current should be maintained for a period two or three times larger than the turn-off time of the thyristor to prevent commutation failure; whereas the duration of the gate turn-off current is only several microseconds. Thus, the electrical power required to turn

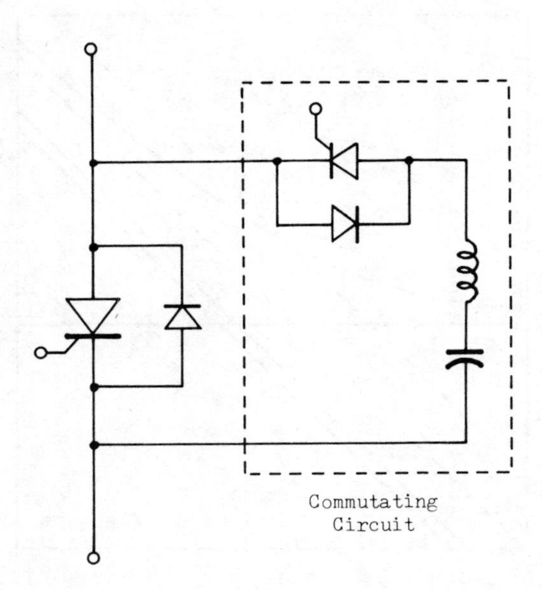

Fig. 17 Fundamental circuit for turning off conventional thyristor by commutating method.

off a GTO is much less than that required for a conventional thyristor. As a consequence, the GTO equipment has a higher efficiency than the thyristor equipment.

As the switching speed of the GTO is faster than that of the conventional thyristor, the GTO can be used at higher frequencies than the thyristor. This allows the use of a smaller capacitance and inductance in the filter circuit, and enables to reduce the weight and the cost of the equipment.

GTO vs. gate-assisted turn-off thyristor (GATT). A GATT is a thyristor that is turned off by the application of a negetive voltage to the anode and the gate simultaneously.[18~21] The use of this device can improve the tradeoff between the blocking voltage and the turn-off time of the usual high speed thyristor. For example, the GATT reported by Shimizu et al[20] is rated at the blocking voltage of 1200 volts and the turn-off time of 6 microseconds, and has a high-frequency (10 kHz) rating of more than 500 amperes. Though the GATT can handle a higher power than the GTO's of the same turn-off time, it requires a more complicated circuitry.

GTO vs. power transistor. It is generally known that the transistor is preferred to the GTO in high frequency applications, because the transistor inherently has a higher switching speed. The recent development[22] of the high power transistors which have collector current ratings of 400 amperes and collector-emitter voltage ratings of 300 volts has made the power handling capability of the transistor to be comparable to that of the high power GTO. However, it was reported that the switching speeds of those transistors were also comparable to that of the GTO. Thus, other considerations will be necessary to compare them in high power regions. One is the duty cycle; that is, the transistor is efficient for short on-duty cycles, while the GTO for long on-duty cycles. This is because the former needs power to hold it on, while the latter does not. For a short-circuit problem, the transistor will come out of saturation and will be led to the secondary breakdown, while the GTO has a higher current capability owing to its regenerative nature.

Some Examples of GTO Application

Many application circuits of the GTO have been proposed in the literatures.[11,12,23] A few examples will be shown here.

DC-AC inverter. Figure 18 shows the inverter circuit reported by Wolley et al.[12] Two GTO's and a center tapped transformer are used to convert 28 volts DC into 120 volts AC (square wave) at 1200 Hz. The inverter rating is approximately 3 kVA and each GTO carries approximately 75 A RMS (106 A peak, 50% duty cycle). The efficiency is approximately 86%. The features of this inverter are (1) simple circuitry due to the elimination of the commutation circuitry, and (2) the improved controllability of the output voltage by the pulse width modulation.

Figure 19 shows the inverter circuit under development in our laboratory. Four GTO's forming a bridge, and a transformer were used. The output voltage was controlled at 100±2 volts by the pulse width modulation, and the output frequency at 50 or 60 ±0.05 Hz.

In the GTO inverter, a higher switching frequency can be used because of its short turn-off time. Consequently, the reduced harmonic components of lower order allows to use a small capacitance and inductance in the filter circuit. In the thyristor inverter, the energy stored in the commutating circuit usually depends upon the load impedance; whereas the GTO inverter has no commutation circuitry. Consequently, the operation of the GTO inverter is more stable for the variation of the load.

In the thyristor inverter, as the charging current for the commutating circuit flows superimposed on the main current, the current rating of the main thyristor would be much larger than that of the GTO to achieve the same output power. The modulation frequency of our inverter was approximately 1 kHz, and the current rating of the GTO was about 60% of that of the thyristor, and the total efficiency was about 85%.

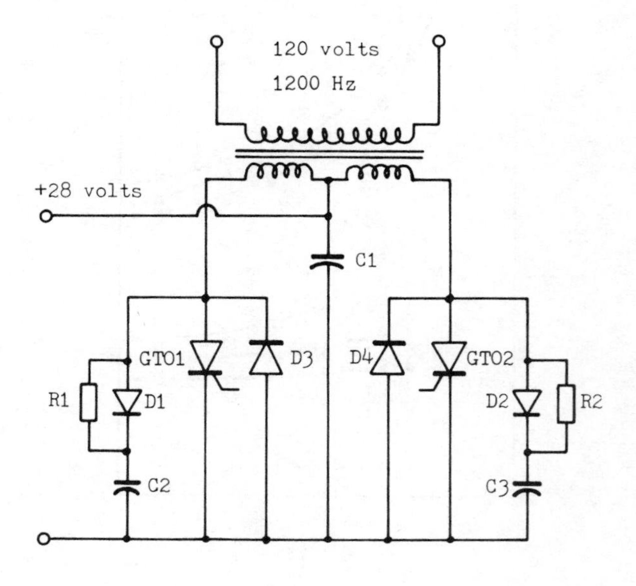

Fig. 18 DC-AC inverter.[12]

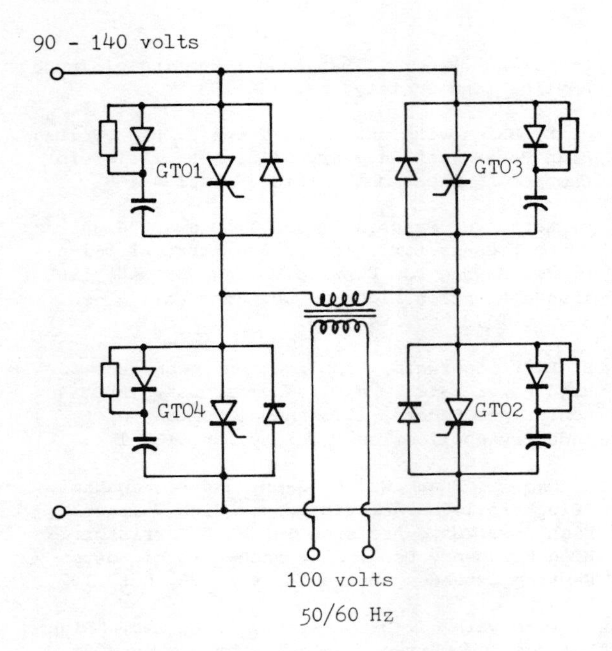

Fig. 19 5 kVA DC-AC inverter.

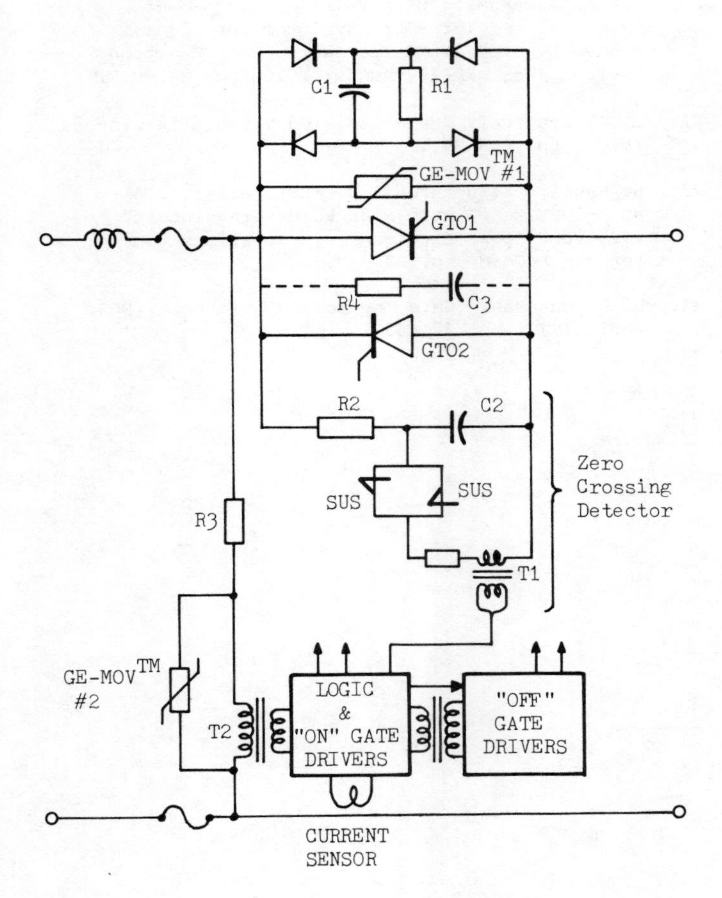

Fig. 20 GTO circuit breaker.[12]

AC circuit breaker. The simplified diagram of the circuit breaker shown in Fig. 20 was reported by Wolley et al.[12] Two GTO's are placed in reverse parallel. When the line current exceeds 200 amperes, the circuit is interrupted by turning off the GTO's. The short turn-off time of the GTO can interrupt the current in a sub-cycle.

Conclusion

Improved understanding of the design and operation has led to the successful developments of a new generation of the GTO. Recent studies have proved the feasibility of high power GTO's which have the rating of several hundred volts and can turn off 100 - 200 amperes in microseconds. Better understandings of the turn-off operation and of the switching performance of the devices have been obtained. These advancements will extend the usefulness of the GTO in power electronics applications, and will help the cost reduction of power converter equipment.

The practical application of GTO's began only recently, and the demand for them is expected to grow in the future. Although the high power GTO's are not commercially available at present, they will be ready for volume production if a sufficient demand is found.

Acknowledgement

The authors would like to thank Dr. E. Kobayashi, S. Kawai, K. Morita, and H. Ikeda for their guidance and encouragement in this work. T. Kamei and Dr. M. Naito are acknowledged for their helpful discussions of the device modeling and characteristics. H. Amano and K. Onda are also thanked for their comment on the device applications. The authors are indebted to J. Koizumi and Y. Kawaguchi for their support in the electrical measurements and to I. Sanpei for his part in the fabrication of the devices.

References

1. R. H. van Ligten and D. Navon, "Base Turn-Off of p-n-p-n switches", 1960 IRE WESCON Conv. Rec., pt. 3, pp. 49-52.

2. A. K. Jonscher, "Notes on the Theory of Four-Layer Semiconductor Switches", Solid-State Electronics, vol. 2, 1961, pp. 143-148.

3. J. M. Goldey, I. M. Mackintosh, and I. M. Ross, "Turn-Off Gain in p-n-p-n Triodes", Solid-State Electronics, vol. 3, September 1961, pp. 119-122.

4. T. A. Longo, M. Miller, A. L. Derek, and J. D. Eknaian, "Planar Epitaxial p-n-p-n Switch with Gate Turn-Off Gain", 1962 IRE WESCON Conv. Rec., pt. 3, pp. 1-6.

5. E. D. Wolley, "Gate Turn-Off in p-n-p-n Devices", IEEE Transactions on Electron Devices, vol. ED-13, July 1966, pp. 590-597.

6. T. C. New, W. D. Frobenius, T. J. Desmond, and D. R. Hamilton, "High Power Gate-Controlled Switch", IEEE Transactions on Electron Devices, vol. ED-17, September 1970, pp. 706-710.

7. M. Kurata, "A New CAD Model of a Gate Turn-Off Thyristor", 1974 IEEE Power Electronics Specialists Conference Record, pp. 125-133.

8. Y. C. Kao and J. B. Brewster, "A Description of the Turn-Off Performance of the Gate Controlled Switches", IEEE Conference Record, 1974 Annual Meeting of Industry Applications Society, pp. 689-693.

9. H. W. Becke and J. M. Neilson, "A New Approach to the Design of a Gate Turn-Off Thyristor", 1975 IEEE Power Electronics Specialists Conference Record, pp. 292-299.

10. Y. C. Kao and J. B. Brewster, "Recent Development in Gate Controlled Switches", 1973 IEEE Power Electronics Specialists Conference Record, pp. 90-96.

11. F. M. Matteson, R. L. Steigerwald, E. D. Wolley, and R. U. Yu, "200 Ampere Gate Controlled Switch", General Electric Co., Technical Report, AFAPL-TR-73-39, August 1973.

12. E. D. Wolley, F. M. Matteson, R. Yu, and R. L. Steigerwald, "Characteristics of a 200 Amp Gate Turn-Off Thyristor", IEEE Conference Record, 1973 Annual Meeting of Industry Applications Society, pp. 251-257.

13. K. S. Tarneja and J. Bartko, "Electron Irradiation - A Method of Producing Fast Switching High Power Diodes", 1975 IEEE Power Electronics Specialists Conference Record, pp. 269-272.

14. C. K. Chu, J. Bartko, and P. E. Felice, "Electron Radiated Fast Switch Power Thyristor", IEEE Conference Record, 1975 Annual Meeting of Industry Applications Society, pp. 180-183.

15. J. D. Balenovich and K. S. Tarneja,"High Voltage, High Power Fast Recovery Silicon Rectifier for Chopper Applications", ibid., pp. 184-188.

16. M. Naito, H. Matsuzaki, and T. Ogawa, "High Current Characteristics of Asymmetrical p-i-n Diodes Having Low Forward Voltage Drops", IEEE Transactions on Electron Devices, vol. ED-23, August 1976, pp. 945-949.

17. R. L. Steigerwald, "Application Techniques for High Power Gate Turn-Off Thyristors", IEEE Conference Record, 1975 Annual Meeting of Industry Applications Society, pp. 165-174.

18. H. Oka, S. Funakawa, H. Gamo, and A. Kawakami, "Electrical Characteristics of High Voltage High Power Gate-Assisted Turn-Off Thyristor for High Frequency Use", Electrochem. Soc. Boston Meeting Extended Abstracts, vol. 73, Oct. 1973.

19. T. P. Nowalk, J. B. Brewster, Y. C. Kao, "High Voltage and Current, Gate-Assisted, Turn-Off Thyristor Development", NASA CR-121161.

20. J. Shimizu, H. Oka, S. Funakawa, H. Gamo, T. Iida, and A. Kawakami, "High-Voltage High-Power Gate-Assisted Turn-Off Thyristor for High-Frequency Use", IEEE Transactions on Electron Devices, vol. ED-23, August 1976, pp. 883-887.

21. E. S. Schlegel, "Gate-Assisted Turnoff Thyristors", ibid., pp. 888-892.

22. S. Saeki, "Structures and Characteristics of 400A-300V Monolithic High Power Transistors", 1975 IEEE Power Electronics Specialists Conference Record, pp. 274-281.

23. R. H. Cushman, "Gate Controlled Switches", EDN, April 1970, pp. 37-43.

INVESTIGATIONS OF GATE TURN-OFF STRUCTURES*

H. W. Becke**, RCA Corporation
R. P. Misra, New Jersey Institute of Technology
Newark, New Jersey 07102

ABSTRACT

R & D was carried out on gate turn-off devices. 1. An optimized, high speed epi-base GTO was developed. Fall times of <200ns and rise times <400ns were simultaneously obtained for average chip current densities >50A/cm^2 @ T_j=125°C. The fast t_r and t_f response was obtained through a controlled gold distribution in the active device volume. A voltage source with series inductance at the gate will establish safe turn-off conditions; a 1.6KW switching capability @ 50kHz is calculated for a chip of 0.15cm^2. 2. The introduction of anode shorts improves turn-off, however, turn-on sensitivity is substantially reduced. Replacing these shorts with Schottky barrier diodes fully restores the turn-on sensitivity. Devices with identical V_T and similar turn-off capability, ≃ 30A @ 125°C, show about an order of magnitude improvement in turn-on sensitivity @ -40°C if Schottky barriers are added. 3. A dynamic ballasting concept was introduced. Through resistive, edge metalized cathodes the operational range for GTO's was extended from -60°C for turn-on (I_{gt}=300µA) to +150°C for turn-off (J>55A/cm^2). The formation of high current density filaments is countered.

INTRODUCTION

To achieve high speed of the order of 10^{-7} seconds, and high current turn-off capability for GTO thyristors it is important to control the minority carrier lifetime in a low range (∿1µs) and to minimize the lateral gate resistance.[1, 2, 3] The introduction of anode shorts is also an effective way to improve turn-off.[1, 4] Further, proper circuit conditions must be chosen to assure a "safe" transition from the on-state to the off-state. Unfortunately, measures taken to obtain good turn-off, generally lead to adverse characteristics in the turn-on behavior and result in an increase of on-state voltage. By far of greatest concern, however, is the sensitivity of GTO's to high current density filament formation causing catastrophic failures during the turn-off phase. This is similar to reverse bias second breakdown in transistors. This paper summarizes work done to improve turn-off capability without compromising turn-on, on-state, and maintaining useful temperature range, as well as implementing safe turn-off.

*This work was done in partial fulfillment of the requirements for the degree of Doctor of Engineering Science at NJIT.

**Present address: Bell Laboratories, MH 2C-446
Murray Hill, New Jersey 07974

HIGH SPEED DEVICE

For a high speed device delay time and storage time are not as critical as are risetime, fall time, and the turn-off "tail", as long as one considers for example, an upper frequency limit of 100 kHz for a power switch. In this section we examine some of the aspects leading to optimum performance of simple four-layer GTO's. These devices feature a low resistivity near-uniform p-base, gold doping, and interdigitated cathodes. Geometries, profile, static data, and switching characteristics were presented previously.[2] We shall turn here to the test conditions and attempt to determine turn-off criteria.

Test Conditions and Turn-off Response

The input and output characteristics for this GTO were determined for an anode load configuration. The device is turned on by a standard square wave pulse. The turn-off, however, is accomplished by applying a negative pulse from a voltage source across a series inductance to the gate. An input current ramp is produced while the inductance is charged up. The discharge of the stored energy into the gate causes an added negative voltage spike which aids in extinguishing the "squeezed" electron-hole plasma. The actual time response for input and output voltages and corresponding currents during turn-off are shown in Figure 1. The gate current I_{gq} rises during the whole storage phase, while the anode current I_A remains constant. I_{gq} reaches a peak value of -2A at 1.5µsec where it intersects the anode current which is now falling rapidly (Figure 1). At the same instant (∿1.5µs) we see in Figure 1b the gate voltage V_{gq}, which remained positive during the storage phase, assuming the open circuit valve V_{goc} =-10V, indicating that the voltage across the inductance is zero. The decaying gate current now produces additional negative voltage until the cathode breakdown voltage of -23V is reached at about 1.6µs. At this instant the anode current is smaller than the gate current,' thus the cathode has shut off. Simultaneously with the fall of I_A the anode voltage V_A rises to the supply voltage of +200V (The overshoot of 100V is due to some inductance associated with the load resistor bank). The anode current tail is extremely small in magnitude (2% I_T) and short in duration (<0.5µs). Since I_{gq} is always larger than I_A in this phase, the cathode will not be retriggered.

Reprinted from *IEEE Int. Electron Devices Meet.*, 1980, pp. 649-653.

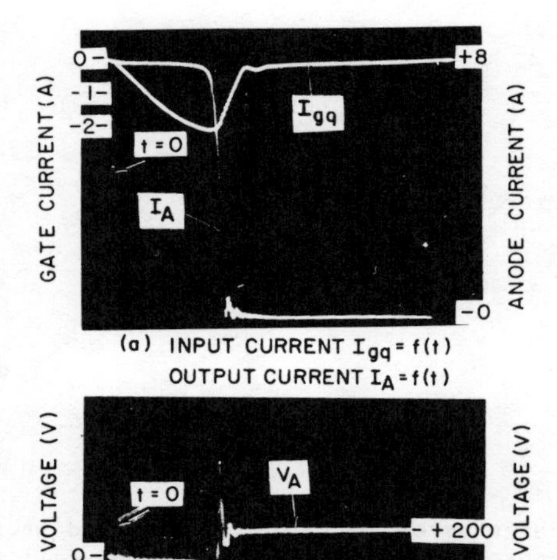

(a) INPUT CURRENT $I_{gq} = f(t)$
OUTPUT CURRENT $I_A = f(t)$

(b) INPUT VOLTAGE $V_{gq} = f(t) = 10V/DIV.$
OUTPUT VOLTAGE $V_A = f(t) = 50V/DIV.$

HORIZONTAL SCALE : $t = 0.5\,\mu s/DIV.$
VERTICAL SCALE : 1A/DIV.
GATE SERIES INDUCTANCES $L_{gs} = 4.6\,\mu H$
DEVICE 3 (4 MIL CATHODE)

92CS-29895

Figure 1. Turn-Off Characteristics for High-Frequency GTO-Thyristor (Example of Waveforms)

Turn-off Criteria

Figure 2 shows the turn-off for devices having the cathode width varied over an order of magnitude, from 2 mil to 20 mil, under identical off-drive and load conditions. Examining the I_A - traces during the fall period we detect that for devices 2, 3, and 4 the slope dI_A/dt increases up to the tail break point. Also, the maximum gate current is reached only after the anode current has substantially decreased. That does not hold for Device 1. Here the slope of I_A goes through an inflection at about 4A decreasing with time, i.e.

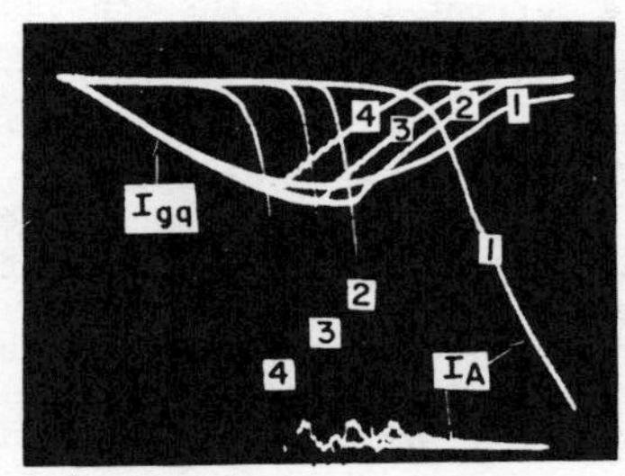

CATHODE WIDTH

1 = 20 MIL 3 = 4 MIL
2 = 8 MIL 4 = 2 MIL

$t = 0.2\,\mu s/DIV.$ T = 25°C
I = 1A/DIV. $V_{goc} = -15V$
$L_{gs} = 4.6\,\mu H$

Figure 2. Turn-Off Behavior of GTO-Thyristor with varying Cathode Width $[I_A, I_{gq}] = f(t)$ $W_K = L_X$ as Parameter

d^2I_A/dt^2 becomes negative. Also, the gate current peak occurs well before the anode current drops. A slight decrease in turn-off voltages or increase in temperature will cause failure to turn-off. The difference between safe turn-off and potential turn-off failure becomes rather clear if we plot the ratio of anode current to gate current dI_A/dI_{gq} against time during turn-off as is shown in Figure 3 using the turn-off waveforms of Figure 2. Normal turn-off without a gate series inductance would commence with infinity at t=0, where the gate current is zero and $I_A = I_T$, and then monotonically decrease to unity at the tail break, from which point anode current and gate current are identical. With the gate series inductance the differential turn-off gain becomes smaller than one and approaches unity as the gate current and anode current go to zero. This behavior constitutes a safe turn-off condition as is exhibited by devices 2, 3, and 4. Device 1 shows an increase of dI/dI_{gq} toward the end of the fall time signaling instability. The peak will grow or move out in time with a slight increase in local temperature or decrease in gate voltage.

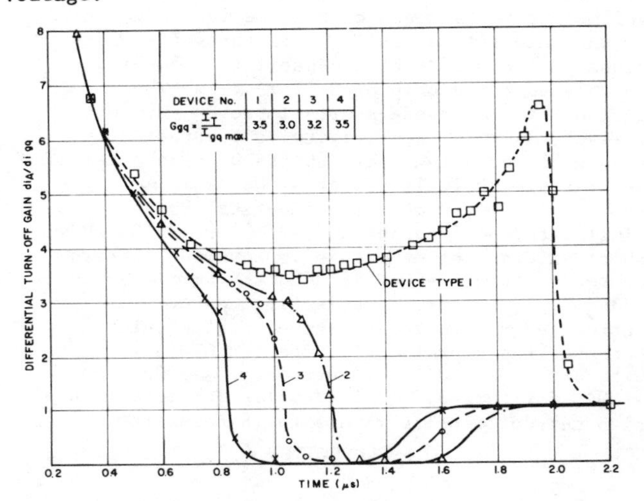

Figure 3. Differential Turn-Off Gain (dI_A/dI_{gq}) vs. Time for GTO Thyristors having different Cathode Widths under Identical Input and Output Conditions; showing safe Turn-Off for Device Types 2, 3, 4 (2, 4 and 8 mil cathode, respectively), and Indicating Onset of Turn-Off Failure for Device Type 1 (20 mil Cathode). V_D=200 V, I_T=8 A, V_{goc}=-15 V, L_{gs}=4.6 μH, T=25°C (for I_A, I_{gq}=f(t), See Figure 3

Switching Performance

Figure 4a shows the storage time t_s and Figure 4b the rise time t_r and fall time t_f for Device D_3 (4 mil cathode) in the temperature range from 25°C to 125°C with the open-circuit source voltage V_{goc} as the parameter. For values of V_{goc}>10V=the fall time remains well below 0.2μs and the tail current is less than 3% as noted before (Figure 2). It is noteworthy that simultaneously a fast rise time of ∿0.4μs was obtained and the gate trigger current of 50mA as well as the on-state voltage of 2.4V could be kept rather low. Usually, a device with such

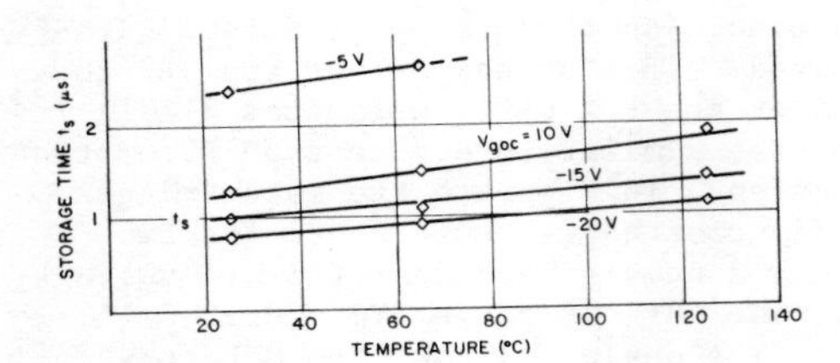

Figure 4a. Storage Time t_s=f(T)

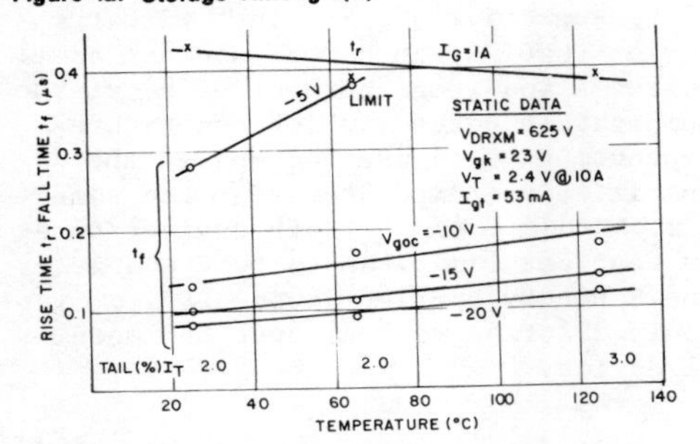

Figure 4b. Rise Time and Fall Time [t_r,t_f]=f(T)

Figure 4. Switching Performance of GTO-Thyristor as Function of Temperature. Device 3 (4 mil Cathode) Conditions: I_T=8 A=constant, V_D=200 V, L_{gs}=4.6 μH

excellent turn-off characteristics would have a rise time of ∿1.5μs, a gate trigger current of ∿300mA, and an on-state voltage of ∿4V. The reason for the desirable combination of fast t_r and t_f, and low I_{gt} and V_T can be found if we analyze the spreading resistance profile (see Figure 5 and 6).

Gold Distribution and Life Time

Figure 5 shows the vertical spreading resistance variation of the device taken through the p^+-gate toward the anode along $\overline{CD}$ (see insert). The center junction is located about 36μ from the surface. The resistivity throughout the n^--region is >1000Ωcm. The entire n-base is upconverted from its initial value of approximately 30Ωcm. Thus, we assume that the Au-concentration is about 1.5×10^{14} corresponding to a life time t_n ∿0.5μs in the n-base.[5] The p-base in this region will have an even lower life time value.

Now we turn to Figure 6 in which the spreading resistance is taken through the n^+ cathode, again vertically toward the anode along pass $\overline{AB}$. The interesting fact here is that in the vicinity of the center junction, the starting resistivity is preserved (30Ωcm) and upconversion becomes noticeable only about 50μ into the n-base with a pronounced increase in direction of the anode. The difference in this spreading resistance profile from that of Figure 5 suggests that Au was gettered by the heavily doped phosphorus emitter above this region. As a consequence the active gold distribution in the device is such that the life time is kept high (>1.0μs) in the p-base and in the n-base adjacent to the center junction while it is low (∿0.5μs) in the vicinity of the anode. The high life time keeps the npn gain reasonably high and accounts for the low I_{gt} and the fast t_r. The low life time limits the anode emitter efficiency and also causes

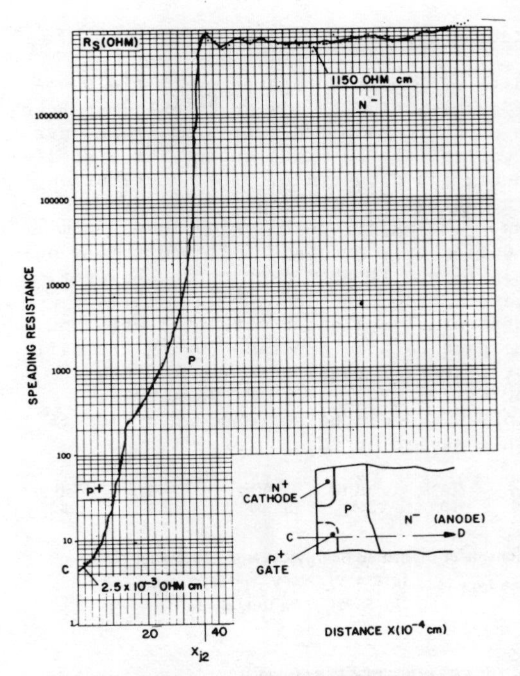

Figure 5. Two-Point Probe Spreading Resistance Profile (C-D) through P⁺ Gate showing Total Up Conversion of N⁻ Region. (Insert shows Schematic of Cross Section and Probe Travel Pass)

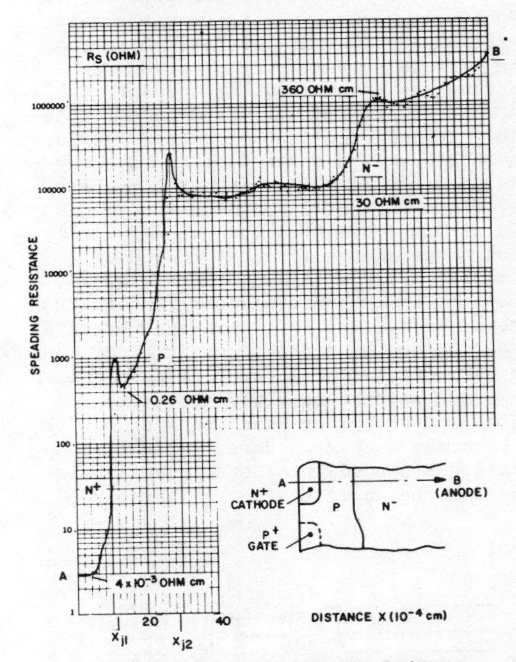

Figure 6. Two-Point Probe Spreading Resistance Profile (A-B) through N⁺ Cathode Showing Partial Up Conversion of N⁻ Region in Direction of P⁺ Anode (Insert shows Schematic of Cross Section and Probe Travel Pass)

a rapid recombination of the stored charge in the n-base, the majority of which is located close to the anode junction.[6] Thus, the tail-effect is minimized.

Power Switching Capability

The data of Figure 4 together with the waveforms of Figures 1 and 2 allow us to calculate the components of the dissipated switching energy per cycle, as well as the on-state energy for a specified frequency and duty cycle. The distribution is shown in Table 1 for the conditions given in the caption. The major contribution comes from the anode during t_{f1}, and t_{f2}, followed by the on-state dissipation. It should be noted, however, that the gate dissipation is not negligible because substantial avalanche current is flowing through the gate cathode junction during the tail phase. For a thermal resistance $\Theta = 1.0°C/Watt$ and $\Delta T = 50°C$; i.e. $T_{case} = 75°C$, an average power of 1550 Watts can be switched at a rate of 50 kHz with a device efficiency of 97%.

$E_{A(t_r)}$	$E_{G(t_r)}$	$E^{*)}(on)$	$E_{A(t_{f1})}$	$E_{A(t_{f2})}$	$E_{G(t_{f1})}$	$E_{G(t_{f2})}$
16	3	188	640	96	2.3	25.7

Table 1. Components of Dissipated Energy/Cycle in μJoules for GTO (Device Type 3). I_A=8 A, V_D=400 V, T_j=125° C, Ind. Load, *) f=50 kHz, 50% Duty Cycle

SERIES SCHOTTKY BARRIER

An effective method of improving turn-off characteristics of GTO's is to introduce anode shorts.[1,4] If these shorts are dimensioned properly and are placed most remote from the gate contact, the excess minority carrier charge in the n-base will be partially removed through those shorting resistances during turn-off. Also, the squeezed plasma will tend to be extinguished in this non-regenerative center section. The anode shorts may be thought of as being by-pass transistors in parallel with the thyristor. Therefore, while turn-off is aided, turn-on is impaired. It was found that this is especially true at low temperatures, i.e. -40°C. A Schottky diode having a similar barrier height as the adjacent pn anode junction does restore the turn-on sensitivity, while the non-regenerative property of the center region, required for effective turn-off, is preserved. Figure 7 is a schematic cross section of a GTO structure featuring an integrated by-pass transistor with a series Schottky barrier. Here, lateral bias currents are not needed in order to cause hole injection from the anode emitter. Therefore, the

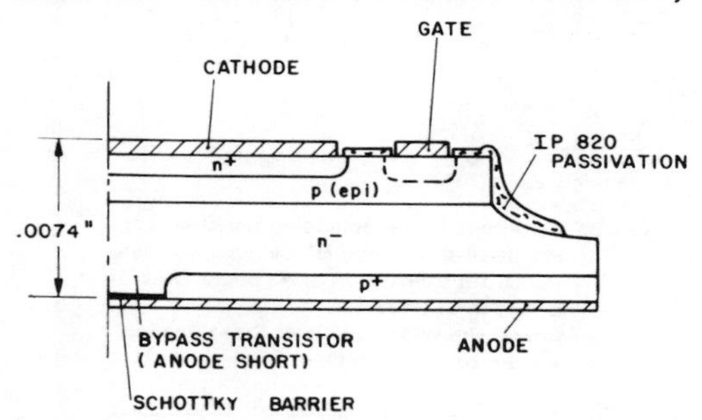

Figure 7. GTO Structure showing Anode Short (By-Pass Transistor) and Series Schottky Barrier aligned with Center of Cathode

restriction placed on the degree of interdigitation is removed. This principle was applied to the geometry of the RCA G400, which has 8-mil cathode fingers, similar to device type D2 of the previous section. In Figure 8 the gate trigger current I_{gt} is plotted vs temperature for two devices having a nearly identical forward voltage drop over a wide current range (V_T = 0.9V @ 1A, 1.61V@ 30A). The device without the Schottky barrier (G400-1) has an I_{gt} of 2.4mA and 65mA @ 25°C and -40°C, respectively. The unit with the Schottky barrier (G400-8) in comparison has an I_{gt} of only 1.0 mA and 10mA, respectively, at the identical temperature points, which constitutes a drastic improvement. Figure 9 compares the turn-off capability vs temperature for the same devices for a cathode load. Both devices turn off 40A (the test set limit) up to 90°C and at T = 125°C the Schottky barrier device $I_{A(max)}$ is reduced by only 25% (25A vs 33A) over the more conventional device.

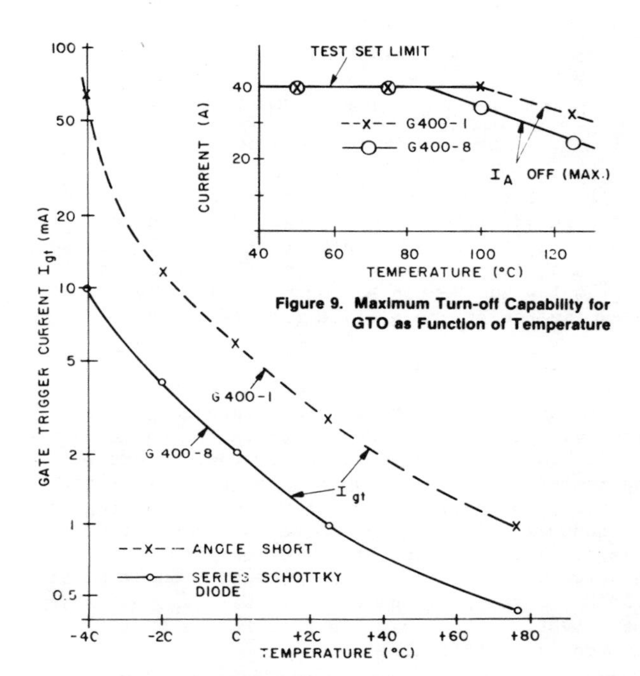

Figure 9. Maximum Turn-off Capability for GTO as Function of Temperature

Figure 8. Turn-on Sensitivity as Function of Temperature for GTO with Anode Short (dashed line) and Series Schottky Diode (solid line)

DYNAMIC BALLASTING (DEFOCUSING)

Figure 10 is a schematic cross section of an epitaxial GTO structure in which the standard, high concentration phosphorus cathode is replaced with a resistive, ion-implanted layer. Also, the center of the cathode is insulated, only the edge is contacted with the metal electrode. Therefore, for turn-on and the on-state while modest current densities exist in the device, we should expect normal thyristor behavior; i.e., fast turn-on with high sensitivity and low forward voltage drop. In the process of turn-off, however, when the electron-hole plasma is being squeezed toward the center [7, 8, 1] the high current

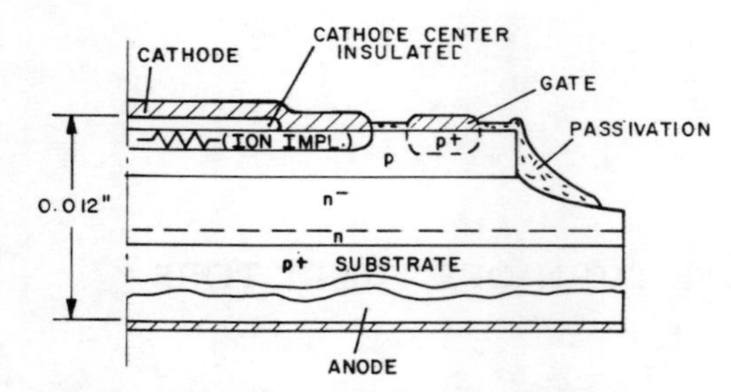

Figure 10. Schematic of Epi-GTO featuring Resistive Cathode with Isolated Center to achieve Dynamic Ballasting (Defocusing)

densities which locally develop will cause lateral voltage drops in the cathode and debias the gate-cathode junction counteracting the formation of filaments. The effect of this dynamic ballasting or defocusing mechanism can be observed in Figure 11 and results are given in Table 2. The upper traces of Figure 11 reveal a long turn-off (∿15µs) and considerable tailing (∿30µs) for this device having a standard cathode at room temperature (Table 2, line 2, col. 4). The lower traces show a remarkable improvement (t_{off} ∿4µs) and the tail is negligible even at 125°C (Table 2, line 4, col. 5). It was possible by further optimization to extend the temperature range for sensitive turn-on down to -60°C and to turn off a maximum current of 8.5A @ 150°C (Table 2, data line 5).

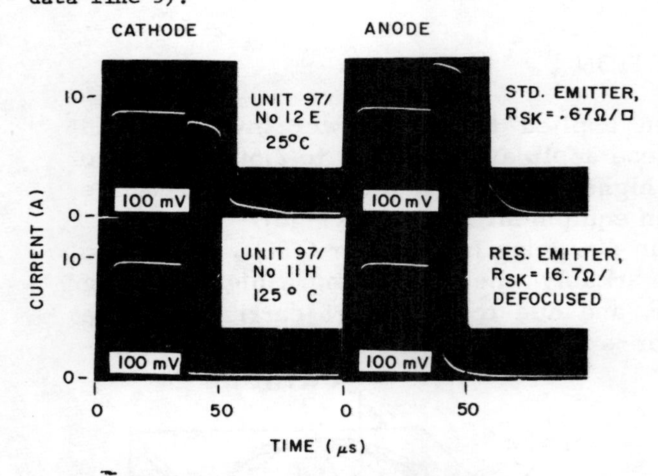

Figure 11. Cathode Load Response I_K=f(t) (Left), and Anode Current I_A=f(t) (Right) for Epitaxial (N)PN⁻ N(P⁺) GTO-Thyristor with Std. Diffused Cathode Emitter (Top) and Resistive, Defocused Cathode Emitter (bottom) Vert.=2 A/Div, Hor.=10 µs/Div

CONCLUSIONS

High speed-high voltage GTO's are feasible. Low t_r and t_f values are simultaneously obtained by a controlled gold distribution resulting in a high life time near the cathode and a low life time around the anode. - Safe turn-off is aided in using for the gate input a voltage source and series inductance. - Anode shorts improve turn-off but adversely effect turn-on. Series

Geometry	I_{gt} (mA)		V_T (V)	Max. $I_{turn-off}$ (A)	
Std. G400 R_{sk}∼.7 Ω/□	12.0	1.8	1.15	25	12
D3, PN⁻N⁺ Epi R_{sk}∼.7 Ω/□	.7	.06	1.22	9	2
D3, PN⁻N⁺ Epi R_{sk}∼20 Ω/□	.75	.10	1.48	15	13
D3, PN⁻N⁺ Epi R_{sk}∼20 Ω/□ + Defocusing	1.3	.22	1.53	18	14
	-40°C	+25°C	+25°C	+25°C	+125°C
D3, PN⁻N⁺ Epi R_{sk}∼7 Ω/□ + Defocusing	.3	.06	1.28	9.3	8.5
	-60°C	+25°C	+25°C	+25°C	+150°C

Table 2. Effect of Resistive Cathode and Dynamic Ballasting (Defocusing) on GTO Temperature Behavior.

Schottky barriers will restore useful turn-on while the turn-off improvement is preserved. - Resistively ballasted cathodes with metallic contact at only the periphery will drastically widen the operational range and improve safe turn-off for GTO's by counteracting the formation of high current density filaments.

ACKNOWLEDGEMENT

The authors acknowledge the assistance of Mr. J. Dean in fabricating the devices. Thanks to Mr. G. Guenther and Mr. E. McKeon for their help with the measurements. Messrs. J. Neilson, R. Amantea, and R. Martinelli are thanked for many valuable discussions as are Messrs. D. Burke, R. Denning, and H. Kressel for their suggestions and support.

REFERENCES

1. H. W. Becke and J. M. Neilson "A New Approach to the Design of a Gate Turn-Off Thyristor", 1975 PESC, pp 292-299.
2. H. W. Becke "A High-Speed High-Voltage Epi Base GTO" 1977 IEDM, pp 46A-46D
3. M. Azuma, K. Takigami, M. Kurata "2500V, 600A Gate Turn-Off Thyristor (GTO)", 1979 IEDM, pp 246-249.
4. T. Nagano, M. Okumara, T. Owaga "A High Power, Low-Forward-Drop Gate Turn-Off Thyristor" 1978 IAS, pp 1003-1006.
5. W. M. Bullis "Properties of Gold in Silicon" SSE, Vol. 9, 1966, pp 143.
6. M. Naito, T. Nagano, H. Fukui, Y. Terasawa "One-dimensional Analysis of Turn-Off Phenomena for a Gate Turn-Off Thyristor." IEEE Trans. ED, VOL. 26, 1979, pp 226-231.
7. E. D. Wolley, "Gate Turn-Off in pnpn Devices" IEEE Trans. ED, 13, 1966, pp 590-597.
8. M. Kurata "A New CAD Model of a Gate Turn-Off Thyristor" 1974 PESC, pp 125-133.

ULTRA HIGH VOLTAGE, HIGH CURRENT GATE TURN-OFF THYRISTORS

Tsutomu YATSUO, Takahiro NAGANO, Hiroshi FUKUI, Masahiro OKAMURA
Hitachi Research Laboratory, Hitachi, Ltd.
AND Shuroku SAKURADA
Hitachi Works, Hitachi, Ltd.
Hitachi, Ibaraki, Japan

Abstract

A high power GTOs with ratings of 2,500 V · 2,000 A has been developed, and 4,500 V · 2,000 A GTO trial fabricated and performance tested, for use in traction motor control equipment. Their low on-state voltage was attained by applying a unique anode emitter shorting structure which does not require doping of a lifetime killer such as gold to obtain suitable GTO characteristics. Their high interrupt current was obtained by introducing a ring shaped gate structure which has uniform operation between many segments in the devices during turn-off process.

1. INTRODUCTION

Gate turn-off thyristors (GTOs) have been applied to high power inverters and choppers for traction motor drive.[1),2),3)] For these applications, 1,200 to 2,500 V, 600 to 1,000 A GTOs were developed.[4),5),6)] However, higher current, higher voltage GTOs are needed in order to realize substantial reductions in equipment size and weight.

There are two major factors to be resolved in designing high power GTOs. One is to reduce the on-state voltage, which increases remarkably when producing a high blocking voltage due to the inherently thick n-base layer, and due to the short carrier lifetime required in conventional GTOs. The second factor is to increase the maximum gate turn-off current without causing device failure.

To resolve these two points, the basic structure consisting of an anode emitter shorting was introduced and its structural parameters were optimized by experimental fabrications and characterizations of small size test samples. As a result, a high voltage, high current GTO with the rating of 2,500 V · 2,000 A was developed, and a 4,500 V · 2,000 A GTO trial fabricated and performance tested.

In this paper, various design considerations, device structures and electrical characteristics of the newly developed high power GTOs are described.

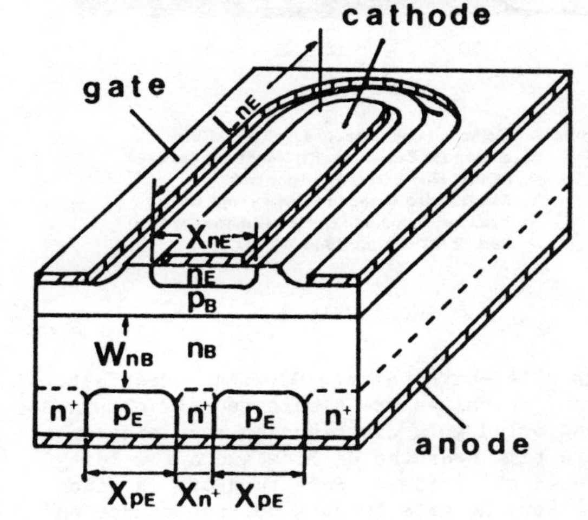

Fig. 1. Schematic structure of the unit-GTO with anode emitter shorting.

Reprinted with permission from *Int. Power Electron. Conf.*, 1983, pp. 65–74.

2. DESIGN CONSIDERATIONS

2.1 Typical GTO Characteristics and the Shorted Emitter Structure

In general, GTO has a multi-emitter structure consisting of a plurality of narrow emitter strip surrounded by the gate electrode, which is called a unit-GTO. It is easy for the gate current to drive out electrons and holes from the current conducting area during the turn-off process.

Figure 1 shows a schematic structure of the unit-GTO with anode emitter shorting.[7] This structure dose not require a particular reduction in carrier lifetime to obtain good turn-off performance, because p-emitter shorting could reduce the carrier concentration in the base layer at the current conducting state, and have an effect of sweeping out carriers during the turn-off phase, similar to a reduction of carrier lifetime by doping with lifetime killer such as gold.[8] Thus it is easier here to decrease the on-state voltage than in a conventional GTO structure.

From a design viewpoint, it was important to estimate the electrical characteristics of the unit-GTO. The relations between turn-off characteristics, as well as on-state voltage and the structural parameters of the unit-GTO were investigated by using a group of small-size test samples.

Figure 2 shows examples of turn-off waveforms for the unit-GTO with different p-emitter shorting structures. The samples consisted of a single n-emitter strip with constant width and length of 0.3 and 5.8mm, respectively, and with various p-emitter shorting factors X_{n^+}/X_{pE}. It can be seen that turn-off time t_{GQ} and the tailing part of the anode current are strongly related to the shorting factor X_{n^+}/X_{pE}.

Figure 3 shows the effect of p-emitter shorting structure on turn-off time t_{GQ} and a turn-off gate charge Q_{GQ}, where Q_{GQ} was defined as the gate current integral during the turn-off time. The sample having $X_{n^+}/X_{pE} = 0$ was a conventional structure without p-emitter shorting. As is evident from Fig. 3, both t_{GQ} and Q_{GQ} decrease with an increase in the value of X_{n^+}/X_{pE}, and a saturating tendency can be seen at higher values of X_{n^+}/X_{pE}.

Figure 4 shows anode currents as a function of time in a late stage of the gate turn-off period for the various unit-GTOs. In the tailing part of the anode current, I_A decays linerly the semi-logarithm plots, so I_A can be expressed as the following relation, $I_A \propto e^{-Rt}$, where R represents a shorting resistance which decreases with an increase in

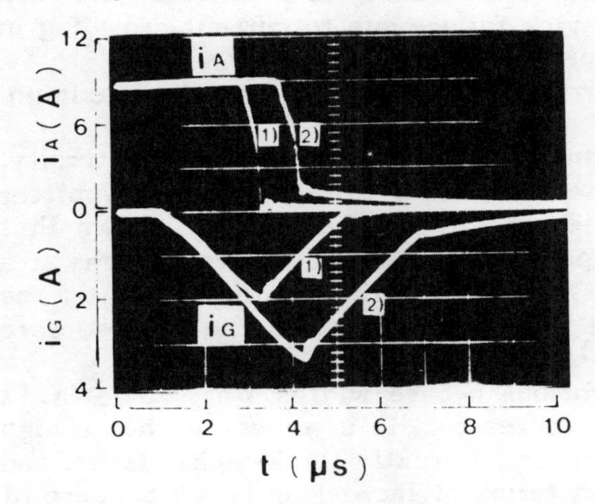

Fig. 2. Turn-off waveforms for the unit-GTOs with different p-emitter shorting structures.

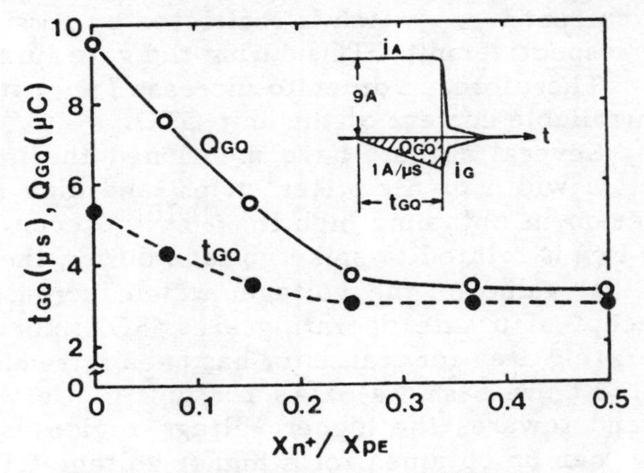

Fig. 3. Effect of shorting factor X_{n^+}/X_{pE} on the turn-off time t_{GQ} and turn-off gate charge Q_{GQ} of the unit-GTOs.

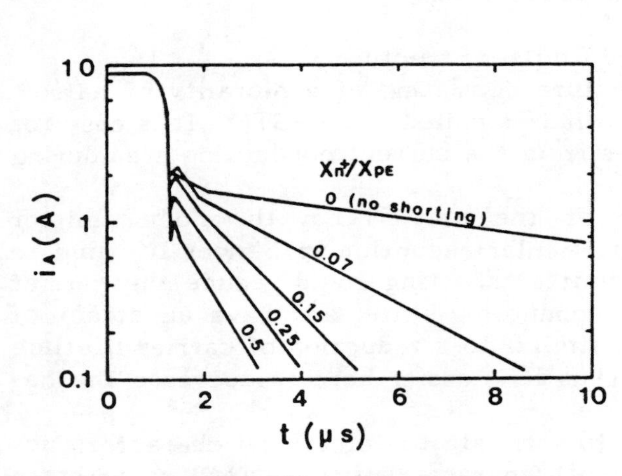

Fig. 4. Anode currents as a function of time in a late stage of the gate turn-off process.

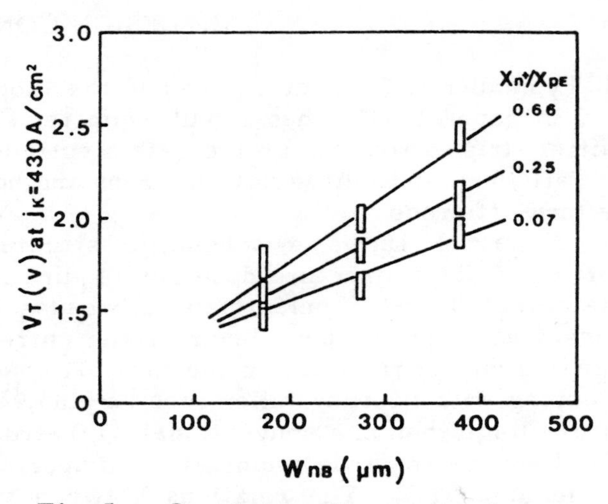

Fig. 5. On-state voltage vs. n-base width W_{nB} for various shorting factors.

X_{n^+}/X_{pE}. From the results, it is seen that power dissipation in the turn off process can be reduced by an increase in the p-emitter shorting of an actual GTO device.

Figure 5 plots the on-state voltage V_T versus n-base width W_{nB} for various shorting factors. The V_T increases with an increase in W_{nB} and with the shorting factor X_{n^+}/X_{pE}. A lowering of V_T below 2.5 volts can be obtained by optimizing the p-emitter shorting structure, even for such thick n-base widths as 0.4 - 0.5 mm or more.

Summarizing the above investigation, it was found that unit-GTO characteristics, such as on-state voltage, turn-off speed and gain, and turn-off power dissipation, depend primarily on the p-emitter shorting structure, and the degree of the shorting can be controlled by the p-emitter width X_{pE} and the n^+-shorting width X_{n^+}.

2.2 Maximum Gate Turn-Off Currents and GTO Structure

2.2.1 Safe operating area of unit-GTOs

In designing high current GTOs, it is important to increase the maximum gate turn off current I_{TCM}, witch is restricted by possible device failure due to current crowding in the respective unit-GTOs during the gate turn-off process.

Therefore, in order to increase I_{TCM}, it is first necessary to increase the maximum controllable current of the unit-GTO.

Several authors have mentioned the importance of a low p-base sheet resistiveity, narrow width of n-emitter strips, and high reverse breakdown voltage of the n-emitter junction in obtaining high I_{TCM}.[9),10)] Recent studies on turn-off failure have shown that the I_{TCM} is related to spike voltage during the fall period,[10)] and that failure happens at a critical value of the voltage, which increases in W_{nB}. According to these results, the concept of a safe operating area (SOA) for GTOs, which is similar to the reverse safe operating area for transistor has been introduced.[11)]

Figure 6 shows SOAs for unit-GTOs with various n-base widths W_{nB}. The SOAs extend towards the higher voltage region as W_{nB} increases. It is expected that a high I_{TCM} can be obtained for a higher voltage GTO with an inherently thick n-base layer, and that the p-emitter shorting structure is superior in terms of increasing I_{TCM}, because of the lower on-state voltage.

Figure 7 shows the effect of the number of unit-GTOs N_{nE}, in parallel operation, on SOA. The SOAs extend towards the higher current region as N_{nE} increases, however, a saturating tendency is seen and the voltage limit does not increase in the lower current region. Therefore, only an increase in the number of unit-GTOs can not lead to hgiher I_{TCM} values.

2.2.2 Arrangement of the unit-GTOs

In designing a large area GTO device, it is important not only to increase the maximum controllable current of the unit-GTOs, but also to provide operation which is as uniform as possible in all of the unit-GTOs during the turn-off process. Uniformity of turn-off currents in the device are strongly affected by unbalance in lateral conductance of the gate electrode on the p-base layer, between the contact of the external gate lead and the respective unit-GTOs.

Figure 8 shows experimental results for arrangement effects on I_{TCM} in the unit-GTOs. All of the sample GTOs had the same diameter of 40 mm, but different gate structures as illustrated in the figure. The ring shaped gate sample shows a higher interrupt current than that of the center gate sample, which is probably caused by

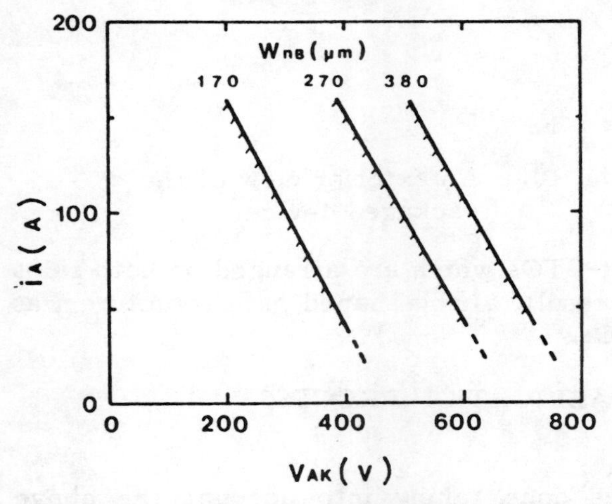

Fig. 6. SOAs for unit-GTOs with various n-base widths.

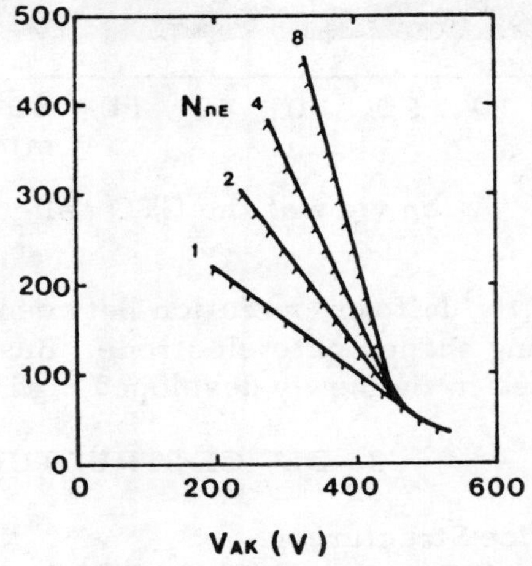

Fig. 7. Effects on SOA by the number N_{NE} of unit-GTOs in parallel operation.

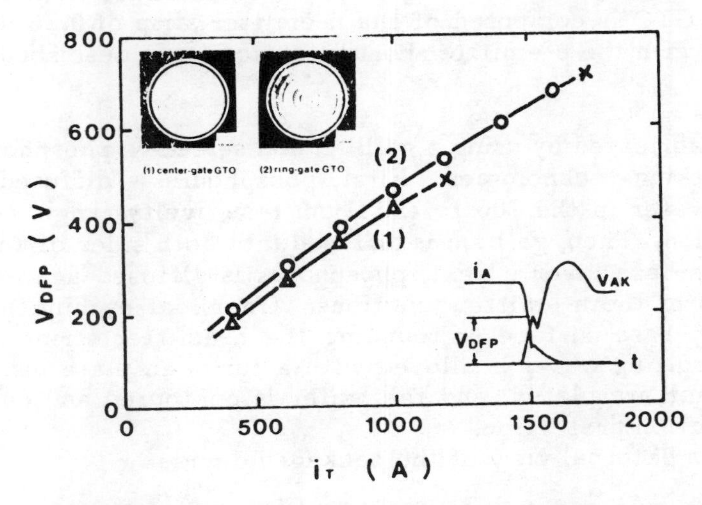

Fig. 8. Comparision of the maximum interrupt current I_{TCM} and of spike voltage V_{DFP} for the different gate structures.

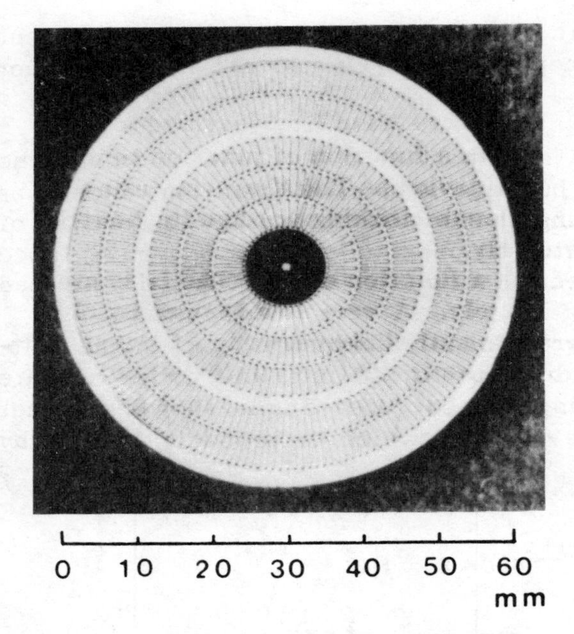

Fig. 9. A top view of the GTO pellet.

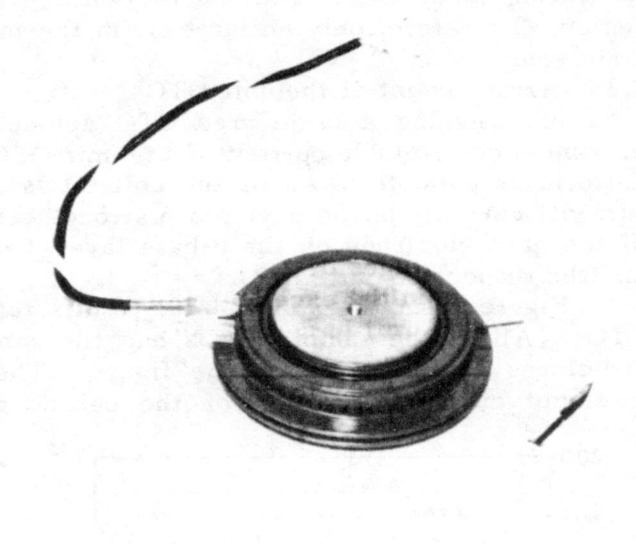

Fig. 10. An exterior view of the packaged device.

sufficiently uniform operation between many unit-GTOs which are arranged on both sides of the ring shaped gate electrode. Based on the result, a ring shaped gate structure was introduced in the newly developed high power GTOs.

3. DEVICE STRUCTURE AND FABRICATION PROCESS

3.1 Device Structure

Design of the 2,000 A GTO structure was done taking into account the above mentioned considerations. Figure 9 shows a top view of the GTO pellet which has a 60 mm diameter. Many unit-GTOs are arranged in five concentric circles on the pellet, and the external gate electrode, which has a ring shape makes contact on the internal circuler area. Eath of the unit-GTO is composed of the n-emitter strip of 0.26 mm in width and 4 mm in length, provided with the p-emitter shorting structure as described in Fig. 1.

3.2 Fabrication Process

The devices are fabricated by using a gallium and selective phosphorus diffusion, and conventional photo masking technologies. First, phosphours is diffused selectively into the anode side of the wafer in the 200 to 250 Ω·cm resistiveity range, to form the anode emitter shorting n^+ region. Then, gallium is diffused into both sides of the wafer, to form the p-emitter and the p-base layer. Next, phosphorus is diffused again into the cathode side of the wafer to form the n-emitter junctions. Chemical etching is applied to etch down the part of the p-base surface surrounding the n-emitter strips to form the gate contacts. After the resulting wafer is alloyed with a tungsten plate using aluminum, the cathode and gate contact are plated, and the wafer is contoured and covered by silicone rubber for the main junction protection.

Figure 10 shows an external view of the packaged device.

4. ELECTRICAL CHARACTERISTICS

4.1 Static Characteristics

4.1.1 Voltage blocking capability

Figure 11 plots forward blocking voltage V_{DRM} as a function of junction temperature T_J for the trial fabricated 2,500 V and 4,500 V GTO devices. The V_{DRM} is defined as the applied voltage at a leakage current of 20 mA. The peak blocking voltage is about 3,000 V and 5,200 V at 125°C, for the two devices respectively.

Figure 12 shows forward leakage current I_R as a function of T_J. The I_R is measured at applied voltages of 2,500 V and 4,500 V respectively. A very low I_R can be obtained, which results in the excellent blocking characteristics stated above.

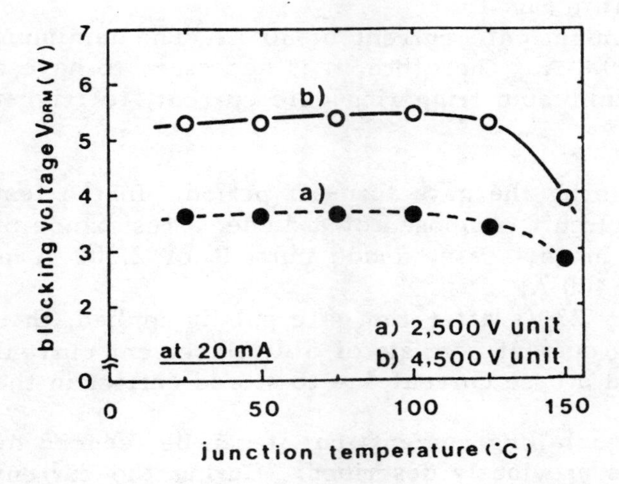

Fig. 11. Forward blocking voltage V_{DRM} vs. junction temperature.

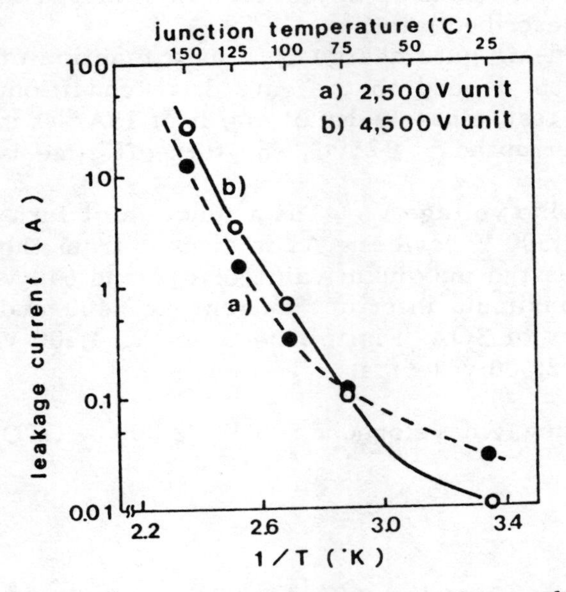

Fig. 12. Leakage current at forward blocking state.

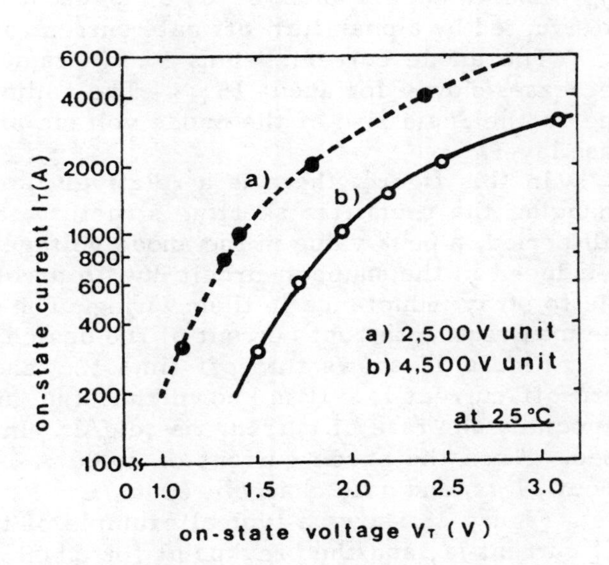

Fig. 13. Forward currnt conducting characteristics.

4.1.2 On-state voltage

Figure 13 shows the forward conducting characteristics, V_T vs. I_T, at the junction temperature of 25°C. At a current of 2,000 A, V_T is less than 1.8 V and 2.5 V, for the devices respectively. These values are lower compared with those of gold doped type GTOs.

From above results, advantages of the p-emitter shorted GTO, such as low on-state voltage and high blocking voltage at high temperature are comfirmed.

4.2 Dynamic Characteristics

4.2.1 Turn-on responce

Figure 14 shows typical turn-on waveforms of the 2,500 V · 2,000 A GTO. In this case, it took a relatively long time, about 6 μs, to turn-on. Measured turn-on time t_{GT} is plotted as a function of triggering gate current in Fig. 15.

The t_{GT} decreases to 4 μs at the maximum gate current of 30 A. The minimum triggering gate current of the GTO is about 0.8 A. Therefore, it is necessary to have a large gate current as much as 40 times the minimum triggering gate current, to trigger the GTO in a sufficiently short time.

4.2.2 Turn-off characteristics

Figure 16 shows turn-off waveforms during the gate turn-off period. In the test circuit, the load is inductive, and a snubber circuit composed of a diode, a resistance of 10 Ω and a capacitance of 6 μF is used. In this case, anode current of 2,000 A is interrupted by a peak turn-off gate current of 400 A.

The anode current begins to fall rapidly 23 μs after the gate puls is applied, then decreases slowly for about 15 μs. The tailing current consists of a displacement current due to the rapid rise of the anode voltage and a base current due to stored carrier in the base layer.

In this figure, there is a relatively large tailing current, but it can be reduced by changing the p-emitter shorting structure, as previously described. During the current fall period, a peak value in the anode voltage can be seen. This is the spike voltage, which is induced in the snubber circuit due to a voltage drop in the diode and capacitance, and due to stray inductance in their wiring. The spike voltage is an important value related to the maximum interrupt current of the device, as described later.

Figure 17 shows turn-off time t_{GQ} and turn-off gate charge Q_{GQ} as a function of turn-off current I_T. It is known that t_{GQ} and Q_{GQ} depend on the gate drive condition, especially the rate of current rise dig/dt. In this test, a gate pulse of dig/dt of 15A/μs is used. When the anode current of 2,000 A is interrupted at 125°C, the turn-off time is about 29 μs, and gate charge is 6000 μC.

Figure 18 shows a typical example of the spike voltage V_{DFP} as a function of turn-off current I_T, and the breakpoint for 2,500 and 4,500 V devices. As is evident from the figure, V_{DFP} increases with increasing I_T, and has the maximum value of 610 and 640 V obtained at the breakpoint, respectively. The maximum interrupt current is 2,400 and 2,600 A. This result agrees well with the results of SOA (Fig. 6), because the 4,500 V device has a thicker n-base width than that of the 2,500 V device.

4.2.3 Ratings and typical characteristics

Ratings and typical characteristics of the newly developed 2,500 V · 2,000 A GTO are described in Table 1.

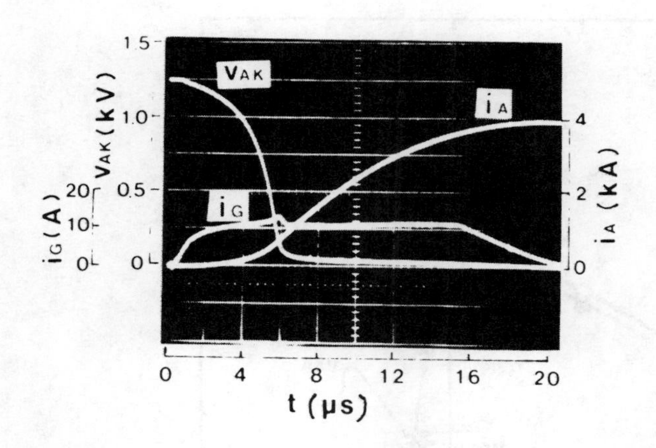

Fig. 14. Typical turn-on responses (T_j = 25°C).

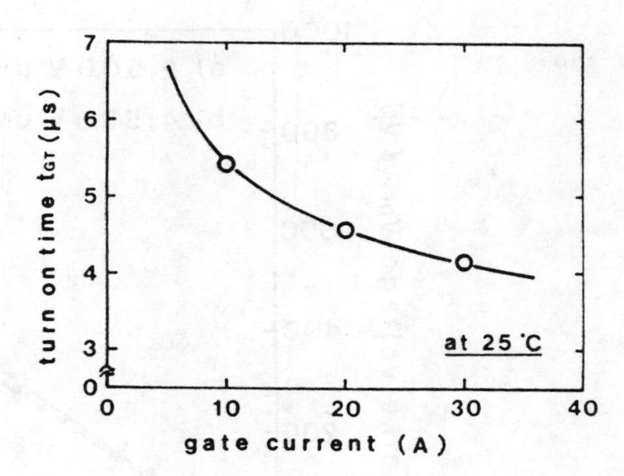

Fig. 15. Turn-on time t_{GT} vs. gate current I_G.

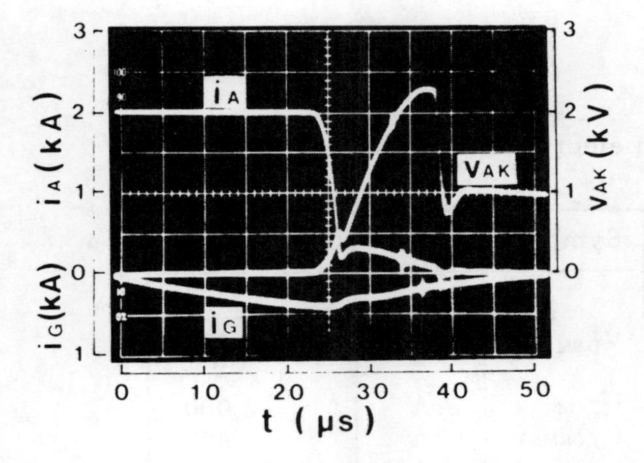

Fig. 16. Waveforms at 2,000 A turn-off (T_j = 125°C).

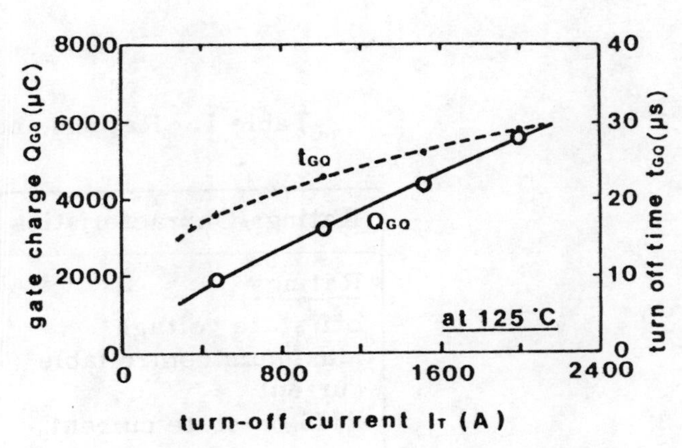

Fig. 17. Turn-off time t_{GQ}, turn-off gate charge Q_{GQ} vs. turn-off current.

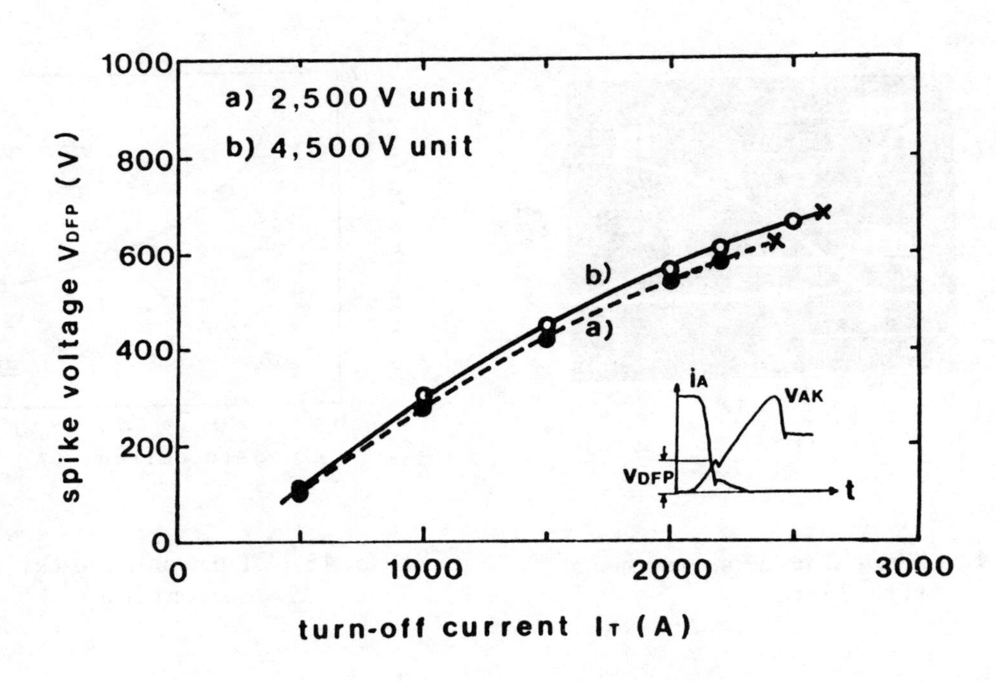

Fig. 18. Spike voltage V_{DFP} as a function of turn-off current.

Table 1. Ratings and electrical characteristics

Ratings/Characteristics	Symbol	Units	GFP2000B25
Ratings			
off-state voltage	V_{DRM}	V	2,500
maximum controllable current	I_{TCM}	A	2,000
RMS on-state current	$I_{T(RMS)}$	A	800
surge on-state current	I_{TSM}	A	14,000
gate reverse voltage	V_{GRM}	V	16
junction temperature	T_J	°C	-40 to 125
Characteristics (typical value)			
on-state voltage	V_{TM}	V	1.8
gate trigger current	I_{GT}	A	0.8
turn-on time	t_{GT}	μs	6
turn-off time	t_{GQ}	μs	30
turn-off gate charge	Q_{GQ}	μC	6,000
thermal impedance	Rth (j-f)	°C/W	0.02

5. CONCLUSION

Using a unique anode emitter shorting structure, which shows good switching performance without particular lifetime control, a high voltage and high current GTOs with ratings of 2,500 V · 2,000 A was developed, and 4,500 V · 2,000 A GTO was trial fabricated. In these devices, electrical charcteristics such as on-state voltage, turn-off time and gain, and switching power dissipation depended upon the degree of the p-emiter shorting structure which could be controlled mainly by the width of the p-emitter and n^+-shorting layer. The new devices had excellent forward blocking characteristics, low on-state voltage, less than 2.5 volts and a peak higher allowable gate turn-off current, 2,400A.

Acknowledgements

The authors wish to express their sincere thanks to Dr. T. Takasuna, Messrs. T. Tsuboi, H. Kawakami for their continuing guidance and encouragement, and Messrs. H. Amano, Y. Ikeda and A. Ueda for many valuable discussions. They also wish to acknowledge the significant roles played by Messrs. S. Oikawa, I. Sampei and S. Okano in the fabrication of the device, and by Messrs. S. Kimura, M. Sato, Y. Sato and K. Koga in the electrical measurments.

References

1) A.Ueda et al. "GTO Inverter for AC Traction Drives" IEEE - IAS '82 Annual meeting Rec. p.645 (1982-10)
2) M.Ohta et al. "Gata Turn-off Thyristor Application Technology for Rolling Stock" IEEE - IAS '82 Annual Meeting Rec. p.239 (1982-10)
3) T.Kanzaki et al. "Inverter Control system for Driving Induction Motors in Papid Transit cars Using High Power Gate Turn-Off Thyristors" IEEE - ISPCC '82 Meeting Rec. p.145 (1982)
4) T.Nagano, T.Yatsuo and M.Okamura "Characteristics of a 3000 V, 1000 A Gate Turn-off Thyristor" IEEE - IAS '81 Annual Meeting Rec. p.750 (1981)
5) M.Azuma, M.Kurata, and K.Takigami "2500-V 600-A Gate Turn-Off Thyristor (GTO)" IEEE Trans. Electron Devices, ED-29, p.270 (1981)
6) A.Tada and H.Hagion "A High-Voltage, High-Power, Fast-Switching Gate Turn-Off Thyristor" IEEE - ISPCC '82 Meeting Rec. p.66 (1982)
7) T.Nagano, M.Okamura, and T.Ogawa "A High-Power, Low-Forward-Drop Gate Turn-Off Thyrsitor" IEEE - IAS '78 Annual Meeting Rec., p.1,003 (1978)
8) Y.Shimizu et al. "Numerical Analysis of Turn-Off Characteristics for a Gate Turn-Off Thyristor with a Shorted Anode Emitter" IEEE Trans. Electron Devices, ED-28, p.1,043 (1981)
9) E.D. Wolley "Gate Turn-Off in p-n-p-n Devices" IEEE Trans. Electron Devices ED-13, p.590 (1966)
10) H.Ohashi and A.Nakagawa "A Study on GTO Turn-Off Failure Mechanism" IEEE-IEDM Conf. Rec. p.414 (1981)
11) T.Nagano et al. "A Snubber-Less GTO" IEEE - PESC '82 Rec. p.383 (1982)

GATING CIRCUIT DEVELOPED FOR HIGH POWER THYRISTORS

N. Seki, Y. Tsuruta and K. Ichikawa

Toshiba Corporation, Fuchu Works, Tokyo, Japan

ABSTRACT

Gate turn off thyristor (GTO) gating circuits, especially off-gating circuits, are the most important for reliable operation of GTO equipment. This paper describes a new gating circuit for high power GTO of 600A class. The off-gating circuit can provide a negative pulse of 200A with its rate of rise of 30A/us. Its power dissipation decreases to 20 percent of a previous type. The on-gating and negative bias circuits are also described.

1 INTRODUCTION

The development of power switching devices is remarkable in recent years. High power GTO, being expected to appear for a long time, have now been available up to 2.5kV 600A. The special feature of GTO is to make it possible to eliminate a commutating circuit from a self-commutated inverter or a chopper.

The GTO gating circuits have a great influence on performance and operation of GTO equipment. The gating circuits are mainly consist of two circuits. One is for turnig on the GTO and the other, for turning it off. The former is hereafter called an on-gating circuit, the latter an off-gating circuit. Sometimes a negative bias circuit is also required to prevent dV/dt (critical rate of rise of off-state voltage) triggering.

Since the on-gating requirements are similar to those of conventional thyristors, a similar circuit may be applicable. The off-gating requirements, however, are quite different from the on-gating ones, and must be taken into special considerations. Various gating circuits previously presented are applicable for relatively small power GTO's, but not for high power GTO's. Because an off-gating circuit for high power GTO's has such a difficulty that it must supply a large current, say 200A, with a large di_g/dt, say 30A/us, through a comparatively low voltage source, say 30V, without causing any backswing. This is too severe for previous circuits.

The circuits shown in the reference 1, were developed for these deveces. But they have a large power dissipation for higher frequency operation and are apt to cause spurious current

Reprinted from *IEEE Power Electron. Specialists Conf.*, 1981, pp. 215–225.

due to backswing.

This paper presents a newly developed gating circuits shown in Fig. 4, which achieves reliable operation with extremely low power dissipation. This circuit is designed to operate from low frequency to mid-frequency of 1000Hz. Experimental results are described in comparison with the previous circuits.

2 HIGH POWER GTO's AND GATING REQUIREMENTS

The major ratings and characteristics of high power GTO family are shown in Table 1. The biggest one is 2.5kV 600A GTO (SG600EX21).

When a GTO is triggered at the time $t\text{-}t_0$ in Fig. 1, GTO's anode-cathode voltage begins to decrease. Higher on-gating current is desirable for a shorter turn on time, say more than 2A, with its rate of rise of more than 1A/us. At time $t=t_2$ illustrated in Fig. 1, an off-gating current is supplied with its rate of rise of 20$\sim$30A/us. After the storage time (T_s), the period of time from t_2 to t_3, the on-state

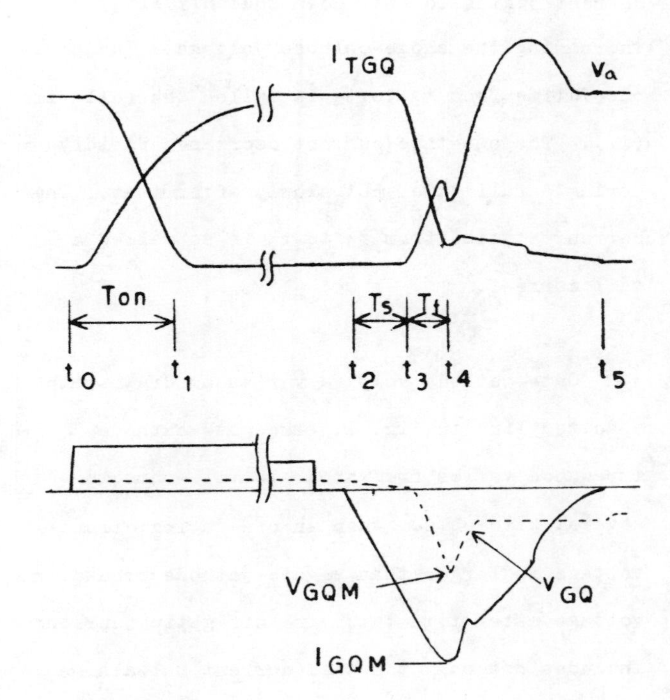

Figure 1 Typical switching waveforms

I_{TGQ} : Peak turn off current

I_{GQM} : Peak gate turn off current

V_{GQM} : Peak gate turn off voltage

V_{GQ} : Gate turn off voltage

Item	Symbol	SG600J21	SG600R21	SG600EX21
Repetitive Peak Off-state Voltage	V_{DRM}	600V	1300V	2500V
Peak Turn-off Current	I_{TGQM}	600A	600A	600A
ON-state Current (RMS)	$I_{T(RMS)}$	400A	400A	400A
Peak One-cycle Surge On-state Current	I_{TSM}	6000A	6000A	5000A
Critical Rate of Rise of On-state Current	di/dt	100A/us	100A/us	100A/us
Minimum DC Gate Trigger Current	I_{GT}	400mA	400mA	400mA
Gate-cathode Breakdown Voltage	$V_{(BR)R}$	14V	14V	14V

Table 1 Major ratings and characteristics of GTO's

current begins to fall down suddenly with increasing the anode-cathode voltage. The period of time from t_3 to t_4 is called the fall time (T_f). The on-state current decreases rapidly during a fall time, but slowly after that. The current at time from t_4 to t_5 is so called a tail current.

Gate-cathode voltage varies as drawn with a dotted line in Fig. 1. The gate-cathode impedance varies from about 10 mill-ohms to several kilo-ohms. When an off-gating circuit voltage is larger than a gate-cathode breakdown voltage after time $t=t_4$, the off-gating current includes not only the tail current but also a current through gate-cathode junction. Thus, when turn off current I_{TGQ} is small, almost of all the off-gating current flows through the junction against its breakdown voltage.

3 USUAL GATING CIRCUITS

3.1 Classifications of Off-gating Circuits

Various types of off-gating circuits have been used up to now. Table 2 shows the classifications of these typical circuits.

Type 1: Power source voltage Ed is chosen less than the gate-cathode breakdown voltage $V_{(BR)R}$ in Type 1(a). The gate current supplied from a power source becomes nearly zero, after the tail current diminishes.

The power dissipation of this type is least. But it is hardly expected that the rate of rise of gate turn off current (di_g/dt) is more than 10A/us and peak gate turn off current I_{GQM} is more than 100A. On the other hand, in

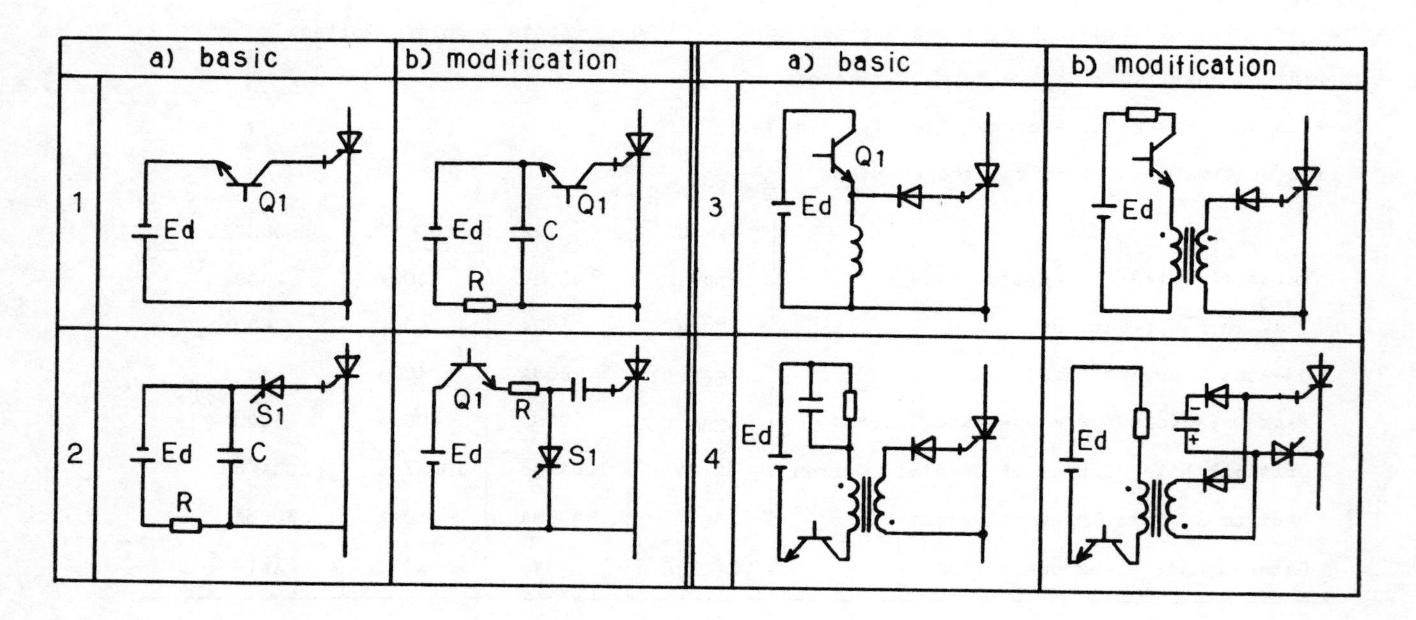

Table 2 Various Off-gating Circuits

Type 1(b), Ed is chosen higher than $V_{(BR)R}$ so as to improve di_g/dt, hence current limiting resistor R is inserted, The di_g/dt increases in proportion to Ed. But the higher Ed is, the larger power dissipation becomes. Because the current $(Ed-V_{(BR)R})/R$ flows during the conduction of the transistor Q1 even after the gate-cathode impedance has become high. Power loss decreases with increasing resistance, but increase in time constant CR limits its upper operating frequency.

Type 2: Both Type 2(a) and 2(b) are designed so that the capacitor charge is sufficient to turn off GTO. In Type 2(a), resistance R should be shosen to limit the thyristor current to less than its holding current after the GTO tail current has decreased Zero, therefore the resistance must be of the order of kilo-ohm. Upper frequency becomes necessarily low. In Type 2(b), a charging circuit, composed of a transistor, low resistance and a negative DC power source, is disconnected while the thyristor S_1 is turned on. Practically capacitor charging current is effectively used as an on-gating current. But larger capacitance needs longer charging time, which restricts minimum conducting period of GTO, or needs larger charging current.

Type 3: Reactor current, which has been built up due to a transistor Q1 conduction for a few ten micro-seconds before turning off a GTO, is transfered to the closed circuit consisting of the reactor, GTO and a diode, as soon as the transistor is switched off. The inductance must be large enough to keep the current flow against increased gate-cathode impedance for several ten micro-seconds.

Type 4: It is similar to a conventional on-gating circuit of thyristor, but it must have a bigger current capability. Type 4(b) is a modified circuit having two current sources, one current source which supplies a high rate of rise but short duration pulse current and the other which supplies a slow rate of rise but long duration pulse.

3.2 Example Using Pulse Transformer

Since each type of circuit listed in Table 2 has merits and demerits, it is selectively used corresponding to its application's requirements or sometimes may be combined each other. In case of a 1000Hz VVVF inverter using GTO listed in Table 1, Type 4 is most suitable regarding insulation between main and control circuits and switching devices to be available. Figure 2 shows a trially manufactured example. During whole conducting period of GTO, the transistor Q2 is switched on and off at 20kHz to supply on-gating current and during its non-conducting period, the transistor Q3, in turn, to supply negative bias. The transistor Q1 is conducted for 100 micro-seconds when turning off GTO. The waveforms are shown in Fig.3.

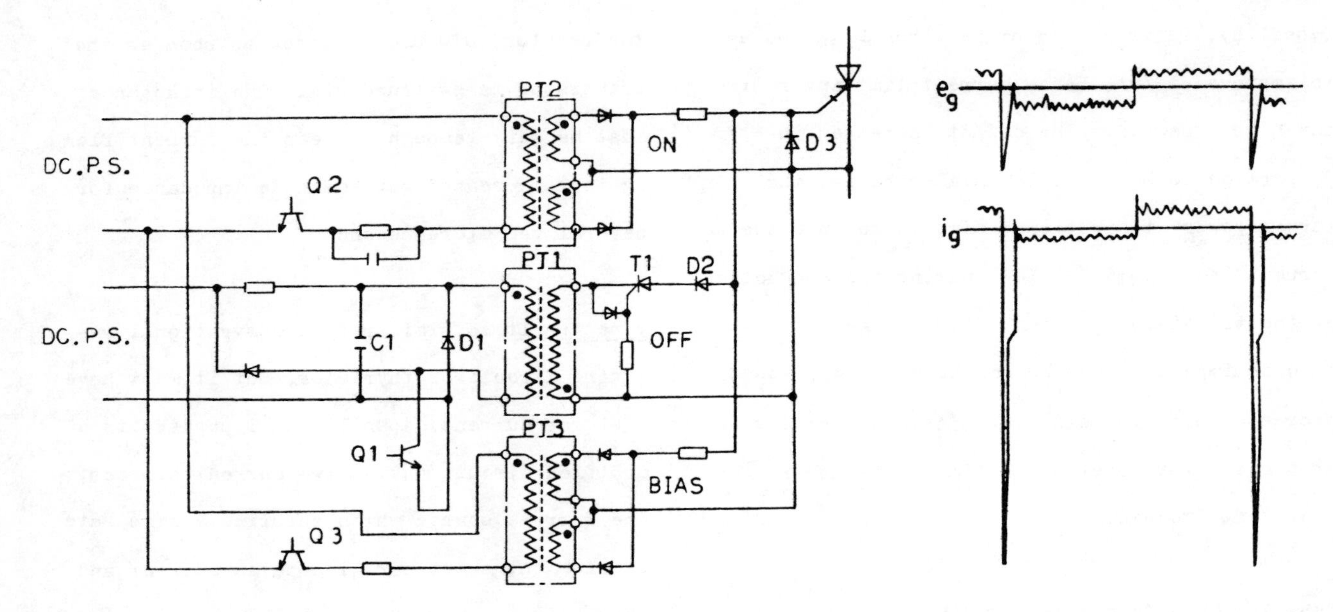

Figure 2 Example using pulse-transformer

Figure 3 Waveforms of Fig. 2

Each circuit seems to be identical to a conventional pulse generating circuit but has such problems listed below.

a) Power dissipation in the off-gating circuit is very large, for example, 200 Watt at 1000 Hz.

b) There is an interaction between off- and on-gating circuits. The stored energy in the pulse transformer PT_2 increases due to a part of the off-gating current's flowing into PT2 secondary windings, even if a diode D_3 exists for bypass. This may cause a GTO to fire faulsely.

c) A sudden change in the voltage polarity of PT_1 at the instant when the transistor Q_1 is switched off, causes a reverse current to flow in PT1 secondary winding. High speed diode D_2 should block this reverse current, but even a small amount of reverse recovery charge may be large enough to fire GTO falsely.

4 NEW CIRCUIT

Figure 4 shows a new gating circuit including on-gating, off-gating and bias circuit.

4.1 Description of On-gating and Bias Circuits

On-gating circuit using pulse-transformer is quite popular to a conventional thyristor. But it was proved to be interfered with off-gating current as described before. Consequently a transistor switch is provided for blocking the flow of off-gating current into the on-gating circuit. The primary winding of a transformer PT_2 is connected to a 20kHz power source. The 20kHz power source would be readily obtainable, for example, by means of slight modification of a switching regulator on the market. Such a power source is rather preferable, because DC voltage of each on-gating and bias

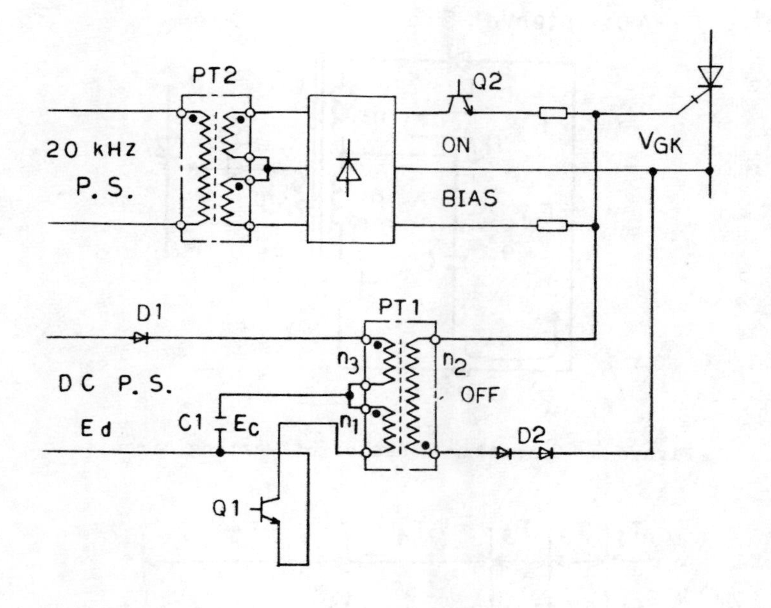

Figure 4 The new gating circuit

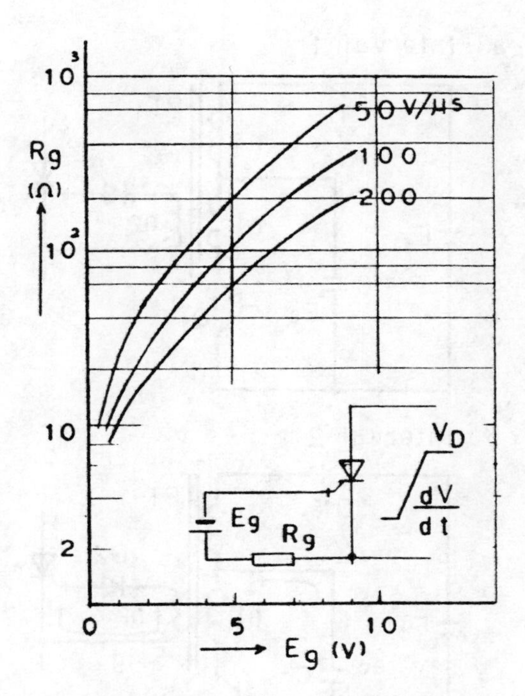

Figure 5 dV/dt capability

circuits would result in stabilization against line voltage fluctuation.

No transistor switch is used in a bias circuit. Figure 5 shows series resistance vs negative bias voltage characteristics of the before-mentioned GTO's with dV/dt capability as a parameter. In a practical use, negative bias current of 0.1 Ampere is sufficient. Commutating dV/dt capability is very high and has no relations with bias current, because a large negative off-gating current is flowing at that time.

4.2 Operating Principle of Off-gating Circuit

The followings characterize the circuit as compared with the previous one.

a) A diode D1 is inserted between a pulse-transformer PT1 and the positive terminal of a DC power source.

b) A capacitor C1 is connected across the connecting point of PT1 primary and tertiary windings and the negative terminal of the power source.

The operation will be described with reference to Figures 6 and 7.

Interval 1: The capacitor C1 is charged up to the voltage Ec, where Ec is higher than power source voltage Ed.

Interval 2: Suppose that number of turns of primary and tertiary windings are equal. When the transistor Q1 is turned on, capacitor voltage Ec appears across the winding n_1, thus off-gating current i_g starts to flow. i_g and its initial rate of rise $di_g/dt|_{t=0}$ are approximately given as follows:

(a) Interval 1

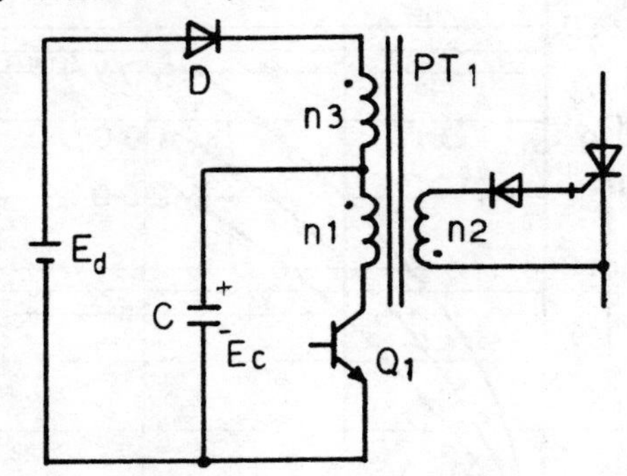

(b) Interval 2

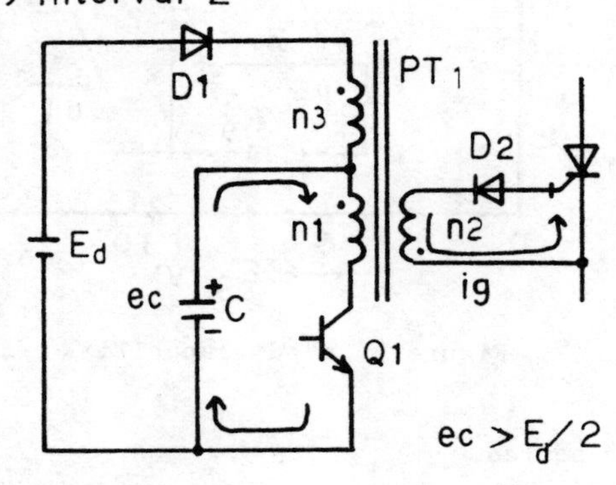

(c) Interval 3

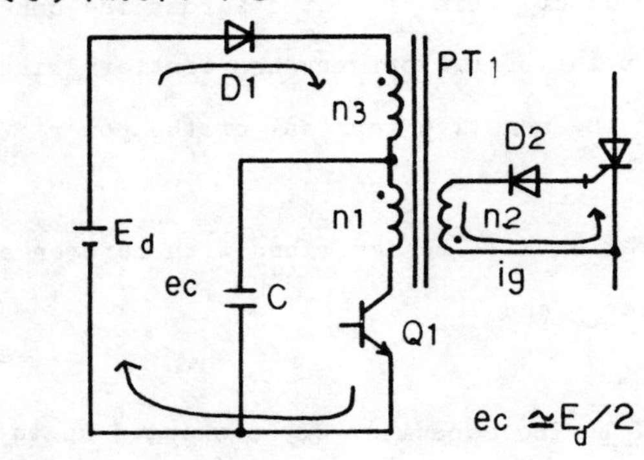

(d) Interval 4

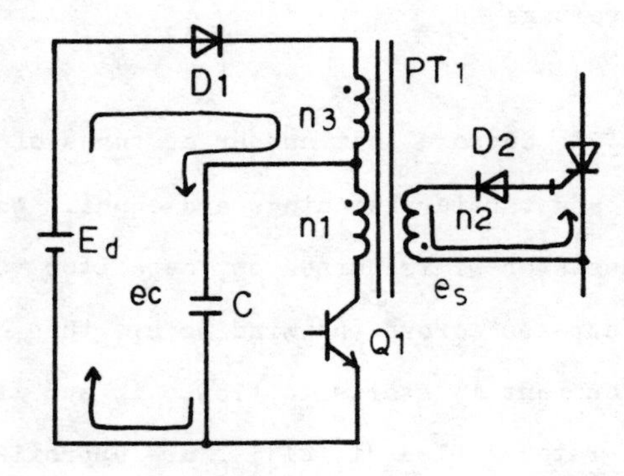

(e) Interval 5

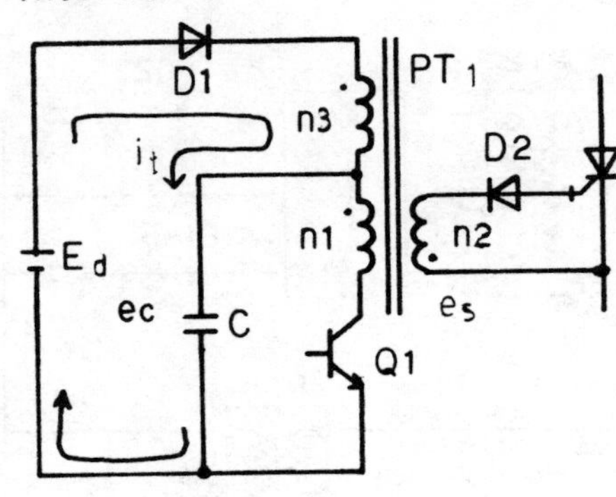

Figure 6 Operating modes of off-gating circuit

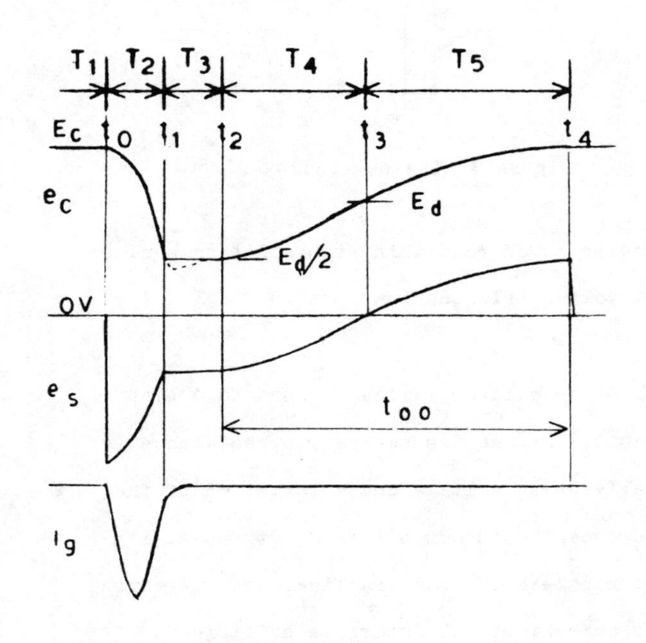

Figure 7 Waveforms of off-gating circuit for one operating cycle

$$i_g = E_c\sqrt{C/l}\,\sin(t/n\sqrt{lC}) \qquad \text{-----(1)}$$

$$di_g/dt\big|_{t=0} = E_c/nl \qquad \text{-----(2)}$$

where l is the total inductance reduced to terms of the secondary and n is n_1/n_2.

Since the voltage equal to the primary voltage induces across the tertially winding, diode D_1 will block current flowing through the tertially until the capacitor discharges to Ed/2. By the end of the interval 2 the gate-cathode junction has reverted to a blocking state, thus the gate-cathode impedance has become high. In case that on-state current is zero, i_g is given by replacing Ec in equation (1) by (Ec - $nV_{(BR)R}$):

$$i_g = (Ec - nV_{(BR)R}) \cdot \sqrt{C/l}\ \sin(t/\sqrt{lC}) \text{---(3)}$$

Ec is chosen so as to meet the requirement of the initial rate of rise of off-gating current and normally $Ec > nV_{(BR)R}$

Interval 3: When the capacitor voltage reaches the value Ed/2, the diode D_1 starts to conduct. Ed is impressed across the primary and tertiary windings in series. The capacitor voltage holds Ed/2 and the secondary voltage e_s is given by:

$$e_s = Ed/2n \qquad \text{---(4)}$$

If e_s is chosen less than $V_{(BR)R}$, then i_g becomes nearly zero after the tail current has reduced to zero.

Interval 4 and 5: When the transistor Q_1 stops conducting, the exciting current, originally flowing in the primary winding, is transfered to the tertiary. Almost of all energy, stored in PT, are used for recharging C_1 to higher value than Ed. It may be comprehensible from the fact that the capacitor will be charged up to Ed even if no exciting current exists.

Practically i_g becomes negligibly small during the intervals, hence the final value Ec of the capacitor voltage can be approximately obtained by solving the following equations:

$$i_t = \left\{(Ed/2X)\sin wt + I_m \cdot \cos wt\right\} \cdot \exp(-wt/2Q) \qquad \text{---(5)}$$

$$e_c = Ed + \left\{XI_m \sin wt - (Ed/2)\cos wt\right\} \cdot \exp(-wt/2Q) \qquad \text{---(6)}$$

where $X = \sqrt{L_m/C}$, $Q = X/R$, $w = 1/\sqrt{L_m C}$,

R is circuit loss in terms of a resistance, I_m, exciting current and L_m, exciting inductance.

The time too until which capacitor C_1 has been charged, is obtained by putting equation (5) equal to zero:

$$wt_{oo} = \tan^{-1}(-2XI_m/Ed) \qquad \text{---(7)}$$

where $\pi/2 < wt_{oo} < \pi$.

Substituting equation (7) to (6) gives Ec:

$$Ec = e_c(t_{oo}) = \sqrt{(XI_m)^2 + (Ed/2)^2}\ \exp(-wt_{oo}/2Q) \qquad \text{---(8)}$$

and $XI_m = w(L_m I_m)$

= w x Constant determined by volt-seconds of pulse-transformer.

The voltage v_s across the secondary winding is given by $-(Ed - e_c)/n$. When e_c= Ed, the interval 4 ends. During the interval 5 e_s becomes positive, but diode D2 blocks current i_g to flow in opposite direction. The rate of decay of v_s at time $t = t_3$ is so slow, and i_g is so small that the diode reverse recovery charge is negligible.

5 EXPERIMENTAL RESULTS

An off-gating circuit on test has the following circuit parameters:

Input voltage	DC100V
Turn ratio of PT_1	$n_1 : n_2 : n_3$ = 5 : 1 : 5
Exciting inductance	3mH
Capacitance C_1	2.2μ F
Transistor Q_1	2SD644(450V30A)

The experiment was made using a GTO chopper circuit. Figure 8(a) shows off-gating current waveforms with turn off current as a parameter. From this figure, it is clear that maximum off-gating current increases as turn off current increases. The di_g/dt is about 25A/us. The capacitor voltage Ec is 180V, which agrees with the caluculated value. With respect to gate-cathode voltage during turn off process, in the previous circuit small but positive voltage spike may appear as drawn in Fig. 3 at the instant when a transistor Ql is turned off. But new circuit has not such a spike as shown in Fig. 8(c). This involves very reliable gate turn off operation.

Figure 9 shows waveforms of on- and off-gating current and gate-cathode voltage at zero turn off current. The current was measured by using a conventional current transformer, due to

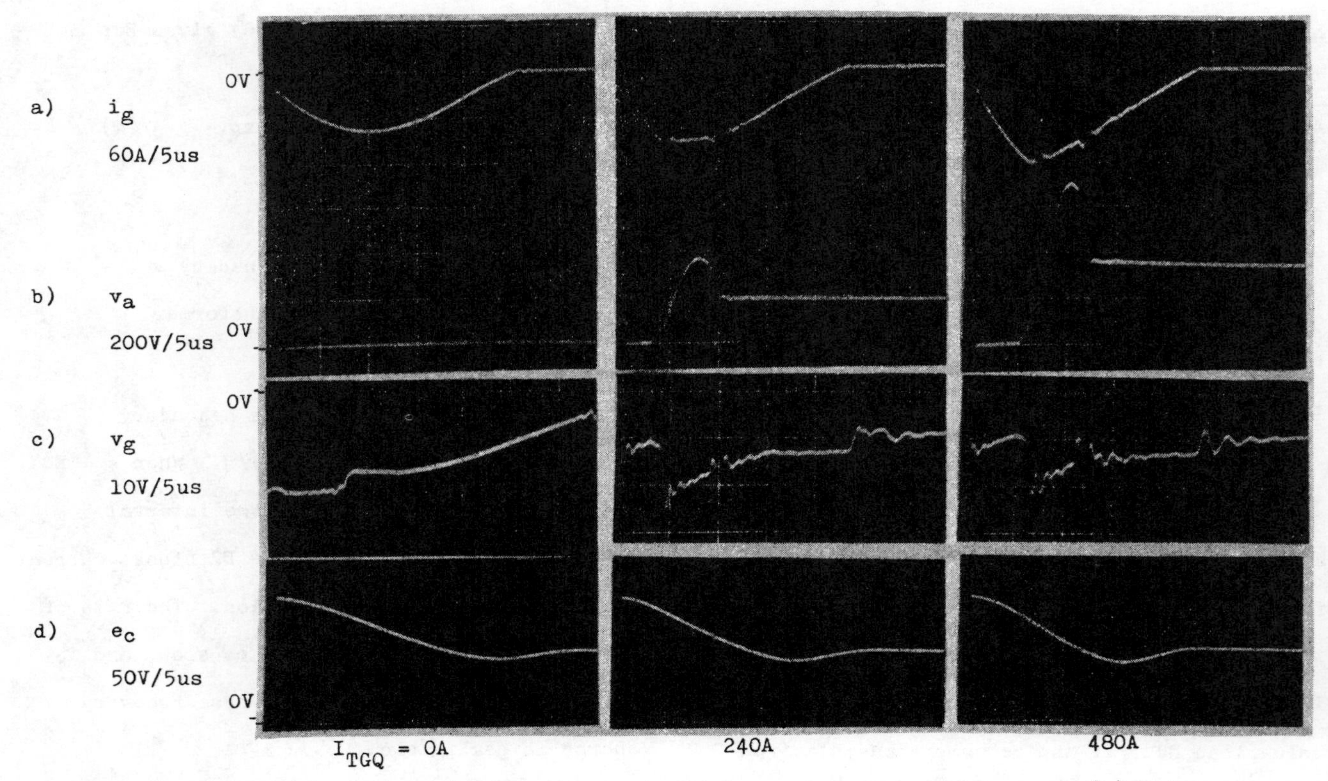

Figure 8 Waveforms of (a) off-gating current i_g, (b) anode-cathode voltage v_a, (c) gate-cathode voltage v_g and (d) capacitor voltage e_c with turn off current I_{TGQ} as a parameter.

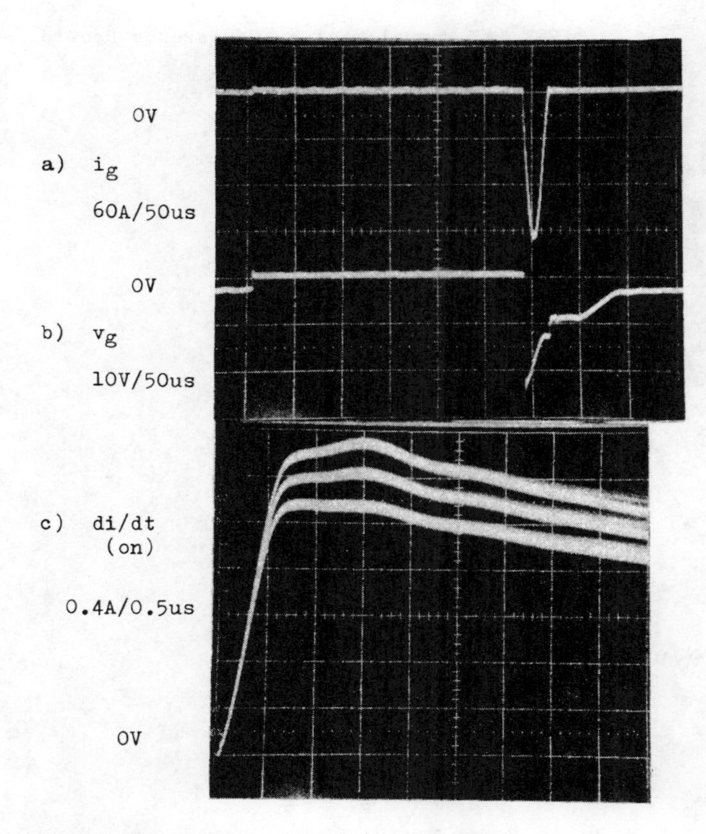

Figure 9 Waveforms of (a) gating current i_g, (b) gate-cathode voltage v_g and the rate of rise of on-gating current di/dt (on) with varing input voltage 100 + 10%

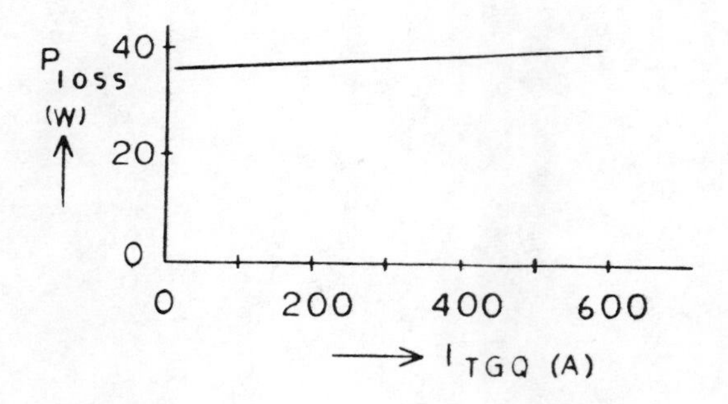

Figure 10 Power dissipation of off-gating circuit with varying I_{TGQ}

Figure 11 A packaged gating unit.

which a small oscillation was obsereved after a large off-gating current finished flowing. The rate of rise of on-gating current was 2A/us. The delay time of on-gating signal from photo-coupler input to transistor Q2 output was less than 1 us. These results meet all the requirements. Figure 10 shows the power dissipation. This is approximately independent of turn off current, but proportional to frequency. The power dissipation is extremely reduced to about 40W at 1000Hz from the previous 200W.

Figure 11 shows a packaged unit of the circuit, whose size is one-thirds of the previous one.

6 CONCLUSION

A new gating circuit for high power GTO has been presented and its performance has been compared to the previous one. An experiment proved that the new circuit operates satisfactorily and power dissipation in the off-gating circuit reduces extremely.

REFERENCE

1) T. Jinzenji et al.: "Three Phase Static Power Supplies for Air-conditioned Electric Coaches Using High Power GTO", IEEE IAS Annual Meeting Conference Record pp.1088-1098,1979.

PRESENT CONDITION OF HIGH POWER GTO APPLICATIONS

N.Seki, K.Ichikawa, Y.Tsuruta, and K.Matsuzaki

Toshiba Corporation, Fuchu Works, Tokyo, Japan

ABSTRACT

High power gate turn-off thyristor(GTO) is now available up to 2.5kV-1.6kA in Japan. The special feature of GTO is to make it possible to eliminate a commutating circuit from a self-commutated inverter or chopper. In this paper, practical applications using high power GTO, that is, constant-frequency power supply from 30kVA to 500kVA series for UPS system, 300kVA mid-frequency(1000Hz) inverter and 1200kVA variable voltage variable frequency inverter, are described. As compared with the conventional thyristor system, these inverters have been proved to obtain higher efficiency, lower size and audible noise, and higher reliability. The total capacity of GTO equipment, which has been already produced in Toshiba, amounted to about 35000kVA including electric railway apllications till May in 1982.

1. INTRODUCTION

High power gate turn-off thyristor(GTO) has now entered into the stage of practical appliciation. Several kinds of GTOs have been developed and available up to 2.5kV-1.6kA. At present, the cost of GTOs is still expensive in comparison with that of the conventional thyristors. However, in the near future, it will become reasonable by improved manufacturing process and mass-production.

In GTO application, some requirements, especially off-gating ones, which are quite different from those of conventional thyristors, must be taken into special considerations. On the other hand, on-gating requirements are similar to those of conventional thyristors. The gating circuit developed for high power GTOs was described in details in reference (1).

This paper presents some practical GTO applications which have achieved high performance and high reliability. The experimental results are described in comparison with the previous equipment employing conventional thyristors.

2. HIGH POWER GTOS AND THEIR REQUIREMENTS

High power GTOs

The ratings (off-state voltage and RMS on-state current) of major GTOs which have been already published in Japan are shown in Fig.1. Each numerical value shows peak turn-off current, that is, the maximum anode current turned off by off-gating current. The low power GTOs (less than 200A class) are omitted. At present, the voltage rating from 600V to 2500V and the peak turn-off current from 200A to 1.6kA GTOs have been developed.

Figure 2 shows the exterior of 2.5kV-800A GTO. 600~800V class GTOs ca. be mainly employed for application of AC 200~220V system. 1200~1600V class

Reprinted with permission from *5th Int. PCI Conf. Power Conversion*, 1982, pp. 391-403.

GTOs for application of AC 400-440V system, and 2500V class for application of DC 750-1500V system.

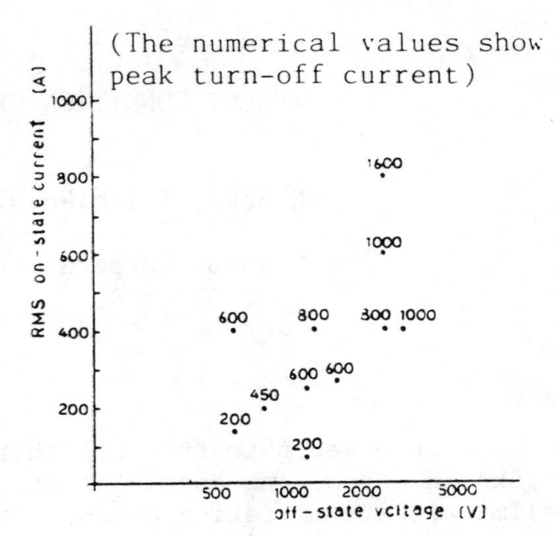

Fig.1 Ratings of major GTOs

Fig.2 Exterior of 2.5kV-800A GTO (SG800EX21)

Gating circuits

The GTO gating circuits have a great influence on performance and operation of GTO equipment. The gating circuit mainly consists of two circuits. One is for turning on the GTO and the other, for turning it off. The former is called an on-gating circuit, the latter an off-gating one. Sometimes a negative bias circuit is also required to prevent dv/dt (critical rate of rise of off-state voltage) triggering. Because dv/dt capability of unshorted-emitter structure GTO is not so high. The off-gating circuit is most important to achieve the full turn-off capability.

Figure 3 shows a gating circuit suitable for high power GTOs including an on-gating, an off-gating and a negative bias circuit. On-gating circuit using pulse transformer is quite popular to a conventional thyristor. This method is applicable to the on-gating circuit of GTOs. In Fig.3, the primary winding of the transformer PT1 is connected to a 20kHz power source. The 20kHz power source would be easily obtainable, for example, by means of slight modification of a switching regulator on the market. Such a power source is rather preferable, because DC voltage of each on-gating and negative bias circuits will result in stabilization against line voltage fluctuation. And also, the transformer PT1 will be decreased in external dimensions. An on-gating signal is isolated by a photo-cuppler, and a transistor switch is provided for blocking the flow of off-gating current into the on-gating circuit.

No transistor switch is used in a negative bias circuit. In a practical use, a bias current of 0.1 ampere will be sufficient. Because a large negative off-gating current is flowing during turn-off operation, the dv/dt capability is very high and has no relations with the bias circuit.

As shown in Fig.3, the off-gating circuit is quite simple. The capacitor C1, which is charged up to 2Ed by the stored energy of PT2, provides a higher off-gating pulse. The features of this circuit are as follows.

(1) Off-gating circuit is theoretically low in power dissipation.

(2) No backswing of PT2 due to its stored energy appears because of using its energy for charging up a capacitor C1.

Figure 4 shows a packaged unit of the gating circuit which can operate up to 1000Hz.

Snubber circuit

The GTO turn-off characteristics will depend on the snubber circuit. The snubber capacitor is used to suppress the voltage rate of rise (dv/dt) of GTO and to improve the turn-off capability. The requirements of snubber

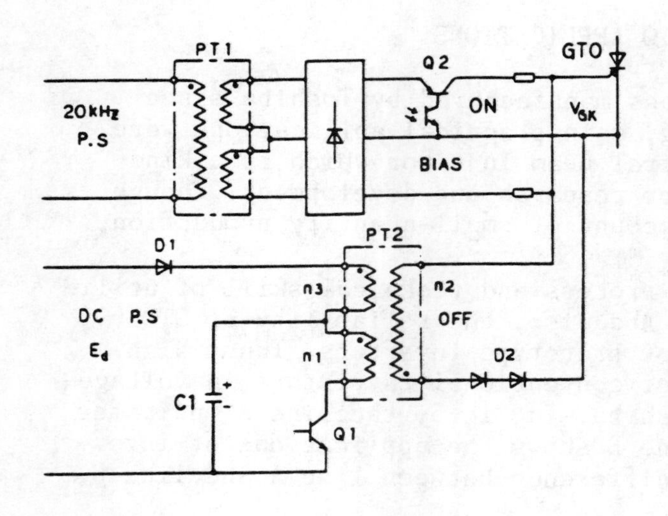

Fig.3 Gating circuit of GTO

Fig.4 Packaged gate unit

circuit are as follows.

(1) The snubber circuit must be wired as short as possible to reduce wiring inductance.

(2) Fast recovery diode is necessary to decrease the dv/dt of GTO and turn-off power dissipation.

(3) Smallest internal inductance of the capacitor will be also desirable.

The power dissipation (Ws) of the snubber circuit is approximately expressed as follows

$$Ws = 1/2 \cdot Cs \cdot Vc^2 \cdot f$$

where, Cs: capacitor
Vc: capacitor voltage
f: operating frequency

The snubber capacitor of GTO should be larger, say 2 μF, than that of a conventional thyristor, say 0.5 μF. Therefore GTO snubber circuit power dissipation will be larger by about 4 times than thyristor one. But a commutation circuit power dissipation of conventional thyristor equipment will be larger by about 50 times than off-gating circuit one of GTO. For instance, at operating frequency of 1000Hz, calculated results are as follows,

	GTO	SCR
snubber + switching	680W	190W
commutating (turn-off)	40W	2140W
on-state	600W	580W
total	1320W	2910W

Therefore, the total efficiency of GTO inverter will increase more than conventional thyristor one.

3. TOTAL CAPACITY OF GTO APPLICATIONS

The total capacity of GTO apllications manufactured by Toshiba since 1976 are shown in Fig.5. In the beginning, main practical apllications were used as a switch of power supply for Neutral Beam Injector which is a kind of plasma heating device and inverters for research and development. Though the cost of GTOs was very expensive on account of small-quantity production, GTO equipment increased gradually.

In recent years, both manufacturing process and technical skill of utilization have been developed and improved. And also, the reliability of GTO device has been proved by long-run test of prototype inverters. Thus, high power GTOs have begun to take the place of conventional thyristors in voltage-sourced type inverter applications in Toshiba. In Fig.5, the line A indicates the transition of total capacity, the line B shows the applications of inverter and chopper for UPS system, and the difference between line A and line B is the use of above mentioned switch.

4. INVERTER FOR UPS STSTEM

New types of constant-voltage constant-frequency (CVCF) inverter using 1300V-800A GTO, called "TOSNIC-1000 series",have been developed for the uninterruptible power supply (UPS) system. "TOSNIC-1000 series" connected with storage batteries will be usually used for the important equipment such as on-line computers, satellite communication stations, airport control tower,radio and TV broadcasting stations, and other instruments of plants. The capacity of this series is lined up from 30kVA to 500kVA.

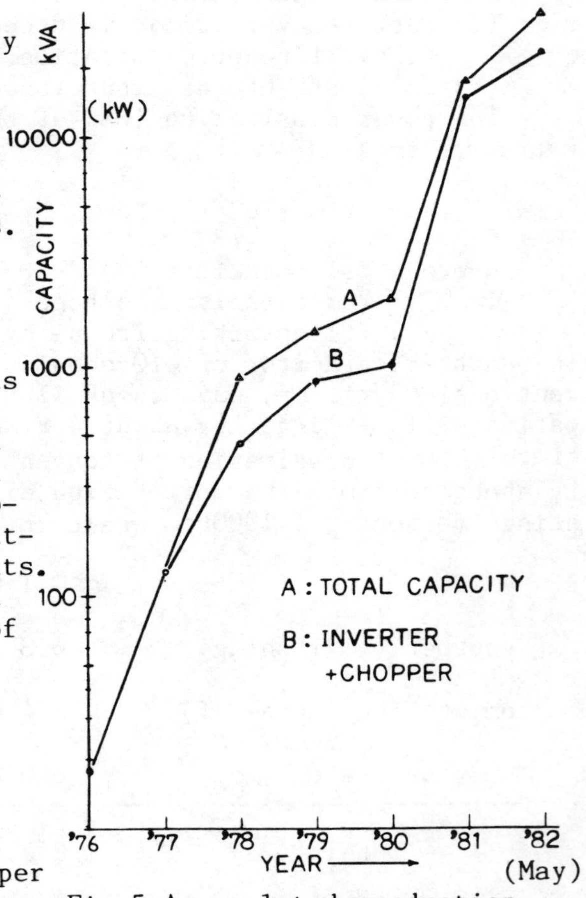

Circuit configuration

The block diagram of the CVCF inverter is shown in Fig.6(a). Special features are that the step-up chopper is employed and an output high-speed thyristor interrupter for parallel operation system is removed. The step-up chopper converts DC input voltage to the most suitable DC voltage,say 650V, for inverter circuits. Figure 6(b) shows the step-up chopper circuit of "TOSNIC-1000 series". The output voltage of the chopper is given as follows,

$$Eout = \frac{Ton + Toff}{Toff} \times Ein$$

where, Eout: output voltage of step-up chopper
Ein : input voltage of step-up chopper
Ton : on-state period of GTO
Toff: off-state period of GTO

Fig.5 Accumulated production records of GTO equipment in Toshiba

The output voltage (Eout) is determined by controlling the on-time (Ton) of GTO when one cycle time (Ton + Toff) is constant. The operating frequency of chopper is set by 12 times of inverter frequency. Therefore, in case of 50Hz inverter, the chopper operating frequency of 600Hz is used. Plural step-up choppers are employed according to the output capacity of equipment and DC input voltage. The unit inverter circuit is shown in Fig. 6(c). The inverter circuit consists of 6 GTOs and 6 diodes which are connected in antiparallel with each GTO in case of less than 200kVA, and from 250kVA to 500kVA, two or three unit inverters will be employed. Their unit inverters are coupled by transformers with zigzag connection.

Control and protection

The output voltage is determined by DC voltage, which is controlled by the step-up chopper. Inverter circuit itself has no voltage controlability. The following design is provided for increasing reliability of the parallel operation system. All functions required for parallel operation are provided in each CVCF inverter. It can operate with little disturbance from another parallel inverter's start and stop.

GTO in combination with fuse serves the parallel operation system without using the high-speed thyristor interrupter, and also a protection of DC short circuit fault. When the short circuit fault occurs, off-gating current is immediately supplied to all GTOs excluding the short circuit arms of the unit inverter.

For the fault analysis in many cases, the digital simulations by means of RUNGE-KUTTA methods were carried out. The digital simulation of short circuit at load terminals is shown in Fig. 7. On account of the differece in rate of rise, protection against AC short circuit is safely done by turning off the GTOs at once.

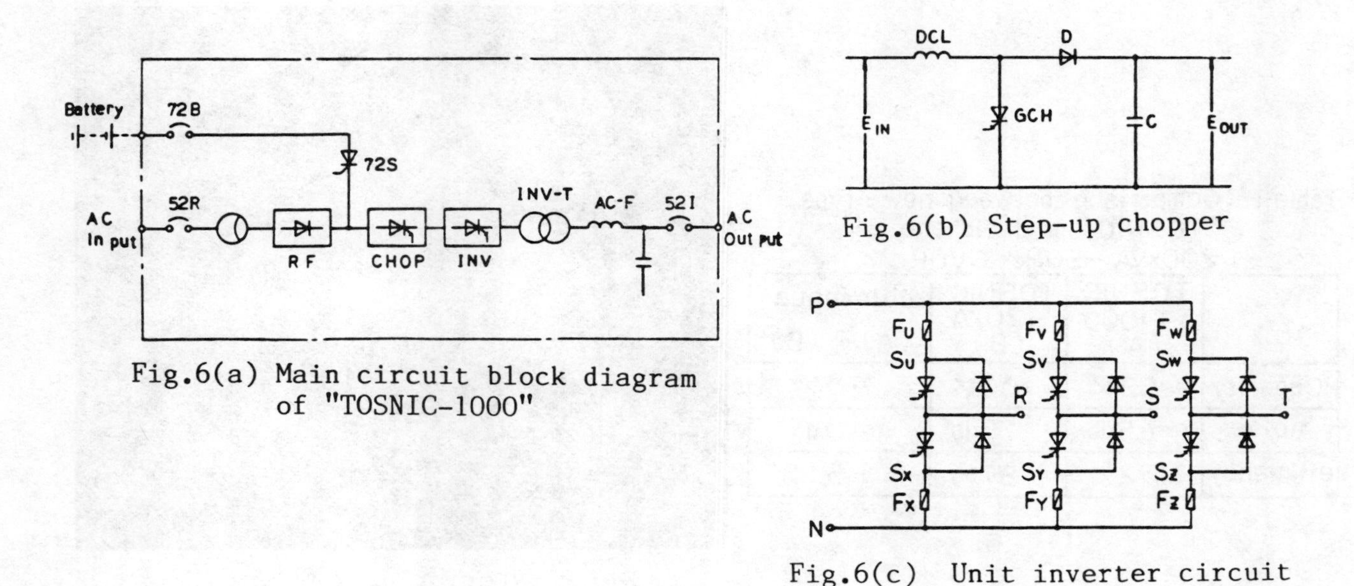

Fig.6(a) Main circuit block diagram of "TOSNIC-1000"

Fig.6(b) Step-up chopper

Fig.6(c) Unit inverter circuit

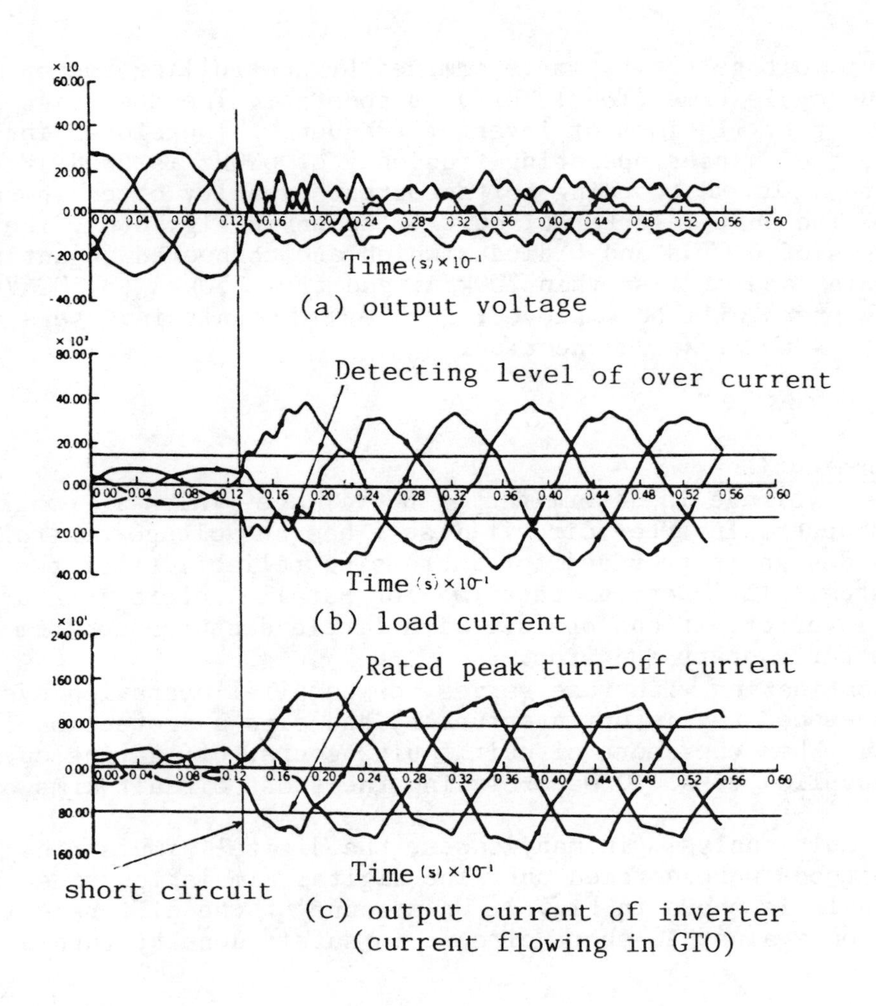

Fig.7 Digital simulation of short circuit at load terminals

Fig.8 Exterior of 200kVA CVCF

Table 1. Comparison between new-type "TOSNIC" and old one 200kVA - 50Hz CVCF

	TOSNIC -1000 (A)	TOSNIC -201A (B)	difference (A) - (B)
Capacity	52%	100%	-48%
noise	65db	75db	-10db
efficiency	92%	87%	+5%

Performance

The special features of 200kVA CVCF inverter are as follows.

(1) Unitized all main circuits and control circuits permit easy inspection and repairing by only front side.

(2) DC step-up chopper by high frequency operation offers a high-speed voltage response required in CVCF inverter

(3) The external dimentions ot 200kVA inverter are reduced to about half size compared with conventional one.

(4) The audible noise is reduced to less than 65 dB.

Table 1 shows the main features of 200kVA GTO CVCF inverter compared with the conventional thyristor one. The exterior of the inverter is shown in Fig. 8.

5. MID - FREQUENCY INVERTER

A mid-frequency power supply, which operates 500Hz~2kHz, is sometimes required, such as for the test of higher harmnoics filters, for the high speed motor drives.

In case of conventional thyristor inverter,there are problems that operating frequency of over 1000Hz is limited in circuit operatĭon, and also that high efficiency can not be achieved because of the commutating circuit power dissipation becoming larger in proportion to operating frequency. In addition, large external dimensions and big audible noise are defects. However, these problems can be solved by employing GTOs. To grasp these effects quantitatively, 1000Hz prototype GTO inverter was produced.

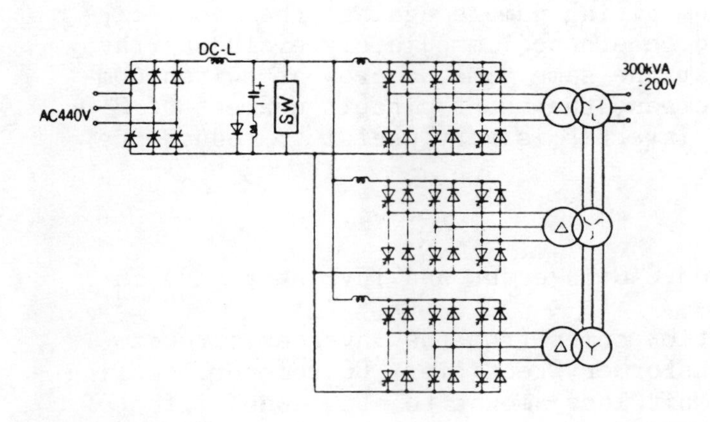

Fig.9 Schematic of 300kVA-1000hz inverter

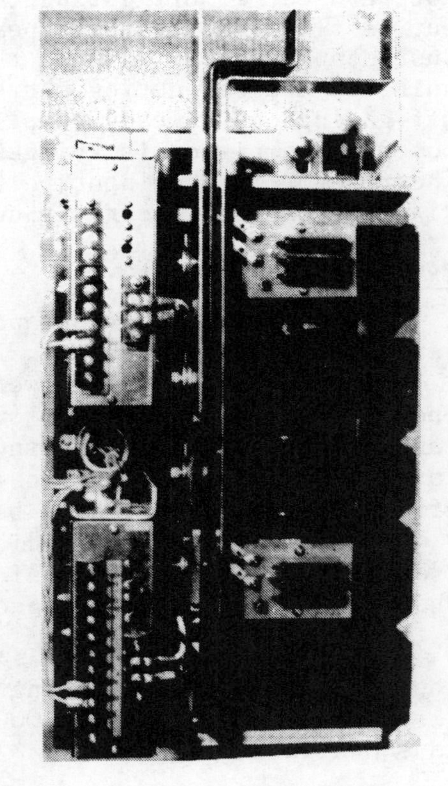

Fig.10 GTO stack

Main Circuit Configuration

Fig.9. shows the schematic of 300kVA-1000Hz prototype GTO inverter developed for the test of higher harmonics filters. And GTO stack configuration is shown in Fig.10. The GTO stack consists of 2 GTOs, 2 diodes and 2 gating units. In the GTO mid-frequency inverter, the output voltage is controlled by a thyristor controlled rectifier. The inverter circuits are composed of three unit inverters and three output transformers with zigzag connection.

The high speed free-wheeling diode which is connected to each GTO in antiparallel is required. Because, when the power factor of load is leaded, the momentary short circuit will occur through a turned-on GTO and storage carrier of a free-wheeling diode. If the storage carrier of diode is larger, the turn-on switching dissipation of GTO will become larger. A reactor, in the input part of each unit inverter, is provided so as to reduce the di/dt of GTO.

Each unit inverter is shifted each other by 20 degrees of electrical angle and connected to three output transformers.

The harmonics of the output voltage is 18n±1 (n=1,2,) and theoretical wave distortion is about 10%, but on load operation, Less than 10% wave distortion is obtained owing to the leakage reactance of output transformers.

Protection method

Since the operating frequency of this GTO inverter is relatively higher than that of normal one, the protection method will be problem when DC short circuit fault occurs. Because in case of 1000Hz inverter, for instance, one of the GTOs turns off every 56 μs in 18 pulse inverter. That is, the off-gating current is supplied to one of the GTOs every 56 μs. So all triggering method, which has been employed on the conventional thyristor inverter, could not be used. If off-gating current is supplied to the GTO through which current is flowing over rated peak turn-off current, the GTO will be destroyed instantaneously. To prevent the GTO from giving damage against the short circuit, on and off-gating current should be stopped immediately excluding the off-gating one already supplied. And at the same time, a crowbar switch composed of thyristor is triggered to decrease the short-circuit current of GTO. The reactor in the input part of unit inverter is very useful to reduce the di/dt of the short circuit current.

Experimental results

Comparison of performances between GTO inverter and Thyristor(SCR) inverter is shown in Table 2.

About 40% of total power dissipation comes from the inverter circuit one. And the rest are from output transformer, rectifier, DC reactor, cooling fans and so on. Since the snubber circuit loss amounts to about one - fifth of total power dissipation because of rather high frequency, total conversion efficiency of inverter can be improved by regenerating the snubber energy.

The main source of audible noise will be commutation reactor and output transformer in case of usual SCR inverter. In 1000Hz GTO inverter, less than 70 dB on light load, and about 72 dB on rated load, have been obtained.

A main source of noise excluding the transformer would be cooling fan. More decreasing of noise can be expected by improving them. Fig. 11 shows the exterior of 1000Hz GTO inverter and its output voltage waveform is shown in Fig. 12.

Table 2. Comparison between GTO inverter and SCR one
300kVA – 1000Hz INVERTER

	GTO (A)	SCR (B)	difference (A) – (B)
Capacity	75%	100%	-25%
noise	72db	83db	-11db
efficiency	88%	83%	+5%

Fig.11 Exterior of 300kVA-1000Hz inverter

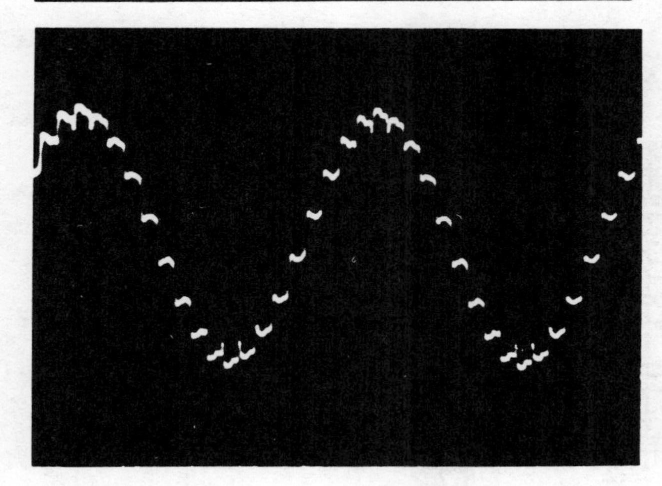

Fig.12 Output voltage waveform

6. VVVF INVERTER

The inverter system for driving AC motors, especially induction motors, has various merits such as, elimination of contacts in motors, maintenance free, as compared with DC motor driving system . And also, speed control of induction motor for power saving system has come into practical use by using variable voltage variable frequency (VVVF) inverter. In this field, GTO inverter will be also available.

Circuit configuration

Figure 13 shows main circuit configuration for driving an induction motor. The electrical ratings are given as follows.

(1) Input voltage / frequency : 1100 V/ 50Hz - 3 phases
(2) DC voltage : 1000 V
(3) Output capacity : 1200kVA - 3 phases
(4) Output voltage / frequency: 3.3 kV/55Hz - 120 V/2 Hz

The main circuits are composed of a 3-phase thyristor controlled rectifier (REC), a surge absorber (ZX), DC filter (DX,FC),thyristor crowbar switch (S), reactor (X1-X9), and nine single-phase unit inverter and three output transformers connected with each other. The output voltage is stepped up to four times by transformers. The single-phase unit inverter is shown in Fig.14. The rectifier consists of 4kV-1500A thyristors, and the unit inverter, 2.5kV-600A GTOs and 2.5kV-400A free-wheeling diodes. The reactors (X1-X9) connected to the input part of each unit inverter and the reactors (X11-X22) included in each unit inverter have effects as already mentioned.

Control System

The output voltage/frequency is linearly controlled from 120V/2Hz to 3300V/55Hz. From 2Hz to 10Hz, the rectifier controls the output voltage with the phase-angle of the unit inverter constant. After that, DC voltage is controlled at constant,and the output voltage is determined by controlling the phase-angle θ in order to obtain a quick response. Figure 15 shows the control pattern of this inverter.

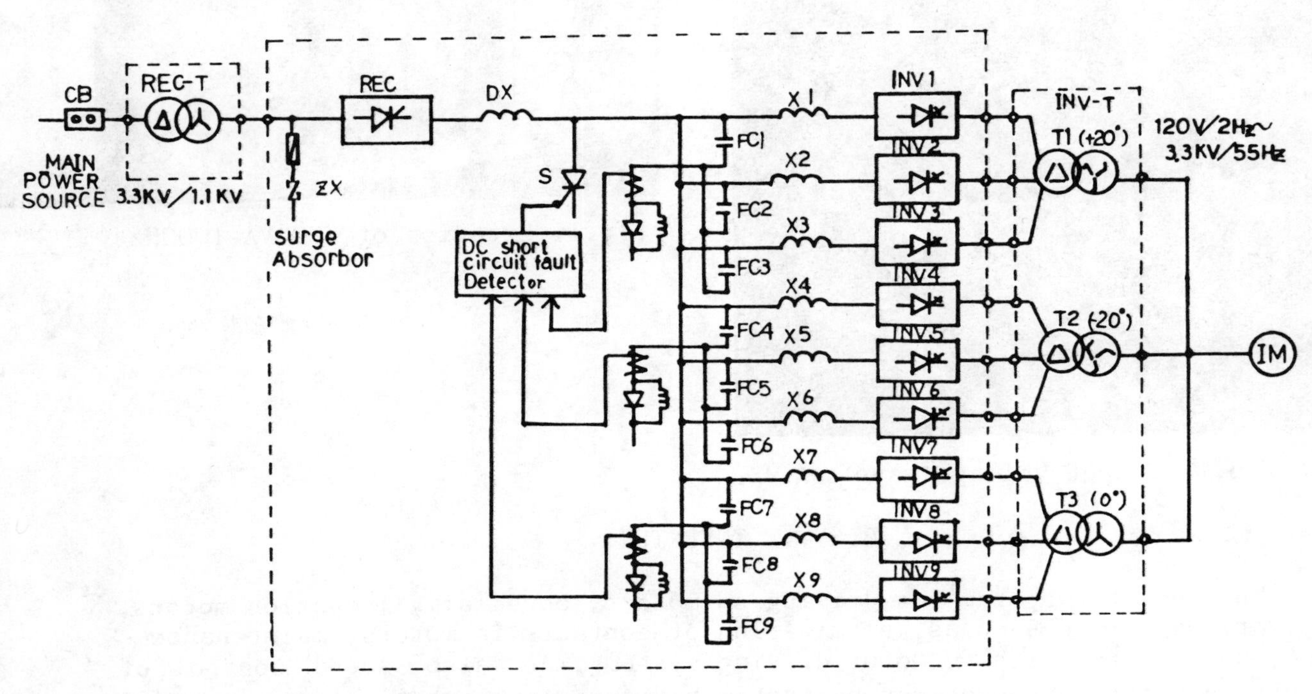

Fig.13 Main circuit configuration of 1200kVA VVVF

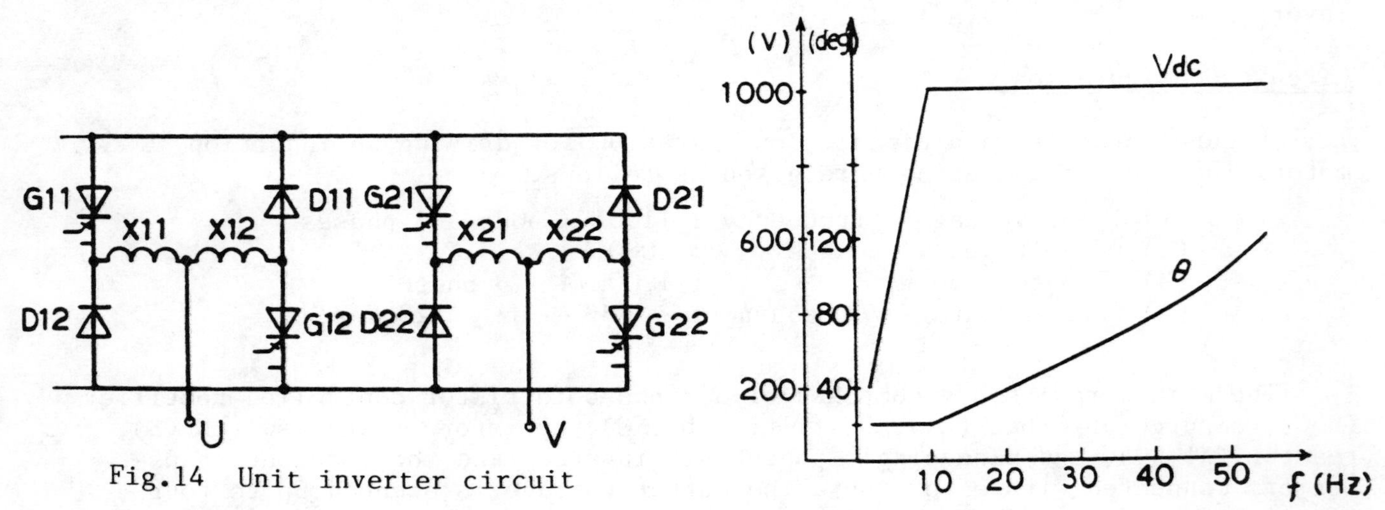

Fig.14 Unit inverter circuit

Fig.15 Control pattern of DC voltage and phase-angle

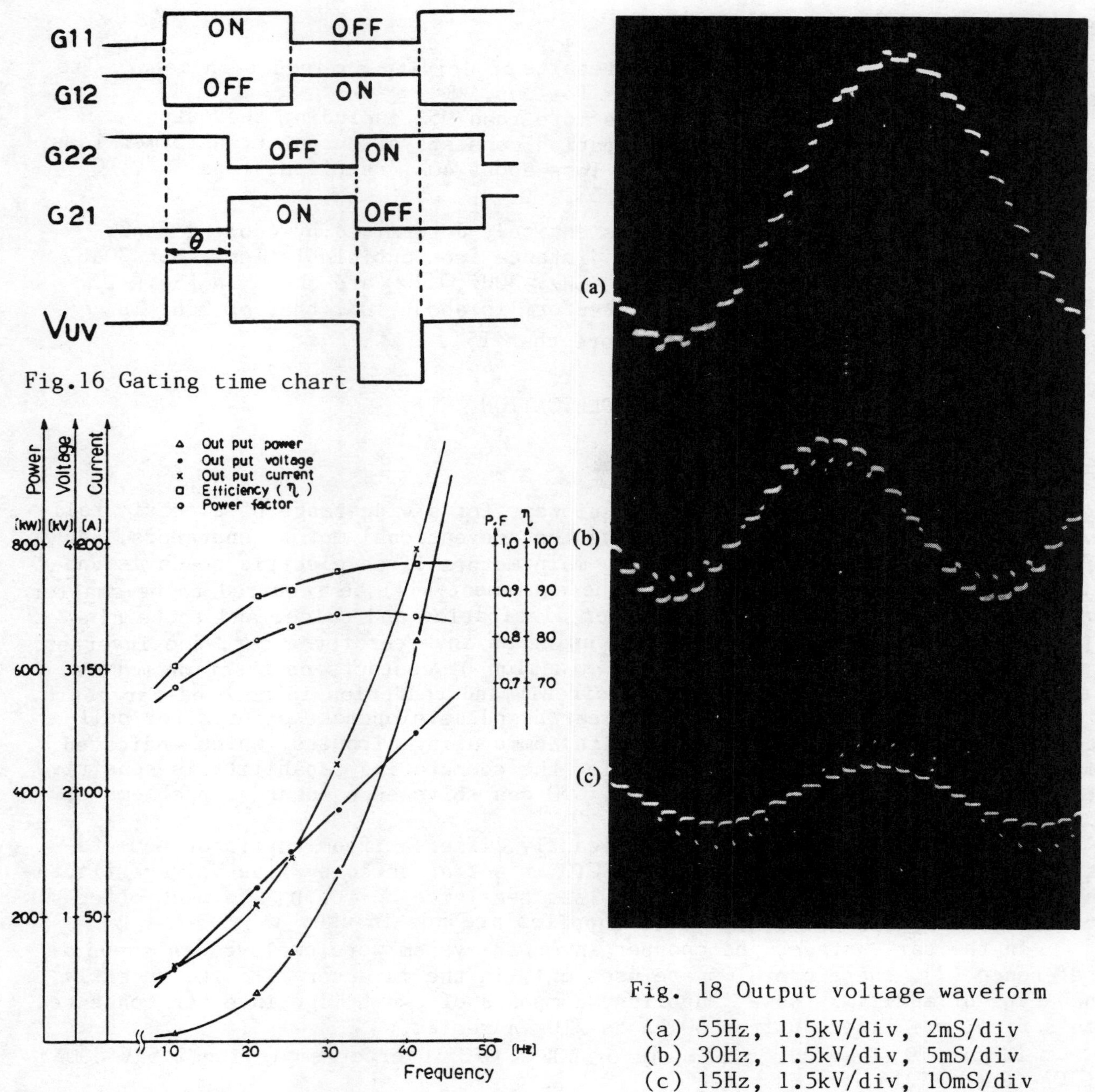

Fig.16 Gating time chart

Fig. 17 Induction motor drived data

Fig. 18 Output voltage waveform

(a) 55Hz, 1.5kV/div, 2mS/div
(b) 30Hz, 1.5kV/div, 5mS/div
(c) 15Hz, 1.5kV/div, 10mS/div

Figure 16 presents the principle of control operation corresponding to the unit inverter circuit shown in Fig.14 . When the GTOs (G11,G12,G21,G22) are controlled like this, the single-phase voltage of phase angle θ is obtained. The response speed of DC voltage control system is designed less than 1/10 as compared with phase angle control one.

Manufactured gating unit is the same configuration as shown in Fig.3 except for insulation voltage of AC 4.5kV.

Experimental Results

Figure 17 shows the experimental results of driving an induction motor. Its performance and characterestics are as follows,

(1) AC to AC conversion effciency is more than 96% including the output transformers. Total power dissipation consists of output transformer loss of about 40%, inverter circuits loss about 40%, rectifier loss 15%, DC reacter loss and the rest about 5%.

(2) Audible noise in the cubicles is entirely determined by cooling fans. Noise level measured at 1 meter distance from cubicle is less than 70dB.

(3) The output voltage waveforms of 55Hz, 30Hz, 15Hz are shown in Fig.18. The distortion factor of 55Hz waveform is about 10%, that of 30Hz is about 15% and that of 15Hz is more than 15%.

7. OTHER APLLICATION

Electric Coach

As the auxiliary power source equipment for a wide range of electric railway, thyristor inverters are replacing the conventional motor generators. And also, the inverter system for driving main motors of an electric coach is under testing. Being installed on a car, the equipment will be required to be smaller and lighter, higher reliability, lower dissipation and noise, and to be dispensed with labor of maintenance. By using an inverter for a car, the inverter system has various merits such as elimination of contacts on traction motors, reduction of contactors in the main circuit,and reduction in motored-car ratio.

Most of the past significant research and developement efforts for self-commutated inverter were concerned with commutating circuits, which indicated many problems to be solved, for example, the commutating capability is sensitive to line voltage fluctuation. However, GTO can solve such inherent problems at once.

GTO turn-off behavior is significantly different from thyristor commutation one. The turn-off capability of GTO is not affected by line voltage fluctuation, because off-gating current is less sensitive to it. Due to much effectiveness, many GTO auxiliary power supplies are now in use.

In the early stage, the chopper inverter system were employed as shown in reference (2). Therefore,GTOs are used only in the inverter circuit. Recently, new-type of auxiliary power supplies, composed of two bridge inverter connected in series, are lined up from 50kVA to 210kVA series.

Figure 19 shows the schematic of 50kVA GTO inverter employing 2.5kV-600A GTOs (SG600EX21).

Power Supply for NBI

Various kinds of power supplies are required for a Newtral Beam Injector (NBI). Though details are omitted here on account of limited space, a power suplly is used for generating source plasma of hydrogen. In this circuit, GTOs are employed as a crowbar switch. After conducting for a short time, 5 GTOs in paralleled turn-off 2000A. A small resistor is used as a current balancing device.

In future, a big GTO switch will replace a high power regulator tube in the acceleration and deceleration power supplies.

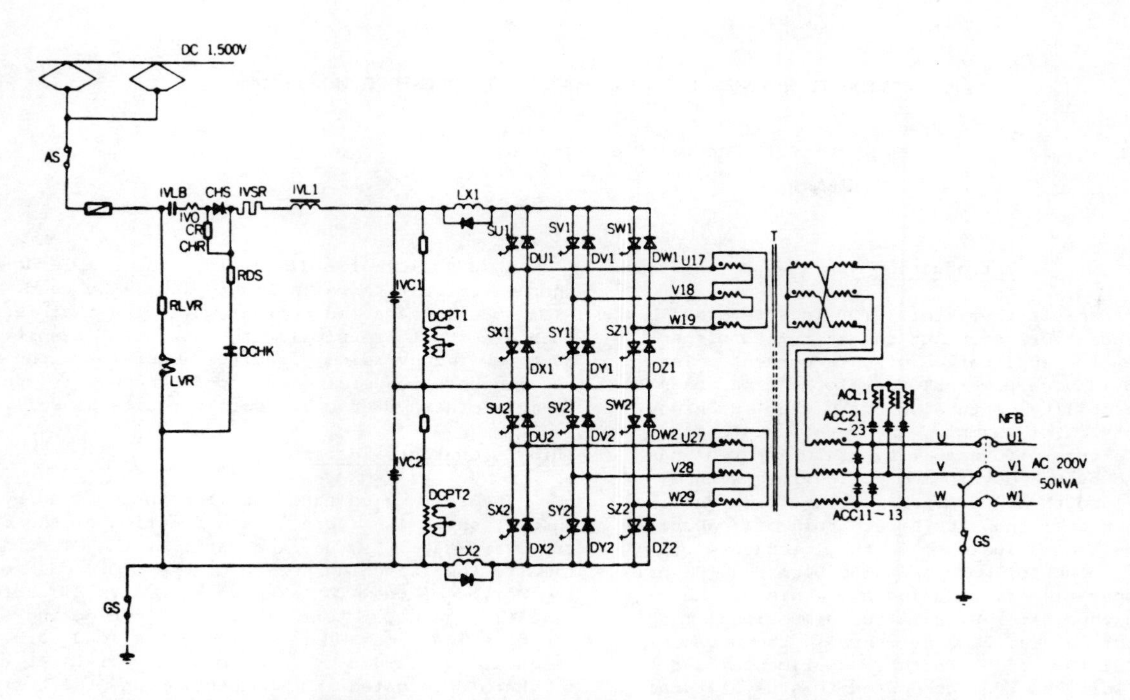

Fig.19 Schematic of 50kVA inverter for electric coach.

8. CONCLUSIONS

Many GTO applications in Toshiba such as UPS, mid-frequency inverter,VVVF inverter etc. have been presented. They are put into practical use and amount to 35000kVA in terms of their output capacity.

Now is the time when GTO plays the leading role in every self-commutated converter.

REFERENCES

(1) N.Seki et al,: " Gating circuit developed for high power GTO thyristor", PESC '81 pp 215-225, 1981

(2) T. Jinzenji et al,: "Three phase static power supplies for air-conditioned electric coaches using high power GTO", IEEE IAS Annual Meeting Conference Record pp.1088-1098, 1979

APPLICATION TECHNIQUES FOR HIGH POWER GATE TURN-OFF THYRISTORS

R.L. Steigerwald
General Electric Company
Schenectady, New York

Introduction

A gate turn-off thyristor (GTO) is a reverse blocking triode thyristor that can be turned off as well as on by the application of gate current. The recent trend in high power GTO development has been to show the feasibility of turning off higher and higher anode currents. For example, the development of a device with a turn-off capability of 50 Amperes[1] has been followed by the development of devices with 100 Amperes[2] and then 200 Ampere[3] turn-off capability. Devices with higher than 200 Ampere turn-off capability have also been alluded to in the literature[4]. A considerable amount of work has also been done describing device operation first using a one dimensional model[5,6] and then later using a two dimensional model[7] to account for lateral base currents in the device which occur at turn-off. Further experimental and theoretical work has been conducted to study lateral device currents at turn-off.[8,9] The study of charge dynamics during GTO switching has been investigated[4, 7, 10] and recently a CAD-model has been developed which included the effect of lateral currents in the base layers, non-linear junction recovery and external circuit conditions.[4] This paper investigates several characteristics of a high power GTO which are important from the circuit designers point of view in order that such a device may be applied properly.

The effect of various anode current and voltage waveforms as well as the effect of various gate circuit waveforms are determined for a high power, highly interdigitated GTO. Practical snubber circuits are analyzed and the GTO dynamic characteristics (di/dt and dv/dt) are related directly to snubber losses. The experimental data given in this paper was gathered using a GTO capable of turning off 200 Amperes anode current in less than 1 μsecond by applying a turn-off gate current of approximately 40 Amperes. A description of this device has been reported previously.[3] It was fabricated on a 33mm diameter wafer using a highly interdigitated cathode-gate pattern composed of spiral elements[12] to obtain a good active area utilization. The device also employed an amplifying gate to reduce the turn-on gate drive requirements.

While higher power GTO's are not generally available at this time, there has been and is active interest in high power device development as evidenced by the literature. This paper illustrates many of the considerations that must be included in a practical circuit design and therefore should allow comparisons between GTO, power transistor, and SCR circuits for a given application.

Switching Characteristics and Gating Requirements

To apply the GTO properly, its switching characteristics must be understood. This section describes the switching characteristics and waveforms one would expect in practical circuits. The necessary auxiliary components (e.g., snubbers, di/dt inductors, etc.) that are needed to properly apply a high power GTO are also discussed in this section.

As a vehicle for discussion, the simple dc-dc chopper circuit shown in Figure 1 will be used. The GTO anode voltage and current waveforms that result in such a circuit are similar to those that result in most power conversion circuits. Therefore, the discussion here concerning the dc-dc chopper is applicable to most other power conversion circuits as well.

Turn-On Characteristics

The average voltage supplied to the load by the chopper shown in Figure 1 is a function of the dc supply voltage, E_d, and the ratio of the GTO conduction time to the switching period of the GTO. One conducting period of the GTO is shown in Figure 2. Before time t_1, the GTO is off (i.e., in the blocking state) and coasting diode D3 is carrying the load current, I_d, which is maintained by load inductance L_2. At time t_1 the GTO is gated into conduction and the current begins to transfer from coasting diode D3 to the GTO. The function of inductance L is to limit the rate of rise of the current (di/dt) in the GTO to safe values. The rate at which the current rises in the GTO (also the rate at which current decreases in diode D3) is given by: $di/dt = E_d/L$.

Because the GTO is a highly interdigitated structure, relatively high di/dt ratings compared to a conventional SCR can be expected. A high di/dt capability (small L) is important because not only is the cost and size of the inductor minimized, but the snubber losses will be reduced due to decreased voltage overshoot occurring at turn-off (discussed in next section). At turn-on the GTO also discharges snubber capacitor C through snubber discharge resistor R_1. The rate of rise of this pulse is not limited. This pulse therefore must be limited in magnitude to safe values. However, a certain amount of snubber discharge current appears to be beneficial if single pulse gating is used for the following reason: If the rate of rise of anode current at turn-on is relatively slow, then it is possible that only a few emitter "fingers" will remain latched following the termination of the gate pulse. As current builds up, it does not spread to nonconducting fingers by the usual plasma spreading phenomena since the emitter fingers are each individual islands. Consequently device failure can occur if the current rises to high values because the majority of emitter fingers are not conducting. Note that the snubber discharge provides a relatively high pulse of anode current at turn-on (Figure 2) which brings the entire GTO into conduction. Thus, it is concluded that if single-pulse gating is used, a relatively high di/dt and/or a snubber discharge current should be available. For the GTO discussed above, a snubber discharge of from 13 to 20 Amperes with the anode current never falling below 10 Amperes after the discharge will result in satisfactory operation (i.e., 200 Amperes can be successfully turned off after the anode current has risen to that value).

If pulse train or DC gating is used, the possible difficulties associated with single pulse gating are eliminated. Pulse train gating or dc gating will latch additional fingers into conduction as the anode

Reprinted from *IEEE Ind. Appl. Soc. Meet.*, 1975, pp. 165-174.

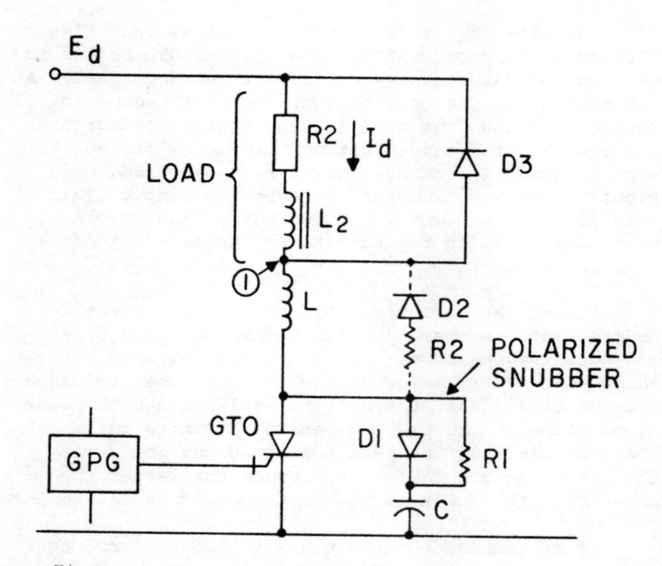

Figure 1 GTO DC-DC Chopper Circuit

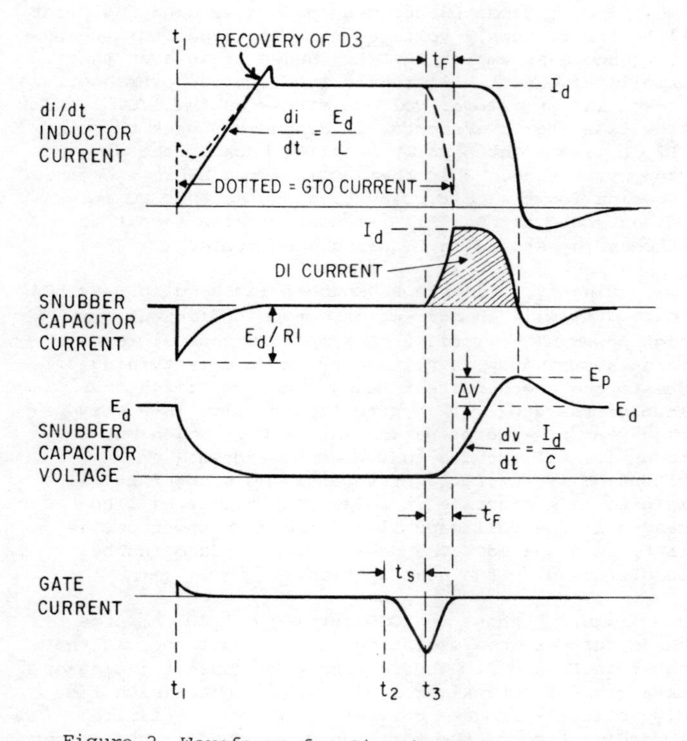

Figure 2 Waveforms for Circuit of Figure 1

current rises, regardless of the rate of the rise of the current. Figure 3 shows the effect of pulse train gating. The device was continuously pulse train gated and the anode current was slowly raised while the forward drop was monitored. The gate pulses used were approximately 25-μs wide at a repetition rate of 10 kHz.

The resulting anode volt-ampere characteristic shown in Figure 3 is a tracing of an oscillogram showing anode current on the horizontal axis with the envelope of forward drop on the vertical axis. The forward drop is multi-valued, because the forward drop is lower with gate current (i.e., when the gate pulse is present) than with no gate current.

The interesting point to be noted relative to Figure 3 is that the forward drop when the gate current is not present initially increases with current and then suddenly decreases as the anode current is raised. This happens several times and then the decrease appears to be more continuous (e.g., in the neighborhood of 24 to 28 Amperes). This phenomenon appears to indicate that as the anode current is raised, more emitter fingers come into conduction and remain latched, due to the pulse train gating.

The forward drop when the gate pulse is present appears essentially continuous, indicating that most of the fingers conduct when driven by gate current but do not necessarily remain conducting when the gate pulse is removed (depending on the anode current available). It is concluded that pulse train or DC gating is sufficient regardless of the anode current shape.

One factor that should be considered when using pulse gating is the shape of the pulse itself. It has been stated[13] that the gate-to-cathode junction capacitance plays a role in both turn-off and turn-on. It has been found experimentally that the falling slope of the gate-pulse current has an effect upon the latching current; that is, the faster the rate at which the trailing edge of the current pulse falls, the higher the latching current will be.

Figure 4 defines the parameters and shows the variation in latching current as a function of the rate of fall of the gate current. The time for which the peak gate current was applied remained constant (120 μs), while only the -di/dt of the gate pulse was varied. These data were taken at relatively low supply voltages, because this represents a more severe condition and is likely to occur in practice. From this curve it can be concluded that if relatively light load currents are to be switched using single-pulse gating, then the rate of fall of the gate current should be relatively slow, to ensure that the device latches and that all the emitter fingers are brought into conduction. The fact that the rate of fall of the gate current has an effect on the latching current has been attributed to the charging and discharging of the gate-cathode junction capacitance.

One important point to be made in this section is that latching current or holding current measured by conventional methods may be a useless parameter. For example, the latching current measured may be only that current necessary to latch on a single emitter finger. Consequently, as the device current rises the device may fail. A more meaningful measure

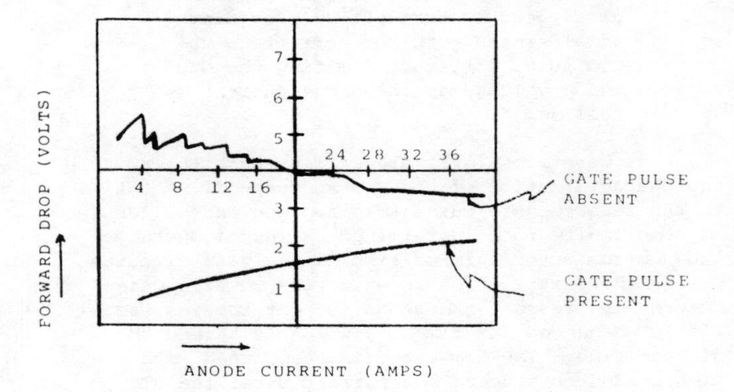

Figure 3 Envelope of Forward Voltage Drop vs. Anode Current for Pulse Train Gating

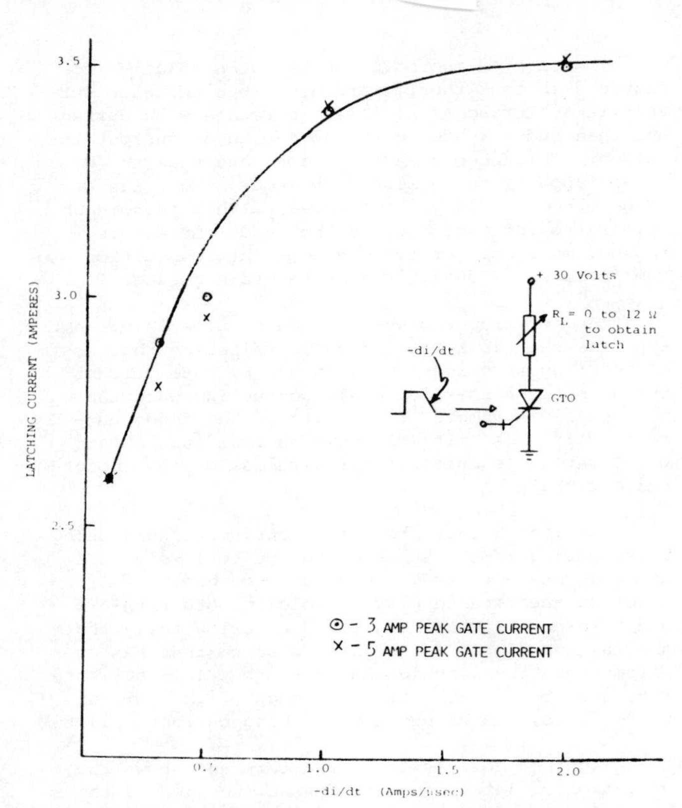

Figure 4 GTO Latching Current as a Function of Rate of Fall of Gate Current Pulse

would be that anode current which must be available at turn-on in order that rated current can be carried and successfully turned off. The forward voltage drop will be a function of the number of emitter fingers conducting.

Turn-off Characteristics

The above comments apply to the turn-on characteristics of the GTO. The following comments apply to the turn-off characteristics. In Figure 1, a snubber comprised of diode D1, capacitor C, and snubber discharge resistor R1 is connected directly across the GTO. The purpose of the snubber is twofold:

- The snubber limits the rate of rise of voltage (dv/dt) during and after GTO turn-off, to prevent retriggering of the GTO.
- The snubber limits the peak power dissipated at turn-off by providing an alternate path for current as the GTO turns off, thus limiting the GTO anode voltage as the anode current is falling.

The waveforms in Figure 2 illustrate the switching characteristics of the GTO at turn-off. At time t_2 the reverse gate current is applied and builds up at some finite rate (determined by lead inductance and semiconductor turn-on time in the gate circuit). During the initial time in which the reverse gate current is present, the anode current remains essentially unchanged for time t_s, which is termed the storage time. The time, t_s, is associated with the time needed to squeeze the current under the GTO emitter fingers to a small dimension and with the time to remove excess carriers from the p-base of the GTO.

The time, t_s, is analogous to the storage time of a power transistor. After time t_s, the GTO begins to come out of saturation and turn off anode current. At the same time, the gate-cathode junction begins to recover and thus the gate current begins to decrease. Subsequently the anode current turns off, falling to zero in time t_F (the fall time, t_F, is defined more precisely as the time for the anode current to fall from 90 to 10 percent). Concurrently, the reverse gate current falls toward zero as the gate cathode junction recovers.

During the fall time, t_F, the anode current is shunted into snubber capcitor C through diode D1 and consequently the anode voltage rises. Because the anode voltage is rising during the time when the anode current is falling, there temporarily is a high power dissipation in the GTO. Capacitor C limits this peak power by limiting the peak voltage occurring during the anode current fall time. Hence the larger the capacitor, the lower is the peak power dissipation.

After the anode current has fallen to zero, the capacitor voltage rises at a constant rate of dv/dt = I_d/C, where I_d is the anode current being turned off. The voltage continues to rise at this rate until the capacitor voltage exceeds the dc supply voltage at which point diode D3 becomes conductive clamping point 1 to the dc supply voltage, E_d. The capacitor then overshoots an amount $I_d\sqrt{L/C}$. Hence it is seen that a small value of L will result in a smaller overshoot, resulting in a lower voltage stress on the GTO. After the capacitor overshoots, it discharges back through R1, L, and D3 eventually settling back to the dc supply voltage. When the GTO is gated back into conduction for the next conduction cycle, the GTO must discharge C through R1. R1 must dissipate most of the energy stored in C during each cycle.

Finally, a simple RC snubber (as used in many SCR circuits) will in most cases not be sufficient for high power GTO circuits. A simple RC snubber results in an abrupt jump in voltage as the GTO is turning off, due to the anode current being shunted through the snubber resistor. This situation results in a very high peak power dissipation in the GTO, which may lead to device failure due to second breakdown. The situation is further aggravated if no snubber at all is used. In practice, a polarized snubber will be needed if the full capability of a high power device is to be attained. The design of polarized snubbers is discussed in the next section.

Figure 5 shows the GTO turning off 200 Amperes anode current when operating in a circuit such as that shown in Figure 1. The reverse gate current increases as a ramp of approximately 30 A/μs. This finite rise time of turn-off gate current is due to the finite switching time of the semiconductors used in the gating circuit, as well as a small amount of lead inductance present.

As can be seen in Figure 5, the storage time with the ramped gate current is approximately 0.5 μs, after which time the anode current begins to fall and the gate current begins to decrease as the gate-cathode junction voltage increases (the gate current is produced from a voltage source).

Due to the snubber, the anode voltage rises as a ramp, thus limiting dv/dt and turn-off dissipation. Approximately an 80-volt spike occurs across the GTO as the anode current is falling. The spike is the result of the high di/dt developed by the falling anode current (≈320A/μs), acting in conjunction with any snubber lead inductance and the turn-on charac-

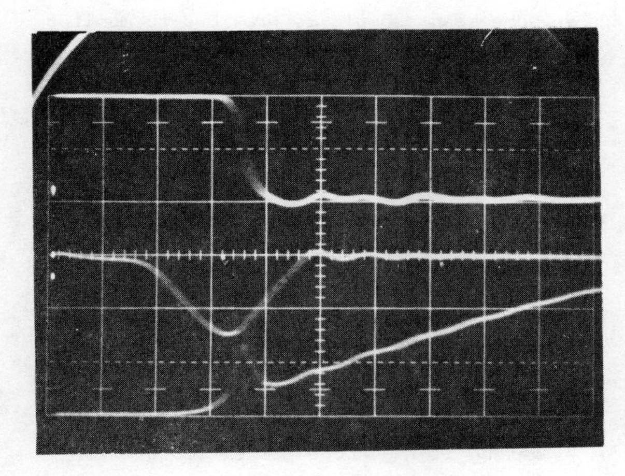

FIGURE 5 GTO TURNING OFF 200 AMPS

TOP: ANODE CURRENT = 100 AMPS/CM
MIDDLE: GATE CURRENT = 20 AMPS/CM
BOTTOM: ANODE VOLTAGE = 50 VOLTS/CM
SWEEP: 1.0 μSEC/CM

teristics of the snubber diode. To reduce this spike and thus decrease turn-off dissipation in the GTO, a fast turn-on diode should be used for D1 and the snubber should be mounted as close as possible to the GTO, to eliminate lead inductance. The power GTO is a relatively fast switching device and is capable of high-frequency operation.

During turn-off of a high power GTO the gate cathode junction should not be avalanched. The gate voltage necessary to establish the turn-off current should be less than the gate-cathode avalanche voltage. Otherwise, the region near the gate contact would avalanche and additional gate current would not contribute to the turn-off process but rather would simply be avalanche current. Proper device design eliminates such a junction avalanche for the range of turn-off gate currents expected. It is also necessary that the gate not be driven into avalanche by the external circuit immediately after turn-off. Experience indicates that if the junction is avalanched immediately after turn-off of currents near the rated value, device failure will occur. A high-power GTO is a highly interdigitated structure, and in practice each emitter finger will not have exactly the same breakdown voltage. Thus avalanching probably occurs in a localized region leading to localized power dissipation and device failure at turn-off. It was found that some gate avalanching could be tolerated while turning off lower anode current but failure would occur if substantial gate avalancing was allowed to occur while turning off currents near the rated value.

As discussed in the above section, the shape of the GTO turn-on gate pulse can affect the GTO latching current, due to the effect of the gate capacitance. In this section, the effect of gate capacitance is investigated as far as it affects GTO turn-off.

Immediately after turn-off, many gating circuits will produce a ramp of gate voltage from a negative value toward zero as the gate turn-off voltage is removed. Such a ramp can actually produce a positive gate current even though the gate voltage is always negative. Thus, there is the danger of inadvertently turning the device back on. As the gate voltage is ramped toward zero, the junction capacitance discharges and a forward gate current is generated, depending on the rate of the rise of the gate voltage. Data showing the dependence of the peak forward gate current generated as a function of the dv/dt at the gate are shown in Figure 6. A linear relationship exists, resulting in a capacitance of approximately .071 μF. This capacitance compares with small signal bridge measurements of from 0.1 to 0.06 μF, depending on the magnitude of the reverse bias.

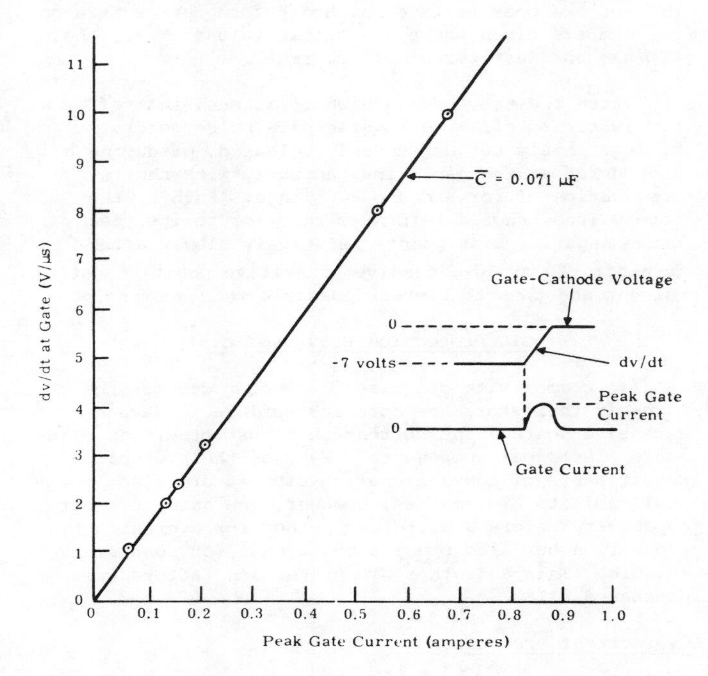

Figure 6 Peak Forward Gate Current Caused by dv/dt at Gate

The next question to be answered is whether this capacitively generated gate current can turn the GTO on with a constant voltage applied to its anode. With an anode voltage of 400 volts, the steepness of the gate voltage rise was increased to obtain approximately 1.2 Ampere peak gate current. Under these conditions, the GTO could not be turned on. Some forward anode current was generated. However, the device could not be made to latch on, due to either insufficient anode current or the shortness of the capacitively generated gate pulse (≈0.5 μs). The next question to be answered is whether or not the simultaneous application of the gate ramp and reapplied forward anode voltage would decrease the dv/dt capability of the device. This is the expected situation that would occur in practice.

To answer this question, a reapplied dv/dt of approximately 750 volts/μsec was applied to the device anode simultaneously with a ramped gate-cathode voltage. Other gate conditions were also tested with the same anode voltage waveform. In each case, the peak voltage at which the GTO turned on was taken as the relative measure of dv/dt capability under the various gate conditions. The results are listed below:

Gate Condition	Peak Voltage (volts)	Relative Voltage
Open	170	1
Shorted	185	1.09
-5-volt bias	240	1.41
Ramped (≈10V/μs from -7 to 0 volts)	160	0.94

Compared to the open circuited gate case, the anode voltage reached was approximately 10 percent higher for the shorted gate and 40 percent higher for the reverse bias case. Note that the voltage reached for the ramped gate case was less than for the open gate case. Thus, though the gate voltage was negative, the fact that it was ramping and producing a forward capacitive gate current (≈1 Ampere in this case) decreased the peak voltage reached before device turn-on. The numbers given above are static values (i.e., the GTO has not just turned off current).

From the above discussion it is seen that after a GTO is turned off, the reverse gate pulse generator voltage should not be suddenly released, producing a high dv/dt at the gate simultaneously with the reapplication of forward anode voltage. Rather, the gate voltage should return to zero (or to its steady state negative bias point) relatively slowly after turn-off, to avoid excessive capacitive positive gate current and thus to prevent possible retriggering.

GTO Protection Requirements

In common with other semiconductor devices, gate turn-off thyristors are more susceptible to damage by excessive voltage and current than most other nonsolid-state electrical components. Many of the protection requirements of conventional thyristors are also applicable to GTO devices; however, the gate turn-off capability offers a different method for overcurrent protection but also opens a new avenue for possible failure. This and other GTO protection factors are discussed below.

Overcurrent Protection

Many electric power converters or other apparatus employing power semiconductors require that the equipment be designed to deliver a certain amount of overcurrent to the load for a certain time and to protect the load against higher currents for longer durations. Such operation must be regarded as normal for the switching devices and must be within their repetitive rating. However, if such an overcurrent control system is not provided or fails to operate properly, or if there is an internal failure in the converter equipment, fault conditions will prevail and the devices must be protected as much as possible. For conventional thyristors, the surge current should be limited to less than the rated capability of the device by suitable surge limiting impedances, eventual natural or forced commutation of the surge, fuses, or fast circuit breakers.

Because GTO devices also have a significant surge capability, they may also be protected in the same way; however, any attempt to extinguish a high surge current by gate turn-off control is almost certain to result in destruction of the device. The limit of safety is the nonrecurrent peak turn-off capability, $I_{ATO(Max)}$. The usual failure mechanism of the GTO under an overcurrent condition is reverse bias second breakdown, just as in a power transistor. For most practical circuits, the reapplied dv/dt will increase in direct proportion to the anode current, because the snubber capacitor is usually a fixed value. For example, in the circuit illustrated in Figure 1, current I_d will build up to large values if the load is short circuited. As the GTO turns off, the reapplied $dv/dt = I_d/C$. Thus the peak power dissipated by the GTO will increase very rapidly with the anode current.

From the above discussion, the following design techniques are recommended:

- Some means should be provided to monitor the anode current in a GTO device and to suppress the generation of any gate pulses if a level exceeding I_{ATO} (nonrepetitive) is detected. The conventional surge limit will then be in effect.

- If a current level between the repetitive and nonrepetitive turn-off rating is sensed, turn-off action should be initiated at once, on a one-shot basis. The current sensing device is already available in many power conversion circuits, to perform other functions. In any event, the current sensor should have a suitably fast response, so corrective action can be taken to gate the GTO off at less than I_{ATO} (nonrepetitive) in the event of a fault or to suppress gating if the anode current is greater than I_{ATO} (nonrepetitive).

Blocking Voltage Protection

As in the case of conventional thyristors, a GTO will breakover into conduction if a critical forward voltage is exceeded and will enter an avalanche state if a critical reverse voltage is exceeded. Turning the device on due to exceeding the critical forward voltage is not a recommended mode of operation, due to the interdigitated nature of the GTO, unless the anode current available is very limited. That is, if the GTO is brought into conduction due to exceeding its forward breakover voltage, the greatest probability is that only one or perhaps a few emitter fingers will be brought into conduction. Thus the entire available current will be in only a few fingers (a small area of the device), leading to device failure. Turning the GTO on by exceeding its breakover voltage therefore should be avoided. Limitation of overvoltages during the dissipation of energy trapped in inductive elements when a GTO is commutated off is an important consideration.

For example, in Figure 1 energy is trapped in inductance L when the GTO is turned off and thus the voltage on capacitor C overshoots the supply voltage as L and C go into a quarter cycle of oscillation. Limitation of peak GTO voltages in such circuits is considered in the next section with regard to snubber circuit design.

Snubber Circuits

As the previous discussion has illustrated, auxiliary circuit components are needed if a given GTO is to reach its full capability. This fact is true for all high power semiconductor devices. This section will consider the snubber circuit in Figure 1 and discuss the effect of GTO parameters on the design of such a circuit.

The function of L is to limit the rate of the rise of the current in the GTO when the device is gated into conduction and also to limit the peak reverse recovery current, in feedback rectifier D3, which otherwise would be substantial and even destructive. Capacitor C serves to limit reapplied dv/dt to the GTO when the device is turned off and thus limits peak power dissipation at turn-off. Three types of snubbers are described in the following paragraphs.

Undamped Snubber

The snubber illustrated in Figure 1 without R2 and D2 is termed an "undamped snubber" because at turn-off the voltage on capacitor C overshoots the dc supply voltage an amount determined by the undamped resonant characteristics of L and C. Figure 2 illustrates the waveforms occurring in such a chopper for one conduction cycle of the GTO.

As the GTO is turned off at time t_3, the voltage applied to the GTO is limited by capacitor C. After a short time (approximately 1 μs for the GTO discussed in this paper), all of the anode current will be shunted through D1 and will be in snubber capacitor C, resulting in a reapplied dv/dt of:

$$\frac{dv}{dt} = I_d/C \tag{1}$$

where I_d is the dc load current which can be assumed constant during the commutation intervals due to the filtering action of the load inductance.

The capacitor voltage continues to rise at the linear rate given by Equation 1 until it reaches the dc supply voltage, E_d. At that instant, point 1 becomes effectively clamped to E_d, and continued voltage rise on C reduces the current in L resonantly to zero. In the design of the snubber components, the amount of overshoot on capacitor C must be determined, because this amount gives the peak voltage applied to the GTO.

The peak overshoot may be determined by simply considering the energy transfer as follows. The energy increase in C is equal to the energy decrease in L plus the energy from the dc supply:

$$\frac{1}{2} C(E_p^2 - E_d^2) = \frac{1}{2} LI_d^2 + E_d \int idt \tag{2}$$

where E_p is the peak capacitor voltage reached. Because i = C dv/dt, where v is the capacitor voltage:

$$\frac{1}{2} C(E_p^2 - E_d^2) = \frac{1}{2} LI_d^2 + E_d \int_{E_d}^{E_p} Cdv \tag{3}$$

Performing the integration leads to:

$$\frac{C}{2} (E_p - E_d)^2 = \frac{1}{2} LI_d^2 \tag{4}$$

which results in a peak capacitor (GTO) voltage of:

$$E_p = E_d + I_d \sqrt{L/C} \tag{5}$$

After the capacitor reaches the peak voltage, E_p, it discharges back through R1 and L to the dc supply, eventually settling out at a voltage equal to the dc supply voltage.

As the capacitor discharges back to the dc supply through R1 and L, the energy dissipated in R1 is the difference between the energy lost from capacitor C and the energy delivered to the dc supply (the energy change in L is zero, because the current in L begins at zero and ends at zero).

Energy dissipated in R1 =

$$\frac{1}{2} C (E_p^2 - E_d^2) - E_d \int (-i)dt \tag{6}$$

Again, because i = C dv/dt:

Energy dissipated in R1 =

$$\frac{1}{2} C (E_p^2 - E_d^2) + E_d \int_{E_p}^{E_d} C\,dv =$$

$$\frac{1}{2} C (E_p - E_d)^2 \tag{7}$$

Substituting Equation (5) into Equation (7) yields:

$$\text{Energy dissipated in R1} = \frac{1}{2} C (I_d \sqrt{L/C})^2$$

$$= \frac{1}{2} LI_d^2$$

= Energy trapped in L just before turn-off (8)

Thus the energy trapped in the di/dt inductance just before turn-off must eventually be dissipated in R1, the snubber discharge resistor. When the GTO turns on for the next conduction cycle, essentially the entire energy stored in C must be dissipated in R1 (= $1/2CE_d^2$). Therefore summing this energy with the energy of Equation 8 and multiplying it by the chopping frequency gives the total power rating of R1:

$$\text{Power rating of R1} = \frac{1}{2} LI_d^2 f + \frac{1}{2} CE_d^2 f \tag{9}$$

Equation 9 should be evaluated for the worst case conditions (i.e., maximum I_d, E_d, and f). The inductance, L, is selected to limit the di/dt when the GTO turns on to safe values for both the GTO and the feedback rectifier (to limit peak reverse current):

$$L = \frac{E_d}{di/dt} \tag{10}$$

Combining equations 1, 9, and 10 results in

Power rating of R1 =

$$\frac{1}{2} E_d I_d f \left[\frac{I_d}{(di/dt)} + \frac{E_d}{(dv/dt)} \right] \tag{11}$$

Equation 11 shows that a GTO with high dv/dt rating (small C) and a high di/dt rating (small L) results in the lowest power dissipation in R1 (i.e., the most efficient chopper operation). The capacitor, inductor, and diode (D1) ratings must be such that the waveforms described in Figure 2 applied at the desired frequency result in operation within the ratings of those devices. Note that it is desirable to minimize the quantity in brackets in Equation (11). Therefore, if a GTO is designed for a low voltage, high current application (low E_d, high I_d) a device design with emphasis on high di/dt capability will result in lower snubber losses. Alternately, a device with high dv/dt capability results in lower snubber losses for high voltage low current applications. The above comments apply to GTO's when operating in chopper circuits and inverter circuits with the type of snubber given in Figure 1. Of course, the switching times of the GTO as well as its forward drop should be minimized to reduce device dissipation to the lowest possible value.

Because in some cases the overshoot occurring on the GTO after turn-off (due to resonant charging of C) may be undesirable, a damping resistor may be added to limit this overshoot as discussed below.

Damped Snubber

Figure 1 shows the addition of a damping resistor, R2, to limit the peak overshoot on capacitor C after the GTO turns off. The peak overshoot voltage can be determined by considering the instant after the GTO has already turned off and the capacitor voltage has charged just to the dc supply voltage, E_d. At this time diodes D2 and D3 become conductive and the equivalent circuit reduces to that shown in Figure 7a, with the initial conditions shown. Remembering that an uncharged capacitor in series with dc voltage is equivalent to a capacitor charged to that same voltage and remembering that an inductor with zero initial current in parallel with a current source is equivalent to an inductor with an initial current equal to that current source, the circuit shown in Figure 7a reduces to that shown in Figure 7c. The peak voltage across resistor R2, which can then be added to E_d to obtain the peak overshoot voltage, is desired. The Laplace transform equation for the voltage across R2 of the damped snubber circuit is with the aid of Figure 7c:

$$v(s) = \frac{I_d L}{LCs^2 + (L/R)s + 1} = \frac{I_d L\omega_o^2}{s^2 + 2\zeta\omega_o s + \omega_o^2} \qquad (12)$$

where:

$$\zeta = \text{Damping factor} = \frac{1}{2(R2)}\sqrt{L/C}$$

$$\omega_o = \text{Undamped resonant frequency} = 1/\sqrt{LC}$$

The time solution of Equation 12 has three forms, depending on the amount of damping

- Underdamped case ($\zeta<1$):

For the case in which $\zeta<1$:

$$v(t) = \frac{I_d L\omega_o}{\sqrt{1-\zeta^2}} \exp(-\zeta\omega_o t)\sin\omega_o\sqrt{1-\zeta^2}\,t \qquad (13)$$

The maximum value is found by differentiating v (t) and setting the result equal to zero, resulting in:

$$\tan\omega_o\sqrt{1-\zeta^2}\,t = \sqrt{1/\zeta^2 - 1} \qquad (14)$$

The maximum value is then given by:

$$v(max) = \frac{I_d L\omega_o}{\sqrt{1-\zeta^2}}\left[\exp\left(-\zeta W^*/\sqrt{1-\zeta^2}\right)\right]\sin W^* \qquad (15)$$

where: $W^* = \omega_o\sqrt{1-\zeta^2}\,t = \arctan\sqrt{1/\zeta^2 - 1}$.

- Critically Damped Case ($\zeta = 1$):

For the critically damped case, Equation 12 reduces to:

$$v(s) = \frac{I_d L\omega_o^2}{(s+\omega_o)^2} \qquad (16)$$

resulting in a time solution of:

$$v(t) = I_d L\omega_o^2 t[\exp(-\omega_o t)] \qquad (17)$$

Differentiating and setting the result equal to zero, gives the maximum value as:

$$v(max) = \frac{I_d L\omega_o}{e} \qquad (18)$$

where e is the base of the natural logarithm.

- Overdamped Case ($\zeta>1$):

For the overdamped case the roots of the denominator of Equation 12 are real. Equation 12 then reduces to a Laplace transform equation of:

$$v(s) = \frac{I_d L\omega_o^2}{b-a}\left[\frac{1}{s+a} - \frac{1}{s+b}\right] \qquad (19)$$

where $a,b = -\omega_o(\zeta \mp \sqrt{\zeta^2 - 1})$. The inverse transform of Equation 19 gives the time response as:

$$v(t) = \frac{I_d L\omega_o^2}{b-a}\left[\exp(-at) - \exp(-bt)\right] \qquad (20)$$

The maximum value is found by differentiating Equation 20 and setting the result equal to zero, resulting in:

$$v(max) = \frac{I_d L\omega_o^2}{b-a}\left[\exp(-aT^*) - \exp(-bT^*)\right] \qquad (21)$$

where $T^* = \dfrac{\ln(b/a)}{b-a}$

To aid in the design of the damped snubber, Equations 15, 18, and 21 are plotted in normalized form in Figure 8. The peak voltage is normalized by dividing the maximum voltage by $I_d\omega_o L = I_d\sqrt{L/C}$. Thus after the di/dt inductor, L, and the snubber capacitor, C, are selected using Equations 1 and 10, the value of the damping resistor can be determined, which results in the desired overshoot using Figure 8 (the peak current being turned off, I_d is assumed known).

Note that from Figure 8 the peak voltage reached by the snubber capacitor is:

$$E_p = E_d + KI_d\sqrt{L/C} \qquad (22)$$

which is the same as Equation 5, with the overshoot reduced by the factor K.

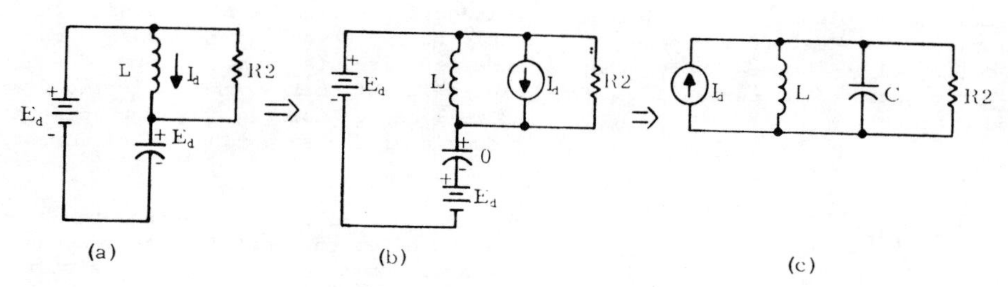

Figure 7 Equivalent Snubber Circuit to Determine Overshoot of Capacitor Voltage at GTO Turnoff

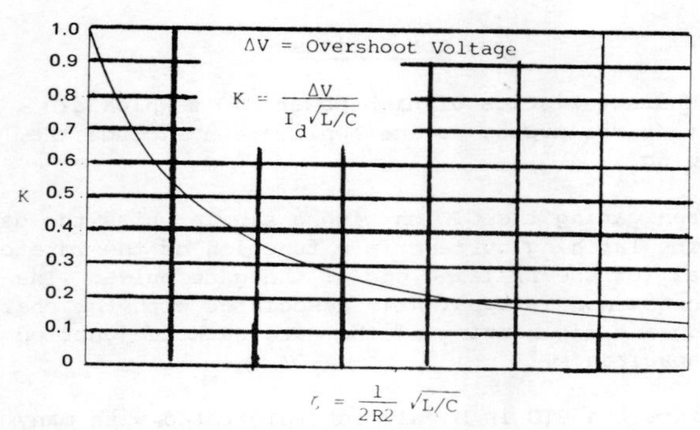

Figure 8 Normalized Overshoot Voltage for Damped Snubber Versus Damping Factor

During the overshoot interval the energy dissipated in R2 may be determined by referring to the energy transfers in the equivalent circuit of Figure 7c. The energy out of the current source, $I_d\int e dt$, must equal the energy stored in L (in the equivalent circuit) at the end of the overcharge interval, $1/2\ LI_d^2$, plus the energy stored in C (in the equivalent circuit) at the end of the overcharge interval, $1/2\ C(E_p-E_d)^2$ plus the energy dissipated in R2. Since $e = Ldi/dt$ where i is the inductor current, the above energy relation can be stated as

$$I_d\int_0^{I_d} L di = \frac{1}{2} LI_d^2 + \frac{1}{2} C\ (E_p-E_d)^2 +$$

energy dissipated in R2 (23)

Using equation (22) for E_p-E_d results in

$$\text{Energy dissipated in R2 during overcharge interval} = \frac{1}{2} LI_d^2(1-K^2) \quad (24)$$

After the capacitor has charged to its peak value, it discharges back through R1 and the parallel combination of L and R2 to the dc supply, eventually settling out at the dc supply voltage. The total energy dissipated in the two resistors, R1 and R2, during this discharge can be determined by considering the energy transfers:

Energy out of C = energy into E_d + energy change in L + energy dissipated in R1 and R2 (25)

$\frac{1}{2}C(E_p^2 - E_d^2) = E_d\int -idt + 0 +$ energy dissipated in R1 and R2. Since

$$i = C\ dv/dt \quad (26)$$

where v is the capacitor voltage:

Energy dissipated in R1 and R2 =

$$\frac{1}{2} C(E_p^2 - E_d^2) + E_dC\int_{E_p}^{E_d} dv =$$

$$\frac{1}{2} C(E_p - E_d)^2 \quad (27)$$

However, because the overshoot is given by Equation 22:

Energy dissipated in R1 and R2 =

$$\frac{1}{2} C(KI_d\sqrt{L/C})^2 = K^2 \frac{1}{2} LI_d^2 \quad (28)$$

which is the same as Equation 8 for the undamped case modified by constant K^2.

Resistor R1 must also dissipate most of the energy stored in C each time the GTO is gated on $(1/2\ CE_d^2)$. Therefore, adding this energy to that dissipated in R2 during the capacitor overshoot, to that given in Equation (28) and multiplying by the chopping frequency results in the total power dissipated in R1 and R2

Power Dissipated in R1 and R2 =

$$\frac{1}{2} CE_d^2\ f + \frac{1}{2} LI_d^2\ f \quad (29)$$

which is the same as that given in Equation 9. The same conclusions are reached concerning the GTO di/dt and dv/dt rating were reached in the discussion following Equation 11. The results obtained above are also applicable to other power circuits. For example, consider one leg of a bridge inverter circuit as shown in Figure 9 which is a basic building block for many inverter circuits.

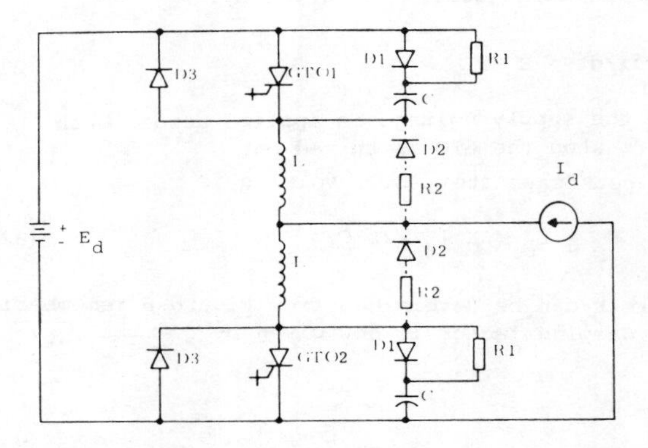

Figure 9 Inverter Leg Showing Snubber Components

The diode and resistor damping networks are shown connected by dotted lines, because they may or may not be needed, depending on the amount of voltage overshoot tolerable following GTO turn-off. The usual case is for an inductive load, thus the load current, I_d, can be considered constant during the commutation interval. The amount of overshoot voltage that the GTO must block can be approximated by considering the instant after GTO2 turns off and its snubber capacitor, C, has charged just to the supply voltage, E_d. The snubber charges to this value at approximately a linear rate, given by:

$$dv/dt = I_d/C \quad (30)$$

When the capacitor has charged to E_d, the condition of the circuit is as given in Figure 10. (The effect of R1 and snubber capacitor C associated with GTO1 are being neglected during the charging time). The inductor, L, associated with GTO1 is carrying zero initial current. The GTO2 snubber capacitor, C, is also charged to the dc supply voltage, E_d. Using these values as initial conditions, the overshoot, E_d, can be determined and thus the peak voltage applied to the GTO can be obtained.

The equivalent circuit can be simplified considerably by remembering that a capacitor with initial voltage E_d is equivalent to an uncharged capacitor in series with a voltage source of voltage E_d. An inductor with initial current I_d is also equivalent to an inductor with zero initial current in parallel with a current source of current I_d. Using these two facts,

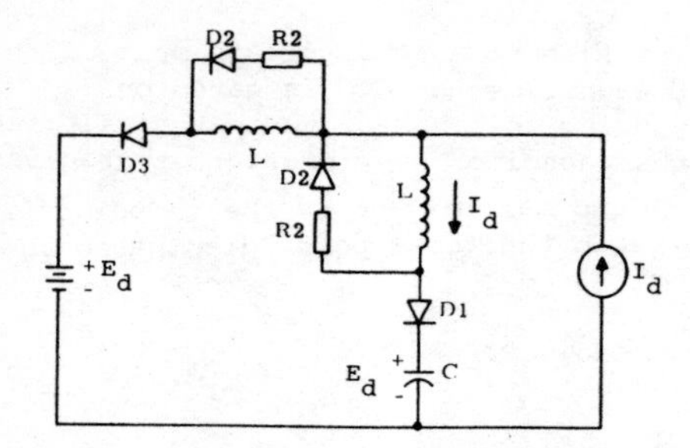

Figure 10 Equivalent Circuit for Determining Overshoot Voltage After GTO2 Turnoff

the equivalent circuit of Figure 10 can be reduced, as shown in Figure 11 to a simple parallel RLC circuit. This circuit is the same as that shown in Figure 7c, except L and R are multiplied by two. Thus the solution of the circuit is the same as that worked out above with R and L replaced by 2R and 2L, respectively.

The di/dt at turn-on is given by:

$$di/dt = E_d/2L \tag{31}$$

because the supply voltage is applied across both inductors when the GTO is turned on.

The peak capacitor (GTO) voltage is

$$E_p = E_d + K\, I_d \sqrt{2L/C} \tag{32}$$

where K can be determined from Figure 8 remembering that the damping factor is now given by $\zeta = \frac{1}{4R2}\sqrt{2L/C}$.

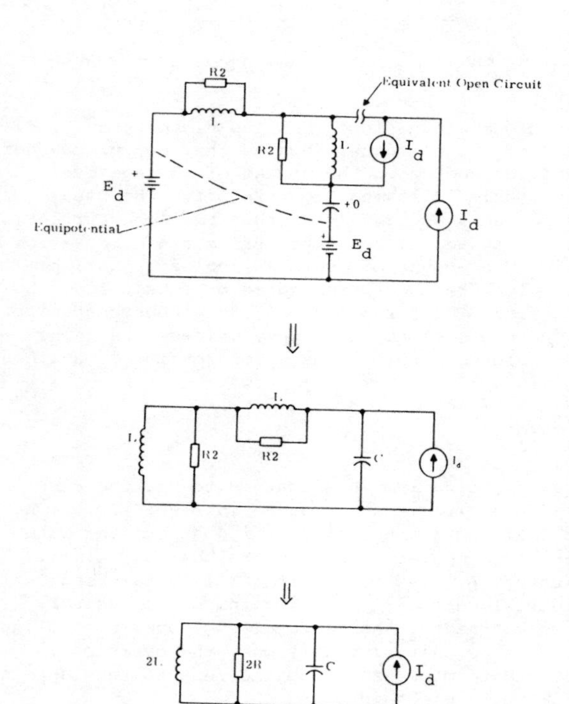

Figure 11 Reduction of Equivalent Circuit of Figure 10

Conclusions

Characteristics of high power GTO's which are important for proper device application include the following:

1. When gating the GTO on with a single pulse the device latching current is a function of the rate of fall of the trailing end of the gate pulse. The phenomenon is apparently associated with the charging and discharging of the gate-cathode junction capacitance.

2. Since the GTO is highly interdigitated with many individual emitter fingers, it is only necessary for a single finger to become conductive for the device to latch. If significant anode current is not available during single pulse turn-on, only a few fingers may come into conduction which could cause device failure if the anode current is subsequently raised to high values. DC or pulse train gating eliminates this problem.

3. To obtain the full turn-off capability of the device, the GTO should be bypassed with a polarized snubber to limit dv/dt at turn off and thus limit turn-off dissipation. Also, a small inductance in series with the GTO is needed to limit turn-on di/dt.

4. The gate characteristics of the GTO at turn-off are similar to those of a power transistor with a small resistance in series with the base.

5. The cathode-gate junction should not be avalanced at turn off if the full turn-off capability of the device is to be obtained.

6. After turn-off, the rate of change of gate voltage from a negative value towards zero can actually produce a positive gate current (even though the gate voltage is negative) due to the reverse biased cathode gate junction capacitance discharging. If this rate of change of gate voltage is rapid enough, the reapplied anode dv/dt capability of the GTO may actually be decreased below the open gate case.

7. Equation 11 shows that, for the type of switching circuits discussed, snubber losses are inversely proportional to the di/dt as well as the dv/dt device ratings. In particular, a low voltage high current application has lower snubber losses with a device having a high di/dt rating while a high voltage low current application has lower snubber losses with a device having a high dv/dt rating.

Acknowledgment

The GTO used to gather data for this paper was developed at the General Electric Semiconductor Products Department, Auburn, New York. This work was partially supported by the Air Force Aero Propulsion Laboratory of the Wright Patterson Air Force Base, Ohio under contract No. F33615-71-C-1729.

References

1. Final Report, June 1969, "50 Ampere Turn-off Switch", Contract F-33615-68-C-1177, Westinghouse Electric Corp., Youngwood, Pa.

2. Y.C. Kao and J.B. Brewster, "Recent Development in Gate Controlled Switches", 1973 Power Electronic Specialists Conference Record, p. 90-96.

3. E.D. Wolley, R. Yu, R.L. Steigerwald, F.M. Matteson, "Characteristics of a 200 Amp Gate Turn-off Thyristor", IEEE Conference Record, 1973 Annual Meeting of Industry Applications Society, p. 251-257.

4. Y.C. Kao, J.B. Brewster, "A Description of the Turn-off Performance of the Gate Controlled Switches", IEEE Conference Record, 1974 Annual Meeting IAS, pp. 689-693.

5. R.H. VanLigten and D. Newon, "Base Turn-off of PNPN Switches", 1960 IRE WESCON Conference Record, Pt. 3, p. 49-52.

6. J.M. Goldey, I.M. MacKintosh and I.M. Ross, "Turn-off of PNPN Switches", 1960 IRE WESCON Conference Record, Pt. 3, p. 49-52.

7. E.D. Wolley, "Gate Turn-off in p-n-p-n Devices", IEEE Transactions on Electron Devices, Vol. ED-13, No. 7, July 1966, p. 590-597.

8. V.A. Gorokhov and G.V. Koshelyayev, "Analysis of Thyristor Turn-off Transients under the Action of the Control Electrode", Telecommunication Radio Engineering, Pt. 2 (USA), No. 12, December 1971, p. 113-117.

9. A.M. Bashirov, Sh.D. Burkhanov, V.V. Garshenin, V.I. Rodov, "Multi-dimensional Nature of Turn-off in a p-n-p-n Structure", Fiz. Tekh. Poluprovodnikov (USSR), Vol. 3, No. 5, p. 633-5 (May 1968), Translation in Soviet Physics-Semiconductors, Vol. 3, No. 5, p. 541-543 (November 1969).

10. T.C. New, W.D. Frobenius, T.J. Desmond, D.R. Hamilton, "High Power Gate-Controlled Switch", IEEE Transactions on Electron Devices, Vol. ED-17, No. 9, September 1970, p. 706-710.

11. M. Kurata, "A New CAD-Model of a Gate Turn-off Thyristor" 1974 IEEE Power Electronics Specialists Conference Record p. 125.

12. H.F. Storm, J. St. Clair, "An Involute Gate-Emitter Configuration for Thyristors and Transistors", IEEE International Electron Devices Meeting, Washington, D.C., December 3-5, 1973, (Technical Digest 73 CHO 781-5ED).
(See also Patent #3,609,476 by H.F. Storm, pat. Sept. 28, 1971).

13. D.R. Grafham, "Now the Gate Turn-off Switch Speeds up DC Switching", Electronics, March 23, 1964, p. 64-71.

Hiroshi Fukui, Hisao Amano
(Hitachi Research Laboratory, Hitachi, Ltd.)
Hisaichi M:ya
(Hitachi Works, Hitachi Ltd.)

ABSTRACT-Direct paralleling of gate turn-off thyristors has been investigated. Gate coupling, which directly connects both gate terminals, determins the current balancing ability of the device itself, and is useful for reducing the transient current unbalance. Any mismatch in the switching time is cancelled by the commutation of the gate current between parallel devices in gate coupling.

Consideration must be given to the mismatch of the cathode wiring resistance and inductance. This is closely related to the transient current unbalance, and symmetrical cathode wiring is necessary for realization of direct paralleling in gate coupling.

1. INTRODUCTION

Paralleling of gate turn-off thyristors (GTO) has been realized using an anode current balancer[1]. Paralleling with no anode current balancers, i.e. direct paralleling, is useful in simplifying large capacity power converters. Direct paralleling also makes it easier to replace a large current GTO with paralleled small current GTO's. For the equipment handling a large current over several hundred amperes, a flat type GTO is used, although it is more expensive than the stud type. By replacing the flat type GTO with the direct-paralleled stud type GTO, costs of a large capacity power can be reduced.

Direct paralleling is an important technique in the application of a GTO, however it is not well-understood.

This paper presents stationally and transient operations for direct paralleling of GTO's. The current balancing ability of GTO's caused by the commutation of the gate current between parallel devices is described.

2. EXPERIMENTAL METHOD

The device (GFF200E) used in the experiments has a blocking voltage of 1200 V and controllable current of 200 A, which is stored in the stud type package.

Current unbalance in the stationally and the tranient state is generally attributed to the mismatch of the on-state voltage and the switching time. However, in case of GTO paralleling, exchange of the gate current between paralleled devices also has much influence on current unbalance, as noted later. In order to investigate this influence, the following two methods of gate connection are used in the experiments ;

1) The external impedances are connected in series with each gate terminal, so that the gate current can be equally divided among the devices, (gate decoupling).
2) The gate terminals are directly connected with each other, so that the gate current can be divided by the gate impedance of the device itself (gate coupling).

Current unbalance is compared for the two methods of gate connections in Fig. 1 to investigate of gate interaction. External impedance is determined as shown in Fig. 1 so that a whole gate current can flow in the same way for the both methods.

Fig. 2 shows typical switching waveforms of both devices. Current unbalance is evaluated using the current value specified by the following. Current unbalance ΔI_p and ΔI_T during turn-on and steady on-state are given by

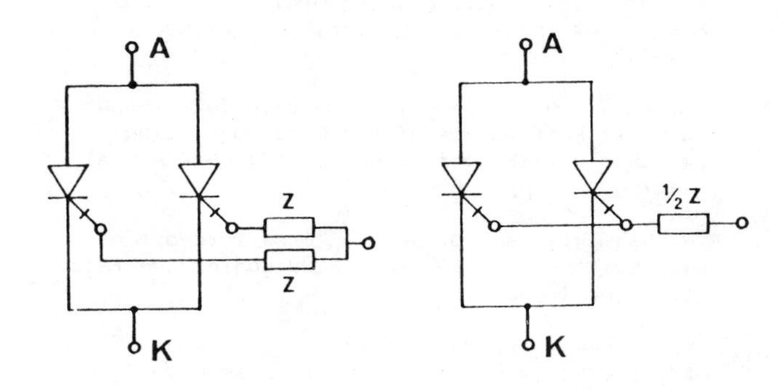

Fig. 1 Connections in direct paralleling of GTO's.

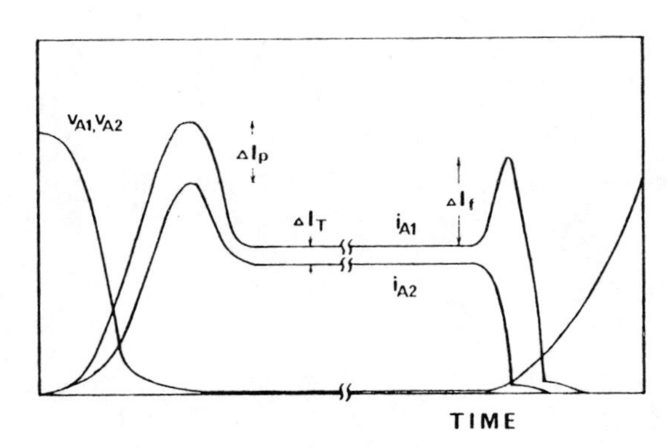

Fig. 2 Switching waveforms of parallel devices.

$$\Delta I_p = I_{p1} - I_{p2} \tag{1}$$

$$\Delta I_T = I_{T1} - I_{T2} \tag{2}$$

where I_p is the peak anode current during turn-on, I_T is the conducting current during the on-state, and subscripts 1, 2 represent the number of the parallel branches, assuming the conducting current of branch 1 is larger than that of branch 2.

Current unbalance during turn-off is evaluated by the incremental current ΔI_f in the transient state, and it is given by

$$\Delta I_f = I_{f1} - I_{T1} \tag{3}$$

where I_{f1} is the peak anode current during turn-off. The whole value of current unbalance during turn-off is the difference between the peak current I_{f1} and the average

Reprinted from *IEEE Ind. Appl. Soc. Meet.*, 1982, pp. 741-746.

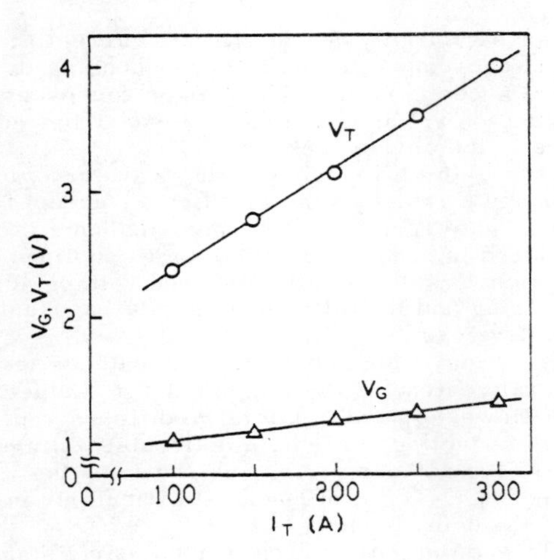

Fig. 3 Dependence of on-state voltage V_T and gate voltage V_G on on-state current.

on-state current. It is the sum of ΔI_f and $\Delta I_T/2$.

The anode and gate currents are measured using a current transformer. The impedance inserted in the test circuit is 0.2 mΩ, which is sufficiently small enough not to produce a disturbance in the parallel operation.

3. STEADY ON-STATE

3.1 On-state operation

In the on-state, current unbalance results from the mismatch ΔV_T of the on-state voltage between parallel devices. Dependence of the on-state voltage on the gate current can not be neglected in a GTO, unlike in a conventional thyristor. Current unbalance in the on-state is derived from not only the mismatch ΔV_T, but also the exchange of the gate current.

The on-state voltage changes linearly with small variations of the gate current and the anode current, and is given by

$$V_T = V_{T0} + R_{TT} \cdot \Delta I_T - R_{TG} \cdot \Delta I_G \tag{4}$$

where V_{T0} = the initial value of the on-state voltage, $R_{TT} = \partial V_T/\partial I_T$ and $R_{TG} = -\partial V_T/\partial I_G$.

Similarly, the gate voltage changes linearly with small variations of both currents, and is given by

$$V_G = V_{G0} + R_{GT} \cdot \Delta I_T + R_{GG} \cdot \Delta I_G \tag{5}$$

where V_{G0} = the initial value of the gate voltage,

$$R_{GT} = \partial V_G/\partial I_T, \text{ and } R_{GG} = \partial V_G/\partial I_G .$$

The values of R_{TT} and R_{GT} are 8 and 15 mΩ, respectively, and correspond to the slopes of V-I curve in Fig. 3. The values of R_{TG} and R_{GG} correspond similarly to the slopes shown in Fig. 4; however, the slopes are different for the negative and positive side of the gate current. When the gate current is not supplied from the external circuit, the values of R_{TG} and R_{GG} are given by the mean slope between both sides, and are 13 and 23 mΩ, respectively.

In gate decoupling, $\Delta I_G \equiv 0$. For the on-state current, $\Delta I_T = -\Delta I_{T2} \equiv \Delta I_T/2$, and for the on-state voltage, $V_{T1} = V_{T2}$. From eq. (4), current unbalance

$$\Delta I_T = \Delta V_T/R_{TT} \tag{6}$$

where $\Delta V_T = V_{T02} - V_{T01}$

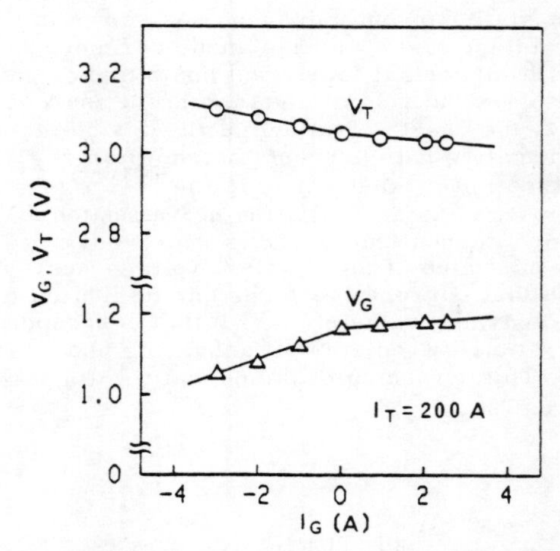

Fig. 4 Dependence of on-state voltage V_T and gate voltage V_G on gate current.

In gate coupling, $\Delta I_{G1} = -\Delta I_{G2} \equiv \Delta I_G/2$, and $V_{G1} = V_{G2}$. The condition with respect to V_T and I_T are the same as gate decoupling. Assuming that the mismatch ΔV_G is negligible small, $V_{G10} = V_{G20}$, current unbalance is given by

$$\Delta I_T = \frac{\Delta V_T}{R_{TT} + R_{GT} \cdot R_{TG}/R_{GG}} . \tag{7}$$

3.2 Current Unbalance

Fig. 5 shows dependence of current unbalcne on the mismatch of the on-state voltage. The experimental and calculated results are in good agreement. As shown in Fig. 5, current unbalance in the gate coupling is smaller than in the gate decoupling. This is also shown in comparison between eqs. (6) and (7).

From a phenomenal viewpoint, the reduction of current unbalance is attributed to the improvement of the mismatch ΔV_T caused by exchange of the gate current. In gate coupling, there is nothing to prevent the exchange of the gate current, except the gate impedance of the GTO itself. Therefore, the exchange of the gate current is

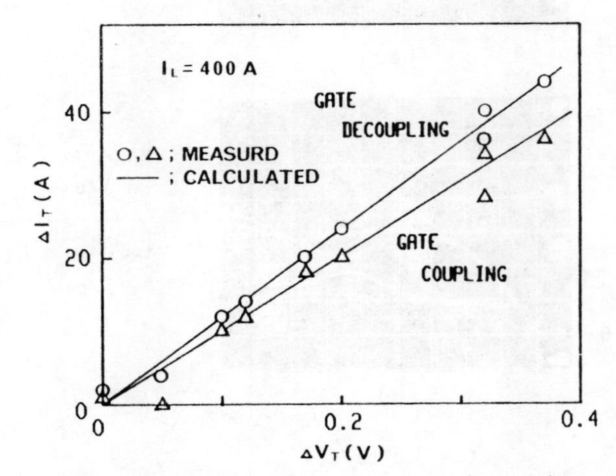

Fig. 5 On-state current unbalance I_T due to the mismatch V_T of on-state voltage.

generated due to the difference of the gate voltage caused by the on-state current unblance. As shown in Fig.3, the gate voltage rises with the anode current, thus the device with larger anode current has a higher gate voltage, and plays the role of a gate current source. Consequently, the on-state voltage of the device increases with the negative gate current, as shown in Fig. 4, while the on-state voltage decreases in the device of small anode current. As a result, the net mismatch $\triangle V_T$ is reduced and current unbalance is improved.

The mismatch of the on-state voltage must be within 0.4 V, so that current unbalance can be limited to less than the desirable value of 40A, which corresponds to 10 % of the ratio between current unbalance and average current. The screening of the on-state voltage is rather easy to realize.

4. TURN-ON

4.1 Turn-on operation

The turn-on operation is measured using a resonant circuit. The load current is a half-sine waveform with the conducting period of about 10 μs. First, the delay time of each device is measured under the following conditions; gate current, 6 A; anode voltage, 800 V; peak anode current, 250 A. Then, the combinations with various mismatch $\triangle t_d$ of the delay time are selected and their turn-on parallel operations investigated.

Fig. 6 shows the turn-on switching waveforms of parallel devices with the mismatch $\triangle t_d$ of 0.34 μs, in gate decoupling. Current unbalance $\triangle I_p$ reaches the value of 155A, which is 30% in the ratio of incremental value to the average peak anode current. The gate current is equally divided due to the external resistor, except in the last stage of the rise period. The gate voltage increases abruptly due to the gate current at first, and then increases gradually corresponding to a rate of rise of anode current. The former represents rise of the junction voltage, and the latter may be caused by inductance in the device package.

Fig. 7 shows the turn-on switching waveforms in the same devices and circuits as the last figure, except for the condition of gate coupling. Current unbalance is remarkably reduced in comprison with gate decoupling. This improvement results from the induced voltage across the cathode wiring and the difference in gate impedance between both devices.

Cathode wiring, which connects both cathode terminals, is several centimeters in length, and its inductance is very low. However, due to the large difference in rate of rise of the conducting current, a noticeable voltage is induced across the cathode wiring, as shown by the waveforms of V_{K12} ($V_{K1} - V_{K2}$) in Fig. 7, particularly in the final stage of the delay period.

The difference in gate current for the early stage results from the difference in gate impedance between parallel devices. The device with a long delay time has low gate impedance; conversely, the device with a short delay time has high gate impedance. The reason for this dependance of gate impedance is not clear.

Due to the induced voltage of the cathode wiring and the difference in gate impedance, a larger gate current flows to the device of longer delay time, while a smaller gate current flows to the device of shorter delay time. As a result, the mismatch of the delay time is reduced, as well as current unbalance is improved.

4.2 Current unbalance

Fig. 8 shows the dependence of current unbalance on the mismatch of the delay time. In gate decoupling, current unblance increases linearly with the mismatch $\triangle t_d$. In gate coupling, current unbalance is small, and about one-tenth compared with gate decoupling. At the maximum

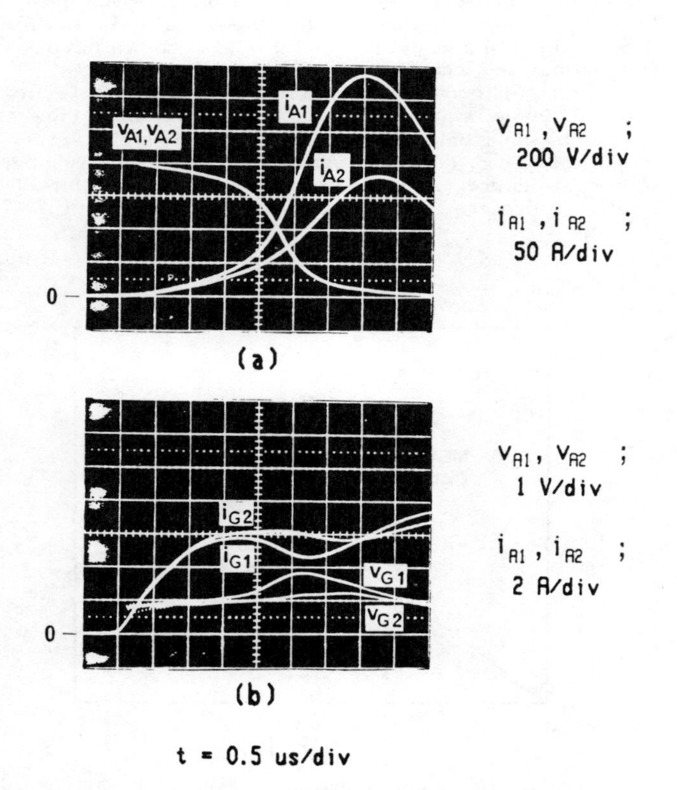

Fig. 6 Turn-on switching waveforms in gate decoupling with the delay time mismatch $\triangle t_d$ of 0.34 μs.

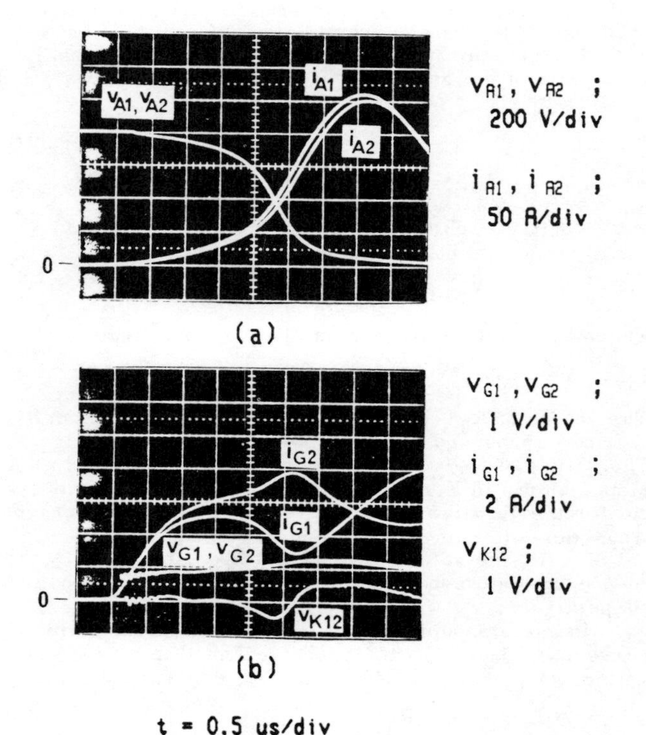

Fig. 7 Turn-on switching waveforms in gate coupling with the same combination as in Fig. 6, where V_{K12} is the induced voltage across the cathode wiring with a zero potential point of cathode terminal 2.

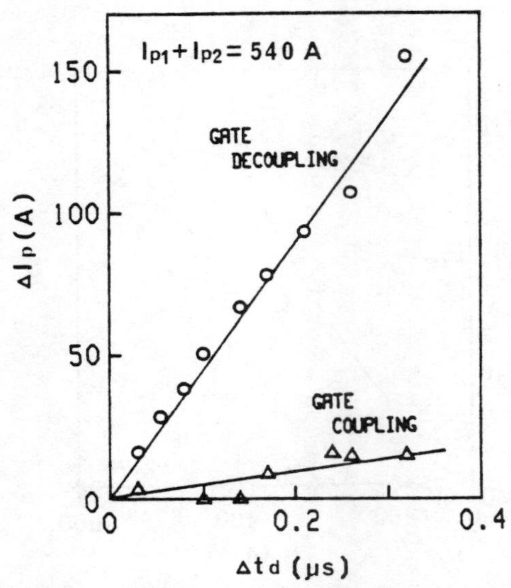

Fig. 8 Relations between turn-on current unbalance ΔI_p and the delay time mismatch Δt_d with peak load current ($I_{p1} + I_{p2}$) of 540 A.

mismatch Δt_d of 0.34 μs, the ratio of the incremental value to the average peak current is about 3 % in gate coupling. Current unbalance is sufficiently small enough not to require screening for the mismatch Δt_d.

Fig. 9 shows the dependence of current unbalance on the inductance L_K of cathode wiring. The difference of L_K between both devices generates a serious current unbalance. Therefore, each device must be connected with the load circuit by cathode wiring of the same length.

In gate coupling, current balance is determined by the induced voltage of the cathode wiring and the difference in gate impedance. Therefore, it depends only slightly on the peak load current and the gate current determined by the external circuit.

The method of gate coupling is useful in achiving the turn-on current balance in direct paralleling provided symmetrical cathode wiring is carried out.

5. TURN-OFF

5.1 Turn-Off Operation

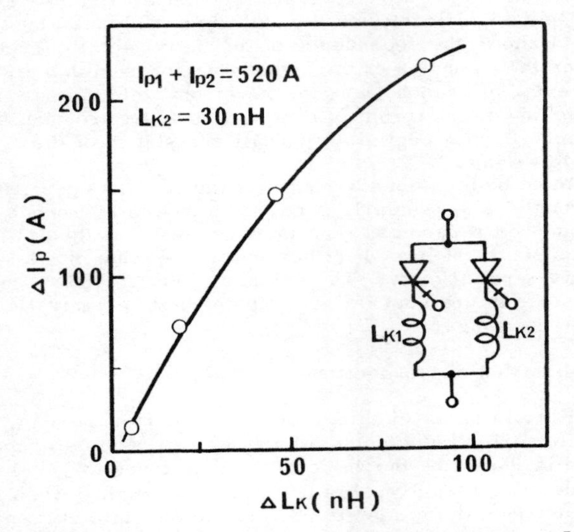

Fig. 9 Dependence of turn-on current unbalance ΔI_p on the mismatch ΔL_K ($L_{K2} - L_{K1}$) of cathode wiring inductance.

Turn-off operation is done in the clamped inductive load, using a polarized snubber, which protects the device from turn-off failure.

In the parallel operation, the mismatch of the storage time is a important parameter affecting current unbalance. The value of the storage time is slightly different between devices, and also changes with the anode current in the on-state. Therfore, the mismatch Δt_s is given by the sum of the difference in storage time for the fixed anode current and the difference caused by current unbalance in the on-state.

Fig. 10 shows the turn-off switching waveforms in gate decoupling with the mismatch Δt_s of 0.4 μs. The operating conditions are as follows : on-state load current, 400 A; rate of rise of the whole negative gate current, 20 A/μs; clamped voltage, 800 V; and snubber capacitance, 0.47 μF per device.

The gate decoupling is carried out using the series gate inductance of 1 μH connected to each gate terminal. As shown in Fig. 10, the gate current is equally divided during the storage period.

When the device with the short storage time is turned off early, the anode current of the other device increases abruptly. In this case, the peak anode current reaches about 320 A from the on-state current of 200 A. Serious current unbalance generates due to even a slight mismatch Δt_s of less than 10 % of the storage time.

Fig. 11 shows the turn-off switching waveforms in the same device and circuit as the last figure, except for th the condition of gate coupling. Current unbalance is remarkably reduced in comparison with gate decoupling. This results from the improvement of the mismatch Δt_s caused by commutation of the gate current.

The storage period of a GTO is the process during which the excess holes in the p-base are removed by the negative gate current, and lasts until the gate charge, or the time-integrated value of the negative gate current reaches a prescribed value [2]. Therefore, the storage time becomes short with an increase of the negative gate current, and longer with its decrease.

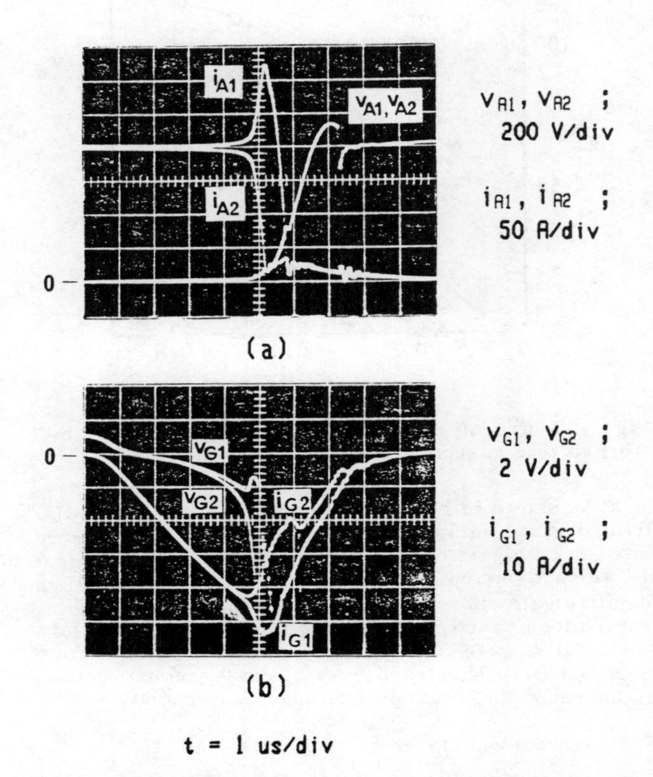

Fig. 10 Turn-off switching waveforms in gate decoupling with the storage time mismatch of 0.4 μs at load current I_L of 400A.

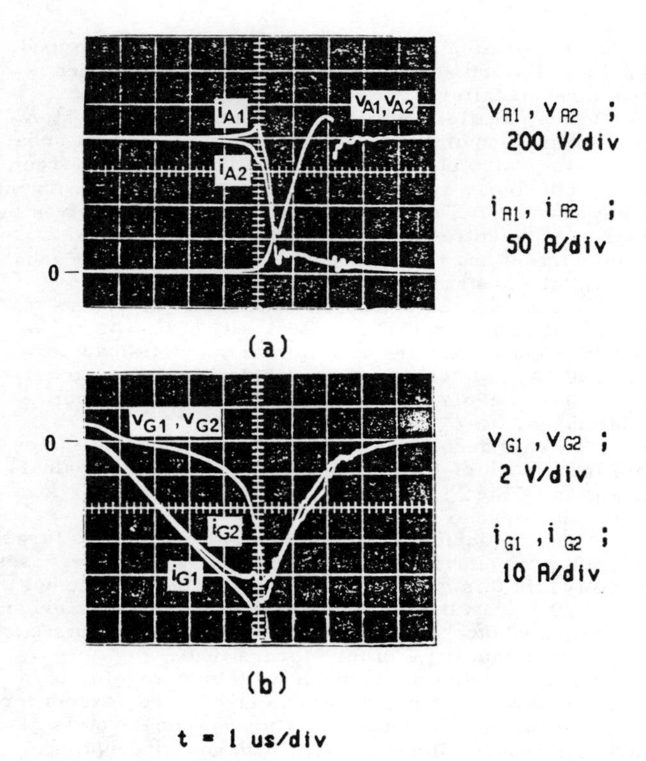

Fig. 11 Turn-off switching waveforms in gate coupling with the same devices and circuit as in Fig. 10.

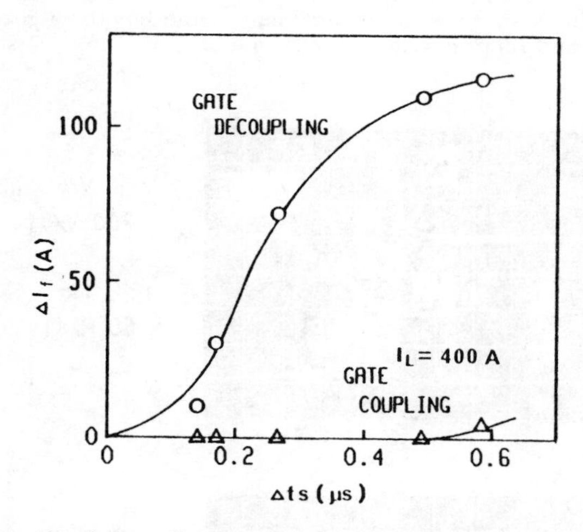

Fig. 12 Turn-off incremental current ΔI_f due to the storage time mismatch Δt_s.

As shown in Fig. 11, the gate current is equally divided in the early stage, however there is a difference between both gate currents generated in the final stage of the storage period. In the device turning off early, the negative gate current decreases so that the storage period may be lengthened, while in the device turning off later, the negative gate current increases so that the storage period may be shortened. As a result, the mismatch Δt_s is improved and current unbalance is reduced.

5.2 Current Unbalance

Fig. 12 shows the dependence of the incremental ΔI_f on the mismatch Δt_s. In gate decoupling, the dependence

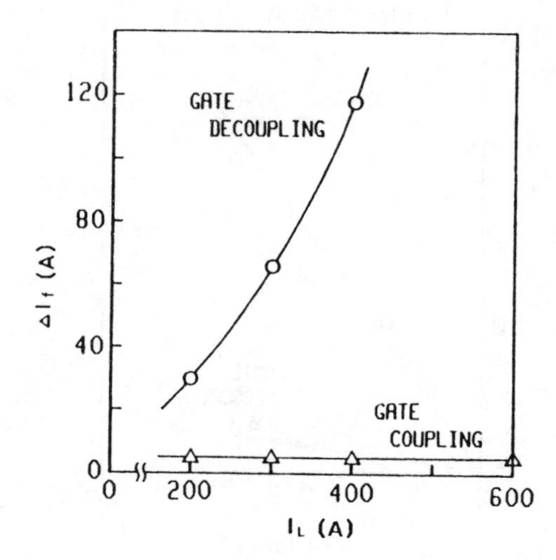

Fig. 13 Dependence of turn-off incremental current ΔI_f on load current I_L.

roughly corresponds to the reverse waveforms of the anode current when the other device turns off early. Basically, the mismatch Δt_s must be sufficiently small enough compared with the fall time of the other device, in order to reduce current unbalance. In this case, the desirable value of Δt_s is within about 0.15 μs, however the realization of the screening is difficult in practicl.

In the gate coupling, the incremental value is remarkably small compared with gate decoupling, and is only slightly dependent on the mismatch Δt_s. In other words, the mismatch Δt_s is reduced within 0.1 μs. If this figure is replotted using the true value of Δt_s, the experimental data in gate coupling would be almost centered on one point near the origin.

Fig. 13 shows the dependence of the incremental current ΔI_f on the load current, using the combination with the mismatch Δt_s of 0.58 μs at load current of 400A. Similar, to the previous figure the value of ΔI_f in gate coupling is small compared with the gate decoupling. Also it is hardly dependent on the controllable current, therfore the ratio of current unbalance decreases with controllable current.

In gate coupling, current balance during turn-off is not affected by the device parameter mismatch; however, it depends significantly on the cathode wiring resistance. Fig. 14 shows the dependence of the incremental current ΔI_f on the mismatch of the cathode wiring resistance. The device of branch 1 has the lower on-state voltage, and the on-state current of branch 1 is 40A more than that of branch 2, at a negligiblely small resistance of the cathode wiring.

When both cathode wiring resistances are equal, the incremental ΔI_f is nearly zero. The on-state current unbalance is reduced with an increase of the cathode wiring resistance of branch 1, however, the value of ΔI_f increases remarkably. The balance of the cathode wiring resistance is the most important point in direct paralleling of gate coupling.

5.3 Mismatch of storage time

Fig. 15 shows the dependence of the mismatch Δt_s on the gate series inductance, where the end of the storage period is judged by the value of the impedance (v_A/i_A) of each device. Usually, this is defined at the time when the anode current decreases to 90 % of the on-state current. However, the usual definition is not applicable for the divice turning off later, in which the anode current

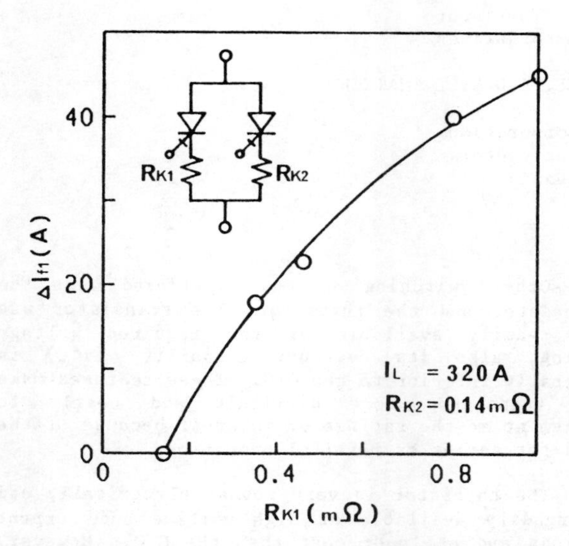

Fig. 14 Dependence of turn-off incremental current ΔI_f on cathode wiring resistance.

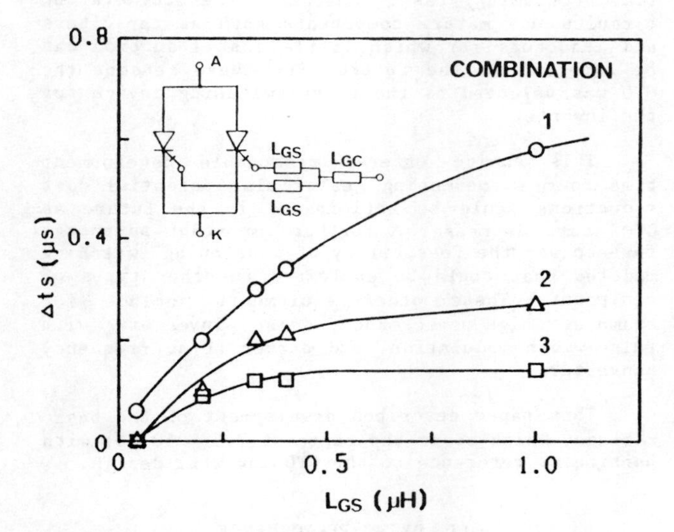

Fig. 15 Dependence of the storage time mismatch Δt_s on the gate series inductance L_G for three different combinations.

increases over the on-state current during the storage period. The value of the impedance is selected so as to be consistent with the usual definition, and is 0.15Ω in this case.

As shown in Fig. 15, the mismatch Δt_s decreases with the gate series inductance L_{GS} connected with each gate terminal. When the value of L_{GS} is changed, the common inductance L_{GC} is also changed so that a rate of rise of the whole negative gate current can be maintained at a fixed value of 20 A/μs. All mismatch Δt_s values in the three different combinations are within less than 0.1 μs for the shortest gate wiring of 0.05 μH in Fig. 15.

The mechanism improving the mismatch Δt_s is considered below. Basically, it results from the time-variation of the gate impedance of the device itself. In the on-state, the gate impedance is low due to conductivity modulation of the p-base. The negative gate current squeezes the modulated area toward the center of the n-emitter. The modulated area changes into the un-modulated area, therefore the gate impedance increases with the reduction of the modulated area. Then, the storage period ends with the disappearance of the modulated area[3].

In parallel operation, a device with a larger modulated area has a lower gate impedance, therefore the modulated area is reduced faster due to the larger gate current corresponding to the gate impedance. In the other device with a smaller modulated area, it is reduced slowly. As a result, the modulated area of both devices would be squeezed in the same area, and would almost simultaneously disappear, i.e., the storage periods of both devices would end at the same time.

6. CONCLUSIONS

Direct paralleling of GTO's was investigated by two methods of gate decoupling and gate coupling. GTO's have essentially the current-balanceing ability based on the gate interaction between parallel devices.

The on-state current unbalance was determined mainly by the mismatch of the on-state voltage, and its improvement in gate coupling was not so large. However, the transient current unbalance was remarkably improved by the reduction of the switching time mismatch in gate coupling. The screening in gate coupling was required for the on-state voltage, however it is not always necessary for the switching time.

An important consideration was the resistance and inductance of the cathode wiring, which are closely related to turn-on and turn-off operation, respectively. The symmetrical cathode wiring was most important in realizing direct paralleling of GTO's.

The authors would like to thank Dr.T.Takasuna, N.Kawakami, K.Oonishi and Dr.M.Okamura for their encouragement and guidance throughout this work.

REFERENCES

[1] A. Ueda et al., "GTO Inverter For AC Traction Drives", in IEEE Conf. Rec., 1982 Annual Meeting of IAS, to be published.

[2] H. Fukui et al., "Switching Oharacteristics Of A Gate Turn-Off Thyristor", PCI'82, pp.256-257.

[3] E.D. Wolley, "Gate Turn-Off in p-np-n Devices", IEEE Trans. Electron Devices, vol. ED-13, pp. 125-135, 1966.

Application of Gate-Turn-Off Thyristors in 460 V, 7.5-250 hp AC Motor Drives

D. A. PAICE, SENIOR MEMBER, IEEE, AND K. E. MATTERN

Westinghouse Electric Corporation
Research and Development Center
Pittsburgh, Pennsylvania

ABSTRACT

A 7.5 - 250 hp range of variable-frequency ac motor drives has been under development at the Westinghouse R & D Center since early in 1981. Design and performance details for prototype equipment rated for 10,20 and 50 hp will be discussed.

The drive system comprises a three-phase thyristor controlled ac/dc converter in conjunction with a dc/ac adjustable-frequency inverter using gate-turn-off thyristors. Output of the inverter is a six-step voltage wave in the range of 6 to 120 Hz suitable for driving three-phase ac induction or synchronous motors.

This paper will especially elaborate upon details of the inverter design, including means to protect the gate-turn-off thyristors from output line-to-line or line-to-ground short circuits.

I. INTRODUCTION

Two major factors have contributed to the accelerated development of variable-frequency, variable-speed motor drive systems in recent years. These are:

1. The need for energy conservation has extended the areas in which variable-speed drives are cost effective; especially in process control and pump applications.
2. New switching devices such as increased rating transistors and gate - turn-off thyristors (GTOs) have become available on a production basis. These devices, especially the GTOs have enabled higher power, higher voltage, simplified inverter circuit designs to be achieved.

Today in the USA there is significant interest in variable-frequency motor drive systems to operate with nominal 460-V, 60-Hz, motors covering the range of 10 - 500 hp. Additional demands exist for 1 - 10 hp at 460 V and 230 V. Possible future applications are envisaged for 4160 V operation at higher horsepowers.

To address these variable-speed drive requirements a line of motor drives covering the range of 7.5 - 50 hp has been developed. Higher horsepower units are also being developed. In the dc/ac inverter section of these equipments the GTO was selected as the switching element because of its appropriate voltage withstand capability (1200 V) and electrical switching and overload capabilities.

Other switching devices considered were the transistor and the thyristor. The transistor was not readily available at the required voltage rating, also its overcurrent ability (I^2t) is generally inferior to the GTO. These features make the transistor more difficult and costly to implement at the ratings of interest because of the need for series or parallel operation.

The thyristor is very robust electrically and is readily available at high voltage and current ratings and at lower cost than the GTO. However, development of reliable commutating circuits for thyristor inverters with variable loading and operation with variable voltage is a difficult and time consuming task. Further, the commutating circuits use mature components such as capacitors and inductors for which little cost reduction can be expected in the future. For these reasons the GTO was selected as the power switching device for the inverter.

This choice offered reasonable development time to meet marketing goals; also potential cost reductions could be anticipated in the future as GTO costs decrease. A further important advantage forseen was the feasibility of developing switching modules that could be exploited in other types of equipment. These other equipments include for example, high-performance dc/ac inverters with pulse-width modulation, and direct ac/ac frequency converters.

This paper describes development of the basic ratings, which cover the range of 7.5 - 50 hp; with particular reference to the GTO inverter design.

II. DRIVE PERFORMANCE

At the outset of the program, technical and cost goals were established for the range of packaged (as opposed to custom designed) drives to be developed. Basic performance features of the design are given below in Table 1.

Table 1: BASIC PERFORMANCE OF ACCUTROL 300 DRIVES

- Input --- 460 V +10%, -5%, Three-Phase 60 Hz
- Output --- Variable frequency 6 to 120 Hz, 460 V maximum, three phase with adjustable Volts/Hertz. Six-step, square-wave voltage for ac induction or synchronous motors. Overload rating to be 150% rated current for one minute. Protected against line-to-line and line-to-ground faults.
- Ambient --- 0 to +40 degrees C, in NEMA 1 enclosure.

Reprinted from *IEEE Ind. Appl. Soc. Meet.*, 1982, pp. 663-669.

In addition to the basic functions given in Table 1 provision was made to incorporate anticipated customer options as required. Most of these options are coordinated and defined by a microprocessor-based control [1]. The system dynamic performance is primarily controlled by analog-based feedback loops which include dc link current, dc link voltage, and transient voltage/frequency damping.

III. BASIC GTO INFORMATION

To be cost effective the GTO applied to an AC induction motor drive inverter must meet specific requirements. For operation on a 460-V supply the devices must have forward blocking capability of about 1200 V. The unavailability of power transistors of significant current handling ability at this voltage has limited transistor inverter development in the 460-V class motor drives. The GTO is suitable for 460-V drives. The lack of significant reverse-blocking capability in some GTOs is not a limitation in the voltage-sourced dc/ac inverter described here.

For 460-V drives rated up to 250 hp, power devices or combinations of devices with turn-off capability up to about 1500 A may be needed, depending on the motor and inverter characteristics. In a GTO, increasing the turn-off capability also tends to increase the forward drop and results in higher conduction losses. Since a six-step type inverter is used, conduction loss will be the predominant loss mechanism and a reasonable compromise in device characteristics must be found.

Keeping snubber circuitry requirements minimal is very important in reducing losses and increasing feasible switching rates. Reapplied dv/dt specifications in the 300 V/μs region or greater are desirable. Earlier available GTOs, in the 1960's-70's, had dv/dt ratings of only 10-50 V/μs, placing excessive burdens on the snubber designs and limiting switching speeds.

GTOs recently available have the desired characteristics for motor-drive inverter circuits [2],[3]. In addition, higher current GTOs are being developed which will enable the present range of horsepower ratings to be increased and other suppliers are also entering the marketplace with suitable devices. Characteristics of the GTOs being used in the range of 7.5 - 100 hp drives are listed in Table 2.

IV. GATE DRIVE

The successful application of the GTO to the dc/ac inverter of the motor drive depends upon exploiting the characteristics previously mentioned; high forward-blocking voltage, high rate-of-rise of anode voltage and current, and high turn-off current capability. These characteristics are strongly impacted by the performance of the gate drive circuitry and the GTO will only perform as well as its gate drive allows. The development of suitable gate drive circuitry is an important starting point in the successful application of GTOs.

The turn-on characteristics of the GTO are similar to those of the conventional thyristor (SCR). To accommodate high rate-of-rise of anode current (di/dt), the gate drive must supply a sharply rising current (t_{rise} < 1 μs) which reaches approximately three times the required dc gate current, I_{GT}. During turn-off of the GTO an appropriate current must be drawn from the gate of the device. This gate current increases with increasing anode current. During the turn-off process the anode current is "squeezed" into small areas in the device. This action causes a momentary high dissipation condition, therefore turn-off should be accomplished as quickly as possible. Typical GTO turn-off behavior is depicted by the waveforms in Figure 1. Reverse gate current in this device rises at a rate of about 5 A/μs to a peak value called I_{GQ}. This value varies from device to device but is usually one-third to one-fifth of I_T, the anode current being turned off. The term "turn-off gain" applied to GTOs is the ratio of I_T to I_{GQ}, and turn-off gains of 3-5 are typical.

Table 2: GTO THYRISTORS USED IN VARIABLE FREQUENCY MOTOR DRIVES

Typical Characteristics		Units	50 A Nom.	90 A Nom.	200 A Nom.	600 A Nom.
Repetitive peak voltage		V	1200	1200	1200	1200
Controllable on-state peak current	Repetitive	A	50	90	200	600
	Non-repetitive	A	80	180	280	600
RMS Current		A	18	30	70	250
On-state voltage drop at repetitive current		V	3.1	2.8	3.8	3.5
Reapplied dv/dt at nonrepetitive current		V/μs	800	800	800	300
di/dt at turn-on		A/μs	100	200	200	2000
Gate trigger current		A	0.3	0.4	0.6	4
Thermal impedance		°C/W	1.0	0.6	0.35	0.075
Turn-on time		us	4	4	4	5
Turn-off time		us	6	6	6	8

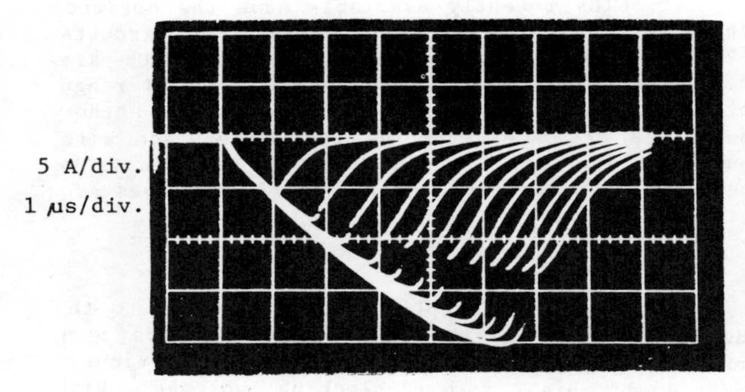

Gate Turn-Off Current for 90 A GTO
with Anode Currents of
10, 20, 30, 40, 50, 60, 70, 80, 90,
100, 110, 120 A

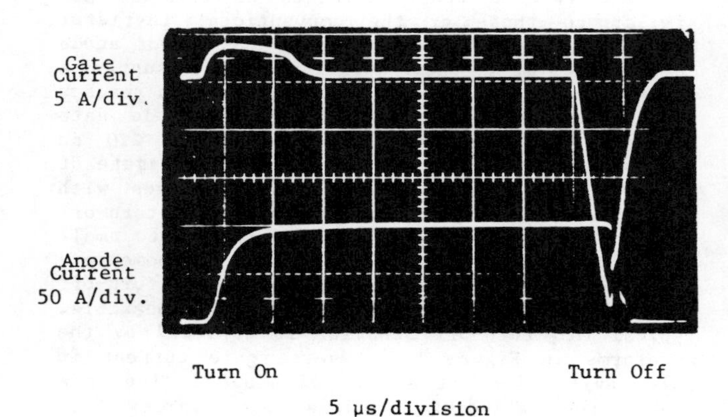

Figure 1. TYPICAL GTO SWITCHING WAVEFORMS

To achieve the high, static dv/dt ratings (>1000 V/µs) and forward blocking capability (1200 V), a reverse bias of approximately 5 V should be applied to the gate when the device is to be in the off state. The rate-of-rise of anode voltage at turn-off or the reapplied dv/dt rating for the GTO varies from 300 to 800 V/µs and is controlled by the snubber networks which are described later.

The GTO, like the SCR, is a latching device. When the anode current reaches a level called the latching current (which tends to be higher than for conventional thyristors), gate drive is no longer needed to maintain conduction. This allows significant reduction in turn-on, gate drive power compared to the power transistor which requires continuous base drive. The GTO gate drive circuits need only supply current when the anode voltage goes positive by greater than 6-7 V and by sensing the GTO anode-cathode voltage, gate current power demands are kept small at operating frequencies up to 120 Hz

The gate drive circuit arrangement used in the 7.5-50 hp range of drives incorporates anode-cathode voltage sensing and the actual printed wiring board contains two gate drive circuits to control two GTO thyristors as shown in Figure 2.

Figure 2. VIEW OF GTO POLE AND GATE DRIVE FOR 20 HP

The dual board was designed to coordinate with the philosophy of a pole module. The pole is treated as a unit with a two-wire, three-state control line. Positive, negative, or zero control current selects upper switch, lower switch, or no switch to be on.

Also incorporated in each gate drive circuit is an overcurrent shutdown circuit. This circuitry will, independent of the status of the control lines, turn-off the GTO if its anode current reaches a maximum limit. As discussed later the rate-of-rise of anode current above the trip level is controlled by an inductor to provide sufficient time for the trip circuit to operate safely.

V. BASIC POWER CIRCUIT

After due consideration of the basic requirements of the packaged drives to be designed, a six-step, voltage-sourced, dc/ac adjustable-frequency power circuit was selected [4],[5]. The block diagram is shown in Figure 3. The 460-V input is rectified by a three-phase bridge converter utilizing six conventional thyristors, to produce a variable dc voltage. In the lower horsepower units dual thyristor packaged modules are used; for the higher horsepowers (50 hp and up), discrete devices are employed.

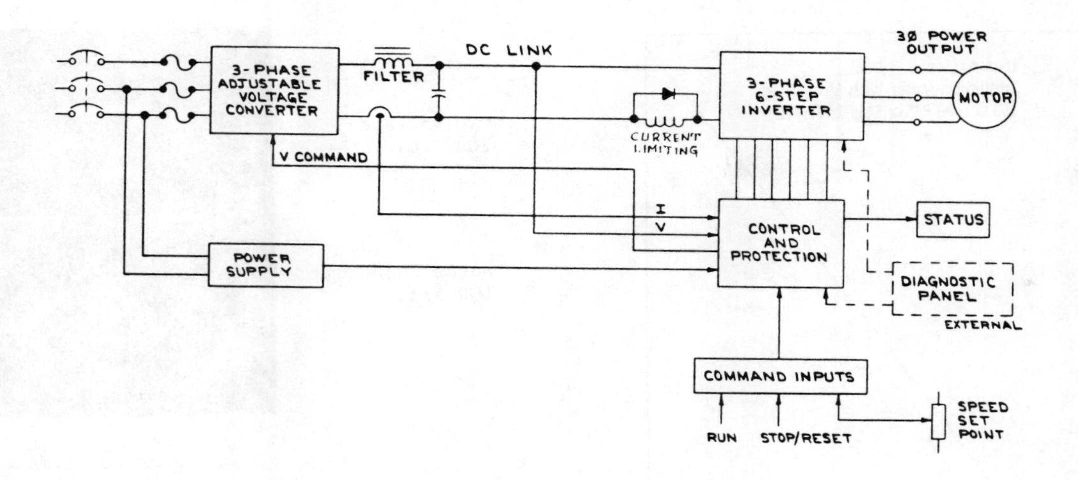

Figure 3. BLOCK DIAGRAM FOR PACKAGED MOTOR DRIVES

The output of the ac/dc converter is filtered by the dc link inductor and capacitor. Selection of the inductor is a compromise between minimizing distortion of the ac supply and providing an economic design. The dc voltage and current are sensed and used by the control to provide appropriate regulation of the drive performance.

The dc voltage (600 V) is applied to the three GTO poles through a current-limiting inductor and diode. The operation of the current-limiting circuit is described later. The GTOs then switch the 600-V link voltage to produce the conventional six-step waveforms which are applied to the motor.

VI. POLE DESIGN

The basic pole schematic is shown in Figure 4. The function of the pole is to control the voltage on the output line which connects to one phase of the motor. Control is accomplished in this six-step scheme by alternately gating GTO1 and GTO2 to connect the output line to the positive dc bus and the negative dc bus respectively. Diodes D1 and D2 which permit reactive power flow, eliminate the need for reverse blocking capability in the GTOs. The remaining components shown in Figure 4 are necessary to provide safe operating conditions for the GTO such as, dv/dt control, di/dt control, and overcurrent protection.

The current which a GTO can turn-off has an upper limit and if an attempt is made to turn-off higher currents, damage will occur. This limit is device dependent. It represents a circuit limitation similar in concept to the peak commutation current capability of impulse commutated SCR circuits which are limited by commutating capacitor charge. It was considered very desirable to prevent the GTO current from exceeding this limit which could occur, for example, from spurious signals which cause both GTOs to turn on simultaneously (shoot-through), output short circuits, or overloads. Measuring shunts 1 and 2 are added to provide feedback signals to the gate drive circuitry to turn-off the GTO before the anode current reaches unacceptable limits. Once the trip level is detected, a period of approximately 20 μs elapses before initiating turn-off of the GTO. This time delay ensures that the snubber circuit capacitors(C1,C2) have discharged to a safe value and are able to prevent high dv/dt at turn-off. During this 20 us period

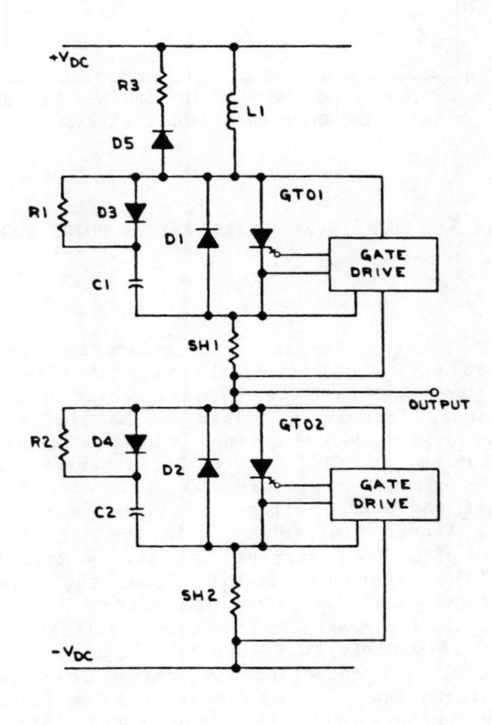

Figure 4. TYPICAL GTO POLE SCHEMATIC

the anode current rate-of-rise must be controlled. This is accomplished by the current-limit inductor shown in Figure 3. Design trade-offs were made to assess the effects of the current-limiting inductor rating, peak turn-off capability of the GTO, and repetitive peak motor currents. The results provided insight into the rating limitations of particular GTOs. Figure 5 shows a graph of typical peak GTO current versus motor horsepower for the six-step induction motor drive. The trip setting of the overcurrent circuitry is set approximately 10-15% above the peak GTO current expected. The relationship between the trip setting, maximum nonrepetitive turn-off capability, and current-limit inductor then determines the inverter horsepower rating for the device.

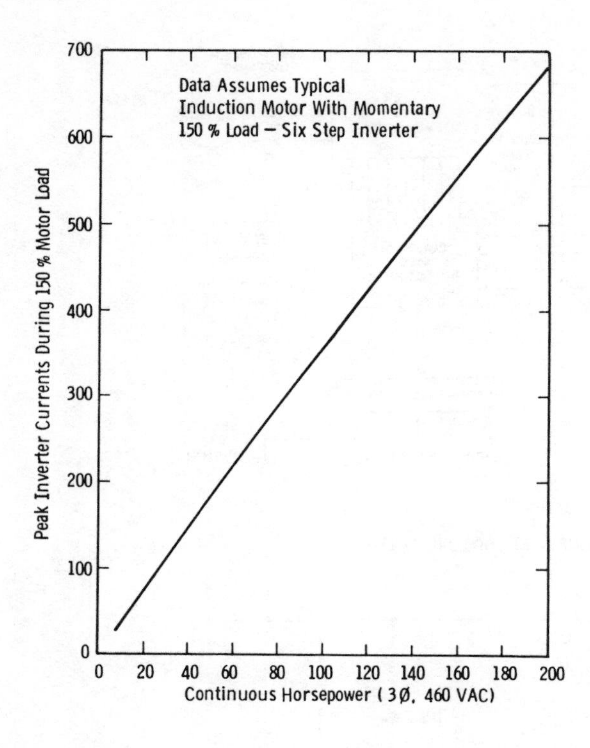

Figure 5. PEAK INVERTER CURRENT vs MOTOR HORSEPOWER

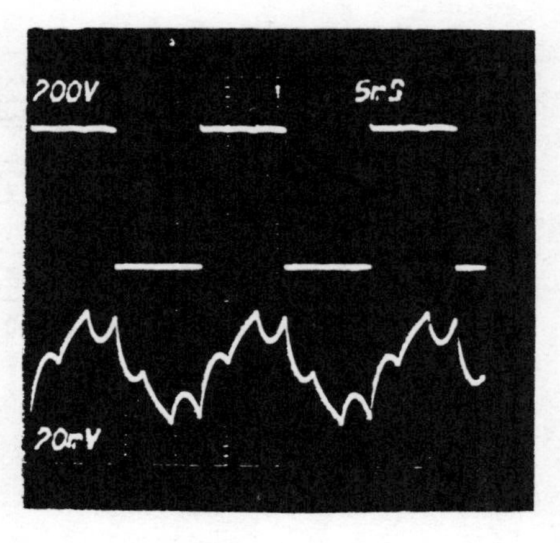

GTO Voltage
200 V/div.

Motor Current
100 A/div.

5 msecs/division

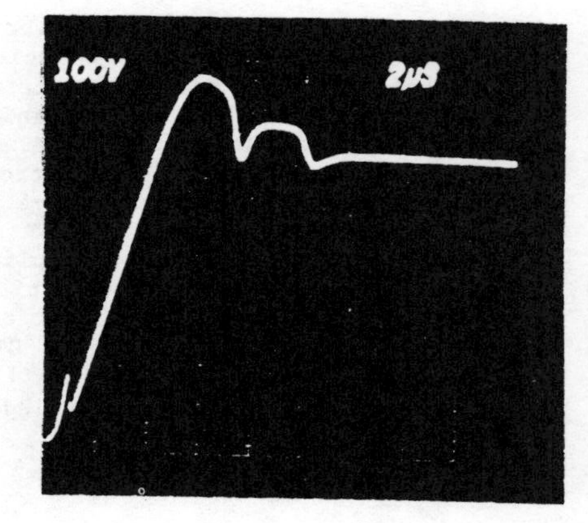

GTO Voltage
100 V/div.

2 μsecs/division

Figure 6. TYPICAL POLE WAVEFORMS at 45-HP, 60-HZ

To prevent excessive peak power dissipation in the GTO during turn-off, the rate of rise of the anode voltage must be limited while the anode current is decreasing. This device limitation is the reapplied dv/dt rating. When a current is interrupted by GTO1 it is transferred to the snubber circuit comprising D3, C1. The rate-of-rise of the anode voltage therefore is controlled by the factor I/C1, where I is the current being turned off. When current is rapidly transferred into the snubber circuit from the GTO any inductance in the snubber circuit causes a transient increase in the anode voltage. This effect is visible in Figure 6. It is undesirable and can be controlled by using low-impedance capacitors, fast turn-on diodes, and low-inductance wiring techniques. The inductances, stray and intentional, cause the voltage on C1 to rise towards twice the dc voltage, however, "clamping" diodes such as D5 were applied and under worst case conditions the GTO voltages are only about 20% greater than the dc link voltage.

When the GTO is gated on, the snubber capacitor is discharged through the GTO with a current which is limited by resistor R1. This resistor dissipates most of the energy which was stored in the snubber capacitor during GTO turn-off; the power is thus approximately $1/2\ CV^2f$, where f is the frequency at which the GTO operates. If the GTO should turn-off before the snubber capacitor has been discharged, the anode voltage will rise "instantaneously" to the capacitor voltage; this is undesirable. The time constant ,R1 C1, of the snubber discharge sets a minimum on-time for the GTO pole. The typical on-time, as discussed for the overcurrent trip circuits, is about 20 μs.

Inductor, L1, acts to limit the maximum rate-of-rise of anode current, di/dt, which the GTO will experience at turn-on. Typical di/dt ratings of GTOs are 100-200 A/μs. Some manufacturers indicate that their devices can successfully operate without di/dt limiting inductors, however circuit inductance does reduce the GTO turn-on losses. Some inductance is necessary between the poles and the dc voltage source such that the snubber capacitors, in conjunction with this inductance can control the GTO rate-of-rise of voltage. The required value is small and is readily available in the stray wiring inductances. When the GTO turns-off, the energy stored in L1 is transferred to the snubber capacitor. When the capacitor voltage equals the dc link voltage diode D5 becomes forward biased, and resistor R3 limits the overshoot voltage which appears on the capacitor by dissipating the excess energy.

VII. PROTECTION ARRANGEMENTS

Five main protective features are included in the GTO power inverter. These are shown in Table 3.

Table 3: PROTECTIVE FEATURES OF ACCUTROL 300

- GTO Instantaneous Overcurrent.
- GTO Heat sink Overtemperature.
- GTO Low-Voltage Gate Drive.
- Output Line-to-Line Short Circuit.
- Output Line-to-Ground Short Circuit.

To illustrate how each of the protective features is obtained the basic power circuit schematic shown in Figure 7 will be used.

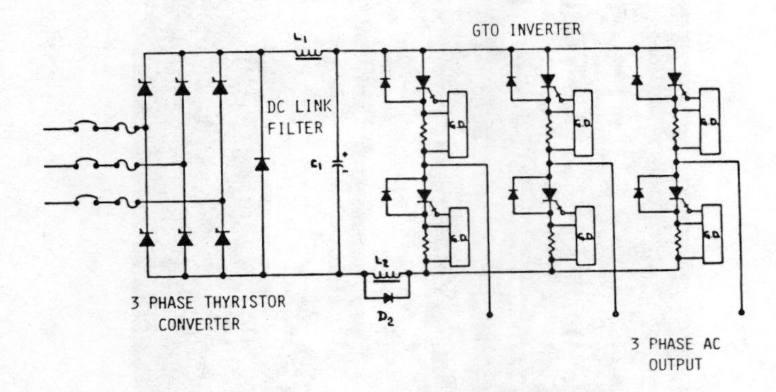

Figure 7. BASIC POWER CIRCUIT SCHEMATIC

Each gate drive circuit has an optically-isolated signal input connected with the main controls. By this means, protection commands fed into the gate drive unit are safely communicated, without dielectric stress problems, to the main controls which are at ground potential.

GTO Instantaneous Overcurrent

Instantaneous overcurrent in any GTO is detected by measuring the voltage drop across a small resistor in series with the GTO cathode lead. When this voltage exceeds a predetermined trip level the gate drive circuit takes immediate action to turn the GTO off. Also a signal is sent via the optocoupler to the microprocessor to provide an electronic shutdown of the drive. An electronic shutdown sequence then ensues, which causes the thyristor converter to phase back to the fully off position. Also all GTOs are turned off and prevented from receiving further ON pulses until a manual reset action is implemented. Inductor L2 ensures a limited rate-of-rise of GTO current under fault conditions and provides time (20 μs) during which the current limiting controls are activated. Under normal operating conditions the current in inductor L2 charges up to the peak of the three-phase ac motor load current and then it circulates as necessary through diode D2. Thus inductor L2 has little influence on normal operation. When sudden increases are required in the GTO currents, inductor L2 limits the rate-of-rise of these so that time is provided for the overcurrent protection circuits to act and safely limit the maximum current.

Low Voltage Gate Drive.

Failure of the GTO thyristors can occur if insufficient reverse gate current is applied at the turn-off condition. This could happen if for any reason the gate-drive circuits had insufficient voltage. An undervoltage sensor on the gate drive board protects against this condition. It prevents starting the inverter until sufficient gate drive voltage is available and also causes an electronic shutdown if the gate-drive voltage falls to an unsafe level.

Overtemperature.

Protection from excessive heat sink temperature is by means of a bimetal switch mounted on the heat sink associated with the lower GTO in each pole. It acts in conjunction with the alarm signal circuit associated with the gate drive for the upper GTO. With this contraposition arrangement there is practically no dielectric stress applied to the bimetal sensor, despite the fact that the heat sink voltage changes very rapidly with respect to ground.

Line-to-Line Short Circuit.

The path for possible line-to-line short circuit currents is shown in Figure 8.

Line-to-line short circuits would cause high currents to flow through two GTOs in series, and in this case both overcurrent detectors may signal to interrupt the circuit. Performance on a 20-hp prototype model was very satisfactory; for example under line-to-line fault conditions the peak current was limited to about 95 A. The duration of fault was such that a slow-acting fuse rated at 3 A and connected in series with the fault did not melt. For comparison it is noted that under normal maximum overload conditions the inverter output current is about 45 A rms.

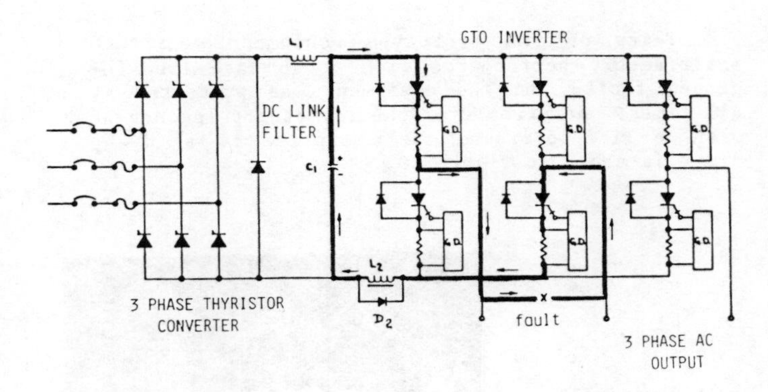

Figure 8. TYPICAL PATH FOR LINE-TO-LINE FAULT

Line-to-Ground Short Circuit.

Performance with line-to-ground faults is dependent upon the instantaneous operating conditions at the time of the fault. If the dc link filter capacitor is charged to about 1/2 voltage or more (i.e. the inverter output voltage is 1/2 rated or more) a line-to-ground fault gives a completely electronic shutdown. The path of a possible line-to-ground fault in which fault current flows through the inverter diodes is shown below.

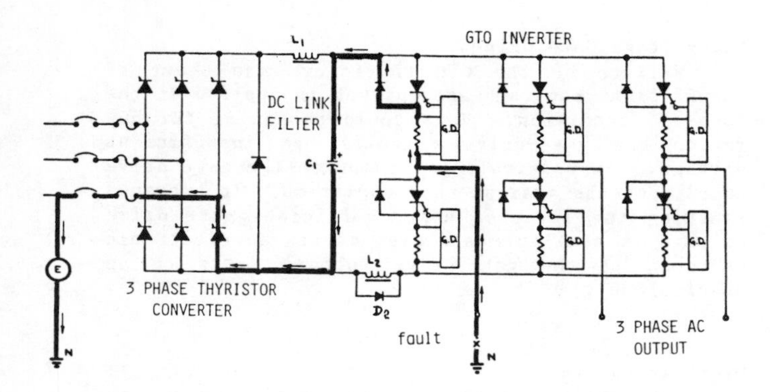

Figure 9. CURRENT PATH OF LINE-TO-GROUND FAULT

It can be noted from Figure 9 that under worst case conditions when the dc link capacitor voltage is discharged below the peak of the line-to-neutral voltage a large fault current may flow through the path comprising the diode associated with the upper GTO and the ac line input fuse. By properly coordinating the capabilities of the upper diodes with the ac input fuses, full protection is provided for all output fault conditions. Full electronic shutdown protection could be achieved for all line-to-ground faults if the dc link inductor is split into two parts, one section in the positive lead of the dc link and one in the negative lead. However this technique was not found to be cost-effective up to 50 hp ratings.

The schematic of Figure 9 shows L1 and L2 in opposite polarity leads of the dc link. This physical arrangement is important; it ensures that the GTOs are always protected against overcurrents caused by line-to-ground faults, whatever the instant of occurrence.

Tests on the prototype equipment confirmed satisfactory performance with worst case line-to-ground faults and the equipment was protected at all times. An example of the inverter pole current with a line-to-ground fault applied to a 50 hp drive is shown in figure 10.

100 A/division
5 ms/div.

Peak Current = 205 A

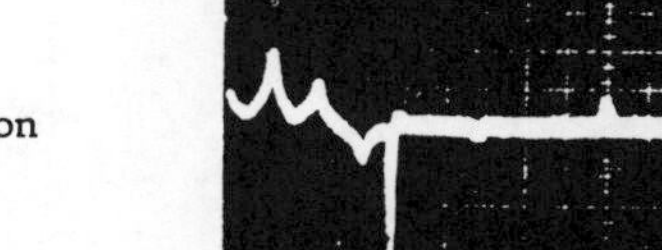

Figure 10. POLE CURRENT INTO LINE-TO-GROUND FAULT
50 HP DRIVE AT FULL SPEED-LIGHT LOAD

VIII. ACKNOWLEDGEMENTS

The work described was carried out as part of a team effort, and the authors wish to acknowledge the contributions to the program from many members of the Power Electronics Laboratory, and the Equipment Design Department, of the Westinghouse Electronics Technology Division. The ACCUTROL 300 motor drives are being manufactured by the Westinghouse Electric Corporation, Vectrol Solid State Control Division in Oldsmar, Florida. A photograph of a 20 hp unit with the front cover removed is shown in Figure 11.

Figure 11. 20 HP ACCUTROL 300 MOTOR DRIVE

REFERENCES

1. C. W. Edwards, "Microprocessor-Based Control of an AC Motor Drive", Conf. Rec. 1982. IEEE/IAS Annual Meeting

2. Hitachi GTO Thyristor Data Book No. SD-E705

3. International Rectifier Data Sheet No. PD-3.064

4. J. M. D. Murphy, Thyristor Control of AC Motors, Pergamon Press Ltd., 1973

5. "Techniques for Energy Conservation in AC Motor-Driven Systems" EPRI Report No. EM-2037, Sept. 1981. Prepared by University of Minnesota, Minneapolis.

Part III
Power MOS Field Effect Transistors

As discussed in the introductory section, the power MOSFET was developed using MOS integrated circuit technology with the objective of providing a very high input impedance power device to the circuit designer. This was realized by using a vertical channel device geometry to allow high current operation. Although the first high voltage laboratory devices were reported in the early 1970's, high voltage commercial devices for power switching applications became available starting in 1975. The reprinted papers in this section provide device design guidance and describe some typical applications for these transistors. The first paper in the section provides an overview of power MOSFET technology and describes the features of these devices that make them attractive for many power switching applications.

In order to design a power MOSFET, it is necessary to begin with the maximum voltage rating. This device parameter must first be used with Figs. A and B in the Appendix to determine the optimum thickness and resistivity of the n-drift region. The next stage in the device design process is to choose the proper MOS cell geometry in order to obtain the desired on-resistance. The on-resistance of the power MOSFET is determined by the resistance of the channel and the resistance of the drift region including spreading resistance effects. It is also dependent upon the type of cell—namely, DMOS, VMOS or LDMOS—used for device fabrication. These issues are dealt with in the second and third papers of this section. These papers demonstrate that an optimum polysilicon overlap distance between the cells exists for each voltage rating. This occurs because the channel resistance becomes high for cells with very large overlap distances and the spreading resistance from the channel to the drift region becomes large for short overlap distances. The drain overlap region also contributes to the input capacitance of the device and designing for minimum on-resistance may not result in minimum input capacitance. A general guideline for device design is to use small overlap distance for low voltage power MOSFET's where channel resistance is dominant and large overlap distance for high voltage power MOSFET's where drift region resistance is dominant. It is also worth pointing out that although these analyses indicate that the VMOS cell will have a lower on-resistance than the DMOS cell for low voltage devices, the high electrical fields at the tip of the VMOS cell and its greater processing complexity have resulted in the preference of the DMOS process by most manufacturers.

The fourth paper in this section deals with second breakdown in power MOSFET's. When the power MOSFET was first introduced, it was claimed to be free from the second breakdown problem that has plagued the power bipolar transistor. However, the power MOSFET structure contains a parasitic bipolar transistor. It is necessary to short the emitter-base junction (i.e., the source and base regions of the power MOSFET) effectively in order to make the parasitic bipolar transistor inactive. This is particularly important in circuits operating at high *dv/dt*'s. The paper by Hu and Chi provides an analysis of second breakdown in power MOSFET's which can be used by the device designer in laying out the source to base shorts in the devices.

One of the interesting characteristics of the power MOSFET structure is the existence of a reverse conducting diode formed between the p-base region and the n-drift region. In certain device applications such as motor drive circuits, it is often desirable to conduct reverse current across the power switching devices. One of the circuit conditions under which this occurs is during regenerative braking of motors. In motor drive circuits, it is customary to use a fast recovery power diode in antiparallel with the power device when using bipolar transistors. In replacing the bipolar transistor with the power MOSFET, it would be attractive use the integral diode of the power MOSFET to eliminate the "flyback" or reverse conducting rectifier. Unfortunately, the integral diodes in as-fabricated power MOSFET's have been found to be too slow in switching response to be used as flyback rectifiers. The fifth paper in this section describes a method for increasing the switching speed of the integral diode without degrading any of the power MOSFET characteristics.

The next two papers deal with device-circuit interactions. Paper six presents a model of power FET's and the results of an analysis of the model. Close form solutions for switching behavior are derived in terms of both circuit and device parameters, including the effect of parasitic elements. Results of this analysis provide physical insight into the FET switching mechanism. Paper seven discusses the causes of spurious *dv/dt* triggering. *dv/dt* false triggering is an important design consideration in a circuit with totem pole configurations such as a bridge inverter. When one of the FET conducts, the other FET in the same totem pole is subjected to large *dv/dt*. A false *dv/dt* triggering of the latter FET would cause a catastrophic short circuit through the two FET's and the power source. It is pointed out in the paper that device gate-drain capacitance and the parasitic BJT are the causes of such a possibility. Practical circuit techniques for minimizing such triggering possibility are also suggested in the paper.

The next two papers deal with FET applications. In both cases, device paralleling is necessary to achieve the desirable current level. Paper eight discusses an experimental low-voltage high-current dc motor drive circuit capable of two-quadrant operation. The integral diodes of FET's are used to provide the current path for both motoring and regenerating modes. The use of fast recovery diodes is essential in such applications because slow diodes may present a temporary short-circuit loop of the FET, the diode and the power supply which cause an unacceptable high current spike. As described in the fifth paper, a method has been reported for increasing the diode recovery speed without degrading FET characteristics. The nineth paper discusses a high-voltage high-frequency induction heating application. A 3-kW 500-kHz production unit is demonstrated using a resonant coverter circuit. Because of the high operating frequency requirements, the use of FET's is absolutely necessary in this case. It can be expected that the high frequency FET resonant converter will become more popular in many application in the future.

POWER MOSFETS - A STATUS REVIEW

Brian R. Pelly
International Rectifier
El Segundo, California

Abstract
This paper gives an overview of the present technical status of power MOSFETs and discusses the features of these devices that are luring more and more designers to use them. A summary of available MOSFET ratings and packages is presented, and information is given on the exacting quality and reliability levels that are now being achieved.

INTRODUCTION

Power MOSFETs were introduced to the marketplace about five years ago, and have now gained a strong foothold in a multitude of applications, at power levels from a few watts to a few kilowatts -- in a few cases, to several hundreds of kilowatts. Typical uses include switching and linear power supplies, speed control of dc and ac motors, stepper motor controllers, relays, lighting controls, solenoid drivers, medical equipment, robotics, appliance controls, induction heating, and instrumentation.

Useage is expected to increase significantly over the next few years, as engineers become more aware of the electrical performance characteristics, quality, and reliability of power MOSFETs. Stimulating the trend will be continuing reductions in prices.

This paper presents a brief technical overview of the present status of the power MOSFET market, and shows -- by reference to key technical comparisons with conventional bipolar transistors -- why power MOSFETs will continue to gain a greater share of the marketplace.

POWER MOSFET TECHNOLOGY

Power MOSFETs differ fundamentally from signal-level field Effect Transistors, in that the current flow is vertical rather than lateral through the silicon. Vertical current flow is vital in a power device in order to utilize the silicon effectively.

The first power MOSFETs to be introduced employed a surface groove technique -- called V-groove, or VMOS. This structure is generally no longer favored, as it is limited in voltage capability, and most manufacturers now use a planar DMOS (double diffused MOS) construction. A DMOS power MOSFET is exemplified by the HEXFET structure shown in Figure 1.

The basic principle of operation of a MOSFET is that application of a potential to a control (gate) electrode alters the conductivity of an adjacent "channel" region, thus allowing the flow of current through the main terminals of the device to be controlled. Because the gate is isolated, the power gain is extremely high; and because the MOSFET is a majority carrier device, switching speeds are very fast, thermal stability is assured, and second breakdown -- a common problem in bipolar transistors -- is absent.

The manufacturing technology needed to achieve the fine geometrical structure of a power MOSFET is borrowed largely from that previously reserved

Reprinted with permission from *Int. Power Electron. Conf.*, 1983, pp. 19–32.

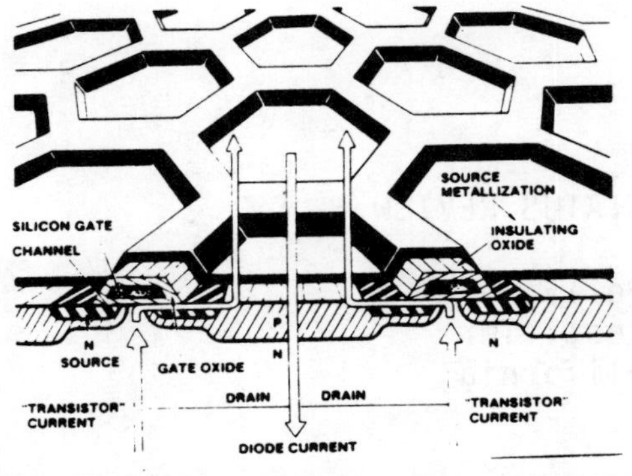

Fig. 1. Basic structure of a HEXFET Power MOSFET

Fig. 2. HEXFET geometry

for Large Scale Integrated circuits. Ion implantation, which allows very precise controlled formation of shallow layers, and very fine-line photolithographic techniques are commonplace in the production of power MOSFETs. The HEXFET structure shown in Figure 1, for example, contains about 1000 hexagonal source cells per square millimeter of silicon. This is the same order of cellular density that is used in today's intricate integrated circuits. Figure 2 shows a magnified view of the HEXFET's surface geometry.

A common (but incorrect) argument when comparing fundamental costs of bipolars and MOSFETs is that since the MOSFET inherently requires greater manufacturing precision and more manufacturing steps, it will inherently be more expensive. This point of view ignores the fact that the manufacture of all MOSFETs, whether high voltage, low voltage, small-chip or large-chip, follows largely the same standard processing steps. Substantially the only variable is the starting silicon material. Real practical economies of scale will therefore be possible as MOSFET manufacturing volumes grow, because of the uniformity with which MOSFETs can be manufactured. By comparison, literally hundreds of different manufacturing processes exist for bipolar transistors, each optimized for the specific mix of parameters needed for the particular application. Bipolar transistor manufacturers are always fine-tuning their processes to optimize the trade-offs between gain, Safe Operating Area, switching speed, and voltage, which inevitably increases manufacturing costs, and narrows any real cost differences with MOSFETs.

ELECTRICAL CHARACTERISTICS -- COMPARISONS WITH BIPOLARS

Power MOSFETs have a number of major performance advantages over bipolar transistors. These are discussed in the following sections.

MOSFETs are voltage controlled

To switch a MOSFET ON, it is necessary simply to apply a voltage -- typically 10 volts for "full enhancement" -- between gate and source. The gate is isolated by silicon oxide from the body of the device, and the dc gain is virtually infinite. Drive power is negligible, and drive circuitry is generally considerably simpler than for a bipolar transistor. A typical comparison between the base drive current of a bipolar transistor and the gate current of a MOSFET in a switching application is illustrated in Figure 3. Note the change of scales between the bipolar and MOSFET oscillograms. Whereas drive current for the bipolar must be supplied continuously during the whole conduction period, gate current for the MOSFET flows only for the short periods needed to charge and discharge the self-capacitance, during the turn-on and turn-off transitions.

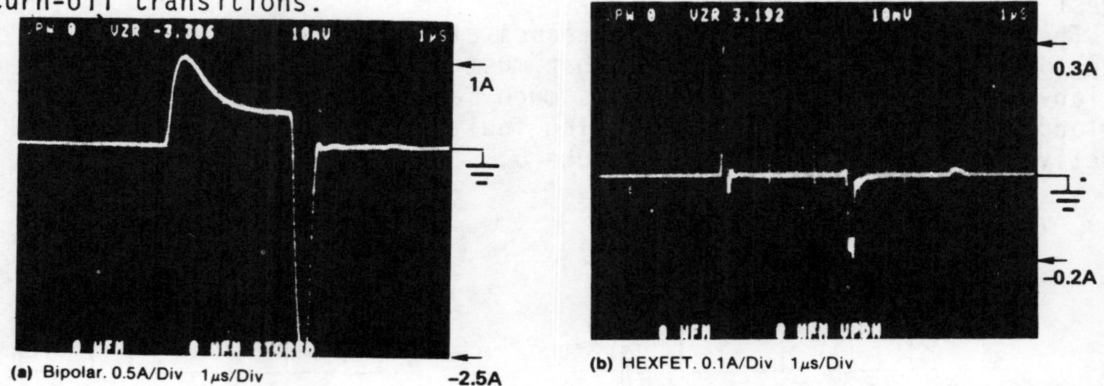

Fig. 3. Drive current waveforms for (a) bipolar transistor and (b) HEXFET
Collector (drain) current = 2.5A
Collector (drain) voltage = 270V
Frequency = 100kHz

A comparison of typical drive circuitry for a bipolar and a power MOSFET is shown in Figure 4. In this example the bipolar uses a "proportional" drive circuit, base current being taken from the collector via a current transformer. A second winding on the current transformer provides electrical isolation to the primary drive circuit. The bipolar's drive transformer is relatively large, because it must handle the relatively large base current, throughout the duration of the collector conduction period. The drive transformer for the HEXFET, by comparison, can be an order of magnitude or more smaller, since it carries current only for the short charging period of the gate-source capacitance (considerably less than a microsecond). Also, in this particular circuit, the size of the drive transformer is <u>independent</u> of the conduction time -- because the MOSFET "holds" its own gate drive voltage, by virtue of its own gate-source self-capacitance (discharge being blocked by the small auxiliary MOSFET drivers).

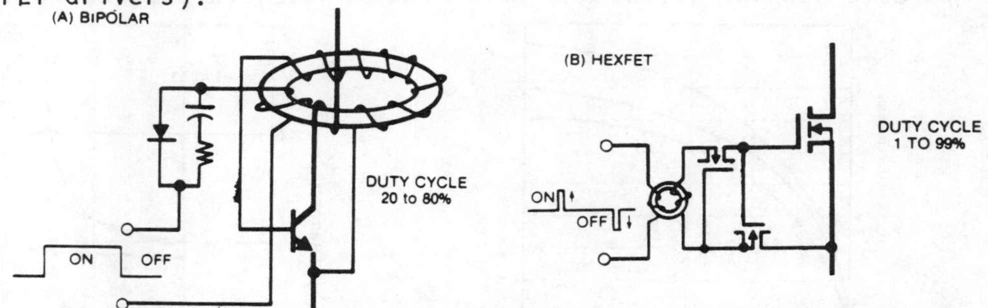

Fig. 4. Typical isolated drive circuits for (a) bipolar and (b) MOSFET

<u>Absense of storage time</u>

The MOSFET has no inherent delay and storage times, though it does, as mentioned above, have self-capacitance, which must be charged and discharged when switching. The time constant formed by the self-capacitance in conjunction with gate circuit and drain and source circuit impedances determines the switching times. Practical switching times range from less than 10 nanoseconds for small MOSFETs rated 1 ampere or less, to 50-100 nanoseconds for the largest ones, rated 10 to 30A.

The absence of storage time increases circuit utilization factors by reducing or eliminating "dead times" that must be built into circuits that use bipolar transistors. It also permits much faster response when reacting to overload and fault conditions, allowing fault current to be arrested much more effectively -- as illustrated in Figure 5.

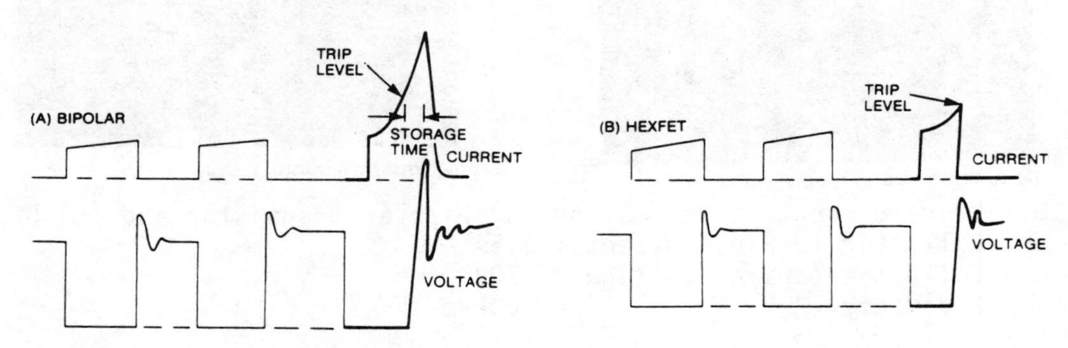

Fig. 5. Response to overload in switching power supply (a) bipolar transistor (b) HEXFET

High peak current capability

The gain of a bipolar transistor decreases with increasing current, but the transconductance of a MOSFET increases with increasing current, as illustrated in Figure 6. High peak current in a bipolar transistor tends to "pull it out of saturation" and destroy it through overheating. The on-resistance of a MOSFET does increase with increasing current -- as shown in Figure 7 -- but the effect is much more benign than with a bipolar, and the MOSFET has a much higher peak current carrying capability. For example a 400V HEXFET with a continuous drain current rating of 5.5A has a useable peak current of 20A. A comparable bipolar transistor would have a useable peak current of perhaps 7.5A.

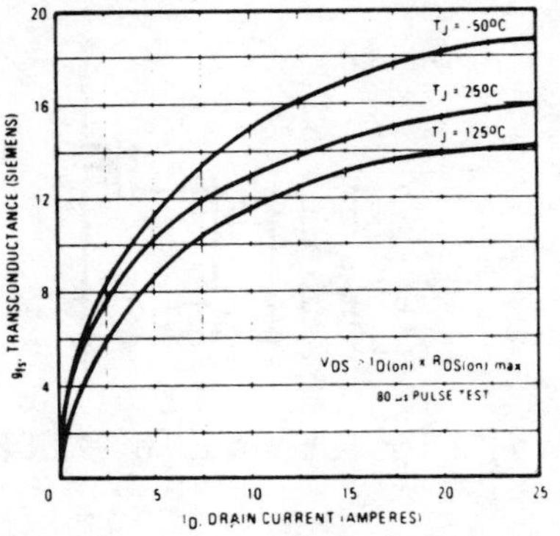

Fig. 6. Typical transconductance vs. drain current IRF450 HEXFET

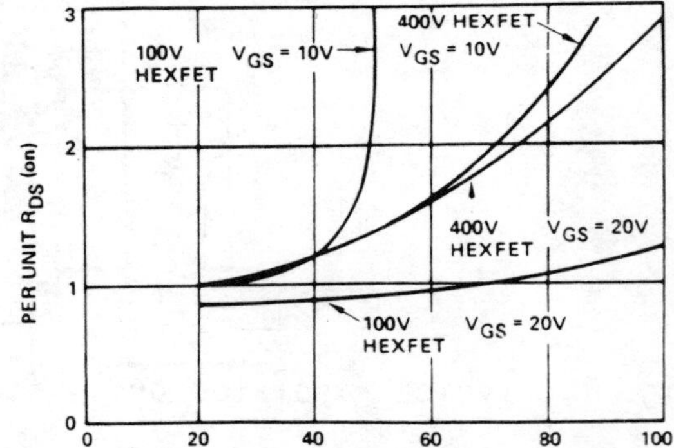

Fig. 7. Typical variation of on-resistance with drain current. 100% $I_{DM} \doteq 4 \times I_D$ @ $T_C = 25^{\circ}C$.

A practical example of the use of the HEXFET's peak current carrying capability is illustrated in Figure 8. In this example, the IRF150 HEXFET, rated 40A continuous, carries a peak capacitive load charging current of 150A, with a time constant of about 10msec.

The underlying limitation on current handling capability of a HEXFET is junction heating. It is able to handle peak current well in excess of its continuous current rating, just so long as the rated maximum junction temperature is not exceeded. This is true both for non-repetitive and repetitive operation.

A comparison of typical die sizes and current ratings of 400V rated bipolar and HEXFET devices is shown in Figure 9.

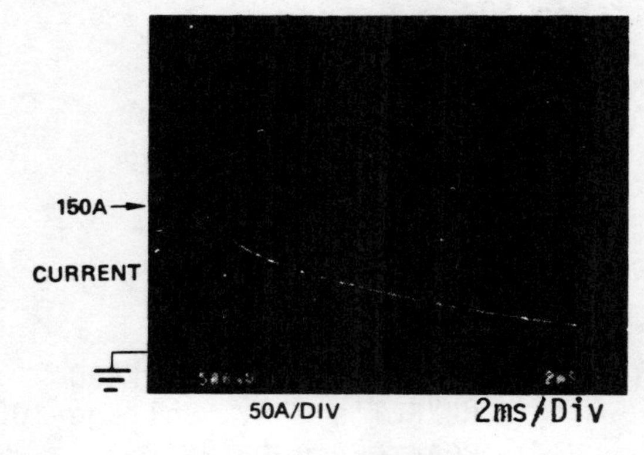

Fig. 8. Single shot 150A peak exponentially decaying pulse applied to the IRF150. Time constant ≑ 8msec.

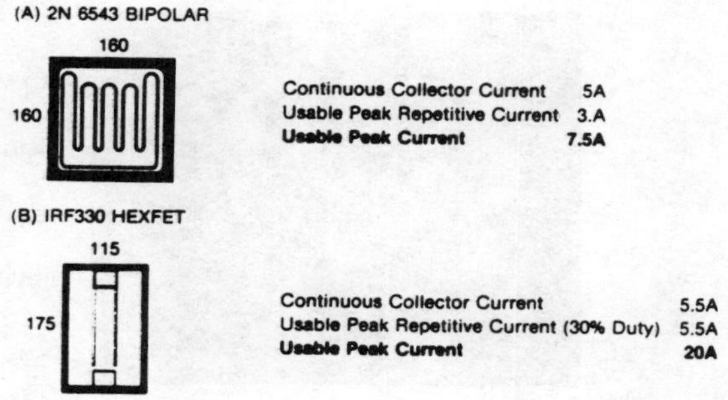

(A) 2N 6543 BIPOLAR

160

160

Continuous Collector Current 5A
Usable Peak Repetitive Current 3.A
Usable Peak Current 7.5A

(B) IRF330 HEXFET

115

175

Continuous Collector Current 5.5A
Usable Peak Repetitive Current (30% Duty) 5.5A
Usable Peak Current 20A

Fig. 9. Comparison of typical bipolar and MOSFET chips

Wide Safe Operating Area

The Safe Operating Area of a power MOSFET is much better than that of a bipolar. The MOSFET, being a majority carrier device, has a positive temperature coefficient of resistance, and is immune from the hotspot formation and second-breakdown phenomena that plague the bipolar transistor. MOSFETs are therefore generally much more rugged than bipolars, and snubber/clamp circuitry can be smaller and less dissipative, as illustrated in Figure 10.

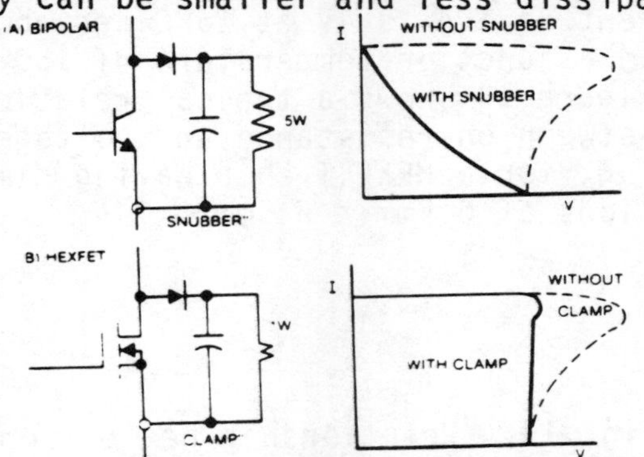

Fig. 10. Comparison of snubber/clamp circuitry for (a) bipolar and (b) MOSFET

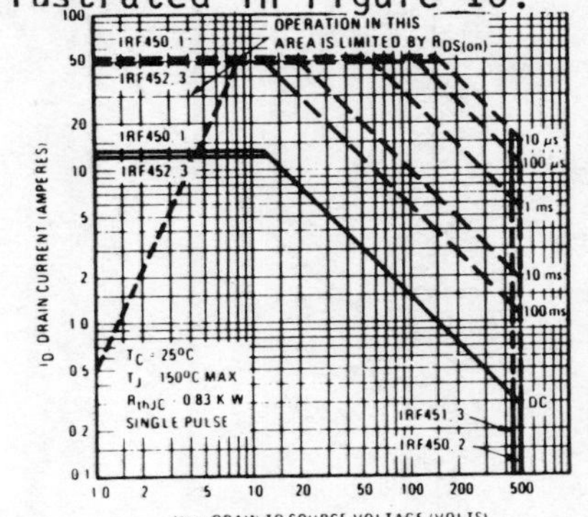

Fig. 11. Maximum Safe Operating Area IRF450 HEXFET

MOSFET data sheets usually show Safe Operating Area curves that cover current and voltage values up the rated I_{DM} and V_{DS} values respectively. Typical SOA data for a HEXFET is shown in Figure 11. This data is based simply on junction temperature rise, and is for a case temperature of 25°C, and an internal dissipation that raises the junction to the rated maximum value of 150°C. SOA curves for each pulse duration follow a line of constant power, and are actually nothing more than a graphical statement of the absence of second breakdown.

A verification of this particular SOA data is shown by the oscillograms in Figure 12. In this example, a 10µs, 50A pulse current is applied to the IRF450, at an applied drain-to-source voltage of 400V. Reference to Figure 11 shows that this point lies beyond the published SOA.

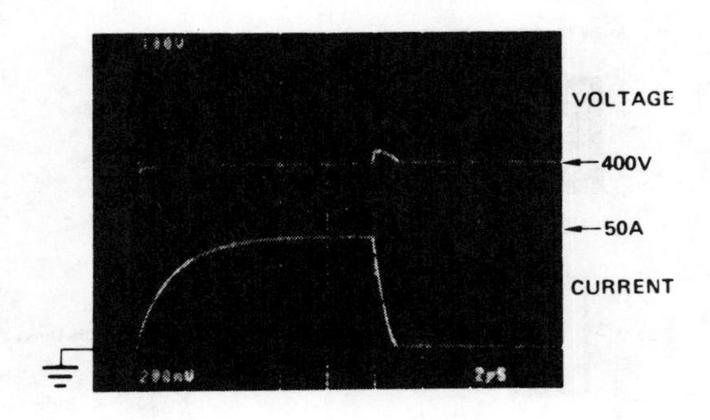

Fig. 12. **Verification of** IRF450 HEXFET's SOA. Voltage=400V. Peak current=50A

MOSFETs are more efficient at high frequency

The on-resistance of a MOSFET is one of its key characteristics, because it determines the device's conduction losses in switching applications. Today's largest 100V MOSFETs have on-resistance in the order of 30-50mOhms at 25°C. The corresponding conduction voltage drop at 30A at 25°C is 1 to 1.5V, and at rated maximum junction temperature of 150°C, it is 2 to 3V. For a given chip size, on-resistance increases with voltage rating. The largest 500V rated power MOSFETs have on-resistance in the order of 0.4 ohms at 25°C. The corresponding conduction voltage drop at rated useable continuous current of 10A is 4V at 25°C and about 8V at a junction temperature of 150°C. Figure 13 shows a typical relationship between on-resistance and voltage rating, for a HEXFET chip having dimensions of 6.5mm X 6.5mm.

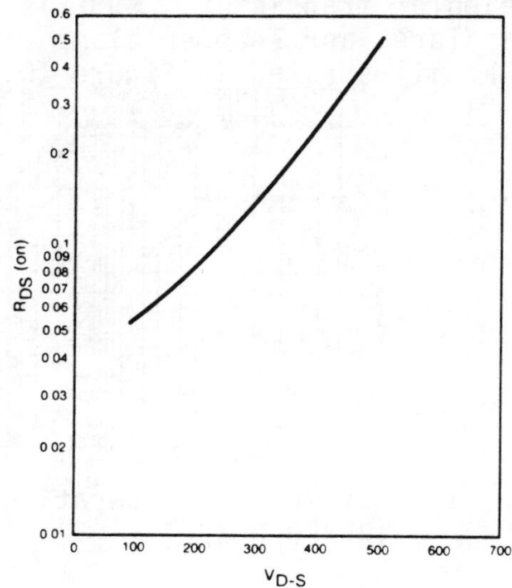

Fig. 13. Relationship between on-resistance and voltage rating. 6.53 X 6.53mm HEXFET

Comparisons between power MOSFETs and bipolars often center around the fact that the conduction voltage of a MOSFET is higher than for a bipolar, and becomes progressively more so as voltage rating increases. Conduction losses of a MOSFET when operating near rated current are therefore generally greater than for a bipolar. Switching losses of a MOSFET, on the other hand, are almost negligible -- while the switching losses of a bipolar are often greater than its conduction losses, particularly at high frequency. The bipolar is therefore usually more energy-efficient at low frequency, while the MOSFET is more energy-efficienct at high frequency. The frequency cross-over point depends upon specific circumstances, but is generally somewhere between 10 and 40kHz.

A comparison between the switching waveforms for a bipolar and a HEXFET is shown in Figure 14; note the sharper waveforms for the HEXFET. Figure 15 shows a comparison between power dissipation for the IRF330 HEXFET, the industry standard 2N6542/3 bipolar, and a recently introduced fast switching bipolar transistor. Power losses were gauged by measuring the case temperature of the device on a calibrated heatsink. The "full" curves represent only the device dissipation. Additional power is dissipated in the external base drive circuit of the bipolar, and the dashed curve includes 1.3W of external base drive power for the fast switching bipolar.

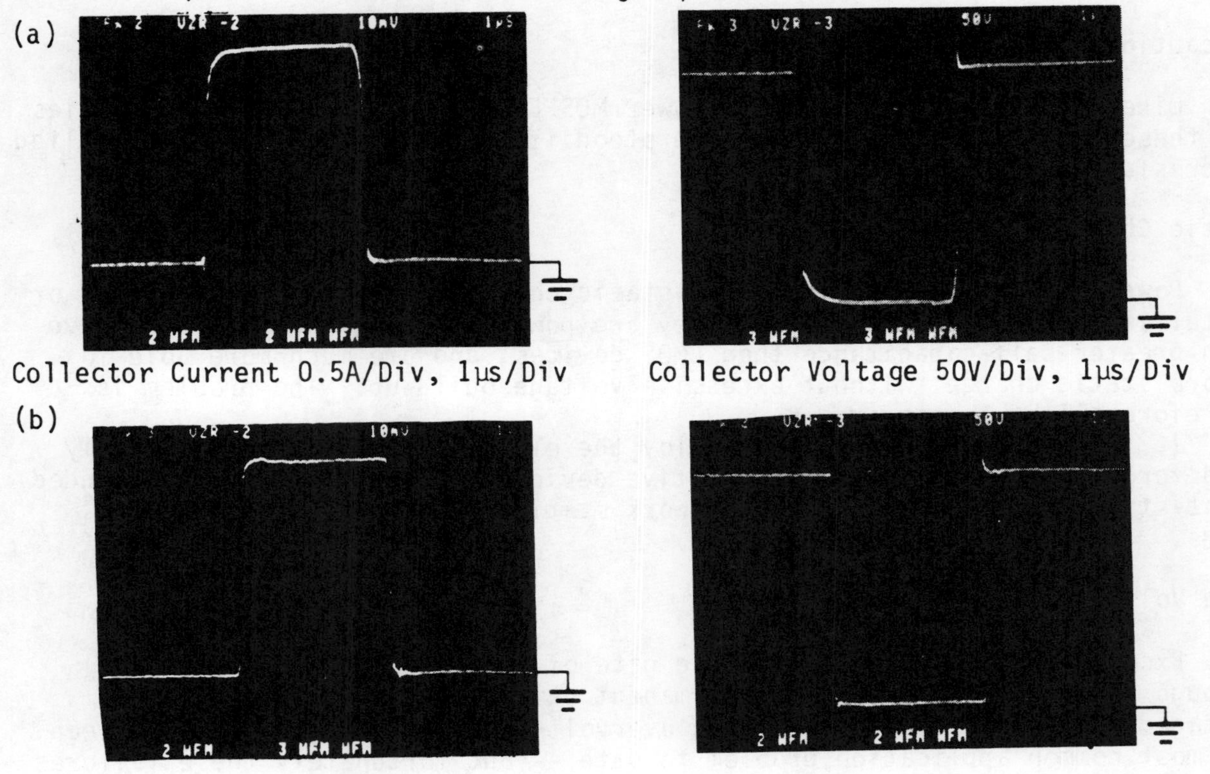

(a) Collector Current 0.5A/Div, 1μs/Div — Collector Voltage 50V/Div, 1μs/Div

(b) Drain Current 0.5A/Div, 1μs/Div — Collector Voltage 50V/Div, 1μs/Div

Fig. 14. Switching waveforms for (a) bipolar transistor and (b) MOSFET operating at 100kHz

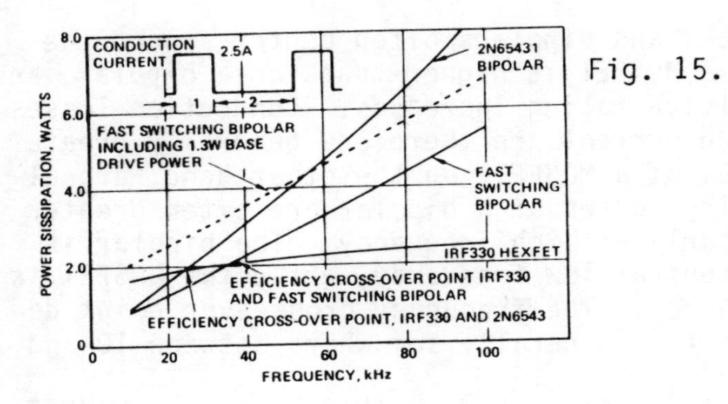

Fig. 15. Power dissipation versus frequency for 2N6542/3, a fast switching bipolar and IRF330 HEXFET. Supply voltage=270V. Conduction duty cycle=0.33 current amplitude =2.5A

It would not be correct to conclude from this discussion that power MOSFETs will find use only in higher frequency applications, and will not also be the preferred choice at lower frequency. Although MOSFETs certainly "shine" at high frequency, their higher conduction losses actually are often inconsequential, when viewed from the standpoint of the overall system design. Considerations such as circuit simplification, ruggedness, cost effectiveness and reliability of the overall system will frequently favor the MOSFET, even in low frequency applications.

Precautions

Like any other semiconductor, power MOSFETs do have their own subtleties, and these must be recognized and understood if these devices are to be applied successfully.

Static Charge

Power MOSFETs can be damaged by static charge when handling, testing, or installing into a circuit. Since they are power devices, however, they have much greater self-capacitance than CMOS devices, and are much more able to absorb static charge, without excessive voltage build-up. The problem is therefore slight by comparison.

It is wise, nonetheless, to employ the elementary precautions commonly used for other types of static-sensitive devices, such as the use of grounded wrist-straps, electrically grounded work stations, and grounded soldering irons.

Gate Voltage Transients

Excessive voltage applied to the gate of a power MOSFET will punch through the gate oxide and cause permanent damage. A typical gate-source voltage rating is ±20V. Gate-source overvoltage failures have probably been the most common application problem to date -- but fortunately these failures can be easily avoided.

The problem usually stems from the fact that designers do not immediately recognize that voltage transients in the drain circuit can be coupled to the gate via the drain-gate self-capacitance. The amplitude of the voltage transient induced at the gate may or may not be dangerous (often it isn't), and it depends upon the impedance of the drive circuit. The simplest solution where gate voltage transients are suspected is to connect a clamping zener diode between the gate and source, physically as close as possible to the terminals.

A more subtle circuit mechanism for producing overvoltage at the gate has been observed. This is due to induced voltage in common source inductance when commutating "freewheeling" current from the internal body-drain diode. If this is a problem (in nearly all applications it isn't), it can be overcome by minimizing common source inductance -- and by keeping this inductance as symmetrical as possible for each MOSFET in the circuit.

Drain Voltage Transients

The extremely fast switching speed of the MOSFET is obviously conducive to self-inflicted overvoltage transients when switching off, due to stray drain circuit inductance -- even though the drain supply voltage may be well within the drain-source voltage rating.

The solution is to minimize stray circuit inductance, and if necessary (usually it is) to use a voltage clamp. Figure 10(b) shows an example.

High Let-Through Fault Current

Because power MOSFETs are not gain-limited to nearly the same degree as bipolar transistors, let-through fault currents can be much higher before self-limiting occurs. If left to flow unabated, or if allowed to reach the very high self-limiting level of the MOSFET, this fault current can be destructive to the MOSFET, as well as to other components in the circuit.

In applications where high let-through currents can occur, the first "line of defense" is to reduce the amplitude of the MOSFET's gate drive voltage, so that the peak let-through current is limited to a more moderate value. A more positive approach, however, is to use current sensing circuitry that turns OFF the MOSFET at a present level of current (as illustrated for example by the waveforms in Figure 5), thereby avoiding operation in a high-dissipation mode for more than a short transient interval.

Body Drain Diode

All power MOSFETs have an integral body-drain diode built into their structure (see Figure 1). In many applications this diode is inactive, and its presence can be ignored. In other circuits the body-drain diode does conduct, and actually performs a necessary circuit function "free of charge", obviating the need for a separate "freewheeling" or "clamping" rectifier.

In other cases, however, the reverse recovery time of the internal diode can be a problem, because this is relatively long by comparison with the switching speed of the MOSFET, and switching losses can become excessive. More subtly, the dv/dt capability of the body diode during its recovery period is limited, and in circuits that apply dv/dt during this period MOSFET failures can occur, if the dv/dt is excessive.

Fortunately, there are relatively simple circuit solutions to these problems when they arise (and in the majority of applications, they don't).

POWER MOSFET RATINGS AND PACKAGES

Power MOSFET ratings and packages cover the spectrum shown in Figure 16.

A typical family of N-Channel MOSFET die, with their associated voltage and current ratings, is illustrated in Figure 17.

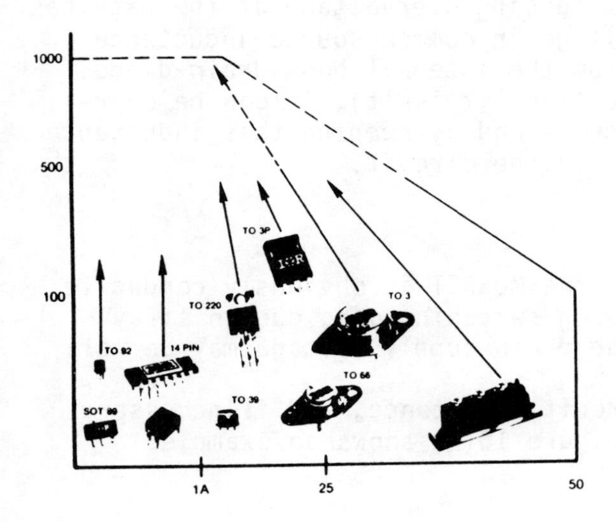

Fig. 16. Summary of power MOSFET ratings and packages

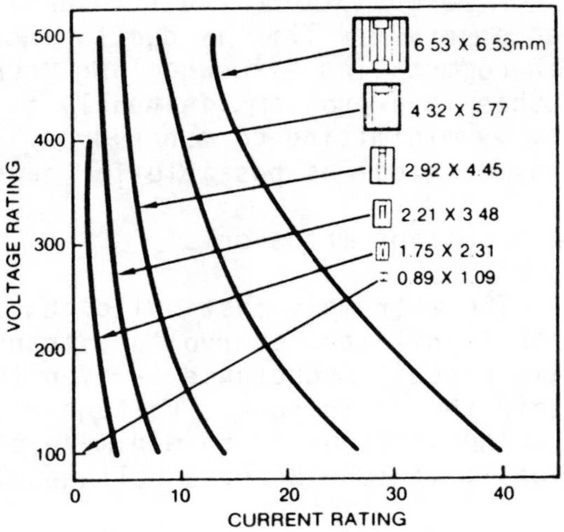

Fig. 17. Voltage versus current ratings. HEXFET family of power MOSFETs

As voltage capability increases above 500V, the power MOSFET requires an increasingly larger silicon die area than a bipolar transistor. Manufacture of MOSFETs above this voltage rating also becomes more difficult, reducing manufacturing yields and further contributing to costs. These are reasons why most of the power MOSFET market is presently at 500V and below.

Two or three years ago, the main packages offered were the industry-standard TO-3, TO-220AB, and TO-92 types. Today's spectrum of packages is broadening considerably, to cover both higher and lower power applications.

Noteable among the packages in Figure 16 are the "HEXPAK", and the 4-pin and 14-pin dual in-lines. The former package style has been popular for several years for SCRs and rectifiers; it is a natural development to put parallel chip HEXFETs in this package, for high power applications such as motor drives and uninterruptible power supplies.

The DIP package -- popular for line drivers, stepper motor controllers, automatic test equipment, instrumentation, and the like -- are automatically insertable, and are ideal where space is at a premium.

Power MOSFETs are being increasingly used in high reliability, space and military applications. Several special packages are now being offered to meet these special needs. Some of these packages are shown in Figure 18.

Hybrids -- containing additional circuitry peripheral to the MOSFET itself -- are also a natural development, and are beginning to appear for special applications. An example is shown in Figure 19.

Fig. 18. Typical packages for high reliability/military applications

Fig. 19. Typical HEXFET hybrid package

QUALITY AND RELIABILITY

The sophisticated processes and modern facilities necessary to the production of power MOSFETs create a manufacturing environment that is ideally conducive to high standards of quality and reliability. As of January, 1983, HEXFETs, for example, are being delivered to an AQL of 0.04% (i.e. a statistical maximum of 400 defective parts per million). Bipolar transistors and related bipolar power semiconductors, by comparison, typically achieve AQL levels that are twenty-five to fifty times inferior -- in the range of 10,000 to 20,000 defective parts per million.

The stringent AQL levels being obtained with power HEXFETs are achieved through rigid in-process controls -- which include numerous computerized monitoring checks during wafer fabrications, inspection of die mount-down, wire bonding, ultrasonic cleaning, and cap welding, pre-cap visual, and 100% computerized testing to electrical specifications.

HEXFETs also undergo an additional 100% batch Reliability Certification Program, that includes accelerated gate stress and high temperature reverse bias, along with tests to destruction, such as Safe Operating Area, gate dielectric breakdown, and avalanche energy failure. These tests determine the ultimate limits of performance, and verify that safety margins lie well beyond published data sheet specifications.

Reliability

Separate from -- but obviously related to -- the very tight quality levels now being achieved are equally impressive reliability specifications. Extensive long term reliability testing of TO-3 and TO-220AB package power HEXFETs, for example, for a total of 680,000 device hours, has demonstrated a relationship between Random Failure Rate and junction operating temperature shown in Figure 20.

This graph gives 23 failures in 10^9 device operating hours at a continuous junction operating temperature of 90°C. This level of reliability matches or exceeds that of the most advanced LSI memories.

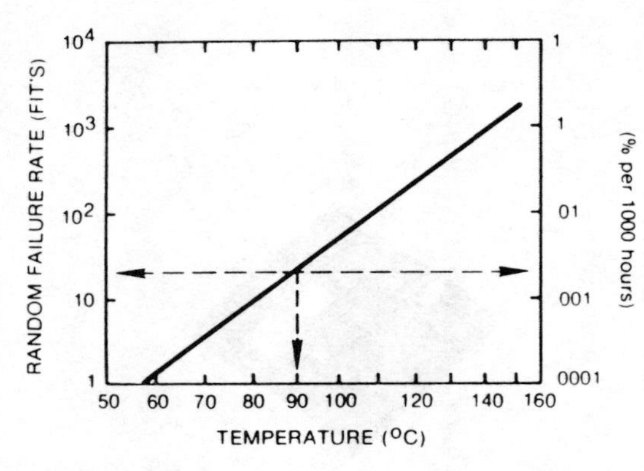

Fig. 20. Random failure rates versus junction operating temperature of HEXFETs

Reliability testing such as this is expected to lead many users to reduce or eliminate their own burn-in test programs, thus accelerating the user trend to power MOSFETs.

Military Approvals

Several power MOSFET types are now qualified to JAN, JANTX, JANTV, and CECC specifications, and many more are in the process of qualification.
Such approvals, for example, have been conferred on all TO-3 package HEX-3 and HEX-5 types (Figure 19).

Radiation Resistance

The effects of radiation are critically important in military and space applications. Testing by various agencies has now demonstrated the suitability of power MOSFETs for use in these environments, provided that the appropriate circuit design features are observed.

The primary effect of ionizing radiation is to cause a decrease in gate threshold voltage of N-Channel types, and an increase for P-Channel types. Typical test results are shown in Figure 21. The change of threshold voltage depends on the gate biasing conditions, particularly at radiation doses that are sufficient to change the polarity of the threshold voltage (N-Channel HEXFETs). A negative bias voltage applied to the gate during the OFF period -- typically -10V -- allows reliable switching operation at doses up to 1 megarad (Si) -- a greater dose than would actually be accumulated in most space applications.

The primary effect of neutron radiation is to cause an increase in on-resistance. This depends on the resistivity of the silicon -- and hence on the voltage rating of the device. The effect is not too pronounced for a 100V rated HEXFET, but can be appreciable for a 400V rated device, as illustrated in Figure 22.

The increase of on-resistance due to neutron radiation produces increased conduction losses in switching applications, but will have little effect in linear applications. It can be allowed for at the design stage, by appropriate choice of MOSFET and heatsink.

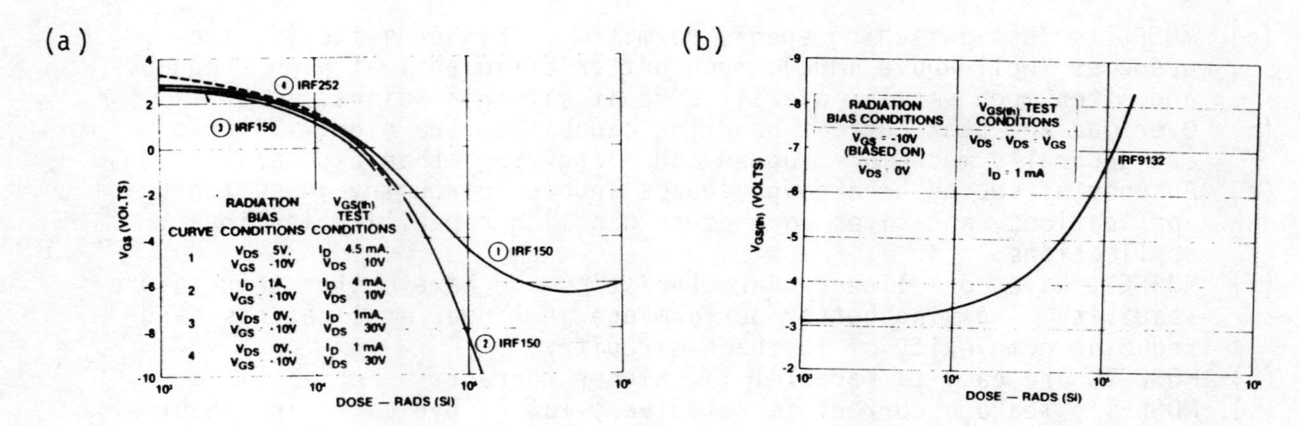

Fig. 21. Typical effects of ionizing radiation on (a) N-Channel and (b) P-Channel HEXFETs

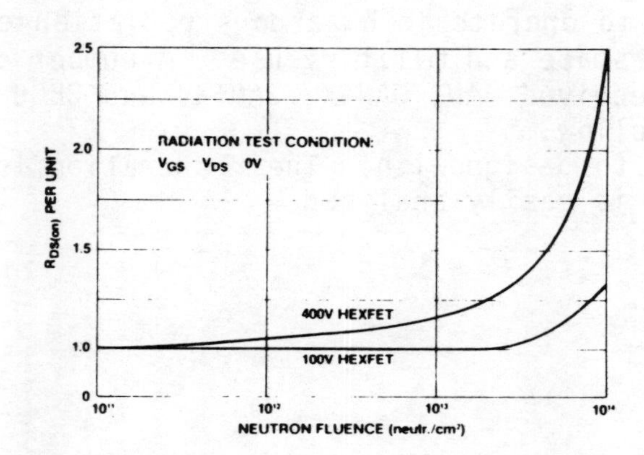

Fig. 22. Typical relationship between on-resistance and neutron fluence level

CONCLUSIONS

MOSFETs are now firmly established, and useage is continuing to increase rapidly. From a fairly modest range of product offerings two or three years ago, users now have at their disposal a variety of different types, in a variety of different packages.

Today MOSFET prices are often quite close to those of bipolars, and overall system costs are frequently lower using MOSFETs, because of the circuit simplifications that result from their use. Price is much less a consideration than it used to be -- and will become even less so in the future. Designers, no longer shackled by overriding cost considerations, are turning in droves to MOSFETs for the technical advantages that they offer. These advantages are summarized as follows:

(a) Quality and reliability of MOSFETs are an order of magnitude better than for bipolar transistors, providing much better system reliability.

(b) Drive circuitry is simpler and cheaper, and often can be standardized for a whole range of products.

(c) MOSFET's fast switching speeds permit much higher switching frequencies (well above 1MHz), much better efficiency at high frequency, and often much smaller overall circuit size and weight.
(d) Overload and peak current handling capability are high. MOSFETs are generally much more rugged and "forgiving" than bipolars.
(e) Absence of second breakdown reduces snubber circuitry in switching applications, and gives more power handling capability in linear applications.
(f) MOSFETs have more linear characteristics and have better temperature stability -- giving better performance in linear applications, and reducing complexity of feedback circuitry.
(g) MOSFETs are easy to parallel for higher current.
(h) MOSFET's leakage current is relatively low -- typically in the order of nanoamperes. This is important in critical applications such as Automatic Test Equipment and relay switching.
(i) Drain-source conduction threshold voltage is absent, eliminating electrical noise in sensitive A-C switching applications.
(j) MOSFETs are able to operate in hazardous radiation environments, and are suitable for space and military use. A number of different types have now received JAN, JANTX, JANTXV and CECC approval, and many more will follow.
(k) MOSFETs are easy to design with. Their operation is "clean" and predictable, and is easily analyzed.

Modeling of the On-Resistance of LDMOS, VDMOS, and VMOS Power Transistors

S. C. SUN AND JAMES D. PLUMMER, MEMBER, IEEE

Abstract—**Power MOS transistors have recently begun to rival bipolar devices in power-handling capability. This new capability has arisen primarily through the use of double-diffusion techniques to achieve short active channels and the incorporation of a lightly doped drift region between the channel and the drain contact, which largely supports the applied voltage. Many different structures have been proposed to implement these new devices. This paper considers three of the most common—LDMOS, VDMOS, and VMOS. Structural differences which result in on-resistance and transconductance differences between the devices are described. Quantitative models, suitable for device design, are developed for the on-resistance of each type of structure. These models are developed directly from the physical structure (geometry and doping profiles) so that they are useful in optimizing a particular device structure or in quantitatively comparing structures for a particular application.**

I. INTRODUCTION

RECENT ADVANCES in processing technology and the introduction of new device structures have allowed dramatic improvements in the current, voltage, and power-handling capabilities of MOSFET devices. The impetus for much of this work is the faster switching ability of majority-carrier devices which do not suffer from the minority-carrier charge-storage problems inherent to bipolar transistors. A second primary motivation is the negative temperature coefficient of carrier mobility which greatly decreases problems of thermal runaway, secondary breakdown, and current hogging, all of which play important roles in the design and application of power bipolar transistors. The recent commercial availability of a variety of discrete-power MOS transistors has made possible a host of new applications for these devices including switching power supplies, linear audio and high-frequency amplifiers, and power control devices.

Two main changes in the basic MOSFET structure have been responsible for these advances. The first of these is the widespread use of double-diffusion techniques to achieve very short channels (1-3 μm) [1], [2] although not all of the new power MOSFET's use double diffusion [3]. Sequential diffusion of p- and n-type impurities in a manner analogous to bipolar-transistor fabrication processes yields channel lengths comparable in dimension to bipolar basewidths. Historically, this process has been difficult to control because the threshold voltage of the device is determined by diffused impurity profiles rather than by bulk substrate doping levels. The widespread use of ion implantation has, however, largely eliminated this difficulty.

Manuscript received June 29, 1979; revised October 1, 1979. This work was supported under an Industrial Grant from Tektronix Inc.

The authors are with Stanford University, Integrated Circuits Laboratory, Stanford, CA 94305.

The second major change in the basic MOSFET structure has been the incorporation of a lightly doped (usually n^-) drift region between the channel and the n^+ drain contact [1]-[4]. This region largely supports the applied drain potential because its doping level is chosen to be much smaller than the p-channel region. These new structures, therefore, effectively separate the active portion of the device (channel) which determines device gain, from the region of the device which supports the applied voltage (drift region). This separation is exactly analogous to modern bipolar transistors in which a lightly doped collector region largely supports the applied potential and a narrow, more heavily doped base region largely determines device gain.

While most of the recently developed power MOS devices are based on these two principles, there exists substantial variation in the structures used to implement them. At least three basic structures have been described, the lateral double-diffused transistor (LDMOS) [4], the vertical double-diffused transistor (VDMOS) [5], and the V-groove double-diffused transistor (VMOS) [2]. Depending upon the voltage, current, power, and speed requirements of a particular application, the choice of which structure to use may vary.

It is the purpose of this paper to compare some aspects of the dc characteristics of these three devices to point out some of the tradeoffs involved in the choice of a particular structure for a given application. Little consideration is given here to layout density since questions of this type are highly technology dependent; rather, the emphasis is on modeling of the devices' dc characteristics through the use of equivalent circuits. To facilitate a direct experimental comparison, all three types of structures have been fabricated side by side on the same wafers, with a simple metal gate technology and identical design rules.

II. THREE DMOS STRUCTURES

The three DMOS structures considered here are shown in cross section in Fig. 1. The LDMOS, VDMOS, and VMOS devices are shown in parts (a), (b), and (c), respectively. The following qualitative comments can be immediately made in comparing the three.

1) Both the LDMOS and the VDMOS depend upon lateral diffusion profiles and sequential diffusion of boron and phosphorus or arsenic through the same oxide cut to achieve their channel regions. The VMOS device, on the other hand, derives its channel from the vertical diffusion profiles. The channel length in the lateral devices, therefore, will be $\approx$85 percent of the vertical channel length due to the two-dimensional effects of lateral diffusion [6]. In addition, the VMOS channel is formed along the side of an anisotropically etched groove

Reprinted from *IEEE Trans. Electron Devices*, vol. ED-27, pp. 356-367, Feb. 1980.

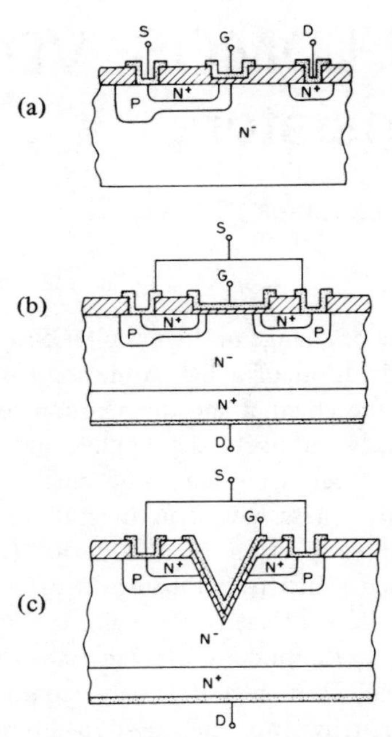

Fig. 1. Cross sections of three high-voltage DMOS devices. (a) LDMOS. (b) VDMOS. (c) VMOS.

which forms a 54.7° angle with the surface. The net result is that for the same diffusion schedule, the VMOS channel length will be approximately $(1/0.85)(\sqrt{3/2})$ or 1.5 times longer than the lateral devices' channel length. For the same diffusion schedule, therefore, the lateral devices would be expected to have about 50 percent lower channel resistance. (This qualitative conclusion must be modified when velocity saturation effects in the short channels are considered as will be described later.)

2) The VDMOS and VMOS structures require only two of the electrodes on the top surface. This would tend to indicate that they should be capable of higher packing density than the LDMOS which has the drain connection on the top surface. In fact, virtually all power MOSFET's manufactured today do use one of the vertical structures for exactly this reason. The LDMOS device, on the other hand, has the advantage of much simpler integration with other components [4], exactly because all three electrodes are on the top surface and readily available for interconnection to other devices. In addition, in lower voltage applications in which the LDMOS gate electrode may be extended all the way to the drain n^+ region over thin oxide [7], the gate potential can substantially reduce the resistance of the n^- drift region by heavily accumulating the surface. This mechanism cannot occur in the vertical structures and it makes the LDMOS more competitive in current per unit area with the vertical devices at low voltages.

3) Both the LDMOS and VDMOS may be readily fabricated on any silicon crystalline orientation. The VMOS is constrained to have its channel along an etched ⟨111⟩ surface. The choice of ⟨100⟩ material for the LDMOS and VDMOS, which is common, provides a 20-percent improvement in electron inversion-layer mobility [8], [9] and a 15-percent improvement in inversion-layer electron scattering-limited velocity [8]. These effects result in lower channel resistance and higher device transconductance per unit width in the LDMOS and VDMOS structures.

4) The fixed oxide charge density Q_{ss} is known to be approximately three times higher on the ⟨111⟩ plane than it is on the ⟨100⟩ plane [10]. For a given device threshold voltage, therefore, the peak channel doping must be higher in the VMOS structure than it is in the LDMOS or VDMOS devices. Since the electron inversion-layer mobility decreases with increasing doping [8], [9] this would again imply higher mobility in the LDMOS and VDMOS devices. In practice, however, this effect is minor in higher voltage structures because channel doping in all three structures is high enough that the bulk charge term dominates device threshold, with little contribution from the Q_{ss} term [11]. The higher Q_{ss} in the VMOS may in itself, however, degrade mobility [12], and also generally implies a higher interface state (N_{st}) density which may impact other device parameters such as noise performance in linear amplification applications [13].

5) The nonplanar VMOS device can present fabrication difficulties in terms of metal coverage and photolithography which do not exist in the planar LDMOS and VDMOS structures. The increasing use of silicon-gate technologies in the fabrication of all three types, however, has largely eliminated this potential problem.

6) All three structures inherently have some overlap of the thin gate oxide over the heavily doped n^+ source region. This contributes directly to gate-to-source capacitance and degrades the speed of the devices. In metal-gate technologies, this capacitance can be substantially reduced by making use of differential oxidation rates over n^+ and n^- regions [14]. Such techniques are more effective on ⟨100⟩ oriented silicon than on ⟨111⟩ substrates. This would tend to favor the LDMOS and VDMOS structures when high-frequency performance is desired. In addition, self-aligned silicon-gate technology is relatively straightforward to implement in the planar devices; it is not as effective in reducing C_{GS} in the VMOS structure.

7) In high-voltage applications, the bulk resistance of the n^- region in each device plays a dominant role in the total device on resistance. (This will be demonstrated quantitatively later.) Each of the three structures inherently has such a bulk resistance, however, quantitatively the values are quite different for all three. Equivalent circuits for the three devices will be described next which will point out the differences in this component for each structure.

III. Equivalent Circuits for the Three Devices

Each of the structures consists basically of an enhancement-mode transistor in series with a parasitic bulk resistance. A more careful look at the equivalent circuits shown in Fig. 2 will reveal some basic differences, however. As in Fig. 1, the LDMOS, VDMOS, and VMOS structures are represented in parts (a), (b), and (c), respectively.

The LDMOS device consists of an enhancement-mode transistor (active channel) in series with a depletion-mode transistor (surface accumulation region), itself in parallel with a bulk resistance R_1, in series with a second bulk resistor R_2 [11]. The middle two components model the gate-controlled accumulation layer in parallel with a bulk resistor; the current through the device will divide between these two in a manner determined by the gate voltage (i.e., the conductance of the accumulation layer). The third series component R_2 models

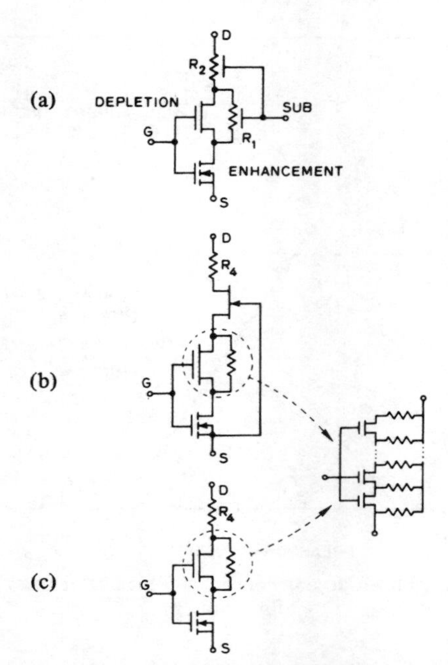

Fig. 2. Equivalent circuits of high-voltage DMOS devices. (a) LDMOS. (b) VDMOS. (c) VMOS.

the region of the device between the gate electrode and the drain contact; such a region would be missing in a low-voltage structure in which the gate metal could extend to the n^+ region [7]. In a junction isolated device, the substrate p-n^- junction modulates the conductivity of R_1 and R_2 as indicated in the figure.

The VDMOS device model (Fig. 2(b)) is distinguished from the LDMOS model in principally two ways. First, the depletion-mode transistor and its parallel bulk resistor must be regarded as distributed devices because the current flow changes in direction from horizontal to vertical along the length of the accumulated surface between the source regions. The distributed model will be simplified here to a single depletion-mode transistor. A correction factor will be introduced into the modeling later to account for the two-dimensional nature of the current flow. The second major factor distinguishing the VDMOS from the LDMOS is the presence of the series JFET in the equivalent circuit. This arises physically because of the pinching of the current between the adjacent p bulk diffusions. The spacing between the p diffusions (channel width of the JFET) can be of great importance in determining the fraction of the device's total on-resistance contributed by the series JFET.

The VMOS equivalent circuit (Fig. 2(c)) is identical to that of the LDMOS except that any influence of a p-type substrate on R_1 and R_2 is absent since the VMOS is regarded here as a nonisolated, discrete device. The physical structure of the VMOS is, however, quite distinct from the LDMOS so that an analysis of the various components in the two models will be different in each case.

The analyses of the three devices will be based on these equivalent circuits. Clearly, approximations are involved in proceeding in this fashion. Specifically, two-dimensional distributed structures are being simplified to lumped equivalent circuits. The motivation for doing this is simply that clearer physical insight results. With minor geometrical corrections to

TABLE I
EXPERIMENTAL PARAMETERS FOR FABRICATED DEVICES

Wafer Resistivity Ωcm	Effective Epi Thickness Between P Diffusion and N^+ Substrate μ	LDMOS, VDMOS L_{eff} μ	VMOS L_{eff} μ
0.5	6.4	1.50	2.17
1.1	7.3	1.62	2.34
3.0	14.6	1.79	2.58
7.0	20.3	2.00	2.88
8.5	26.8	2.13	3.06

account for this approximation, the models in Fig. 2 will be shown to provide good agreement with a variety of experimental structures.

Even with the simplifications apparent in Fig. 2, modeling of these devices is not straightforward. A few of the reasons for this are listed below.

1) Each of the enhancement-mode devices in Fig. 2 has a nonuniformly doped channel. Electron inversion-layer mobility and scattering-limited velocity variations with doping must, therefore, be accurately known as must their variation with gate field.

2) Depletion-mode devices in each model (representing surface accumulation layers) are present. Electron (majority-carrier) mobility in such regions must, therefore, be accurately known, including its dependence on gate field.

3) The bulk resistors in each model are clearly two dimensional. Realistic geometries for these regions must be found.

IV. DEVICE DESIGN AND FABRICATION

In order to investigate experimentally the accuracy of the models in Fig. 2 and the analytic expressions to be presented later, all three types of devices were fabricated side by side on the same wafers. A variety of n^- drain region resistivities ranging from 0.5 to 8.5 Ω · cm were used. All wafers were ⟨100⟩ oriented. Highly doped Sb, n^+ substrates were used; the n^- layers were grown with conventional epitaxial techniques. Following epitaxial growth, the fabrication process followed one previously described [4] and included a p^+ source-contact diffusion, p^- channel implant and drive-in, n^+ source and drain (for the LDMOS) diffusion, V-groove etching (for the VMOS), gate oxidation, contact-hole opening, metal deposition and definition, and passivation. Design rules for each of the devices were 10-μm minimum feature sizes and 5-μm alignment tolerances. A summary of relevant structural parameters is given in Table I for the fabricated devices.

Photomicrographs of three of the resulting devices (from the same die) are shown in Fig. 3. Each of the devices (LDMOS in Fig. 3(a), VDMOS in Fig. 3(b), and VMOS in Fig. 3(c)) has a channel width of 3440 μm. Channel lengths in the three devices are listed in Table I. Gate-oxide thickness was found to be 950 Å in the LDMOS and VDMOS and 1350 Å in the VMOS because of the different growth rates on ⟨100⟩ and ⟨111⟩ surfaces [10]. The diffused guard rings around the perimeter of the two vertical devices were employed to reduce surface effects degrading the device breakdown voltage. Such techniques are not as easily employed in the LDMOS because of its

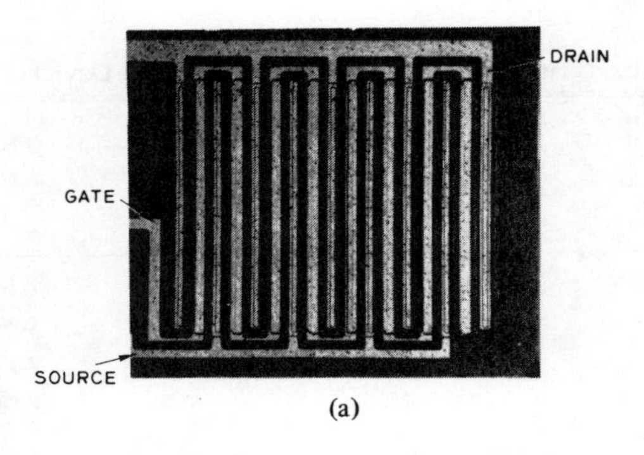

(a)

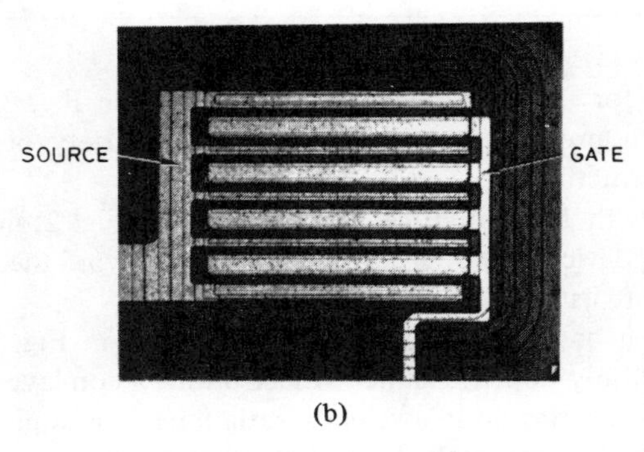

(b)

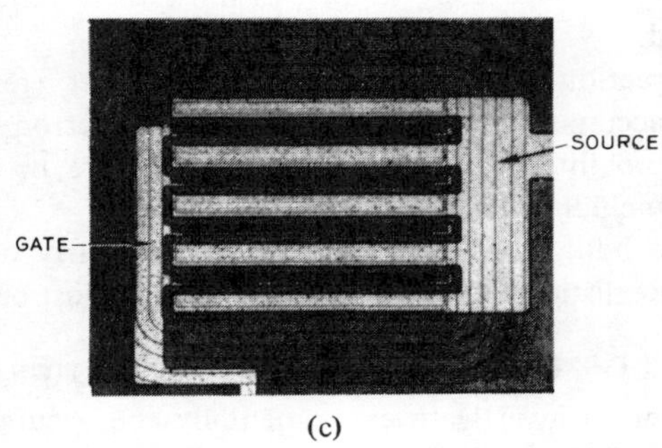

(c)

Fig. 3. Photomicrographys of three devices with identical channel widths fabricated on the same wafers. (a) LDMOS device. (b) VDMOS device. (c) VMOS device.

topside drain contact, as a result they were not used in the structure of Fig. 3(a).

While the area occupied by each of the devices is highly dependent on design rules and desired voltage capability, it is instructive to compare the structures in Fig. 3. Each device was designed with the same design rules and with a voltage capability of >300 V in the highest resistivity epitaxial material (8.5 $\Omega \cdot$ cm). For the same total channel width, the relative surface areas excluding guard rings are LDMOS : VDMOS : VMOS = 1.62 : 1.08 : 1.0. A more important criterion than simply surface area for a given channel width, however, is surface area for a given on-resistance or current capability. A relative discussion of the three devices on this basis will be given later in light of the equivalent circuits of Fig. 2.

V. Epitaxial-Layer Resistivity

Consideration of the models of Fig. 2 in the next section will clearly show that in high-voltage structures, the on-resistance of each device is dominated by the drift region bulk resistance. For this reason, the epitaxial-layer thickness and resistivity must be properly chosen for optimum device performance. This section briefly considers the choice of these parameters.

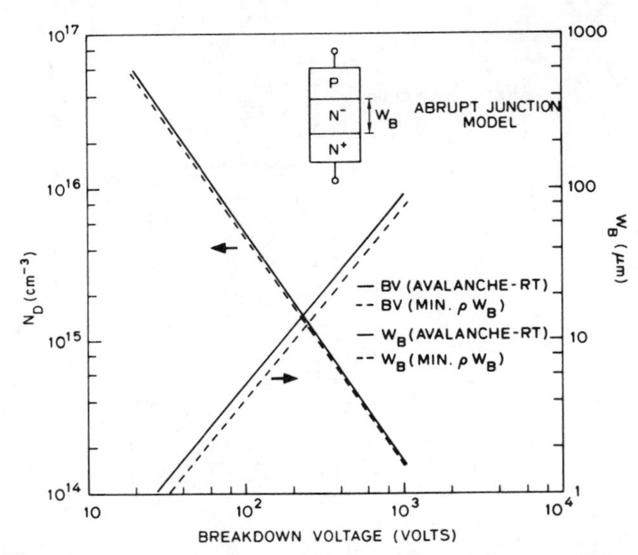

Fig. 4. Epitaxial-layer doping concentration and thickness versus breakdown voltage.

A simple one-dimensional model shown in the inset of Fig. 4 is employed in which the n^- region represents the epitaxial layer with thickness W_B between the topside p diffusion and the n^+ substrate, and doping N_D.

Minimum on-resistance is achieved when the source-drain depletion region is allowed to spread completely through the epitaxial layer. Under this condition (reachthrough) the device voltage capability is limited by avalanche breakdown when the field in the depletion region reaches the critical electric field ϵ_{crit}. For abrupt junctions, the doping dependence of ϵ_{crit} can be approximated by the following expression [16]:

$$\epsilon_{crit} = \left(\frac{2qN_D BV}{\epsilon_{Si}}\right)^{1/2} \tag{1}$$

where the various terms have their usual meanings. Using the data in [16], an approximate empirical expression for the dependence of the avalanche breakdown on doping may be obtained as follows:

$$BV \cong 2.93 \times 10^{12} N_D^{-0.666} \text{ V} \tag{2}$$

where N_D is the doping density/cm^3. The epitaxial-layer thickness W_B can then be obtained from [17]

$$BV = \epsilon_{crit} W_B - \frac{qN_D W_B^2}{2\epsilon_{Si}} \tag{3}$$

under the reachthrough condition. Simultaneous solution of the above three equations yields the solid lines in Fig. 4 which give the required epitaxial-layer thickness W_B and doping density N_D as a function of the desired reachthrough-limited breakdown voltage.

The formulation given thus far does not absolutely minimize the device on-resistance. A more appropriate criterion is the minimization of the product ρW_B which is proportional to the bulk series resistance of each device. That is, we wish

$$\frac{\partial}{\partial N_D}(\rho W_B) = 0 \tag{4}$$

where ρ is the epitaxial-layer resistivity.

This resistivity is given simply by

$$\rho = \frac{1}{qN_D\mu(N_D)} \tag{5}$$

where $\mu(N_D)$ is the bulk electron mobility.

The formulation of Caughey and Thomas for $\mu(N_D)$ is [18]

$$\mu(N_D) = \frac{\mu_{MAX} - \mu_{MIN}}{1 + (N_D/N_{REF})^{\alpha}} + \mu_{MIN} \tag{6}$$

where μ_{MAX} and μ_{MIN} are the limiting values of bulk electron mobility in lightly and heavily doped material, respectively ($\mu_{MAX} = 1330$ cm^2/V · s, $\mu_{MIN} = 65$ cm^2/V · s), $N_{REF} = 8.5 \times 10^{16}$, and $\alpha = 0.72$.

Using (5) and (6) for ρ and (2) and (3) for W_B in (4), optimum values of N_D and W_B may be obtained which minimize the epitaxial bulk resistance. The results are shown by the dashed lines in Fig. 4. The optimum values are not substantially different from those shown in the solid lines; they do indicate, however, that a slightly lower doping and smaller thickness in the epitaxial layer minimize the on-resistance as compared to a direct calculation based only on (1)-(3). In actual practice, the degree of control which may be achieved over epitaxial-layer thickness and doping, and the need to allow for process tolerances make the differences between the solid and dashed curves a second-order consideration.

VI. Modeling of Device On-Resistance

A. The LDMOS Device

The LDMOS equivalent circuit shown in Fig. 2(a) contains a total of four components. However, in high-voltage structures such as those being considered here, the lateral extent of the thin gate oxide over the n$^-$ drift region and the lateral extent of the metal gate after it steps up over the thick oxide over the n$^-$ drift region (see Fig. 1(a)), are both generally small compared to the total extent of the n$^-$ region [4]. This configuration is necessary in order to maximize the device voltage capability since the gate metal acts as a field plate [4], [7]. Because of this, the contributions of the depletion-mode device and the bulk resistor R_1 to the total device on-resistance are generally small. We shall, therefore, neglect the contribution of the depletion-mode device and include R_1 in the calculation of R_2. In lower voltage devices in which this is not a valid approximation, the formulation given later for the VDMOS and VMOS devices can be directly extended to the LDMOS structure. With this simplifying assumption, the equivalent circuit becomes simply an enhancement-mode transistor in series with a bulk resistor.

The enhancement-mode channel resistance is given simply by

$$R_E = \frac{1}{(W/L_{eff})C_o\mu_E(V_G)(V_G - V_{TE})} \tag{7}$$

where

W	channel width
L_{eff}	effective channel length
C_o	gate oxide capacitance/unit area
$\mu_E(V_G)$	electron inversion-layer mobility
V_{TE}	threshold voltage of the enhancement-mode device.

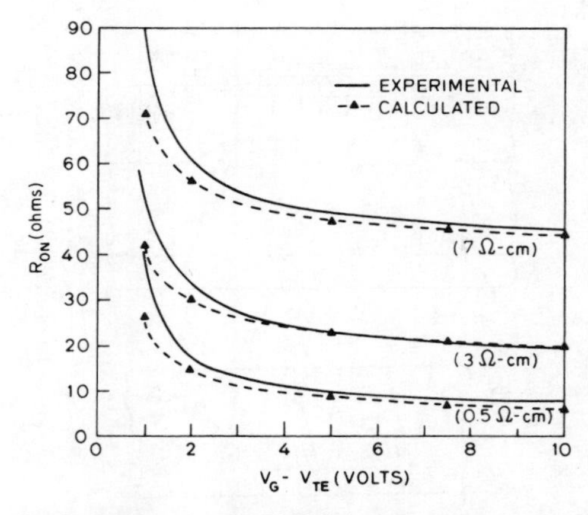

Fig. 5. On-resistance of LDMOS devices versus $V_G - V_{TE}$.

For the device shown in the photomicrograph of Fig. 3(a), $W = 3440$ μm and L_{eff} is ≈85 percent of the vertical difference in the n$^+$ and p diffusion depths. For the same diffusion schedules, L_{eff} will depend upon the epitaxial-layer concentration as shown in Table I, primarily because the p-n$^-$ junction depth depends on the n$^-$ doping. C_o is obtained directly from the measured gate oxide thickness (950 Å). V_{TE} was measured to be ≈2 V. The electron inversion-layer mobility depends upon the channel doping. For the devices considered here, $N_{A_{MAX}} = 5 \times 10^{16}$/cm^3 and, therefore, $\mu_{E_{MAX}} \cong 560$ cm^2/V · s [8], [9]. As gate bias is applied to the device, the effective mobility will be reduced because of the vertical component of the electric field existing in the channel. In the calculation of on-resistance, the data of [9], [19] were used to account for this reduction.

To model the bulk resistor R_2 in the LDMOS equivalent circuit, the current is considered to flow from a line source of radius r_1, at the end of the channel, to line sink with radius r_2 at the n$^+$ contact. This model is analogous to the electrostatic problem of an image charge in a conducting plane [20] and has been treated in detail by Pocha [15]. From [15], the resulting resistance of the bulk epitaxial region is

$$R_2 = \frac{\rho}{W\pi}\left[\ln\left(\frac{L' - r_1}{r_1}\right) + \ln\left(\frac{L' - r_2}{r_2}\right)\right] \tag{8}$$

where L' is the effective length of the n$^-$ epitaxial resistor, r_1 is the effective radius of the current source at the end of the channel, and r_2 is the effective radius of the current sink at the n$^+$ contact.

L' is found directly from the mask dimensions of the device and the extent of lateral diffusion of the p-channel and n$^+$ diffusions. For the devices considered here it is ≈24 μm, r_1 is related to the effective "emitting" area at the end of the channel; a value of 0.5 μm provided reasonable agreement with experiment. r_2 is related to the effective "collecting" area of the n$^+$ diffusion; a value of 2 μm, close to the n$^+$ junction depth, provided good agreement in the devices considered here.

Using the above formulation for R_E and R_2, the calculated results shown in Fig. 5 were obtained. These are compared with experimental results for three different epitaxial-layer resistivities. Epitaxial-layer thicknesses and L_{eff} values for

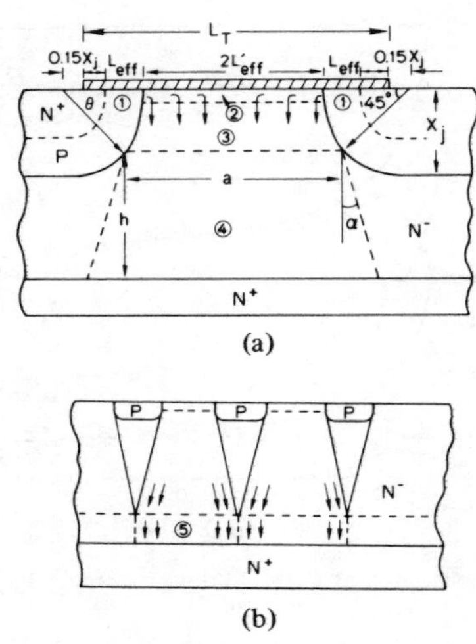

Fig. 6. Model of VDMOS device used to calculate on-resistance. (a) Separation of device into four series components with parameter definitions. (b) Modification to model to account for overlapping of bulk epitaxial-layer trapezoids.

each device are given in Table I. In all cases, the drain voltage was small (<100 mV) to keep the devices in their linear region of operation.

The agreement between the calculated and experimental results is seen to be quite good except for values of gate voltage close to the threshold. There are principally two reasons for this. First, in this region, the channel resistance R_E dominates and inaccuracies in the simple formulation of (7) become important. This equation, for example, does not take into account the nonuniform channel doping profile. More accurate models such as that contained in [21] could be used to reduce this error. Second, under moderate to high gate biases, the thin-oxide overlap of the n^- region creates a surface accumulation layer which helps to avoid current-crowding effects at the end of the channel. As the gate bias is reduced, the crowding becomes more severe (r_1 decreases), leading to higher on-resistance than that predicted by the models. In any case, the small errors present for low V_G are not of much consequence in many applications of power MOS device because gate voltages substantially above V_{TE} would be used. The relative contributions of R_E and R_1 to the total device on-resistance will be considered specifically for the VDMOS device in the next section.

B. The VDMOS Device

Calculation of on-resistance for the VDMOS device is considered in Fig. 6(a) in which several dimensions to be used below are defined. In this device, the thin gate oxide normally extends all the way between adjacent source and channel diffusions. As a result, the surface accumulation layer (region 2 in the figure) plays an important role in total device on-resistance and cannot be neglected. Physically, this region collects the channel current and distributes it over the region between adjacent p diffusions. It, therefore, reduces current crowding at the end of the channel and helps to reduce total device on-resistance. This effect would be expected to be most efficient at high gate voltages which maximize the conductivity of the accumulation layer.

In addition to this factor, a junction field-effect transistor exists between the adjacent p diffusions as shown in the equivalent circuit in Fig. 2(b). Depending on the ratio of channel width (a in the figure) to the p diffusion depth in this JFET, it can play an important role in overall device on-resistance.

Neglecting contact resistances and the resistance of the bulk n^+ substrate, the VDMOS on-resistance is thus given by

$$R_{ON_{VDMOS}} = R_E + R_D + R_{JFET} + R_4 \tag{9}$$

where

- R_E enhancement-mode device on-resistance
- R_D depletion-mode device on-resistance (surface accumulation layer)
- R_{JFET} JFET on-resistance
- R_4 bulk epitaxial-layer resistance.

R_E is given by the same expression [see (7)] as was used in the modeling of the LDMOS device. The same parameters discussed in connection with that device apply in the present case as well.

The depletion-mode device on-resistance is given by an expression analogous to (7). That is

$$R_D = \frac{1}{3} \frac{1}{\frac{W}{L'_{eff}} C_o \mu_D(V_G)(V_G - V_{TD})} \tag{10}$$

where $\mu_D(V_G)$ is the electron (majority-carrier) accumulation-layer mobility and L'_{eff} is the effective depletion-mode device channel length.

The factor of $\frac{1}{3}$ in the expression for R_D appears because of the two-dimensional nature of the current flow from the accumulation layer into the bulk n^- region [22], [23]. In essence, it models the conversion of current flow from lateral to vertical. More accurately, as shown in Fig. 2, the depletion-mode device should be modeled as a distributed structure; the factor of $\frac{1}{3}$ has, however, provided reasonable agreement with experiment.

For the device shown in the photomicrograph in Fig. 3(b), $W = 3440\ \mu m$ as before, L_T is defined in Fig. 6 and is $\approx 20\ \mu m$, C_o is determined from the gate oxide thickness of 950 Å, and V_{TD} depends upon the epitaxial-layer doping concentration. The factor of two in the L'_{eff} definition in Fig. 6(a) does not appear in (10) because two devices effectively act in parallel.

The majority-carrier electron mobility in the surface accumulation layer is a parameter not nearly as well characterized in the literature as inversion-layer mobility. Based upon the work of Reddi [24] and additional data taken in the authors' laboratory [9], a value of $\mu_D = 1050\ cm^2/V \cdot s$ for the ⟨100⟩ surface at $V_G = 0$ has been used in the present work. Note that this is not the limiting value of μ_D for $V_G - V_{TD} \approx 0$ since V_{TD} is ≈ -2 V. A formulation similar to that for μ_E [19] was used to account for the reduction in μ_D as the gate field is increased.

The JFET component of the VDMOS equivalent circuit in Fig. 2 is modeled with the geometry shown in Fig. 6(a). The p-n^- junction is modeled as a portion of a circle with the origin offset from the original mask edge by 0.15 X_j to account for the difference in lateral and vertical diffusion. The current is assumed to originate uniformly from the surface accumulation layer. If X is positive downward from the surface, then the spacing between the two P diffusions is given by

$$L(X) = L - 2\sqrt{r^2 - X^2} \tag{11}$$

where $L = L_T + 0.3\,X_j$ and r is the radius of the circle shown in Fig. 6(a) ($r = X_j$).

The current density as a function of X is given by

$$J(X) = \frac{I}{WL(X)} = \frac{1}{\rho}\frac{dv}{dx}. \tag{12}$$

Substituting (11) into (12) and integrating we find the total resistance of the JFET region to be

$$R_{\text{JFET}} = \int_0^V \frac{dv}{I} = \frac{\rho}{W}\int_0^{X_A} \frac{dx}{L - 2\sqrt{r^2 - X^2}} \tag{13}$$

where X_A is equivalent to $\theta = 45°$.

Conversion to polar coordinates and straightforward integration yields

$$R_{\text{JFET}} = \frac{2\rho}{W}\left[\frac{1}{\sqrt{1 - (2X_j/L)^2}}\tan^{-1}(0.414)\sqrt{\frac{L + 2X_j}{L - 2X_j}} - \frac{\pi}{8}\right]. \tag{14}$$

Note that the formulation given here does not take into account any depletion regions which may exist in the n^- region. As expressed in (14), then, it applies to the case of low V_{DS}.

The JFET region of the VDMOS model is considered to end at the point where $\theta = 45°$. Below this boundary in the n^- material, a bulk resistor bounded by the trapezoidal geometry in Fig. 6(a) represents the series resistance of the epitaxial layer. Following the procedure outlined above for the JFET region, the resistance of this trapezoidal volume is found to be

$$R_4 = \frac{\rho}{W}\frac{1}{\tan\alpha}\ln\left(1 + 2\frac{h}{a}\tan\alpha\right). \tag{15}$$

a and h are easily found from device geometry. The spreading angle α is analogous to the thermal impedance spreading angle that has been calculated by David [25]. His exact calculations can be approximated within ±5 percent by the simple expressions shown below. These were used in the modeling here.

$$\alpha = \begin{cases} 28° - \dfrac{h}{a}, & \text{if } h \geqslant a \\ 28° - \dfrac{a}{h}, & \text{if } h < a. \end{cases}$$

If the epitaxial layer is very thick or if the surface packing density is very high, adjacent trapezoids may overlap as shown in Fig. 6(b). In this case, the resistance of region 5 in the figure is easily found by assuming uniform current flow in the merged region.

The model for on-resistance described above was applied to the structure of Fig. 3(b) with the results shown in Fig. 7. Epitaxial-layer thicknesses and L_{eff} values are given in Table I. In each case, the device channel width (Fig. 3(b)) was 3440 μm.

Over a wide range of epitaxial-layer resistivities, excellent

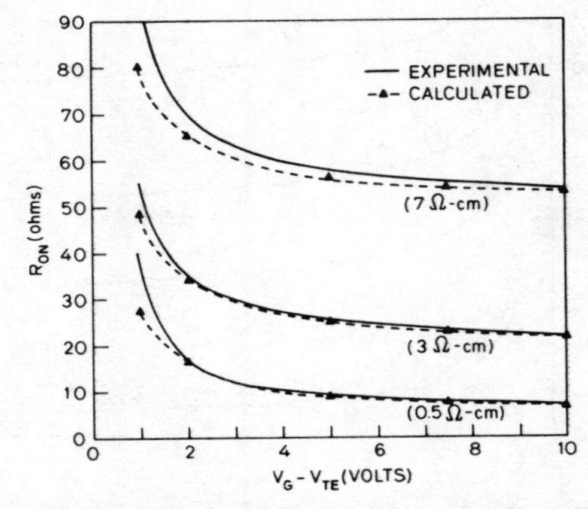

Fig. 7. On-resistance of VDMOS devices versus $V_G - V_{TE}$.

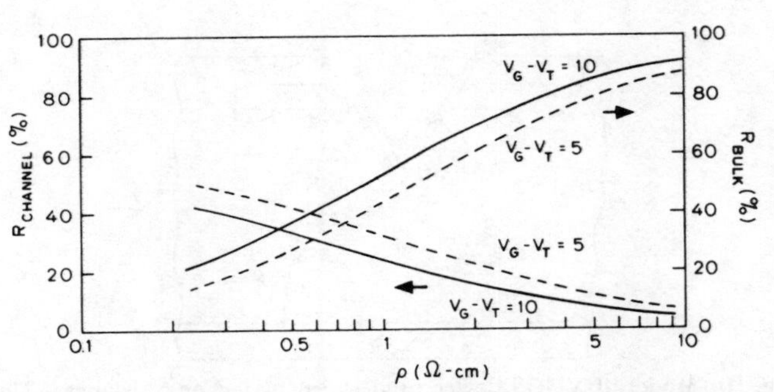

Fig. 8. Calculated percentage of total VDMOS device on-resistance due to channel resistance and to the JFET plus the bulk epitaxial-layer resistances, as functions of epitaxial resistivity.

agreement with experimental measurements is obtained. As with the LDMOS model previously described, the discrepancy between experiment and theory is most pronounced at low vales of $V_G - V_{TE}$. The physical reason for this is likely the relative ineffectivenss of the surface accumulation layer in preventing current crowding at the end of the channel, when $V_G - V_{TE}$ is small.

Using the model described above, it is possible to evaluate the relative contributions of the various parts of the device equivalent circuit to the total on-resistance. A particular example of this is shown in Fig. 8. For the purposes of this calculation, the p-channel junction depth was assumed to be 4 μm and the n^+ junction depth 2 μm. This gives an L_{eff} of $\approx$1.7 μm. L_T was assumed to be 20 μm. For each epitaxial-layer resistivity, an appropriate choice for epitaxial-layer thickness was determined from Fig. 4.

Fig. 8 plots the percentage of total device on-resistance due to the enhancement channel (R_E) and due to the sum of the JFET and bulk resistance terms ($R_{\text{JFET}} + R_4$). The contribution of the depletion-mode device is simply the difference between 100 percent and the sum of the other two terms.

The calculations produce not unexpected results. For low-voltage devices ($\rho < 1\ \Omega\cdot$cm), channel resistance dominates while in high-voltage devices, the bulk contributions of the JFET and the epitaxial layer dominate. In fact, for epitaxial layer resistivities greater than about 8-10 $\Omega\cdot$cm, the total device on-resistance can be reasonably calculated from just the bulk terms $R_{\text{JFET}} + R_4$.

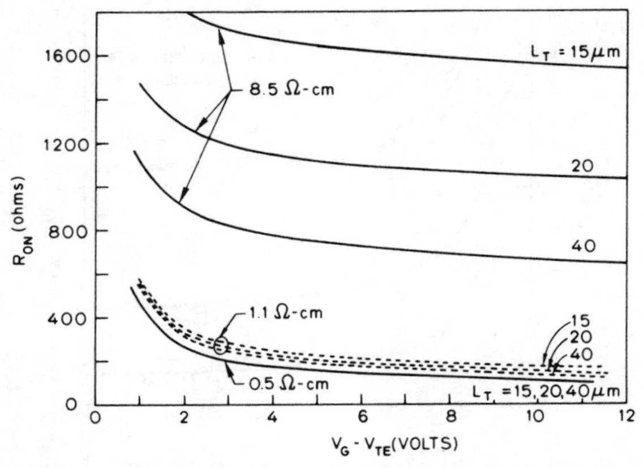

Fig. 9. Experimental results of the effect of p-channel diffusion separation L_T on VDMOS device on-resistance.

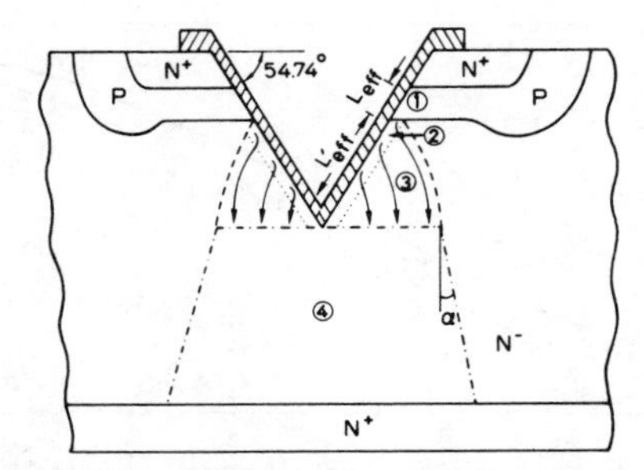

Fig. 10. Model of VMOS device used to calculated on-resistance. The device is separated into four series components.

The mask spacing between the p-type diffusion (L_T in Fig. 6) plays an important role in determining the on-resistance of the JFET and the bulk epitaxial resistor portions of the VDMOS equivalent circuit. Particularly in high-voltage structures in which these two components dominate, L_T is a critical parameter in optimizing device performance. An example of this is shown in the experimental results of Fig. 9. These measurements were made on smaller geometry devices (W = 200 μm in all cases) which accounts for the higher values of on resistance. All other parameters are the same as Table I.

It is clear from the figure that the mask dimension L_T is very important in higher voltage devices (8.5-Ω · cm curves). It is not so important in lower voltage devices (0.5 and 1.1 Ω · cm) in which channel resistance rather than bulk resistance dominates. It is not desirable to arbitrarily increase L_T in high-voltage devices, however, since for a given chip area this will reduce total channel width (and, therefore, increase on-resistance). For a given epitaxial-layer resistivity, there will, as a result, be an optimum value for L_T which will minimize the device on-resistance for a given chip area.

C. The VMOS Device

Calculation of on-resistance for the VMOS device is considered in Fig. 10 in which several dimensions to be used below are defined. In this device, thin gate oxide normally extends down to the bottom of the groove, resulting in a surface accumulation layer under the gate where the groove extends into the n^- region. As in the VDMOS device, this property tends to minimize current-crowding effects at the end of the enhancement channel and, therefore, reduces on-resistance. This region should properly be modeled as a distributed network of depletion-mode transistors in series with bulk resistors as indicated in Fig. 2(c); however, for simplicity, the same single-transistor series-resistor model used in the VDMOS device, will be used here as well.

The total on-resistance of the VMOS structure is, therefore, given by

$$R_{ON_{VMOS}} = R_E + R_D + R_3 + R_4 \tag{16}$$

where

- R_E enhancement-mode device on-resistance
- R_D depletion-mode device on-resistance (surface accumulation layer)
- R_3 bulk resistance of region 3 in Fig. 10
- R_4 bulk resistance of trapezoidal shaped region 4 in Fig. 10.

R_E and R_D are given by the same expressions that were used in the modeling of the VDMOS device, i.e., (7) and (10). All parameters in these equations are obtained as previously described except for L_{eff} and μ. The channel length in the VMOS device is ≈1.5 times longer than the values previously used in modeling the LDMOS and VDMOS structures. Both the inversion-layer mobility μ_E and the majority-carrier accumulation layer mobility are smaller on the ⟨111⟩ surface of the VMOS device than they are on the ⟨100⟩ surfaces of the other two devices. Values of 80 percent of the corresponding numbers on the ⟨100⟩ plane (previously given) were used for the VMOS device [8], [9] with the same reduction factors to account for gate electric field used in the two cases.

The bulk resistance of region 3 in Fig. 10 can be calculated by assuming that the geometry is a sector of a circle with an included angle of 54.7°. The current is assumed to uniformly enter this region from the surface accumulation layer under the gate. Straightforward integration in a manner analogous to that described for the VDMOS device yields

$$R_3 = 0.477 \frac{\rho}{W}. \tag{17}$$

Region 4 in the VMOS device (R_4) is identical to that previously described for the VDMOS device. Thus the resistance of this region is given by (15). The same considerations regarding overlapping of adjacent trapezoids that were described in connection with Fig. 6(b) apply here as well.

Comparison of the predictions of the above model with experiment are shown in Fig. 11 for a variety of epitaxial-layer resistivities. In each case, the width of the mask opening for the V-groove etch was 10 μm. Other structural parameters are given in Table I. In all cases, the agreement for high gate voltages is excellent. As $V_G - V_{TE}$ approaches zero, a divergence of theory and experiment is apparent. This is in exactly the same direction as previously observed on the LDMOS and VDMOS devices. Again, it is believed to be due to a failure of the simple models presented here to account for current-crowding effects at the end of the enhancement channel, particularly at low V_G when the surface accumulation layer is not so effective in reducing this crowding.

The effect of V-groove depth for a given p-channel junction depth on the on-resistance is shown in Fig. 12. The calcula-

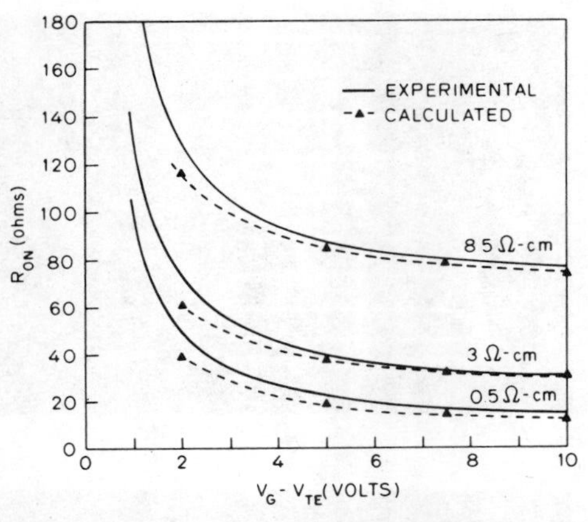

Fig. 11. On-resistance of VMOS devices versus $V_G - V_{TE}$.

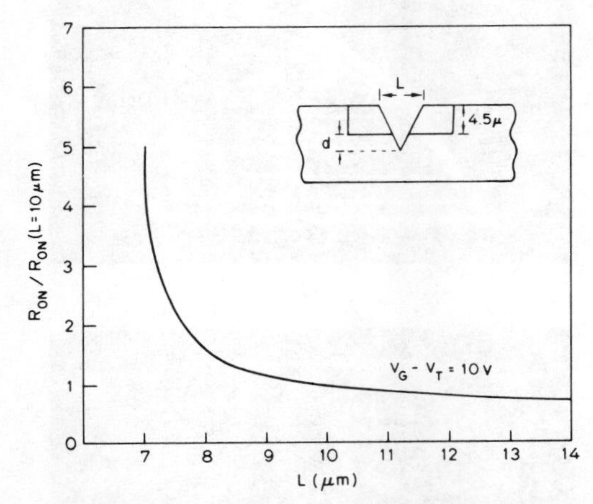

Fig. 12. Calculated effect of V-groove depth on VMOS on-resistance.

Fig. 13. SEM photomicrograph of truncated V-groove structure.

tions in this figure are for the 8.5-Ω · cm, 26.8-μm epitaxial-layer device of Fig. 11. The total on-resistance (vertical scale) is normalized to the value achieved for a nominal 10-μm mask dimension. As L decreases below 10 μm, the on-resistance increases because of more severe current crowding at the end of the enhancement channel. That is, the surface accumulation layer becomes less and less effective in reducing this crowding as its length is reduced. For L less than about 7 μm, the V-groove does not completely penetrate the channel diffusion resulting in infinite on-resistance. This dependence of R_{ON} on L should be less pronounced in low-voltage devices in which R_E tends to dominate; it should also be less pronounced at lower gate voltages for the same reason.

Another form of the VMOS device is the truncated VMOS structure [2]. As the SEM photomicrograph in Fig. 13 indi-

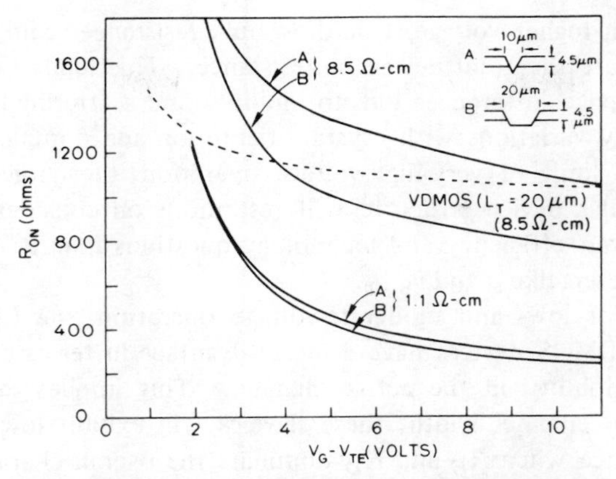

Fig. 14. Comparison of the on-resistance of VMOS devices with A fully etched V-grooves and B truncated V-grooves versus $V_G - V_{TE}$.

cates, such a device is fabricated by terminating the V-groove etch prior to reaching the apex of the groove. This requires a larger surface mask dimension and hence a lower packing density, and much better process control than the standard VMOS structure. The advantage of doing this is that the flat bottom portion of the V-groove behaves very much like the surface accumulation region in the VDMOS device. It acts to spread out the current distribution, reduces current crowding, and, therefore, reduces on-resistance.

This can be seen in the experimental data of Fig. 14. Devices with 10- and 20-μm mask openings for the V-groove etch are compared in terms of on-resistance. The etching was terminated shortly after the 10-μm opening reached the bottom of its groove. This resulted in the 20-μm opening being etched ≈7-8 μm deep with a flat bottom length of ≈8-10 μm. In each case, the channel width was 200 μm. Other relevant data are contained in Table I. It is clear from the figure that in lower voltage devices (e.g., 1.1-Ω · cm curves), the truncated structure offers little advantage. This is to be expected because R_E dominates in such devices. In higher voltage structures, however (e.g., 8.5-Ω · cm curves), the truncated geometry makes a significant difference. This is also to be expected because in these structures bulk resistances dominate; the spreading out of the current distribution by the accumulation layer at the groove bottom should be helpful here.

For comparison purposes, the on-resistance of a VDMOS device also built in 8.5-Ω · cm epi with W = 200 μm and L_T = 20 μm is also shown in Fig. 14. Its shorter channel and higher electron mobility give it a clear advantage in on-resistance at low gate voltages. At higher gate voltages, for which bulk resistances dominate, the truncated VMOS is slightly better. This is because the current in the VDMOS device flows through bulk material all the way from the surface down to the n^+ substrate; in the truncated VMOS, this distance is reduced by the nonplanar configuration.

D. General Comments Regarding the Three Devices

The models described above allow calculation of on-resistance for a variety of planar and nonplanar power MOS devices. General conclusions regarding the relative merits of these structures are difficult to state because they are highly technology and design rule (layout) dependent. Some statements can, however, be made based upon the above modeling.

1) In higher voltage structures, bulk resistances completely dominate the total device on-resistance. This tends to mask any device differences due to mobility and scattering-limited velocity variations with crystal orientation and doping levels. In the limit of very-high-voltage operation, the choice of a particular device structure will rest more on considerations of layout efficiency and technology questions than it will on parameters like μ and v_{SAT}.

2) For low- and moderate-voltage operation, the LDMOS and VDMOS devices have a clear advantage in terms of electron mobility in the active channel. This implies that for a given channel width, these devices will exhibit lower on-resistance when R_E and R_D dominate the overall characteristics. This advantage may be overridden, however, by layout considerations since the overall goal is to achieve a specified on-resistance in minimum chip area.

VII. Device Transconductance

This paper has concentrated on the dc modeling of the low drain voltage on-resistance of various power MOS device structures. This is the parameter of principal importance in most switching applications of these devices. In other situations, however, particularly in linear applications, device transconductance in the saturation region is of interest. We very briefly consider this question here in order to provide additional perspective on the preceding modeling.

Under scattering-limited velocity operation, which is typical of these devices in their saturation regions because of their very short channels, device transconductance is given to first order simply by [1]

$$g_{m\text{MAX}} = C_o W v_{\text{SAT}} \tag{18}$$

where v_{SAT} is the electron scattering limited velocity (6.5×10^6 cm/s in ⟨100⟩ silicon inversion layers).

For gate voltages less than

$$\frac{V_G - V_T}{L_{\text{eff}}} < \epsilon_{\text{crit}} \tag{19}$$

where ϵ_{crit} is the critical electric field to cause velocity saturation ($\approx 1.5 \times 10^4$ V/cm), the device transconductance is directly related to the low-field electron mobility and is given by standard equations. Typically, transconductance in these short-channel structures initially increases as $V_G - V_{TE}$ increases, with a slope proportional to μ_E, and gradually reaches a limiting value given by (18) as velocity saturation is achieved.

The curve tracer photographs in Fig. 15 illustrate this behavior. The devices whose I-V characteristics are displayed here are the VDMOS device shown in Fig. 3(b). The photographs in Fig. 15(a), (b), and (c) are for 0.5-, 3-, and 7-Ω · cm epitaxial-layer resistivities, respectively. All three devices show maximum values of transconductance of approximately 50-60 mmhos, somewhat less than the ≈75 mmhos predicted by (18). There are two principal reasons for this. First, it is apparent that the drift region bulk resistance limits the device current substantially, particularly in the 7-Ω · cm device, but also in the 3-Ω · cm structure. (Note the different horizontal scale in Fig. 15(c).) Because of this, much higher drain voltages are required to maintain operation in the saturation region

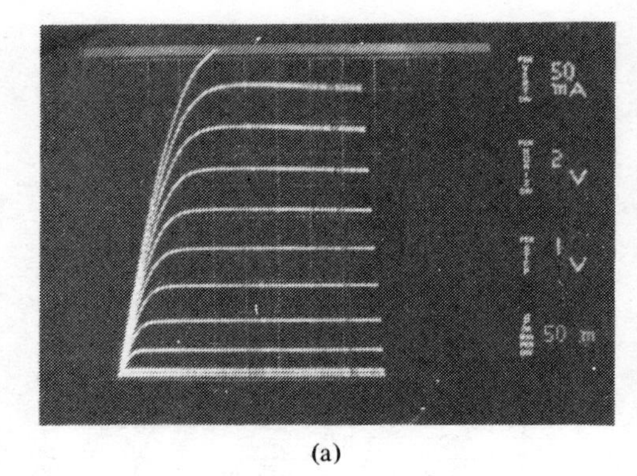

(a)

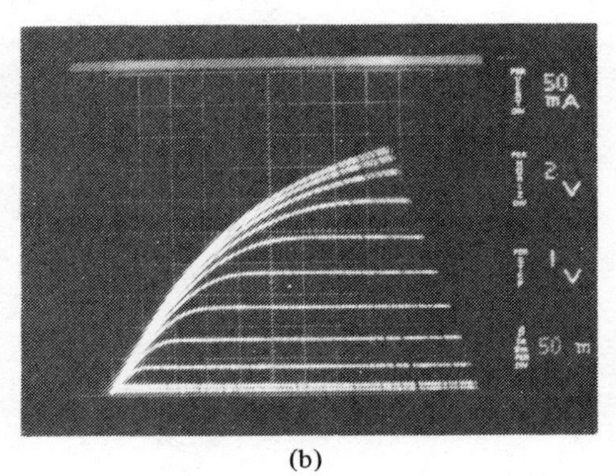

(b)

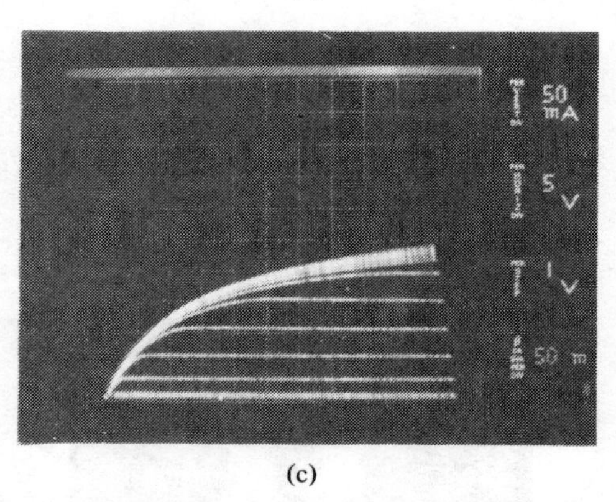

(c)

Fig. 15. Experimental I-V characteristics of VDMOS devices fabricated in various epitaxial resistivities. (a) 0.5-Ω · cm epi. (b) 3-Ω · cm epi. (c) 7-Ω · cm epi.

as the current increases. The inevitable result of this is that power dissipation must rise and, therefore, device temperature as well. Since both the low-field mobility and the scattering-limited velocity decrease with increasing temperature [8], device gain would be expected to decrease. Both the steady-state chip temperature due to average power dissipation and the "instantaneous" temperature under transient power dissipation conditions are important in determining the effective values for μ_E and v_{SAT} [26].

These effects result typically in the g_m versus $V_G - V_{TE}$ experimental curves shown in Fig. 16. These measurements were made on three VDMOS and a single VMOS device ($W = 3440$ μm in all cases), using 80-μs pulsed measurements on a standard curve tracer. For each gate voltage, the drain voltage was adjusted to assure operation in the saturation region; this typically requires higher drain voltages in the higher epitaxial resistivity devices. The 0.5-Ω · cm VMOS device achieves a maximum g_m close to that predicted by (18); the higher

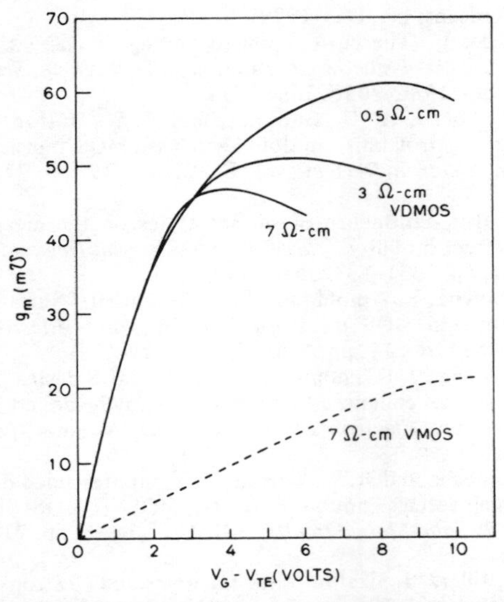

Fig. 16. Experimental transconductance measurements on VDMOS and VMOS large-geometry devices.

epitaxial resistivity devices do not, because of heating effects associated with the higher drain voltages required.

The VMOS structure has a longer effective channel length (≈1.5 times longer for the simultaneously fabricated devices considered here), and in addition a lower low-field mobility and scattering-limited velocity due to its ⟨111⟩ surface orientation. As a result, the VMOS device in Fig. 16 shows much lower transconductance at low gate voltages and requires much higher gate voltages to achieve scattering-velocity-limited operation. Thermal effects are so dominant in the VMOS structure that it does not achieve a maximum g_m comparable to the VDMOS devices.

Fig. 17 compares the I-V characteristics of the 7-Ω · cm large-geometry ($W = 3440$ μm) VDMOS and VMOS devices. The difference in transconductance is apparent as is the higher gate voltage in the VMOS case required to achieve scattering-limited velocity operation.

Using shorter duration pulses to reduce the device heating improves the apparent device transconductance as Fig. 18 illustrates. These measurements are on the 3-Ω · cm small-geometry ($W = 200$ μm) VDMOS device. Even with 1-μs pulses, however, the device does not achieve a transconductance equal to that predicted by (18). This has been postulated to be a result of the very fast response time of the temperature in the active channel region of the device [26].

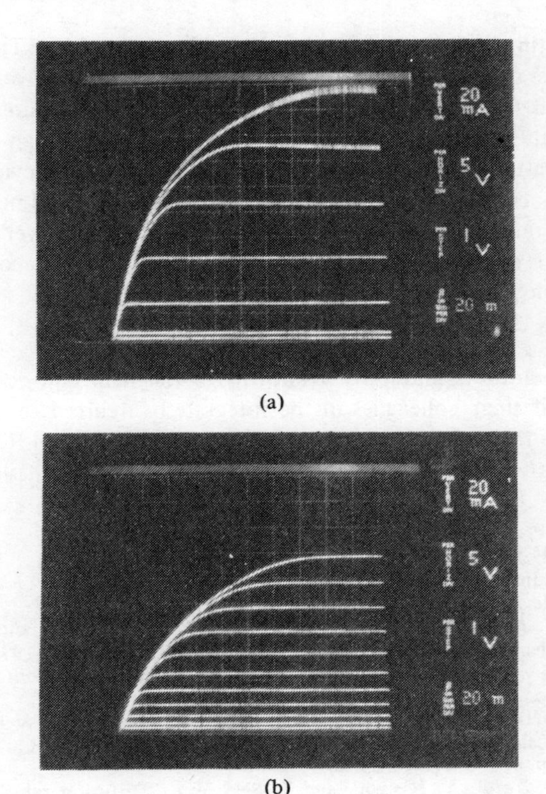

(a)

(b)

Fig. 17. Comparison of the experimental I-V characteristics of VDMOS and VMOS devices with identical channel width. (a) VDMOS, 7-Ω · cm epi. (b) VMOS, 7-Ω · cm epi.

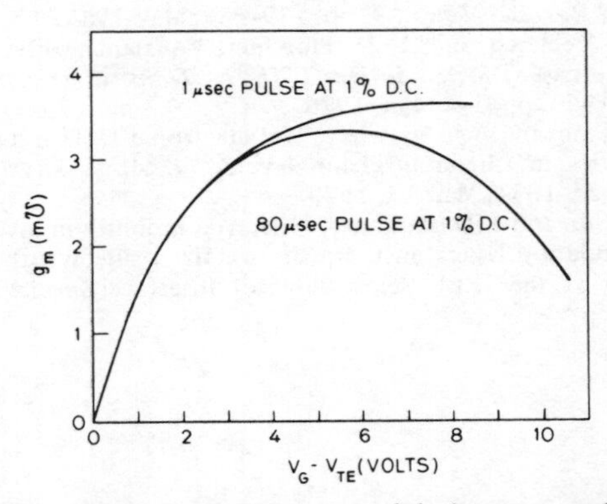

Fig. 18. Effect of pulsewidth on measured device transconductance for the 3-Ω · cm VDMOS.

VIII. Conclusions

This paper has considered several planar and nonplanar DMOS structures suitable for power MOS transistor applications. Three possibilities in particular were considered—the LDMOS, VDMOS, and VMOS structures—both from a theoretical point of view and by direct experimental comparison between devices fabricated side by side on the same wafers. Qualitative comparisons between these devices indicated differences in on-resistance and transconductance are to be expected because of differences in the physical structures and in basic material parameters such as mobility and scattering-limited electron velocity appropriate to each device. Quantitative models relating on-resistance to the physical structures were developed for each device and shown to agree well with experimental data. These models should allow direct quantitative comparison of device structures for a given application, when technological and design rule parameters are specified.

For low-voltage applications, the VDMOS and LDMOS devices show lower on-resistance per unit device width than the VMOS device. At high voltages, all three devices are approximately equal. Technological and layout efficiency considerations may well be more important, however, in determining which device produces lower total on-resistance for a given chip area. Transconductance is generally higher in LDMOS and VDMOS devices than in VMOS devices for equal channel lengths, because of higher electron mobilities and scattering-limited velocities on the ⟨100⟩ crystal plane. Thermal effects make achieving room-temperature scattering-velocity-limited transconductance difficult in all these structures, particularly in higher voltage designs. In fact, in high-power applications, thermal effects may well determine device on-resistance and current capability. It is precisely this negative temperature coefficient, however, that makes power MOS devices relatively immune to thermal runaway and secondary breakdown problems.

Acknowledgment

The authors gratefully acknowledge the help and cooperation of their colleagues in the Integrated Circuits Laboratory at Stanford. In addition, J. Willard of Tektronix and R. Blanchard of Supertex and Stanford provided helpful discussions.

References

[1] H. J. Sigg, G. D. Vendelin, T. P. Cauge, and J. Kocsis, "D-MOS transistor for microwave applications," *IEEE Trans. Electron Devices*, vol. ED-19, pp. 45-53, Jan. 1972.

[2] V. A. K. Temple and R. P. Love, "A 600 volt MOSFET with near ideal on resistance," in *IEDM Conf. Dig.*, pp. 664-666, 1978.

[3] M. Nagata, "High-power MOSFETs," *Inst. Phys. Conf.*, Ser. no. 32, pp. 101-110, 1977.

[4] J. D. Plummer and J. D. Meindl, "A monolithic 200-V CMOS analog switch," *IEEE J. Solid-State Circuits*, vol. SC-11, no. 6, pp. 809-817, Dec. 1976.

[5] Y. Tarui, Y. Hayashi, and T. Sekigawa, "Diffusion self-aligned MOST: A new approach for high speed device," in *Proc. 1st Conf. on Solid State Devices*, Tokyo, 1969, Suppl. to *J. Japan. Soc. of Appl. Phys.*, vol. 39, pp. 105-110, 1970.

[6] D. P. Kennedy and R. R. O'Brien, "Analysis of the impurity atom distributon near the diffusion mask for a planar p-n junction," *IBM J. Res. Devel.*, vol. 9, pp. 179-186, May 1965.

[7] M. J. Declercq and J. D. Plummer, "Avalanche breakdown in high-voltage DMOS devices," *IEEE Trans. Electron Devices*, vol. ED-23, pp. 1-4, Jan. 1976.

[8] F. F. Fang and A. B. Fowler, "Hot electron effects and saturation velocities in silicon inversion layers," *J. Appl. Phys.*, vol. 41, pp. 1825-1831, Mar. 15, 1970.

[9] S. C. Sun and J. D. Plummer, "Electron mobility in inversion and accumulation layers on thermally oxidized silicon surfaces," presented at the IEEE Semiconductor Interface Specialists Conf., New Orleans, LA, Dec. 1979.

[10] B. E. Deal, "The current understanding of charges in the thermally oxidized silicon structure," *J. Electrochem. Soc.*, vol. 121, no. 6, pp. 198C-205C, June 1974.

[11] M. D. Pocha, A. G. Gonzalez, and R. W. Dutton, "Threshold voltage controlability in double-diffused-MOS transistors," *IEEE Trans. Electron Devices*, vol. ED-21, no. 12, pp. 778-784, Dec. 1974.

[12] S. T. Hsu, 'Influence of surface states on the measurement of field-effect mobility," *IEEE Trans. Electron Devices*, vol. ED-25, no. 11, pp. 1331-1332, Nov. 1978.

[13] G. Abowitz, E. Arnold, and E. A. Leventhal, "Surface states and 1/f noise in MOS transistors," *IEEE Trans. Electron Devices*, vol. ED-14, no. 11, pp. 775-777, Nov. 1967.

[14] C. P. Ho and J. D. Plummer, "Improved MOS device performance through the enhanced oxidation of heavily doped n^+ silicon," *IEEE Trans. Electron Devices*, vol. ED-26, no. 4, pp. 623-630, Apr. 1979.

[15] M. D. Pocha and R. W. Dutton, "A computer-aided design model for high-voltage double diffused MOS (DMOS) transistors," *IEEE J. Solid-State Circuits*, vol. SC-11, no. 5, pp. 718-726, Oct. 1976.

[16] S. L. Miller, "Ionization rates for holes and electrons in silicon," *Phys. Rev.*, vol. 105, no. 4, pp. 1246-1249, 1957.

[17] A. S. Grove, *Physics and Technology of Semiconductor Devices.* New York: Wiley, 1967.

[18] D. M. Caughey and R. E. Thomas, "Carrier mobilities in silicon empirically related to doping and field," *Proc. IEEE* (Lett.), vol. 55, pp. 2192-2193, 1967.

[19] D. Frohman-Benchkowsky, "On the effect of mobility variation on MOS device characteristics," *Proc. IEEE* (Lett.), vol. 56, pp. 217-218, Feb. 1968.

[20] S. Ramo, J. R. Whinnery, and T. Van Duzer, *Fields and Waves in Communication Electronics.* New York: Wiley, 1967, p. 98.

[21] B. W. Scharf and J. D. Plummer, "Insulated-gate planar thyristors: II–Quantitative modeling," this issue, pp. 387-394.

[22] A. B. Phillips, *Transistor Engineering.* New York: McGraw-Hill, 1962, p. 216.

[23] H. C. Lin, *Integrated Electronics.* San Francisco, CA: Holden-Day, 1967, p. 110.

[24] V. G. C. Reddi, "Majority carrier surface mobilities in thermally oxidized silicon," *IEEE Trans. Electron Devices*, vol. ED-15, no. 3, pp. 151-160, 1968.

[25] R. F. David, "Computerized thermal analysis of hybrid circuits," *ECC*, pp. 324-332, 1977.

[26] D. Sharma, J. Gautier, and G. Merckel, "Negative dynamic resistance in MOS devices," *IEEE J. Solid-State Circuits*, vol. SC-13, no. 3, pp. 378-380, June 1978.

A PARAMETRIC STUDY OF POWER MOSFETS

Chenming Hu

Department of Electrical Engineering
and Computer Sciences
University of California, Berkeley, California 94720

ABSTRACT

The theoretical limitations of V-grooved and double-diffused power MOSFETs are studied using several design parameters as variables. The results are used to gauge the performance of currently available power MOSFETs and to project the capabilities of future devices. With proper design, the channel resistance can be negligible so that the on-state resistance is that of the bulk material ($\sim 8.3 \times 10^{-9}\, V_{dsB}^{2.5}\ \Omega\cdot cm^2$) plus the lead resistance. A transmission line effect associated with long resistive gate electrodes could limit the speed of certain devices. Devices with the capability of switching 10 kW per cm^2 of chip area and hundred kilowatt per package are theoretically possible over a very wide voltage range. A new structure is proposed that could raise the power capability by an order of magnitude in very high voltage devices.

I. INTRODUCTION

Power MOSFET is the most significant development in power semiconductor devices since the introduction of thyristers. It offers performances unavailable from bipolar transistors and thyristors, thus promises to not only offer better replacements for present-day devices but also lead to new circuit and systems concepts and bring power electronics into new areas of applications.

The main feature that separates power MOSFET from the ordinary MOSFET is the vertical current flow, which greatly facilitates the packaging as shown in Fig. 1. Figure 1a is the VMOS design [1,2,3]. An epitaxial layer is grown on a 100 cut n^+ silicon substrate. Consecutive diffusions create the p and n^+ layers at the surface. Through an oxide mask whose lines are aligned with one of two particular crystalgraphical directions V shaped grooves are etched using an anisotropic etchant. In Fig. 1a, the grooves have not been etched to the bottom of the V but only extend slightly beyond the p layer. This calls for more difficult process control but results in better device performances [2,3]. This design is sometimes referred to as UMOS [2]. After the etching, oxide is thermally grown. Gate and source electrodes may be deposited in a straightforward manner. The device is normally off. When a positive gate voltage is applied, electrons are attracted to the surface and form a very thin n^+ layer as indicated by the heavy line under one gate. (All gates are in fact electrically inter-connected.) This thin n^+ layer in the p region is called the channel and in the n region the accumulation layer. With a positive drain to source bias, electrons flow from the source through the channel into the accumulation layer. From there, they spread into the epitaxial layer and flow to the drain.

Figure 1b shows the DMOS (double diffused MOSFET) design [4]. Its p and n^+ regions are made by consecutive diffusions through an oxide diffusion mask. The gate electrode may be made of poly-crystalline silicon and patterned with the diffusion mask at the same time. Thermal oxide may be grown on the poly-silicon gate so that the source electrode may overlap the gate. This simplifies lead bonding to the source metal in high current devices. In fact the flat pack design popular with high current rectifiers and thyristers may be applied. The same effect may be achieved in VMOS at the expense of more processing steps. The most significant difference between VMOS and DMOS is that the latter can be made with simpler process and therefore probably at higher yield and lower cost. Another difference is that p and n^+ diffusion lines do not have to run in certain special directions. The recently announced HEXFET from International Rectifier is a DMOS with diffusion lines tracing out a hexagonal grid pattern.

Neither VMOS nor DMOS were originally studied for the kind of power applications being considered now [1]. VMOS is a significant technology in very large scale integrated circuits. It has been studied extensively as power devices in recent years [2,3]. Siliconix is believed to be the first company to produce it. DMOS was studied for high power (100W) microwave frequency use [6,7] with the drain electrode also deposited on the top surface. To our knowledge, there has been no scientific papers on DMOS with vertical current flow. International Rectifier is believed to be the first company to produce power DMOS.

Compared with bipolar trnasistors, MOSFETs offer these advantages. Their speed is 10 to 100 times faster because it is not limited by minority carrier storage. MOSFETs is essentially a resistor in on-state and the temperature coefficient of its on-state resistance is negative. This means that localized hot spot does not occur and the second breakdown mechanism is nonexistent. The same negative temperature coefficient tends to even out the

Reprinted from *IEEE Power Electron. Specialists Conf.*, 1979, pp. 385–395.

current distribution on a chip or among paralleled devices. This means that very high current units would be feasible. MOSFETs have very high input impedance, so they require simpler drive circuitry and facilitate the manufacture of large area, high current units.

Compared with the thyristors, MOSFETs have 10^2 to 10^4 times the speed. They can be turned off with control voltage. However, MOSFETs probably cannot match the thyristors in power ratings.

In this paper, we shall examine the limitations of power MOSTs as switching elements, although they may be used as linear elements. It is impractical to control large quantities of power with linear devices. Even in applications involving relatively low power and traditionally using linear devices such as power supplies and audio amplifiers, pulse width modulation is being adopted to reduce power dissipation. Power MOSFETs should hasten this trend.

II. ON-STATE RESISTANCE

Until recently power MOSFETs have considerably higher on-state resistance than bipolar transistors of the same size. Indeed, the on-state resistance may be the major parameter by which MOSFETs will be judged. In this section, this parameter is analyzed in some details.

In Fig. 1b are shown the components of the on-state resistance. R_ℓ is the lead resistance which includes the resistances due to the source diffusion, the n^+ substrate, the contact resistance and any leads that are in close thermal contact with the junction. R_{ch} is the resistance of the channel. R_a is the resistance of the accumulation layer, and R_{epi} is the resistance of the epitaxial layer including any spreading resistance. All resistances will be expressed in ohms for 1 cm^2 of chip area, $\Omega \cdot cm^2$.

In order to analyze R_{epi} we first examine the ohmic resistance of the n region in a pnn^+ diode such as shown in Fig. 2. The resistance is

$$R_n = \frac{W}{q\mu_n N_d} \quad (1)$$

In order to minimize R_n, one should decrease W and increase N_d. However doing either may reduce the breakdown voltage of the diode. The electric field in the n region under three reverse voltages are shown in the figure. The slope is determined by Poisson's equation

$$\frac{dE}{dx} = -\frac{\rho}{\varepsilon} = \frac{-qN_d}{\varepsilon} \quad (2)$$

where ε is the permittivity of silicon. Under a certain reverse bias, the field at the junction reaches a critical field E_c, which is approximately 2×10^5V/cm and avalanche breakdown occurs [8]. This breakdown voltage is simply the area under the field distribution curve.

$$V_B = \frac{W}{2} \left(E_c + E_c - \frac{qN_d W}{\varepsilon}\right) \quad (3)$$

which leads to

$$N_d = \frac{2\varepsilon}{qW} \left(E_c - \frac{V_B}{W}\right) \quad (4)$$

The series resistance of the n region when it is not depleted is

$$R_n = \frac{W}{q\mu_n N_d} = \frac{W^3}{2\mu_n \varepsilon (E_c W - V_B)} \quad (5)$$

Setting the derivative of R_n with respect to W to zero and one obtains this condition for minimum R_n.

$$W = 3V_B/2E_c \quad (6)$$

From Eqs. 4, 5, and 6

$$N_d = \frac{4}{9} \frac{E_c^2 \varepsilon}{qV_B} \quad (7)$$

$$R_n = 27V_B^2/8\varepsilon\mu_n E_c^3 \quad (8)$$

From both theory and experiments, it is known that E_c is really a slowly varying function of V_B such as $E_c \approx 8.2 \times 10^5 \ V_B^{-0.20}$V/cm [8]. The electron mobility is also a slowly varying function of N_d and hence of V_B such as $\mu_n \approx 710 \times V_B^{0.1}$. These values are most reliable in the breakdown voltage range between 200 and 2000 volts. Using them, Eq. 8 becomes

$$R_n = 8.3 \times 10^{-9} \ V_B^{2.5} \quad (9)$$

More careful analysis [9] shows that the optimum doping profile is not a uniform one but rather as shown in Fig. 3, and the minimum n region resistance is $3V_B^2/\varepsilon\mu_n E_c^3$. This value is only 11% lower than the value in Eq. 9. Figure 3 shows that the constant doping profile is not very different from the optimum profile. The difference is even smaller when one considers the complimentary error function dopant profile at the nn^+ interface due to the diffusion of impurities from the n^+ substrate into the epitaxial layer [10].

In the following analysis the higher resistance value given in Eqs. 8 and 9 will be used. This should partially compensate for the fact that breakdown may occur at the edge or surface of the device because of higher local field or lower breakdown field at the surface. Many techniques, including ion implantation can bring the breakdown voltage to between 85 and 95% of the bulk breakdown voltage considered in the above analysis [3].

Comparing the structures in Fig. 1 with the pnn^+ diode in Fig. 2, three differences may be noted. One is that the p layer is not continuous and junction edges as well as the Si/SiO_2 interface are present. Fortunately, this geometry with the gate electrode and oxide overlapping the junction edge is an effective way of supressing the breakdown at the junction edges as long as the oxide is thin enough and the gate is shorted to or bias negative relative to the p layer [11]. Since the permittivity in the oxide is three times less than in the silicon, the field strength in the oxide will be three times higher than in the silicon. The breakdown field, however is at least 20 times higher than E_c. Therefore, the oxide will not breakdown before the silicon underneath it. In addition to this, the n region thicknesses under the p region and under the oxide are different. For simplicity, we shall neglect this difference.

The second difference is that the voltage is applied across the source and drain. If the p region is electrically grounded to the source, the breakdown voltage would be just that of the pnn^+ diode and the previous analysis would still apply. If the p region is left floating, the breakdown voltage would be the open base breakdown voltage of the n^+pnn^+ transistor. It is well known that the open base breakdown voltage is $(\beta+1)^{-1/m}$ times the base collector junction breakdown voltage, where β is the transistor gain and m is often taken to be 4 for Si [12]. This is a severe reduction in device voltage. Therefore, it is important to short the source and the p region together at the expense of extra processing steps.

Finally, the resistance between the accumulation regions and the drain is not that of a semiconductor layer between uniform top and bottom electrodes. Rather, the situation is as illustrated in the inset of Fig. 4. The top electrode is an array of long fingers of width a and separated from the neighboring fingers by s. The resistance between the upper and lower electrodes is larger than R_n by a spreading resistance factor so that

$$R_{epi} = 8.3\times10^{-9} V_B^{2.5} \left[1 + \frac{a+s}{\pi W} \ln \frac{a+\pi W}{a+\pi W \sin(\pi a/2(a+s))}\right] \quad (10)$$

where R_n is taken from Eq. 9 and the spreading resistance factor is a composite of two approximate factors for $a+s \gg W$ [13] and $a+s \ll W$ [14] respectively. The accuracy of this factor has not been verified. It is plotted in Fig. 4. The spreading resistance effect disappears when $s \to 0$ or $a \to \infty$ as expected. As a rough rule, the spreading resistance becomes rather small when $s < W$ and $a > W$.

To calculate R_{epi} in this manner and later add to it an R_a, the accumulation layer resistance is valid only if R_a is much smaller than R_{epi}. Since current is fed from both ends of the accumulation layer, the resistance for each finger of the accumulation layer 1 cm long and of width a is $a/8\sigma_a$ where σ_a is the sheet conductivity of the accumulation layer. Over an area of 1×1 cm^2 there are $1/(a+s)$ such fingers in parallel. Therefore,

$$R_a = a(a+s)/8\sigma_a = a(a+s)/8qN_{as}\mu_{ns} = a(a+s)/8\varepsilon_{ox}E_{ox}\mu_{ns} \quad (11)$$

where qN_{as} is the surface charge density in the accumulation layer, which is equal to $\varepsilon_{ox}E_{ox}$ according to Gauss' law. E_{ox} should be the highest field that may be applied to SiO_2 without causing reliability problem. In other MOS devices 2×10^6 V/cm is often encountered. We shall take this value as the maximum safe oxide field, although the intrinsic breakdown field of SiO_2 is close to 10^7 V/cm. The surface mobility of electrons μ_{ns} is about 700 cm^2/s-V at low oxide field but decreases to about 350 cm^2/s-V at the high field we assume [7]. Using these numbers, Eq. 11 becomes

$$R_a = 517\, a(a+s)\ \Omega\cdot\text{cm}^2 \quad (12)$$

The channel resistance R_{ch} needs some special consideration. The inset in Fig. 5 shows the dopant densities in the n epitaxial layer, and the p and n^+ diffused layers. The diffused dopant profiles are usually assumed to be gaussian but may be approximated by the following equations if L is smaller than the distance between the surface and the $x=0$ point [15].

$$N_a(x) = N\left(\frac{N}{N_d}\right)^{-x/L} \quad (13)$$

$$N_d^+(x) = N\left(\frac{N}{N_d}\right)^{-x/\ell} \quad (14)$$

If $\bar{N}_a$ is the average net doping density in the p layer

$$\bar{N}_a L \equiv \int_0^L [N_a(x) - N_d^+(x) - N_d]dx = \frac{N}{\ln(N_d/N)}\left(\frac{N_d}{N} - 1\right)(L-\ell) - N_d L \approx \frac{N_d L}{\ln(N_d/N)} \quad (15)$$

In order to avoid the punch-through breakdown, $\bar{N}_a L$ must be larger than εE_c. This sets a lower limit on the channel length and/or the approximate peak channel doping density, N.

$$\frac{N}{\ln(N/N_d)} > \frac{\varepsilon E_c}{qL} \approx 1.3\times10^{12}/L \quad (16)$$

The minimum peak channel doping density is plotted in Fig. 5.

In order to invert the surface, the surface potenital must be raised by about 1 volt. This creates a depletion region under the inverted surface. In the limit of very small L, the space charge density in the depletion layer can be calculated on the basis of the average channel doping density $\bar{N}_a$.

$$Q_B \approx \sqrt{2q\varepsilon\bar{N}_a \times 1} \geq 6.5\times10^{-10}/\sqrt{L}\ \text{coul/cm}^2 \quad (17)$$

where Eqs. 15 and 16 have been used to eliminate $\bar{N}_a$.

For each $1 \times 1\ cm^2$ chip area, there can be $2/(a+s)$ 1 cm wide channels in parallel, therefore

$$R_{ch} = \frac{a+s}{2} \cdot L/\mu_{ns} (\varepsilon_{ox} E_{ox} - Q_B)$$

$$= \frac{a+s}{2} [L/(2.24 \times 10^{-4} - 2.3 \times 10^{-7}/\sqrt{L})] \Omega\text{-}cm^2 \quad (18)$$

The quantity in the square brackets is the resistance of one 1cm wide channel and is plotted in Fig. 5. As the channel length L decreases the channel resistance decreases proportionally at first. At the same time the channel doping density must increase in order to prevent punch through. This raises the channel turn-on threshold voltage. At very small L's, the threshold voltage becomes so high that the channel resistance at the maximum safe oxide field increases with decreasing L. However the minimum channel resistance occurs at a very small L, where many of the concepts used in the analysis may be invalid. Nevertheless, it is much smaller than commonly used at the present. The optimum channel length should be the smallest that can be manufactured reproducibly.

The sum of R_{epi}, R_a, and R_{ch} according to Eqs. 10, 12 and 18 are plotted in Figs. 6 and 7 for a 100 volt device and a 1000 volt device respectively. In all devices, there is a value of a at which the resistance is minimum. At smaller a's the spreading resistance is too large. At larger a's the channel and accumulation layer resistances are larger because there are fewer channels and accumulation layers in parallel. It can be seen that the minimum R_{on} occurs at about $a \simeq 0.5W$. As expected from Eqs. 10, 12, and 18, the smaller s is the smaller R_{on} can be. Even at $s = 7\mu m$, minimum R_{on} is about 40% higher than R_n (see Eq. 9) for the 100 volt device. For the 1000 volt device, R_{on} is only 2% higher than R_n.

The minimum R_{on} as a function of the source-drain breakdown voltage V_{dsB} is shown in Fig. 8, with and without a $1m\Omega\text{-}cm^2$ lead resistance added. The lead resistance includes all resistances due to the source diffusion, the n^+ substrate, the contact interfaces and any leads in close thermal contact with the junction. Even with a modest lead resistance of $R_\ell = 1\Omega m \cdot cm^2$, R_{on} can be closely approximated by

$$R_{on} \approx R_n + R_\ell = 8.3 \times 10^{-9} V_{dsB}^{\ 2.5} + R_\ell \quad (19)$$

In other words, with L, a, and s optimized, the channel and accumulation layer resistances and the spreading resistance can be neglected.

III. CURRENT-VOLTAGE TRADE-OFF

Since power MOSFET is a linear resistor in the on-state. The average power dissipation per unit area is $J_{dRMS}^2 \cdot R_{on}$, where J_{dRMS} is the rms drain current. If T_j and T_c are the junction and case temperature and θ_{jc} is the junction to case thermal resistance for one unit area of the chip, the maximum drain current is limited to

$$J_{dRMS} = \sqrt{\frac{T_j - T_c}{\theta_{jc} R_{on}}} \quad (20)$$

J_{dRMS} is also the maximum continuous drain current. It is plotted in Fig. 8. $T_j - T_c$ is assumed to be 100°C and θ_{jc} is taken to be $0.6°C \cdot cm^2/W$ as the average to good value for a large range of device packages including the flat-pack types. In calculating J_{dRMS}, R_{on} at 150°C is taken to be 2.7 times the room temperature R_{on} shown in Fig. 8, because the electron mobility in lightly doped silicon is 2.7 times lower at 150°C [16]. With these values,

$$J_{dRMS} \approx \left(\frac{167}{2.2 \times 10^{-8} V_{dsB}^{\ 2.5} + R_\ell} \right)^{1/2}$$

$$\approx 8.5 \times 10^4 V_{dsB}^{\ -1.25}\ A/cm^2 \quad (21)$$

The room-temperature on-state voltage, V_{on} is the product of J_{dRMS} and R_{on} and is plotted in Fig. 8. V_{on} is approximately 0.5% of V_{dsB} for $V_{dsB} > 100V$. At 150°C, the on-state voltage would be 2.7 times higher. From Eqs. 19 and 21, for high voltage devices,

$$V_{on} \approx 7.1 \times 10^{-4} V_{dsB}^{\ 1.25}\ \text{Volt} \quad (22)$$

Figure 9 shows the continuous switched power $V_{dsB} \times J_{dRMS}$ as a function of V_{dsB}. According to Eq. 21, the power that a 1 cm^2 device can control is proportional to V_{dsB} at low V_{dsB}'s and decreases as $V_{dsB}^{-0.25}$ at high V_{dsB}'s. The V_{dsB} at which the controlled-power density peaks depends on the lead resistance. At $R_\ell = 2.7\ m\Omega \cdot cm^2$ as assumed in Fig. 9, the peak occurs at about 300V.

Figure 9 also shows the power ratings of the best commercial power MOSFET in the TO-3 package. The data points represent products from three manufacturers. It appears that the data points follows the general shape of the theoretical curve, but the power ratings are lower than the theoretical values by a factor of three or more.

IV. SWITCHING SPEED

If an ideal step voltage is applied to the gate-source to turn the device on, V_{ds} would ideally drop to zero instantaneously while I_{ds} rises with time as determined by the load circuit. During turn-off, I_{ds} would ideally drop to zero instantaneously while V_{ds} rises as determined by the load circuit.

Let us trace the events that follow the application of an ideal step voltage signal and examine what prevents V_{ds} from reaching V_{on} instantaneously. First, it takes a finite time for the applied gate voltage to propagate down the transmission line formed by the gate electrode and the channel. If R_g' is the resistance per unit length (in the width direction of the channel) of the gate electrode and C_g' is the gate capacitance per unit length of gate electrode, the transmission line equations are

$$\frac{\partial v_g(y,t)}{\partial y} = -R'_g\, i_g(y,t) \qquad (23)$$

$$\frac{\partial i_g(y,t)}{\partial y} = -C'_g \frac{\partial v_g(y,t)}{\partial t} \qquad (24)$$

After eliminating i_g from Eqs. 23 and 24

$$\frac{\partial^2 v_g}{\partial y^2} = R'_g C'_g \frac{\partial v_g}{\partial t} \qquad (25)$$

The solution of Eq. 25 for the initial condition $v_g(y,o) = 0$ and the boundary condition $v_g(o,t) = V_g$ is

$$v_g(y,t) = V_g \text{ erfc } y/2\sqrt{t/R'_g C'_g} \qquad (26)$$

Note that erfc 1 ≈ 0.15. Assuming that the channel is turned on when $v_g = 0.15\ V_g$, the width of on-state channel then increases with time as

$$y = 2\sqrt{t/R'_g C'_g} \qquad (27)$$

For example, if polysilicon gate is used, the minimum gate electrode sheet resistance would be about 20Ω/□, and $R'_g = 20/b$ Ω/cm, where b is the gate electrode width. If the gate oxide thickness is 1000 Å, $C'_g = \varepsilon_{ox} \cdot b/1000$ Å f/cm. $R'_g C'_g = 7\times10^{-7}$ t/cm^2. Using this number in Eq. 27, it can be seen that it takes 2ns for a 1mm long gate to turn on and 200ns for a 1cm long gate to turn on. It takes four times longer for the whole gate to reach 0.5 V_g, which may be taken as the condition of full turn-on. If aluminum gate is used, the transmission line delay is naturally shorter. For a perfectly conducting gate electrode, it can be shown that $y = 1.5\times10^{10}$t. This spreading of the conducting area is reminiscent of the plasma spreading phenomena in thyristors. In principal, this calls for a di/dt rating to protect the MOSFET from destruction by excessive transient current density. In practice, the safe di/dt would probably be higher than may be encountered in most circuits, since the spreading delay time is several hundred times shorter than in thyristors. Nevertheless, this gate voltage delay could limit the speed of large area devices using polysilicon gates unless care is taken in designing the gate bus-bar.

After v_g has reached the threshold voltage, electrons flow from source through channel and across the depletion region so that the depletion layer can collapse and V_{ds} can drop. This transit time is about $(L + W)/10^7$, where 10^7 cm/s is the electron saturation velocity in the depletion region and is close to the saturation velocity in the surface channel (6×10^6 cm/s). This introduces a 0.1 to 1 ns delay. The accumulation layer is established in a time period comparable to $a/10^7$.

In the above analysis, the p region is assumed to be shorted to the source. It would simplify the device processing if the p region is left floating. We have already seen earlier that a floating p region would reduce the source-drain breakdown voltage. Here we examine its effects on the switching turn-on speed. In order for the pn junction depletion region to collapse, holes must be supplied to the p region. Without such a supply, the p-region potential will be capacitively pulled down. This would increase the threshold voltage, V_t -- a phenomenon known as the substrate bias effect. The change in V_t is usually a fraction of the p region potential. However, the p region to source potential difference cannot be larger than the breakdown voltage of that junction. Therefore, the p region may be left floating if the n^+p junction breakdown voltage is much smaller than V_t. Otherwise, it would take a long time for thermal generation in the depletion region to produce enough holes to bring the p region potential back to zero. This time is approximately $\sqrt{NN_d/n_i^2}\ \tau_0$, where N and N_d are defined in Fig. 5, n_i is the intrinsic carrier concentration; and τ_0 is the carrier generation lifetime, which may be between 0.1 to 1 μs.

During turn-off, the transmission line effect is similarly present. The dielectric relaxation time ε/σ, usually less than 0.1 ns, determines how rapidly the accumulation layer is removed. The depletion region is established by a drain current that continues after the channel is shut off and the time required is

$$t = \varepsilon E_c/J_d \approx 2\times10^{-7}/J_d \qquad (28)$$

where $\varepsilon E_c \approx 2\times10^{-7}$ coul/cm^2 is the charge to be stored in the depletion region (assuming $V_{ds} = V_{dsB}$ in the off-state) and J_d is the drain current density. For $J_d = 1$A/cm^2, this component of the fall time would be about 200 ns. Both the absolute value and the $1/I_d$ dependence of this fall time agree quite well with the reported data [17]. Clearly, this fall time will be particularly long in high voltage units which can handle only limited J_d.

The effect of a floating p region on the turn-off time is also different. As the depletion region is being formed, holes must be removed from the p region -- by injection into the source. This would forward bias the n^+p junction and starts a transistor action. Now, in order to widen the depletion region by removing one electron from it, β+1 electrons must be drawn out by the drain current, where β is the transistor current gain. Equation 28 becomes

$$t \approx 2\times10^{-7}(\beta+1)/J_d \qquad (29)$$

So far we have considered the intrinsic speed limits set by the device. Most of them are neglected in previous analyses [17], which correctly argued that in many circuits the circuit (not device) speeds are limited by the RC time constants of the gate drive circuitry. Due to the finite resistance in the drive circuit and the gate input capacitance, the voltage present at the gate electrode bus-bar has finite rise and fall time as shown in the inset of Fig. 10.

The effects of the rise and fall times of $V_g(t)$ can be minimized by proper device design. Figure 10 shows the device current as a function of V_g.

At small values of $V_g - V_t$, the drain current is limited by the device. In this regime, the drain current is, in the limit of zero voltage drop in the channel (small L), $(V_g-V_t)C_{ox}'v_s/(a+s)=2.2x10^{-6}$ $(V_g-V_t)/(a+s)d_{ox}A/cm^2$, where v_s is th channel electron saturation velocity and d_{ox} is the oxide thickness [7]. At large V_g's, the current is limited by the load. The transition occurs at a fairly low voltage V_{g1}, if L, a, and s are small. In that case the turn-on rise time is the relatively short time needed for V_g to rise from V_t to V_{g1}. The rise period ends at $V_g = V_{g1}$ for I_d will be limited by the load; and it (and V_{ds} in the case of resistive load) will stop changing. For a device with larger L, a, and s, the rise period does not end until V_g reaches V_{g2}. Therefore, the rise and fall time could be shorter than the gate drive time constant.

Finally, the gate input capacitance may be minimized by either increasing the gate oxide thickness or reducing the a/(a+s) ratio. In the previous analysis of on-state resistance, no mention was made of the gate oxide thickness. Therefore, the gate oxide thickness may be chosen to minimize the gate capacitance (this would also reduce the transmission line delay) as long as the required magnitude of the gate voltage can be generated. The a/(a+s) ratio may be made small without introducing a spreading resistance factor (see Eq. 10) greater than 110% only if s/W is kept less than 2a/(a+s) (130% if $s/W < 6a/(a+s)$). For example, $a = 0.1\ (a+s)$ is acceptable if $s < 0.2W$ (or 0.6W). Therefore obtaining very small gate input capacitance in this manner would either require very fine line technology or increase the on-state resistance by more than 10-30%. The optimum design seems to be to use the smallest line width for a and let $s = 1.4\sqrt{aW}$ (or $2.5\sqrt{aW}$).

Table 1 summarizes the events and time scales involved in the turn-on and turn-off transients. Besides the drive circuit RC time constant, which is not intrinsic to the device, the depletion layer forming time and, for certain devices, the gate voltage spreading time are the most important.

V. NEW STRUCTURES FOR HIGH VOLTAGE HIGH POWER MOSFETs

Looking into the future of very high voltage and high current density power MOSFETs as replacements for high power thyristors, one quickly realizes that the designs shown in Fig. 1 are limited by $J_d \propto V_{dsB}^{-1.25}$ and $P \propto V_{dsB}^{-0.25}$. The limit is imposed by the high $R_{on} \propto V_{dsB}^{2.5}$ at high V_{dsB}'s. Figure 11 considers certain alternative designs all based on the fact that the accumulation layer is a high conductance path. Figure 11a extends the V-groove down to the n^+ region. Now, in the on-state, electrons flow from the source through the channel and the surface accumulation region to the n^+ substrate. The on-state resistance is then

$$R_{on} = \frac{\lambda W/\cos 35°}{2\varepsilon_{ox}E_{ox}\mu_{ns}} > \frac{2W^2/\cos 35° \cdot \sin 35°}{2\varepsilon_{ox}E_{ox}\mu_{ns}}$$

$$= 8.8 \times 10^3\ W^2$$

$$= 3 \times 10^{-8}\ V_B^{2.4} \qquad (30)$$

	Turn-on	Turn-off
RC time of gate drive circuit	Less than 1 ns and up, depending on drive circuit impedance. High V_{dsB} devices may be faster than low V_{dsB} devices.	
Voltage spreading on resistive gate	Less than 1 ns to 1 μs for large area poly-Si gate devices	
channel formation or removal	10 ps (μs and up if p region is floating and $V_{n+pB} > V_t$)	10 ps
accumulation layer formation or removal	50 ps - 1 ns	less than 0.1 ns
depletion layer collapse or transit time	50 ps - 2 ns	not applicable
depletion layer formation	not applicable	$2 \times 10^{-7}/J_d$ 1 ns - 1 μs (β+1 times longer if p region is floating; β is the n^+pn transistor gain)

Table 1. Compilation of mechanisms affecting turn-on and turn-off times.

Equation 6 and $E_c = 8.2 \times 10^5 V_B^{-0.2}$ have been used in the last step. Comparing Eq. 30 with Eq. 9, one notices a net increase in the resistance. This is because λ in Fig. 11a must increase with increasing V_{dsB}.

Figure 11b shows a more favorable structure. Here, deep vertical holes (not grooves for higher mechanical strength) are etched into a 110 cut silicon wafer. λ is only limited by lithography and the etching technique. It can be 20 μm and independent of W. The resistance is

$$R_{on} = \frac{\lambda W}{2\varepsilon_{ox} E_{ox} \mu_{ns}} = 4.1W$$

$$= 7.6 \times 10^{-6} \, V_{dsB}^{1.2} \;\; \Omega \cdot cm^2 \qquad (31)$$

Comparing Eq. 31 with Eq. 6, we see that the proposed design has a lower R_{on} for $V_{dsB} > 200V$. At $V_{dsB} = 500V$, R_{on} is 3.5 times lower in the proposed design than the conventional design. At $V_{dsB} = 3000V$, R_{on} is 36 times lower, maximum J_d and controlled power are 6 times higher. Figure 11c shows another structure, it can block voltages of both polarities and conduct currents in both directions.

The proposed structure must have a very thick oxide so that breakdown does not occur. This in turn requires a very high gate voltage to turn on the device. Fabrication method has to be developed. These are some of the disadvantages of these structures.

From Eqs. 31 and 21, $J_d \propto V_{dsB}^{-0.6}$, which is a significant improvement over the $J_d \propto V_{dsB}^{-1.25}$ in Eq. 21. However it is still inferior to thyristors, whose current density rating is independent of the voltage rating, i.e., $J \propto V_B^0$. The tradeoff in thyristors is between voltage and speed. Therefore, power MOSFETs are not expected to match the high power ratings of thyristors.

VI. CONCLUSIONS

Power MOSFETs are expected to be easier to use than bipolar transistors because they are well behaved without extraneous effects such as the second breakdown. For this same reason they lend themselves to simple theoretical analysis. This paper has examined the basic limitations of the voltage, current, power, and speed ratings, and the trade-offs among them. In the process, the optimization of device design is considered.

With optimized channel and gate-epi overlap design, the channel and accumulation layer resistances are negligible. R_{on} is limited by the epitaxial layer resistance for high voltage devices and decreases with decreasing voltage rating as $8.3 \times 10^{-9} V_{dsB}^{2.5}$ until it becomes limited by the lead resistance. Maximum power rating is about 20kW per cm^2 of chip area at around $v_{dsB} = 300V$. Over $10kW/cm^2$ can be obtained up to 5000V V_{dsB}. In low voltage devices, power ratings should be limited by lead resistances and linearly proportional to V_{dsB}. The best available commercial power MOSFETs have power ratings within a factor of four to the theoretical limit, which is calculated on the basis of commonly used packaging technologies.

In some circuits the turn-on and turn-off times will be limited by the RC time constants of the gate drive circuits. In that case, the gate input capacitance may be reduced without significantly incrasing R_{on} by using fine line technology. The rise and fall times may also be reduced by minimizing the channel and the accumulation layer resistances, even if their contributions to the steady state R_{on} is not important. From both these approaches, one may expect to obtain faster operations in higher voltage devices than lower voltage devices of the same chip area.

There are intrinsic device speed limitations that cannot be helped by speeding up the gate drive. One is the finite speed of gate voltage spreading along a long resistive polysilicon gate electrode. The result is a gradual spreading of conducting areas similar to the plasma spreading phenomenon in thyristors. This time constant may be reduced by careful gate electrode design. The other is a fall time associated with the charging of the depletion region and is inversely proportional to the drain current density. Both time constants could bein the order of hundred nanoseconds. They are longer in large area, low current density (i.e., high voltage) devices.

With all time constants considered, the speed should be slower in higher current units but relatively insensitive to the voltage ratings. The p region should be shorted to the source at the expense of additional processing steps. If it is left floating, both V_{dsB} and the fall time will suffer because of the transistor action of the n^+pn structure.

A new power MOSFET geometry is proposed for high voltage high current units. At $V_{dsB} = 3000V$, the new geometry leads to an R_{on} that is 36 times smaller than the minimum R_{on} of a device having the conventional geometry. Another new geometry can block voltage of both polarities and conduct currents in both directions. Although power MOSFETs of megawatt rating are conceivable, MOSFETs are not expected to match the highest power ratings of the thyristors on a per package bases. Ease of paralleling MOSFETs may compensate for this shortcoming to some degree.

REFERENCES

[1] T.M.S. Heng, et.al., IEEE Trans. Microwave Theory and Tech., vol. MTT-24, no. 6, p. 305, June 1976.

[2] C.A.T. Salama, Technical Digest of IEEE Intl. Electron Devices Meeting for 1977, p. 412.

[3] V.A.K. Temple, and R.P. Love, Tech. Digest of IEEE Intl. Elec. Dev. Meeting for 1978, p. 664.

[4] C. Hu, et.al., Notes for a course on Switching Power Supplies, Univ. of California Extension, February, 1979.

[5] H.W. Collins and B. Pelly, Electronic Design, Vol. 27, No. 12, p. 36, June, 1979.
[6] Y. Tarui, et.al., Proc. 1st Conf. Solid-State Devices, supplement to J. Japan Soc. Appl. Phys., Vol. 39, p. 105, 1970.
[7] H.J. Sigg, et.al., IEEE Trans. Electron Dev., Vol. ED-19, p. 45, January, 1972.
[8] M.S. Sze and G. Gibbons, Appl. Phys. Lett., Vol. 8, p. 111, 1966.
[9] C. Hu, IEEE Trans. Elec. Devices, Vol. ED-26, p. 243, March, 1979.
[10] A.S. Grove, A. Roder, and C.T. Sah, J. Appl. Phys., Vol. 36, p. 802, 1965.
[11] A.S. Grove, O. Leistiko, and W.W. Hooper, IEEE Trans. Elec. Devices, Vol. ED-14, p. 157, March, 1967.
[12] A.S. Grove, "Physics & Technology of Semiconductor Devices," John Wiley, p. 233, 1967.
[13] R. Plonsey, R.F. Collin, "Electromagnetic Fields," McGraw-Hill, 1961, Sec. 4.4.
[14] G.A. Coquin, H.F. Tiersten, J. Acoustical Soc. of America, Vol. 41, No. 4, p. 921, 1967.
[15] H.C. Lin and W.N. Jones, IEEE Trans. Elec. Dev., Vol. ED-20, p. 275, March, 1973.
[16] M.S. Sze, "Physics of Semiconductor Devices," Wiley, Ch. 1, 1969.
[17] I Review, International Rectifier, Spring 1979.

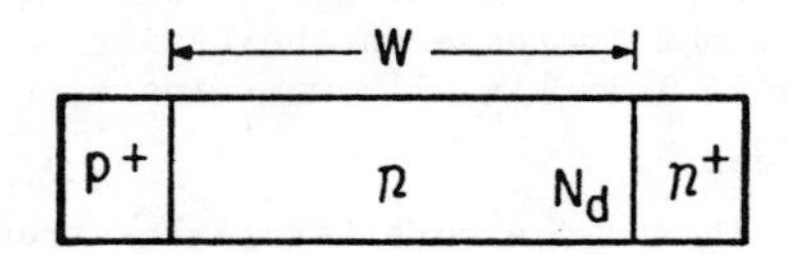
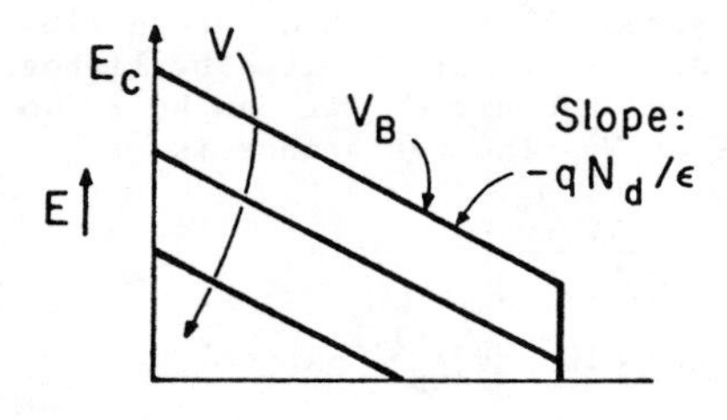

Fig. 2. Electric field distribution in a p^+nn^+ diode under three reverse voltages.

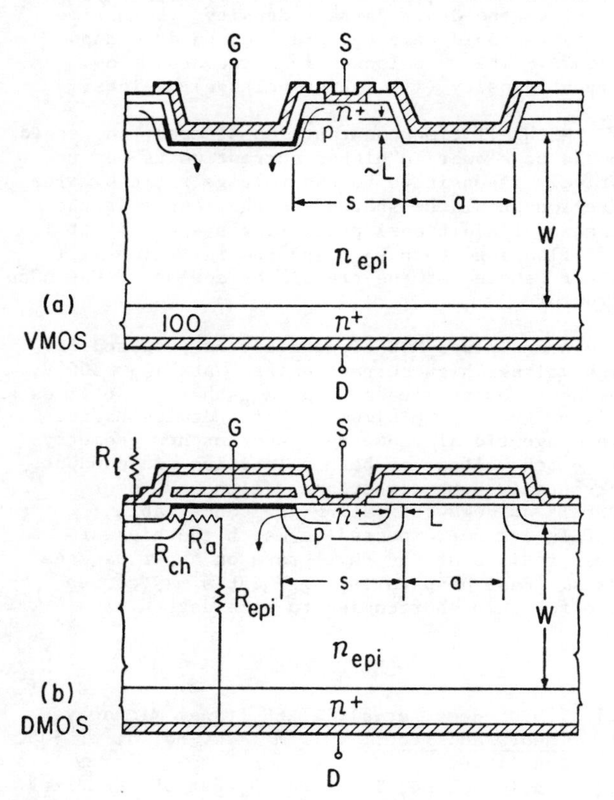

Fig. 1. (a) VMOS (V-groove MOSFET) and (b) DMOS (double-diffused MOSFET). All gate electrodes are electrically connected. The heavy lines show the thin n^+ surface layer induced by the gate voltage. The arrows show the paths of electron flow. The components of on-state resistance are defined in Fig. 1b.

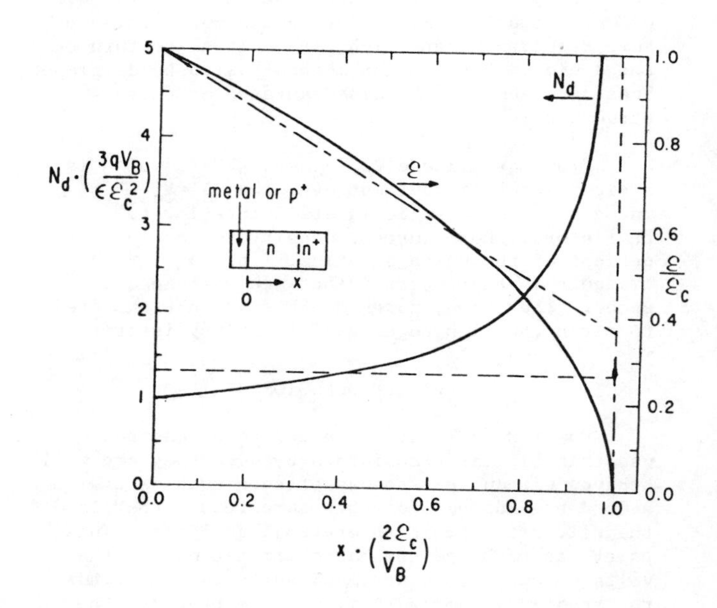

Fig. 3. Normalized optimum doping profile and electric field distribution (solid curves) which result in the minimum series resistance, $3V_B^2/\epsilon\mu_n E_c^3$. The broken lines are the best doping density and field distribution of a uniformly doped n layer; in this case the series resistance is 12.5% higher than the absolute minimum given above.

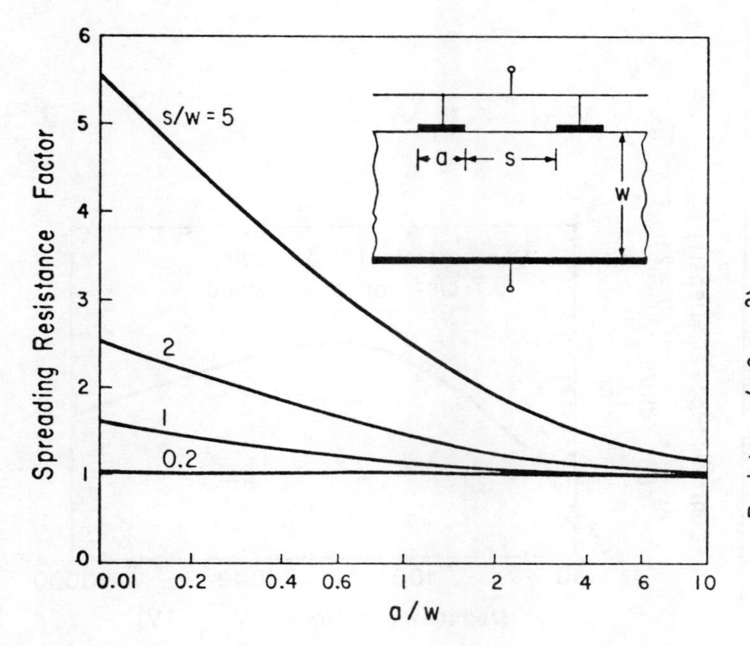

Fig. 4. The resistance between a top electrode long thin contacts consisting of an array and the bottom electrode is larger than that between continuous top and bottom electrodes by the spreading resistance factor.

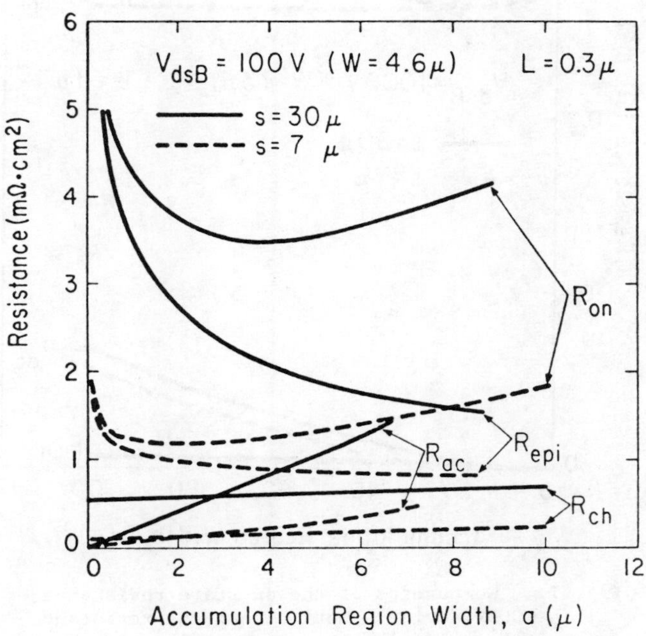

Fig. 6. Components of the on-state resistance of a 100V device, all for 1 cm^2 of chip area. Channel and accumulation layer resistances increase with increasing a and s because the number of channels and accumulation regions in parallel become smaller.

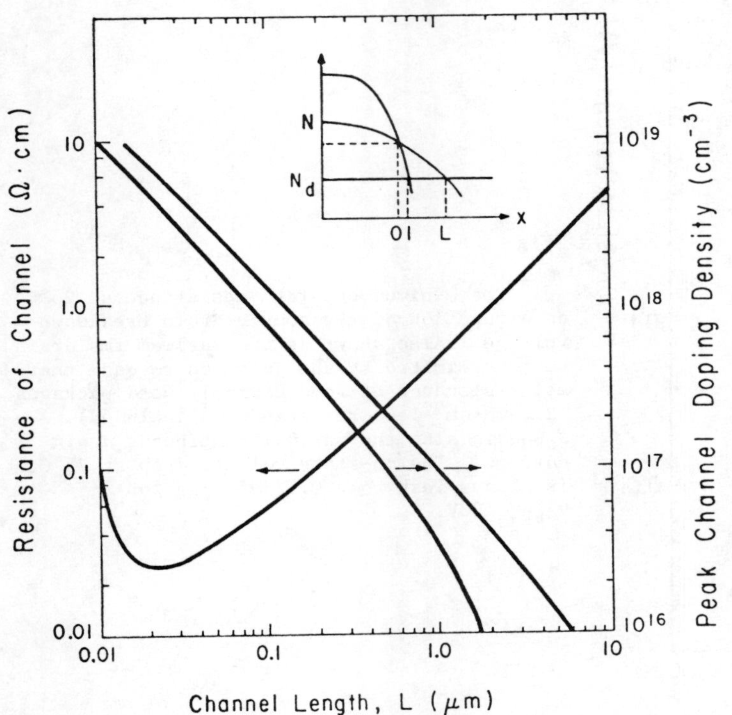

Fig. 5.

The minimum resistance of one 1 cm wide channel and the minimum "peak" channel doping density as functions of the channel length L. The channel doping density is the difference between the p and n^+ gaussian diffusion profiles.

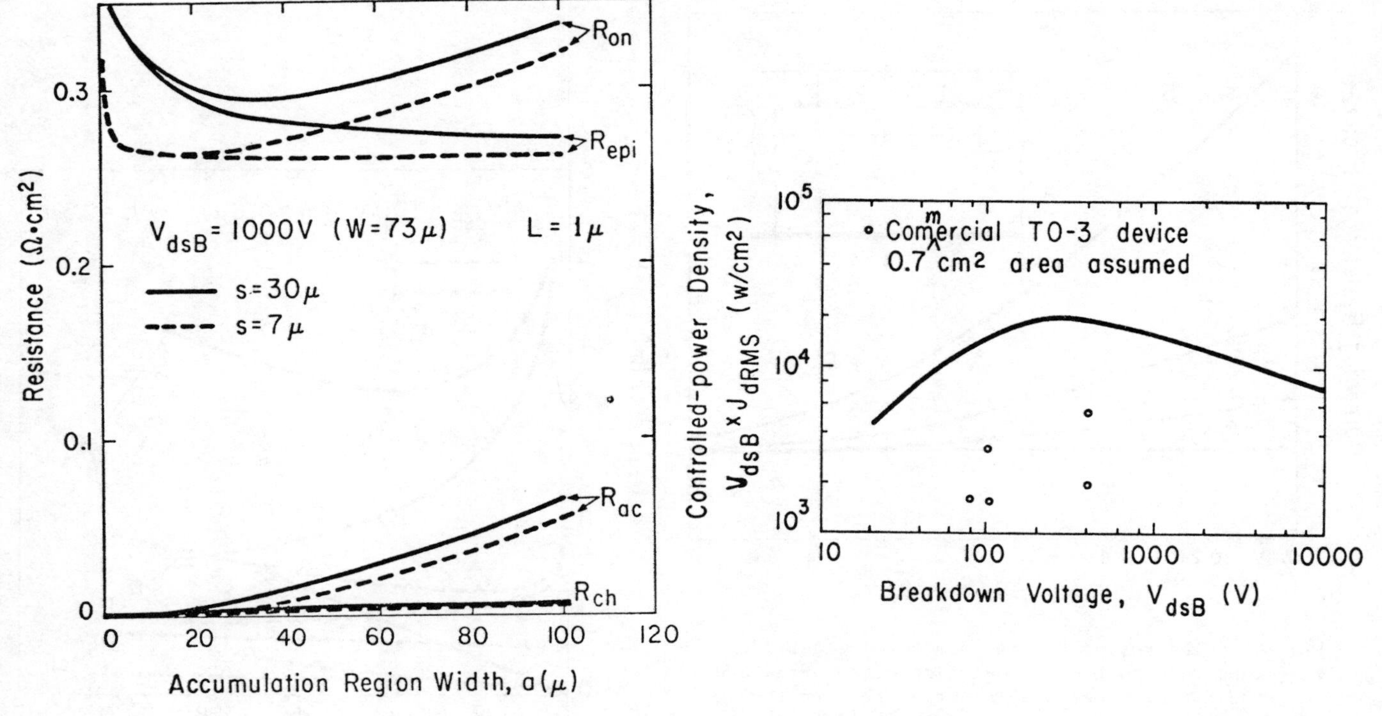

Fig. 7. Components of the on-state resistance of a 1000V device. Notice that the resistance of the epitaxial layer overwhelms the channel resistance.

Fig. 9. Maximum switching controlled power (V_{dsB} x continuous drain current) per cm^2 of chip area. In the low V_{dsB} range, the theoretical curve is limited by lead resistance.

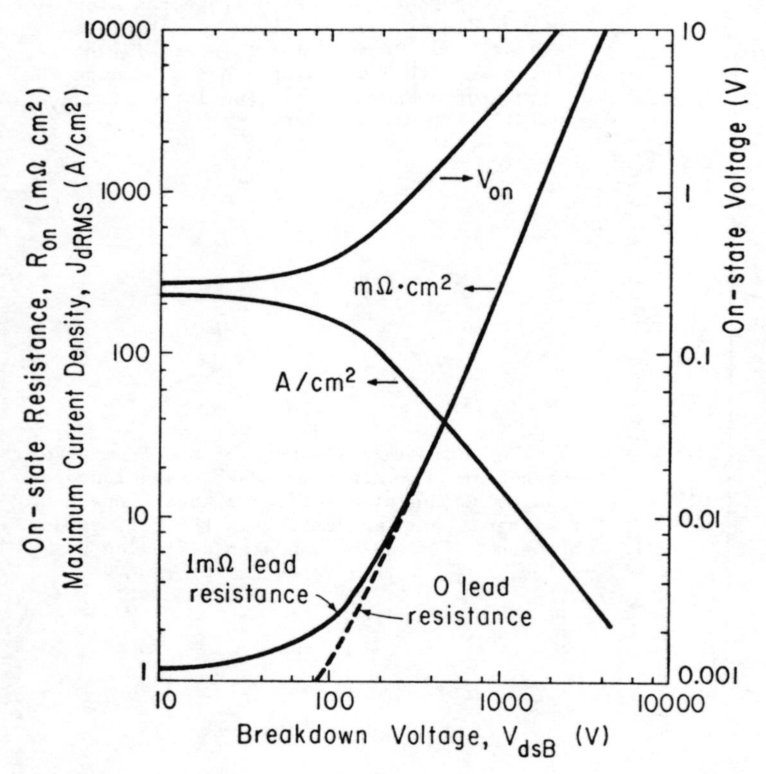

Fig. 8.

The minimum on-state resistance at 25°C as a function of the source-drain breakdown voltage. Also shown is the maximum rms drain current limited by the junction to case thermal resistance of good commonly used packages (2.7 mΩ·cm^2 lead resistance is included). J_{dRMS} is also the maximum continuous drain current. The on-state voltage drop at 25°C is always less than 0.5% of V_{dsB} for $V_{dsB} > 100V$.

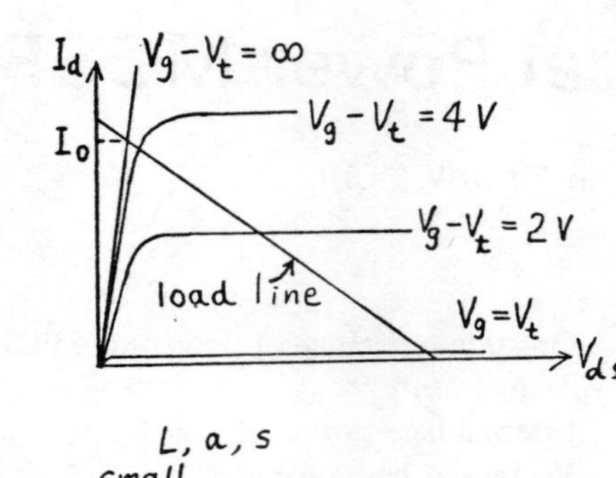

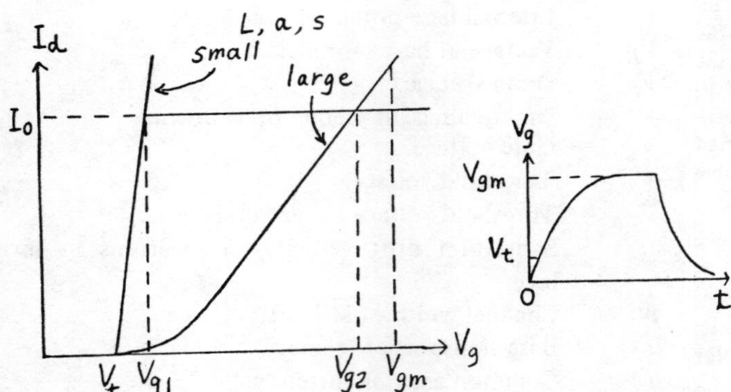

Fig. 10. Schematic plot of drain current versus V_g. For a given V_g waveform, the device rise and fall times are shorter if L, a, and s are smaller.

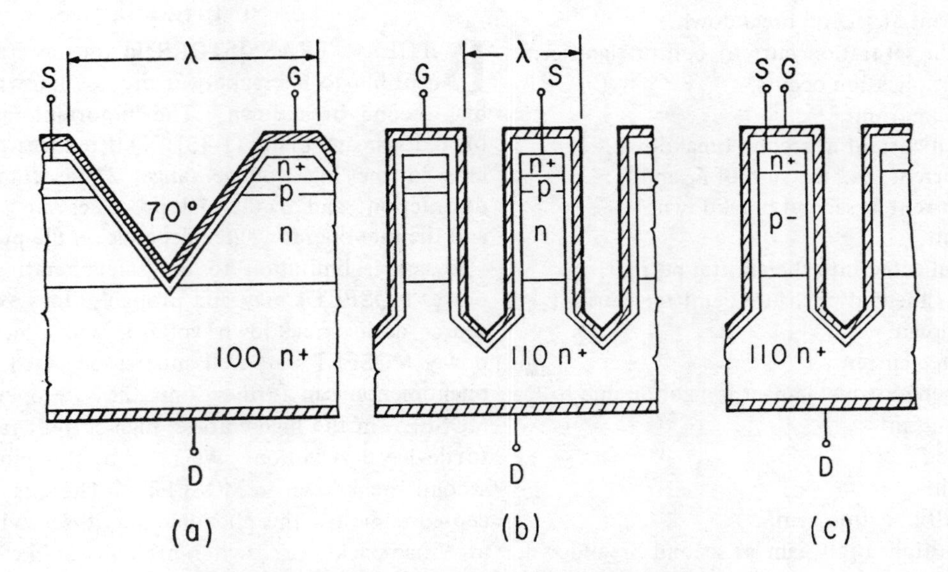

Fig. 11. Proposed new structures for high voltage low R_{on} MOSFETs. Structure in Fig. a offers no improvement. Structure in Fig. b cuts R_{on} by a factor of 36 at $V_{dsB} = 3000V$. Structure in Fig. c has electrically symmetrical source and drain.

Second Breakdown of Vertical Power MOSFET's

CHENMING HU, MEMBER, IEEE, AND MIN-HWA CHI

Abstract—**It is shown that a phenomenon of second breakdown similar to that in bipolar transistors can occur in vertical power MOSFET's. A model for the phenomenon of second breakdown involving the avalanche multiplication of the channel current, the parasitic bipolar transistor, and base resistance is proposed. After presenting the theory, this model is compared with experiments on four-terminal V-groove test devices in which the substrate can be accessed independently. Good agreement is achieved between calculated and measured boundaries of the safe operating area. The model should be applicable to DMOS devices as well.**

List of Symbols

BV	Breakdown voltage in the expression of M (10).
BV_{CE0}	Open-base breakdown voltage of the parasitic bipolar transistor.
C'_{ox}	Gate oxide capacitance per unit surface area.
G	Geometry dependent parameter defined in (B-3).
I_b	Internal base current.
$I_{b,SB}$	Internal base current at second breakdown.
I_{BC0}	Leakage current between base and collector.
I_c	Collector current.
$I_{c,SB}$	Collector current at second breakdown.
I_{d0}	MOSFET drain saturation current before significant avalanche multiplication occurs.
I_d	MOSFET drain current.
$I_{d,SB}$	MOSFET drain current at second breakdown.
I_D	Total drain current, i.e., the sum of I_d and I_c.
$I_{D,SB}$	Total drain current at second breakdown.
I_e	Emitter current.
I_{ep}	Hole current injected into the emitter current.
$I_{ep,SB}$	Hole current injected into the emitter current at second breakdown.
I_s	MOSFET source current.
I_p	Hole current generated by impact ionization due to I_c.
k	Boltzmann constant.
K	Ratio of I_{ep} to I_p.
L	Channel length.
M	Avalanche multiplication gain.
M_{SB}	Avalanche multiplication gain at second breakdown.
n	Fitting parameter in the expression of M (10).
N_a	Doping level in p-base.
q	Electron charge.
r_b	Internal base resistance.
R_B	External base resistance.
R_{on}	On-state resistance of power MOSFET.
T	Temperature.
V_b	Internal base potential.
V_B	Voltage at base terminal.
V_d	Drain voltage.
$V_{d,SB}$	Drain voltage at second breakdown.
V_G	Gate voltage.
V_T	Threshold voltage.
V_{T0}	Threshold voltage at zero body bias.
v_s	Saturation drift velocity of electrons in inversion layer.
W	Channel width of MOSFET.
α_T	Base transport efficiency.
β	Common emitter-current gain.
γ	Large-signal body-bias coefficient for MOSFET (see Appendix I).
γ_E	Emitter efficiency of the bipolar transistor.
ϵ_s	Dielectric constant of silicon.
μ	Mobility of electrons in the inversion layer.

Manuscript received September 15, 1981; revised March 4, 1982. This work was supported by the U.S. Naval Avionics Center under Grant N00163-80-C-0450.

The authors are with the Department of Electrical Engineering and Computer Sciences and the Electronics Research Laboratory, University of California, Berkeley, CA 94720.

I. Introduction

BIPOLAR TRANSISTORS in the on state are often susceptible to a breakdown process known as the forward-bias second breakdown. The important facts about second breakdown are that [1]–[3]: 1) it reduces the safe operating area in the high-voltage range; 2) it often leads to device destruction; and 3) thermal processes are probably involved and the safe operating area depends on the pulse width.

A similar limitation to the safe operating area of vertical power MOSFET's may and probably does exist. The current, source–drain breakdown voltage, and power limitations of power MOSFET's are well understood [4], [5]. An additional phenomenon can further limit the safe operating area of the MOSFET in the high-voltage, high-current range and can lead to device destruction. We refer to this phenomenon as the second breakdown in MOSFET's. There is no universally accepted name for this phenomenon, it is also informally known as "snap-back" or "switch-back." Unlike the forward-bias second breakdown in bipolar transistors, the onset of the second breakdown of MOSFET's does not seem to involve thermal processes and the safe operating area boundary due to second breakdown is independent of the pulse width from dc down to at least 1 μs.

The explanation for the second breakdown phenomenon of MOSFET's, as some have informally suggested, involves the parasitic bipolar transistor contained in all power MOSFET's [4], [5]. The collector and emitter of the bipolar transistor are also the drain and source of the MOSFET as shown in Fig. 2. When the MOSFET current and voltage are high, the

Reprinted from *IEEE Trans. Electron Devices*, vol. ED-29, pp. 1287–1293, Aug. 1982.

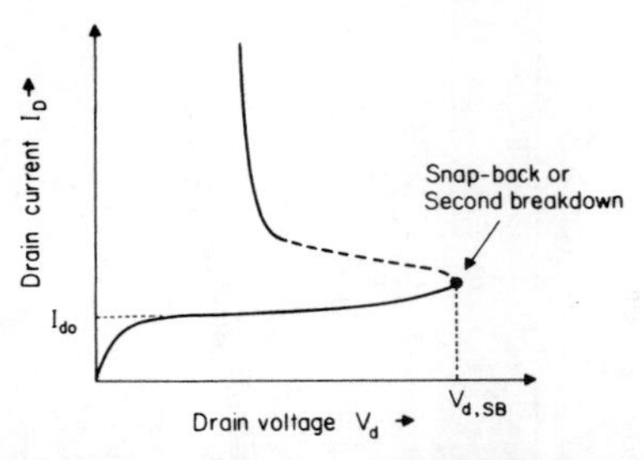

Fig. 1. A typical *I-V* characteristic of n-channel power VMOS with the observed phenomenon of "second breakdown."

(a)

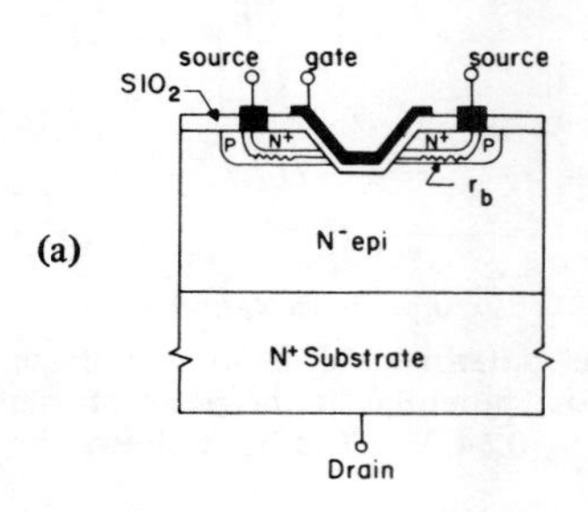

(b)

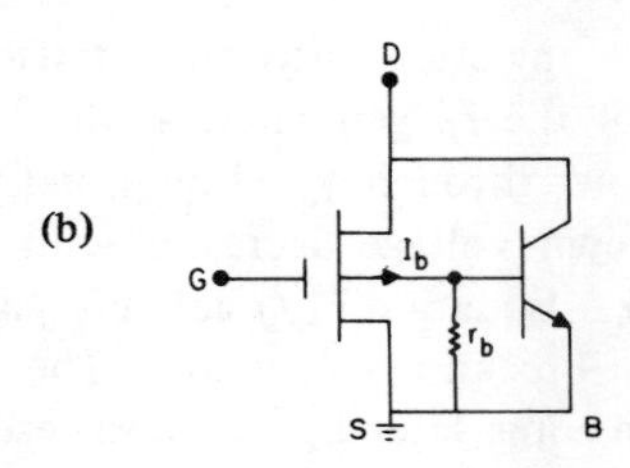

(c)

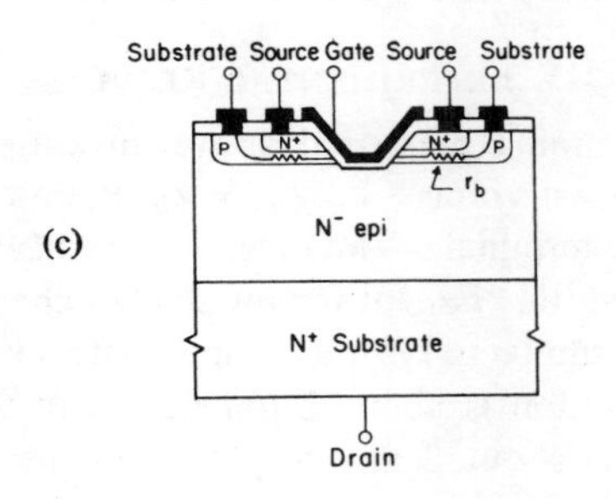

(d)

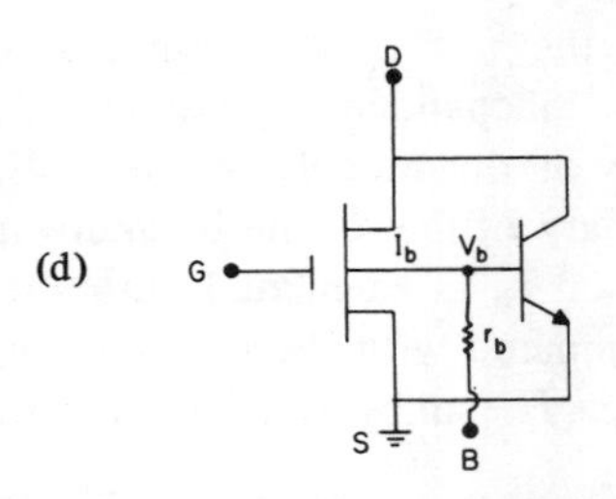

Fig. 2. (a) Structure of U-groove VMOS (n-channel) with substrate shorted to the source. (b) The circuit model of the structure in (a) includes a parasitic n-p-n bipolar transistor and a base resistance. (c) Structure of VMOS (n-channel) with separate contacts to source and substrate. (d) Circuit model of the structure in (c).

bipolar transistor can be turned on and the device voltage snaps back to or below the open-base breakdown voltage BV_{CE} of the transistor—a relatively low voltage in a high gain bipolar transistor. Such a qualitative model has long been reported to explain the "snap-back" phenomenon similar to Fig. 1 and observed in surface IC-type MOSFET's [4], [6]. Quantitative analysis, however, has been lacking because the two-dimensional nature of the planar MOSFET's complicates the analysis. The vertical power MOSFET, however, is quasi-one-dimensional and much more amenable to analysis.

After presenting the theory, the model will be compared with experiments.

II. Theory

A typical structure of a power MOSFET has a short channel between source and drain in order to minimize the "on-state" resistance. The source, drain, and the narrow p-base form a parasitic n-p-n transistor (for an n-MOSFET) as shown in Fig. 2(a), (b). The breakdown behavior of a power MOSFET will be described by using an equivalent circuit which is composed of a MOSFET in parallel with a bipolar transistor (Fig. 2(b)). One connection between the MOSFET and the parasitic transistor is through the common p-base with an internal base resistance (r_b). Various capacitances are not shown in Fig. 2 since only dc breakdown behavior of the device will be discussed in this paper.

At high I_d and V_d, impact ionization multiplication [7] generates hole current which flows to ground (source) through r_b. This raises the internal base potential V_b. The phenomenon "snap-back" is believed to occur when the internal base potential V_b is high enough to turn on the parasitic bipolar transistor. Immediately after the bipolar transistor is turned on, the collector (drain voltage) snaps back to or below BV_{CE} and thus a negative-resistance region is traced out. A typical curve depicting this behavior is sketched in Fig. 1.

A simple model is presented here to help predict or interpret the snap-back behavior of a power MOSFET. For simplicity, two assumptions are used at the onset of snap-back: 1) $V_b \approx 0.6$ V; and 2) $V_b = I_b \cdot r_b$. At a given gate voltage V_G, the source current is

$$I_s = I_{do} + \gamma V_b$$
$$= I_{do} + \gamma r_b I_b \quad (1)$$

where I_{do} is the saturation drain current at small-drain voltage and zero substrate bias and γ is a large-signal body-bias coefficient which is defined as

$$\lambda \equiv \frac{\Delta I_d}{\Delta V_b}. \quad (2)$$

An explicit expression of γ is discussed in Appendix I. The electric field peaks at or near the metallurgical drain junction. At high V_d, the large electric field at the drain causes avalanche multiplication of the electron channel current. The resultant holes are driven into the base. If the multiplication gain is M, the base current I_b is

$$I_b = I_d - I_s$$
$$= MI_s - I_s$$
$$= (M - 1)(I_{do} + \gamma r_b I_b). \quad (3)$$

Therefore

$$I_b = \frac{(M-1)I_{do}}{1 - \gamma r_b(M-1)} \quad (4)$$

$$I_s = \frac{I_{do}}{1 - \gamma r_b(M-1)} \quad (5)$$

$$I_d = \frac{MI_{do}}{1 - \gamma r_b(M - 1)}. \tag{6}$$

In Appendix II, it is shown that the bipolar transistor snaps back to BV_{CEo} when

$$I_b r_b \cong 0.6. \tag{7}$$

Equations (3) and (7) yield

$$M_{SB} - 1 = \frac{0.6}{r_b(I_{do} + 0.6\gamma)}. \tag{8}$$

Equations (3), (6), and (7) yield

$$1 - \frac{1}{M_{SB}} = \frac{0.6}{r_b I_{d,SB}} \tag{9}$$

where subscript SB indicates values at snap-back. M is often expressed empirically as [7]

$$M = \frac{1}{1 - (V_d/BV)^n} \tag{10}$$

where n is a constant and BV is the breakdown voltage at drain and p-base. From (4), (6), and (10) one obtains

$$1 + \gamma r_b + \frac{I_{do}}{I_b} = \left(\frac{BV}{V_d}\right)^n \tag{11}$$

$$\frac{I_d}{I_b} = \left(\frac{BV}{V_d}\right)^n \tag{12}$$

$$1 - \frac{I_{do}}{I_d} = (1 + \gamma r_b)\left(\frac{V_d}{BV}\right)^n. \tag{13}$$

At snap-back, from (12) and (13)

$$I_{d,SB} - I_{do} = (1 + \gamma r_b) I_{b,SB} = \frac{0.6}{r_b} + 0.6\gamma. \tag{14}$$

From (8) and (9)

$$V_{d,SB} = BV[1 + r_b(\gamma + I_{do}/0.6)]^{-1/n} \tag{15}$$

$$V_{d,SB} = BV(0.6/r_b I_{d,SB})^{1/n}. \tag{16}$$

From (16)

$$I_{D,SB} = I_{c,SB} + I_{d,SB} = I_{c,SB} + \frac{0.6}{r_b}\left(\frac{BV}{V_{d,SB}}\right)^n = I_{c,SB} + I_{b,SB} \cdot \left(\frac{BV}{V_{d,SB}}\right)^n \tag{17}$$

where $I_{D,SB}$ includes both "drain current" and "collector current" at second breakdown. From Appendix II

$$I_{c,SB} \approx \frac{kT}{q\gamma_E G}\left(\frac{I_{do}}{0.6} + \gamma\right) = \frac{kT}{q\gamma_E G r_b}\left[1 + \left(\frac{BV}{V_{d,SB}}\right)^n\right]. \tag{18}$$

γ_E is the emitter efficiency of the bipolar transistor and G is geometry dependent and probably less than unity as described in Appendix II. $I_{c,SB}$ may be neglected in (17). The smallness of $I_{c,SB}$ and the high gain of the bipolar transistor indicate that at the time of second breakdown the recombination in the base and hole injection into the emitter consume a negligible portion of the I_b generated at the MOSFET drain region and all I_b flows through γ_b. Equation (15) shows that the second breakdown voltage decreases with increasing I_{do} and/or increasing r_b. Equation (17) actually gives the boundary of the second breakdown region. The shape of this boundary, shown in some later figures, also resembles that of the second breakdown of the bipolar transistor.

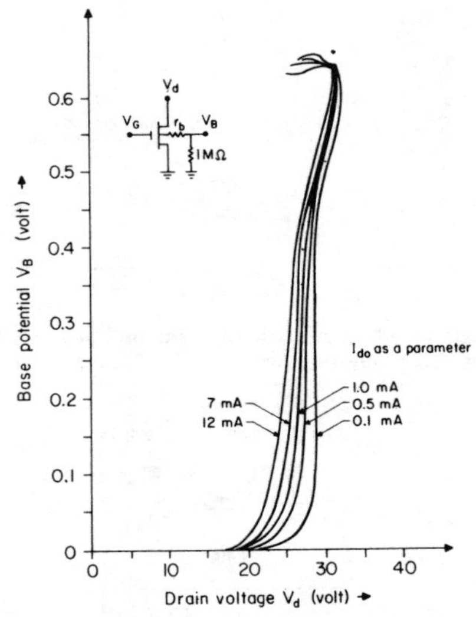

Fig. 3. Internal base potential versus drain voltage with I_{do} as a parameter. The base potential at the onset of snap-back, when V_d collapses, is close to 0.64 V. The inset shows the circuit used for measurements.

III. Experimental Results

To test the model presented above, measurements of the second breakdown voltage $V_{d,SB}$, γ, r_b, n, and BV are made on a small four-terminal n-VMOS test device having 400 μm of total channel width. Except for the smaller channel width, all dimensions are similar to typical commercial power MOSFET's. The source junction is about 2 μm deep and the p-base and n-epi-junction is about 2 μm deep. From the geometry and the p-base resistivity, r_b is estimated to be about 1 kΩ. The special feature of the test device is that a fourth terminal accesses the p-base independently as shown in Fig. 2(c), (d). All salient features of the model are checked against experiments. The boundary of the second breakdown region as well as its dependence on r_b is calculated with the use of device parameters and compared with the measured values. Only dc tests are conducted. The ambient is kept at room temperature during the tests.

The potential of the internal base V_b at the onset of the snap-back is measured with the circuit shown in the inset of Fig. 3. V_G is first adjusted to achieve a particular I_{do}. As V_d is increased, $V_B (\approx V_b)$ also increases due to increasing I_b. The trace ends at snap-back whereupon V_d suddenly collapses. The measurements are repeated for different I_{do} ranging from 0.25 to 30 μA per μm width of channel. The upper value is comparable to the highest current density in commercial power MOSFET's.

Over the whole range of I_{do} tested, snap-back occurs when V_B is approximately 0.64 V. This provides strong evidence that turning-on of the parasitic bipolar transistor is associated with second breakdown, at least in this case of floating sub-

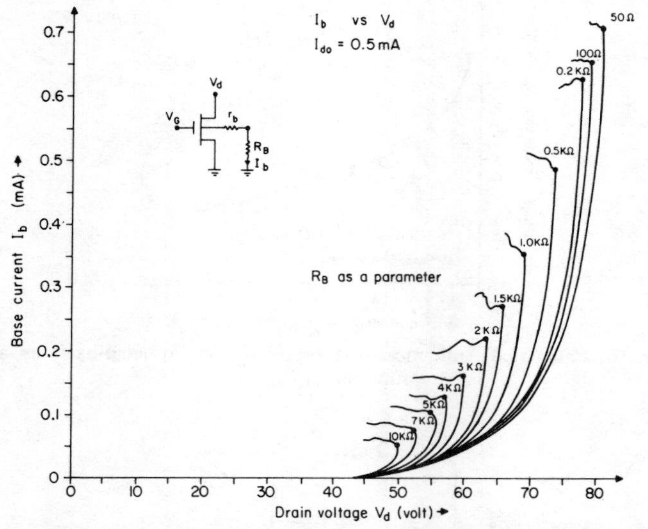

Fig. 4. Base current versus drain voltage with external resistance (R_B) as a parameter and at a constant I_{do}. The inset shows the circuit used for measurements.

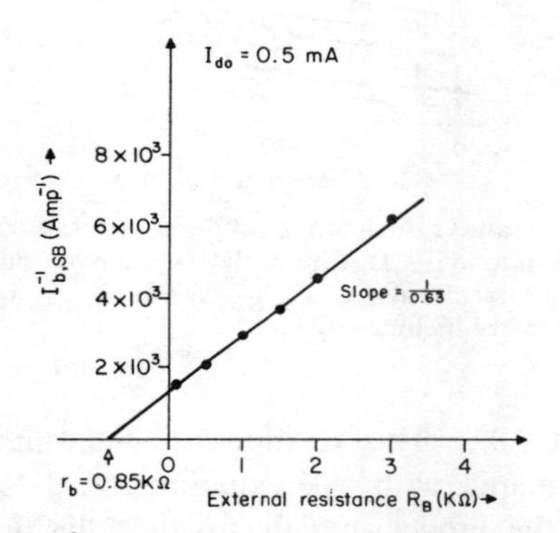

Fig. 5. A plot of $I^{-1}_{b,SB}$ versus the external resistance (R_B) at a constant I_{do}. The base resistance r_b is determined from the intercept of the extrapolated line at the axis of R_B.

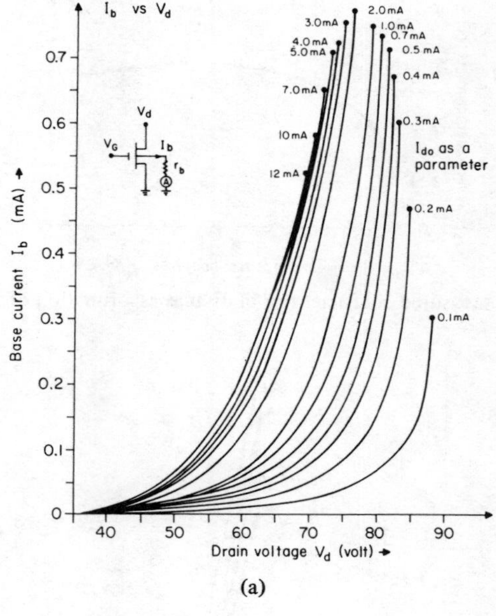

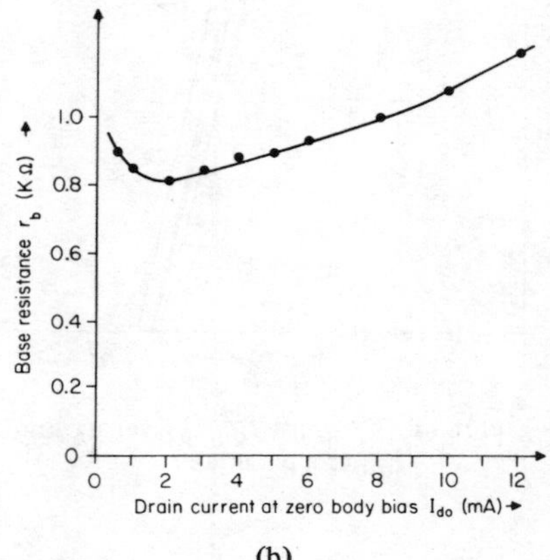

Fig. 6. (a) The base current versus drain voltage with I_{do} as a parameter and zero external resistance. The inset shows the circuit for measurement. (b) Base resistance versus I_{do} calculated from data shown in Figs. 3 and 6(a).

strate. The constancy of V_B at snap-back also suggests that the junction temperature did not vary much over the range of I_{do} in this study.

r_b can be determined in the following manner. At snap-back, $I_{b,SB} \cdot (R_B + r_b) = 0.64$. Therefore

$$I^{-1}_{b,SB} = \frac{1}{0.64}(R_B + r_b) \tag{19}$$

where R_B is the external resistance connected to the p-base. The base current I_b at snap-back is shown in Fig. 4 with different R_B and a constant I_{do}. A plot of $I^{-1}_{b,SB}$ versus R_B is shown in Fig. 5 where r_b is determined from the intercept of the extrapolated line at the R_B axis. This plot follows (19) quite well indicating that V_b is 0.63 V at the onset of snap-back regardless of R_B. In other words, snap-back is associated with the turn-on of the bipolar transistor even when $R_B = 0$. The value of 0.63 V agrees with Fig. 3 and r_b is determined to be 0.85 kΩ for this device. This value of r_b is consistent with the dimensions and the estimated resistivity of the p-layer in Fig. 1(c). r_b, however, is a variable parameter. I_b is expected to originate from a small area at or near the drain junction. r_b depends on the size of the area of origination which could affect the spreading resistance. As a result, r_b may be a function of V_G or I_{do}. Fig. 6(a) shows the I_b at snap-back for different I_{do} and without R_B. From Fig. 6(a) and Fig. 3, r_b as a function of I_{do} is calculated and shown in Fig. 6(b).

The large-signal body-bias coefficient is determined in accordance with (2) at the onset of second breakdown, i.e.,

$$\gamma = \frac{I_d(V_B = 0.6 \text{ V}) - I_{do}}{0.6}. \tag{20}$$

Currents are measured at low V_d but in the saturation region. γ as a function of I_{do} is shown in Fig. 7, where γ is small at low I_{do} but reaches a constant value when I_{do} is large. This behavior is consistent with the constant transconductance at high I_{do} observed in power MOSFET's as discussed in Appendix I.

Using (11) and data from Fig. 6(a), a log-log plot of $(1 + \gamma r_b + I_{do}/I_b)$ versus V_d yields information of n (from the slope) and BV (from the intercept with $\log[1 + \gamma r_b + I_{do}/I_b] = 0$). Fig. 8 shows some of such plots with I_{do} as a parameter. It is seen that n varies between 3.5 and 4.5 while a range of 2 to 6 is most often cited for the multiplication of

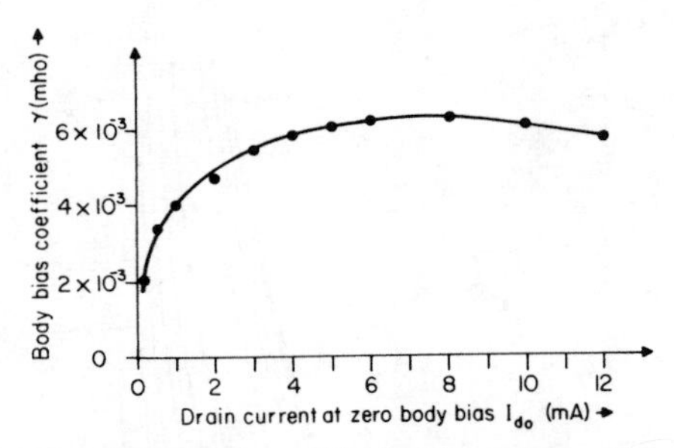

Fig. 7. Measured coefficient of body bias as a function of I_{do}.

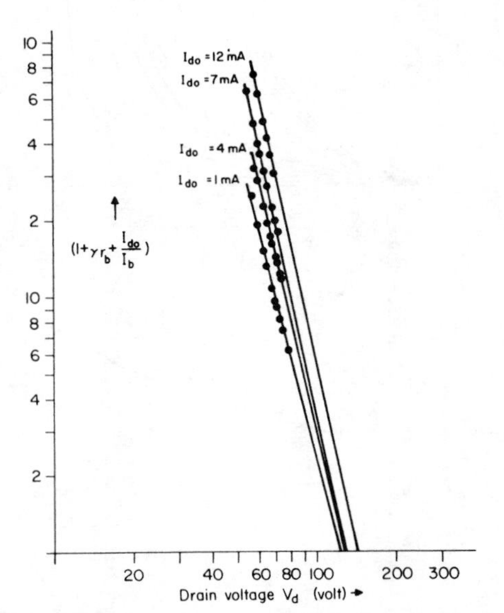

Fig. 8. A log–log plot of $(1 + \gamma r_b + I_{do}/I_b)$ versus drain voltage with I_{do} as a parameter.

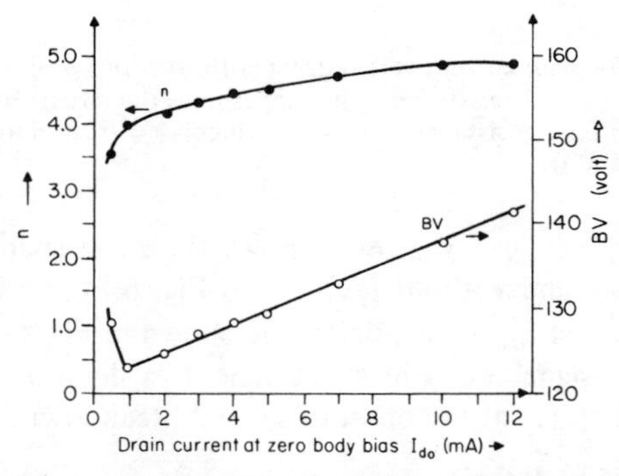

Fig. 9. The values of n and BV are found from Fig. 8 from the slopes and the intercepts with the line $\log(1 + \gamma r_b + I_{do}/I_b) = 0$.

electrons [7]. Both n and BV are weak functions of I_{do} as shown in Fig. 9. The reason is believed to be the finite electron density in the drain depletion region that is necessary for the flow of and approximately proportional to the drain current [8]. The electrons partially compensate the space charge due to ionized donors and hence raise the junction breakdown voltage. Other relationships, e.g., (12) and (13), may also be used to find n and BV from drain-current measurement if the "collector" current is negligible or can be determined independently and substracted from I_D in order to reveal I_d.

Equation (15) suggests that the second breakdown voltage is

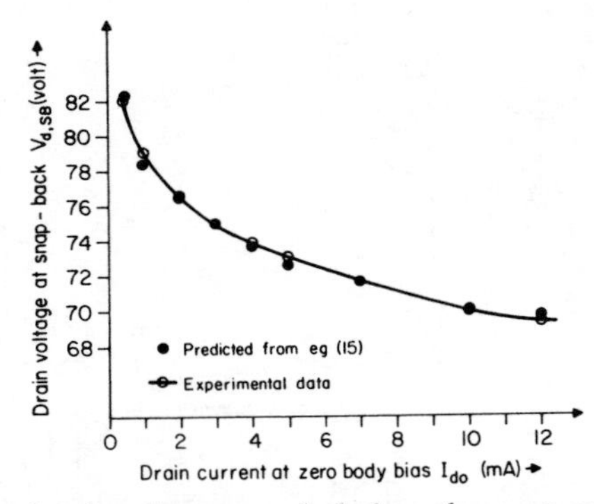

Fig. 10. Calculated and measured drain voltage at snap-back as a function of I_{do}.

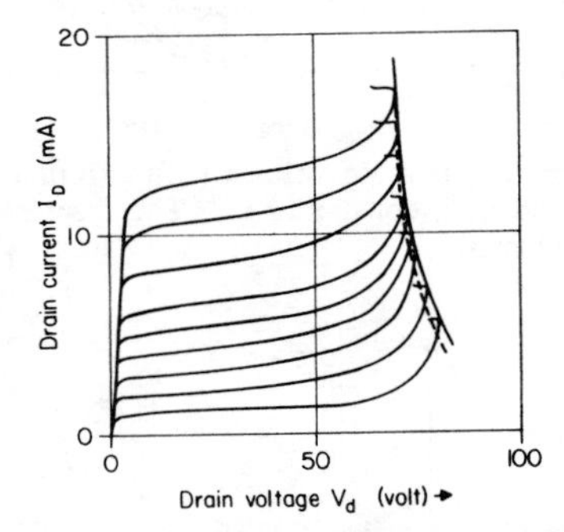

Fig. 11. Measured and calculated boundaries of safe operating area. The measured boundary is where the I-V curves snap back. The dotted curve plots calculated $V_{d,SB}$ versus $I_{d,SB}$ ignoring $I_{c,SB}$, while the solid curve includes $I_{c,SB}$.

a function of I_{do}. This relationship determined from the model is compared with the experimental data in Fig. 10. Interestingly the model actually predicts the boundary of a second breakdown or a safe operation region, i.e., the I_d-V_d contour at which second breakdown occurs (16). Such a calculated SOA boundary is compared with experiments in Fig. 11. The dotted curve plots the $V_{d,SB}$ versus $I_{d,SB}$ ignoring $I_{c,SB}$ in (16); and the solid curve includes $I_{c,SB}$. The "collector" current at snap-back is given in (18) and Appendix II. In our case, $\gamma_E G \approx 0.25$ is needed in order to fit the solid curve to the measured SOA boundary. $I_{c,SB}$, the difference between the dotted and the solid curves is about 1.5 mA. This value agrees with the I_c measured at $V_B = 0.64$ V and $V_G = 0$ V. If this component of I_D is ignored, the calculated SOA region would be somewhat more conservative, as shown in Fig. 11 by the dotted curve.

The multiplication gain at snap-back may be rather small. This and some other numbers of interest at $I_{do} = 5.0$ mA are

$$M_{SB} \approx \left(1 - \left(\frac{73}{128.3}\right)^{4.46-1}\right) = 1.085$$

$$I_{D,SB} - I_{do} = 5.6 \text{ mA}$$

$$I_{c,SB} = 1.4 \text{ mA}$$

$$I_{b,SB} = 0.71 \text{ mA}. \tag{21}$$

IV. Discussion

Let us discuss the model from the viewpoint of device optimization. n is a material constant. BV can be increased

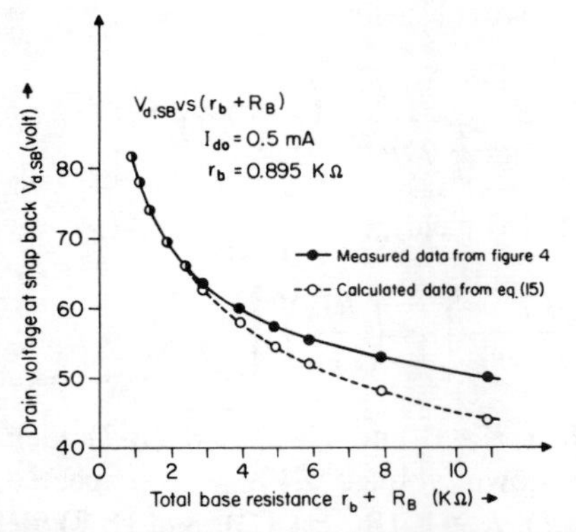

Fig. 12. Measured and calculated drain voltage at snap-back versus total base resistance $(r_b + R_B)$.

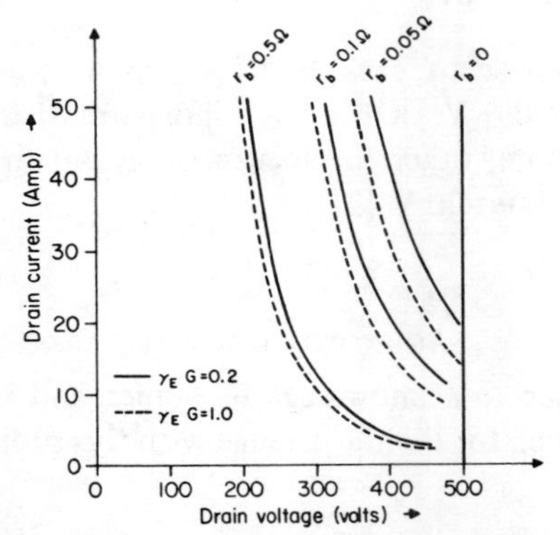

Fig. 13. Calculated boundaries of safe operating area for hypothetical power MOSFET's with BV = 500 V, n = 4.0, r_b = 0.05, 0.1 or 0.5 Ω, and $\gamma_E G$ = 0.2 or 1.0.

by increasing the thickness and resistivity of the n-epitaxial layers, but this is the usual voltage-R_{on} trade-off [9]. Some improvements may be gained by enhancing and exploiting the increase of BV with increasing I_{do} (Fig. 9 and Section III).

The behavior of γ in Fig. 7 is consistent with the analytic expression (A-4). γ increases with I_{do} at small I_{do} and approaches a constant value when I_{do} is large. The p-base doping N_A and channel length L in (A-4) are most likely determined by other more important considerations. The product γr_b, which is the form γ appears in all model equations, is not sensitive to the channel width W, since r_b is inversely proportional to it. Besides, γ does not enter the most important equation at all—the boundary of second breakdown (16) or (17).

$I_{c,SB}$ may be increased by reducing $\gamma_E G$ such as when $\gamma_E = 0$, for example, $I_{D,SB}$ rises to infinity or when the snap-back phenomenon disappears (see (15) and (16)).

The one parameter that may have the most room for improvement is r_b, which depends on the geometry and the sheet resistance of the p-base.

Fig. 4 also shows how r_b (or $r_b + R_B$) affects the second breakdown voltage. The relationship is plotted in Fig. 12 and is compared with (15).

Only dc measurements taken on small V-groove MOSFET's are presented here, but similar behavior has been observed on three-terminal high current/voltage devices tested under pulse conditions. It is believed that the model presented here applies to power MOSFET's of all current and voltage ratings as well as different designs (VMOS and DMOS). Fig. 13 shows the calculated boundaries of the second breakdown of a hypothetical power MOSFET with BV = 500 V and n = 4.0. Here, the effects of r_b are clearly demonstrated.

A standard power MOSFET has three terminals. There is no access to the substrate and it is impossible to measure I_b, I_c, γ, and r_b, however, it is still possible to characterize them in light of the model presented here. One possible set of procedures is outlined below. First, use (A-4) to estimate γ (at high gate voltage, or I_{do}). Then, measure I_d versus V_d at a given I_{do}, where (14) yields r_b. Equation (13) can be used to find n and BV after r_b is determined. There is no way to measure I_c, nevertheless, I_c can be neglected if the device operates at high I_{do} or is estimated with (18).

The measurements in this study were performed at room temperature. Both BV and n are expected to increase with temperature. Both BV and r_b are expected to increase with down voltage, however, larger r_b tends to reduce the drain breakdown voltage. In any event, the second breakdown voltage and current are only weakly dependent on temperature from our high-temperature (25–120°C) measurements.

p-channel power MOSFET's are not expected to exhibit significant snap-back behavior from the experience with surface MOSFET's [4], [10]. The reasons are: 1) lower r_b due to higher electron mobility; and 2) lower hole impact-ionization coefficient [11].

Finally during the turn-off transient, large dV_d/dt can induce a sufficient I_b through the drain-base capacitance (omitted in Fig. 2(b)) to satisfy $I_b r_b \geq 0.6$ V and cause breakdown [12]. Since this phenomenon occurs only during the turn-off transient, it may be thought of as a counterpart to the bipolar transistors' reverse second breakdown.

Appendix I
Analytical Expression for γ

For an n-channel MOSFET, the saturation drain current is usually expressed as

$$I_d = \frac{W}{2L} \mu C'_{ox} [V_G - V_T]^2 \tag{A-1}$$

where

$$V_T = V_{T0} + \frac{\sqrt{2\epsilon_s q N_A}}{C'_{ox}} [\sqrt{2|\phi_p| - V_B} - \sqrt{2|\phi_p|}] \tag{A-2}$$

and all parameters have their usual meanings. Standard power MOSFET's are short-channel devices; and (A-1) is only good in a small V_G (or I_d) regime. At higher V_G (or I_d), I_d for a short-channel device is known to be linearly proportional to $V_G - V_T$ [13]

$$I_d = \frac{W}{L} \cdot C'_{ox} \cdot v_s \cdot (V_{GS} - V_T) \tag{A-3}$$

where v_s is the saturation drift velocity of charge carriers in the inversion layer.

Analytical expression of γ defined by (2) or (20) may be derived from (A-1), (A-2), and (A-3)

$$\gamma \equiv \int_0^{0.6} \left(\frac{\partial I_d}{\partial V_B}\right) dV_B$$

$$\approx \begin{cases} 2(\sqrt{2|\phi_p|} - \sqrt{2|\phi_p| - 0.6}) \cdot \sqrt{\dfrac{W}{2L}\mu I_d \cdot \dfrac{2\epsilon_s q N_A}{C'_{ox}}} & \text{(small } I_d) \\ (\sqrt{2|\phi_p|} - \sqrt{2|\phi_p| - 0.6}) \cdot \dfrac{W}{L} \cdot v_s \cdot \sqrt{2\epsilon_s q N_A} & \text{(large } I_d). \end{cases}$$

Once W and L are fixed, γ becomes nearly fixed, and dependent only on the substrate doping density.

Appendix II
V_b and I_c at the Breakdown of the Bipolar Transistor

If avalanche multiplication is considered, the relation between I_e and I_c is

$$I_c = (\gamma_E \alpha_T I_e + I_{BCo}) M' \tag{B-1}$$

where γ_E is the emitter efficiency, α_T is the base transport efficiency, and

$$M' = \frac{1}{1 - (V/BV')^n} \tag{B-2}$$

where BV' is the breakdown voltage of the planar portion of the base-collector junction [7] and may be higher than BV of the MOSFET. n may also be different than that in (10). At the same V, using (10) and (B-2)

$$M' - 1 \approx \left(\frac{BV}{BV'}\right)^n (M - 1) \equiv G(M - 1) \tag{B-3}$$

$$I_e = \frac{I_{ep}}{1 - \gamma_E} = \frac{1}{1 - \gamma_E} I_c \cdot \left(1 - \frac{1}{M'}\right) K. \tag{B-4}$$

I_{ep} is the hole current injected into the emitter, and

$$K \equiv \frac{I_{ep}}{I_P} \approx \frac{qI_e/kT}{qI_e/kT + 1/r_b}. \tag{B-5}$$

I_P is the total hole current generated at the collector junction by impact ionization and is divided between the r_b path and the emitter diode whose admittance is qI_e/kT. A combination of (B-1) and (B-4) yields

$$I_c = \frac{M' I_{BCo}}{1 - (\gamma_E \alpha_T K / 1 - \gamma_E)(M' - 1)}. \tag{B-6}$$

Clearly I_c approaches infinity when

$$M' - 1 = \frac{1 - \gamma_E}{\gamma_E \alpha_T K}. \tag{B-7}$$

From (B-7), (8), and (B-3), K at snap-back is approximated to be

$$K \approx \frac{1 - \gamma_E}{\gamma_E \alpha_T G} \left(\frac{r_b I_{do}}{0.6} + r_b \gamma\right). \tag{B-8}$$

(B-8) and (B-4) would yield

$$I_{c,SB} \approx \frac{I_{ep,SB}}{1 - \gamma_E} \alpha_T \frac{kT}{q\gamma_E G} \cdot \left(\frac{I_{do}}{0.6} + \gamma\right) \tag{B-9}$$

and (B-9) and (15) yield

$$I_{c,SB} \approx \frac{kT}{q\gamma_E G r_b} \left[1 + \left(\frac{BV}{V_{SB}}\right)^n\right]. \tag{B-10}$$

If $r_b = 0$, then $K = 0$ and (B-2) and (B-7) yield the open-emitter breakdown voltage BV_{CBo} as expected. At large I_e ($V_b > 0.6$ V), $K = 1$, (B-5), (B-2), and (B-7) yield the open-base breakdown voltage

$$BV_{CEo} = BV' / \sqrt[n]{\beta}.$$

This is why V_d snaps back to BV_{CEo} as soon as V_b reaches approximately 0.6 V. Once the bipolar transistor enters the high-current state, other mechanisms may set in and further reduce V_d to about 20 V [2].

Acknowledgment

We would like to acknowledge R. Mullen and Dr. K. Lisiak of Siliconix, Inc., for their assistance with this project.

References

[1] H. A. Schafft, "Second breakdown—A comprehensive review," *Proc. IEEE*, vol. 55, no. 8, p. 1272, Aug. 1967.

[2] P. L. Hower and V.G.K. Reddi, "Avalanche injection and second breakdown in transistors," *IEEE Trans. Electron Devices*, vol. ED-17, no. 4, p. 320, Apr. 1970.

[3] B. A. Beatty, S. Krishna, and M. S. Adler, "Second breakdown in power transistors due to avalanche injection," *IEEE Trans. Electron Devices*, vol. ED-23, no. 8, p. 851, Aug. 1976.

[4] J. Yoshida, T. Okabe, M. Katsueda, S. Ochi, and M. Nagata, "Thermal stability and secondary breakdown in planar power MOSFET's," *IEEE Trans. Electron Devices*, vol. ED-27, no. 2, p. 395, Feb. 1980.

[5] R. W. Coen, D. W. Tsang, and K. P. Lisiak, "A high-performance planar power MOSFET," *IEEE Trans. Electron Devices*, vol. ED-27, no. 2, p. 340, Feb. 1980.

[6] E. Sun, J. Moll, J. Berger, and B. Alders, "Breakdown mechanism in short-channel MOS transistors," in *IEDM Tech. Dig.*, pp. 478–482, 1978.

[7] R. S. Muller and T. I. Kamins, *Device Electronics for Integrated Circuits*. New York: Wiley, 1977, pp. 128–133.

[8] A. W. Wieder, C. Werner, and J. Tihanyi, "2-D analysis of the negative resistance region of vertical power MOS-transistors," in *IEDM Tech. Dig.*, pp. 95–99, 1980.

[9] C. Hu, "A parametric study of power MOSFETs," in *Rec. of IEEE Power Electronics Specialists Conf.*, pp. 385–395, June 1979.

[10] R. D. Josephy, "MOS transistors for power amplification in the HF band," *Philips Tech. Review*, vol. 31, no. 7/8/9, p. 251, 1970.

[11] T. Toyabe, K. Yamaguchi, S. Asai, and M. Mock, "A numerical model of avalanche breakdown in MOSFET's," *IEEE Trans. Electron Devices*, vol. ED-25, p. 825, 1978.

[12] R. Severns, "The power MOSFET, a breakthrough in power device technology," Intersil Inc., Application Bull. A033, 1980.

[13] R. H. Dennard, F. H. Gaensslen, E. J. Walker, and P. W. Cook, "1-μm MOSFET VLSI technology: Part II—device designs and characteristics for high performance logic applications," *IEEE Trans. Electron Devices*, vol. ED-26, no. 4, pp. 325–333, 1979.

UNDERSTANDING POWER MOSFET SWITCHING PERFORMANCE

S. Clemente, B. R. Pelly
International Rectifier Corporation, El Segundo, California
A. Isidori
University of Rome, Italy

Abstract - A simple analytical technique for predicting the switching performance of the power MOSFET is presented.

Closed-form solutions for the gate voltage, drain current, and drain voltage during the switching interval, in terms of each of the relevant device and circuit parameters, are derived.

A specific design example is considered, in which the effects are demonstrated of the drive circuit resistance, drain circuit inductance, and drive voltage, on the switching time and switching energy.

I. INTRODUCTION

The power MOSFET is an almost ideal switch, which is characterized by very high gain and extremely fast switching characteristics. Whilst users often ignore the intricacies of the switching operation, on the assumption that this is not critical to the overall design, the fact is that a clear understanding of the factors that effect switching can have a profound effect upon the system performance, particularly in high frequency circuits, and is therefore of vital interest to the user who needs to optimize his design.

Another reason why many users have a rather incomplete understanding of the MOSFET's switching operation is that the device is still relatively new, and MOSFET circuit design know-how has not yet matured. Users also tend to relate to their experience with bipolar transistors. The switching operation of bipolars is very difficult to analyze, and hence an empirical "try it and see" approach has generally held sway over more rigorous analytical techniques.

One of the major "incidental" benefits of the MOSFET - in addition to its very real operating advantages - is that it lends itself rather well to analytical modeling; its operation can therefore be predicted rather easily at the design stage.

The primary objective of this paper is to show how, starting with a simple model of the MOSFET, and using logical reasoning, the principles that govern the MOSFET's operation in a switching circuit can be readily predicted, and approximate mathematical relationships that describe these waveforms can be readily derived. Emphasis will be placed up on an understanding of basic principles.

II. THE MOSFET MODEL

The electrical model for the MOSFET is shown in Figure 1. The self capacitances are actually nonlinear functions of the applied voltage - also, to some extent, - of the drain current. For purposes of analysis, however, these capacitances will be assumed to have fixed values; this does not detract from our basic objective, which is to understand fundamental principles.

This simple model of the MOSFET is assumed to have a linear transfer characteristic, with slope g_{fs} and gate threshold voltage V_T. The external drain current is assumed to be instantaneously responsive to the gate voltage, for operation in the active region.

Under transient switching conditions, charging and discharging currents flow through the various self-capacitive elements. The paths for components of these currents is through the drain-to-source terminals. The presence of these internal capacitive currents is assumed not to effect the transfer characteristic between the gate voltage and the external drain current.

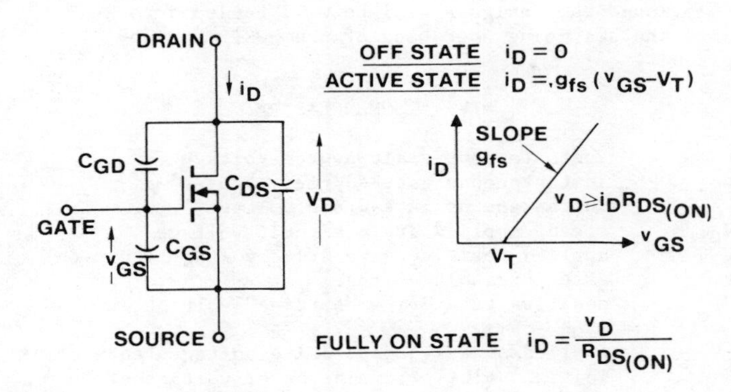

Figure 1: Electrical Model For MOSFET

The presence of C_{DS} will also generally be ignored for operation in the active region. This is valid because the effect of the gate-to-drain capacitance C_{GD} - providing, as it does, a coupling path from the drain circuit to the relatively sensitive gate circuit - generally "swamps" the effect of C_{DS}.

III. THE CIRCUIT MODEL

The circuit model considered in this paper is shown in Figure 2. The clamped load is assumed to have sufficient inductance that the current flowing in it has a constant value I_0 throughout the switching interval. The inductance L_ℓ represents "unclamped" stray circuit inductance.

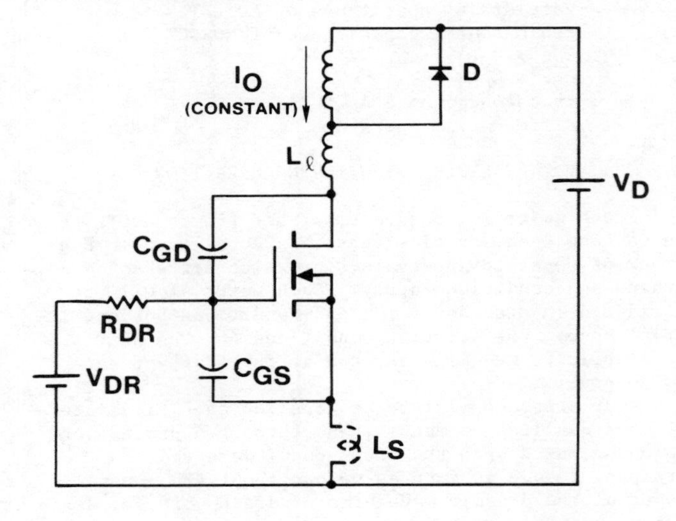

Figure 2: General Circuit Model

The effect of the common source inductance L_S, shown dashed in Figure 2, will generally be neglected. This is not because it is necessarily negligible, but because to include it in a general analysis complicates the issue, making clarity of presentation and a grasp of fundamental principles more difficult. We prefer instead to consider the modifying effect of this inductance once the basic analysis is complete.

A number of switching circuits can be resolved

Reprinted with permission from *Solid-State Electron.*, vol. 26, no. 12, pp. 1133-1141, Dec. 1983.

into the equivalent circuit shown in Figure 2, or variants thereof, and in this sense the analysis is fairly general. The main point, however, is that the chosen circuit serves as a vehicle for obtaining an understanding of basic principles; once this has been accomplished the designer will be well equipped to deal with the switching operating of the MOSFET in any circuit.

IV. NOMENCLATURE

v_D	instantaneous drain-source voltage
v_{GS}	instantaneous gate-source voltage
v_{GD}	instantaneous gate-drain voltage
V_D	steady applied drain circuit voltage
V_{DR}	applied positive gate drive voltage (turn-on)
V_T	gate threshold voltage
V_F	positive gate drive "forcing" voltage, $(V_{DR} - V_T)$
$-V_2$	applied negative gate drive voltage (turn-off)
V_D*	initial value of drain-source voltage at start of interval
$V_{GS}*$	initial value of gate-source voltage at start of interval
V_{CLAMP}	drain-source clamping voltage
i_D	instantaneous current flowing into drain terminal
i_{GS}	instantaneous current in C_{GS}
i_{GD}	instantaneous current in C_{GD}
I_O	steady current in clamped inductive load
I_D*	initial value of current flowing into drain terminal at start of interval
R_{DR}	gate drive circuit resistance
$R_{DS(ON)}$	on-state resistance of MOSFET
R_ℓ	stray drain circuit resistance
L_ℓ	stray drain circuit inductance
L_S	inductance in series with source that is common to gate circuit
C_{GS}	gate-source capacitance of MOSFET
C_{GD}	gate-drain capacitance of MOSFET
C_{DS}	drain-source capacitance of MOSFET
C_G	$C_{GS} + C_{GD}$
C_D	$C_{DS} + C_{GD}$
g_{fs}	transconductance of MOSFET
p	differential operator

V. ANALYSIS OF SWITCHING OPERATION

Each switching sequence, either from the OFF to the ON condition, or vice versa, is subdivided into a number of separate intervals, for which different constrains and conditions apply. Each interval will be considered in sequence. The end-conditions for one interval become the starting conditions for the next. For simplicity we will take t=0 at the start of each new interval.

Our approach will be to consider each time interval in a qualitative manner, and through a process of reasoning based upon the known conditions and constraints, deduce as much as we can about the general shapes of the dynamic waveforms of drain voltage, drain current and gate voltage.

For certain time intervals this qualitative reasoning leads directly to the parametric analytic solution for that interval; for other time intervals, however, the analytic solutions are not so quickly obtained, except for parametric extremes at each end of the possible spectrum of external circuit conditions; a wide middle range of conditions remains for which derivation of the parametric solutions is rather too lengthy to be presented in its entirety, and in these cases we will simply state the final solutions.

A. TURN-ON

Turn-On Delay - Interval 1

The circuit model for this interval is shown in Figure 3, and operating waveforms are shown in Figure 4. The applied drive voltage is assumed to rise instantaneously to its full value; however the voltage actually appearing between the gate and source terminals, which directly controls the external drain current, rises at a finite rate determined by the gate-to-source and drain-to-source self-capacitances. No drain current flows so long as the gate voltage is less than the threshold voltage, V_T. The end of the turn-on delay period is defined as the point at which the gate-to-source voltage becomes equal to the threshold voltage.

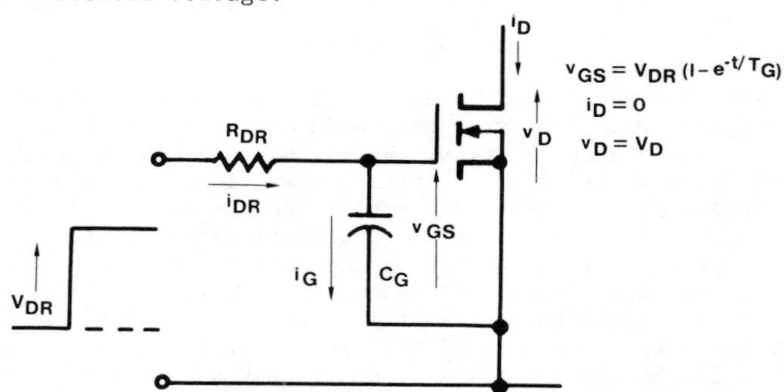

Figure 3: Circuit Model For Turn-On Delay Interval

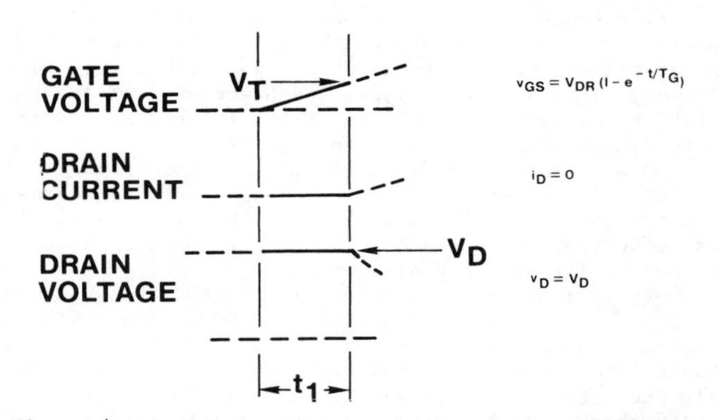

Figure 4: Waveforms For Turn-On Delay Interval (t_1)

The analytic solution for the turn-on delay is almost trivial. Since no drain current flows, the drain voltage remains at V_D. Both the "drain" terminal of C_{GD} and the "source" terminal of C_{GS} sensibly do not change their potentials. The drive source voltage, V_{DR}, 'sees' the parallel combination of $C_{GD} + C_{GS} = C_G$, through the series resistor R_{DR}. The gate-to-source voltage v_{GS} follows a classical exponential:

$$v_{GS} = V_{DR} (1 - e^{-t/T_G}) \tag{1}$$

where $T_G = R_{DR} C_G$ (2)

Turn-On Interval 2

The general circuit model for this interval is shown in Figure 5. The drain current now rises as the drain voltage falls. Which of these events is completed first depends upon the external circuit parameters. When one of these events is completed (or both simultaneously) the interval ends.

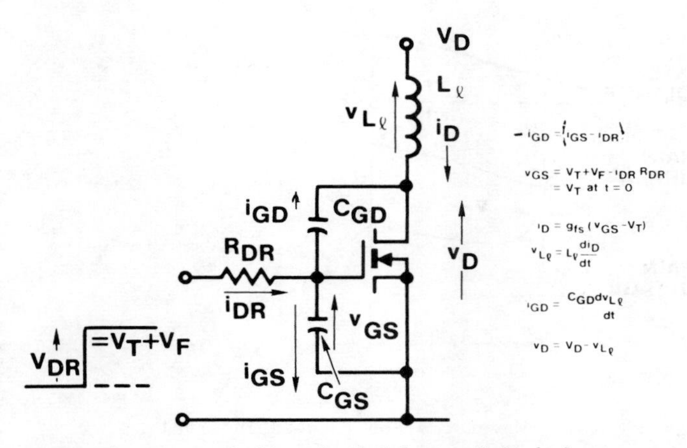

Figure 5: General Circuit Model For Switch-On Interval 2

Since the drain current i_D is less than the current I_O throughout this period, the difference between I_O and i_D must continue to circulate in the freewheeling rectifier D, forcing this diode to stay in conduction. This keeps the potential at the "top" of L_ℓ virtually constant at V_D.

As the gate-to-source voltage rises above the threshold level, the drain current starts to increase, since drain current is proportional to gate voltage. The drain voltage also starts to fall, because the increasing drain current induces a voltage across L_ℓ. As the drain voltage falls, current i_{GD} flows out of the "Miller" capacitance C_{GD}; this current is drawn from the drive source, and deprives the gate-source capacitance C_{GS} of a portion of the charging current it would otherwise have received. This in turn reduces the rate of change of gate voltage, and hence also of drain current.

A dynamically "intertwined" situation obviously exists, by virtue of the "negative feedback" effect that couples the drain circuit to the gate circuit via the "Miller" capacitance C_{GD}. The "strength" of this feedback depends upon the ratio of the external circuit parameters L_ℓ to R_{DR}, as we will now see.

Large L_ℓ means large impedance to the rate of change of drain current, whilst small R_{DR} means fast gate circuit response, and hence potentially fast rate of change of drain current. With a high ratio of L_ℓ to R_{DR} the reactance of the drain circuit will therefore be high, the voltage drop across L_ℓ will be high, the "Miller" effect will predominate, and the rate of change of drain current will be unable to match the applied gate circuit stimulus. High L_ℓ/R_{DR} therefore means that the switching speed is severely limited by the constraints of the drain circuit; the drive circuit is "too fast" for the drain circuit.

Small L_ℓ/R_{DR} ratio means just the opposite; the potential rate of change of drain current is now much faster than the drive circuit actually allows. The voltage drop across L_ℓ is small, the "Miller" effect is small, and the gate circuit largely controls the switching time, virtually unimpeded by the drain circuit. Both of these extreme conditions are rather easy to analyze.

For intermediate L_ℓ/R_{DR}, the drain circuit and gate circuit responses can be envisioned as being reasonably "compatible" with one another. From a purist's viewpoint, compatibility of the gate and drain circuit responses might be considered to be the "correct" design point, because the gate circuit is neither too fast nor too slow for the drain circuit.

Small L_ℓ/R_{DR}

We will start the analysis by considering the situation when L_ℓ/R_{DR} is small. The circuit model is shown in Figure 6, and switching waveforms are shown in Figure 7. Since there is very little voltage developed across L_ℓ, the drain voltage v_D stays virtually at the circuit voltage V_D, until the drain current has risen to its full load value, I_O.

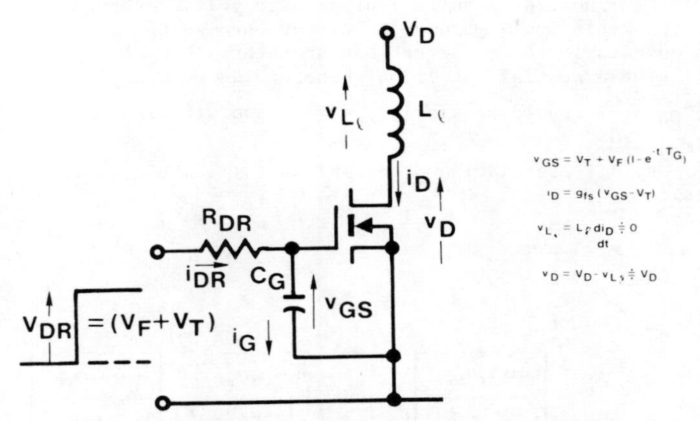

Figure 6: Circuit Model For Switch-On Interval 2 Small L_ℓ/R_{DR}

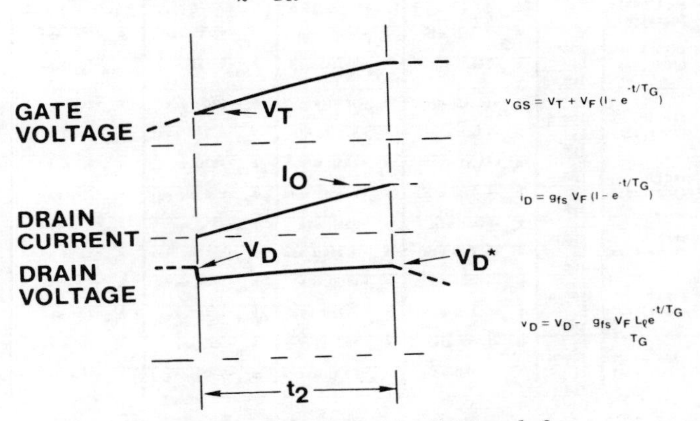

Figure 7: Waveforms For Turn-On Interval 2 Small L_ℓ/R_{DR}

$$\frac{L_\ell}{R_{DR}} < \frac{C_{GS}^2}{10 C_{GD}\, g_{fs}}$$

Because the rate of change of drain voltage is small (almost zero), virtually no current flows through C_{GD}, and the drive circuit continues to see the simple parallel combination of C_{GD} and C_{GS} (as it did during the turn-on delay period). The gate-to-source voltage, v_{GS} therefore continues to rise exponentially:

$$v_{GS} = V_F\,(1 - e^{-t/T_G}) \tag{3}$$

The drain current rises in sympathy with the gate voltage:

$$i_D = g_{fs}\, V_F\,(1 - e^{-t/T_G}) \tag{4}$$

The drain voltage is equal to the circuit voltage V_D, less the small (almost negligible) voltage drop across L_ℓ:

$$v_D = V_D - \frac{g_{fs} V_F L_\ell e^{-t/T_G}}{T_G} \tag{5}$$

The period ends when $i_D = I_O$.

It remains to quantify how small the ratio L_ℓ/R_{DR} must be for equations (3) through (5) to remain valid. The essential condition is that the rise of drain current must, for all practical purposes, be exclusively under the influence of the applied drive voltage. This means that whatever voltage change occurs across L_ℓ should not be noticed in the gate circuit. The current through C_{GD} will therefore be small by comparison with the current through C_{GS} (C_{GS} is typically about 10 X C_{GD}; however a sufficiently large voltage change at the drain would produce a current through C_{GD} which is comparable to or larger than that through C_{GS})

The essential condition therefore is that i_{GD} ($\doteq C_{GD}\frac{dv_D}{dt}$) should be small by comparison with i_{GS} ($\doteq C_{GS}\frac{dv_{GS}}{dt}$).

By differentiation of equations (3) and (5), this yields:

$$\frac{L_\ell}{R_{DR}} << \frac{C_{GS}^2}{g_{fs}\,C_{GD}} \qquad (6)$$

		Small L_ℓ/R_{DR}	Intermediate L_ℓ/R_{DR}		Large L_ℓ/R_{DR}
		$\frac{L_\ell}{R_{DR}} < \frac{C_{GS}^2}{10C_{GD}g_{fs}}$	$\frac{L_\ell}{R_{DR}} < \frac{C_{GS}^2}{4C_{GD}g_{fs}}$	$\frac{L_\ell}{R_{DR}} > \frac{C_{GS}^2}{4C_{GD}g_{fs}}$	$\frac{L_\ell}{R_{DR}} > \frac{10C_{GS}^2}{C_{GD}g_{fs}}$
Applicable Equations		3-5, 37-39	8-10,40-42	14-16,43-45	24-26-46-48
IRF510 (100V,3A)	L_ℓ 100nH	R_{DR}=2.4kΩ min	R_{DR}=960Ω min	R_{DR}=960Ω max	R_{DR}=40Ω max
	L_ℓ 1μH	R_{DR}=24kΩ min	R_{DR}=9.6kΩ min	R_{DR}=9.6kΩ max	R_{DR}=400Ω max
IRF130 (100V,12A)	L_ℓ 100nH	R_{DR}=1.3kΩ min	R_{DR}=520Ω min	R_{DR}=520Ω max	R_{DR}=20Ω max
	L_ℓ 1μH	R_{DR}=13kΩ min	R_{DR}=5.2kΩ min	R_{DR}=5.2kΩ max	R_{DR}=200Ω max
IRF150 (100V,28A)	L_ℓ 100nH	R_{DR}=410Ω min	R_{DR}=165Ω min	R_{DR}=165Ω max	R_{DR}=6Ω max
	L_ℓ 1μH	R_{DR}=4.1kΩ min	R_{DR}=1.65kΩ min	R_{DR}=1.65kΩ max	R_{DR}=60Ω max
IRF710 (400V,1.2A)	L_ℓ 100nH	R_{DR}=820Ω min	R_{DR}=325Ω min	R_{DR}=325Ω max	R_{DR}=13Ω max
	L_ℓ 1μH	R_{DR}=8.2kΩ min	R_{DR}=3.25kΩ min	R_{DR}=3.25kΩ max	R_{DR}=130Ω max
IRF330 (400V,4A)	L_ℓ 100nH	R_{DR}=420Ω min	R_{DR}=170Ω min	R_{DR}=170Ω max	R_{DR}=7Ω max
	L_ℓ 1μH	R_{DR}=4.2kΩ min	R_{DR}=1.7kΩ min	R_{DR}=1.7kΩ max	R_{DR}=70Ω max
IRF350 (400V,11A)	L_ℓ 100nH	R_{DR}=120Ω min	R_{DR}=50Ω min	R_{DR}=50Ω max	R_{DR}=2Ω max
	L_ℓ 1μH	R_{DR}=1.2kΩ min	R_{DR}=500Ω min	R_{DR}=500Ω max	R_{DR}=20Ω max

Table 1: Limiting values of R_{DR} that define which equations(turn-on interval 2, and turn-off interval 3) are applicable, for various HEXFET's.

Table 1 puts the above criterion into perspective, and shows typical values of L_ℓ and the corresponding "minimum" values of R_{DR}, for various HEXFETs. Clearly the values of R_{DR} needed to satisfy this condition are very high relative to most normal application requirements. This condition will not therefore be frequently met in practice; its consideration here is useful, however, because it helps to introduce the overall problem.

Intermediate L_ℓ/R_{DR}

We will now consider the situation when the ratio of L_ℓ/R_{DR} is not small, but has some intermediate value; the voltage drop across L_ℓ due to the increasing drain current becomes significant, and the current through C_{GD} cannot be neglected. The general circuit model of Figure 5 applies, and typical switching waves are illustrated in Figures 8 (a) and (b).

The mathematical analysis is a little too lengthy to keep touch with physical realities. We will therefore confine ourselves to a simple statement of the results.

There are two possible sets of solutions, depending upon whether or not the system is critically damped. If overdamped, then:

$$\frac{L_\ell}{R_{DR}} < \frac{C_{GS}^2}{4C_{GD}g_{fs}} \qquad (7)$$

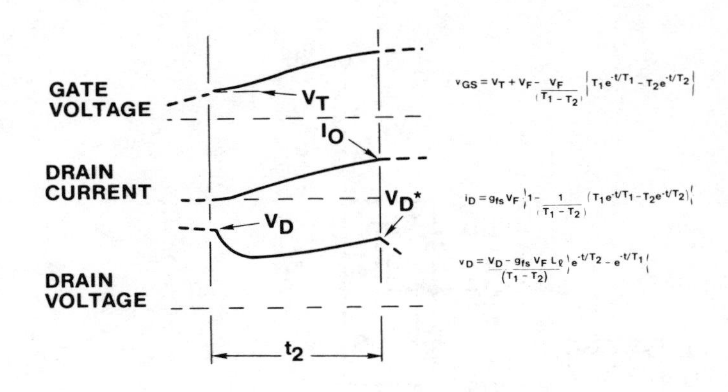

Figure 8 (a): Waveforms For Turn-On Interval 2, Intermediate L_ℓ/R_{DR}

$$\frac{C_{GS}^2}{10C_{GD}g_{fs}} < \frac{L_\ell}{R_{DR}} < \frac{C_{GS}^2}{4C_{GD}g_{fs}}$$

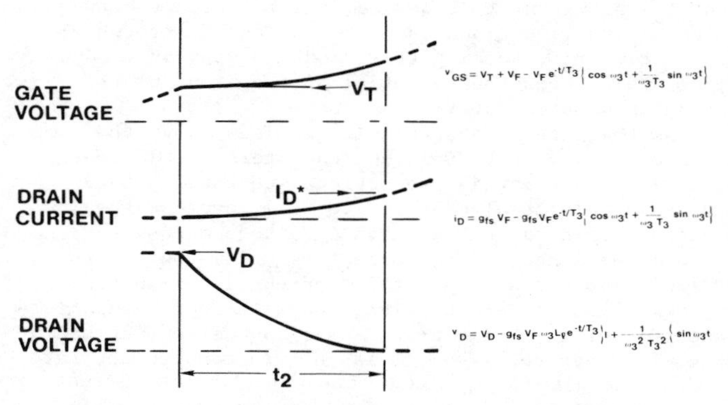

Figure 8 (b): Waveforms For Turn-On Interval 2, Intermediate L_ℓ/R_{DR}

$$\frac{C_{GS}^2}{4C_{GD}g_{fs}} < \frac{L_\ell}{R_{DR}} < 10\,\frac{C_{GS}^2}{C_{GD}g_{fs}}$$

Note the similarity of condition (7) to (6). Table 1 also shows typical values of L_ℓ and corresponding minimum values of R_{DR} that satisfy equation (7). This condition is certainly more likely to be encountered than condition (6), though once again it is generally not representative of most typical practical situations.

The gate voltage, v_{GS}, the drain current, i_D, and the drain voltage v_D, are:

$$v_{GS} = V_T + V_F - \frac{V_F}{(T_1 - T_2)}\left\{T_1\, e^{-t/T_1} - T_2 e^{-t/T_2}\right\} \qquad (8)$$

$$i_D = g_{fs}V_F\left\{1 - \frac{1}{(T_1 - T_2)}\right\}\left\{T_1 e^{-t/T_1} - T_2 e^{-t/T_2}\right\} \qquad (9)$$

$$v_D = V_D - \frac{g_{fs}V_F L_\ell}{(T_1 - T_2)}\left\{e^{-t/T_2} - e^{-t/T_1}\right\} \qquad (10)$$

where

$$T_1 = \frac{2L_\ell C_{GD}R_{DR}g_{fs}}{R_{DR}C_{GS} + \sqrt{R_{DR}^2 C_{GS}^2 - 4L_\ell C_{GD}R_{DR}g_{fs}}} \qquad (11)$$

$$T_2 = \frac{2LC_{GD}R_{DR}g_{fs}}{R_{DR}C_{GS} - \sqrt{R_{DR}{}^2C_{GS}{}^2 - 4L_\ell C_{GD}R_{DR}g_{fs}}} \quad (12)$$

The end of the time interval will generally be marked by the drain voltage having fallen all the way to $i_D \times R_{DS(ON)}$, with the drain current not having completed its rise.

For an "underdamped" system, the converse of (7) applies:

$$\frac{L_\ell}{R_{DR}} > \frac{C_{GS}{}^2}{4C_{GD}g_{fs}} \quad (13)$$

The minimum values of R_{DR} shown in Table 1 that satisfy equation (7) now become the maximum values that satisfy equation (13). Generally, most practical situations will be covered by equation (13).

The gate voltage v_{GS}, the drain current i_D, and the drain voltage v_D, are:

$$v_{GS} = (V_T+V_F) - V_F e^{-t/T_3}\left\{\cos \omega_3 t + \frac{\sin \omega_3 t}{\omega_3 T_3}\right\} \quad (14)$$

$$i_D = g_{fs}V_F - g_{fs}V_F e^{-t/T_3}\left\{\cos \omega_3 t + \frac{\sin \omega_3 t}{\omega_3 T_3}\right\} \quad (15)$$

$$v_D = V_D - g_{fs}V_F\omega_3 L_\ell e^{-t/T_3}\left\{1 + \frac{1}{\omega_3^2 T_3^2}\right\} \sin \omega_3 t \quad (16)$$

$$\text{where } T_3 = \frac{2L_\ell C_{GD}g_{fs}}{C_{GS}} \quad (17)$$

$$\omega_3 = \frac{\sqrt{4L_\ell C_{GD}R_{DR}g_{fs} - R_{DR}{}^2C_{GS}{}^2}}{2L_\ell C_{GD}R_{DR}g_{fs}} \quad (18)$$

The end of the time interval will be marked either by the drain circuit i_D reaching I_O, or the drain voltage v_D collapsing to $i_D \times R_{DS(ON)}$, whichever occurs first.

Large L_ℓ/R_{DR}

Now consider the situation when L_ℓ/R_{DR} has a large value - representing a "fast drive" circuit with a "slow" drain circuit. The equivalent circuit model is shown in Figure 9, and switching waves are illustrated in Figure 10. Note that we are ignoring the gate-to-source capacitance C_{GS}. This is valid because with large L_ℓ/R_{DR} ratio, the "Miller" effect predominates, and current through C_{GS} is small by comparison with that through C_{GD}.

The inductance L_ℓ now presents such a high impedance that the increase of drain current "requested" by the drive circuit cannot be satisfied; the drive circuit is largely impotent to bring about the drain current that it asks for.

The drain voltage now collapses relatively quickly - generally well before the current rise is completed. The end of the period is marked by the MOSFET reaching the essential condition of a "closed switch" - the voltage across it having collapsed completely.

The mathematics are rather simple; in order to gain insight, it is useful to proceed through the analysis step by step:

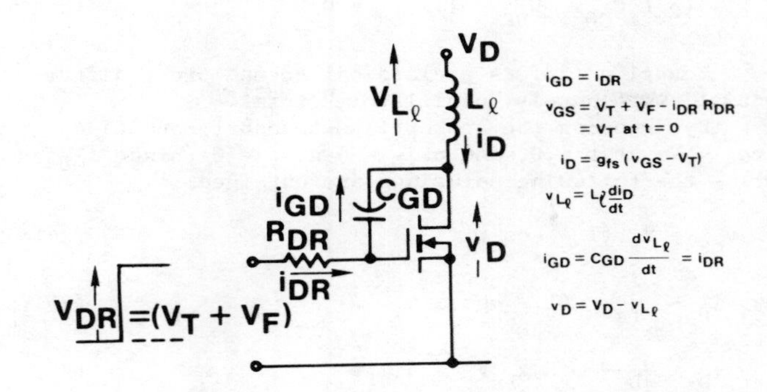

Figure 9: Circuit Model For Turn-On Interval 2, High L_ℓ/R_{DR}

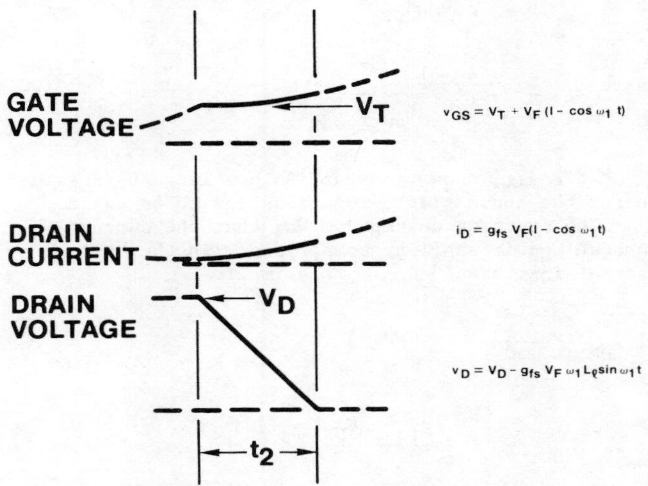

Figure 10: Waveforms For Turn-On Interval 2 Large L_ℓ/R_{DR}

$$\frac{L_\ell}{R_{DR}} > \frac{10C_{GS}{}^2}{C_{GD}g_{fs}}$$

$$v_{GS} = (V_T + V_F) - i_{DR}R_{DR} \quad (19)$$

$$i_D = g_{fs}\,(v_{GS} - V_T)$$

Therefore, from (19):

$$i_D = g_{fs}\,(v_F - i_{DR}R_{DR}) \quad (20)$$

$$\therefore\ pL_\ell i_D = -pL_\ell g_{fs}R_{DR}i_{DR} \quad (21)$$

$$v_D = V_D - pL_\ell i_D$$

Therefore, from (21):

$$v_D = V_D + pL_\ell g_{fs}R_{DR}i_{DR} \quad (22)$$

$$i_{DR} = -pC_{GD}v_D$$

Therefore from (22):

$$i_{DR} = -p^2L_\ell C_{GD}g_{fs}R_{DR}i_{DR}$$

$$\therefore \; (p^2 L_\ell C_{GD} g_{fs} R_{DR} + 1)\, i_{DR} = 0 \tag{23}$$

Equation (23) is a classical second order differential, with purely "oscillatory" terms.

By imposing the appropriate boundary conditions ($v_{GS} = V_T$ at $t = 0$, and $pLi_D = 0$ at $t = 0$ (since $i_{GD} \neq \infty$)) - the following solutions are obtained:

$$v_{GS} = V_F\,(1 - \cos \omega_1 t) \tag{24}$$

$$i_D = g_{fs} V_F\,(1 - \cos \omega_1 t) \tag{25}$$

$$v_D = V_D - \omega_1 L_\ell g_{fs} V_F \sin \omega_1 t \tag{26}$$

$$= V_D - \sqrt{\frac{g_{fs} L_\ell}{C_{GD} R_{DR}}}\, \sin \omega_1 t \tag{27}$$

$$\text{where } \omega_1 = \frac{1}{\sqrt{g_{fs} C_{GD} R_{DR} L_\ell}} \tag{28}$$

It remains now to establish how large L_ℓ/R_{DR} must be for the above simple relationships to be valid.

The starting assumption was that the current through C_{GS} is small by comparison with the "Miller" current i_{GD} through C_{GD}. This implies:

$$R_{DR} << \frac{1}{\omega_1 C_{GS}}$$

$$\therefore \; R_{DR} << \frac{\sqrt{g_{fs} C_{GD} R_{DR} L_\ell}}{C_{GS}}$$

$$\therefore \; \frac{L_\ell}{R_{DR}} >> \frac{C_{GS}^2}{C_{GD} g_{fs}} \tag{29}$$

Table 1 shows maximum values of R_{DR} for various HEXFET's for different values of L_ℓ that satisfy the above condition. It is clear that this condition, and hence expressions (24) through (26), will generally apply only to relatively low impedance drive circuits.

Simple qualitative checks on the above relationships will prove their validity. From equation (28), ω_1 increases as R_{DR} or L_ℓ decrease. The rate of rise of drain current therefore increases as either of these parameters decrease, which is to be expected. From equation (27), the voltage across L_ℓ is proportional to $\sqrt{L_\ell/R_{DR}}$. Thus increasing L_ℓ or decreasing R_{DR} gives increasing voltage across L_ℓ - again, to be expected.

The end of the interval occurs when either the drain current i_D reaches I_0, or the drain voltage collapses to zero (more precisely, when it becomes equal to $i_D \times R_{DS(ON)}$). If I_0 or V_F, or both, are small, i_D could reach I_0 before the collapse of drain voltage is complete. In practice, the voltage collapse will generally occur well before the current has risen to I_0. To take an example, with the IRF150 HEXFET (rated 28A continuous) operating in a 60V circuit, with a gate forcing voltage V_F of 7V, $L_\ell = 1\mu H$, and $R_{DR} = 2\Omega$, the voltage collapse will be completed by the time the drain current has risen to 0.25A (i.e. about 1% of rated current).

This result is to be expected; we have already reasoned that for large L_ℓ/R_{DR} ratio, the MOSFET essentially acts as a closed switch, the voltage across it collapsing quickly, with the current rising much more slowly, at a rate determined by the external circuit inductance.

Turn-On Interval 3

The second time interval ends at the completion either of the drain current rise or the drain voltage fall. The completion of the remaining event - voltage fall, or current rise - whichever it is, takes place during the third time interval.

Fortunately, since only the drain voltage or the drain current are now still changing, the analysis is easy, and is independent of the ratio of L_ℓ/R_{DR}. If the drain current is no longer changing, then L_ℓ is irrelevant, since there is no voltage drop across it, whilst if the drain voltage is no longer changing, the MOSFET already acts as a closed switch, and R_{DR} is irrelevant.

Consider first the situation when the voltage completes its fall during the third interval. The equivalent circuit model is shown in Figure 11. At the start of the period the drain voltage is V_D^*. Since the drain current is constant, v_{GS} must also be constant:

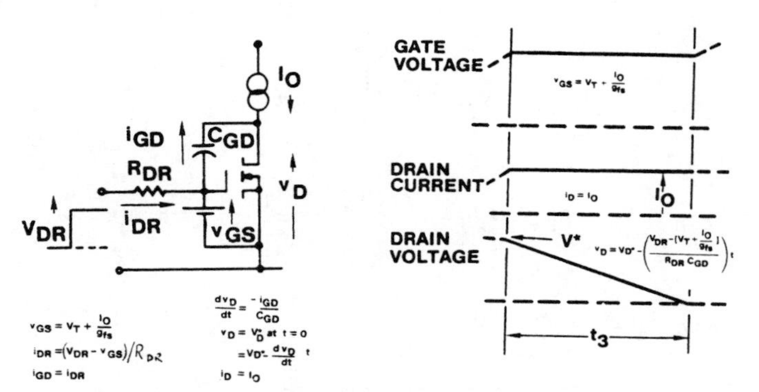

Figures 11 & 12: Circuit Model and Waveforms For Turn-On Interval 3, i_D is Already Equal to I_0. v_D Collapses

$$v_{GS} = V_T + \frac{I_0}{g_{fs}} \tag{30}$$

Therefore i_{DR} is also constant:

$$i_{DR} = \frac{1}{R_{DR}}(V_{DR} - v_{GS}) = \frac{(V_{DR} - (V_T + I_0/g_{fs}))}{R_{DR}} \tag{31}$$

Since v_{GS} is constant, no current flows in C_{GS}, and all of i_{DR} flows in C_{GD}. The rate of change of voltage across C_{GD} is therefore:

$$\frac{dv_{GD}}{dt} = \frac{i_{DR}}{C_{GD}} = \frac{V_{DR} - (V_T + I_0/g_{fs})}{R_{DR} C_{GD}} \tag{32}$$

The rate of change of drain-source voltage is equal to the rate of change of drain-gate voltage, since v_{GS} is constant. Therefore the drain voltage is:

$$v_D = V_D^* - \frac{V_{DR} - (V_T + I_0/g_{fs})}{R_{DR} C_{GD}} \tag{33}$$

We will now consider the situation when the current completes its rise during the third time interval, the drain voltage having already collapsed.

The equivalent circuit model is shown in Figure 13,

and switching waveforms are shown in Figure 14. The drain current i_D is:

$$i_D = I_D^* + \frac{V_D}{L_\ell} t \tag{34}$$

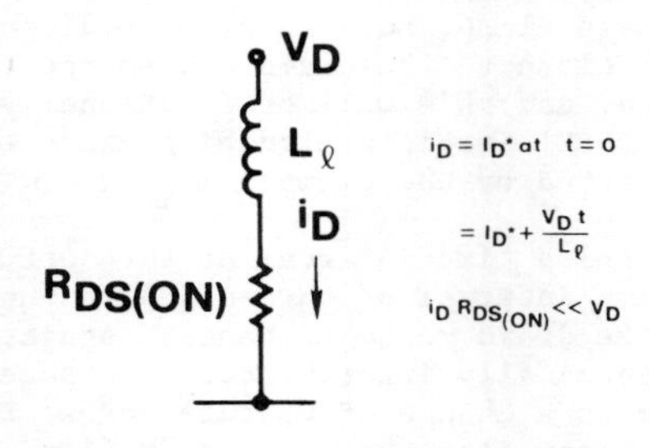

Figure 13: Circuit Model For Turn-On Interval 3, v_D Has Already Collapsed. i_D rises to I_0

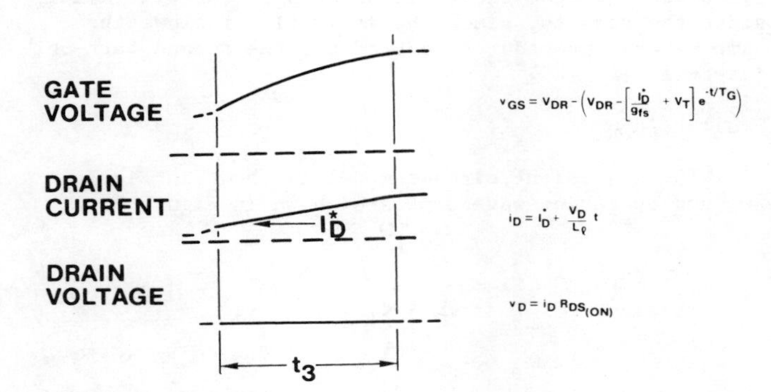

Figure 14: Waveforms For Turn-On Interval 3 V_D Has Already Collapsed. i_D Rises To I_0

The gate voltage continues to increase exponentially during the third interval, at time constant T_G (equation (2)). This however has no influence over the drain current or voltage, since the MOSFET is already "fully on".

Turn-On Interval 4

The gate voltage completes its exponential charge, at time constant T_G, to the level of the applied drive voltage V_{DR}. This has no influence over the drain current or voltage, since the switching sequence in the drain circuit has already been completed.

B. TURN-OFF

Turn-Off Delay - Interval 1

The equivalent circuit model is shown in Figure 15, and operating waveforms are shown in Figure 16. The applied drive voltage V_{DR} is assumed to fall instantaneously to a negative voltage $-V_2$ (this could of course be zero, or even positive, representing a small residual positive drive voltage). The voltage appearing between the gate and source terminals falls at a rate determined by the time constant $R_{DR}C_G$, and nothing happens in the drain circuit until the gate voltage falls to $V_T + \frac{I_0}{g_{fs}}$, which corresponds to the gate voltage needed to sustain the drain current I_0. This point marks the end of the turn-off delay period. The gate voltage during the turn-off delay interval is given by:

$$v_{GS} = (V_{DR} + V_2)\, e^{-t/T_G} - V_2 \tag{35}$$

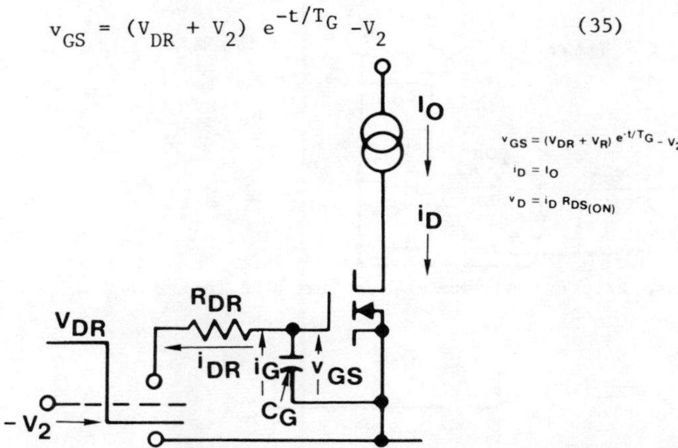

Figure 15: Circuit Model For Turn-Off Delay Interval

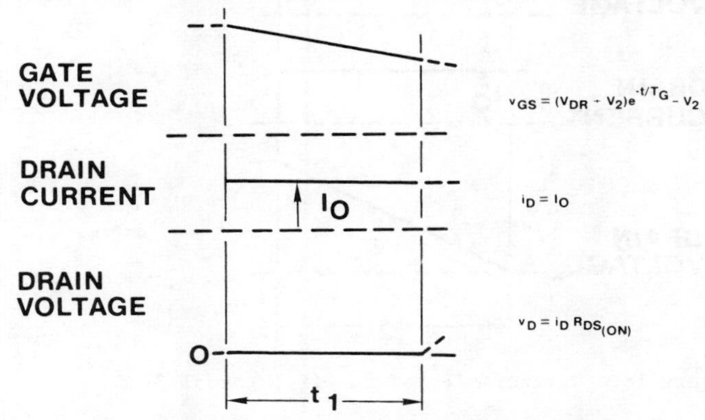

Figure 16: Waveforms For Turn-Off Delay Interval (t_1)

Turn-Off Interval 2

The equivalent circuit model is shown in Figure 17, and typical switching waveforms are shown in Figure 18. The drain voltage rises to V_D whilst the drain current remains constant at I_0, and the gate voltage remains constant at $(V_T + I_0/g_{fs})$. At first sight this may be surprising; a moment's thought shows it has to be so. Until the drain voltage just exceeds the circuit voltage, V_D, the freewheeling rectifier D (Figure 2) remains reverse biased; the whole of I_0 must therefore continue to flow into the drain of the MOSFET. So long as the drain current is constant, the gate voltage will also be constant, (since these two parameters are inextricably tied to one another by the MOSFET's transfer characteristic), and the current flowing "out of" the resistor R_{DR} is drawn exclusively from the gate-to-drain capacitance.

Since the drain current is constant, the ratio of L_ℓ / R_{DR} has no bearing upon the operation during this period. By similar reasoning used to analyze the voltage fall during the third interval of switch-on, the following relationship is derived:

$$v_D = \frac{(I_0/g_{fs} + V_T + V_2)\ t}{C_{GD}R_{DR}} \tag{36}$$

Turn-Off Interval 3

The general circuit model for this interval is shown in Figure 19. At the end of the second interval, the drain voltage is just equal to the supply voltage V_D, whilst the current is equal to the full load value, I_0. The freewheeling rectifier D (Figure 2) is now poised at the point of conduction, ready to receive

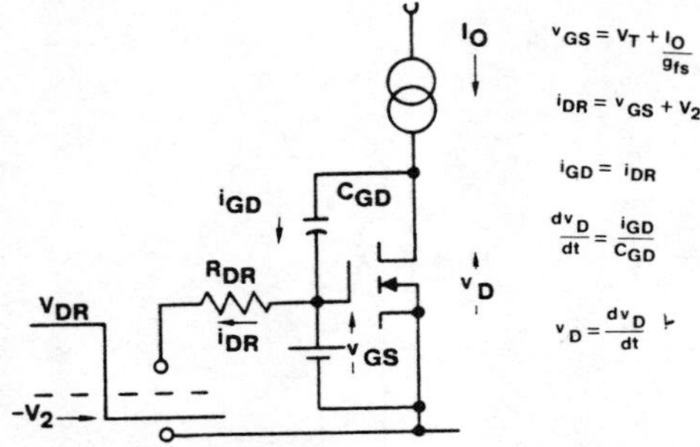

Figure 17: Circuit Model For Turn-Off Interval 2

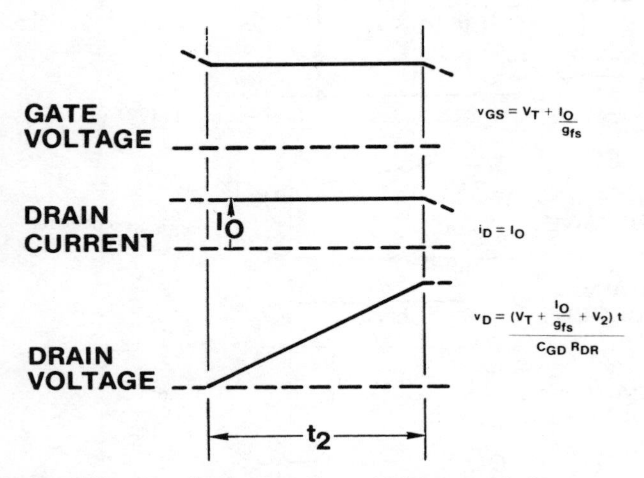

Figure 18: Waveforms For Turn-Off Interval 2

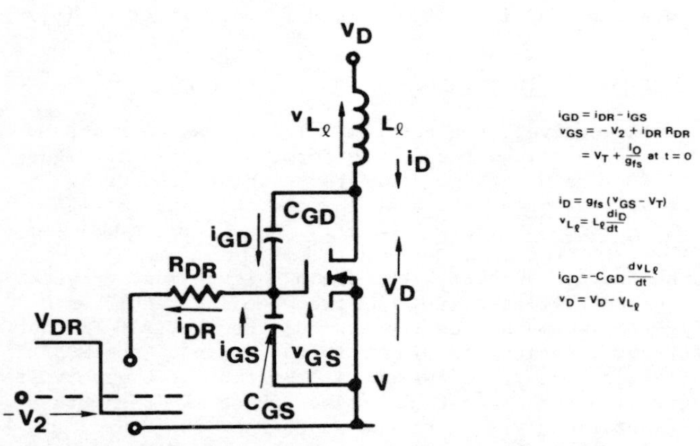

Figure 19: General Circuit Model For Turn-Off Interval 3

the load current I_O, and the potential at the "top" of L_ℓ is now fixed essentially at V_D. In order for the drain current to be commutated into the freewheeling rectifier, it is axiomatic that the drain voltage must increase above V_D. This reflects the fundamental property of inductance L_ℓ; the voltage across it must reverse in order for the current in it to reduce; a voltage-time integral must be developed, equal to I_O X L_ℓ, for the drain current to be returned to zero. This fundamental consideration relates directly to the inductance L_ℓ, and is quite independent of any other circuit considerations. The magnitude of the peak over-voltage developed across the MOSFET will be proporitonal to the size of the inductance L_ℓ, the magnitude of the current I_O, and the speed of switching.

In most practical circuits, the voltage transient at the drain can easily exceed the voltage rating of the MOSFET. In the absence of an externally connected local voltage clamp, the MOSFET will likely be driven into avalanche, acting, in effect, as its own voltage clamp, and preventing further substantial increase of voltage. This may or may not be permissible, depending upon whether the MOSFET is rated to handle the avalanche energy. If it cannot do so, then a local external voltage clamp, such as a zener diode, connected physically close to the drain and source terminals will be needed, and this will be functionally equivalent to the MOSFET itself avalanching, save that the energy is absorbed by the clamp, rather than by the MOSFET.

In this third time interval of turn-off, as during the second time interval of turn-on, both the drain current and the drain voltage change. Again, these two events are dynamically intertwined. A change of drain current produces a change of voltage across L_ℓ; this produces a current flow through the "Miller" capacitance C_{GD}; this restrains the rate of decrease of gate voltage, which in turn restrains the original rate of change of drain current.

As we would expect, the form of the analytic solutions depends upon the ratio of L_ℓ/R_{DR}. We will simply state the results, since the derivation follows the same general procedures covered for the second turn-on interval.

Small L_ℓ/R_{DR}

The equivalent circuit model is shown in Figure 20, and operating waveforms are shown in Figure 21.

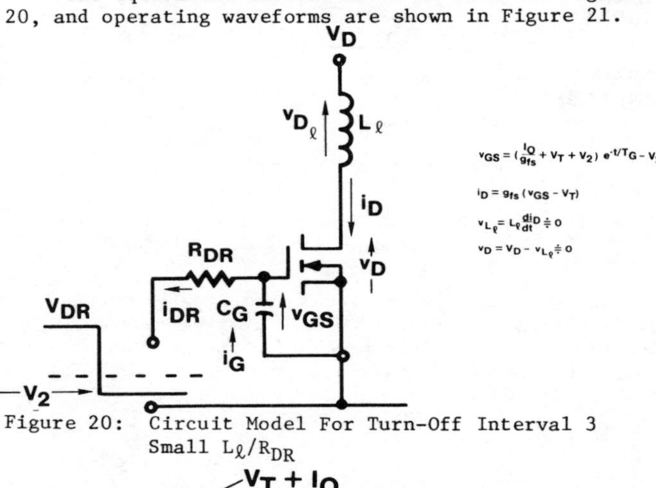

Figure 20: Circuit Model For Turn-Off Interval 3 Small L_ℓ/R_{DR}

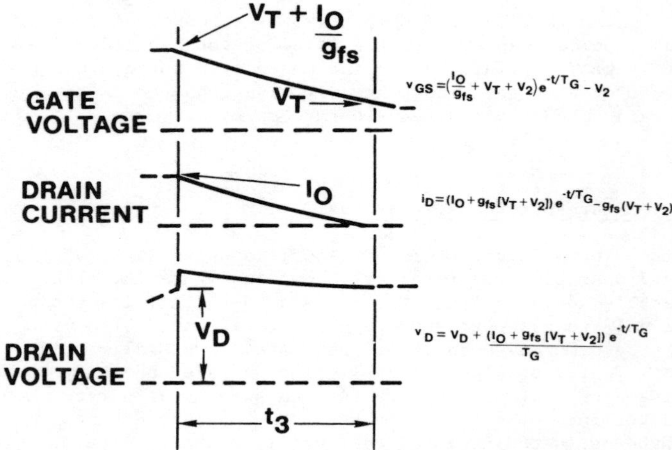

Figure 21: Waveforms For Turn-Off Interval 3 Small L_ℓ/R_{DR}

$$\frac{L_\ell}{R_{DR}} < \frac{C_{GS}^2}{10 C_{GD} g_{fs}}$$

For small L_ℓ/R_{DR}, equation (6) must be satisfied:

$$v_{GS} = (I_0/g_{fs} + V_T + V_2)\ e^{-t/T_G} - V_2 \quad (37)$$

$$i_D = (I_0 + g_{fs}\ [V_T + V_2])\ e^{-t/T_G} - g_{fs}[V_T + V_2] \quad (38)$$

$$v_D = V_D + \frac{(I_0 + g_{fs}[V_T + V_2])\ e^{-t/T_G}}{T_G} \quad (39)$$

The interval ends when the drain current i_D falls to zero.

Intermediate L_ℓ/R_{DR}

The general circuit model shown in Figure 19 applies. Either equation (7) or (13) must be satisfied. Operating waveforms for L_ℓ/R_{DR} that satisfy equation (7) are shown in Figure 22 (a). Expressions for the gate voltage v_{GS}, the drain current i_D, and the drain voltage v_D are as follows:

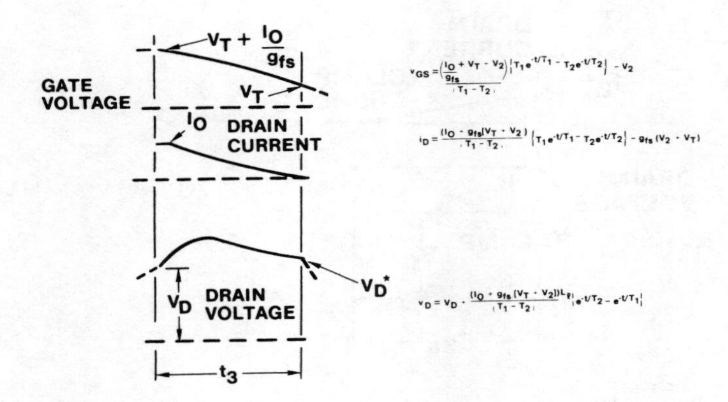

Fiugre 22 (a): Waveforms for Turn-Off Interval 3 Intermediate L_ℓ/R_{DR}

$$\frac{C_{GS}^2}{10C_{GD}g_{fs}} < \frac{L}{R_{DR}} < \frac{C_{GS}^2}{4C_{GD}g_{fs}}$$

$$v_{GS} = \frac{(I_0/g_{fs}+V_T+V_2)}{(T_1-T_2)}\left\{T_1e^{-t/T_1}-T_2e^{-t/T_2}\right\}-V_2 \quad (40)$$

$$i_D = \frac{(I_0+g_{fs}[V_T+V_2])}{(T_1-T_2)}\left\{T_1e^{-t/T_1}-T_2e^{-t/T_2}\right\}\ldots$$

$$\ldots -g_{fs}[V_T+V_2] \quad (41)$$

$$v_D = V_D + \frac{(I_0+g_{fs}[V_T+V_2])\ L_\ell}{(T_1-T_2)}\left\{e^{-t/T_2}-e^{-t/T_1}\right\} \quad (42)$$

where T_1 and T_2 are given by equations (11) and (12) respectively.

Operating waveforms for L_ℓ/R_{DR} given by equation (13) are shown in Figure 22 (b). Expressions for the gate voltage v_{GS}, the drain current i_D, and the drain voltage v_D are as follows:

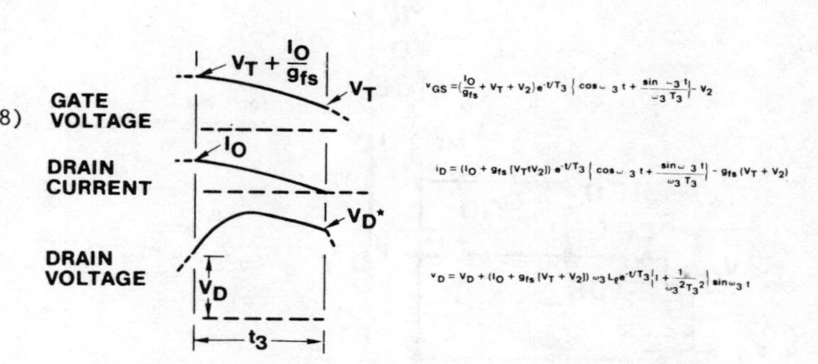

Figure 22 (b): Waveforms For Turn-Off Interval 3 Intermediate L_ℓ/R_{DR}

$$\frac{C_{GS}^2}{4C_{GD}g_{fs}} < \frac{L_\ell}{R_{DR}} < \frac{10C_{GS}^2}{C_{GD}g_{fs}}$$

$$v_{GS} = (I_0/g_{fs}+V_T+V_2)e^{-t/T_3}\left\{\cos\omega_3 t + \frac{\sin\omega_3 t}{\omega_3 T_3}\right\}-V_2 \quad (43)$$

$$i_D = (I_0+g_{fs}[V_T+V_2])e^{-t/T_3}\left\{\cos\omega_3 t + \frac{\sin\omega_3 t}{\omega_3 T_3}\right\}\ldots$$

$$\ldots -g_{fs}[V_T+V_2] \quad (44)$$

$$v_D = V_D + (I_0+g_{fs}[V_T+V_2])\omega_3 L_\ell e^{-t/T_3}$$

$$\ldots \times\left\{1+\frac{1}{\omega_3^2T_3^2}\right\}\sin\omega_3 t \quad (45)$$

where T_3 and ω_3 are given by equations (17) and (18) respectively.

Large L_ℓ/R_{DR}

Large L_ℓ/R_{DR} is defined by equation (29). The circuit model is shown in Figure 23, and operating waveforms are shown in Figure 24. Expressions for the gate voltage v_{GS}, the drain current i_D, and the drain voltage v_D, are as follows:

$$v_{GS} = (I_0/g_{fs}+V_T+V_2)\ \cos\omega_1 t - V_2 \quad (46)$$

$$i_D = (I_0+g_{fs}[V_T+V_2])\ \cos\omega_1 t - g_{fs}(V_T+V_2) \quad (47)$$

$$v_D = V_D + (I_0+g_{fs}[V_T+V_2])\ \omega_1 L_\ell\ \sin\omega_1 t \quad (48)$$

where ω_1 is given by equation (28).

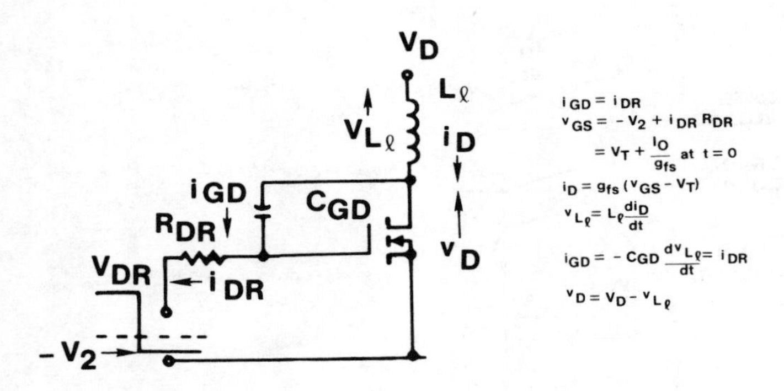

Figure 23: Circuit Model For Turn-Off Interval 3 Large L_ℓ/R_{DR}

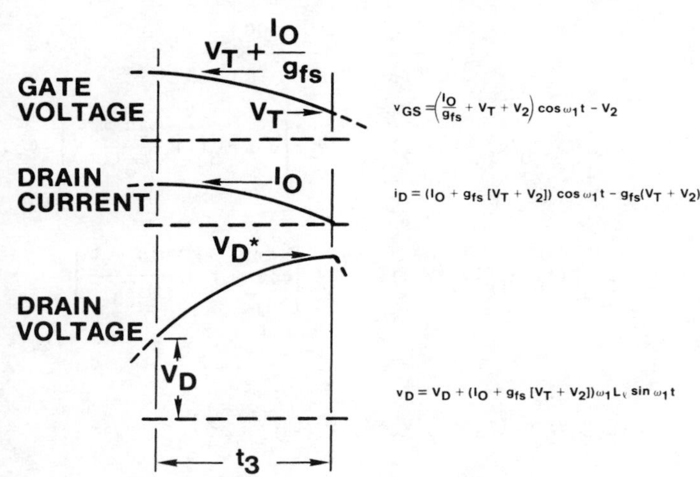

Figure 24: Waveforms For Turn-Off Interval 3. Large L_ℓ/R_{DR}

$$\frac{L_\ell}{R_{DR}} > \frac{10C_{GS}^2}{C_{GD}g_{fs}}$$

Turn-Off Interval 3a (Clamping of the Drain Voltage)

The expressions just derived assume that the drain voltage will increase to whatever extent the circuit operation dictates. In practice, as already stated, the instantaneous drain voltage is likely to exceed the voltage rating of the MOSFET; this is particularly true for high L_ℓ/R_{DR} ratio.

In this event, either the MOSFET will be driven into avalanche - in effect acting as its own "voltage clamp" and limiting further increase of voltage - or, if the MOSFET is unable to handle this, an external local voltage clamping device would have to be connected.

In either event, at the instant at which the drain voltage becomes equal to the "clamp" voltage, interval 3, as given by the previous equations, comes to an end, and interval 3 a - the clamping interval - starts.

Figure 25 shows the equivalent circuit for the "clamping" interval, with an external clamp, and operating waveforms are shown in Figure 26. The drain voltage is assumed to stay constant at the "clamp" level, V_{CLAMP}, while the drain circuit current decays linearly to zero:

$$i_D = I_D^* - \frac{(V_{CLAMP} - V_D)}{L_\ell}t \qquad (49)$$

The period ends when $i_D = 0$. Note that if the MOSFET acts as its own clamp and is driven into avalanche, then equation (49) applies to the MOSFET's drain current; if an external clamp is used, drain current can be assumed to stop flowing at the start of this interval, and equation (49) then applies to the current in the external clamp.

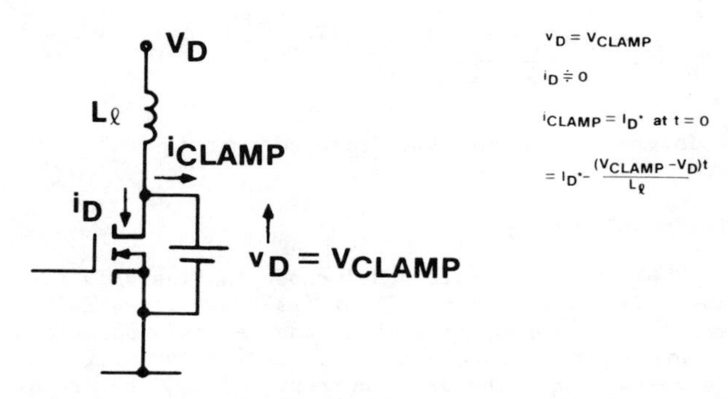

Fiugre 25: Circuit For Clamping Turn-Off Interval 3a

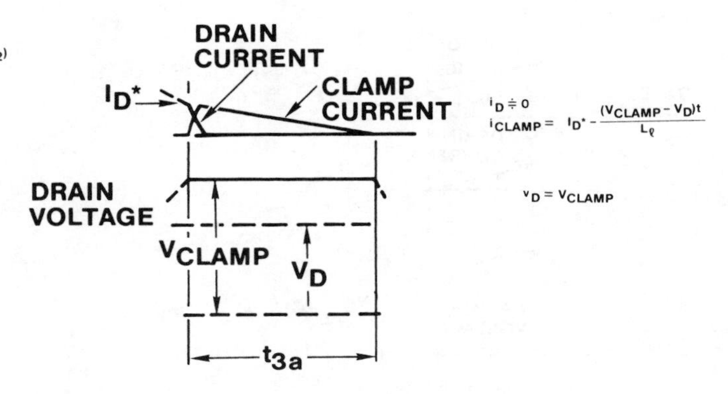

Figure 26: Waveforms For Clamping Turn-Off Interval 3a

Turn-Off Interval 4

At the end of interval 3 (or 3 a) the drain current has fallen to zero, but the drain voltage V_D^*, is greater than the circuit voltage V_D. The drain capacitance C_D then "rings" with the stray circuit inductance L_ℓ, the oscillation being damped by the stray circuit resistance R_ℓ. Figure 27 shows the equivalent circuit for this interval, and Figure 28 shows a typical drain voltage waveform.

$$v_D = V_D + (V_D^* - V_D)e^{-t/T_4}\cos\omega_4 t \qquad (50)$$

$$\text{where } T_4 = \frac{2L_\ell}{R_\ell} \qquad (51)$$

$$\omega_4 = \frac{\sqrt{4L_\ell C_D - C_D^2 R_{e\ell}^2}}{2L_\ell C_D} \qquad (52)$$

During this interval the gate voltage discharges exponentially with time constant T_G towards a final value of $-V_2$.

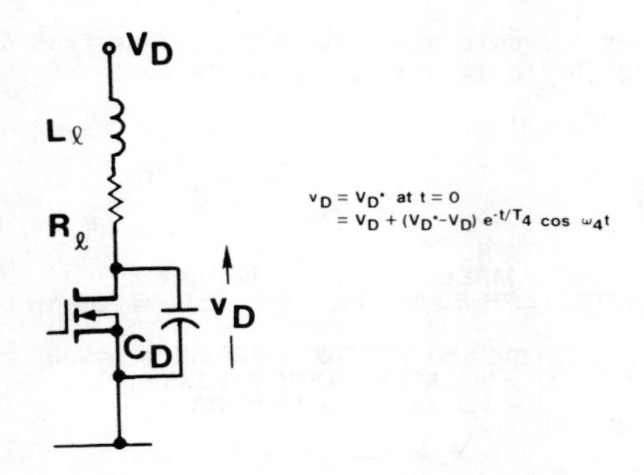

Figure 27: Circuit Model For Turn-Off Interval 4

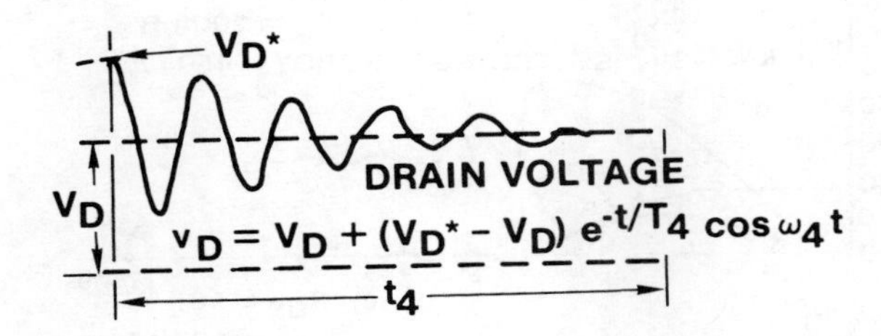

Figure 28: Drain Voltage For Turn-Off Interval 4

VI. A WORKED DESIGN EXAMPLE

Figures 29 through 32 show switching waveforms for a specific design example, obtained from the analytic expressions presented in this paper. Various combinations of L_ℓ/R_{DR}, and amplitude of drive voltage, are considered in order to illustrate the effects of these parameters on the switching performance. The following data is used:

MOSFET type - IRF150

C_{GS} - 2650pF
C_{GD} - 350pF
V_T - 3V
g_{fs} - 8A/V
V_D - 50V
I_O - 35A

Figure 29 shows waveforms calculated for the turn-on interval for (a) R_{DR} = 5 ohms, L_ℓ = 200nH, (b) R_{DR} = 50 ohms, L_ℓ = 200nH, and (c) R_{DR} = 50 ohms, L_ℓ = 1μH. The drive voltage V_{DR} is 10 volts.

Condition (a) is representative of a fast drive circuit, and a relatively high impedance of L_ℓ. The drain voltage falls rapidly, and most of the current rise time occurs subsequent to the collapse of drain voltage. The switching energy is almost negligible - a mere 0.12μJ. In Figure 29 (b), the inductance is the same, but the drive resistance has increased to 50 ohms. The gate drive circuit is now much slower, and the drain voltage collapses much less rapidly; in fact the drain current now completes its rise before the drain voltage collapses completely. The total switching time (current rise + voltage fall) increases from 150ns in Figure 29 (a) to 360ns in Figure 29 (b). More significantly, the switching energy increases from 0.12μJ to 55μJ.

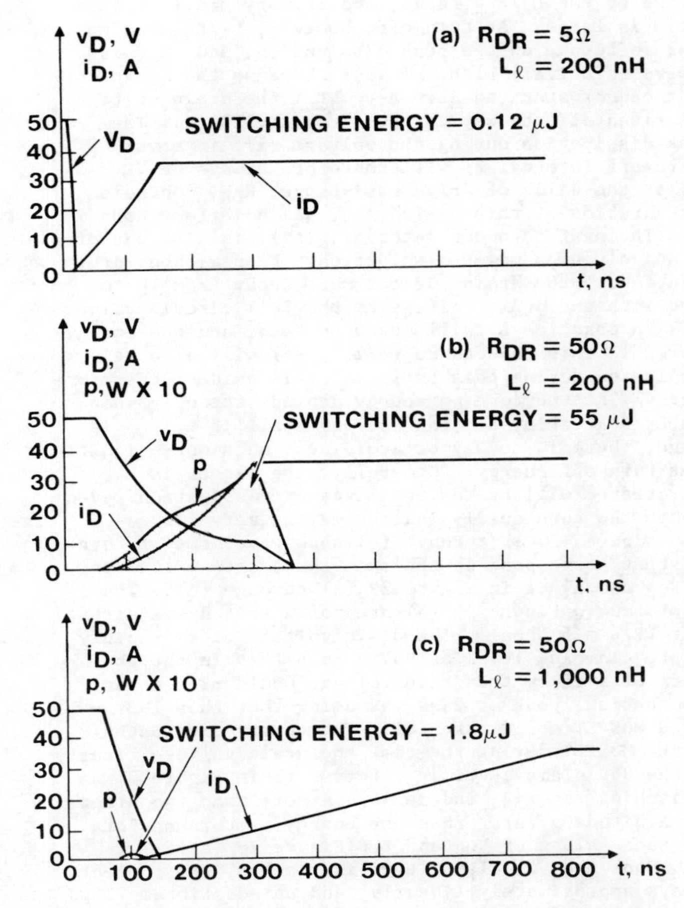

Figure 29: Turn-On Waveforms For Various Circuit Conditions
(a) R_{DR} =Ω5 , L_S = 200nH
(b) R_{DR} =Ω50 , L_S = 200nH
(c) R_{DR} =Ω50 , L_S = 1000nH
IRF 150, V_{DR} = 10V
V_2 = 0

In Figure 29 (c), the drive circuit resistance is still 50 ohms, while the drain inductance L_ℓ has increased from 200nH to 1μH. The speed of the drive circuit is therefore the same as in Figure 29 (b), but the impedance of L_ℓ increases by a factor of 5. The voltage drop in the drain circuit is therefore once again very significant, and the drain voltage collapses much more rapidly. Because of the increased inductance, however, the current rise time is much longer. The switching energy decreases from 55μJ in Figure 29 (b) to 1.8μJ in Figure 29 (c), because of the much faster voltage collapse. It would be wrong to believe, however, that the overall switching losses can be decreased by increasing L_ℓ. As we will see, the energy saved during turn-on by increasing L is more than offset by increased energy at turn-off. Increasing L_ℓ to reduce the turn-on losses is counterproductive; it simply postpones the "day of reckoning" to the turn-off interval.

Before studying the details of the turn-off waveforms in Figure 30, it will be instructive to make some basic comparisons between the operation during the turn-on and turn-off intervals.

At turn-on the peak dissipation is drastically effected by the L_ℓ/R_{DR} ratio, and is very small if this ratio is large. At turn-off, however, L_ℓ/R_{DR} has no real influence on the peak dissipation, and this is always relatively high. This is because the drain current cannot start to decrease until the drain voltage has risen all the way to the circuit voltage. The peak dissipation during the voltage rise interval (turn-off interval 2) will therefore always be $V_D \times I_0$. Whilst the value of drive resistance, R_{DR}, controls the duration of this period, L_ℓ, has no effect upon it.

The next turn-off interval, (t_3), is also one of relatively high power dissipation. Even with no drain inductance, the drain current must decay from I_0 to zero with the drain voltage at the full circuit value, V_D. In practice L_ℓ will never be zero, and the energy stored in this inductance ($1/2\ L_\ell I_0^2$) will also be dissipated during this period. It is evident therefore that whilst the turn-on energy depends strongly upon the L_ℓ/R_{DR} ratio, and can be very small if L_ℓ/R_{DR} is large, there is no way of avoiding a much more significant turn-off energy. Generally, the larger is L_ℓ, the greater will be the total energy dissipation, even though the turn-on dissipation may be very low.

Figures 30(a) through (c) show waveforms at turn-off that correspond to the same three sets of values of R_{DR} and L_ℓ as in Figure 29 (a) through (c). The waveforms in Figure 30 (a) are for a fast drive circuit (R_{DR} = 5 ohms). The drain voltage rises rapidly to the clamping level of 95V. Note that in the absence of a clamp the drain voltage would rise to a hypothetical peak of 235V (assuming that this 100V rated MOSFET would take it!). The energy dissipated in the MOSFET during the time the drain voltage rises to the 95V clamp level is referred to in Figure 31 as "switching" energy, and is 45µJ - more than two orders of magnitude greater than the energy at turn-on for the same values of R_{DR} and L_ℓ (Figure 29 (a)).

Once the 95V clamp level is reached, the current decays approximately linearly, and an additional 235µJ of energy is dissipated during the clamping period. This energy would be dissipated either in an external clamp, if this is used, or in the MOSFET itself - assuming that it is capable of operating in its avalanche mode.

Note that the energy stored in L_ℓ, $1/2\ L_\ell I_0^2$µJ = 122µJ, is about half the total energy dissipated during the clamping period. Simple physical reasoning confirms the correctness of this; not only must the energy stored in L_ℓ be dissipated, but since the supply voltage V_D continues to feed energy to the circuit (i_D continues to be drawn from V_D), this energy also must end up being dissipated during this period.

Figure 30 (b) shows waveforms for R_{DR} = 50 ohms, with L_ℓ the same as for Figure 30 (a). The response of the gate drive circuit is much slower, and hence the rate of rise of drain voltage is also much slower - so slow, in fact, that the drain voltage never reaches the clamping level of 95V. In this case, all the switching energy must be dissipated in the device itself, and there is no opportunity for shunting some of this into an external clamp. The total switching time increases from 175ns (Figure 30 (a)) to 400ns, and the total switching energy increases from 280 to 450µJ. Once again, the turn-off energy of 450µJ is much greater than the turn-on energy of 55µJ for the same value of L_ℓ and R_{DR} (Figure 29 (b)).

Figure 30 (c) shows turn-off waveforms for R_{DR} = 50 ohms, but with L_ℓ increased to 1µH. As would be expected, the intial rate of change of drain voltage is the same as in Figure 30 (b); until the drain voltage becomes equal to the circuit voltage of 50V, the drain current remains constant at I_0, and L_ℓ has no effect. Thereafter, however, the drain voltage moves much more rapidly upwards, and has no difficulty in reaching the clamp level of 95V. The total switching time increases to 950ns, because of the increased value of L_ℓ, and the total switching energy increases from 450µJ in Figure 30 (b) to 1435µJ in Figure 30 (c).

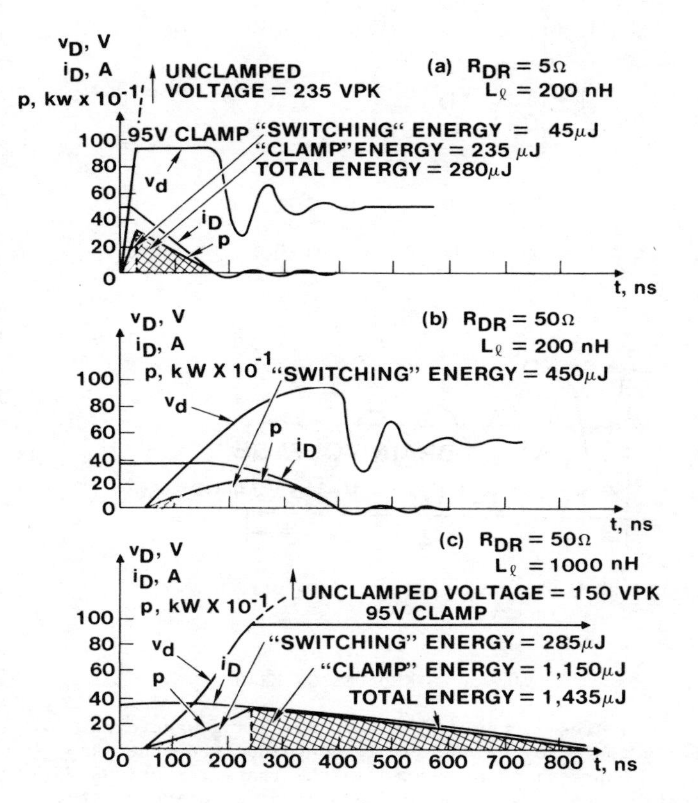

Figure 30: Turn-Off Waveforms For Various Circuit Conditions
(a) R_{DR} = 5Ω, L_S = 200nH
(b) R_{DR} = 50Ω, L_S = 200nH
(c) R_{DR} = 50Ω, L_S = 1000nH

IRF 150, V_{DR} = 10V
V_2 = 0

It is interesting to compare the energy reduction at turn-on when L_ℓ is increased from 200nH to 1µH, Figures 29 (b) and (c), versus the energy increase at turn-off (Figures 30 (b) and (c)). The energy reduction at turn-on is (55 - 1.8) = 53.2µJ, while the energy increase at turn-off is (1435 - 450) = 985µJ. The net effect of increasing drain circuit inductance is a very substantial increase in the total energy dissipation.

The waveforms in Figure 31 show the effect of increasing the applied drive voltage from 10V to 15V, for R_{DR} = 50 ohms and L_ℓ = 200nH. The total switching time decreases from 360ns to 160ns, and the switching energy decreases from 55µJ to 6µJ.

Figure 32 shows the same comparison for the turn-off interval. The waveforms in Figure 32 (a) are for no applied drive voltage during the turn-off interval, while those in Figure 32 (b) are for a negative drive voltage of -15V. The total switching time decreases from 400 to 250ns, and the switching energy from 450 to 305µJ. The negative gate drive voltage not only reduces the total switching energy, but also, because it forces the drain voltage to reach the 95V clamping level, it offers the possibility for "dumping" 195µJ of energy which would otherwise be dissipated in the MOSFET, into an external clamp.

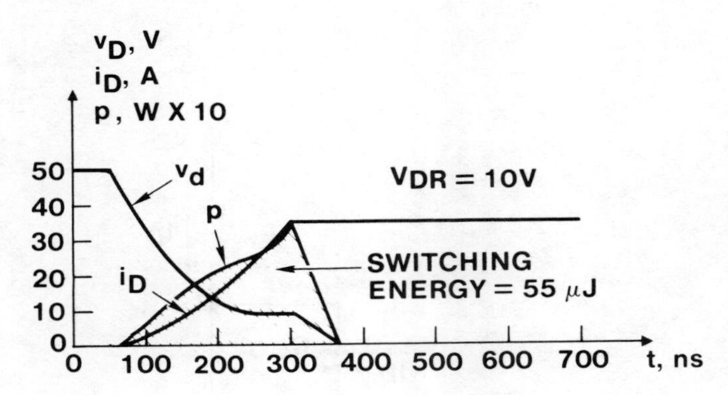

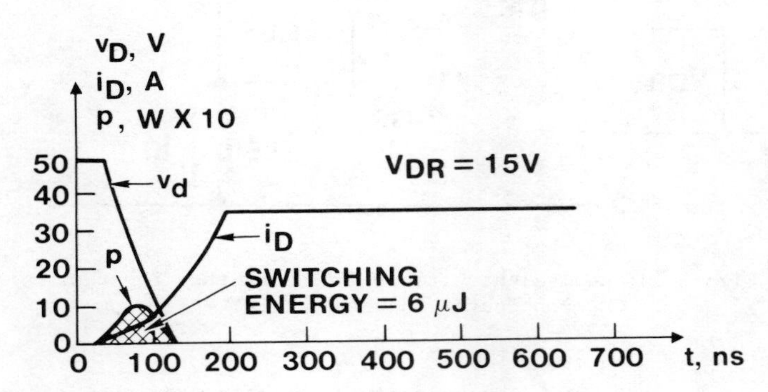

Figure 31: Turn-On Waveforms For
(a) Drive Voltage V_{DR} = 10V
(b) Drive Voltage V_{DR} = 15V

IRF 150 R_{DR} = 50Ω
L_S = 200nH

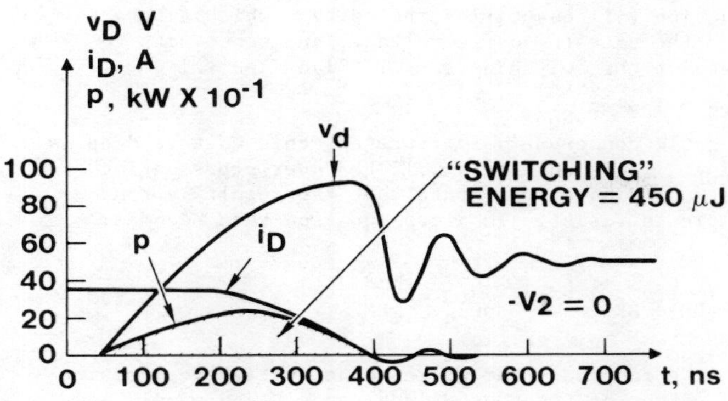

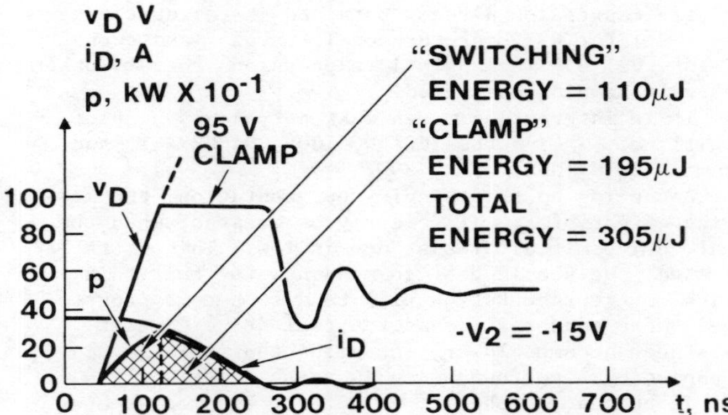

Figure 32: Turn-Off Waveforms For
(a) "Negative Drive Voltage" $(-V_2)$ = 0
(b) "Negative Voltage" $(-V_2)$ = -15V

IRF 150 R_{DR} = 50Ω
L_S = 200nH

VII. THE EFFECT OF COMMON SOURCE INDUCTANCE

So far we have ignored the effect of the common source inductance L_S, shown dashed in Figure 2. This inductance will always be present to some extent, even with careful circuit layout, the user will have to accept, at a minimum, the internal lead inductance within the package of the device. For a TO-3 package, this inductance is in the order of 5 to 10nH. We will now consider briefly the modifying effect of L_S on the switching operation.

Figure 33 shows the general equivalent circuit which includes L_S. As the drain current i_D starts to increase at turn-on, a voltage will be developed across L_S due to the rate of change of drain current. This voltage is common to the gate circuit, and its polarity is such to reduce the net voltage appearing between the gate and source terminals. Like the "Miller" effect, which provides a negative feedback from the drain to the gate, slowing down the rate of change of current, so the common source inductance also provides a negative feedback, from the source circuit to the gate, also slowing down the change of drain current.

A complete analysis of the switching operation that includes the effect of the common source inductance can be accomplished by means of the procedures already presented. This is beyond the scope of this paper. We will content ourselves instead with an approximate analysis, the main benefit of which is the extreme simplicity of the result.

Referring to the equivalent circuit in Figure 33, it is evident that L_S only has an effect when the drain current is changing, and the MOSFET is in its active region. This restricts the analysis to interval 2 during turn-on, and interval 3 during turn-off.

The loop equation for the gate circuit is:

$$i_{DR}R_{DR} + \frac{i_{GS}}{pC_{GS}} + pL_S i_{GS} + pL_S i_D = V_{DR} \tag{53}$$

By making the approximation (valid for practical operating conditions) $pL_S i_D >> pL_S i_{GS}$, equation (53) becomes:

$$i_{DR}R_{DR} + \frac{i_{GS}}{pC_{GS}} + pL_S i_D = V_{DR} \tag{54}$$

Now $i_D = g_{fs} v_{GS}$

$$= \frac{g_{fs} i_{GS}}{pC_{GS}} \tag{55}$$

Substituting for i_D into equation (54) gives:

$$i_{DR}R_{DR} + \frac{i_{GS}}{pC_{GS}} + \frac{g_{fs} L_S i_{GS}}{C_{GS}} = V_{DR} \tag{56}$$

It is seen from equation (56) that the effect of L_S has been to add a resistive voltage drop, equal to $R' i_{GS}$, into the loop equation, where:

$$R' = \frac{g_{fs} L_S}{C_{GS}} \tag{57}$$

Equation (57) quantifies the voltage which subtracts from the gate-to-source voltage, and hence acts to restrict the switching speed. Since $i_{GS} \leq i_{DR}$ ($i_{GS} \rightarrow i_{DR}$ as $\frac{L_\ell}{R_{DR}} \rightarrow 0$) we could conservatively represent this voltage drop as $R' i_{DR}$. To be sure this will be "over representing" the modifying effect of L_S, but the result is too simple to resist. The loop equation then becomes:

$$i_{DR}\,(R_{DR} + R') + \frac{i_{GS}}{pC_{GS}} = V_{DR} \tag{58}$$

We conclude that we can conservatively represent the effect of the source inductance L_S by simply adding an equivalent resistance R', given by equation (57) to the "real" drive circuit resistance, R_{DR}.

The only modification we then have to make to the analytic expressions already obtained is to substitute $(R_{DR} + R')$ for R_{DR}, for turn-on interval 2 and turn-off interval 3. The analytic expressions for all other intervals remains unaltered.

It is interesting to see what a typical value of R' will be. Taking the IRF150, 100V 28A HEXFET, and assuming L_S = 10nH, $R' \doteqdot 30\Omega$.

According to this simple representation, the modifying effect of L_S will clearly be greatest when the "real" drive circuit resistance is low. This is to be expected. We should bear in mind however that this equivalent representation will tend to become progressively more conservative as the real drive circuit resistance becomes lower, (assuming that this is accompanied by a corresponding increase in L_ℓ/R_{DR}). This is because as the "Miller" effect becomes greater i_{GS} becomes smaller relative to i_{DR} and hence the approximation $i_{GS}R' \doteqdot i_{DR}R'$ becomes more conservative.

VIII. CONCLUSIONS

Our major objective has been to convey an understanding of the switching operation of the MOSFET, by giving a physical insight into the switching mechanism, and into the interaction of the MOSFET with the external circuit, and to provide analytical tools that enable the switching performance to be quantified at the design stage. The major conclusions are as follows:

1. The shapes of the switching waves during the intervals that the drain current and drain voltage simultaneously change are profoundly influenced by the external circuit parameters - specifically the drain circuit inductance and the gate circuit resistance. A given ratio of these parameters substantially determines the geometry of the switching waves, but the actual switching time depends upon specific values.

2. At turn-on, the current rises before the voltage fall is completed, if the ratio of L_ℓ/R_{DR} is small (assuming a clamped inductive load). The converse is true, if the ratio of L_ℓ/R_{DR} is large.

3. At turn-off, the drain voltage always rises to the circuit voltage while the drain current remains constant (assuming an inductive load). The rate of rise of drain voltage is controlled by the drive circuit resistance, but it is independent of drain circuit inductance. Once the drain voltage reaches the circuit voltage, the waveshapes then depend upon the ratio of L_ℓ/R_{DR}.

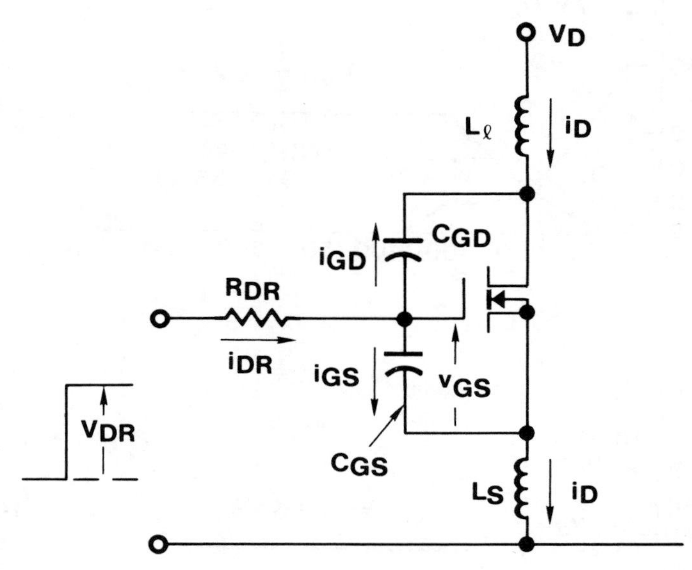

Figure 33: Equivalent Circuit Including the Effect of Common Source Inductance, L_S

4. The turn-off energy is always greater than the turn-on energy. Increasing the drain circuit inductance decreases turn-on losses, but increases the turn-off losses by a far greater amount. The smaller this inductance, the faster the switching times, and the lower the overall energy losses. It is good design practice therefore to keep this inductance as small as possible, by appropriate attention to circuit layout, transformer design, etc.

5. Turn-on time and turn-on energy can be decreased by increasing the applied drive voltage. Turn-off time and turn-off energy can be decreased by applying an increasing negative drive voltage.

6. Common source inductance reduces the switching speed and increases switching energy. It effects the operation during those intervals in which the drain current changes, and the MOSFET is in its active mode (turn-on interval 2, and turn-off interval 3). For these intervals it can be conservatively represented by an equivalent resistance added in series with the drive circuit resistance.

References

1. "Paralleling of Power MOSFETs For Higher Power Output", James B. Forsythe, 1981 IAS - IEEE

2. "Switching Transients in High Frequency,High Power Converter Using Power MOSFET's", T. Sloane, H. Owen, T. Wilson, International Rectifier Application Note AN933.

IMPROVING THE REVERSE RECOVERY OF POWER MOSFET INTEGRAL DIODES BY ELECTRON IRRADIATION

B. JAYANT BALIGA and JOHN P. WALDEN
General Electric Company, Corporate Research and Development Center, Schenectady, NY 12345, U.S.A.

(*Received* 18 *March* 1983; *in revised form* 11 *May* 1983)

Abstract—This paper demonstrates that controlled electron irradiation of silicon power MOSFET devices can be used to significantly improve the reverse recovery characteristics of their integral reverse conducting diodes without adversely affecting the MOSFET characteristics. By using 3 MeV electron irradiation at room temperature it was found that the reverse recovery charge in the integral diode could be continuously reduced in a well controlled manner from over 500 nC to less than 100 nC without any significant increase in the forward voltage drop of the integral diode under typical operating peak currents. The reverse recovery time was also observed to decrease from 3 microseconds to less than 200 nsec when the radiation dose was increased from 0 to 16 Megarads. The damage produced in gate oxide of the MOSFET due to the electron radiation damage was found to cause an undersirable decrease in the gate threshold voltage. This resulted in excessive channel leakage current flow in the MOSFET at zero gate bias. It was found that this channel leakage current was substantially reduced by annealing the devices at 140°C without influencing the integral diode reverse recovery speed. Thus, the electron irradiation technique was found to be effective in controlling the integral diode reverse recovery characteristics without any degradation of the power MOSFET characteristics.

INTRODUCTION

In some power switching circuits it is necessary to allow for reverse current flow across the power switching devices. Examples of these circuits are d.c. to a.c. inverters for adjustable speed motor drives, switching power supplies, and d.c. to d.c. choppers for motor speed control with regenerative braking. In conventional power switching circuits using bipolar transistors this reverse current flow is achieved by using an antiparallel diode connected across the device terminals. This diode must exhibit good reverse recovery characteristics with a small reverse recovery charge to keep the power dissipation low and a reverse recovery current waveform without abrupt changes so as to prevent high voltage transients across the power devices resulting from high di/dt in inductive loads.

Recently, the power MOSFET has been commercially introduced with voltage ratings above 500 V and current handling capability in excess of 5 A. These devices are attractive components for the above power switching circuits due to their insulated gate structure which greatly simplifies the gate drive circuitry and due to their substantially better safe-operating area. A typical DMOS power FET structure is shown in Fig. 1 together with its equivalent circuit. In this device, forward current flow in the MOSFET occurs for positive drain voltages when a sufficient positive gate bias voltage is applied to create an inversion layer at the surface of the p-base region under the gate electrode. This inversion layer connects the N+ source region to the N− drift layer and provides a continuous path for electronic current flow from the source to the drain. In high voltage device structures, the N− drift layer is designed to support the high forward blocking voltages. This drift layer is then found to limit the on-resistance of the device and to control the maximum current handling capability[1].

As illustrated in Fig. 1, the power MOSFET structure also contains a "parasitic" diode formed between the p-base and the N− drift region. This diode will conduct current whenever negative voltages are applied to the drain. It would be desirable to use this integral power MOSFET to conduct reverse currents in the device so as to eliminate the need for the external flyback or freewheeling diode. However, it has been found that the integral diode in the as-fabricated power MOSFET is substantially slower in switching speed than the MOSFET. This is due to bipolar conduction in the integral diode with minority carrier (holes here) injection into the N− base region. These minority carriers must be removed during the diode reverse recovery by recombination leading to a slow switching speed. In contrast the power MOSFET is inherently a high speed switching device because it is a unipolar device in which no minority carrier current flow is involved. This paper discusses a technique developed to improve the switching speed of the power MOSFET integral diode by using electron irradiation. The impact of the electron irradiation upon both the integral diode and the power MOSFET characteristics are described below.

TYPICAL DIODE REVERSE RECOVERY CHARACTERISTICS

Before discussing the impact of electron irradiation, typical diode reverse recovery characteristics desirable for the above applications will be discussed here for reference. In order to illustrate the desirable diode reverse recovery characteristics, a typical set of reverse recovery current and voltage transients for a good discrete diode are shown in Fig. 2. To obtain these waveforms, the Motorola MR854 diode was connected across the power MOSFET in both the upper and lower leg of a half bridge circuit which was used to supply current to an inductive load representive of a motor. The gate drive voltage to the power MOSFETs was then adjusted so as to limit the peak reverse recovery current in the diode (in

Reprinted from *IEEE Ind. Appl. Soc. Meet.*, 1981, pp. 763–776.

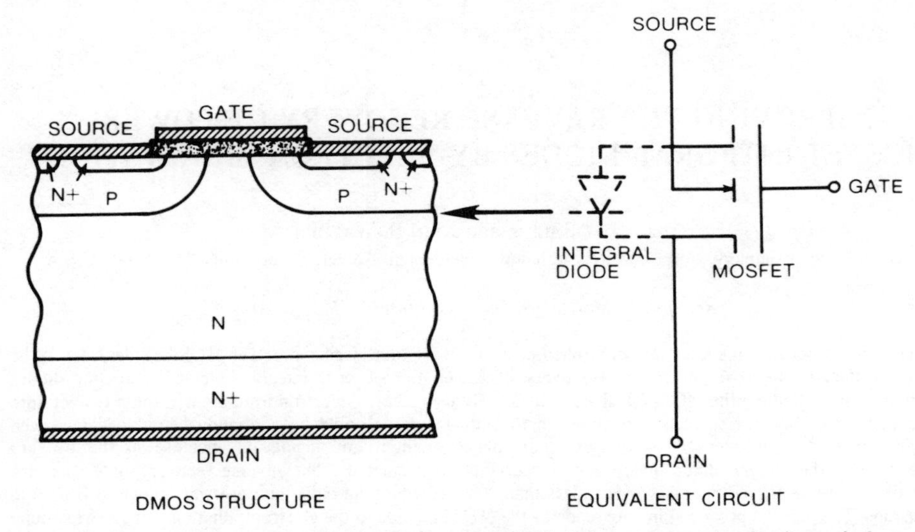

Fig. 1. Basic cross-section of the vertical channel power DMOS device structure showing its equivalent circuit containing the reverse conducting integral diode formed between the p-base region and the N− drift layer.

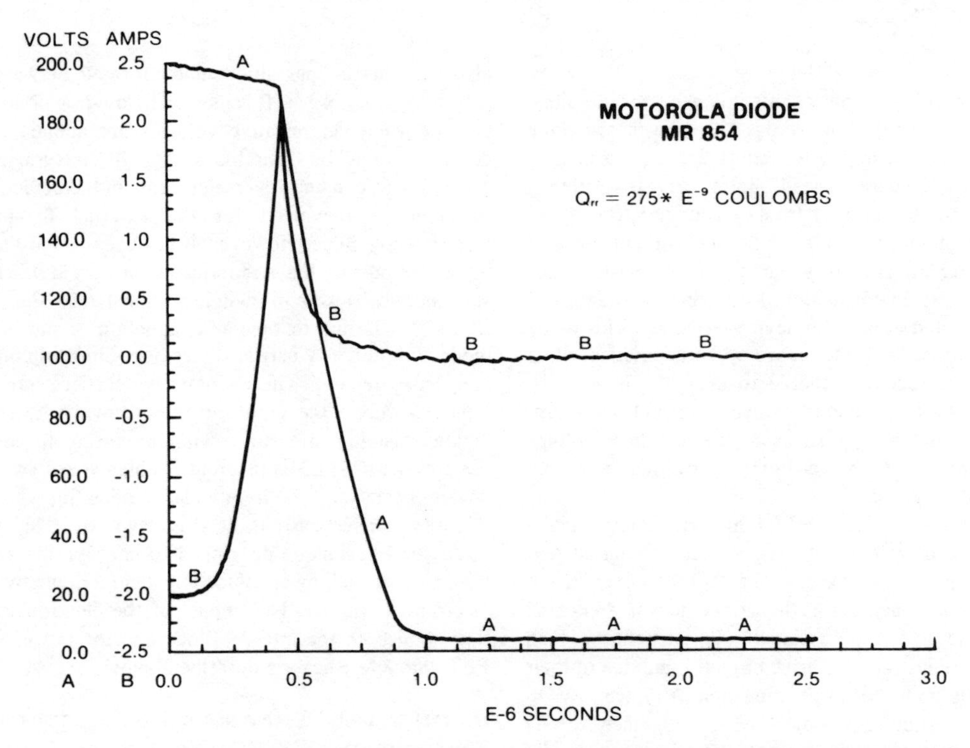

Fig. 2. Reverse recovery switching transients for a good discrete diode (Motorola MR 854). Note that the waveform A is the voltage across the power MOSFET connected to the upper link and waveform B is the current flowing in the diode. The current flowing through the power MOSFET is also represented by waveform B if the current zero is shifted down by 2 A.

this case 2 A) so as not to exceed the maximum current handling capability of the power MOSFETs (4 A for the devices used here). As discussed elsewhere[2], in these circuits, the power MOSFET must handle a peak current consisting of the sum of the motor current and the peak reverse recovery current flowing through the diode during its reverse recovery transient. As can be seen from Fig. 2, during this reverse recovery transient the voltage across the power MOSFET is equal to the link voltage resulting in simultaneous high current and high voltage on the power switch. This transient, therefore, controls the power dissipation in the devices. It is obviously desirable to minimize the reverse recovery time while limiting the peak reverse diode current. The discrete diode used here was adequate for keeping the switching times below 1 μsec while maintaining the peak reverse current at 2 A. This provided an acceptable power loss in the power MOSFET.

When the discrete diode was removed and the integral diode of the power MOSFET used to provide the reverse current flow, it was found that the gate drive voltage had to be substantially slowed down in order to keep the peak reverse recovery current at 2 A. The current and voltage waveforms in this case are shown in Fig. 3. It can be seen that the turn-off now takes place in 3 μsec instead of the less than 1 microsecond turn-off time observed with the discrete diode. This produces a corresponding (about 3 fold) undesirable increase in the power dissipation in the power MOSFET. This increase in switching time is directly related to the much large stored charge (Q_{rr}) of 530 nC in the power MOSFET integral diode when compared with the 275 nC stored charge of the discrete diode. As a result of this, the integral diode of the as-fabricated power MOSFET cannot be used to provide the reverse current conduction function in the above applications.

DIODE SPEED CONTROL

It has been previous demonstrated that the reverse recovery of power rectifiers can be controlled by the introduction of deep level recombination centers in their N− base region. This can be accomplished by either doping the silicon with gold and platinum, or by creating lattice damage via electron irradiation [3, 4]. In this study, electron irradiation was chosen as the lifetime control process because it can be performed in a highly controlled manner after complete device fabrication. The energy for the electron irradiation was chosen as 3 MeV because it has been found that the radiation process is efficient for creating recombination centers when the energy exceeds 1 MeV [5]. The radiations were performed at room temperature with doses up to 16 Mrad. It is worth pointing out here that although this process has been applied in the past for controlling lifetime in bipolar devices, this is the first time that it has been applied to control lifetime in a power MOS device structure. Consequently, in addition to the integral diode characteristics, all the power MOSFET characteristics were also monitored as a function of the radiation dose to look for any undesirable effects.

INTEGRAL DIODE CHARACTERISTICS

It was found that the electron irradiation was indeed very effective in reducing the stored charge (Q_{rr}) in the integral diode of the power MOSFET. This resulted in a greatly improved turn-off switching transients. An example for the turn-off transients observed for the case of a power MOSFET radiated with a dose of 4 Megarads is shown in Fig. 4. It can be seen that the stored charge in the integral diode has been reduced from 530 to 210 nC. This resulted in the turn-off time being reduced to less than 1 μsec. This makes the integral diode compare very favorably with the discrete diode. Increasing the radiation dose further resulted in even superior turn-off transients until the dose was raised to 16 Mrad. An example of the switching transients observed after the diodes had been radiated with a dose of 16 Mrad is shown in Fig. 5. It can be seen that the turn-off time has now been shortened to about 200 nsec. At this point, the switching transients are so fast that some undesirable oscillations are starting to appear in the waveforms.

During these experiments, it was found that the electron irradiation provided a very precise and controlled

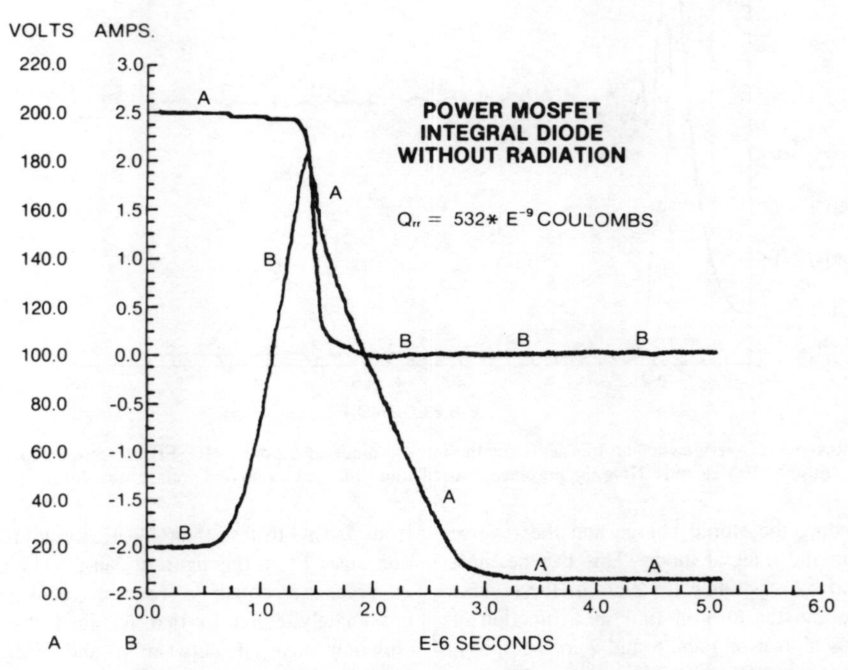

Fig. 3. Reverse recovery transients for the integral diode of an as-fabricated power MOSFET. Note the much longer turn-off time when compared to the discrete diode.

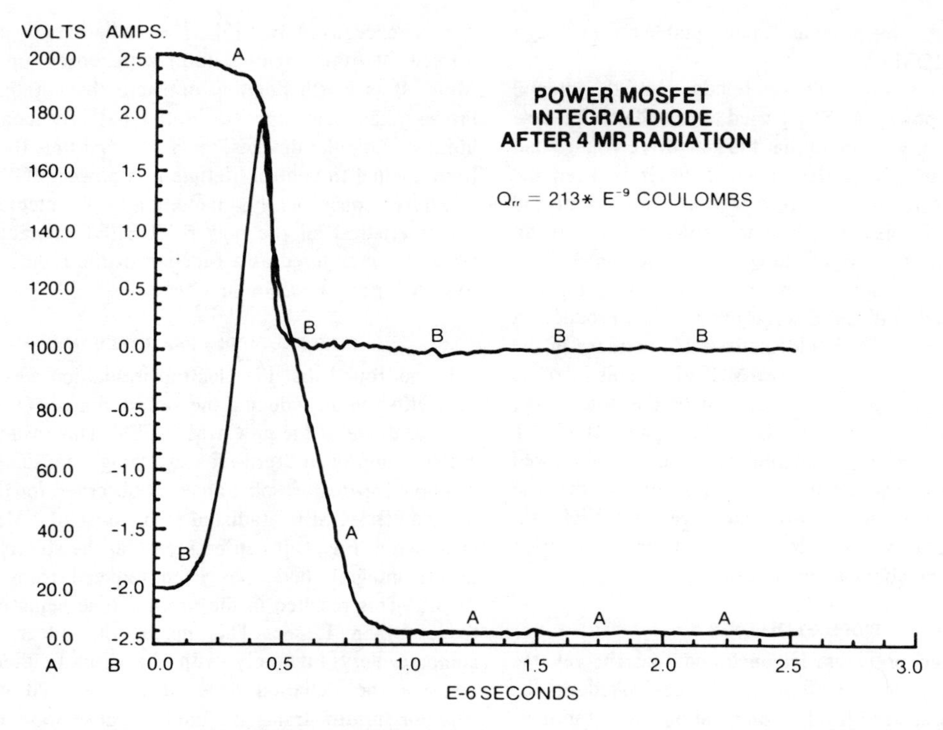

Fig. 4. Reverse recovery switching transients for the integral diode of a power MOSFET after radiation with a dose of 4 Megarands. Note that turn-off now occurs in the same time scale as for the discrete diode.

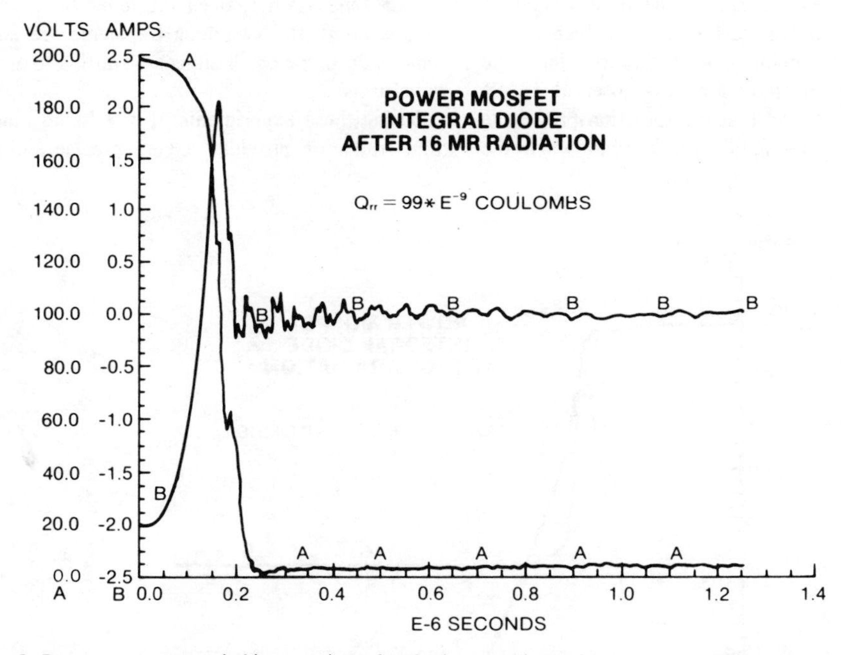

Fig. 5. Reverse recovery switching transients for the integral diode of a power MOSFET after radiation with a dose of 16 Megarads. Note the presence of oscillations on the current and voltage waveforms.

means for adjusting the stored charge and the reverse recovery time in the integral diode. This can be most clearly illustrated by examining the plots of the reverse recovery charge and the turn-off time as a function of radiation dose as shown in Figs. 6 and 7, respectively. From Fig. 6 it can be seen that the reverse recovery charge in the integral diode can be controllably reduced from 530 nC to less than 100 nC by adjusting the radiation dose. From this figure it can also be concluded that in order to obtain a reverse recovery charge approximately equal to that of good discrete diode, a radiation dose of between 1 and 2 Mrad should be used. Similar conclusions can also be derived from Fig. 7 by examining the reverse recovery time as a function of

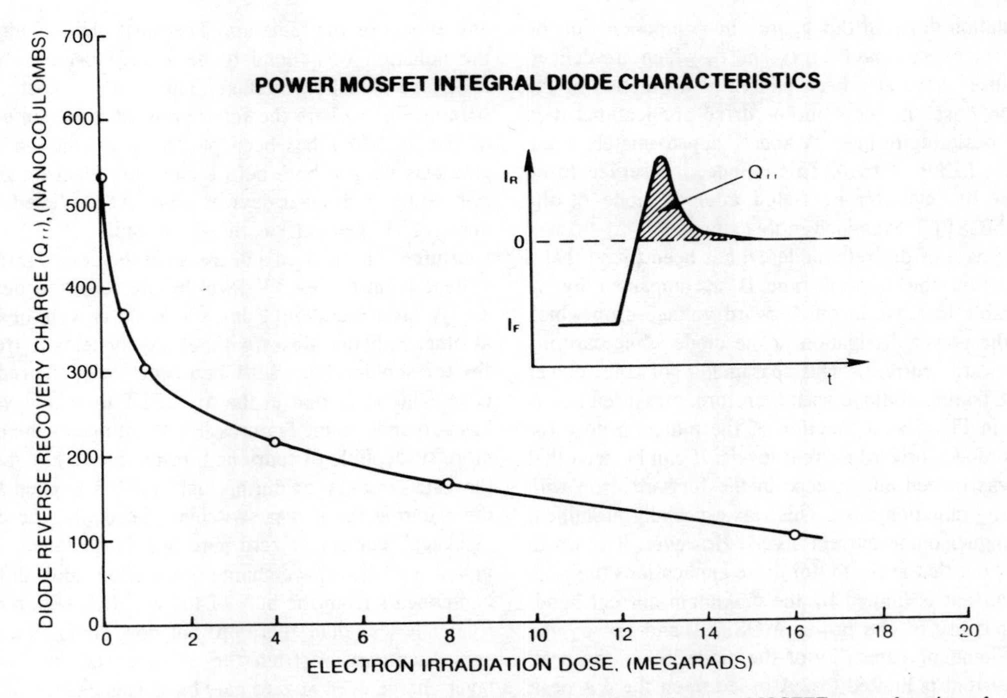

Fig. 6. Reduction in the reverse recovery charge of the integral diode of power MOSFETs as a function of increasing radiation dose. Note the well controlled dependence upon the radiation dose. The inset indicates how the reverse recovery charge was measured by integration of the diode reverse recovery current waveform.

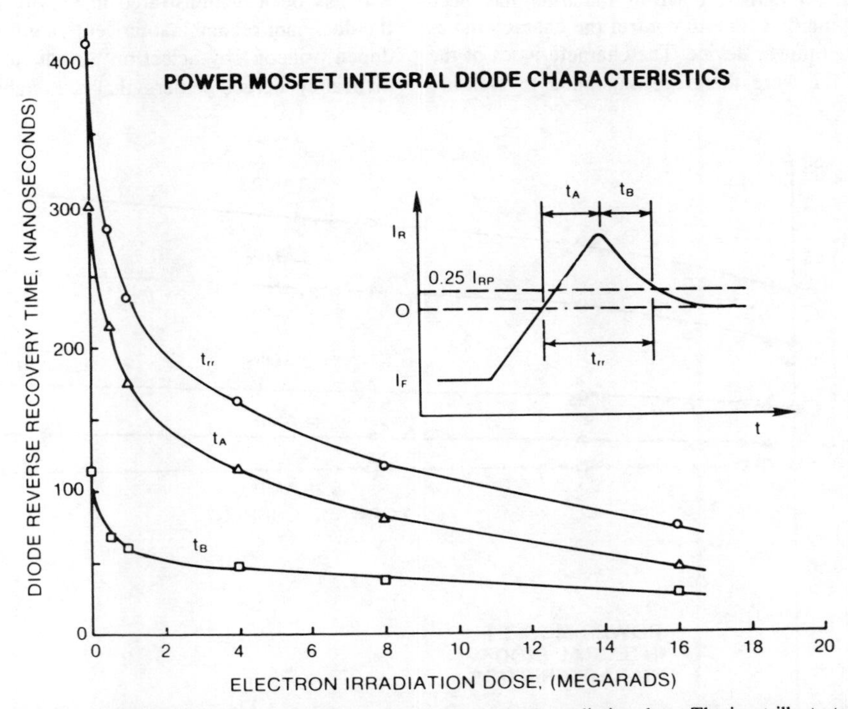

Fig. 7. Dependence of the integral diode reverse recovery time upon the radiation dose. The inset illustrates how the turn-off time t_{rr} and is components t_A and t_B have been defined here. Note that both t_A and t_B are reduced in equal proportion with increasing radiation dose, thus, maintaining an approximately constant ratio t_A/t_B).

the radiation dose. In this figure, the components of the reverse recovery time (t_{rr}), t_A and t_B which are defined in the inset, have also been plotted as a function of the radiation dose. In these motor drive applications it is usually desirable to keep t_A and t_B approximately equal (within a factor of two). This is indeed observed to be true for the electron irradiated integral diode of the power MOSFET examined in this study.

In the case of discrete diodes, it has been found that a reduction in the turn-off time is accompanied by an undesirable increase in the forward voltage drop which raises the power dissipation in the diode while carrying the forward current. This parameter of the power MOSFET integral diode was, therefore, measured and is plotted in Fig. 8 as a function of the radiation dose for various diode forward current levels. It can be seen that there was indeed an increase in the forward drop with increasing radiation dose. This was especially prominent at the higher diode current levels. However, it is worth pointing out that in the motor drive applications the peak diode current is limited by the maximum current handling capability of the power MOSFET and not by the current handling capability of the diode. Since the peak diode current is limited to 2 A based upon the 4 A peak current rating of power MOSFETs used here, the radiation was found to have a very small impact upon the forward drop of the integral diode.

POWER MOSFET CHARACTERISTICS

As pointed out earlier, electron radiation has been used here for the first time to control the characteristics of a MOS gated power device. The characteristics of the power MOSFET were therefore examined to measure any effects of the radiation. The most obvious effect of the radiation was found to be a large decrease in the MOSFET threshold voltage. This can be most clearly seen in Fig. 9 where the square root of the drain current of the MOSFET has been plotted as a function of the gate bias voltage both before and after electron irradiation. In this particular device, which was radiated with a dose of 1 Megarad, a threshold shift of 3.2 V was measured. This caused a decrease in the device threshold voltage from the +4.3 V level before radiation down to +1.1 V after radiation. Comparable shifts were observed at other radiation doses without any systematic trend in the threshold voltage shift as a function of the radiation dose. This reduction in the MOSFET threshold voltage has two undesirable features. Firstly, it makes the device more susceptible to spurious turn-on due to any noise in the gate circuitry or during high dV/dt's applied to the drain during the power switching. Secondly, the device "leakage" current at zero gate bias is no longer determined by the space-charge generation and diffusion components from the bulk of the N− drift region region (typically less than 10 μamp) but now contains a larger component arising from the presence of an inversion layer charge even at zero gate bias. This channel leakage current was found to range from a few milliamperes on up to even several hundred milliamperes after the radiation. This magnitude of the leakage current causes unacceptably high power dissipation in the MOSFET when blocking current flow.

It has been demonstrated in previous studies[6] that the dominant recombination centers introduced in lightly doped silicon by electron irradiation are due to divacancy defect centers that are stable up to 300°C.

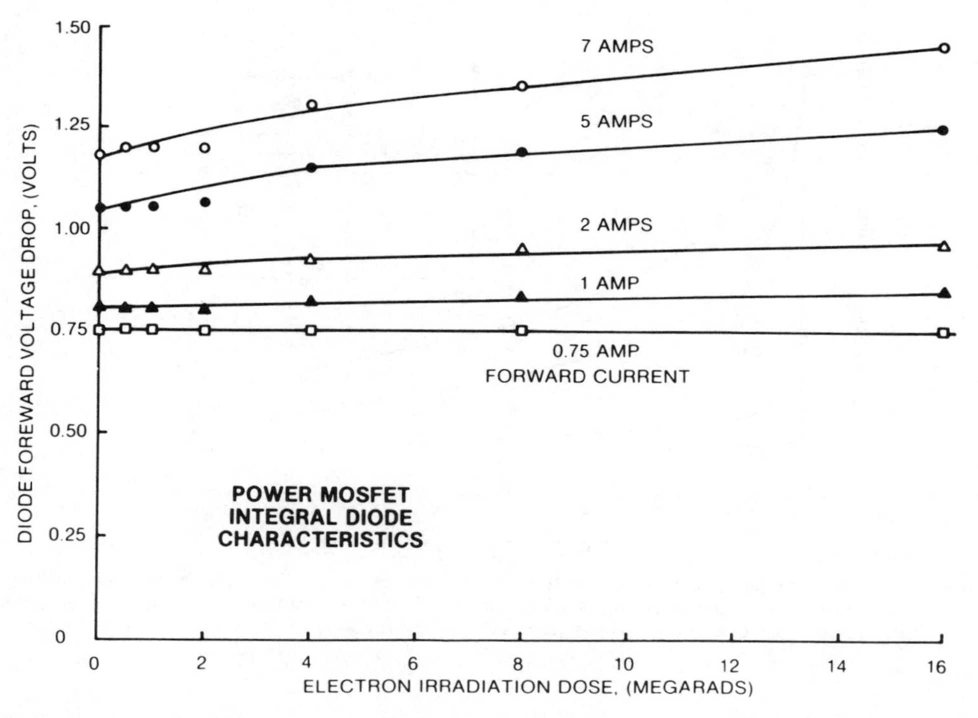

Fig. 8. Impact of the electron radiation upon the foreard voltage drop of the integral diode. A slight increase in the forward voltage drop is observed with increasing radiation dose.

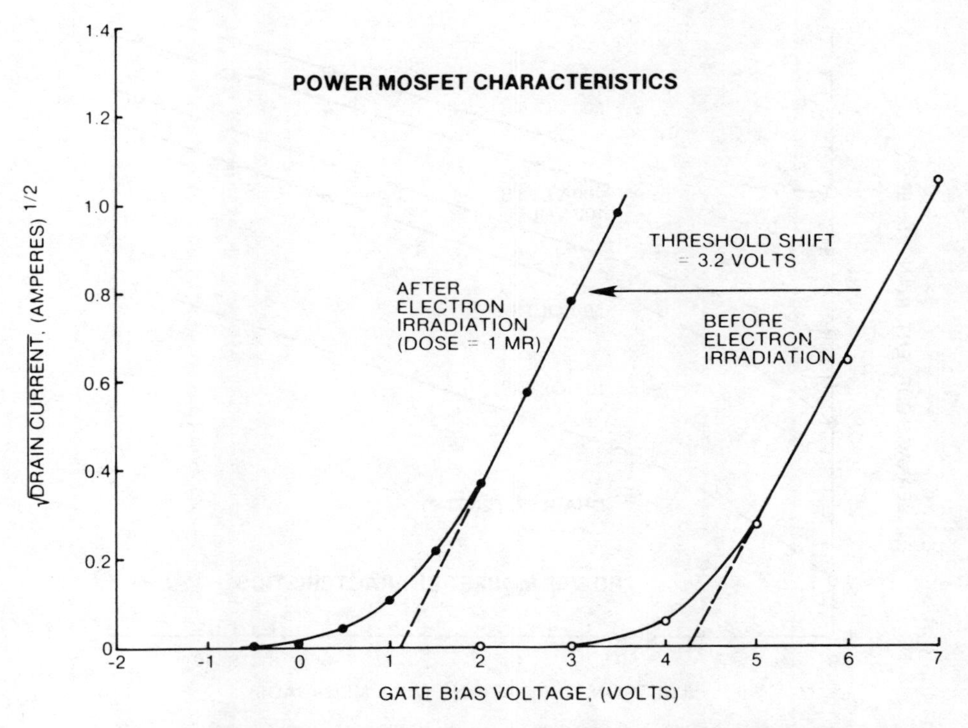

Fig. 9. Threshold voltage reduction of the power MOSFET as a result of the electron irradiation.

These centers are consequently stable during device operation as experience has shown in the case of power rectifiers and thyristors. A reduction of the threshold shift arising from the introduction of positive charge in the MOS gate oxide by the electron radiation was, therefore, attempted at 140°C so as to not affect the reverse recovery characteristics of the integral diode. It was found that annealing the devices at 140°C in a nitrogen ambient was very effective in reducing the channel leakage current as illustrated in Fig. 10. From this figure, it can be seen that the leakage current of the MOSFET is reduced from over 1 mA immediately after radiation down to less than 10 μA after annealing for 7 hr. This leakage current is sufficiently low for using the power MOSFET in its blocking mode without significant power dissipation being incurred. It is worth pointing out here that even after the 7 hr anneal, there is still some channel leakage current contribution because the device threshold voltage increased from the 1 V range to the 3 V range but did not fully recover to the preradiation value of about 4 V. This was verified by the application of a negative gate bias to the gate electrode of the MOSFET during the leakage current measurements. It was found that the negative gate bias reduced the leakage current from over 1 mA at zero gate bias to less than 100 nA. In all these measurements, made over a broad range of blocking voltages and for various radiation doses, it was found that beyond a certain negative gate bias voltage, the leakage current became invariant. This leakage current is representative of the bulk leakage current after the channel leakage contribution had been eliminated by the negative gate bias. This bulk leakage current is plotted in Fig. 11 as function of the radiation dose using

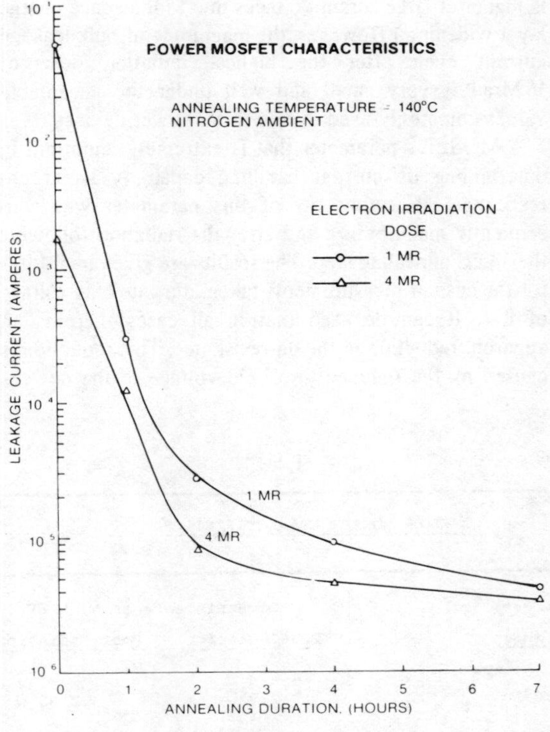

Fig. 10. Reduction in the channel leakage current of the power MOSFET by low temperature annealing.

the drain voltage as a parameter. From this data it can be seen that the radiation induced deep levels in the bulk cause a progressive increase in the device leakage current with increasing radiation dose. (This leakage current

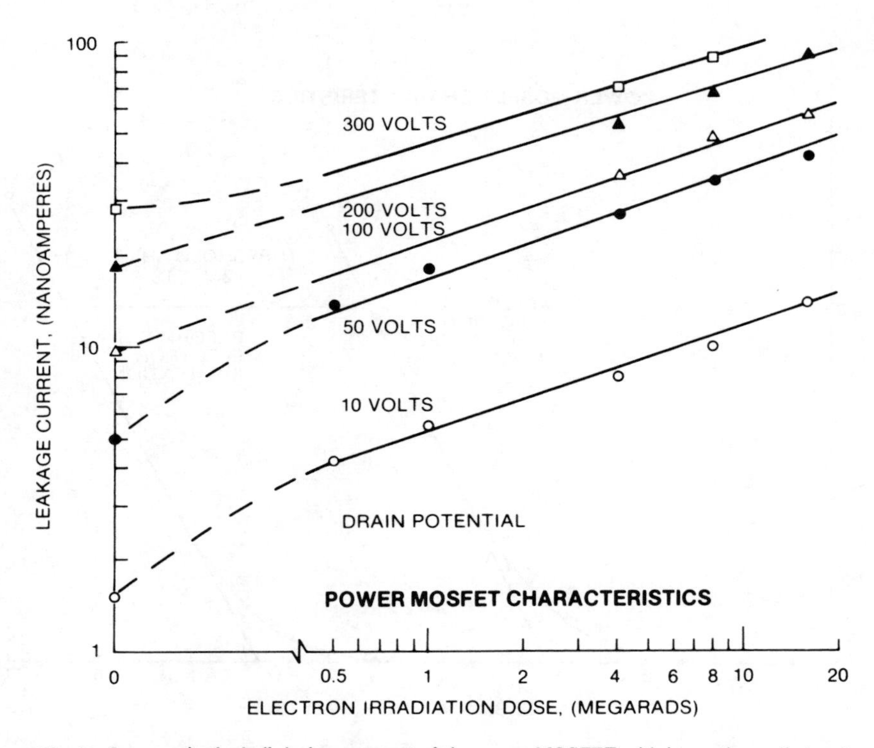

Fig. 11. Increase in the bulk leakage current of the power MOSFET with increasing radiation dose.

is higher at larger drain voltages due to the space charge layer widening.) However, the magnitude of bulk leakage current even after the highest radiation dose of 16 Mrad is very small and well under the acceptable values commonly used to rate these power devices.

A MOSFET parameter that is extremely important in determining its current handling capability is its on-resistance. Measurements of this parameter was consequently made before and after the radiation following the 140°C annealing step. The results are given in Table 1 for the case of measurements taken at a gate bias voltage of 8 V. It can be seen that in all cases there is an apparent reduction in the on-resistance. This reduction is caused by the reduced threshold voltage of the devices even after the anneal. This is a desirable feature in terms of reducing the power losses in the MOSFET during forward current conduction. It is also indicative of no significant reduction in the electron mobility in either the inversion layer or the N− drift region due to radiation induced defect centers. This can be most conclusively demonstrated by plotting the MOSFET transconductance as a function of the gate bias voltage for devices that have been irradiated to various doses and annealed at 140°C for various lengths of time as shown in Fig. 12. In Fig. 12 it can be observed that the slope of the plots are all equal even though the curves are shifted with respect to each other due to differences in the threshold voltage among the devices. Since the transconductance (g_m) is proportional to the inversion layer mobility and the slopes of the curves in Fig. 12 are equal, it is clear that there is no change in the electron mobility.

Table 1.

POWER MOSFET CHARACTERISTICS

Dose (MR)	On-Resistance at V_G = 8V Pre-Radiation	Post-Radiation
0.5	2.9 Ω	2.35 Ω
1	2.8 Ω	2.26 Ω
2	2.9 Ω	2.46 Ω
4	3.5 Ω	2.82 Ω
8	3.2 Ω	2.87 Ω
16	3.2 Ω	2.96 Ω

CONCLUSIONS

It has been determined that the integral diode in as-fabricated power MOSFETs has too large a reverse recovery charge which causes long turn-off times in motor drive circuits. This makes the integral diode a poor component for use in these circuits to carry the reverse current across the power device due to excessive power dissipation caused by the prolonged turn-off. The use of 3 MeV electron irradiation has been demonstrated to be a very effective technique for reducing the reverse recovery charge in a controlled manner by monitoring the radiation dose. For the power MOSFETs used in this study a radiation dose of between 1 and 8 Megarads was found to produce the optimum diode reverse recovery characteristics. Faster turn-off speeds with lower switch-

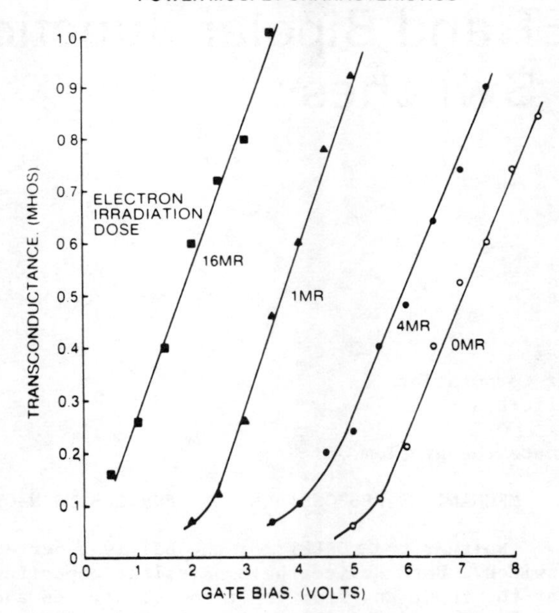

Fig. 12. Tranconducgance of power MOSFETs electron irradiated to various doses and annealeld at low temperatures. Note that the slope of all the curves are equal indicating no effect of the radiation upon the electron mobility.

ing power losses were achieved with progressive increase in the radiation dose. When a radiation dose of 16 Mrad was used, the reverse recovery because so rapid that circuit oscillation began to show up in the current and voltage waveforms.

The most prominent effect of the electron radiation upon the power MOSFET characteristics was a decrease in the device threshold voltage. This resulted in an undesirable channel leakage current flow when the devices were required to block current flow at zero gate bias. It also made the device more susceptible to undesirable turn-on due to noise in the gate circuitry. A reduction in this threshold shift was achieved by low temperature (140°C) annealing in nitrogen. This annealing process was successful in reducing the channel leakage current in these devices to less than 10 μA. Although this leakage current is satisfactory for efficient device performance, a channel leakage current component was still present after the annealing as evidenced by further reduction in the device leakage current when a negative gate bias was applied. The electron radiation damage in the bulk did cause an increase in the leakage current due to contributions to the generation and diffusion currents. This leakage current was observed to be less than 100 nA at room temperature. While the annealing process improved the power MOSFET leakage current it did not alter the integral diode reverse recovery characteristics as measured right after the radiation. Other MOSFET parameters remained unaffected by the radiation.

REFERENCES

1. B. J. Baliga, Silicon power field controlled devices and integrated circuits, In *Silicon Integrated Circuits*, Applied Solid State Science Series, Supplement 2B, (Edited by D. Kahng) Academic Press (1981).
2. S. Clemente and B. Pelly, A chopper using parallel connected power MOSFETs, *Power Conversion International*, pp. 22–45, April 1981.
3. B. J. Baliga and E. Sun, *IEEE Trans. Electron Dev.* **ED-24**, 685–688 (1977).
4. K. S. Tarneja and J. E. Johnson, Tailoring the recovered charge in power diodes using 2 MeV electron irradiation, Electrochem. Society Meeting, Paper 261 RNP (1975).
5. R. O. Carlson, Y. S. Sun and H. B. Assalit, *IEEE Trans. Electron Dev.* **ED-24**, 1103–1108 (1977).
6. A. O. Evwaraye and B. J. Baliga, *J. Electrochem. Soc.* **124** 913–916 (1977).

dv/dt Effects in MOSFET and Bipolar Junction Transistor Switches

R. Severns

International Rectifier Corporation
El Segundo, California

ABSTRACT

Spurious turn-on due to dV/dt triggering is a real possibility in high speed switching circuits using MOSFETs or bipolar junction transistors (BJTs). This paper discusses the mechanisms leading to spurious turn-on, test methods to determine dV/dt limits, the effect of dV/dt turn-on circuit operation, and methods to minimize dV/dt triggering in practical circuits.

INTRODUCTION

Most designers of power switching circuits are well aware of the dV/dt limitations of SCRs which, if exceeded, can cause these devices to turn on in the absence of a normal trigger pulse. Despite an earlier discussion of the problem (1) it is not generally appreciated that a similar, and equally detrimental phenomenon, can appear in both BJTs and MOSFETs. Until the last few years very few BJTs were available which could switch fast enough to induce false triggering in most circuits. In the special case of a resonant circuit being driven off resonance, it was possible to see dV/dt triggering with even relatively slow switching times but very few designers ever encountered this application so that dV/dt triggering remained a matter of academic rather than practical interest. In the last three to four years however, BJTs have shown dramatic improvements in switching times. It is now readily possible to obtain 450V devices that can turn on in 50-70 nsec. In addition, the power MOSFET has become a practical reality and a 450V MOSFET can be switched on in 4-5 nsec generating transitions exceeding 100V/nsec.

Off line converters operating with switching frequencies well above 100kHz are becoming very common. For a variety of practical reasons resonant converter topologies are frequently used in high frequency regulators. Many designers who formerly used only switchmode circuits are increasingly turning to resonant circuits.

Because of these device changes and the changes in circuit practice, spurious turn-on due to false triggering is no longer an academic possibility, it has become a practical problem. The following discussion is intended to illuminate the mechanisms causing the problem, to show test circuits determining the dV/dt capability of a device and to present some methods to circumvent or eliminate the problem.

MECHANISMS RESPONSIBLE FOR SPURIOUS TURN-ON

Neither the MOSFET nor the BJT is a perfect switch. Both devices have parasitic capacitances at the input and from input to output. In addition any practical drive circuit will have some residual resistance when the switch is off.

Equivalent circuits for the MOSFET and BJT taking into account these parasitic elements are shown in Figure 1. If a positive going voltage ramp is

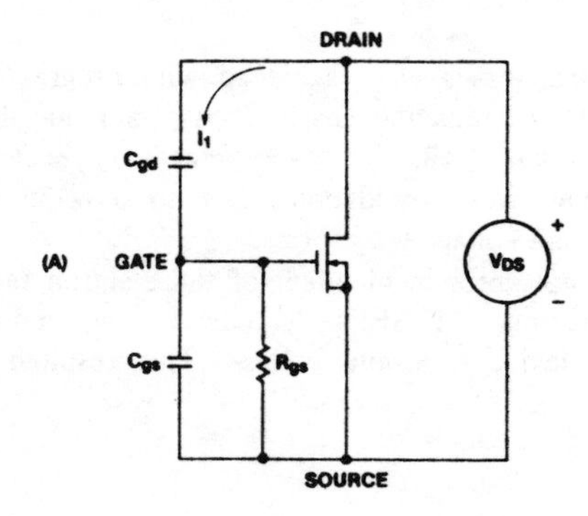

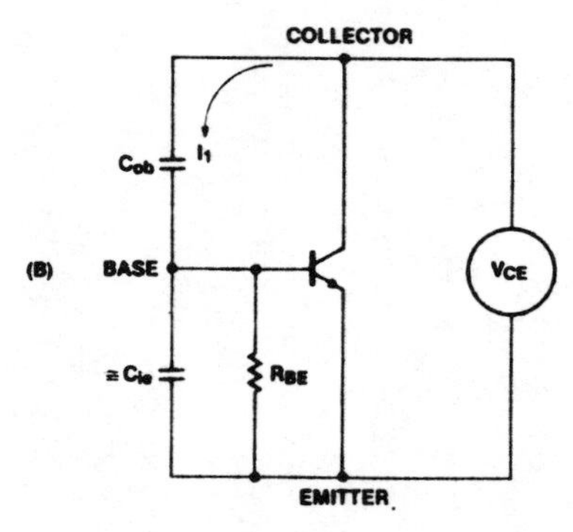

Figure 1

applied either drain-to-source or collector-to emitter a current, I_1, will flow through the feedback capacity (C_{gd} or C_{ob}). If I_1 is sufficiently large the voltage at the gate or base will exceed the turn-on threshold and the switch will turn on, even though the switch is nominally in the off state.

Reprinted from *IEEE Power Electron. Specialists Conf.*, 1981, pp. 258-264.

The important question is: does V_{gs} or V_{BE} exceed the device turn-on threshold or not? To answer this question the equivalent circuit in Figure 2 can be used.

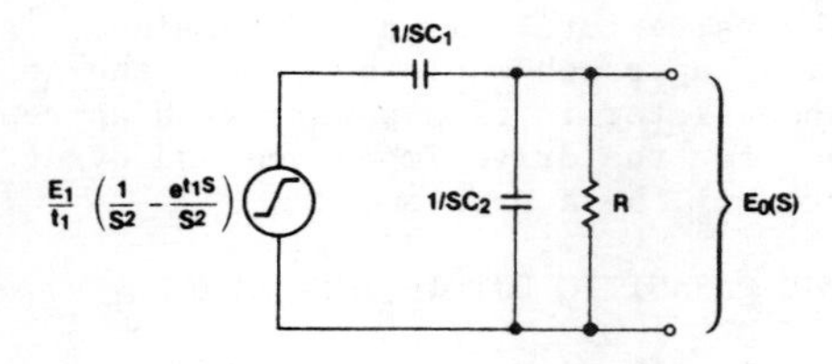

Figure 2

As is shown in appendix 1 the output voltage of the network is:

$$e_o(t)=\frac{(E_1)(RC_1)}{(t_1)}\left(c^{-t/R(C_1+C_2)}\right)\left(c^{t_1/R(C_1+C_2)}\right) \quad (1)$$

The waveform corresponding to this equation is shown in Figure 3.

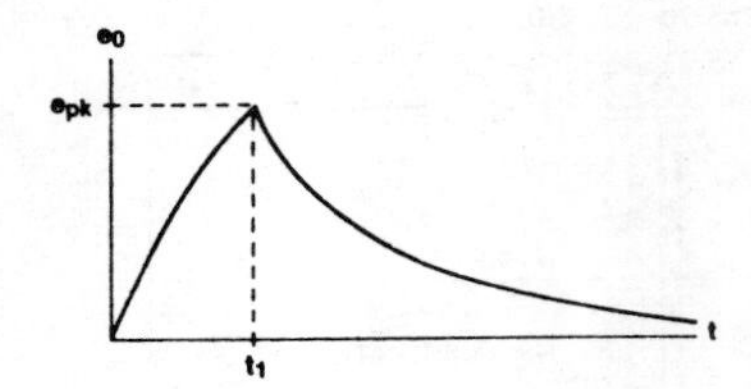

Figure 3

The waveform in Figure 3 applies to V_{GS} for a MOSFET whether below or above $V_{GS(th)}$ presuming that the drain impedance is zero or some small value. Obviously the waveform does not apply when the threshold voltage of the BJT base-emitter diode is exceeded. In that case the waveform would be clamped by the diode. This distinction is really of no great importance. What is important is whether or not e_{pk} is greater than the threshold voltage of the device. The expression for e_{pk} is:

$$e_{pk}=\left(\frac{dV}{dt}\right)(\tau_1)\left(1-e^{-\frac{E_1}{dV/dt}\frac{1}{\tau_2}}\right) \quad (2)$$

where: $\tau_1 = RC_1$

$\tau_2 = R(C_1 + C_2)$

The exponential term of equation 2 is usually quite small so for practical purposes equation 2 may be reduced to:

$$e_{pk} \cong \left(\frac{dV}{dt}\right) RC_1 \quad (3)$$

For a MOSFET where C_1 = 10pF, V_{th} = 3.5V and dV/dt = 50V/ns, the maximum gate to source resistance permissible to avoid false triggering is:

$$R_{gs} \leq \frac{V_{GS(th)}}{(dV/dt)(C_1)} \quad (4)$$

$$R_{GS} \leq 7\Omega$$

The internal gate resistance for such a device could well be as high as 1 to 2Ω. This means that the external resistance due to the drive circuit may need to be less than 5Ω. Strictly speaking this too is an approximation since the internal capacitance and gate resistance form a distributed network more complex than the simple RC network in Figure 2, but the point of this example is that the impedance of the driver must be quite low when the switch is in the off state and some allowance for the internal gate resistance is required.

COMPONENT VALUES

In order to perform the foregoing calculations it is necessary to know the values for C_1, C_2 and V_{th}. For a MOSFET C_1=C_{gd} and C_2=C_{gs}. As shown in Figure 4, C_{gs} changes little with variations in V_{DS}. In a BJT C_{ob} will also show a similar strong dependence on V_{CE}. If the calculation is to be performed then the designer must have a curve similar to that shown in Figure 4. If this information is not provided in the data sheet then the test procedure and circuits outlined in appendix 2 can be used to generate the necessary data.

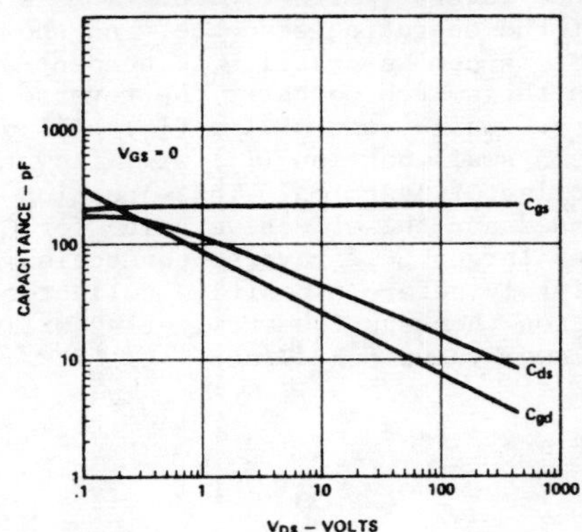

Figure 4

For a silicon BJT $V_{th} \cong$ .7V with a temperature coefficient of about -2 mV/°C. In a MOSFET, V_{th} can have a wide range of values ranging from 1.5 to 7 volts depending on the design of the device. The range of values for V_{th} are usually given in the data sheet but if a more exact value is required V_{th} may easily be measured on a transistor curve tracer. The wise designer will use the worst case low value for V_{th} as given in the data sheet rather than rely on measurements of a few devices which may or may not be typical. The temperature coefficient of V_{th} for the MOSFET can have a wide range of variance, -4 to -8 mV/°C is typical of present power MOSFETS.

Note that for both types of devices the threshold decreases as the temperature increases. At high temperatures the dV/dt capability will be reduced. Another temperature effect is the increase in internal gate resistance, R_G, with temperature in a MOSFET. For a silicon gate device the temperature coefficient is typically .7%/°C. If R_G is a significant portion of the total gate to source impedance then the dV/dt capability will again be decreased at elevated temperatures.

When a BJT is triggered on the time of conduction will be extended by the storage time. Since it is unlikely that during dV/dt triggering the device will be turned on hard enough to allow V_{CE} to fall below V_{BE}, the storage time will be relatively brief. The storage time is a strong function of temperature and at high junction temperatures the conduction time may well be increased significantly.

EFFECT OF REVERSE CURRENT CONDUCTION

Up to this point we have been discussing a model which presumes that immediately before a positive dV/dt is applied that the device is in the quiescent state with at least a small positive potential from drain-to-source or collector-to-emitter. Under these conditions C_1 will be small because the junction with which C_1 is associated will be reversed biased. This assumption is not always valid however.

In a resonant converter or a switching circuit with a reactive load it is possible for the switch to experience reverse current conduction as a normal part of the operating sequence. As shown in Figure 5, it is common practice to connect a diode (D_1) across the switch to carry the reverse load current, I_{LR}. While most of I_{LR} (I_1) will pass through D_1, a small portion (I_2) will flow through the base-collector junction. This junction is now forward biased and the effective value for C_1 can be 100 times larger. If reverse conduction is present immediately before a positive collector-emitter transition then spurious triggering will occur at a much lower value for dV/dt.

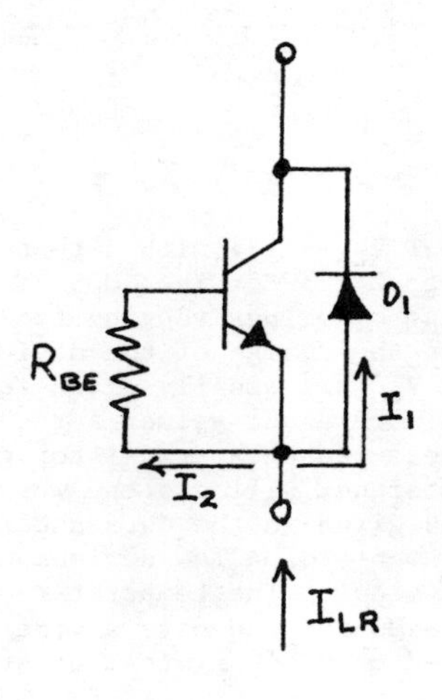

Figure 5

If this is a normal part of the circuit operation then it may be necessary to prevent reverse conduction entirely by using a series diode in the collector. It can be very frustrating to measure the value for C_1, when this junction is in forward conduction, with any accuracy. This is particularly so if the value for I_2 is not determined. A more practical approach would be to use the test circuit, shown later in this paper, with appropriate values for the drive impedance and dV/dt to see if there will be a problem.

THE PARASITIC BJT IN THE MOSFET

Inherent in all present power MOSFET structures there is a parasitic BJT. The presence of this BJT can be seen from Figure 6 which shows the structure of a typical vertical D-MOS device. The N^- epi region corresponds to collector of the BJT. The p diffusion corresponds to the base of the BJT and the N+ diffusion is the emitter. Notice that the aluminum source contact connects the N+ and p diffusions together effectively shorting the base emitter junction. Unfortunately, the p region will still have some resistance so that the parasitic BJT is not completely deactivated.

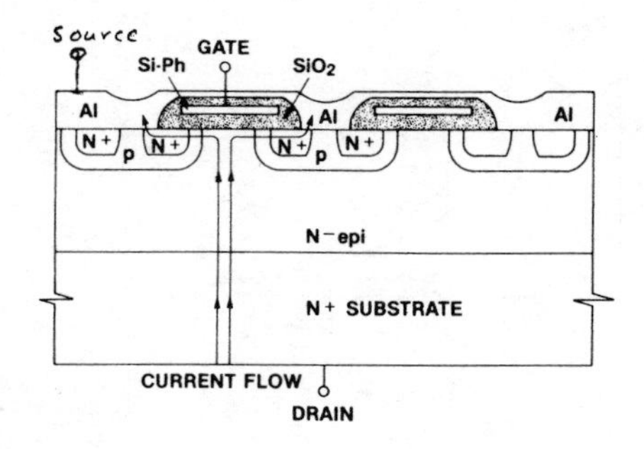

Figure 6

A more complete model for the power MOSFET including the gate resistance, interterminal capacitances and parasitic BJT is shown in Figure 7. In this type of device either the MOSFET or the BJT can be triggered on. The element which conducts first depends on the external gate to source resistance, R_{GS}. If R_{GS} is small (<1Ω) then it is the BJT that will trigger first. A typical 400V MOSFET

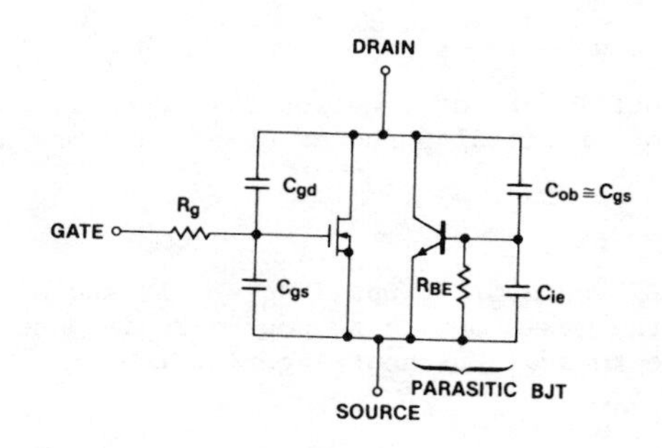

Figure 7

will have a dV/dt threshold of >75V/nsec which is more than adequate for most applications. In fact it is quite difficult to generate transitions that will produce false triggering in devices introduced since 1979. The earlier devices (1976-1978) are for the most part relatively small (V_{DS}=80,$R_{DS(ON)}$= 2.5Ω) and many of these can be turned on by transitions of 5 to 20 V/nsec.

dV/dt TEST CIRCUITS

If it is suspected that there may be a dV/dt problem in a particular design, the best course of action would be to measure the dV/dt capability of the switches being used. One possible circuit is given in Figure 8.

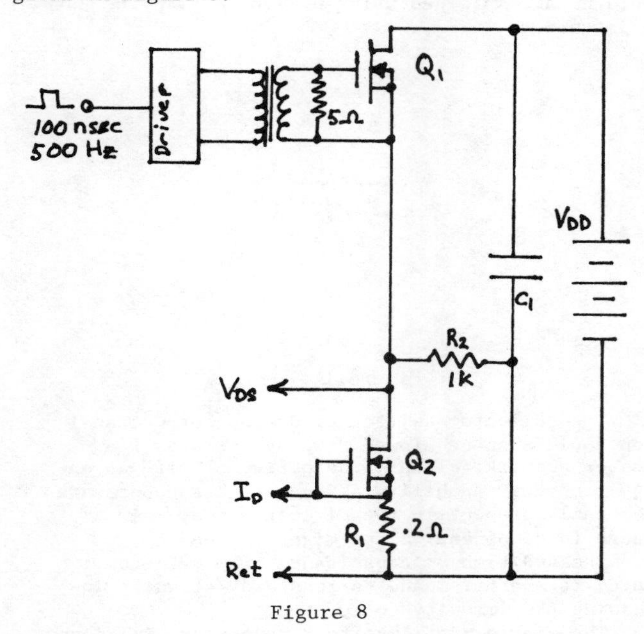

Figure 8

The device under test is Q_2. Q_1 is a very fast switch driven from a pulse generator through T_1. By driving the gate of Q_1 with a very fast pulse and an amplitude of 18 to 20V, Q_1 can be turned on in 5 nsec. From a 400V source this will provide a 80V/nsec transition across Q_2. The current through Q_2 is measured by looking at the voltage across R_1. R_1 is used for current viewing in preferance to a current probe for several reasons:

1. The upper frequency response of most probes is not adequate for the speeds present in this test circuit.

2. Most current probes will have delay times in excess of 10nsec. This makes it difficult to correlate the voltage and current waveforms properly in time.

3. The insertion of the current probe into the circuit will introduce an additional parasitic series inductance which will slow down the charging of C_{gs} and C_{gd} of Q_2. This is turn reduces the dV/dt of the transition from drain-to-source.

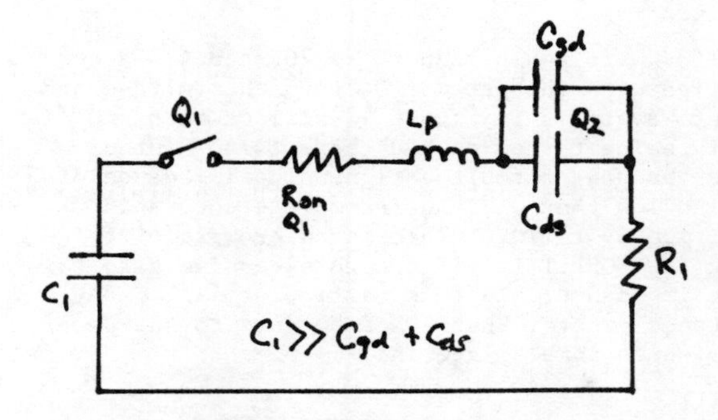

Figure 9 is an equivalent circuit. When Q_1 is turned on the current in the loop will ring as a damped sinusoid because the circuit is a series RLC. To achieve a rapid rise in voltage across Q_2, it is imperative that the parasitic inductance, L_p, be as small as possible.

Actual test waveforms for a power MOSFET are shown in Figures 10, 11 and 12 for an IVN5200 (.5Ω, 80V). In Figure 10 a 70V, 6 nsec transition is applied to Q_2. The current through Q_2 is, as predicted, a damped sinusoid, symetrical about zero. The transition voltage is next increased to 80V and, as shown in Figure 11, spurious triggering occurs. The current waveform is no longer a damped sinusoid. After an initial one cycle of ringing a DC current of approximately 1 A flows for 60nsec. This is due to triggering of the parasitic BJT.

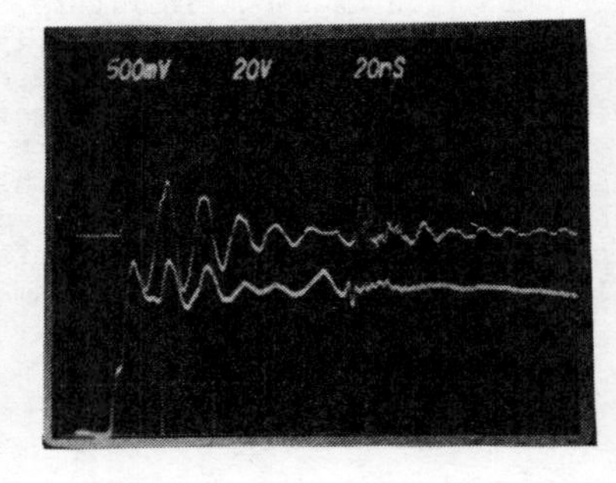

Figure 10

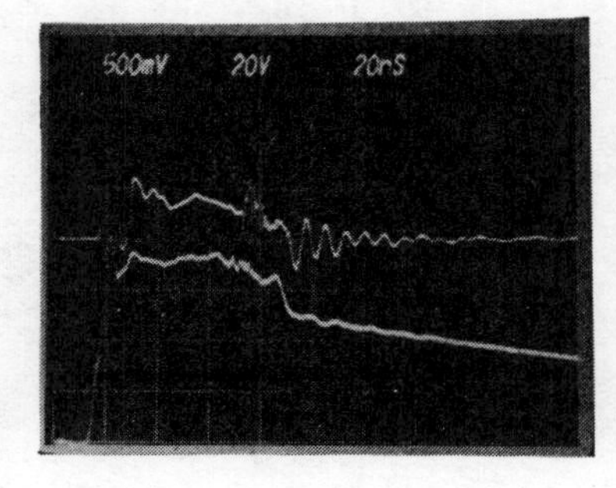

Figure 11

If V_{DD} is again reduced to 70V and 47Ω added in series with the gate of Q_2 then the voltage and current waveforms in Figure 12 will be seen. Q_2 is clearly being turned on and there is no hint of a series resonant circuit action. The impedance of Q_2 is so low that Q_1 cannot complete the rapid transition up to 70V. This is an example of triggering the MOSFET itself due to excessive gate to source impedance. In this particular device any resistance greater than 10Ω from gate to source would induce some triggering.

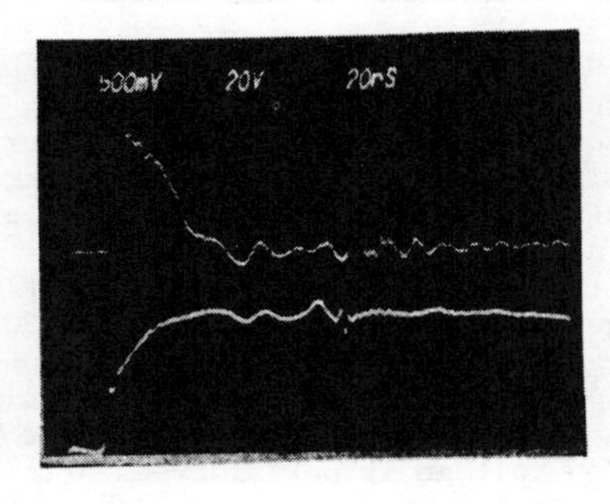

Figure 12

To examine the behavior of a device when reverse conduction is present, the test circuit in Figure 8 can be modified as shown in Figure 13 to include an inductor, L_1. A double pulse drive is now employed so that during the first pulse energy is stored in L_1. During the interpulse interval L1 discharges through Q_2 in the reverse direction until the beginning of the second pulse. This allows the dV/dt capabilities to be measured when reverse conduction is present. Again the repitition rate is kept low to limit dissipation to a small value.

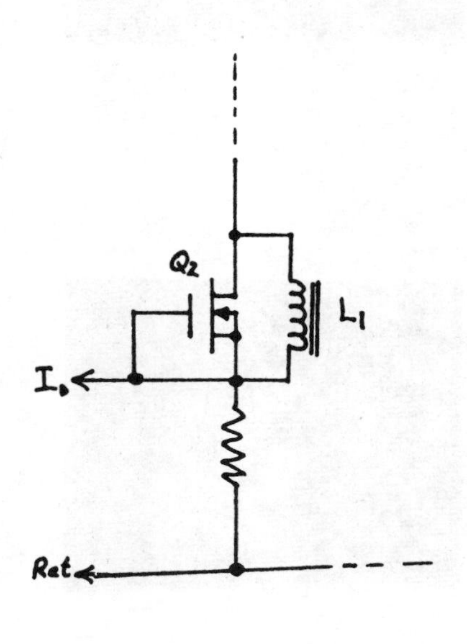

Figure 13

MEANS FOR REDUCING dV/dt TRIGGERING

There are a number of things that can be done to reduce dV/dt turn-on. The simplest and most obvious means would be to slow down the switching time. In the case of the power MOSFET a series resistance at the gate terminal can be used to produce any desired transition time. As was pointed out earlier however, any increase in the gate-to-source resistance decreases the dV/dt capability of the MOSFET proper. The addition of gate resistance will reduce the chances of turning on the parasitic bipolar but does nothing for the MOSFET itself unless the drive resistance is asymetrical, i.e. the resistance during the off state is smaller than during the on state. Figure 14 shows how this could be accomplished using a diode across the gate resistance.

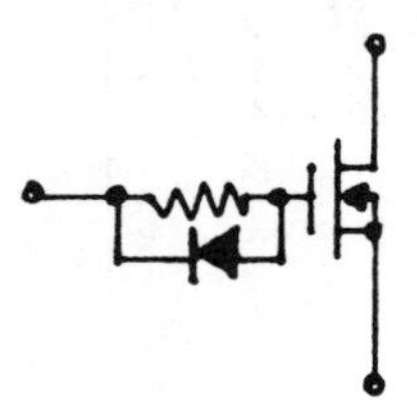

Figure 14

The collector-emitter or drain-source transition could also be slowed down by the use of a snubber network, either dissipative or with an energy recovery capability. Since such snubbers are frequently present in the circuit anyway only a change in component values might be needed.

The addition of capacitance from gate-to-source is another means to reduce dV/dt while increasing the immunity to turn-on.

The single most important means for limiting spurious turn-on is a low value for R_{GS} or R_{BE} during the off state.

Another possibility would be to apply a negative bias to the base or gate during the off state effectively requiring a large value for dV/dt in order to reach V_{th}.

If, however, there is a problem with reverse conduction then the foregoing methods may not be adequate. The most effective method in that case would be to eliminate the reverse conduction entirely by inserting a series diode in the collector or drain as shown in Figure 15. D_1 can be a low

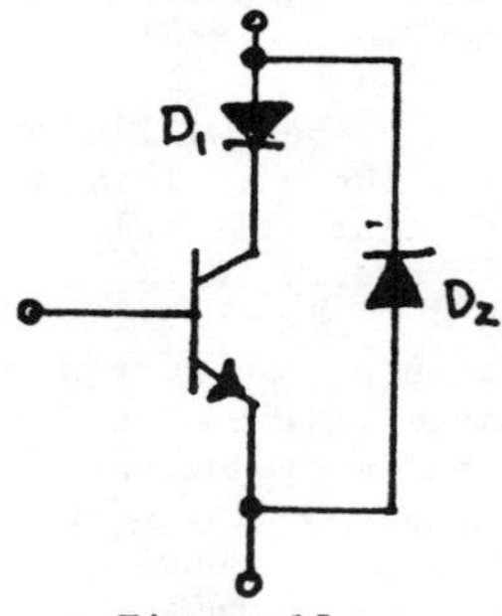

Figure 15

voltage Schottky diode to achieve low forward losses since the reverse voltage is limited to about one volt by D_2. D_2 must be rated to carry the full reverse current and have the same breakdown voltage as the switch.

It should be pointed out that BJT switches, using regenerative proportional base drive, are particularly susceptible to this type of false triggering. Even a slight degree of turn-on may be sufficient to trigger the drive circuit to turn-on the switch fully which will normally destroy the switch and may also destroy other switches in the circuit.

CONSLUSIONS

Spurious turn-on due to dV/dt triggering is a real possibility in high speed switching circuits. A simplified model has been examined which provides a first order explanation of the problem. Triggering of the parasitic BJT in the MOSFET is not a significant problem but triggering of the MOSFET itself or of a normal BJT is possible expecially if reverse condition is present in the switch. The designer however, has a number of means available to eliminate this problem in most circuits. What is most important is that the designer recognize the possibility of the problem and take steps to minimize it.

BIBLIOGRAPHY

(1) "The Power Transistor in its Environment", Thompson-CSF Semiconductor Division, Pages 165 to 180, October 1978.

(2) R. Severns, "Using the Power MOSFET as a Switch", Intersil Applications Note AO-36, January 1981.

APPENDIX 1

DERIVATION OF EQUIVALENT CIRCUIT TRANSIENT RESPONSE

For the purposes of calculating the voltage at the gate or base, the equivalent circuit shown in Figure 2 can be used with the appropriate component values. The ramp function is provided by the combination of a positive until ramp starting at t=0 and a negative unit ramp at $t=t_1$. The slope of the ramp, dV/dt, is:

$$\frac{dV}{dt} = \frac{E_1}{t_1} \qquad \text{(A1-1)}$$

The output voltage, $E_o(s)$, can be shown to be:

$$E_o(S) = \left(\frac{E_1}{t_1}\right)\left(\frac{C_1}{C_1+C_2}\right) \bullet \left[\frac{1}{S(S+1/R(C_1+C_2))} - \frac{e^{-t_1 s}}{S(S+1/R(C_1+C_2))}\right] \qquad \text{(A1-2)}$$

By applying the following inverse Laplace transformations to equation A1-2:

$$\frac{1}{S(S+\alpha))} \Longrightarrow \frac{1}{\alpha}\ (1-e^{-\alpha t}) \qquad \text{(A1-3)}$$

$$F(S)e^{-t_1 S} \Longrightarrow F(t-t_1) \qquad \text{(A1-4)}$$

The time domain response for $e_o(t)$ can be obtained as:

for: $t \leq t_1$

$$e_o(t) = \frac{E_1 RC_1}{t_1}\left(1-e^{-t/R(C_1+C_2)}\right) \qquad \text{(A1-5)}$$

and for $t \geq t_1$

$$e_o(t) = \frac{E_1}{t_1}\ (RC_1)\ \left\{e^{-t/R(C_1+C_2)}\right\} \bullet \left\{e^{t_1/R(C_1+C_2)}\right\} \qquad \text{(A1-6)}$$

The waveform represented by equations (A1-5) and (A1-6) was shown in Figure 3. For practical purposes, the only feature of the waveform that is of interest is the peak value of e_o. By combining equations (A1-1) and (A105) and setting $t=t_1$, the expression for e_{pk} becomes:

$$e_{pk} = \left(\frac{dV}{dt}\right)\ (\tau_1)\ \left(1-e^{-\frac{E_1}{dV/dt}}\right)\left(\frac{1}{\tau_2}\right) \qquad \text{(A1-7)}$$

where $\tau_1 = RC_1$ (A1-8)

$\tau_2 = R(C_1+C_2)$ (A1-9)

APPENDIX 2

POWER MOSFET SMALL SIGNAL CAPACITANCE MEASUREMENTS

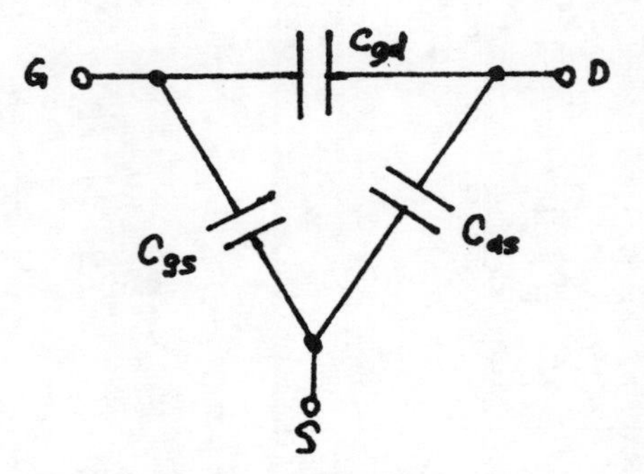

Figure A2-1

All of the capacitances are functions of V_{DS}.

Definitions

$C_{iss} \equiv C_{gs} + C_{gd}$, C_{ds} = short circuit

$C_{oss} \equiv C_{ds} + \frac{C_{gs}\ C_{gd}}{C_{gs} + C_{gd}}$

$C_{oss} \simeq C_{ds} + C_{gd}$; C_{gs} = short circuit

$C_{rss} \equiv C_{gd}$

Measurement Techniques

The measurement is performed by shorting out one capacitance (C_{gs}, C_{gd} or C_{ds}) at a time and measuring the parallel capacitance of the remaining two with a 1Mhz capacitance bridge.

Three capacitances are measured:

$$C_1 = C_{gs} + C_{gd}$$

$$C_2 = C_{gd} + C_{ds}$$

$$C_3 + C_{ds} + C_{gs}$$

From the values of C_1, C_2 and C_3 the values for C_{iss}, C_{oss} and C_{rss} can be derived:

$$C_{iss} = C_1$$

$$C_{oss} \cong C_2$$

$$C_{rss} = \frac{C_1 + C_2 - C_3}{2}$$

Or if you prefer the values for C_{gd}, C_{gs} and C_{ds} may be calculated:

$$C_{gd} = \frac{C_1 + C_2 - C_3}{2}$$

$$C_{gs} = \frac{C_1 - C_2 + C_3}{2}$$

$$C_{ds} = \frac{-C_1 + C_2 + C_3}{2}$$

The actual test circuits are shown in Figure A2-7.

DERIVATION OF EQUATIONS

(1) $C_1 = C_{gs} + C_{gd}$

(2) $C_2 = C_{gd} + C_{ds}$

(3) $C_3 = C_{gs} + C_{ds}$

(1-2) $C_{gs} - C_{ds} + C_1 - C_2$

(3) $C_{gs} + C_{ds} = C_3$

$$C_{gs} = \frac{C_1 - C_2 + C_3}{2}$$

(2-1) $-C_{gs} + C_{ds} = -C_1 + C_2$

(3) $C_{gs} + C_{ds} = C_3$

$$C_{ds} = \frac{-C_1 + C_2 + C_3}{2}$$

(2-3) $C_{gd} - C_{gs} = C_2 - C_3$

(1) $C_{gd} + C_{gs} = C_1$

$$C_{gd} = \frac{C_1 + C_2 - C_3}{2}$$

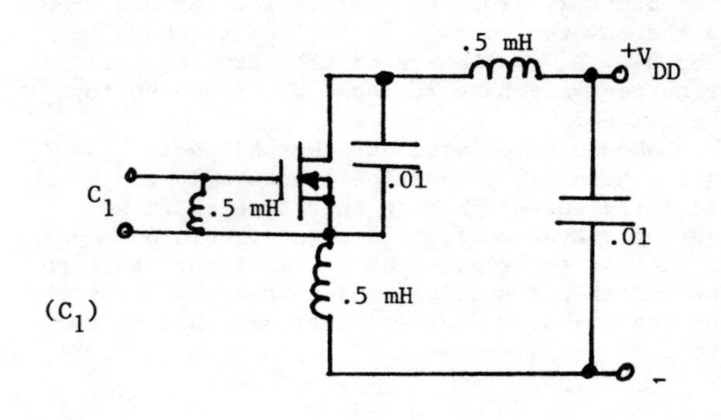

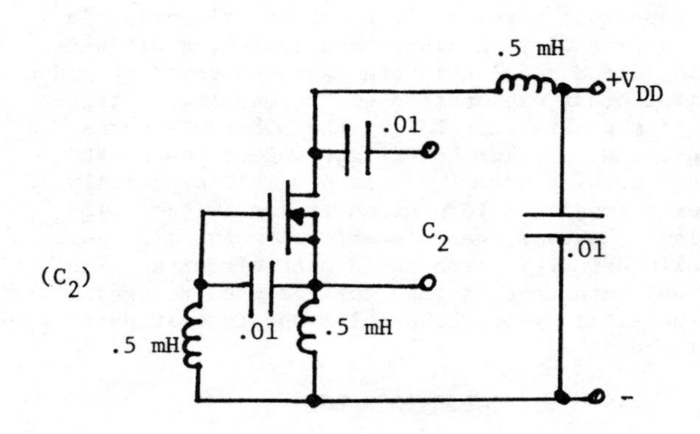

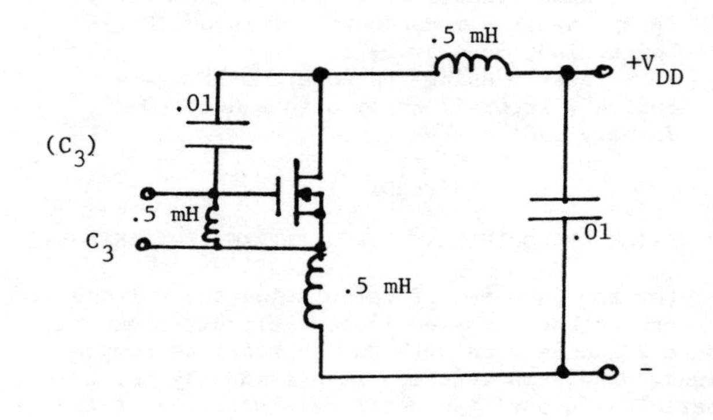

Figure A2-2

A 48V, 200A CHOPPER FOR MOTOR SPEED CONTROL, WITH REGENERATIVE BRAKING CAPABILITY, USING POWER HEXFETS

By S. Clemente, B. Pelly

International Rectifier
El Segundo, California

Abstract

An experimental DC to DC chopper circuit using parallel connected power HEXFETs for speed control of a separately excited DC motor is described. The circuit operates from a 48V battery and provides two quadrant operation, with maximum motoring and regenerating currents of 200A and 140A respectively.

A particular feature is the use of the integral body-drain diode of the HEXFET as a circuit component in its own right, to provide the "freewheeling" and "flyback" functions for the "motoring" and "regenerating" modes respectively.

Particular focus is made on detailed questions that relate to the successful operation of the HEXFETs in this type of system.

Introduction

Efficient speed control of DC motors operating from DC supplies is conventionally accomplished with switching chopper circuits using forced commutated thyristors or bipolar transistors. Battery operated systems rated at hundreds of amperes are in common use in forklift truck and electrical vehicle controllers. Larger thyristor choppers rated at thousands of amperes, at DC voltages up to 1500V, are in use in high power railway traction applications.

An exciting new power semiconductor switching device is the power MOSFET. Today's largest MOSFETs having ratings of 28A continuous and 70A peak at 100V, whilst 400V rated devices have continuous and peak current ratings of 11A and 25A respectively. Power MOSFETs lend themselves readily to paralleling (providing the proper precautions are observed) and therefore a power MOSFET chopper operating at currents of several hundred amperes is technically within grasp.

The major objective of the work described in this paper has been to demonstrate the technical feasibility of a HEXFET chopper circuit using parallel devices to provide a 200A, 48V output for motor speed control. An ancillary objective has been to demonstrate the use of the integral body-drain diode of the HEXFET as a circuit element in its own right, to provide a regenerative braking capability.

The Power HEXFET

The HEXFET is a power MOSFET that is characterized by very high gain, rugged safe operating area, and fast switching speed. The basic structure is illustrated in Figure 1, and the electrical symbol is shown in Figure 2. Current flows from the drain region vertically through the silicon, then horizontally through the channel, then vertically out through the source.

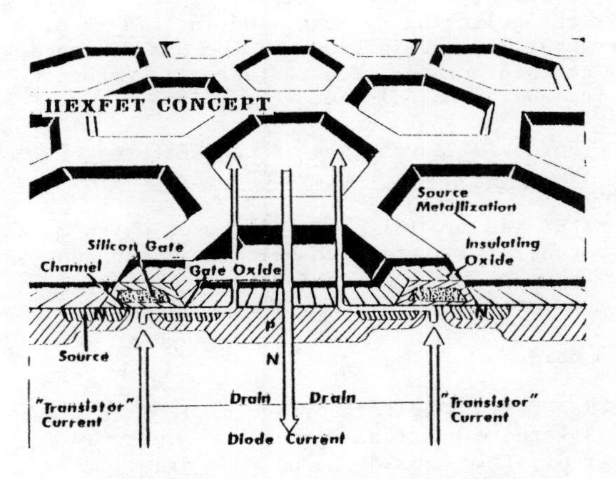

FIGURE 1. BASIC HEXFET STRUCTURE

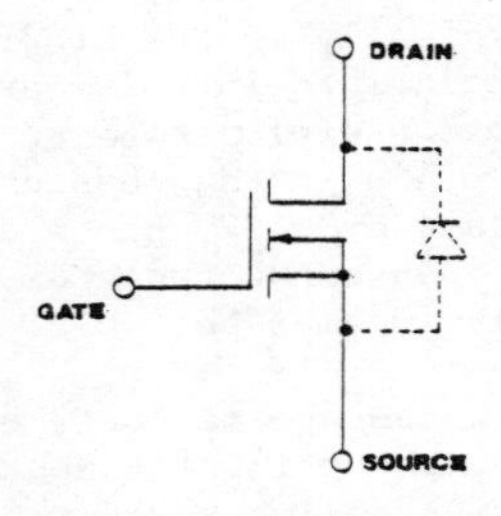

FIGURE 2. ELECTRICAL SYMBOL FOR POWER HEXFET (MOSFET) INCLUDING THE BODY-DRAIN DIODE

The HEXFET design is based upon vertical D-MOS technology. The closed hexagonal cellular structure with the buried silicon gate allows for optimum utilization of silicon, and yields a rugged, highly reliable device.

Reprinted from *IEEE Ind. Appl. Soc. Meet.*, 1980, pp. 664–675.

A feature of the HEXFET (actually, of all power MOSFETs) is that it inherently has built into it an integral reverse "body-drain" diode. The full electrical symbol for the power MOSFET includes the reverse parallel rectifier shown dashed in Figure 2. The existence of this integral reverse rectifier is explained by reference to Figure 1. Current is free to flow through the middle of each source cell across a forward biased P-N junction, and out of the drain. The path for this "reverse" current flow is at least comparable in cross section to that of the "forward" current "transistor" channel. Far from being an inconsequential "parasitic" component, the integral reverse body-drain diode is therefore a real circuit element, with a current handling capability as high as that of the transistor.

The integral reverse body-drain diode may or may not be important in a practical circuit. In some circuits it is irrelevant, because the circuit operation is such that the voltage across the switching device never changes polarity, and the forward conduction characteristic of the body-drain diode never comes into play. This is the case, for example, in a simple DC to DC chopper circuit for motor control which is not configured for regenerative energy flow, and in which the motor voltage never exceeds the source voltage.

A DC to DC chopper circuit for motor speed control that provides a regenerative braking capability would, however, require rectifiers to be connected across the switching devices, and in this case, the reverse body-drain diode of the HEXFET can be used for this purpose, and in fact eliminates the need for additional discrete rectifiers.

Potential Advantages of the HEXFET for a Motor Drive Chopper

The HEXFET has several unique features which make it a potentially attractive switching component for a chopper drive. These features are briefly discussed below:

High Gain

The HEXFET is a voltage driven device. The gate is isolated electrically from the source by a layer of silicon oxide. The gate draws only minute leakage current, in the order of nano amperes, and the DC gain - in the conventional sense used for a bipolar transistor - is extremely high, typically in the order of 10^9. Actually, the gain defined in this way is rather meaningless. A more useful parameter is the transconductance. This is the change of drain current brought about by a 1V change of voltage on the gate. The transconductance of the IRF150 HEXFET is typically 10 amps per volt.

Another important advantage is that, unlike the bipolar transistor, the gain of the HEXFET does not decrease with increasing current. This means that the HEXFET is able to handle high peak current, without showing the bipolar transistor's tendency to "pull out of saturation".

Because the gain of the HEXFET is very high, the drive circuitry required is relatively simple. It should be clearly recognized, however, that although the gate consumes virtually no current under "steady" conditions, this is not so under transitional switching conditions. The gate-to-source and gate-to-drain self-capacitances must be charged and discharged appropriately to obtain the desired switching speed, and the drive circuit must have a sufficiently low output impedance to supply the required charging and discharging current. Even once these requirements have been catered for, the fact remains that the drive circuitry required for a HEXFET is considerably simpler than that required for a bipolar transistor.

Ruggedness

One of the outstanding features of the HEXFET is that it does not display the second breakdown phenomenon of the bipolar transistor, and as a result it has an extremely rugged switching performance.

A simple physical explanation accounts for this superiority. If localized, potentially destructive heating occurs within a HEXFET, the carrier mobility in that area decreases. As a result, the device has a positive temperature coefficient, and acts in a self-protective manner to force currents to be uniformly distributed throughout the silicon.

Ease of Paralleling

Power HEXFETs are potentially easy to parallel, because the positive temperature coefficient forces current sharing among parallel devices. They therefore lend themselves well to the construction of a chopper rated at several hundred amperes, and the problems of paralleling will be much less than those associated with bipolar transistors.

The Basic HEXFET Two Quadrant Chopper Circuit

Figure 3 shows the basic circuit of a DC to DC chopper that provides continuous speed control in the "motoring" mode of operation (i.e., with the motor receiving power from the DC source), and also provides the facility for the motor to return regenerative energy to the DC source, over the whole speed range. Theoretical waveforms that describe the operation are shown in Figure 4, whilst Figure 5 defines the two operating quadrants of the circuit.

In the "motoring" mode of operation, HEXFET 1 is switched ON and OFF, at an appropriate repetition rate, and provides control of the average voltage applied to the motor. HEXFET 2 is OFF, but its integral reverse body-drain diode acts as the conventional freewheeling rectifier, and carries the freewheeling motor current during the periods when HEXFET 1 is OFF. When the motor is required to act as a generator and return energy to the DC source, HEXFET 2 is chopped ON and OFF, and controls the current fed back from the motor to the supply. In this operating mode, HEXFET 1 is OFF, but its integral reverse rectifier carries the motor current back to the DC source during the intervals when HEXFET 2 is OFF.

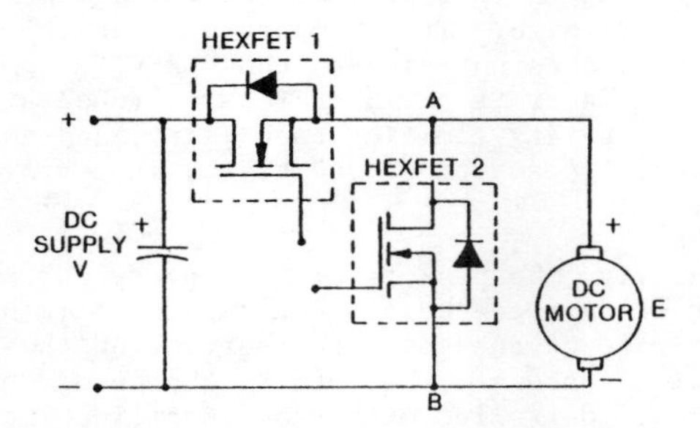

FIGURE 3. BASIC TWO QUADRANT CHOPPER CIRCUIT USING HEXFETS

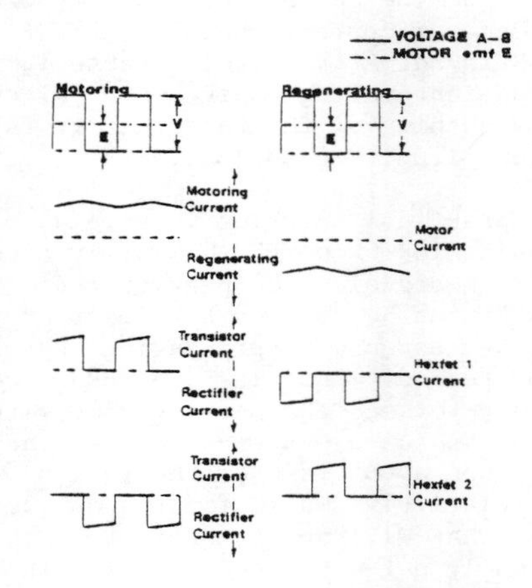

FIGURE 4. THEORETICAL OPERATING WAVEFORMS FOR FIGURE 3

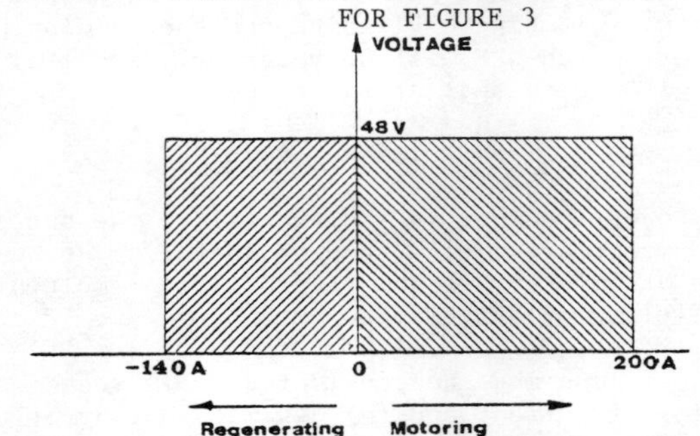

FIGURE 5. MOTORING & REGENERATING OPERATING QUADRANTS FOR THE CIRCUIT OF FIG. 3

In order for the motor to "regenerate", it is necessary for it to have either a shunt or a separately excited field. A series connected field is not feasible, unless the connections to it are reversed for the regenerative mode of operation, which is not practically convenient.

The major objectives of the work described in this paper have been to demonstrate the feasibility of operating a group of parallel connected HEXFETS at currents in the order of hundreds of amperes, and of using the reverse body-drain diode of the HEXFET as a circuit element in its own right, in the basic two quadrant chopper circuit shown in Figure 3.

To achieve these objectives, it is necessary to consider certain detailed aspects of the operation of the HEXFETs. We do this in the following section.

Use of the HEXFET's Body-Drain Diode

An important consideration when using the HEXFET's integral body-drain diode is its reverse recovery characteristic. This rectifier is a conventional P-N junction device, and therefore it exhibits a classical reverse recovery charge. That is to say, when the rectifier switches OFF, the current through it reverses for a short period. The reverse recovery time depends upon the operating conditions. For the IRF150 HEXFET, rated 28A continuous at 100V, (the type used here), the reverse recovery time is about 400ns at maximum operating temperature, and about 260ns at 25°C, for an initial peak forward current of 70A, and a di/dt of 100A/µs.

Reverse recovery presents a potential problem when switching any rectifier OFF. The slower the rectifier, the greater the problem. Although the HEXFET's body-drain diode is relatively fast - not as fast as the fastest discrete rectifiers available, but considerably faster than comparably rated general purpose rectifiers - by comparison with the HEXFET itself, it is rather slow. This presents a potential problem in a chopper circuit, as we will now see.

To illustrate the problem, we will consider the motoring mode of operation. The operating condition that is troublesome is when freewheeling current is commutated from the body-drain diode of HEXFET 2 to the transistor of HEXFET 1. The operating sequence is illustrated by the theoretical waveforms shown in Figure 6.

Throughout the commutating sequence which, of course, is short by comparison with the overall fundamental operating cycle of the circuit, a constant current, I_M, is assumed to flow through the motor. During the operating period, t_0, the current, I_M, is freewheeling through the rectifier of HEXFET 2. At the start of the operating period, t_1, HEXFET 1 is turned ON, and the load current starts to transfer to the transistor of HEXFET 1. The current, i_1, in HEXFET 1 increases, whilst the current, i_2, flowing in the rectifier of HEXFET 2 decreases. The sum of i_1 and i_2 is equal to I_M. At the end of period t_1, the current flowing in HEXFET 1 is equal to the motor current, I_M, and the current flowing in the rectifier of HEXFET 2 is instantaneously zero.

Note that during period t_1 (also during the subsequent period t_a) the voltage across HEXFET 1 theoretically is virtually the full source voltage. This is because so long as the rectifier of

HEXFET 2 remains conducting, the voltage across it can be only its conduction voltage; the difference between this relatively small voltage and the total source voltage is developed across HEXFET 1.

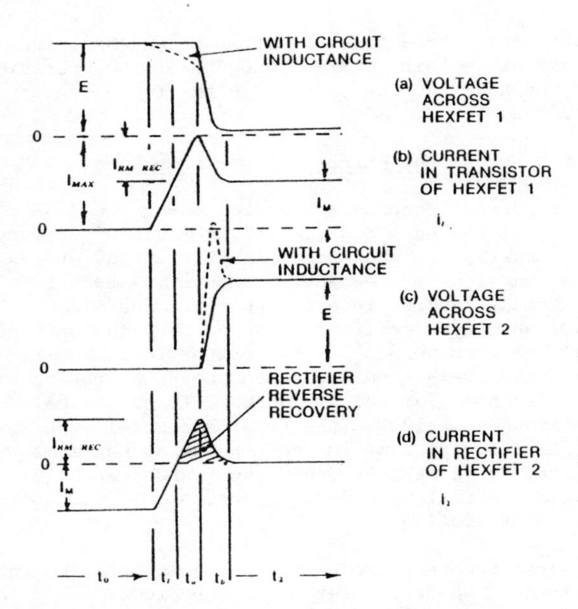

FIGURE 6. THEORETICAL OPERATING WAVEFORMS FOR THE CIRCUIT OF FIGURE 3 WHEN SWITCHING-ON HEXFET 1

This ignores the effect of circuit inductance. In practice, some of the source voltage will be dropped across circuit inductance, and the voltage across HEXFET 1 will be less than the source voltage, by the voltage drop across this inductance. A typical voltage across HEXFET 1 that takes account of the voltage drop across circuit inductance is represented by the dashed wave in Figure 6.

If the rectifier was "perfect", with no recovered charge, the commutation process would be complete at the end of period t_1. In practice, the rectifier current reverses during the recovery periods t_a and t_b. During the period t_a, the reverse current i_2 increases until it reaches its peak value, $I_{RM(rec)}$. The current, i_1, through HEXFET 1 is now the sum of the rectifier reverse current i_2, and the motor current I_M, and its peak value, I_{MAX}, is the sum of I_M and $I_{RM(rec)}$. The voltage across HEXFET 1 still theoretically remains high because the voltage across the rectifier of HEXFET 2 is still relatively low.

During the second part of the recovery period t_b, the rectifier of HEXFET 2 begins to support reverse voltage. The rectifier recovery current i_2 decreases, and the voltage across HEXFET 1 falls to its final conduction level. Note the effect that circuit inductance has in producing an overvoltage transient across the rectifier, as illustrated by the dashed wave in Figure 6.

Certain important points are evident. First, t_1, t_a, and to a lesser extent, t_b, are high dissipation periods. Second, the peak current in HEXFET 1 is the sum of the motor current and the rectifier reverse recovery current, and this peak current occurs at an instant when the voltage across the HEXFET is high. It is important that this peak current does not violate the HEXFET's I_{DM} rating. In fact, if the HEXFET is switched at a speed close to its limiting capability, and no other special precautions are taken, it certainly will do.

Fundamentally, the peak reverse recovery current of the rectifier can be reduced only by slowing down the rate of change of current during the commutation process. The rate of change of current can be controlled either by inserting inductance into the circuit, or by purposefully slowing down the rate of rise of the gate pulse that drives HEXFET 1. A linear inductor inserted in the circuit for the purpose of slowing down the rate of change of current when the HEXFET is switched ON is not attractive, because it produces a transient voltage spike when switching OFF, to say nothing of the fact that it is an added "power circuit" component.

The better practical solution is simply to slow down the switching-ON of the HEXFET by slowing down the drive signal. The peak current carried by the HEXFET can be reduced to almost any desired extent, at the expense of prolonging the high dissipation period. This is a necessary compromise in order to keep the peak current within safe limits, and as a practical matter the switching losses when averaged over the full operating cycle are relatively small, for the operating frequencies that will be of interest in this application - normally a few hundred to a few thousand Hz.

Note that it is not necessary (nor desirable), to slow the switching-OFF; hence the energy dissipation at switch-OFF will be relatively small by comparison with that at switch-ON.

Paralleling of HEXFETs

A key question that is fundamental to the successful demonstration of a chopper operating at hundreds of amperes, is the feasibility of multiple paralleling of HEXFETs.

Two questions must be considered: (a) "steady state" sharing of current and (b) dynamic sharing of current, under the transitional switching conditions.

Steady State Sharing of Current

During the periods outside of the switching transitions, the current in a parallel group of HEXFETs will distribute itself in the individual devices in inverse proportion to their ON resistances. The device with the lowest ON resistance will carry the highest current. This will, to an extent, be self-compensating because the power loss in this device will be highest. It will run hottest, and the increase in ON resistance due to heating will be more than that of the other devices, which will tend to equalize the current.

An analysis of the "worst case" device current in a group of "N" parallel connected devices can be based on the simplifying assumption that (N - 1) devices have the highest limiting value of ON resistance, whilst just one lone device has the lowest limiting value of ON resistance. The analysis can then be concentrated on the current in this one device.

Using this approach for a large group of paralleled IRF150's, it can be calculated that the "worst case" maximum value of device current would be 27A for the hypothetical situation where all devices but one have high limiting ON resistance of 0.045Ω, and carry 20A each, whereas the remaining one has low limiting ON resistance. This is a practically acceptable result.

Dynamic Sharing of Current Under Switching Conditions

Turn-On

It is necessary to take positive steps to ensure that the current is distributed properly between a group of parallel connected devices during the switching transition. Since the HEXFETs will not all have identical threshold and gain characteristics, some will tend to switch sooner than others, and attempt to take more than their share of the current. Adding to the problem is the fact that circuit inductance associated with each device may be different, and this will also contribute to unbalancing the current under switching conditions.

The problem will be introduced by considering the switching waveforms for the basic chopper circuit, shown in Figure 3, which contains a single HEXFET in each of the "motoring" and "regenerating" positions. We will consider the motoring mode of operation, under which HEXFET 1 is switched ON and OFF, whilst the motor current (assumed to be smooth, due to the motor inductance) alternates between this HEXFET and the body-drain diode of HEXFET 2, which acts as a freewheeling rectifier.

Figure 7 shows waveforms of drain current, drain-to-source voltage, and gate voltage during the turn-ON interval. We have already seen that in order to limit the peak recovery current of the body-drain diode of HEXFET 2, the gate drive voltage for HEXFET 1 must be applied at a controlled rate. This is the reason that the applied drive pulse is shown increasing at a relatively slow rate.

At time, t_0, the drive pulse starts its rise. At t_1 it reaches the threshold voltage of the HEXFET, and the drain current starts to increase. At this point, two things happen which make the gate-source voltage waveform deviate from its original "path". First, inductance in series with the source which is common to the gate circuit develops an induced voltage, as a result of the increasing source current. This voltage counteracts the applied gate drive voltage, and slows down the rate of rise of voltage appearing directly across the gate and source terminals; this in turn slows down the rate of rise of the source current. This is a negative feedback effect; increasing current in the source produces a counteractive voltage at the gate, which tends to resist the change of current.

The second factor that influences the gate-source voltage is the so-called "Miller" effect. During the period t_1 to t_2 some voltage is dropped across circuit inductance in series with the drain, and the drain-source voltage starts to fall. The decreasing drain-source voltage is reflected across the drain-gate capacitance, pulling a discharge current through it, and increasing the effective capacitance load on the drive circuit. This in turn increases the voltage drop across the source impedance of the drive circuit, and decreases the rate of rise of voltage appearing between the gate and source terminals. This also is a negative feedback effect; increasing current in the drain results in a fall of drain-to-source voltage, which in turn slows down the rise of gate-source voltage, and tends to resist the increase of drain current. These effects are illustrated diagrammatically in Figure 8.

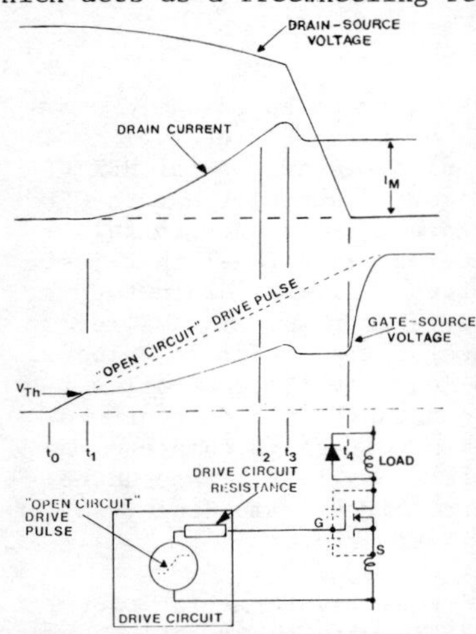

FIGURE 7. WAVEFORMS AT TURN-ON SINGLE HEXFET

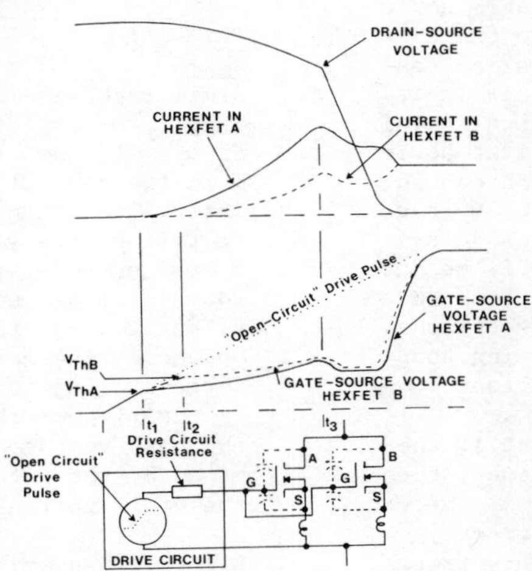

FIGURE 9. WAVEFORMS AT TURN-ON 2 HEXFETS WITH GATES TIED TOGETHER

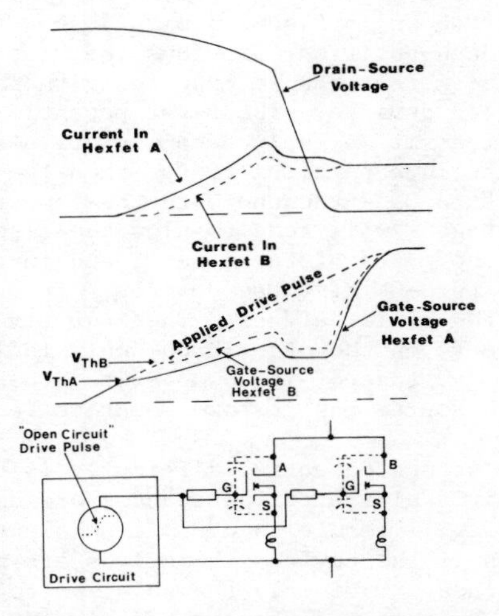

FIGURE 10. WAVEFORMS AT TURN-ON 2 HEXFETS WITH DECOUPLED GATES

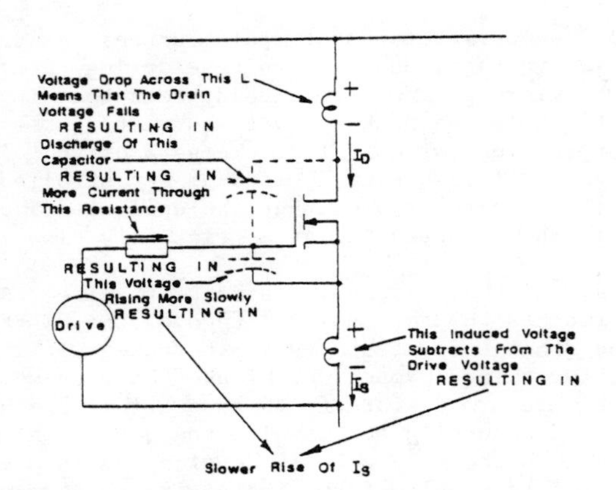

FIGURE 8. DIAGRAMMATIC REPRESENTATION OF EFFECTS WHEN SWITCHING-ON

This state of affairs continues throughout the period t_1 to t_2, whilst the current in the HEXFET rises to the level of the current, I_M, already flowing in the freewheeling rectifier, and it continues into the next period, t_2 to t_3, whilst the current increases further, due to the reverse recovery of the freewheeling rectifier.

At time t_3 the freewheeling rectifier starts to support voltage, whilst the drain current and the drain voltage start to fall. The rate of fall of drain voltage is now governed by the Miller effect, and an equilibrium condition is reached, under which the drain voltage falls at just the rate necessary for the voltage between gate and source terminals to satisfy the level of drain current established by the load. This is why the gate-to-source voltage falls as the recovery current of the freewheeling rectifier falls, then stays constant at a level corresponding to the motor current, whilst the drain voltage is falling.

Finally, at time t_4, the HEXFET is switched fully ON, and the gate-to-source voltage rises rapidly towards the applied "open circuit" value.

The gate-to-source voltage waveform for the circuit shown in Figure 3 with just a single device in each position, provides the clue to the difficulties that can be expected with parallel connected devices. The first potential difficulty is that if we apply a common drive signal to all gates in a parallel group, then the first device to turn ON - the one with the lowest threshold voltage - will tend to slow the rise of voltage on the gates of the others, and further delay the turn-ON of these devices. This will be due to the Miller effect. The inductive feedback effect, on the other hand, only influences the gate voltage of its own device, (assuming that each source has its own separate inductance).

The second potential difficulty is that if the individual source inductances are unequal, then this will result in dynamic unbalance of current, even if the devices themselves are perfectly matched. Obviously the solution to this is to ensure that inductances associated with the individual devices are as nearly equal as possible. This can be done by proper attention to the circuit layout.

Figure 9 illustrates the turn-ON of two parallel connected HEXFETs, A and B, having different threshold voltages, V_{thA} and V_{thB}, with their gates connected directly together. V_{thB} is greater than V_{thA}, and HEXFET A switches ON first, at time t_1. The voltage between gate and source of HEXFET B is now "deflected" slightly as shown, due to the Miller effect and the turn-ON of HEXFET B is delayed until time t_2. During the period t_2 to t_3 the gate-source voltages "drift" towards one another (being "kept apart" only by the individual source inductances), and this produces a separation of the drain currents, with a significant resulting unbalance.

This basic problem stems from the fact that the two gates are coupled together by the common impedance of the drive source. If, on the other hand, the drive circuit is designed to have a low output impedance, and the gates are decoupled from one another by individual gate resistors, the dynamic unbalance can be alleviated.

Figure 10 illustrates theoretical waveforms for two parallel connected HEXFETs, A and B, having different threshold voltages, V_{thA} and V_{thB} as assumed in Figure 9, driven from a low impedance circuit with individual gate resistors. The dynamic unbalance of currents in the two HEXFETs is now substantially reduced.

In summary, attention must be paid to the following points, in order to ensure satisfactory sharing of current between parallel HEXFETs at turn-ON:

(a) Threshold voltages should be within determined limits.
(b) Inductance that is common to the gate and source circuits of the individual devices must be matched within determined limits.
(c) Gates should be decoupled with individual resistors, and the output impedance of the drive circuit must be low by comparison.

Turn-Off

Similar considerations apply to the dynamic sharing of current during the turn-OFF interval. Figure 11 shows theoretical waveforms for HEXFET 1 in the circuit of Figure 3 during the turn-OFF interval. At t_0 the gate drive starts to fall. At t_1 the gate voltage reaches a level that just sustains the drain current, I. The drain-to-source voltage now starts to rise. The Miller effect governs the rate of rise of drain voltage, and holds the gate-to-source voltage at a level corresponding to the constant drain current. At t_2 the rise of the drain voltage is complete, and the gate voltage starts to fall at a rate determined by the gate-source circuit impedance, whilst the drain current falls to zero.

Figure 12 shows theoretical waveforms for two parallel connected HEXFETs with their gates

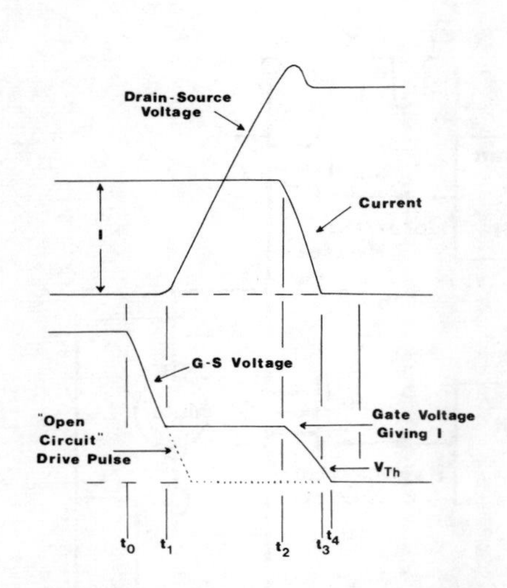

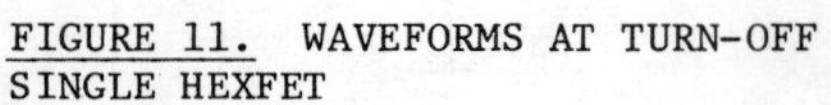

FIGURE 11. WAVEFORMS AT TURN-OFF SINGLE HEXFET

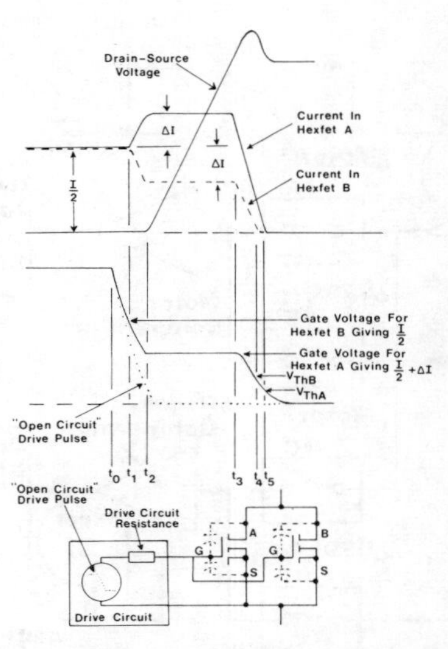

FIGURE 12. WAVEFORMS AT TURN OFF 2 HEXFETS WITH COMMON GATES AND COMMON SOURCES

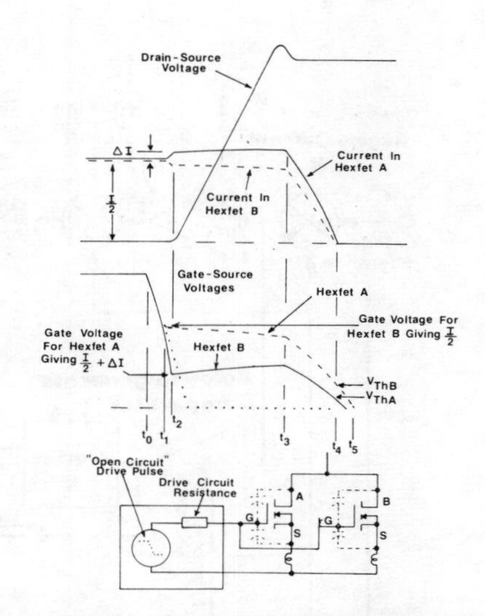

FIGURE 13. WAVEFORMS AT TURN-OFF 2 HEXFETS WITH COMMON GATES AND DECOUPLED SOURCES

connected directly together. For purposes of discussion, the source inductance is assumed to be zero. At t_1 the gate voltage reaches the point at which HEXFET B can no longer sustain its drain current. The load current now redistributes; current in HEXFET B decreases, whilst that in HEXFET A increases. At t_2 HEXFET B can no longer sustain its current; both HEXFETs now operate in their "linear" region, and the drain voltage starts to rise. The gate-to-source voltage is kept practically constant by the Miller effect, whilst the currents in the two HEXFETs remain at their separate levels. Clearly, the unbalance of current in this example is significant.

Figure 13 shows waveforms for two HEXFETs having their gates tied directly together, but with individual inductances in series with each source. The negative feedback effect of these inductances has a pronounced balancing effect upon the individual drain currents. Whilst the drain voltage is rising, however, the voltages on the individual gates, being "separated" only by the source inductances, drift towards one another, whilst the two drain currents drift apart.

Finally, Figure 14 shows waveforms for two parallel connected HEXFETs, driven from a low impedance circuit, with their gates decoupled by individual resistors, and with inductance in series with each source. Because the gates are now separated by decoupling resistors, the gate voltages no longer drift together, and the drain currents no longer drift apart during the period when the voltage is rising. This minimizes the unbalance of current.

In summary, to achieve good sharing of current when switching-OFF, the following points are important:

1. Threshold voltage and transconductance of the HEXFETs should be within determined limits.
2. Source inductance is a key factor in keeping the individual currents balanced.

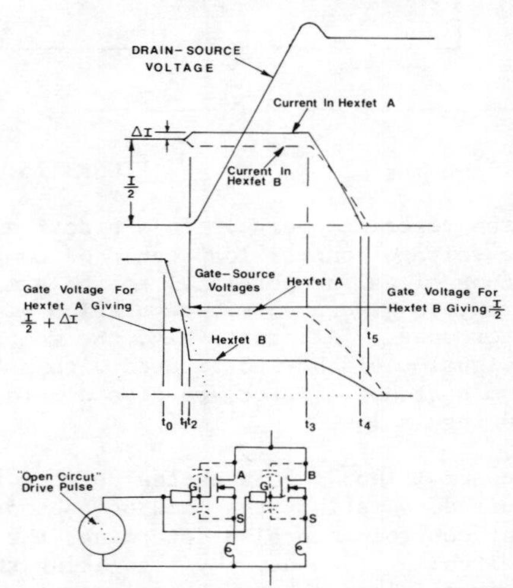

FIGURE 14. WAVEFORMS AT TURN-OFF. 2 HEXFETS WITH DECOUPLED GATES & DECOUPLED SOURCES

3. Gates should be decoupled with individual resistors. The output impedance of the drive circuit must be low.

A Complete Functional Control Scheme

A simplified functional diagram of the control and drive circuitry is shown in Figure 15; this is intended to demonstrate the basic operating principle of the overall chopper system, and differs in some minor details from the actual practical circuitry presented later (Figure 19).

The control system has an outer voltage feedback loop, which compares the motor voltage with a reference voltage, and processes the resulting "error" signal to keep the motor voltage essentially

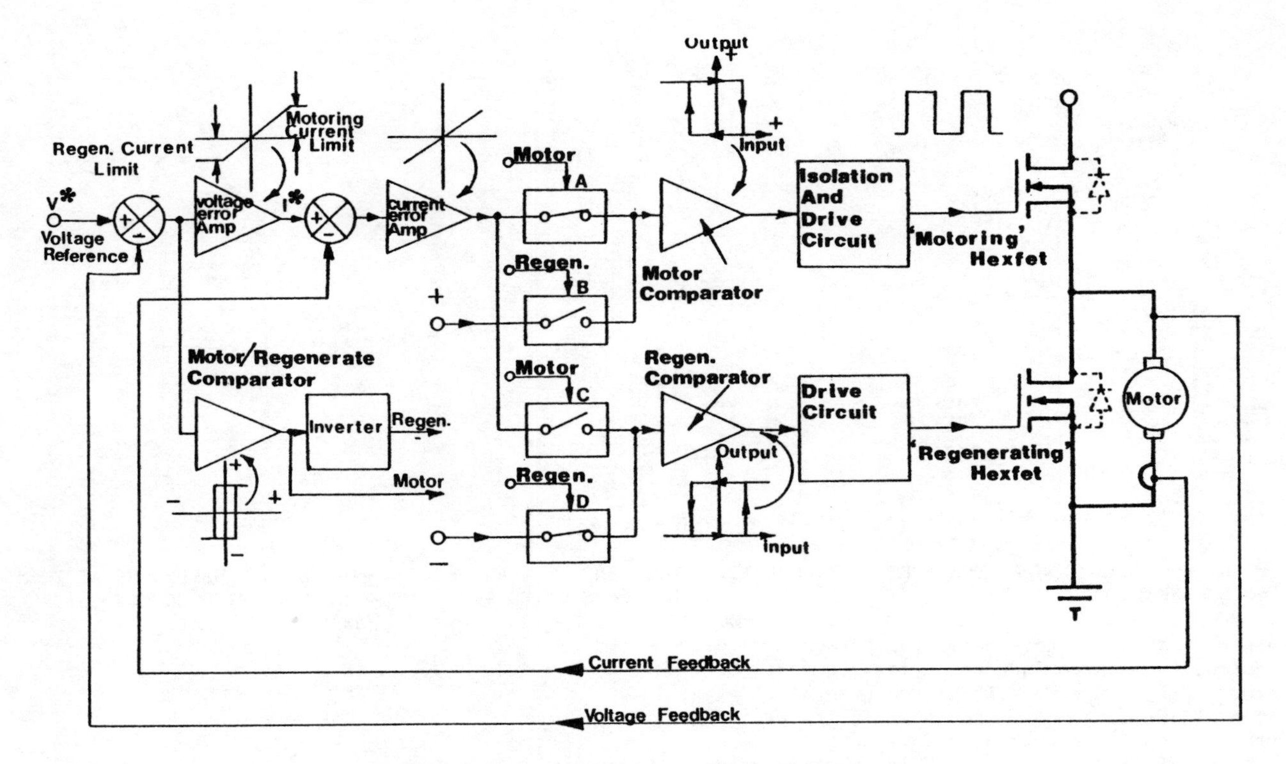

FIGURE 15. SIMPLIFIED CONTROL SYSTEM

equal to the reference value. In a practical system the voltage control loop could be complemented with a signal proportional to the armature voltage drop, to give a closer regulation of actual motor speed. Alternatively, the voltage feedback signal could be substituted with a signal from a tachogenerator, to give a more precise speed regulation.

An inner control loop regulates the current to the level required to satisfy the load on the motor. The current control loop also determines the chopper switching frequency, by regulating the peak-to-peak ripple current between preset upper and lower limits. This it does by switching the HEXFET ON whenever the current falls a given amount below the reference value, and switching the HEXFET OFF whenever the current rises a given amount above it.

The current control loop also provides instantaneous limiting of the peak HEXFET and motor current. This is accomplished simply by setting a maximum limit on the current reference signal, and clamping it to this level. Whenever the instantaneous motor current attempts to exceed the maximum current reference by more than the preset peak ripple current, the HEXFET is immediately switched OFF. Thus the system is completely self-protecting against overcurrent.

Referring to the functional diagram in Figure 15, the voltage reference is compared with the voltage across the motor, and the error signal is amplified through the voltage error amplifier.

The output of the voltage error amplifier is the current reference signal. The voltage error signal is also fed into the MOTOR/REGENERATE logic comparator. When the voltage error is positive, the current reference is also positive, and the control circuit is demanding "motoring" current. The output of the MOTOR/REGENERATE logic comparator is high, the MOTOR signal has a logic "1" value, whilst the REGEN signal has a logic "0" value. Switches A and D are closed, whilst switches B and C are open.

When the current reference is negative, the control circuit is demanding "regenerating" current. The output of the MOTOR/REGENERATE logic comparator is low, the REGEN signal has a logic "1" value, whilst the motor signal has a logic "0" value. Switches B and C are closed, whilst A and D are open.

The MOTOR/REGENERATE logic comparator has a built-in hysteresis to prevent unwanted "bouncing" back and forth between the "regeneration" and "motoring" modes of operation at low current levels.

Consider the "motoring" mode of operation. The positive current reference signal is compared with a signal representing the actual motor current; the difference is amplified through the current error amplifier. The output of this amplifier is fed through switch A, which is closed, to the MOTOR comparator. This comparator produces a "0" output signal in response to a positive input signal above a preset threshold level, and a "1" output signal in response to a negative input signal below a certain preset level.

The output signal of the MOTOR COMPARATOR is isolated and shaped to become the gate drive signal for the "motoring" HEXFET. The "motoring" HEXFET is thereby switched ON when the motor current falls a predetermined amount below the reference value, and OFF whenever the motor current rises a predetermined amount above the reference value, whilst the switching frequency automatically adjusts itself to keep the peak-to-peak ripple current constant. The peak-to-peak ripple current and operating frequency can be adjusted by adjusting the hysteresis of the MOTOR comparator.

Note that in the motoring mode, switch D is closed, applying a steady negative input to the REGEN comparator, and shutting OFF the gate drive signal to the "regenerating" HEXFET. Theoretical waveforms which illustrate the operation of this scheme in the motoring mode are illustrated in Figure 16.

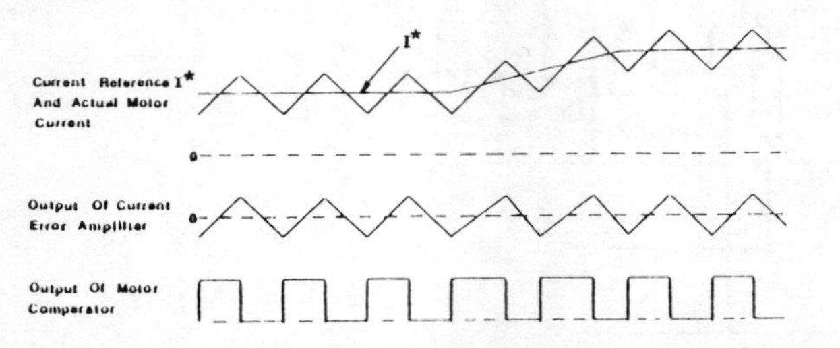

FIGURE 16. THEORETICAL WAVEFORMS FOR THE MOTORING MODE OF OPERATION

In the regenerating mode of operation, switches B and C are closed. A continuous positive signal is applied to the input of the MOTOR comparator, shutting OFF the drive to the "motoring" HEXFET. The current reference is negative, and the current error signal is fed to the input of the REGEN comparator. This comparator produces a "1" output signal in response to a positive input signal above a certain preset level, and a "0" output signal in response to a negative input signal below a given preset level. The "regenerating" HEXFET is now switched ON whenever the regenerative current from the motor falls a preset amount below the reference value, and OFF whenever the motor current rises a preset amount above the reference value. Theoretical waveforms which illustrate the operation in the regenerating mode are shown in Figure 17.

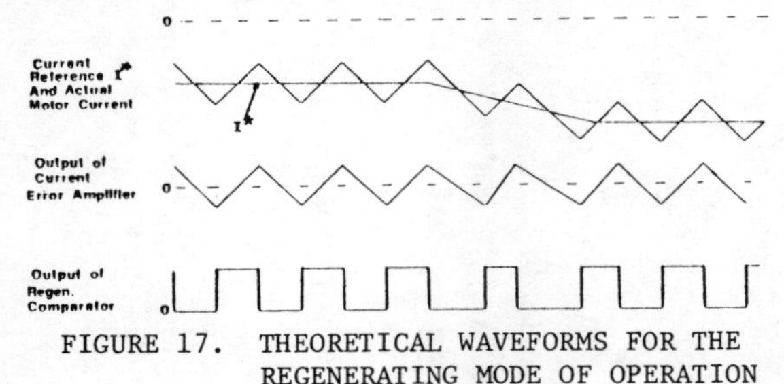

FIGURE 17. THEORETICAL WAVEFORMS FOR THE REGENERATING MODE OF OPERATION

A 48V, 200A Laboratory Chopper

Power Circuit

A schematic diagram of the power circuit of an experimental laboratory chopper is shown in Figure 18. This employs a total of ten IRF150 HEXFETs connected in parallel for the "motoring" switch, and five IRF150 HEXFETs connected in parallel for the "regenerating" switch.

All HEXFETs are mounted on a 22-inch length of aluminum heatsink extrusion, with outside dimensions of 5 inches by 3 inches, with the regenerating HEXFETs being isolated electrically from the heatsink.

The assembly was tested at 200A forward "motoring" current, and 140A of "regenerating" current. The "motoring" HEXFETs by themselves actually are capable of carrying about 300A of output current; the 200A limit was conservatively set by the current carrying capacity of the five freewheeling body-drain diodes of the "regenerating" HEXFETs.

Individual zener diode and snubber circuits are connected across the drain-to-source terminals of each HEXFET, to provide individual protection against overvoltage transients. A total of five local decoupling capacitors, each of 22µF, are distributed at intervals along the "positive" and "negative" bus bars of the assembly, to prevent unwanted oscillation of the local DC source voltage.

Control and Drive Circuitry

Figure 19 shows a diagram of the control and drive circuitry. This is based upon the functional circuit shown in Figure 15, and requires no additional explanation, other than to point out that for practical reasons some of the signal polarities are opposite to those assumed for the simplified functional circuit of Figure 15.

Test Results

Practical test results are shown in Figures 20 through 26.

Figure 20 shows the current when the motor accelerates from standstill to about half speed. The current limit circuit keeps the peak motor current to just over 200A. The chopping frequency is about 2kHz.

Figure 21 shows motor current and voltage waveforms when accelerating from half speed to almost full speed, then decelerating back to half speed. The current limit holds the peak motoring current to about 205A, and the peak regenerating current to about 140A.

Figure 22 shows the output voltage and current of the chopper with a passive inductive load. Note the classical linear rise and fall of the current associated with an inductive load.

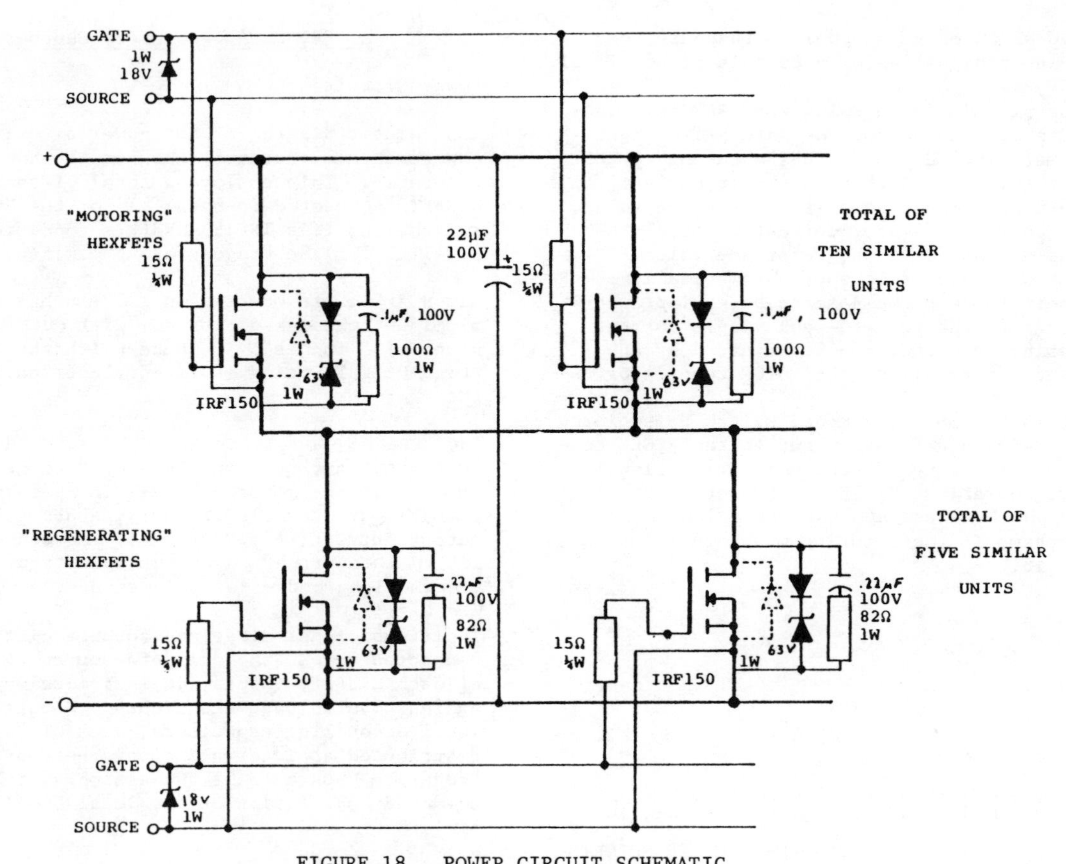

FIGURE 18. POWER CIRCUIT SCHEMATIC

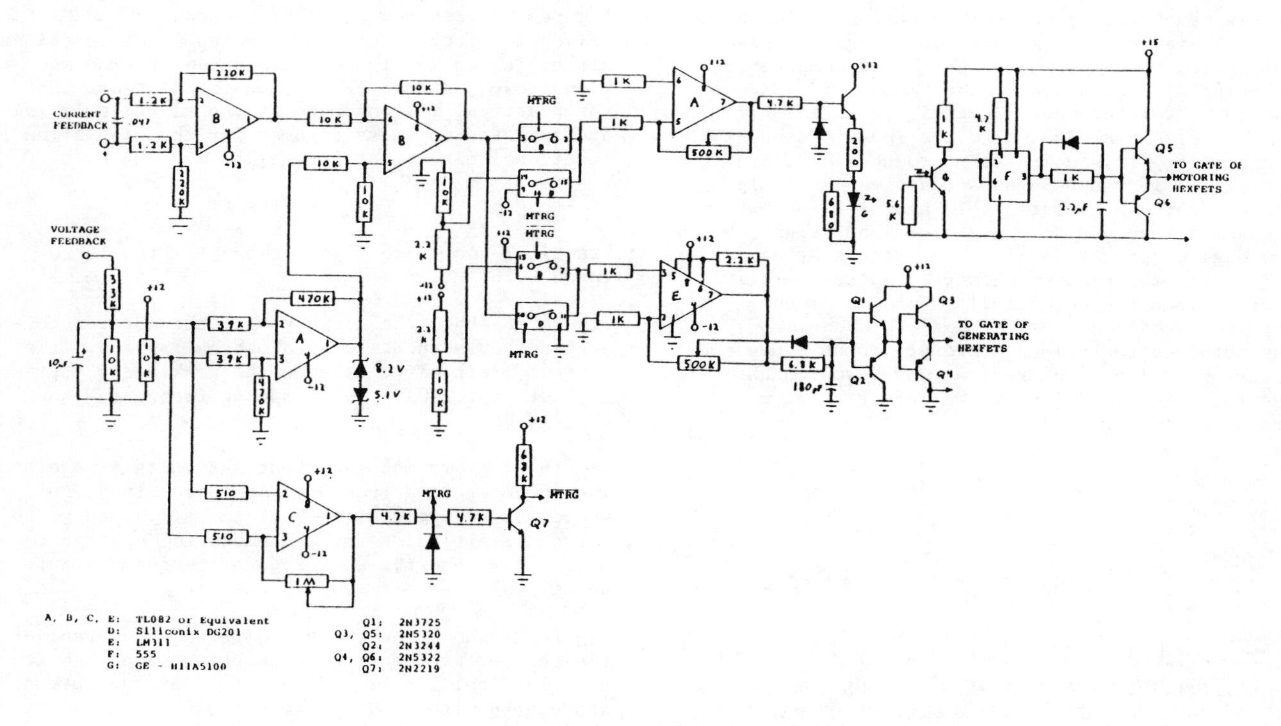

FIGURE 19. CONTROL AND DRIVE CIRCUIT SCHEMATIC

Figures 23 and 24 show turn-ON and turn-OFF oscillograms respectively for one HEXFET, with the chopper operating with a passive inductive load of 120A. These waveforms generally agree with the foregoing theoretical discussion. Note, however, in Figure 24 that the gate voltage reverses at switch-OFF; this is due to resonance between the gate capacitance and circuit inductance.

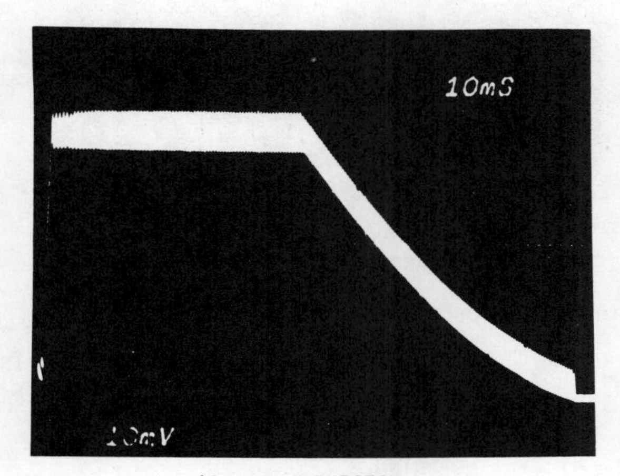

40A PER DIVISION
10ns PER DIVISION

FIGURE 20. MOTOR CURRENT UNDER ACCELERATION
PEAK MOTOR CURRENT = 220A

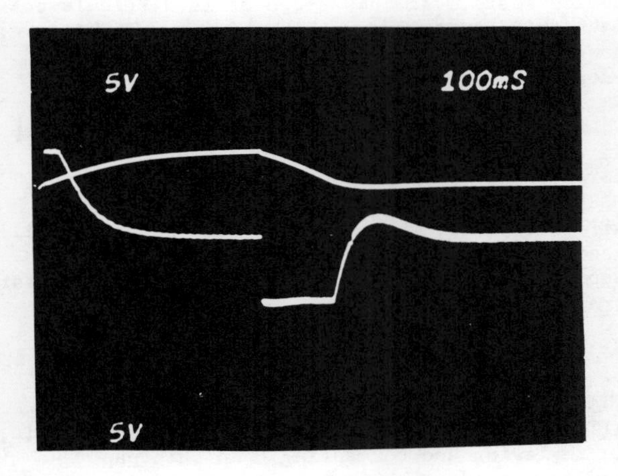

TOP TRACE: VOLTAGE 25V PER DIVISION
BOTTOM TRACE: CURRENT 115A PER DIVISION
100 ms PER DIVISION

FIGURE 21. MOTOR VOLTAGE AND CURRENT WHEN ACCELERATING AND DECELERATING

PEAK MOTORING CURRENT = 205A
PEAK REGENERATING CURRENT = 140A

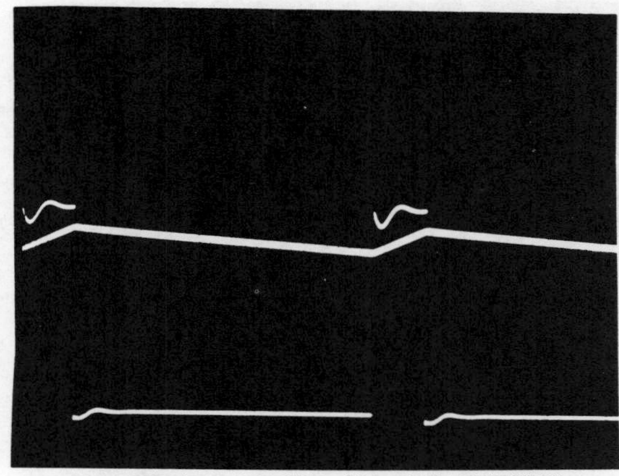

TOP TRACE: CURRENT 40A PER DIVISION
LOWER TRACE: VOLTAGE 10V PER DIVISION

FIGURE 22. OUTPUT VOLTAGE & CURRENT OF CHOPPER INTO PASSIVE INDUCTIVE LOAD - 200μs PER DIVISION

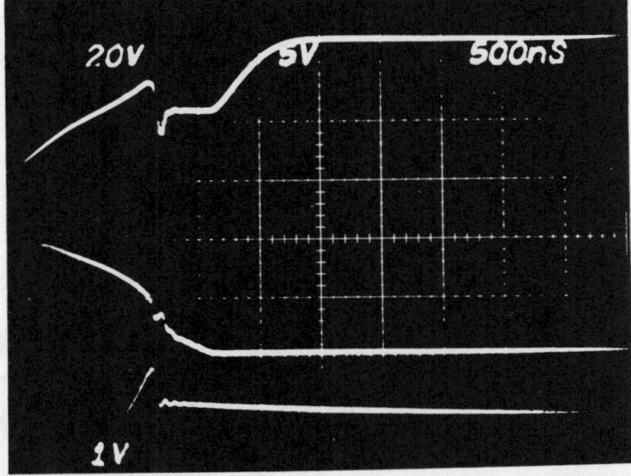

TOP TRACE: GATE-SOURCE VOLTAGE 5V PER DIVISION
MIDDLE TRACE: DRAIN-SOURCE VOLTAGE 20V PER DIV.
LOWER TRACE: DRAIN CURRENT 10A PER DIVISION

FIGURE 23. TURN-ON OSCILLOGRAMS FOR ONE HEXFET
TOTAL OUTPUT CURRENT = 120A. 500ns PER DIVISION

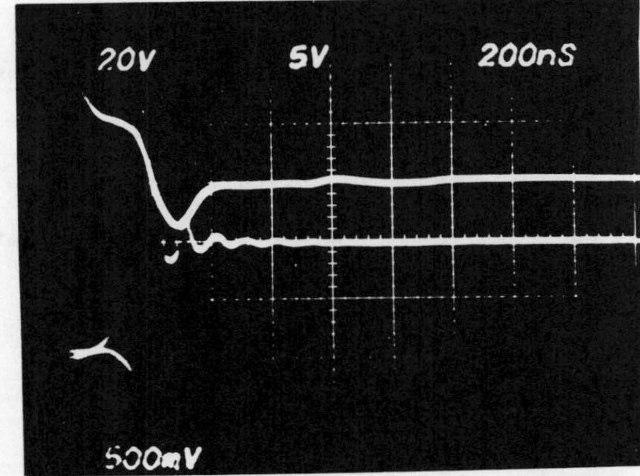

TOP TRACE: GATE-SOURCE VOLTAGE 5V PER DIVISION
MIDDLE TRACE: DRAIN-SOURCE VOLTAGE 20V PER DIV.
LOWER TRACE: DRAIN CURRENT 5A PER DIVISION
200 NANOSECONDS PER DIVISION

FIGURE 24. TURN-OFF OSCILLOGRAMS FOR ONE HEXFET
TOTAL OUTPUT CURRENT = 120A

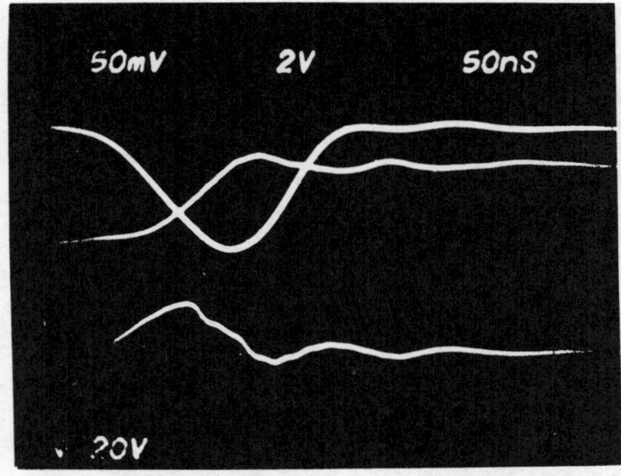

TOP TRACE: GATE VOLTAGE 2V PER DIVISION
MIDDLE TRACE: CURRENT 20A PER DIVISION
LOWER TRACE: VOLTAGE 20V PER DIVISION
50 NANOSECONDS PER DIVISION

FIGURE 25. OSCILLOGRAM FOR "REGENERATING" HEXFET WHEN ITS BODY-DRAIN DIODE IS USED AS FREEWHEELING RECTIFIER CURRENT IS BEING COMMUTATED FROM THE MOTORING HEXFET INTO FREEWHEELING RECTIFIER

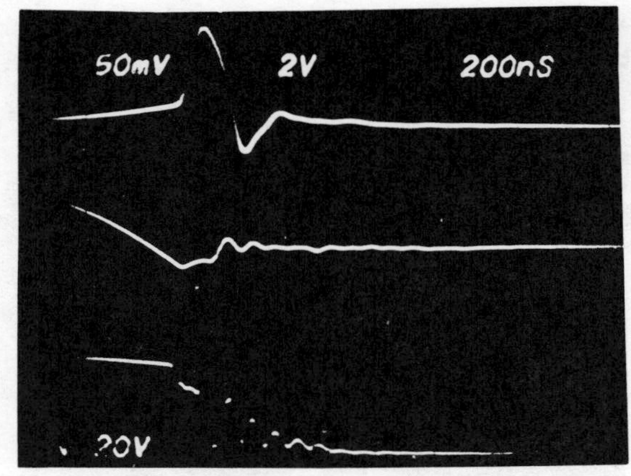

TOP TRACE: GATE VOLTAGE 2V PER DIVISION
MIDDLE TRACE: CURRENT 20A PER DIVISION
LOWER TRACE: VOLTAGE 20V PER DIVISION
200 NANOSECONDS PER DIVISION

FIGURE 26. OSCILLOGRAMS FOR "REGENERATING" HEXFET WHEN ITS BODY-DRAIN DIODE IS IN USE AS A FREEWHEELING RECTIFIER IN THE MOTORING MODE OF OPERATION. CURRENT IS BEING COMMUTATED FROM THE FREEWHEELING RECTIFIER INTO THE MOTORING HEXFET.

Conclusions

We have demonstrated the technical feasibility of a DC to DC chopper using parallel connected HEXFETs for motor speed control, operating at the 200A level, and the use of the HEXFET's body-drain diode to provide the freewheeling and flyback functions needed for two quadrant operation. The potential attractions of using HEXFETs are simplicity of drive circuitry, ruggedness, speed of response, ease of paralleling and overall compactness

The limiting current capability of the system has not been definitively established at the time of writing; it is probably close to 300A for the motoring quadrant of operation. We have not addressed perhaps the most pertinent question - a hard cost comparison with existing systems using bipolar transistors or thyristors. We believe at this stage that such a comparison would be misleading. A comparison of this immature first generation HEXFET circuit, based upon today's device prices, would not be favorable to the MOSFET. This is largely irrelevant, however. What is of interest is the cost of a future mature MOSFET system, based upon future mature device costs. At this stage it is difficult to make this comparison with precision. We believe, however, that the MOSFET system will become economically competitive.

What we have done here is to take the first step in what we see as an evolutionary process. We foresee that as improvements in circuit design, MOSFET technology, packaging, and device costs all take place, the type of system described will become economically as well as technically superior to today's chopper systems using bipolar transistors or thyristors.

Acknowledgements

The assistance of H. Murphy and T. Gilmore of Allis Chalmers, and of F. Stich of Siemens Allis, in reviewing and commenting upon this paper, and in providing the motor, is gratefully acknowledged.

SOLID STATE RF GENERATORS FOR INDUCTION HEATING APPLICATIONS

W.E. Frank, Member IEEE & C.F. Der, Member IEEE
Westinghouse Electric Corporation
Industrial Equipment Division
7301 Sykesville Road
Sykesville, Maryland 21784

Abstract - Radio frequency power for induction heating has traditionally been generated by vacuum tube oscillators. These generators use high voltage tubes and generally operate in the frequency range of 100 - 500 kHz, 1 KW - 400 power output at approximately 50 - 60% efficiency. Although thyristor SCR's and bipolar transistors are marketed at this time, they are unsuitable due to either slow switching or low power performance. Until just recently, solid state devices with both high-frequency and high-power capabilities were not available.

A new power semiconductor - the power MOSFET - a highspeed field effect transistor, can now meet these criteria. The power MOSFET switching time is in the order of 50 - 100 nanoseconds and can generate many kilowatts of power at frequencies to 500 kHz.

This paper describes a solid state RF generator using MOSFET transistors for the power semiconductors. The RF generator is a load tracking, resonant inverter capable of full power output over a frequency range of 100 kHz to 500 kHz. Output power levels are in the kilowatt range with induction coil KVA's to 200 KVA (400 V and 500 A) and more. This solid state RF generator has similar characteristics to the more familiar low frequency power converters presently used in the industry; namely, all solid state, high efficiency (approaching 90%), small and compact, and has no vacuum tubes or moving parts.

I. INTRODUCTION

The art of induction heating is to uniformly heat metallic workpieces at specified temperature and times. For surface heating and non-magnetic parts, high frequency power is required. Therefore, the power source must be capable of supplying a minimum of several kilowatts of power in the radio frequency range. In addition, the power supply must have adequate control, protection and monitoring circuits.

Over the past 50 years, the high voltage vacuum tube was the only device available for radio frequency power generation. Semiconductors, such as SCR's and bipolar and darlington transistors have appeared in the last decade. These components work well in the frequency range up to 10 kHz, but they do not have the characteristics that are required for a radio frequency power source necessary for induction heating applications. In addition, these solid state devices are limited by slow switching speeds and/or low power ratings.

A new switching device, the power field effect transistor, is capable of both higher power outputs and significantly higher switching speeds. Some various market names for these transistors are VMOS, MOSFET, HEXFET and DMOS (these are the manufacturer's distinct trade names). The availability of such components have been disclosed and described in increasing detail since 1979 or so. [1,2,3,4] As a result, Westinghouse Electric Corporation has developed and manufactured a solid state, radio frequency power supply using the power MOSFET in the inverter circuit design of the RF generator.

II. POWER MOSFET TRANSISTOR

The power MOSFET is a field effect transistor with ratings presently ranging from 50V/60A to 1000V/3A among the various manufacturers. Figure 1 depicts an N channel and a P channel MOSFET transistor showing polarity of the terminal potentials and current flow for normal transistor action. The MOSFET contains an internal, reverse P-N junction diode (shown dotted in Figure 1) which has the same current rating as the transistor.[5] This diode is a viable circuit element that can be used, just as any externally connected discrete diode, for a reverse current path around the transistor.

The MOSFET transistor is a voltage controlled device with an insulated gate (G) which controls current flow between the drain (D) and source (S) terminals. A positive gate-source potential applied to the N channel MOSFET causes current to flow from the drain to the source. For the P channel MOSFET, the opposite is true: a negative gate-source potential causes current flow from the source to the drain. The drain current can be controlled linearly by the gate-source voltage or can be "switched on" by gate overdrive to a level determined by the impedance of the external circuit. With sufficient gate overdrive, the drain to source characteristic appears resistive with "ON" resistances for high current MOSFETS being less than a ohm. This "ON" resistance has a positive temperature coefficient which forces current sharing among paralleled devices eliminating the need for external current balancing components and making paralleling relatively easy.

Drain-source switching speeds can be very fast, as low as 20 - 100 nanoseconds, depending on drive circuit source impedance and gate capacitance. MOSFET transistors do not exhibit minority carrier storage effects as do bipolar transistors and switching times are determined primarily by how fast the gate capacitance can be charged and discharged. The fast switching speeds result in a low and thermally manageable switching energy loss per cycle which is requisite for RF operation.

The steady-state gate-source input impedance is very high being essentially capacitive. But dynamically the gate must be sourced and sinked by a high current source to produce fast switching speeds. For example, a MOSFET having an input capacitance C_{iss} = 1000 pf, passes a gate current of 1A when driven with a gate-source voltage rise time of 20V/20 ns. At RF frequencies, these drive requirements can become formidable.

A limiting factor with the MOSFET is the recovery time of the reverse diode. This internal diode is relatively slow, having a recovery time which can be in the order of 200 ns or more compared to FET transistor switching times of 20 - 100 ns. These differences in speeds must be recognized and dealt with in certain switching circuits.

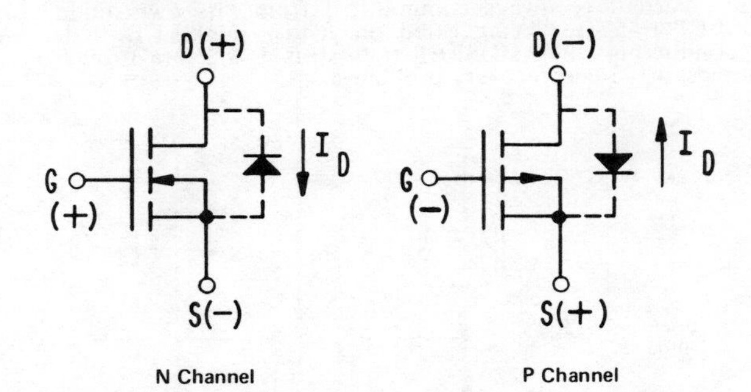

Figure 1
Power MOSFET Symbol Shown With Internal Diode & Polarities of Voltage & Current

III. SOLID STATE RF GENERATOR

A. Basic RF Inverter

Figure 2 shows the basic circuit of a 1 KW resonant inverter developed for induction heating. MOSFET transistors Q1-Q4 are configured as a full bridge, voltage fed inverter with the internal diode of each providing the return path to the DC source for reactive currents. For higher powers, paralleled MOSFET transistors are used; for example, 4 paralleled 6A/450V devices in each leg for a 3 KW rating. The DC terminals of the inverter are tightly coupled to RF bypass capacitor (RFBC) whose capacitance is sufficient to pass the AC component of inverter input without substantially changing its DC potential. The AC terminals of the inverter drive the RF load circuit which is essentially a high Q series resonant circuit formed by tuning capacitor, C_t, and induction load coil, L_t. Transormer RFT matches the load impedance to the VA capability of the inverter, while coupling capacitor RFCC prevents any DC current from flowing in the primary winding and saturating the core. These two components are sized for minimal effect on the resonant circuit parameters, namely:

$$\text{RFCC} \gg \left(\frac{N_s}{N_p}\right)^2 C_t \qquad (1)$$

$$\text{Leakage Inductance of RFT} \ll \left(\frac{N_p}{N_s}\right)^2 L_t \qquad (2)$$

$$\text{Primary Inductance of RFT} \gg \frac{1}{w_{min}} \cdot \frac{E_o}{I_o} \qquad (3)$$

Where:

$\frac{N_p}{N_s}$ Primary - secondary turns ratio of RFT

E_o Rated RMS output voltage of inverter

I_o Rated RMS output current of inverter

w_{min} Minimum output frequency of inverter

In operation, the MOSFET transistors are switched as diagonal pairs: Q1 & Q2 alternating each half cycle with Q3 & Q4 to provide a square wave voltage output at the AC terminals of the inverter. The waveform of the output current depends on the inverter output fre-

Reprinted from *IEEE Ind. Appl. Soc. Meet.*, 1982, pp. 939–944.

quency which is the switching rate of the MOSFET transistors. Driving the series resonant load off resonance - i.e. at a MOSFET switching frequency differing from the natural resonant frequency of C_t & L_t - results in low output current, while driving it at resonance results in maximum power to the load coil. In fact the output current is controlled in a closed loop by varying the driving frequency.

The differences in switching speeds between the MOSFET transistor and its relatively slow reverse diode imposes an operational limitation on the inverter. Had MOSFET diodes 3 & 4 been conducting at the time MOSFET transistors Q1 and Q2 were switched on, they would have failed to recover in time to prevent short circuit currents from flowing between the positive bus and negative bus through the oncoming transistors and offgoing diodes. Commutation of current from a diode to a FET in adjacent legs of the inverter occurs whenever the output current leads the output voltage. One simple solution to the problem is to avoid leading currents and assure that the output current always lags the output voltage (lagging power factor). This is accomplished by always gating the MOSFETS at a frequency greater than the resonant frequency of C_t and L_t. With such a lagging power factor, current is always commutated from "its own diode" to the MOSFET transistor being gated on. Thus, MOSFET diodes 1 & 2 are conducting when MOSFET transistors 1 & 2 are turned on eliminating short circuit recovery problems.

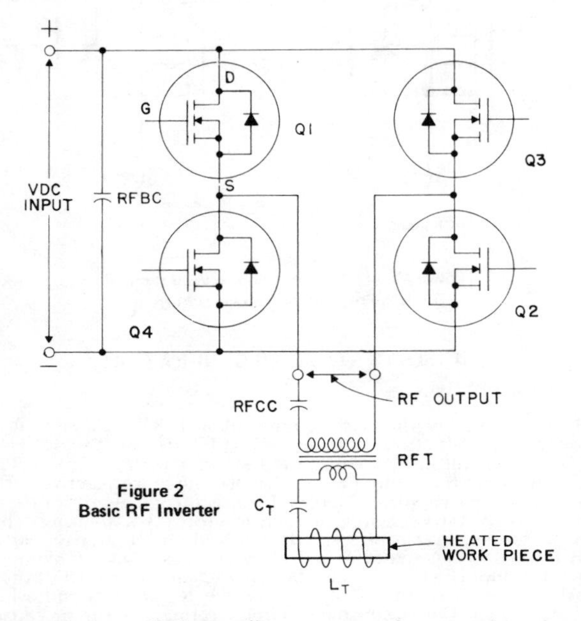

Figure 2
Basic RF Inverter

Figure 3 shows "light load" inverter waveforms resulting from driving of the series load at a frequency $f \gg \frac{1}{2\pi\sqrt{L_t C_t}}$.

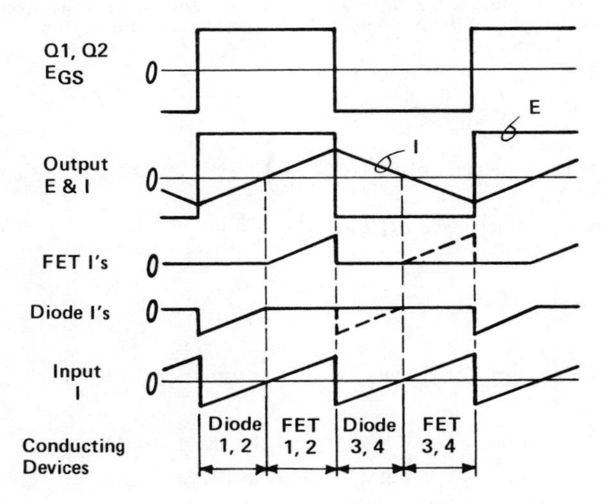

Figure 3 Inverter Waveforms for Light Load

The output current takes the form of an "inductive ramp" with positive current from the DC source flowing through MOSFET transistors and a nearly equal negative current being returned to the source by the MOSFET diodes. The net DC current is nearly zero and very little power is delivered to the highly reactive load (PF $\approx$ 0). Figure 4 shows the waveforms for the same series load being driven at nearly resonance by decreasing the inverter output frequency. The output current is a sinusoid with a high amplitude. In this situation a mostly positive current flows from the DC source through the MOSFET transistors with very little negative current being returned by the MOSFET diodes. The net DC current is large and "heavy power" is delivered to the highly resistive load (PF $\approx$ 1). In both extremes, the commutation of current from a MOSFET diode to "its own MOSFET transistor" is illustrated.

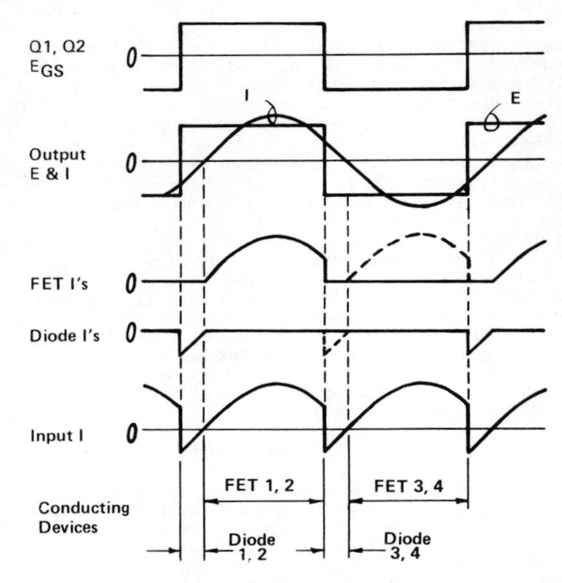

Figure 4 Inverter Waveforms for Heavy Load

B. Gate Drive Circuits

The gate drive circuitry shown in Figure 5 is designed to perform the following functions:

1. Charge and discharge the input capacitances (C_{iss}) of the inverter MOSFETS for drain to source switching times of 100 ns or less.
2. Provide a dead time between turn on and turn off of alternate firing MOSFETS serially connected in one pole of the inverter to prevent short circuit "shoot through" currents.
3. Provide a low impedance to the gate and source of each MOSFET during steady state inverting operation to prevent possible misfiring by the switching of the adjacent device.[6]
4. Prevent misfiring of the MOSFETS during shutdown.

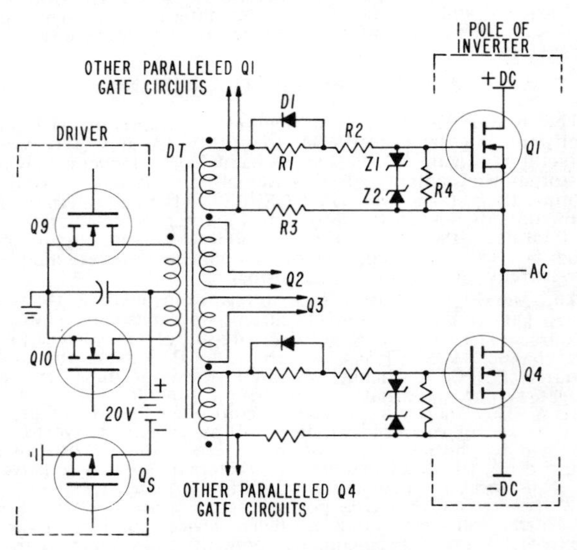

FIGURE 5 - BASIC GATE DRIVE CIRCUITS

MOSFETS Q9 and Q10 and transformer DT form a push-pull driver with four isolated secondaries for gating the inverter power devices. The turns ratio from ½ primary to each secondary is 1:1 being specially wound for low leakage inductance in the order of 0.4 uhy. This unavoidable transformer leakage inductance plus stray inductance in the same order of magnitude as the leakage forms a RLC circuit to charge the input capacitance of the MOSFET. Resistors Rl,

R2 & R3 are in the MOSFET "turn on" charging path being dimensioned for a circuit $Q \approx 1$. The "turn off" discharge path includes only R_2 & R_3 (D1 shorts R1) increasing the circuit $Q \approx 2$ resulting in a faster discharge than charge of the input capacitance. This provides an asymmetrical gate drive signal with relatively slower turn on time and relatively faster turn off time. The result as shown in Figure 6 is that the "off going MOSFET" reaches its gate threshold voltage before the "on coming MOSFET" attains its threshold, providing a gating dead time of about 50 ns for the pair. Figure 7 shows G-S turn on tracking waveforms of diagonally fired MOSFETS.

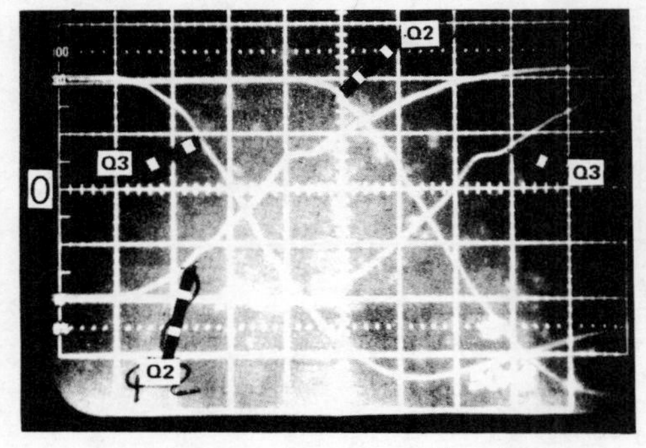

10 V/div
50 ns/div

Photograph is double exposed to show on and offgoing dead times for the devices

Figure 6

MOSFET G-S Turn On-Turn Off Waveforms for Alternate Firing Devices

Since the gate drive is an alternating source, a significant negative potential (see Figure 6) is on the gate of the "off going MOSFET" when the "on going MOSFET" begins to switch. Thus, transient D-G currents caused by switching of the adjacent MOSFET should not cause the "off going" MOSFET to turn back on spuriously.

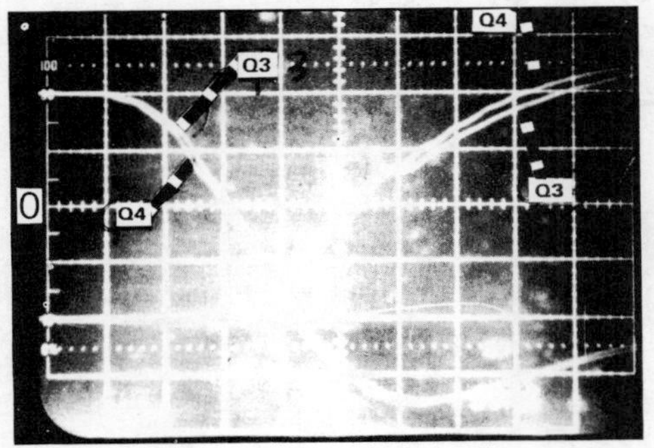

10 V/div
50 ns/div

Photograph is double exposed to show on and offgoing tracking for the devices

Figure 7

MOSFET G-S Turn On- Turn Off Waveforms for Series Fired Devices

If the driver were stopped by removing pulses from driver MOSFET transistors Q9 and Q10, the current established in the primary of DT would freewheel through the other ½ primary and reverse diode of the device that was off. This would result in possible unscheduled gating of the inverter MOSFETS until the energy in DT transformer was dissipated and then result in an open circuited G-S drive connection thereafter. If the tuned load on the inverter was still ringing down through the inverter MOSFET diodes to RFBC bypass capacitor, the switching of the diodes - i.e., rapid charge of D-S voltage - might now gate on the open circuited adjacent MOSFET causing a "shoot through" and failure. To alleviate this condition, driver MOSFETS Q9 and Q10 are both turned on at the STOP command while series MOSFET Q_S is turned off. Thus, the driver transformer is shorted at turn off and secondary drive voltage is reduced to nearly zero immediately. This presents a low impedance to the G-S drive circuits diverting any transient D-G currents from the gate of inverter MOSFET and preventing abnormal turn on.

C. 3 KW RF Generator Circuits

The basic schematic of the overall 3 KW RF generator is shown on Figure 8. A 3 phase full wave diode rectifier is used to effectively utilize the voltage rating of the MOSFET transistors in the inverter. This is so because a single phase full wave rectifier would stress the inverter transistors with a 1.57/1 peak to average voltage (about 393V peak for a 250 Vdc supply) unless an elaborate DC filtering and lightload bleeder were used. The peak to average stress with the 250Vdc 3 phase supply is 1.049/1 or only 263V peak. Since induction heating loads rarely have waveform modulation restrictions (especially when 3 phase diode rectifiers are used) filtering is quite unnecessary. Thus, about 50% more power can be handled by an inverter powered from 3 phase than from single phase given the same voltage stress on the transistors.

Radio frequency filtering by RFL choke and RFFC feed through capacitor minimizes RF feedback (at twice the output frequency) to the AC power lines. The DC surge suppressor holds down any surge voltage across bypass capacitor RFBC during the following situations:

1. Turn on of the main circuit breaker and LC resonant charging of RFBC.
2. Turn off of the inverter gating under heavy load and transfer of energy from the RFL choke to RFBC capacitor.
3. Turn off of the inverter gating under light load and transfer of energy from the high Q output circuit to RFBC.

The inverter bridge has 4 paralleled MOSFET transistors per leg mounted on a common water cooled heat sink. The gate circuits of each transistor is as shown in Figure 5. Four such circuits are mounted on a FET drive board and assembled to the heat sink for each leg. Paralleling of the four gate drive circuits is done by one set of leads from the appropriate driver transformer secondary. Each inverter leg assembly contains a snubber network (RN, CN), tranzorb voltage suppressor network (VZ, DZ) and DC equalizing resistor (RD).

The four inverter leg assemblies (Q1 - Q4) are interconnected in a symmetrical layout by multiple sandwiched, flat bus such that equal and opposite currents are flowing to minimize inductances. This low inductance technique is used from RFBC bypass capacitor, through each drain-source tie, to the RF output. The total inductance from DC input to RF output is in the order of 0.3 uhy.

Three RF current transformers provide control and protection signals to the controller:

- RFCT provides a closed loop feedback signal for control of the RF output current and for load overcurrent trip.
- ϕ CT provides a signal for output phase measurement and limiting.
- DCCT provides a signal for instantaneous trip of the gating for an inverter fault.

The output rating of the 3 KW RF generator is nominally a 250V peak square wave at 18.7 Arms (sinusoid). The output voltage remains nearly constant, except for 60 hz line variations, whereas the current will vary in amplitude as required. The fundamental of the square wave (E_1 = .9 x 250 = 225 Vrms) is the exciting voltage for the series resonant load circuit. RF induction load coils exhibit Q's in the order of 20-60. Thus quite high potentials (4500V-13,500V ideally) would be developed across a coil and tuning capacitor connected directly to the RF output terminals when driven at the resonant frequency for maximum power. In general, RF load coils at 3 KW or so, can have much less impedance than 4500V/18.7A = 240 ohm and generally require impedance matching.

The usual RF matching transformer is air core with considerable exciting current and poor primary to secondary coupling. Such a transformer is not suitable for use with an inverter because of the reactive current and voltage requirements.

A universal, core type transformer (RFT) was developed to suit a 54:1 load impedance matching range.[7] Split primary and secondary windings on a 2 mil strip wound core are seriesed or paralleled to provide output voltages and currents from 5.1V/823A to 37.5/112A in approximately 15 steps. Thus induction coils with impedances ranging from 102V/823A to 2250V/112A ($20 \leq Q \leq 60$) can be matched at the 3 KW level. The transformer is designed to operate from 100 kHz to 500 kHz with a maximum primary exciting current of 1 amp and maximum loss of 94 watts.

D. Control

Basic control requirements for the RF generator are:

1. Control the RF output current (and necessarily the output power) by a closed loop system which effects changes in output frequency to excite the series load nearer to or further from its natural resonance.

2. Detect the phase angle (ϕ) of output current relative to output voltage and "switch in" limiting phase control to maintain a worst case lagging phase nearly zero. This is requisite for two situations: (1) light load where the "asked for current" by the reference is higher than can be achieved at resonance. This would result in an out-of-control situation with the frequency sliding past resonance to the low frequency limit. (2) to prevent improper commutation of current from the slow reverse connected MOSFET diodes to the fast transistors.
3. Detect voltage on the series tuning capacitor (C_t) and "switch in" limiting voltage control where excessive potential would occur as with higher than expected circuit Q.
4. Provide high frequency and low frequency limiting.

Figure 9 shows the basic control scheme which implements the aforementioned requirements. It is a straight forward closed loop system with the operators control potentiometer PT1 providing the current reference I_c* and RFCT/REC1 providing the current feedback signal I_{fb} to OA1 error integrator. The output of OA1 decreases the frequency of pulses from the VCO (which is twice the output frequency) until zero error is achieved; i.e. the magnitude of output current is satisfied. Pulses from the VCO clock the timing flip flop whose Q and $\overline{Q}$ outputs drive OR gates IC12A,B and IC12C,D alter-

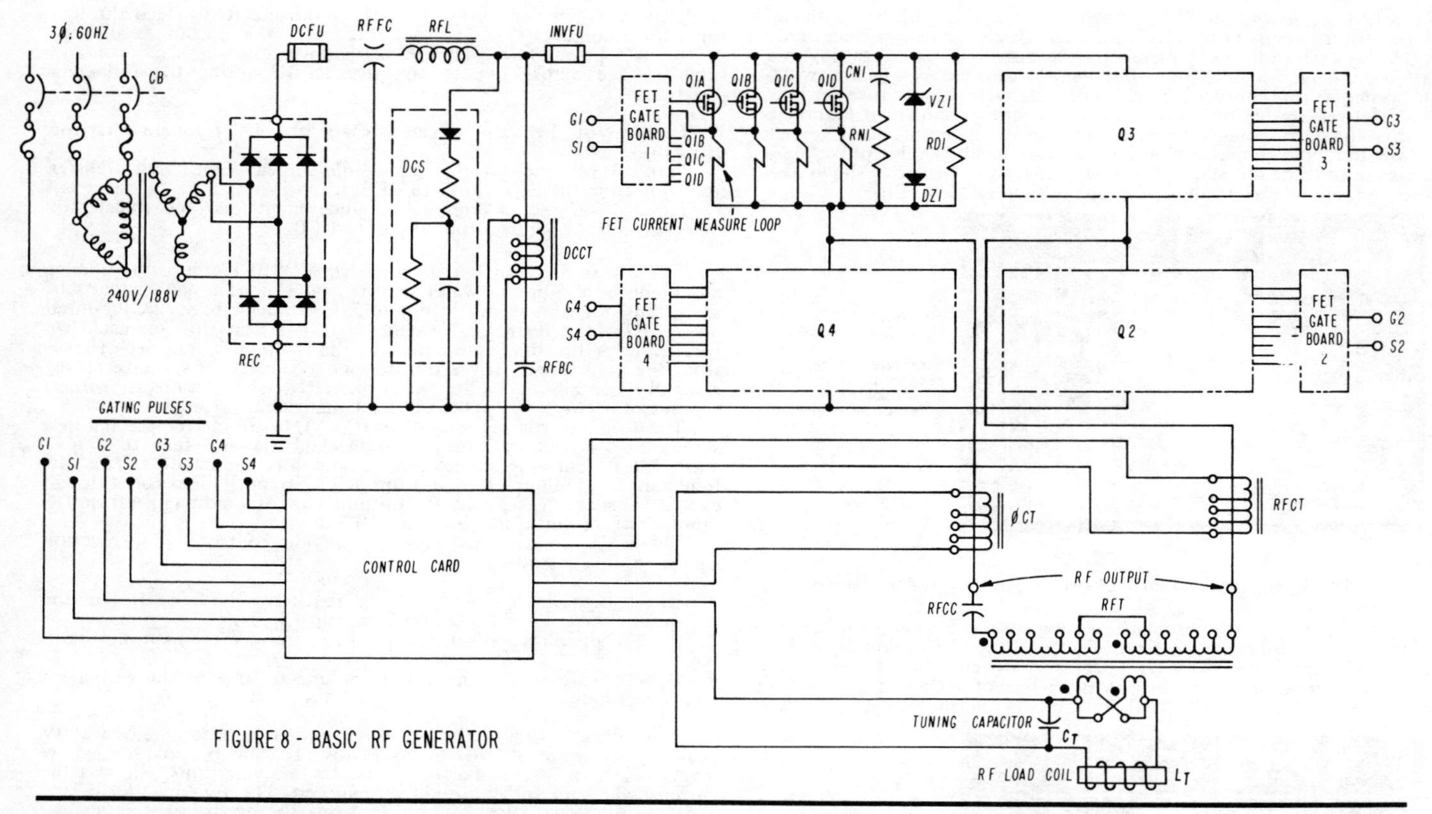

FIGURE 8 - BASIC RF GENERATOR

nately at ½ the VCO frequency. The OR gates drive emitter follower EFT1 and EFT2 which, in turn, provides low impedance gating signals to the driver N channel MOSFETS Q9 and Q10.

Rise time and fall time at the gates of Q9 and Q10 are about 10 ns with drain-source switching times in the order of 20 ns. The IC12A,B and IC12C,D paralleled pairs of OR gates are paralleled within one quad IC to increase current sinking and sourcing to enhance their switching times. All logic elements are CMOS operated at 15 volts. This allows direct interfacing with bipolar and FET transistors in the controller without need for level conversion.

A high STOP or FAULT signal from the sequencing and protective logic circuits simultaneously make the outputs of OR gates IC12A,B and IC12C,D a high. Driver MOSFETS Q9 and Q10 are both turned on shorting the primary of the driver transformer (DT) for low impedance stopping of the inverter gate drive pulses. At the same time the -15V gate drive signal to the P channel series clamp MOSFET Q_S is removed, disconnecting the drive power supply from the shorted driver.

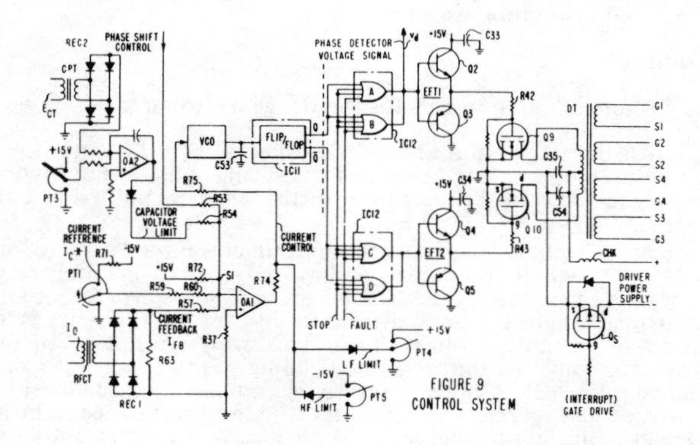

FIGURE 9 CONTROL SYSTEM

Series capacitor voltage limiting occurs when the voltage sensed by transformer CPT and rectified by REC2 tends to exceed the maximum level desired. Capacitor voltage limit pot PT3 sets the maximum level at V_c*. When rectified voltage signal V_c is less than V_c*, integrator 0A2,, has a negative output back biasing D15 and disconnecting 0A2 from summing node S1 to 0A1. Should V_c tend to exceed V_c*, the output of 0A2 goes positive and is connected to 0A1 summing node. This constitutes additional negative feedback to 0A1 which "makes up" the existing difference between I_c* and I_{fb} limiting the VCO frequency from further reduction and holding the capacitor voltage at the limit level.

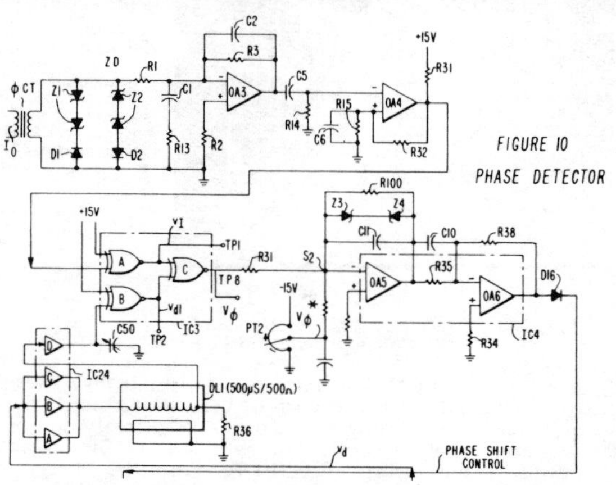

FIGURE 10 PHASE DETECTOR

The phase detector circuitry is shown in Fig. 10. Basic phase detection occurs with the EXCLUSIVE NOR gate IC3C. The inputs are square waves representative of the RF output current (V_I) and RF output voltage (V_{dl}). The truth table for EXCLUSIVE NOR gate IC3C is as follows:

Input		Output
V_I	V_{dl}	V_ϕ
0	0	1
1	0	0
0	1	0
1	1	1

Thus, an output occurs from the EXCLUSIVE NOR gate whenever two 1's or two 0's overlap, resulting in an average output voltage versus phase of the two square waves as shown in Fig. 11. This "phase output" after processing by OA5 integration, OA6 inverter and diode D16 provides negative feedback to S1 summing node of 0A1 whenever it exceeds the minimum desired phase determined by the setting of PT2 potentiometer. The VCO frequency is limited from further reduction holding the phase angle between voltage and current at the limit level.

The interesting part of the phase detector lies in the development of the V_I and V_{dl} input signals to the EXCLUSIVE NOR gate IC3C.

First it should be recognized that the phase characteristic shown in

Figure 11 is really not suitable for stable closed loop control. This is because the phase detector average output voltage does not monotonically increase with variation of ϕ from 90° lag to 90° lead. At ϕ = 0°, the slope changes and the average output decreases for further increasing leading phase. Therefore, detection at ϕ = 0° is hazardous since the phase control loop would become unstable with each incremental increase in ϕ incrementally decreasing the error signal to OA5. This results in increasing error to OA1 which necessarily drives the frequency of the VCO to its lower limit giving nil current output at ϕ = 90° lead. The solution to this problem is to phase shift the

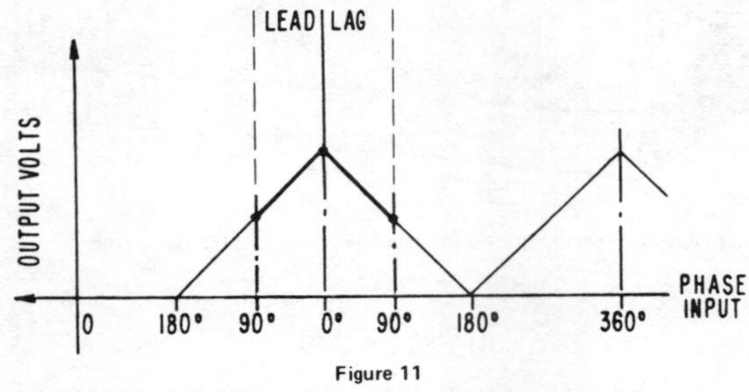

Figure 11

Average Output of EXCLUSIVE NOR Phase Detector

current 90° lagging and change the characteristic to that shown in Figure 12. Thus, at ϕ = 0°, should the phase tend to go leading, the phase error would tend to increase resulting in more (not less) feedback to hold the phase at the desired level. This is accomplished by wideband amplifier OA3 connected as an integrator. OA3 integrates the current signal from ϕCT which is squared by ZD clamp network into a ramp (see Figure 13). The integrated current waveform from OA3 is AC coupled to comparator OA4 which generates a square wave at 90° lagging (almost) from the original current signal. The lagging square wave is further "squared up" by EXCLUSIVE NOR gate IC3A into V_I.

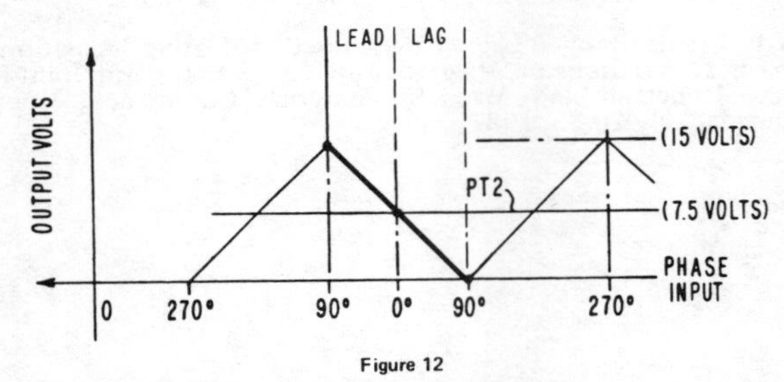

Figure 12

Average Output of EXCLUSIVE NOR Phase Detector with 90° Lag

The second problem is in the "almost" 90° lag generated by OA4 comparator. It happens that there is a constant, added delay time (t) of about 200 ns between the actual RF current and output of O4A - mostly caused by OA4 (an LM211 high speed comparator). This fixed time delay is formidable since it represents a variable phase angle delay of 7.2° - 36° when the frequency is varied from 100 kHz to 500 kHz. The problem is further compounded by a constant lead time (t´) of V_d voltage signal from IC12A,B which is a digital representation of the RF output voltage. V_d leads the actual RF output voltage by about 300 ns. The solution to this problem is to make the voltage signal V_{dl} lag the actual RF output voltage by the same fixed time as the current signal V_I lags the output current. Since the EXCLUSIVE NOR phase detector measures the relative phase of V_I and V_{dl}, the absolute phase of each is unimportant as long as they are equal. To accomplish this, delay line DL1 and adjustable vernier capacitor C50 on the output of buffer gate IC24D provide most of the required time delay (t´ + t). The remaining is provided by the propogation times of the squaring gate IC3B and paralleled delay line buffers IC24 A,B,C.

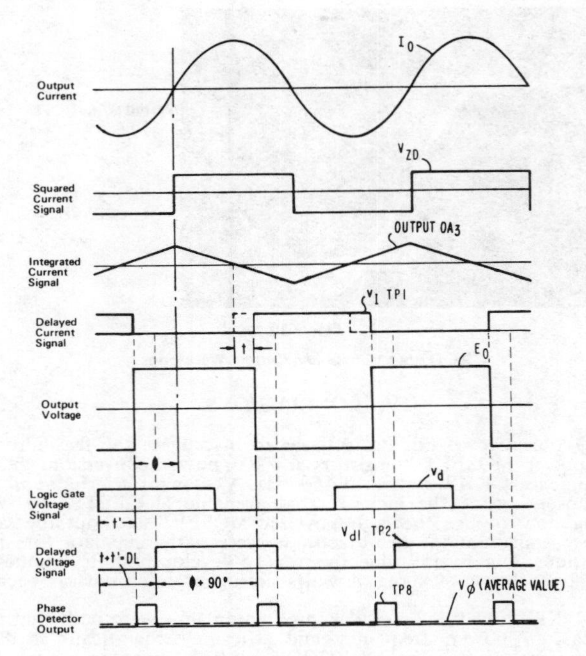

Figure 13 Phase Detector Waveforms

E. Test Results

The RF generator was tested from a 1 KW output level using 4 HPWR-6501 MOSFETS to over 3 KW using 16 MOSFETS, 4 paralleled per leg. Some pertinent data and statements about those results follow:

1. Waveforms of voltage and current (see Figures 14 and 15) were as expected, except that MHz ringing at current commutations were significant. Worst case D-S ringing occurred off resonance with maximum E_{DS} = 370V at 250Vdc.

2. Current sharing among the paralleled MOSFETS was excellent with a worst case unbalance of 13.5% from the mean and a typical unbalance of less than 10% measured. The only precaution taken for paralleling was symmetry in circuit layouts and matching of gate threshold voltages (V_t) as follows:

V_t within ½V for devices in one inverter leg.
V_t within ½V for groups in alternate firing legs.
V_t within 1V for groups in diagonal firing legs.

3. Over 3 KW of output power was achieved with a typical set of recorded data tabulated below:

Edc Volts dc	Idc Amps dc	E_{01} Volts rms	I_{01} Amps rms	f kHz	ϕ Degrees	Po Watts (Calculated)	DC-RF Efficiency % (Calculated)
236	16.6	198	18.74	387.2	20.9	3466	88.5

(1) Above data for RLC load - no transformer

L_t= 30 uhy C_t= 0.0055 uf R = 9.87 ohm

(2) MOSFET TO3 case temperature rise above inlet water < 5°C

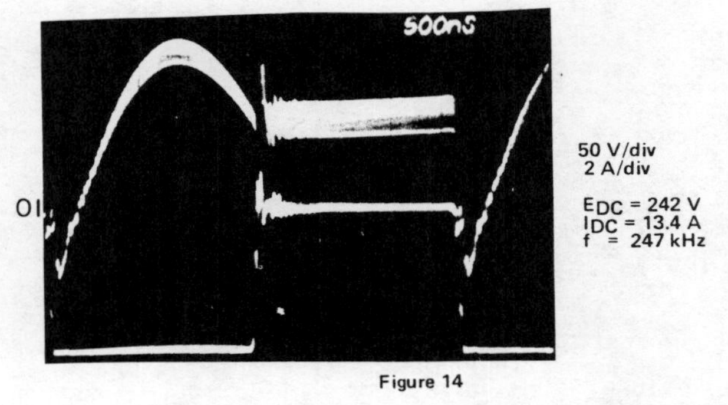

Figure 14

MOSFET Voltage and Current Waveforms

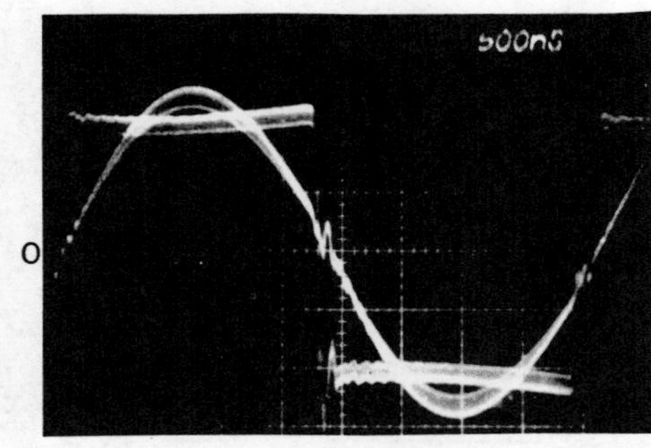

Figure 15

RF Output Voltage and Current Waveforms

IV. CONCLUSIONS

The objective of this project was to investigate the feasibility of using power MOSFET transistors for RF power conversion and to achieve at least 1 KW output at 450 kHz. A resonant inverter approach was chosen using the highest rated available MOSFETS. We were learning how to use the high powered MOSFET at about the same time as semiconductor manufacturers were gathering data for their applications. Our overall objective was to develop basic techniques to eventually achieve tens of kilowatts as higher rated devices became available.

A 1 KW model and a 3 KW production unit were built that met the goals of power, frequency and efficiency. Significant ancillary developments were megahertz CMOS controls, nanosecond risetime drivers, iron-core RF load matching transformers and circuit measuring techniques for nanosecond transitions.

More work needs to be done to "harden" the design for rugged field applications. The low surge current capability of the MOSFET (like the bipolar transistor, but unlike the SCR) makes it difficult to protect even with high speed gate suppression. We would recommend that compatible protective devices and/or techniques be developed to match the speed and surge capability of the MOSFET. We recommend also, that semiconductor manufacturers develop higher rated MOSFETS in more suitable packages for higher power designs.

V. ACKNOWLEDGEMENTS

The authors wish to acknowledge the assistances of Dave Hoffman, Siliconix; Victor Li, Hewlett-Packard; and Brian Pelly, International Rectifier in application of the power MOSFET to this project. We are indebted to Reuben Lee, retired from Westinghouse, for design of the RF matching transformer and for consultative help in design of other magnetics.

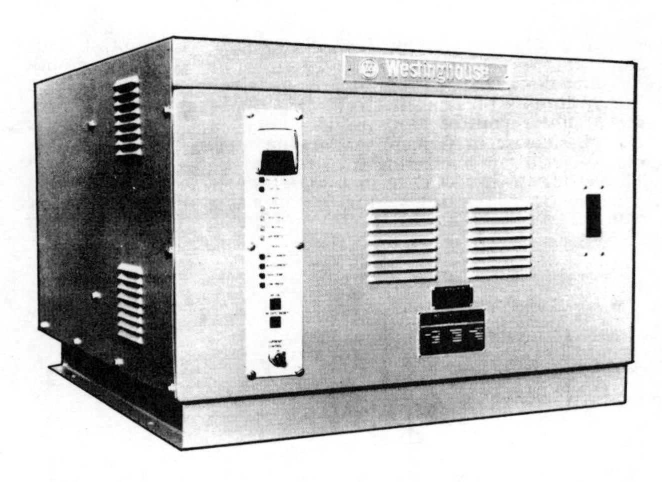

Figure 16 Solid State 3 kW, 100 kHz - 450 kHz RF Generator

REFERENCES

1. B.R. Pelly, "Applying International Rectifiers Power MOSFETS," IR Application Note AN-930.
2. D. Hoffman, "VMOS - Key to the Advancement of SMPS Technology," Power Conversion International - March/April, 1980, Volume 6, Number 2, PP 37-42.
3. S. Davis, "Switching - Supply Frequency to Rise; Power FETS Challenge Bipolars," EDN, January 20, 1979, PP 44-50.
4. R. Severns, "The Power MOSFET, A Breakthrough in Power Device Technology," Intersil Application Bulletin A033.
5. W. Fragle, B.R. Pelly, B. Smith; "The HEXFET's Integral Reverse Rectifier - A Hidden Bonus for the Circuit Designer," Power Conversion International - March/April, 1980, Volume 6, Number 2, PP 17-36.
6. S. Clemente, "Gate Drive Characteristics and Requirements for Power HEXFET's," IR Application Note AN-937.
7. W.E. Frank, Reuben Lee; "New Induction Heating Transformers," IEEE Transactions on Magnetics presented at the 3rd Joint Intermag-Magnetism and Magnetic Materials Conference; Montreal, Canada, July 20-23, 1982.

Part IV
Emerging Transistor Technology

In the last 10 years, a plethora of new transistor types have been reported in the literature. Most of these devices have been laboratory curiosities and some have become commercially available. In this section, reprinted papers dealing with power junction gate field effect transistors (JFET's), power field controlled thyristors (FCT's), and some new power devices based upon combining MOS and bipolar technologies are provided to acquaint the reader with transistor technology that may become increasingly important in the future.

The section begins with three papers that discuss the design and characteristics of high voltage power JFET's. As mentioned in the introduction, the primary application for these devices is in high performance audio and RF amplifiers where their unique high frequency characteristics are utilized and their normally on characteristics are acceptable. Since the buried gate structure for the JFET has a poor frequency response due to its high series gate resistance, most of the interest lies in high voltage, surface gate, power JFET devices. The major difficulty in the development of these devices has been achieving high voltage performance with sufficiently large blocking gains. The recessed gate structure described in the first paper provides the best method for obtaining adequate blocking gains and devices with the highest forward blocking capability have been fabricated using this technology. The second paper provides a more detailed discussion of the design of the gate structure of the high voltage JFET. Although the analysis in the paper was performed using a buried gate structure, the conclusions derived by the authors are equally valid for surface gate devices. Here design of the channel length and width to achieve either triodelike or pentodelike characteristics are discussed in detail. It is worth pointing out that an optimally designed high voltage power JFET should have mixed pentode-triode characteristics because its channel width must be greater than the zero-bias depletion width in order to achieve high current handling capability. The characteristics of these devices are triodelike at high drain voltage and low currents and pentodelike at high currents and low drain voltages. The third paper, as well as the first paper, demonstrate that by using highly interdigitated source and gate fingers, 300-600 V devices can be fabricated with maximum frequency of oscillation in excess of 500 MHz.

The current handling capability of the power JFET is limited by the high resistance of the drift layer. This problem is overcome in the power FCT by the modulation of the drift layer resistance by the injected minority carriers from an anode region. Since these devices usually have normally on characteristics, a high blocking gain is essential to achieving high voltage performance. The best characteristics in those devices have been obtained in the asymmetrical FCT structure described in the fourth paper in this section. Devices with 1000 V forward blocking capability and direct current blocking gains of over 50 have been successfully developed. The next paper also deals with these devices and describes a method for varying the forced gate turnoff time in these devices from over 20 microseconds to less than 0.25 microseconds. The sixth paper then describes a recently developed gating technique for these devices using a power MOSFET in series with the cathode of the FCT. This gating circuit results in the operation of the FCT with an extremely high input impedance while achieving high speed forced gate turnoff at effectively a unity current gain condition. It also makes the overall circuit perform as a normally off device and eliminates this serious drawback of the FCT. However, this gating technique requires the use of a separate high current power MOSFET which can add to circuit cost. The normally off characteristic has also been accomplished in the FCT by using VLSI technology in the seventh paper. By using high resolution lithography, the authors have reduced the spacing between gate regions until the built-in zero-bias potential of the gate junction creates a large potential barrier in the channel. This has allowed the fabrication of 500 V devices. However, these devices require large gate drive currents during forced gate turnoff.

The insulated gate transistor (IGT) described in the tenth paper in this section is likely to be the most promising new transistor development. In these devices the best characteristics of power MOSFET's and bipolar devices have been combined. The high input impedance of the MOS gate structure, the very efficient current carrying capability, the tailorable switching speed characteristics and the reverse blocking capability in these devices make them ideal for high voltage dc and ac circuits operating at frequencies up to 100 kHz. Since the announcement of the first devices in the technical literature, there has been a rapid growth of interest in these devices which can be expected to lead to the publication of many papers in the future dealing with device physics, design, and circuit applications.

The use of a BJT-FET combination in an emitter-open configuration, as described in the last paper, has great appeal for high voltage applications. Because of emitter-open turnoff of the BJT, the voltage blocking capability of the BJT can be raised from BV_{CEO} to BV_{CBO}, which is often an improvement of more than 50 percent. Turnoff time can also be greatly reduced. Furthermore, to turn off the BJT, all it requires is to gate off the FET, which is much easier to do from the circuit point of view. The price one has to pay for this scheme is the need for a low voltage high-current power MOSFET and a relatively high conduction voltage drop across in BJT-FET power switch. However, for high voltage high frequency applications, the reduction in turnoff switching losses more than offsets the increase in conduction losses. Techniques for integration of the BJT and FET are discussed in the ninth paper.

High-Voltage Junction-Gate Field-Effect Transistor with Recessed Gates

B. JAYANT BALIGA, SENIOR MEMBER, IEEE

Abstract–**A new recessed-gate structure for vertical-channel junction field-effect transistors (JFET's) is described together with a self-aligned gate–source process developed to fabricate these devices. Using this technology, devices with groove depths ranging from 8 to 18 μm have been fabricated. The characteristics of these devices is described as a function of the groove depth. It has been found that the devices display pentode-like characteristics at low gate voltages and triode-like characteristics at high gate voltages. The blocking gain has been found to increase with groove depth. However, this is accompanied by an increase in the on-resistance and a decrease in the saturated drain current. Devices with gate breakdown voltages of up to 600 V have been fabricated with the recessed-gate structure. These high-voltage field-effect transistors (FET's) have a unity power gain cutoff frequency of 600 MHz and gate turn-off times of less than 25 ns.**

I. Introduction

ALTHOUGH the junction gate field-effect transistor (FET) was invented over twenty years ago, it is only recently that high-voltage devices have been developed which are suitable for power switching applications. Among the device structures that have been developed, the vertical-channel structure with source and drain on opposite surfaces of the wafer has been found to be the best for achieving a high breakdown voltage and a low on-resistance. These vertical-channel devices have been made using either a buried-gate structure [1]-[3] or a surface-gate structure [4]-[8]. Although the buried-gate devices can be fabricated with large active areas, their frequency response is poor due to the high-input gate resistance of the buried-gate fingers which require remote gate contacts. This problem is overcome in the surface-gate device where the gate metallization is placed directly upon the diffused-gate fingers. However, the surface-gate devices made using planar diffusion techniques [4]-[7] have been found to exhibit low blocking gains due to their open-channel structure and a poor yield due to the several critical photolithographic steps required to fabricate the interdigitated gate and source regions. The low blocking gain in these devices has also restricted their operation to drain voltages of less than 150 V. Recently, a vertically walled gate structure has been developed by using preferentially etching techniques to fabricate vertically walled grooves in (110) oriented silicon followed by epitaxial refill of the grooves with p-type silicon to obtain a planar structure [8]. This device structure allows the fabrication of high-voltage devices with high blocking gains. Devices have been fabricated

Manuscript received September 10, 1981; revised May 12, 1982.

The author is with the Device Physics Unit, General Electric Corporate Research and Development Center, Schenectady, NY 12345.

Reprinted from *IEEE Trans. Electron Devices*, vol. ED-29, pp. 1560–1570, Oct. 1982.

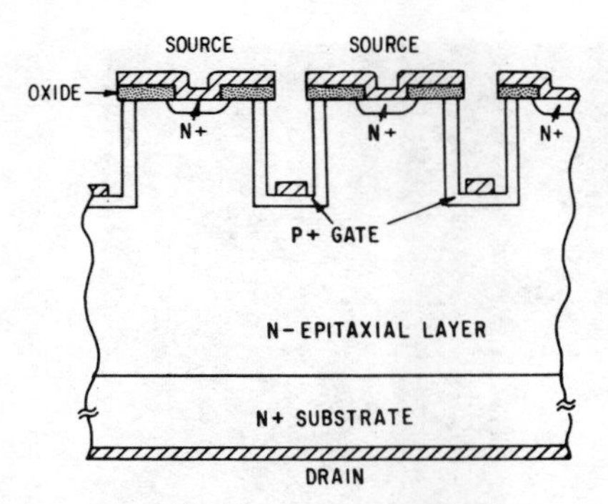

Fig. 1. Recessed-gate JFET structure.

with operating voltages of up to 400 V and blocking gains of more than 20. However, the epitaxial refill process is complex and results in poor yield, unless extreme care is taken during the refill to obtain a planar surface with no voids in the grooves. This paper describes a nonplanar junction field-effect transistors (JFET) device structure which also contains vertically walled gate regions. However, this structure can be fabricated using a self-aligned source-gate process without the photolithographic problems usually encountered due to the interdigitation of the source and gate regions. With this recessed-gate process, a significantly better device yield has been observed and device breakdown voltages of up to 600 V have been achieved. This is the highest breakdown voltage reported for devices of this type. In this paper, the technology developed to fabricate these devices as well as their static and high-frequency characteristics will be described.

II. Device Structure and Fabrication

The recessed-gate structure of the FET is shown in Fig. 1. This device structure has vertically walled gate junctions which have previously been demonstrated to provide field-controlled devices with very high blocking gains [8]-[10]. However, instead of achieving a planar device surface by performing the epitaxial refill process, in the recessed-gate structure the nonplanar nature of the structure is taken advantage of to obtain separation of the gate and source metallization without the need for extra photolithography steps which normally reduce the yield. The device fabrication procedure for the recessed-gate structure is described below.

In order to fabricate high-voltage JFET's, high-resistivity n-type epitaxial layers were grown on 0.01-$\Omega \cdot$ cm antimony-doped substrates. The substrates were chosen to have (110) surface orientation because the (111) crystallographic planes lie perpendicular to this orientation. This feature is essential to achieving the vertically walled gate surfaces. The epitaxial layers were grown in this study by using dichlorosilane under conditions which provided epitaxial layers with a very low hillock and stacking fault density [11]. The doping profiles of the epitaxial layers were controlled by adjusting the flow of the phosphine dopant gas during the growth. The doping concentration and the thickness of the epitaxial layers were determined by using spreading resistance measurements. Typically, the epitaxial layers had a doping level of 2×10^{14} per cc and a thickness of 50 μm.

After the growth of the epitaxial layer, the wafers were thermally oxidized to grow a 10 000-Å masking layer on the surface. This layer was photolithographically patterned to open a window surrounding the active area of the device. A boron diffusion was then performed in this region with a junction depth of 10 μm to serve as the device termination. This device termination region is indicated in Fig. 2 which shows a photomicrograph of the completed device. The recessed-gate structure was then fabricated in the active area of the device by using the following process which will be described with the aid of Fig. 3. First, windows were opened in the oxide at both the source and the gate locations by using a single mask (Fig. 3(b)). This was done to ensure self-alignment between the source fingers and the gate fingers. This self-alignment is important for these field-controlled devices because any misalignment of the source region with respect to the gate region results in a degradation of the gate-source breakdown voltage. After opening these windows in the oxide, the wafers were coated with layer of silicon nitride deposited by the reaction of silane with ammonia [12], [13]. The silicon nitride layer was in turn covered with a layer of silicon dioxide grown by the oxidation of silane [14], [15]. This silicon dioxide layer was then patterned using conventional photolithography to open windows in the source region of the device and the silicon nitride was removed in these areas by etching it in phosphoric acid at 180°C using the pyrolytic silicon dioxide as a mask. As illustrated in Fig. 3(c), some degree of misalignment can be tolerated during this photolithography step. A phosphorus diffusion was then performed in the source windows and a protective oxide layer was simultaneously grown in this area. The silicon nitride layer was then selectively etched away by using hot phosphoric acid to expose the silicon in the gate areas as shown in Fig. 3(d). A preferential etch comprising a mixture of potassium hydroxide and isopropanol was then used to etch deep grooves into the silicon [16]. The composition of the etch was selected such that an overhang was deliberately created in the oxide due to undercutting as illustrated in Fig. 3(e). This was followed by a boron diffusion in the grooves to create the junction-gate regions. The oxide in the source windows as well as any grown in the gate areas during the boron diffusion was then removed by using buffered hydroflouric acid. Aluminum metallization was then evaporated on the wafer surface. Due to the shadowing created by the overhang in the oxide at the top of the grooves, the aluminum in the gate and source areas was automatically separated as illustrated in Fig. 3(f) without the necessity of a critically aligned photolithographic metal definition step that would normally be required at this stage of device fabrication. The separation of the aluminum in the recessed-gate area from the aluminum connecting the source fingers together on the top surface can be clearly seen in scanning electron micrographs shown in Fig. 4. It is worth pointing out that the device has been designed with a recessed-gate contact pad to which each of the gate fingers is connected. This gate bonding pad is visible in the lower section of the photograph shown in Fig. 4(a) together with the gate lead bonded to the gate metallization. The separation of the source and gate metallization can be clearly seen in Fig. 4(b). The above process, thus provides continuous contact metallization over not only all the

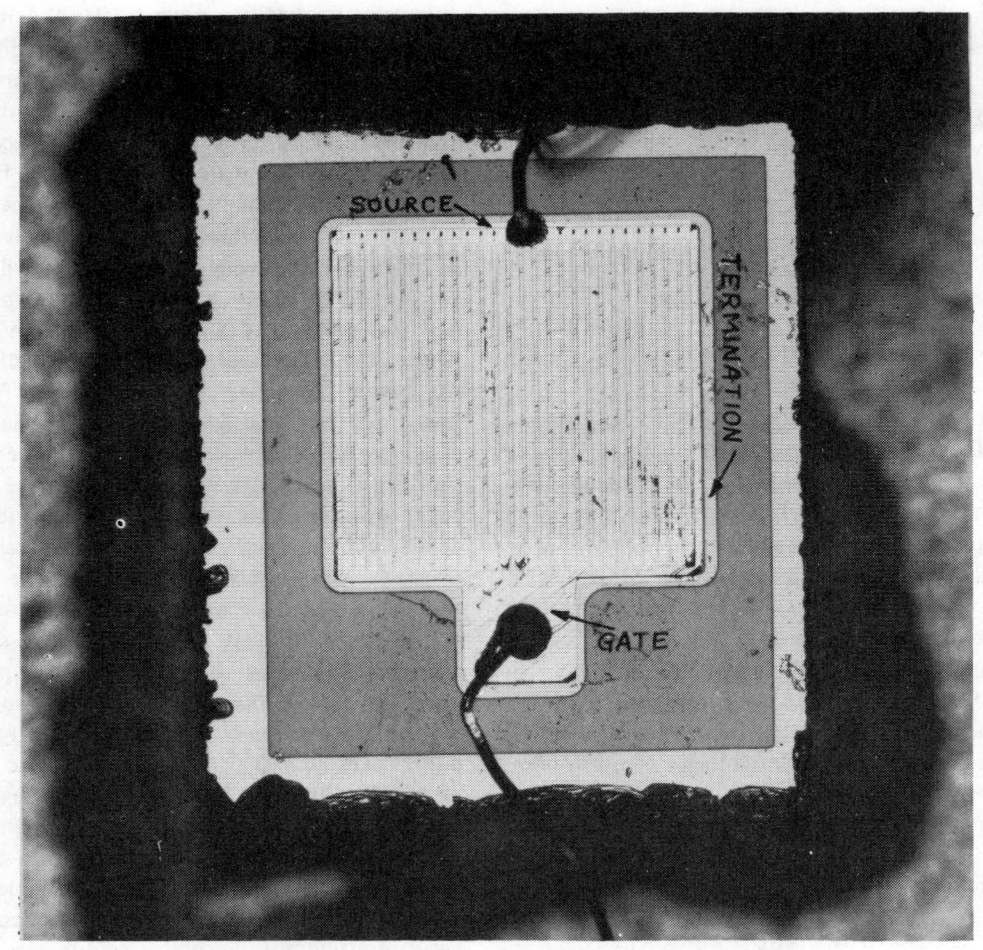

Fig. 2. Photomicrograph of high-voltage JFET device fabricated with the recessed-gate device structure. The active area of the device is 800 μm × 800 μm.

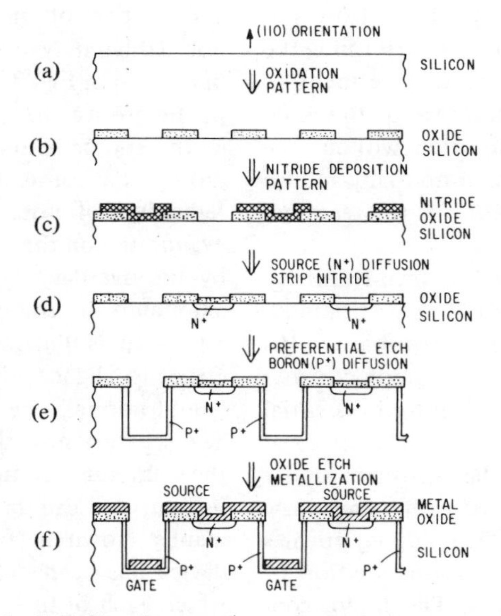

Fig. 3. Self-aligned recessed-gate device fabrication process.

source fingers but also over all the gate fingers. Further, the above process maximizes the width of the source-finger metallization, thus decreasing the current density in the aluminum used to carry the source current. This results in a low contact resistance to both the source and the gate regions. Since the above process can be performed without the use of any critical photolithographic alignment steps, improved device yield can be simultaneously achieved.

(a)

(b)

Fig. 4. Scanning electron micrographs of the recessed-gate structure. (a) Gate pad area. (b) Ends of source finger.

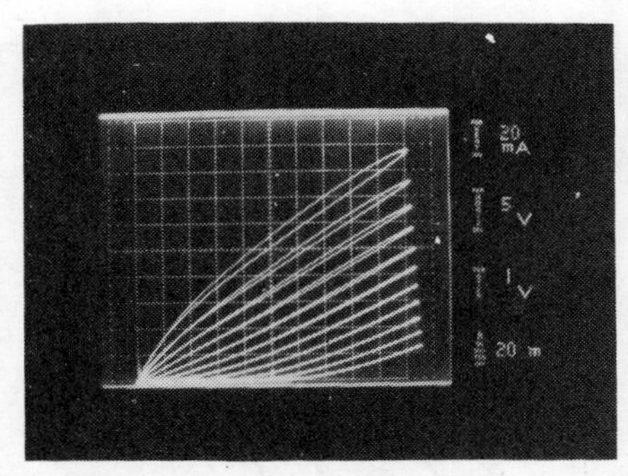

Fig. 5. Typical pentode-like characteristics of the recessed-gate JFET devices observed at low gate bias voltages. Upper trace obtained at zero gate bias (gate shorted to source). Gate bias steps are –1 V per step.

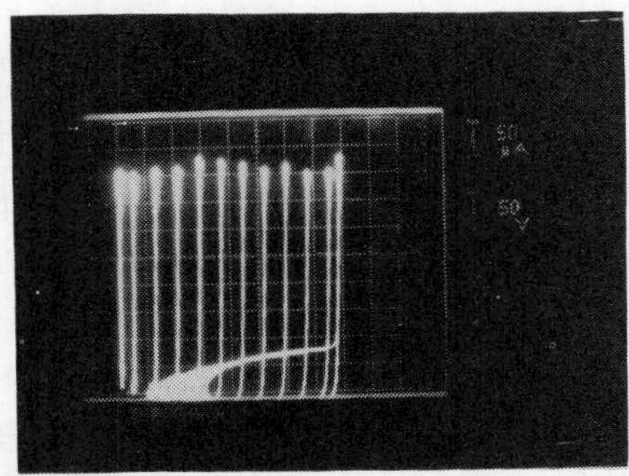

Fig. 6. Typical triode-like characteristics of the recessed-gate JFET devices observed at large gate bias voltages. Gate bias has been increased in 11 steps of –5 V each.

III. Device Characteristics

The devices fabricated in this study were designed to obtain breakdown voltages of over 500 V with a minimum input gate capacitance. The active area of the chip was 800 μm × 800 μm within which 27 source fingers were located interdigitated between 28 gate fingers. Each source finger had a length of 675 μm and a width of 5 μm, thus providing the device with a total source area of 9.1×10^{-4} cm^2. The depth of groove etched in the gate area was varied to examine its influence upon the blocking gain of the devices. The spacing between the grooves (channel width) was maintained at 20 μm for all the devices.

A. Drain-Output Characteristics

In the JFET's fabricated using the recessed-gate structure the half width of the channel exceeded the zero bias depletion layer width of the gate junction. Consequently, these devices were expected to exhibit pentode-like forward conduction characteristics at zero gate bias, as well as at the lower gate voltages when the gate depletion layers did not completely pinchoff the channel. A typical set of pentode-like characteristics observed for the recessed-gate JFET's are shown in Fig. 5. In this region of device operation, the drain–source characteristics exhibit ohmic behavior at low currents where the voltage drop in the channel region is small. At larger drain currents, the voltage drop along the channel causes an increase in the reverse-bias potential on the gate towards the bottom of the grooves. This causes an extension of the gate depletion layer into the channel and reduces its cross-sectional area. The pentode-like characteristics resulting from this phenomena can be adequately modeled by using the gradual channel approximation developed by Schockley [17]. A detailed analysis of the device characteristics, including the effect of the large drift region of these high-voltage JFET's, has been published [18].

As the reverse-bias gate voltage is increased, the gate depletion layers will eventually intersect in the channel region. Once this occurs, a potential barrier is formed between the drain and the source. This potential barrier prevents the injection of electrons from the source to the drain and allows the device to block drain current flow. The height of the potential barrier is determined not only by the gate bias but is also dependent upon the drain potential. Two-dimensional numerical analysis of the current flow in these devices has demonstrated that the potential barrier height will decrease with increasing drain potential [6], [19]. This decrease in the potential barrier allows the drain current flow to commence when the drain voltage is increased. The injection of the electrons over the potential barrier also results in triode-like characteristics. Typical triode-like characteristics observed for the recessed-gate JFET devices are shown in Fig. 6.

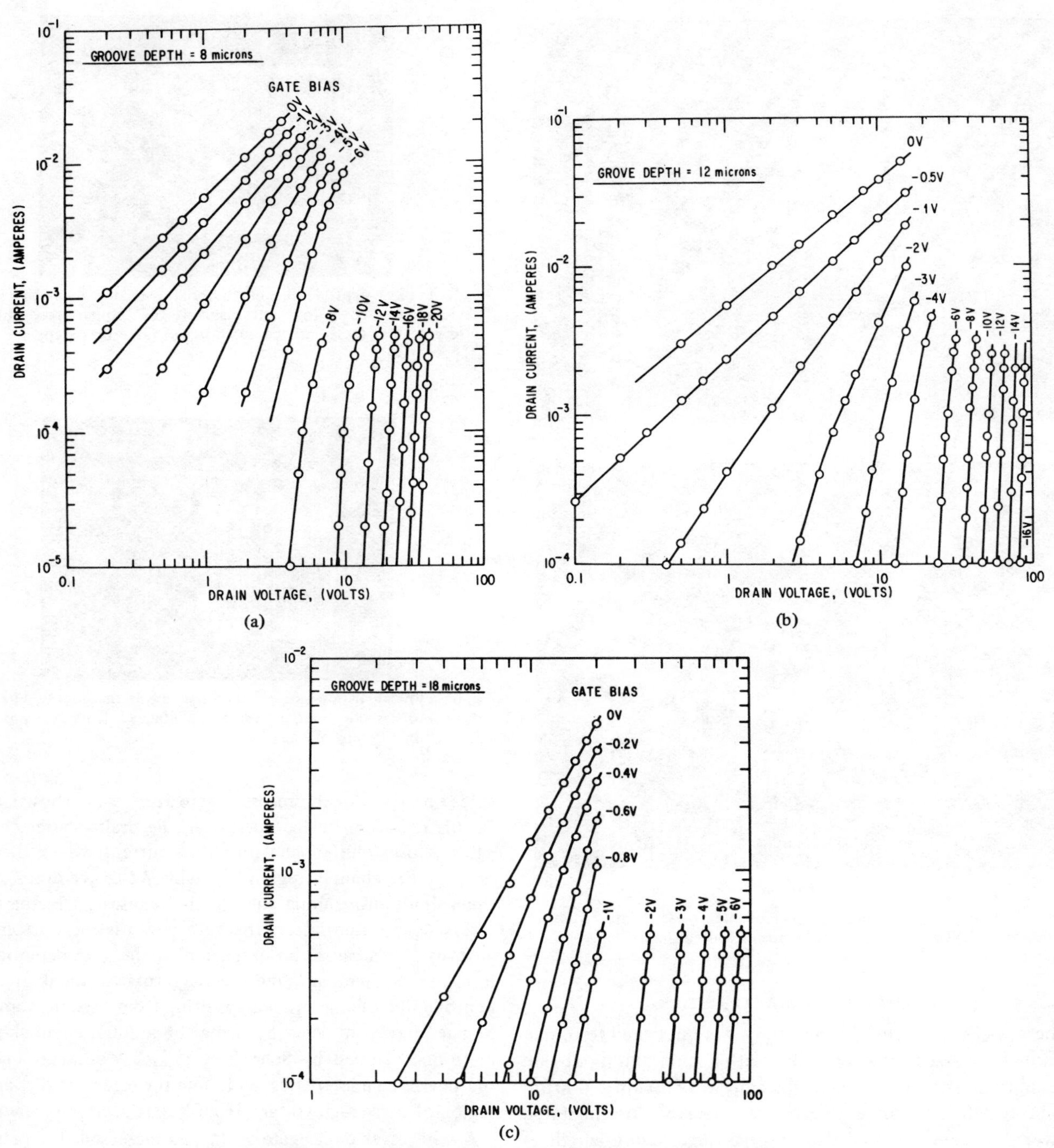

Fig. 7. Drain output characteristics of recessed-gate JFET devices fabricated with groove depths. (a) 8 μm. (b) 12 μm. (c) 18 μm. The reverse-gate bias has been indicated next to each curve.

The transition in the device characteristics from pentode-like to triode-like is determined by the depth of groove in the recessed-gate structure. Fig. 7 shows the drain output characteristics as a function of the gate bias for devices fabricated with groove depths of 8, 12, and 18 μm. In this figure, the transition from pentode-like characteristics to triode-like characteristics is indicated by an increase in the slope of the drain output characteristics. Once in the triode-regime, the drain current increases exponentially with increasing drain voltage and the slope of the drain characteristic becomes independent of the gate bias voltage. This is consistent with the following equation:

$$I_D = I_0 \exp \frac{q}{kT} [\alpha V_D + \beta V_G] \tag{1}$$

used to describe the output drain current flow in earlier devices [1], [8] where α and β are constants.[1] From Fig. 7, it is evident that the transition occurs at smaller gate voltages as the groove depth increases. For example, the transition takes

[1] Note that V_D is positive and V_G is negative.

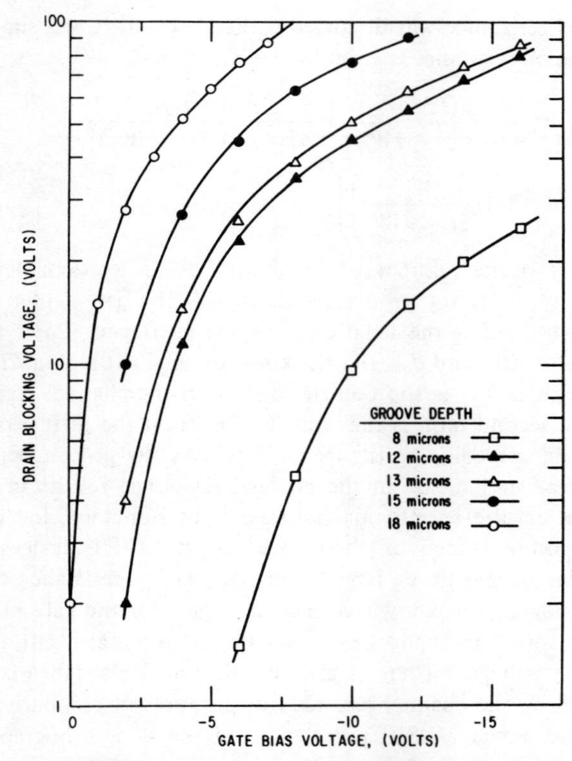

Fig. 8. Blocking characteristics of recessed-gate JFET devices fabricated with groove depths ranging from 8 to 18 μm.

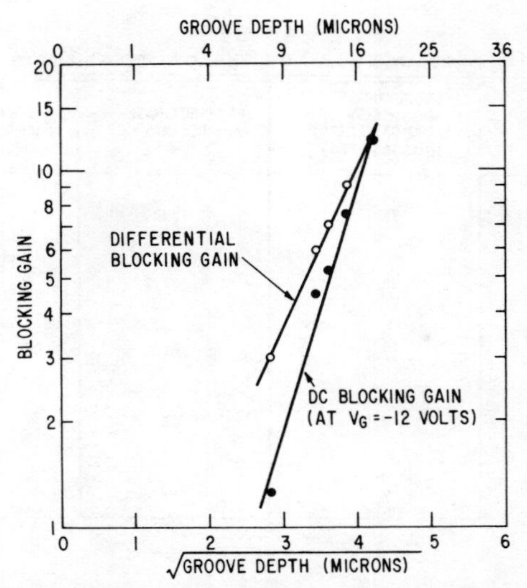

Fig. 9. Increase in the blocking gain with increasing groove depth. Both dc and differential blocking gain were measured at $V_G = -12$ V.

place at about 8 V for the groove depth of 8 μm, 5 V for the groove depth of 12 μm, and at only 1 V for the groove depth of 18 μm. This change in the characteristics has a strong effect on the maximum output drain current of these devices as will be discussed in a later section in the paper.

B. Blocking Characteristics

The blocking characteristics of the JFET determine its maximum operating voltage. Since these devices have normally-on characteristics, a reverse gate bias is necessary to hold these devices in their blocking state. For each gate bias, there is a drain voltage above which significant drain current begins to flow due to the injection of electrons over the channel potential barrier. The ratio of this drain voltage to the applied gate voltage is called the blocking gain. In order to develop devices capable of operating at high voltages, it is essential to achieve large blocking gains. In the recessed-gate device structure, the blocking gain can be increased by increasing the groove depth.

The blocking characteristics of recessed-gate JFET devices fabricated with groove depths of 8, 12, 13, 15, and 18 μm are shown in Fig. 8. As expected, an increase in the blocking voltage capability is observed with increasing groove depth. This is consistent with theoretical analyses of current flow in these devices which have demonstrated that the blocking gain should increase with channel aspect ratio [19], [20]. The channel aspect ratio is the ratio of the depth of the channel region to its width. In these recessed-gate devices, the channel aspect ratio increases in proportion to the groove depth since the channel width is the same for all the devices. The theory [20] predicts that the blocking gain depends upon the channel aspect ratio (A) and the normalized gate bias (V = gate bias voltage/channel pinchoff voltage) according to the following equation:

$$G = \zeta e^{\gamma\sqrt{A}} [V - 1]^2 \tag{2}$$

where ζ and γ are constants. The differential and dc blocking gain of the recessed-gate devices have, therefore, been plotted as a function of the square root of the groove depth in Fig. 9. In this figure, the dc blocking gain is the ratio of the drain voltage at which a drain current of 100 microamperes is observed for a reverse gate bias voltage of -12 V. The differential blocking gain is the change in the drain blocking voltage capability for a 1-V change in the gate bias voltage. In both cases, the blocking gain increases exponentially as the square root of the groove depth, in excellent agreement with the theoretical analysis.

For power switching applications, a high blocking gain is desirable in order to reduce the necessary gate drive voltage required to maintain the devices in their blocking state. It is clear from Figs. 8 and 9 that this can be achieved by increasing the groove depth. For device operation at drain potentials as high as 500 V, a blocking gain of 10 would require the application of a gate voltage of -50 V. This can be achieved by using recessed-gate devices with groove depths of about 15 μm. Although higher blocking gains can be achieved by increasing the groove depth, this has an adverse effect on the on-resistance of these devices as discussed in the next section.

C. On-Resistance

For power JFET's, the on-resistance has been commonly used to characterize their forward conduction capability. A low on-resistance is desirable to reduce the forward voltage drop and maintain a low power dissipation. The ideal specific on-resistance (on-resistance for 1 cm^2 area) can be easily calculated for each breakdown voltage by using the doping level and depletion width corresponding to this breakdown voltage. For a 500-V device, the specific on-resistance can be shown to be 0.04 $\Omega \cdot$ cm^2. In practice, this ideal on-resistance cannot be

TABLE I
ON-RESISTANCE AS A FUNCTION OF GROOVE DEPTH

GROOVE DEPTH (MICRONS)	DRAIN CURRENT AT V_D = 5V (ZERO GATE BIAS) (MILLIAMPERES)	APPROXIMATE ON-RESISTANCE (OHMS)	SPECIFIC ON-RESISTANCE (OHM-CM²)
8	28	179	1.1
12	22	227	1.1
13	17.2	291	1.5
15	9.5	526	2.4
18	0.38	13200	--

Note 1: Specific On-Resistance Has Been Measured at Low I_D to Avoid Channel Pinch-Off Effects

Note 2: Specific On-Resistance Has Been Calculated Based Upon an Active Device Area of 6.4 X 10^{-3} cm² (800μm X 800μm)

achieved because the actual device breakdown voltages are less than the ideal case and some loss in the active area is incurred due to the device structure. With careful device termination, as well as proper internal design to avoid field crowding within the structure, device breakdown voltages of about 80 percent of the ideal values can be achieved. The loss in active area is again dependent upon the device design. It can be minimized by using higher resolution lithography. For the recessed-gate devices fabricated in this study, the minimum feature size was limited to 5 μm. This reduced the area of the channel to about 30 percent of the active area. The combination of this loss in area and the reduced breakdown voltage due to the device edge termination was expected to increase the on-resistance by an order of magnitude when compared to the ideal case.

The specific on-resistance of the recessed-gate devices were obtained by measurement of the on-resistance of the devices at low drain currents where channel pinchoff effects are small. For these calculations a device active area of 800 μm X 800 μm was used. Table I gives a list of the calculated specific resistance for devices fabricated with groove depths ranging from 8 to 18 μm. As the groove depth increases, an increase in the specific on-resistance is also observed. This increase in the specific on-resistance arises from the resistance of the channel region. In these recessed-gate devices, the total on-resistance is comprised of two contributions, namely, that of the channel region and that of the drift region. Further the gate junction of these devices must support not only the drain potential but also the additional reverse bias gate potential. Thus device design based upon a blocking gain of 10, requires the doping level and width of the n-base region to be chosen so as to allow the gate junction to support 600 V if the device is to be capable of operating at drain potentials of up to 500 V. This results in an increase in the on-resistance of the drift region by 40 percent from the ideal on-resistance corresponding to 500 V. In addition to this, the resistance of the channel regions increases the on-resistance even further. The effect of the channel resistance was discussed earlier [8] and it was shown that the on-resistance is given by the expression

$$R_{ON} = \frac{1}{n}\left[\frac{\rho(L + W_D)}{Z(2a - 2W_D)} + \frac{\rho(d - W_D)}{Z[W_b - (2a - 2W_D)]} \cdot \ln\left[\frac{W_b}{2a - 2W_D}\right]\right] \tag{3}$$

where n is the number of channels, ρ is the epitaxial-layer resistivity, L is the groove depth, W_D is the gate-depletion-layer width, Z is the length of the source fingers, $2a$ is the channel width, and d is the thickness of the drift region. The first term in (3) is the contribution from the channel region and the second term is the contribution from the drift region including spreading resistance effects. As the groove depth increases, an increase in the channel resistance results in an increase in the specific on-resistance. In summary, the observed on-resistance of these recessed-gate JFET devices is an order to magnitude larger than the ideal on-resistance due to the higher breakdown voltage necessary for the gate junction, in order to support not only the drain voltage, but also the gate voltage. There is also an additional resistance contributed by the channel regions. For purposes of comparison, a typical power MOSFET design using the same lithography would have a specific on-resistance of about 0.2 $\Omega \cdot cm^2$, which is five times larger than the ideal case and about five times lower than that observed in the recessed-gate JFET devices fabricated in this study.

In addition to the increase in the on-resistance with increasing groove depth, the channel pinchoff effect due to the channel voltage drop at higher drain currents also become increasingly pronounced. This results in a saturation of the drain current at lower values for devices fabricated with deeper grooves. Further, in the case of devices fabricated with the 18-μm deep grooves, the triode-like characteristic persists even at the larger drain currents. In this case, the drain characteristic was found to be nonlinear even at low drain currents and the specific on-resistance could not be measured. From Table I it is clear that increasing the groove depth beyond 15 μm can cause a significant deterioration in the current carrying capability of the recessed-gate JFET's.

D. Transconductance

The transconductance of JFET's has been commonly used to characterize its gain and high-frequency operating capability. In the pentode region of device operation, the transconductance of the recessed-gate JFET devices can be expected to be similar to that of low voltage JFET's used for small-signal amplification. This transconductance is related to the drain voltage (V_D) and the gate voltage (V_G) by the equation

$$g_m = \frac{2Z\mu}{L}\sqrt{2\epsilon\epsilon_0 q N_D}\,[\sqrt{V_D + V_G + V_{bi}} - \sqrt{V_{bi} + V_G}] \tag{4}$$

where Z is the channel width, L is the channel length, μ is the electron mobility, N_D is the channel doping density, and V_{bi} is the built-in diffusion potential of the gate junction. According to this equation, the transconductance should increase as

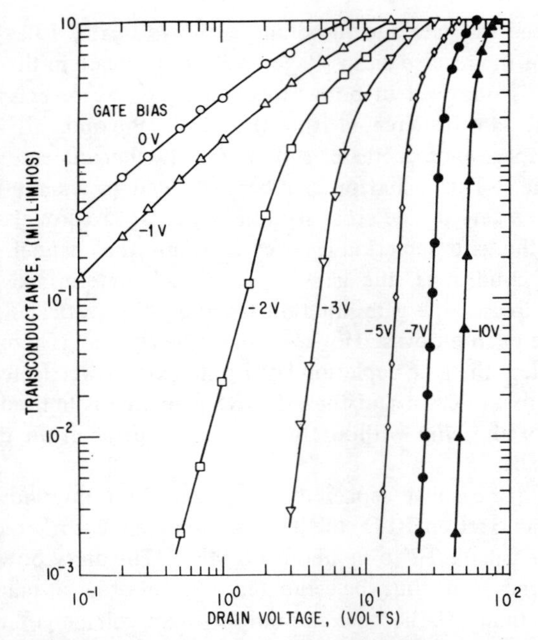

Fig. 10. Dependence of the transconductance of a recessed-gate JFET device with a groove depth of 12 μm upon the drain voltage.

the square root of the drain voltage. In Fig. 10, the measured transconductance of a recessed-gate JFET device fabricated with a groove depth of 12 μm has been plotted as a function of the drain voltage. At low gate voltages and low drain currents, the device operates in the ohmic regime where the drain current increases in proportion to the drain voltage

$$I_D = \frac{V_D}{R_{on}(V_G)} \tag{5}$$

where the on-resistance has been indicated to be a function of the gate voltage but independent of the drain voltage. In this case, from (5) it is obvious that the transconductance (dI_D/dV_G) will also increase linearly in porportion to the drain voltage. A linear increase in the transconductance with drain voltage is indeed observed at low drain currents for the drain characteristics obtained with $V_G = 0$ and $V_G = -1$ V. Further, at low gate bias voltages and higher drain currents where the channel pinchoff effects become dominant, the transconductance should increase as the square root of the drain voltage as indicated by (4). This is observed in the data shown in Fig. 10 at the higher drain currents. This behavior is, thus similar to that of low-voltage small-signal JFET devices.

However, at the higher gate bias voltages, the high-voltage recessed-gate devices exhibit triode-like characteristics because the channel potential barrier controls drain current flow. As discussed earlier in Section III-A, the drain current then increases exponentially with increasing drain voltage and decreasing gate voltage as indicated by (1). The experimental observation of the exponential variation of the drain current with increasing drain voltage in the triode region was discussed in Section III-A with reference to Fig. 7. Consequently, in this region of operation, the transconductance also varies exponentially with drain voltage as observed at the higher gate bias voltages in Fig. 10. The exponential variation of the drain

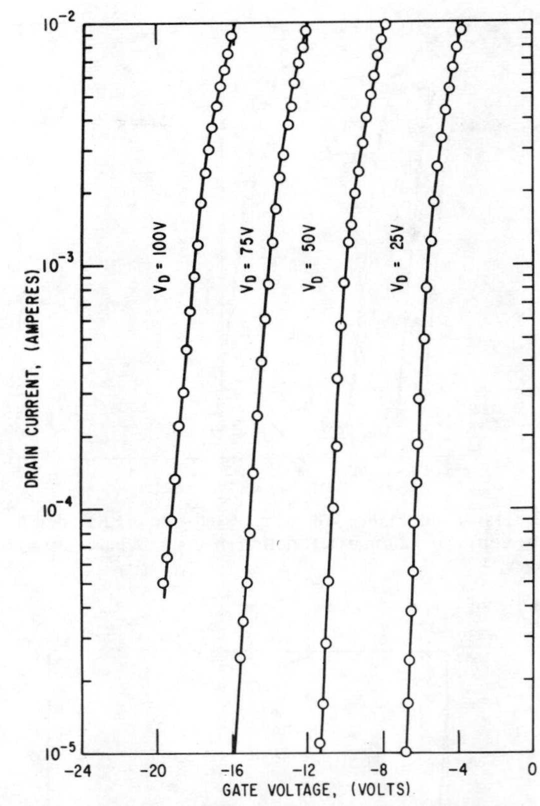

Fig. 11. Exponential variation of the drain current with gate bias voltage for a recessed-gate JFET with a groove depth of 12 μm in the triode region.

current with gate voltage predicted by (1) has also been observed in these devices. A typical example of this is shown in Fig. 11 for a recessed-gate JFET device fabricated with a groove depth of 12 μm. The nonexponential behavior observed in these characteristics at higher drain currents arises from the transition of the drain characteristics from the triode to the pentode regime. Using (1), the transconductance of these devices in the triode region can be derived

$$g_m = -\frac{q\beta}{kT} I_0 \exp \frac{q}{kT}(\alpha V_D - \beta V_G). \tag{6}$$

From this equation it can be expected that the transconductance should vary exponentially with the gate bias voltage in the triode region. This behavior has been observed in the recessed-gate devices. A typical set of transconductance curves for a recessed-gate device fabricated with a groove depth of 12 μm are shown in Fig. 12. Again, in these characteristics a transition from the triode regime to the pentode regime is observed when the drain currents exceeds 3 mA.

E. Frequency Response

The frequency response of JFET's can be limited by several device parameters such as the input capacitance and the transit time. As the breakdown voltage of the FET's is increased, the width of the drift region must also be increased in order to support the higher voltage. This results in a increase in the transit time for the carriers. In addition for the recessed-gate devices with deep grooves, the transit time through the

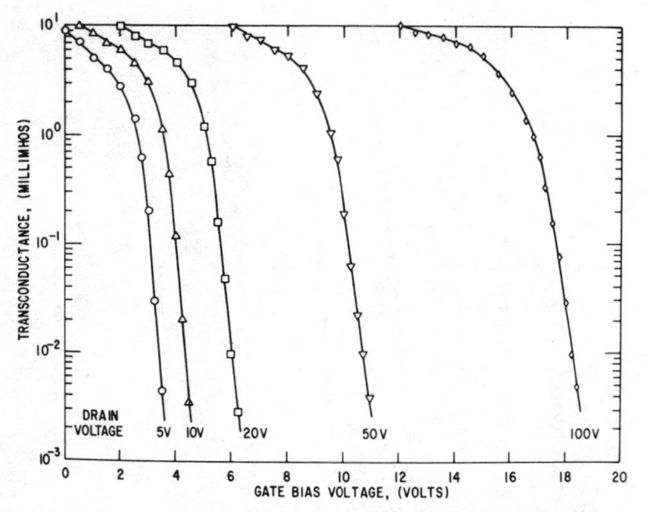

Fig. 12. Transconductance of a recessed-gate JFET device with a groove depth of 12 μm as a function of the gate bias voltage (negative voltages).

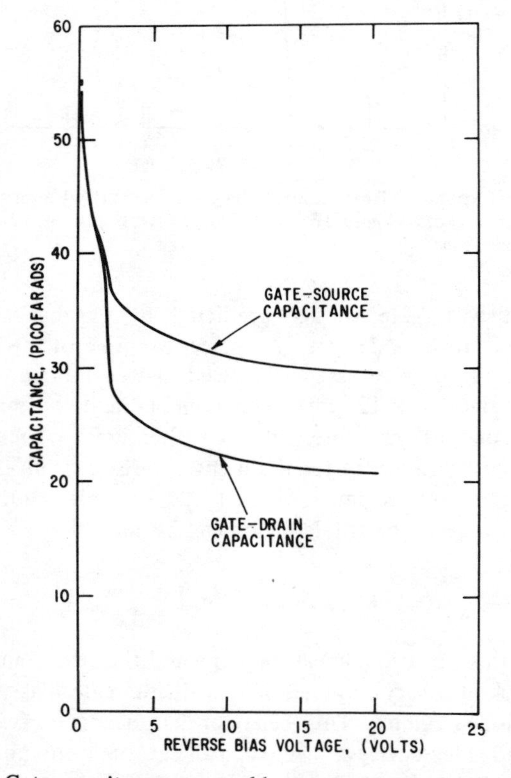

Fig. 13. Gate capacitance measured between source and drain terminals as a function of the reverse bias voltage.

channel region causes a further degradation in the frequency response. Using a saturated drift velocity for electrons in silicon of 10^7 cm/s, the transit time through the channel region and the drift region are calculated to be 1.5×10^{-10} s and 4×10^{-10} s, respectively.

Another limit on the frequency response of these devices can arise from the time taken to charge the gate input capacitance. In order to obtain a low-input capacitance the recessed-gate devices were fabricated with small active areas. The capacitance measured between the gate and the source, as well as between the gate and the drain, is shown in Fig. 13 as a function of the reverse bias voltage. A sharp decrease in the capacitance is observed in both cases when the reverse bias voltage on the gate is increased from 0 to 3 V. Beyond this voltage, the capacitance decreases slowly with further increase in gate voltage. This behavior can be explained by examining the recessed-gate device cross section in Fig. 1. At low gate voltages, the gate depletion layer extends into the channel. Under these conditions, the gate capacitance is determined by the total area of the gate junction including the vertical sidewalls of the deep grooves. However, once the channel is completely depleted, the gate depletion layer only extends vertically down towards the drain and the effective gate area is limited to the area of the chip without further contribution from the gate sidewalls.

Using the above capacitance data and the transconductance given in Section III-D, the unity current gain cutoff frequency can be calculated to be about 60 MHz. The unity power gain frequency can, thus, be estimated to be an order of magnitude larger than this because the output drain voltage is ten times larger than the input gate voltage. To confirm this, the power amplification capability of these high-voltage recessed-gate JFET devices was measured as a function of frequency for devices mounted on both TO5 headers and microwave packages. The results are shown in Fig. 14. From this data, a unity power gain cutoff frequency of 600 MHz can be measured. This value is in good agreement with the estimates based upon the measured transconductance and capacitance.

In addition to the cutoff frequency, the power efficiency was also measured at 300 MHz. Both the drain efficiency and the overall efficiency are shown in Fig. 15 as a function of the input RF power. Power efficiencies of over 30 percent have been obtained in these devices at this frequency. These values are comparable to those reported for static induction transistors [21] despite the much higher operating voltages of the recessed-gate JFET devices. As mentioned earlier, the performance of the recessed-gate JFET devices was limited by the 5-μm lithography used to fabricate the devices. It is estimated that an increase in frequency response by a factor of 4 can be achieved by improving the device structure with the use of 1- to 2-μm resolution lithography. This can be expected to allow these high-voltage JFET devices to operate at lower microwave frequencies.

F. Power Switching Performances

In addition to high-frequency power amplification, these devices are potentially useful for power switching applications such as in switching power supplies and high-frequency lamp ballasts. In these applications, the gate turn-off time has been found to be the limiting factor because this determines the switching losses. The gate turn-off time of the recessed-gate JFET devices was consequently measured as a function of the reverse gate bias. During these tests, a steady-state drain current of 100 mA was turned off with the drain voltage rising to 100 V after turn-off. Typical turn-off and fall times are shown in Fig. 16 as a function of the reverse gate bias. Here, the turn-off time has been defined as the time taken for the drain current to drop to 10 percent of its steady-state value

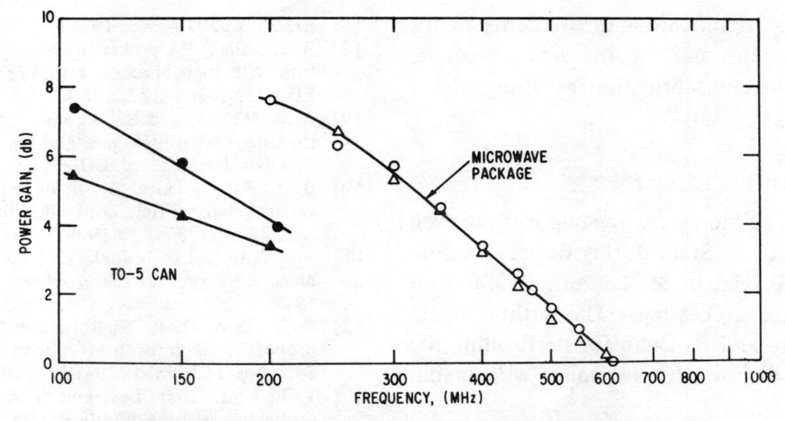

Fig. 14. Dependence of the RF power gain as a function of frequency for typical recessed-gate JFET devices. The circles and triangles represent data taken on different devices.

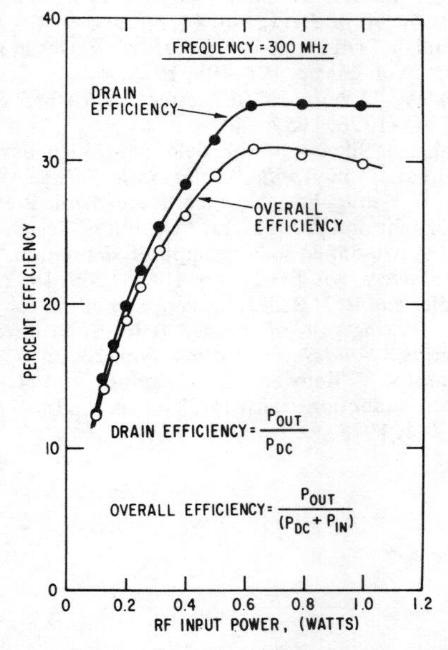

Fig. 15. Power efficiency measured as a function of the RF input power at 300 MHz.

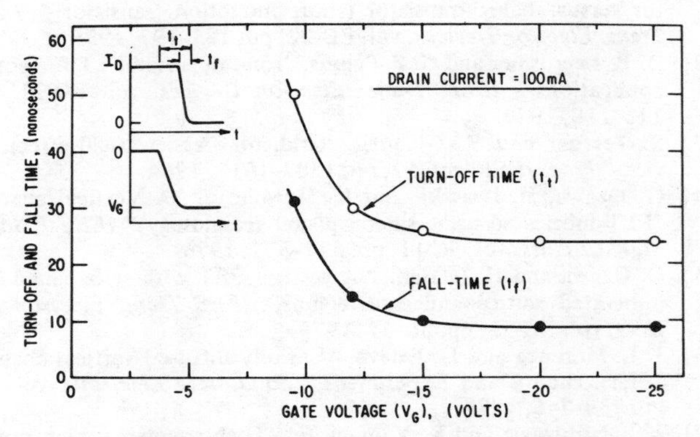

Fig. 16. Gate turn-off time of a recessed-gate JFET device with a groove depth of 12 μm as a function of the gate bias voltage.

after the application of the reverse bias to the gate, and the fall time has been defined as the time taken for the drain current to decrease to 10 percent of its steady-state value. The inset in Fig. 16 shows typical gate turn-off waveforms and pictorially defines the turn-off and fall times. From Fig. 16 it can be seen that the turn-off time decreases with increasing gate bias. This reduction stems from a reduction in the fall time with increasing gate bias. Since these recessed-gate JFET devices are operated without minority-carrier injection, the fall time is determined by the time taken to sweep out the charge in the channel and drift region and set up the gate depletion layer which can support the drain voltage. In these tests, the rate of removal of the charge via the gate was controlled by the peak gate turn-off current. This gate current was proportional to the gate bias voltage due to the 50-Ω internal impedance of the gate bias supply. Thus although these tests indicate that gate turn-off times of about 25 ns can be obtained with a 50-Ω gate source impedance, even faster turn-off speeds can be expected by using a lower impedance gate bias source albeit at the expense of larger gate-drive power requirements.

IV. Summary

A new method for the fabrication of high-voltage vertical-channel JFET's has been developed. In this process, vertically walled recessed-gate regions are formed which are self-aligned to the source contacts. The process eliminates several critical photolithographic alignment steps which have limited the yield and the breakdown voltage of earlier surface-gate JFET structures. With this recessed-gate structure, the highest breakdown voltage reported for surface-gate devices has been achieved with gate-to-drain breakdown voltages of up to 600 V, and drain blocking voltages of up to 500 V.

In these recessed-gate devices, the gate groove depth plays a critical role in determining the device characteristics. As that groove depth is increased, the blocking gain increases while the saturated output drain current decreases because of an increase in the on-resistance as well as an increase in the channel pinchoff effect. With a groove depth of 15 μm, devices with blocking gains of about 10 have been fabricated with on-resistances of 1.5 Ω · cm^2. These devices have been found to have a unity power gain cutoff frequency of 600 MHz and a forced gate turn-off switching time of less than 25 ns. These devices were fabricated by using 5-μm lithography. It is estimated that significant improvements in the on-resistance, the blocking gain, and the frequency response can be obtained with improved fabrication by using higher resolution lithography.

Acknowledgment

The author would like to acknowledge the support received from Dr. A. P. Ferro and L. L. Stahl during device development, and from B. Love, B. Hatch, M. Lazzeri, P. Menditto and G. Gidley during device processing. The author would also like to thank N. Lavoo and R. Dehn for performing the RF measurements, and N. Waldron for assistance with manuscript preparation.

References

[1] J. I. Nishizawa, T. Terasaki, and J. Shibata, "Field effect transistor versus analog transistor (static induction transistor)," *IEEE Trans. Electron Devices*, vol. ED-22, pp. 185-197, 1975.

[2] D. P. Lecrosnier and G. P. Pelous, "Ion-implanted FET for power applications," *IEEE Trans. Electron Devices*, vol. ED-21, pp. 113-118, 1974.

[3] S. Teszner and R. Gicquel, "Gridistor–A new field-effect device," *Proc. IEEE*, vol. 52, pp. 1502-1513, 1964.

[4] O. Ozawa, H. Iwasaki, and K. Muramoto, "A Vertical channel JFET fabricated using silicon planar technology," *IEEE J. Solid-State Circuits*, vol. SC-11, pp. 511-517, 1976.

[5] O. Ozawa and H. Iwasaki, "A vertical FET with self-aligned ion-implanted source and gate regions," *IEEE Trans. Electron Devices*, vol. ED-25, pp. 56-57, 1978.

[6] J. L. Morenza and D. Esteve, "Entirely diffused vertical channel JFET–Theory and Experiment," *Solid-State Electron.*, vol. 21, pp. 739-746, 1978.

[7] J. I. Nishizawa and K. Yamamoto, "High frequency high power static induction transistor," *IEEE Trans. Electron Devices*, vol. ED-25, pp. 314-322, 1978.

[8] B. J. Baliga, "A power junction gate field effect transistor structure with high blocking gain," *IEEE Trans. Electron Devices*, vol. ED-27, pp. 368-373, 1980.

[9] B. W. Wessels and B. J. Baliga, "Vertical channel field controlled thyristors with high gain and fast switching speeds," *IEEE Trans. Electron Devices*, vol. ED-25, pp. 1261-1265, 1978.

[10] B. J. Baliga, "Grid depth dependence of the characteristics of vertical channel field controlled thyristors," *Solid-State Electron.*, vol. 22, pp. 237-239 1979.

[11] —, "Control of hillocks and stacking faults during silicon vapor phase epitaxy," in *Electrochem. Soc. Mtg.*, Abs. no. 222, Oct. 1978.

[12] V. Y. Doo, D. R. Nichols, and G. A. Silvey, "Preparation and properties of amorphous silicon nitride films," *J. Electrochem. Soc.*, vol. 113, pp. 1279-1281, 1966.

[13] T. L. Chu, C. H. Lee, and G. A. Gruber, "The Preparation and properties of amorphous silicon nitride films," *J. Electrochem. Soc.*, vol. 114, pp. 717-722, 1967.

[14] N. Goldsmith and W. Kern, "The deposition of vitreous silicon dioxide films from silane," *RCA Rev.*, vol. 28, pp. 153-165, 1967.

[15] B. J. Baliga and S. K. Ghandhi, "Growth of silica and phosphosilicate films," *J. Appl. Phys.*, vol. 44, pp. 990-994, 1973.

[16] D. L. Kendall, "On etching very narrow grooves in silicon," *Appl. Phys. Lett.*, vol. 26, pp. 195-198, 1975.

[17] W. Shockley, "A unipolar field effect transistor," *Proc. IRE*, vol. 40, pp. 1365-1376, 1952.

[18] B. J. Baliga, "Silicon power field controlled devices and integrated circuits," in *Applied Solid State Science Series*, Supplement 2B, D. Kahng, Ed. New York: Academic Press, 1981.

[19] K. Yamaguchi and Y. Kodera, "Optimum design of triode-like JFET's by two-dimensional computer simulation," *IEEE Trans. Electron Devices*, vol. ED-24, pp. 1061-1069, 1977.

[20] M. S. Adler and B. J. Baliga, "A simple method for predicting the forward blocking gain of gridded field effect devices with rectangular grids," *Solid-State Electron.*, vol. 23, pp. 735-740, 1980.

[21] Y. Yokimoto, Y. Kajiwara, G. Nakamura, and M. Aiga, "1 GHz 20 W static induction transistor," *Japan. J. Appl. Phys.*, vol. 17, pp. 241-244, 1978.

Optimum Design of Triode-Like JFET's by Two-Dimensional Computer Simulation

KEN YAMAGUCHI, MEMBER, IEEE, AND HIROSHI KODERA, MEMBER, IEEE

***Abstract*—Design criteria of triode-like JFET's are studied by fully utilizing two-dimensional numerical analysis. The current is caused by the carriers injected over a potential barrier in a depleted channel. In contrast to normal pentode-like FET's, the drain field plays an important role reducing the barrier height and thus causing triode-like I–V characteristics.**

Triode-like characteristics depend strongly on device geometry. This operation can be realized only in short gate devices. The channel thickness a is an essential parameter in determining the operational mode. The devices operate as triodes or pentodes corresponding to thin or thick channels, respectively. If applied to low-resistance load direct-drive circuits, the mixed characteristics situated between the triode- and pentode-like ones, are more desirable when compared to pure triode-like ones. This is because of their low on-resistance and high ac power efficiency. The gate-drain distance l_{gd} is also essential in determining breakdown voltage.

The design criteria are discussed and an optimum design specified on the N_D (channel doping)-a and $N_D - l_{gd}$ planes with respect to triode-like characteristics, circuit application and breakdown phenomena. Calculated results are compared with experiments and good agreement is found without using any adjustable parameters. The present design criteria will be useful for designing triode-like JFET's.

I. INTRODUCTION

JUNCTION-gate field-effect transistors (JFET's) applications have recently been extended to various branches of electronics. The main features of JFET's are low noise; absence of minority carrier storage effect, high output voltage, and thermal stability due to the negative temperature coefficient of majority carrier mobility. These features, especially the latter two, are of great advantage for high power application. Recently, vertical structure JFET's with multigates have been reported for power applications [1], [2].

Usual JFET's exhibit current-saturating I–V curves, so-called pentode-like characteristics. However, current-nonsaturating "triode-like" JFET's with channels constructed of an intrinsic high-resistivity material (i- or ν-layer) were proposed by Nishizawa [3] and Shockley [4]. The operational mechanism of these FET's is analogous to that of a vacuum triode, and the current flowing through the i-layer is considered to be space-charge-limited. Triode-like JFET's are more favorable for applications in low-resistance load direct-drive circuits, for example, audio power amplifiers [5], because of their low output impedance, low distortion, and high output voltage [6].

The space-charge-limited current was studied by an analytic approach [6]–[8], in which the existence of the i- or ν-layers corresponding to the "vacuum" in solid-state devices is essential for triode-like operation. The theories are based on individual treatment of each variable, that is, the interaction between the fields parallel and perpendicular to the current flow direction is not fully taken into account. The physical picture of the triode-like operation for JFET's was clarified by two-dimensional numerical analysis [9]. The triode-like current is derived from the drain field penetrated into the depleted channel region under the gate, that is, it results from two-dimensional interaction. Such operation is possible only in short gate devices and the existence of an i- or ν-layer is not absolutely essential. Moreover, the basic characteristics of triode-like JFET's, including breakdown phenomena, were analyzed theoretically [10].

In this paper, the advantages of two-dimensional numerical analysis are fully utilized for obtaining design criteria on triode-like Si JFET's. Current–voltage characteristics, avalanche breakdown voltage, voltage ampli-

Manuscript received October 20, 1976.
The authors are with Central Research Laboratory, Hitachi, Ltd., Higashi-koigakubo 1-280, Kokubunji-shi, Tokyo 185, Japan.

Reprinted from *IEEE Trans. Electron Devices*, vol. ED-24, pp. 1061–1069, Aug. 1977.

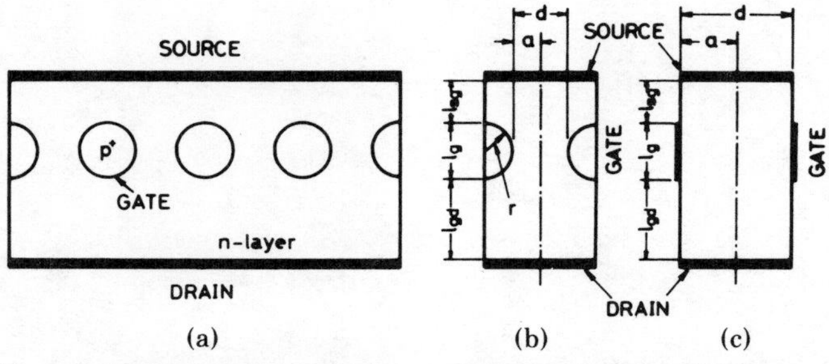

Fig. 1. Schematic representations of JFET's. (a) Multigate FET model. (b) Cylindrical-gate structure. (c) Planar-gate structure.

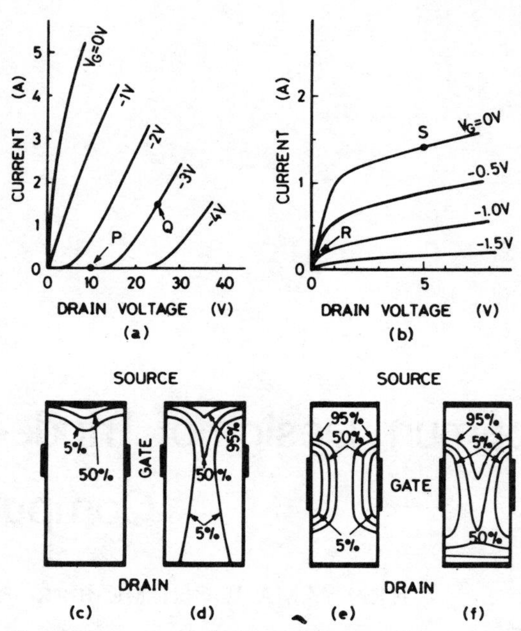

Fig. 2. Comparison of a triode-like operation with a pentode-like one by two-dimensional computer simulation. Typical triode-like and pentode-like I–V characteristics are shown in (a) and (b), respectively, where the assumed parameters are $N_D = 1 \times 10^{15}$ cm^{-3}, $l_g = 1$ μm, and $a = 1.2$ μm in (a), and $N_D = 1 \times 10^{16}$ cm^{-3}, $l_g = 1$ μm, and $a = 0.5$ μm in (b). Contour-line maps of the electron distributions in the devices corresponding at points P, Q, R, and S in (a) and (b) are shown in (c), (d), (e), and (d), respectively.

fication factor, pinch-off characteristics, etc., are calculated from physical and geometrical parameters. The calculations are evaluated with experimental results. Based on the physical insight revealed through two-dimensional numerical analysis and device characteristics, an optimum triode-like JFET design is discussed.

II. Analysis Method of a Triode-Like Operation

The multigate construction of a JFET, the common structure for vertical FET's [2], [6], [11], is shown schematically in Fig. 1(a). Since this structure is a repetition of single channels, the present analysis is carried out for only one channel. Two models were employed for the calculations, i.e., the cylindrical-gate structure shown in Fig. 1(b) and the planar-gate structure shown in Fig. 1(c). The latter is a simplified model of a vertical FET. The coordinate axes and the notations to be used in the following analysis are also defined in this figure.

A. Operational Mechanism and I–V Characteristics

Fundamental equations describing the operation of FET's are Poisson's equation and the current continuity equation:

$$\nabla^2 \psi = -\frac{q}{\epsilon}(N_D - n) \tag{1}$$

$$q\frac{\partial n}{\partial t} = \nabla \cdot \mathbf{J} \tag{2}$$

where ψ and n are the potential and electron density, respectively. These partial-differential equations are solved numerically under the boundary conditions mathematically described which correspond to the device structure and electrostatic conditions [12]. Firstly, ideally ohmic and rectifying contact conditions are imposed at the source and drain electrodes and the gate electrode, respectively. As a barrier height for the rectifying contact, 0.8 eV is assumed. Then, through the surface except for the electrodes, no current must flow, i.e., normal derivatives of the potential and carrier distributions are zero. For simplicity, the planar-gate structure shown in Fig. 1(c) is assumed to solve both (1) and (2).

Since the drift velocity in the solids is saturated in the high field region, the following formula [13] is used to simulate the velocity-field relation

$$v(E) = \frac{\mu E}{1 + \mu E/v_s} \tag{3}$$

where μ and v_s are the low-field mobility and saturation velocity, respectively.

The typical triode- and pentode-like I–V characteristics are depicted in Figs. 2(a) and (b), respectively. The results of two-dimensional numerical analysis show that a triode-like current is caused by the carriers injected into the depleted region, as shown in Figs. 2(c) and (d). In other words, the number of mobile carriers increases with increasing drain bias and an enhancement or carrier induced type operation takes place. This operation is different from the usual or normal (pentode-like) FET operation, as illustrated in Figs. 2(e) and (f). The channel becomes narrower with increasing drain bias, i.e., depletion type operation.

B. Pinch-Off and Avalanche Breakdown Characteristics

The main features of triode-like operation are 1) the channel is depleted due to the gate potential, and 2) the carriers are injected into the depleted region due to the drain field. Since the depleted region reaches the drain, the gate potential gives rise to a potential barrier [9] which prevents carrier flow under low drain bias conditions. When the drain bias is high enough, the potential barrier is diminished and injection of carriers into the depleted region begins. Thus the actual current-pinch-off voltage

V^*_P is obtained as follows:

$$V^*_P = V_G|_{\psi_h=0} \tag{4}$$

where ψ_h is the potential barrier height in the conducting channel. It is defined by a potential difference between the peak potential in the conducting channel and the source. Another important factor characterizing triode-like operation is a voltage amplification factor μ^*, which is given by

$$\mu^* = \left.\frac{\partial V_D}{\partial V_G}\right|_{I_D=\text{const}}. \tag{5}$$

Since μ^* is almost independent of I_D as shown in Fig. 2(a), μ^* is obtained as follows:

$$\mu^* = V_D|_{V_G=V^*_P} - V_D|_{V_G=V^*_P+1}. \tag{6}$$

When high voltage is applied between the gate and drain electrodes or the gate and source electrodes, avalanche breakdown occurs limiting the device operation. The breakdown voltage (BV) is determined from the electric-field distribution and the breakdown condition is given by

$$\int_0^W \alpha_n \exp\left[-\int_0^x (\alpha_n - \alpha_p)dx'\right] dx = 1 \tag{7}$$

where W is the width of the depleted region, and α_n and α_p are the electron and hole ionization rates, respectively. The α_n and α_p are functions of the electric field E, and are given by [14]

$$\alpha_n = 7.0 \times 10^5 \exp\left(-\frac{1.4 \times 10^6}{E}\right) \tag{8}$$

$$\alpha_p = \begin{cases} 4.4 \times 10^5 \exp\left(-\dfrac{1.4 \times 10^6}{E}\right), & \text{for } E > 6.07 \times 10^5 \text{ V/cm} \\ 1.3 \times 10^6 \exp\left(-\dfrac{2.09 \times 10^6}{E}\right), & \text{for } E < 6.07 \times 10^5 \text{ V/cm.} \end{cases} \tag{9}$$

Since the channel is completely depleted before carrier injection or avalanche breakdown occurs, the electric-field distribution and ψ_h can be calculated using (1), assuming $n = 0$. The potential or electric-field distribution strongly depends on the device structure, thus, the cylindrical-gate structure shown in Fig. 1(b) is used for solving (1). However, because $n = 0$, equation (2) is ignored when analyzing V^*_P, μ^* and BV.

III. Results

Two-dimensional numerical analysis is carried out and the relation between device parameters and electric characteristics investigated. Calculated and experimental results are compared and design criteria of triode-like JFET are examined. Throughout the calculations, the electron mobility given in a standard textbook [15] is used for a given impurity concentration, and v_s is assumed to be 1×10^7 cm/s.

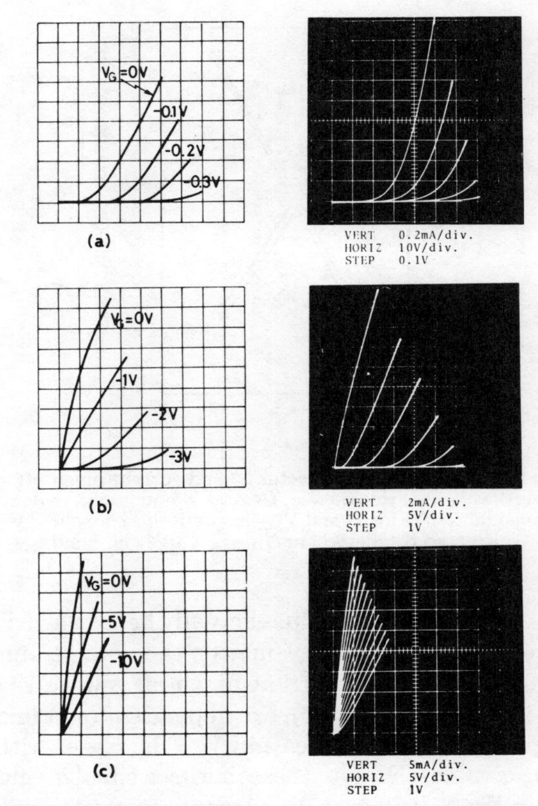

Fig. 3. Effect of channel thickness a (or gate-spacing, $d = 2a$) on JFET operation. The graphs on the left-hand side represent the calculated results and the right-hand side the experimental results. The channel doping N_D is 2×10^{14} cm^{-3}. The gate length is 2.7 μm in (a), and 3 μm in (b) and (c). (a) $a = 1.35$ μm. (b) $a = 3$ μm. (c) $a = 7.5$ μm.

A. *Effect of Channel Thickness on JFET Operation*

Current–voltage characteristics of JFET's with $N_D = 2 \times 10^{14}$ cm^{-3} are shown in Fig. 3 when the channel thickness a (or gate-spacing $d(= 2a)$) is variable. The graphs on the left are the calculated results and those on the right are the experimental results. Agreement between the calculations and experiments is good. As the channel thickness a becomes larger, a transition from triode- to pentode-like characteristics is observed. The transitional characteristics shown in Fig. 3(b) are hereafter called "mixed characteristics" in contrast to the triode- or pentode-like characteristics.

Similar I–V characteristics can be also found for other N_D values. Therefore, it can be concluded that the channel thickness is an essential parameter for realizing triode-like characteristics. Physically speaking, the depletion of the channel by the gate potential is important for triode-like operation. The depletion, i.e., no mobile carrier in the channel, is equivalent to a vacuum for the device operation and the triode-like operation, as illustrated in Fig. 2, results from the carrier injection into the depletion region.

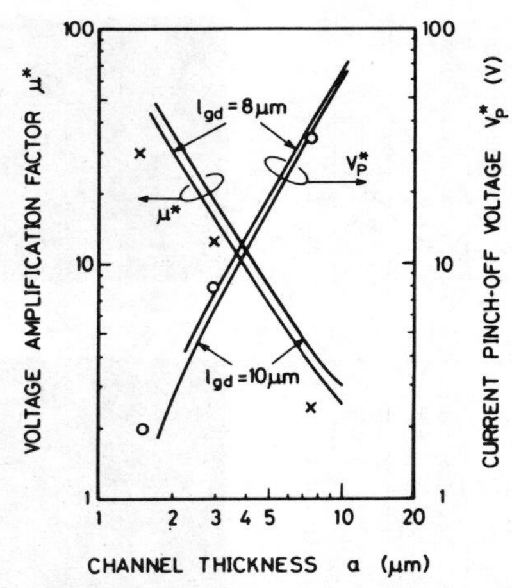

Fig. 4. Voltage amplification factor μ^* and current pinch-off voltage V^*_P versus channel thickness a. Crosses × and circles ○ denote the experimental results for μ^* and V^*_P, respectively. The solid lines stand for the calculated results, where $N_D = 2 \times 10^{14}$ cm^{-3} and $l_g = 2$ μm.

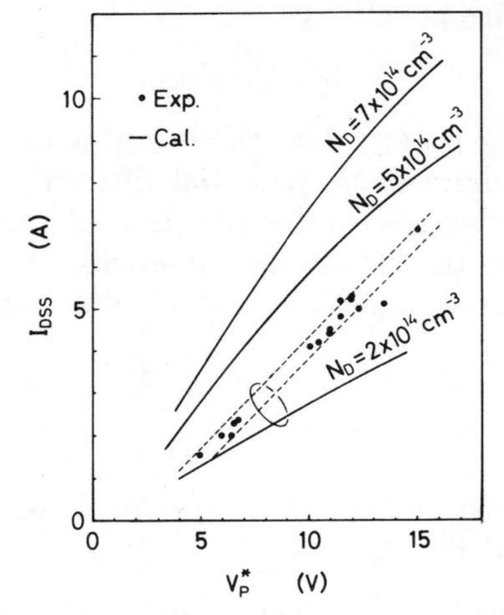

Fig. 5. I_{DSS}, defined in footnote 1, versus current pinch-off voltage V^*_P. The gate length is 4 μm.

FET's with smaller a values actually behave as triodes. Let's consider two figure-of-merits; the voltage amplification factor μ^* and the current-pinch-off voltage V^*_P. As shown in Fig. 4, μ^* and V^*_P are dependent on a and l_{gd}. For a given l_{gd}, μ^* increases and V^*_P decreases with decreasing a values. Despite these merits, a small a value, as shown in Fig. 3, brings about a large "on-resistance" R_{on}, which is the resistance between the source and drain at $V_G = 0$ V and $V_D \approx 0$ V.

For practical application in an amplifier, it is desirable that the on-resistance R_{on} be small and I_{DSS}[1] large. I_{DSS} increases with increasing V^*_P (or increasing a) as shown in Fig. 5. Therefore, it is found that an optimum channel thickness a exists. Although the use of a high doping material is one method of improving the $I_{DSS} - V^*_P$ relation (see Fig. 5), the breakdown voltage, discussed later, is lowered. Thus the optimization must be carefully accomplished taking into account various device design parameters.

It is noted that V^*_P is a different quantity than the channel "pinch-off" voltage V_P, at which the gate depletions expanded from each gate are joined at the channel center. This is a characteristic of triode-like operations contrary to pentode-like operations. The previous characteristics (shown in Figs. 3, 4, and 5) are calculated on the basis of physical and geometrical parameters and no adjustable parameter is included. Considering this, agreement between the calculated and experimental results is remarkably good.

B. Effect of Gate Length on Triode-Like Operation

Triode-like operation results from the two-dimensional effect, that is, the drain field (the source-to-drain direction) penetrates into the depleted region which is expanded from the gate (perpendicular to the drain field). This suggests that gate length has great influence on triode-like operation. Current–voltage characteristics when the gate length is variable are shown in Fig. 6. The FET's with the shortest gate ($l_g \approx a$) behave as triodes, as shown in Fig. 6(a). When the gate length becomes large, I–V characteristics become pentode-like, as shown in Figs. 6(b) and (c).

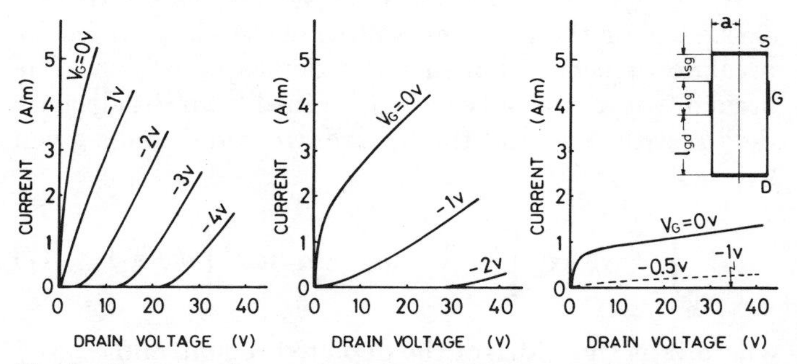

Fig. 6. Current–voltage characteristics when the gate length l_g is variable. The common parameters are $N_D = 1 \times 10^{15}$ cm^{-3}, $a = 1.2$ μm, $l_{sg} = 1.2$ μm, and $l_{gd} = 2.6$μm. The gate length assumed in (a), (b) and (c) is 1, 2, and 4 μm, respectively.

Such transitions can be physically interpreted by Fig. 7 which illustrates the electrostatic potentials along the symmetric line of the devices. As the drain bias is raised, the drain field increases and penetrates into the gate depletion region. If the gate length is short ($l_g \lesssim a$), the potential barrier which prevents the carrier flow rapidly vanishes as shown in Fig. 7(a) and the triode-like current begins to flow. However, if the gate length is long, as shown in Fig. 7(b), the speed, at which the potential barriers vanish, as a function of the applied drain bias is slower than that of Fig. 7(a). Therefore, the output resistance in Fig. 6(b) becomes large compared to that in Fig. 6(a).

If the gate length is long enough, the potential barrier hardly vanishes, as shown in Fig. 7(c). The drain current is completely pinched-off under small reverse gate bias conditions and the FET behaves as a pentode, as shown in Fig. 6(c). From these results, the gate length is restricted within narrow limits for realizing triode-like operation.

Although the small l_g/a ratio is necessary for triode-like operation, the above condition is not sufficient. For ex-

[1] In this paper, I_{DSS} is defined by the drain current at $V_G = 0$ V and $V_D = 5$ V.

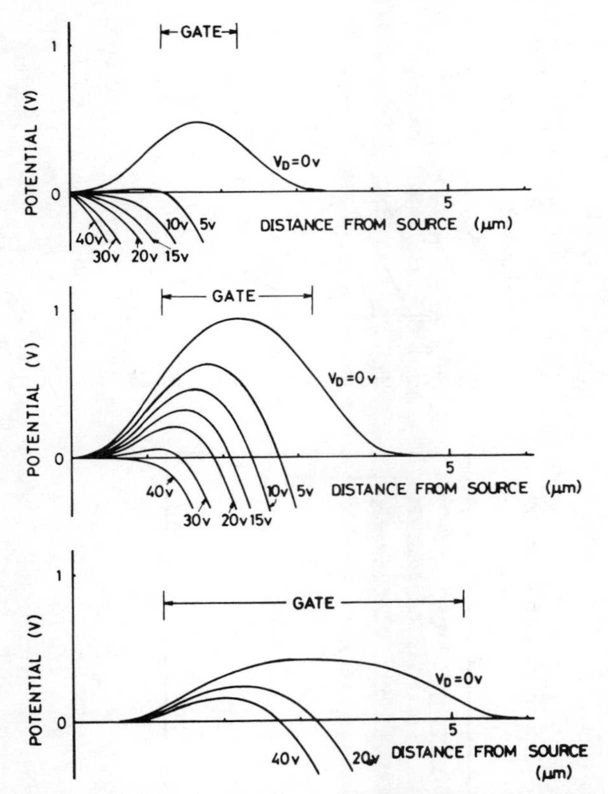

Fig. 7. The electrostatic potential along the symmetric line in the device. The device parameters commonly used are identical with those of Fig. 6. (a) $l_g = 1$ μm and $V_G = -2$ V. (b) $l_g = 2$ μm and $V_G = -2$ V. (c) $l_g = 4$ μm and $V_G = -1$ V.

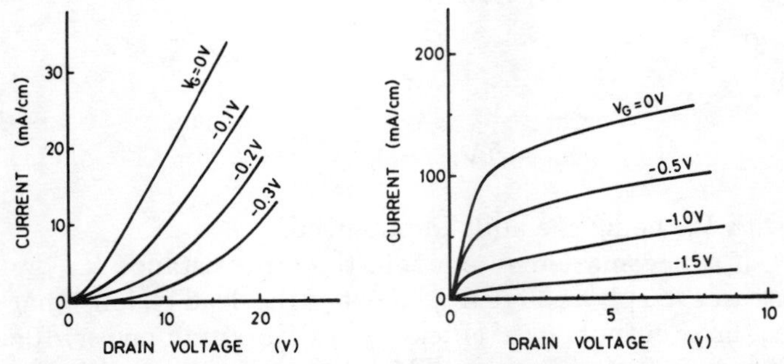

Fig. 8. *I–V* characteristics of JFET's with equal ratios $l_g/a = l_{sg}/a = l_{gd}/a = 2$ when $N_D = 1 \times 10^{16}$ cm^{-3}. (a) $a = 0.25$ μm. (b) $a = 0.5$ μm.

ample, Fig. 8 shows current–voltage characteristics of JFET's with the ratio $l_g/a = 2$ for $N_D = 1 \times 10^{16}$ cm^{-3}. Actually, triode-like characteristics are obtained in FET's with small a values (Fig. 8(a)). However, the other FET's behave as pentodes (Fig. 8(b)). Since the zero-bias depletion layer width is about 0.3 μm for $N_D = 1 \times 10^{16}$ cm^{-3}, the FET's with $a = 0.5$ μm are the normally-on type and the FET's with $a = 0.25$ μm are normally-off type in traditional appellation. FET's with thick channels operate as pentodes even if the ratio l_g/a is small.

C. Effect of Gate–Drain Distance on Triode-Like Operation

When an FET operates as a triode, the depletion reaches the drain as shown in Fig. 2(c). Therefore, it is understood

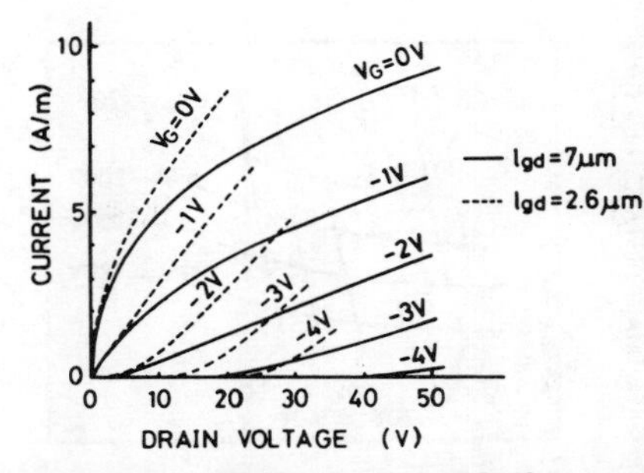

Fig. 9. Comparison of two FET's in which only the gate–drain distances l_{gd} are different. Common parameters of the FET's are $N_D = 1 \times 10^{15}$ cm^{-3}, $l_{sg} = 1.2$ μm, $a = 1.2$ μm, $l_g = 1$ μm, and $\mu = 1500$ cm^2/V · s.

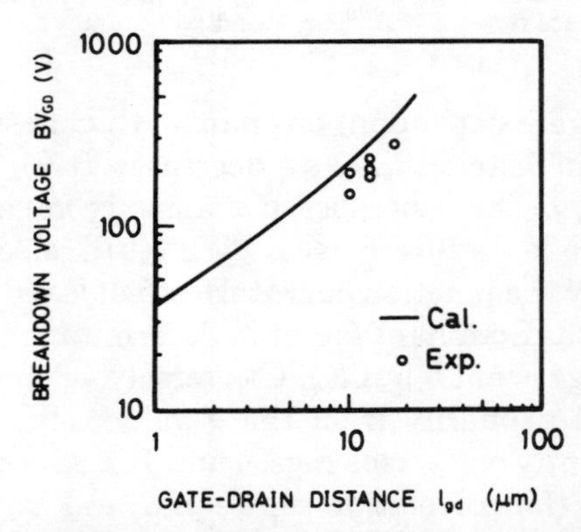

Fig. 10. Breakdown voltage BV_{GD} versus gate-drain distance l_{gd}. The assumed parameters are $N_D = 2 \times 10^{14}$ cm^{-3}, $l_g = 2$ μm and $a = 2.5$ μm in the calculation.

that the region between the gate and drain plays a dominant role on the operation. *I–V* characteristics of two FET's in which only the gate–drain distances are different from each other, are shown in Fig. 9. When the gate–drain distance l_{gd} is short, triode-like operation occurs. However, the output resistance of the FET with the longer l_{gd} is large and the device does not behave like a triode. The gate depletion does not reach the drain at a moderate drain voltage and the drain field does not attain sufficient strength to inject the carriers into the depleted region if l_{gd} is long. This suggests that an upper limit for l_{gd} exists.

The gate–drain breakdown voltage BV_{GD} as a function of l_{gd} is shown in Fig. 10. The breakdown voltage increases with increasing l_{gd}, and reasonable agreement between the calculated and experimental results is found. It is noted that the BV_{GD} is small compared to the breakdown voltage of a one-dimensional p-n junction with the same doping density. This is because of the electric field crowding around the cylindrical junction and short l_{gd}. In order to obtain a high breakdown voltage, a long l_{gd} is required, that is, there exists a lower limit for l_{gd}. As a whole, these analyses indicate the existence of an optimum length for l_{gd}.

According to the present analysis, the basic relation between the device parameters and characteristics of triode-like JFET's are clarified. The correlations between several design parameters are also considered. First, the BV_{GD}, μ^*, and V^*_P are depicted by several contour lines on the l_{gd} and a plane in Fig. 11. It is seen that the break-

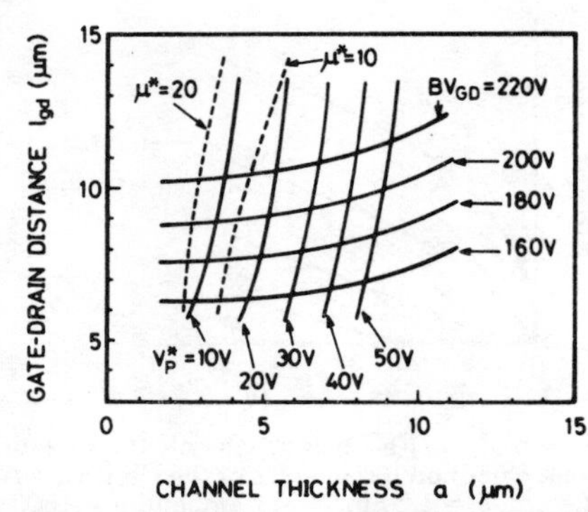

Fig. 11. Breakdown voltage BV_{GD}, voltage amplification factor μ^* and current pinch-off voltage V^*_P on the $l_{gd} - a$ plane, where the assumed parameters are $N_D = 2 \times 10^{14}$ cm^{-3} and $l_g = 2$ μm.

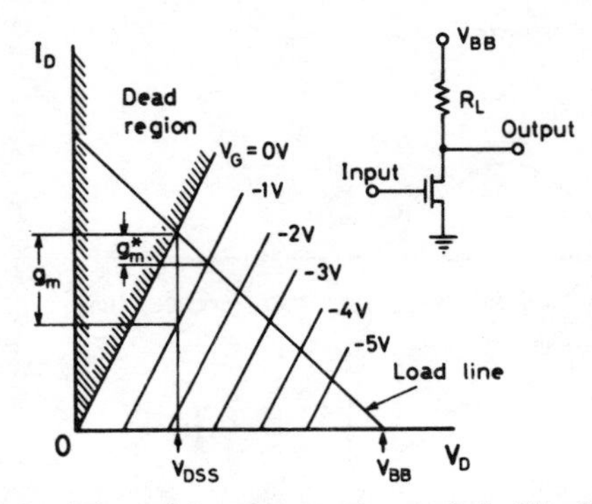

Fig. 12. Direct drive circuit using a triode-like JFET.

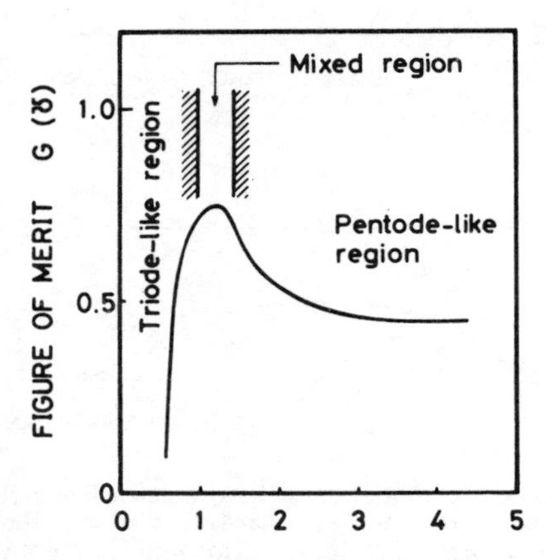

Fig. 13. Figure of merit G given by (12) under the circuit conditions shown in Fig. 12, where $R_L = 8\ \Omega$ and $V_{BB} = 40$ V.

down voltage is dependent on channel thickness a as well as gate-drain distance l_{gd}. As a decreases, BV_{GD} increases for a given l_{gd}. This is because, if a is small compared to the gate electrode radius r (see Fig. 1(b)), electric field crowding by the junction curvature effect is reduced [10]. Of course, the dominant factor in determining the breakdown voltage is not a but l_{gd}. Conversely, μ^* and V^*_P are determined primarily from the a value, although they depend slightly on l_{gd} (see Figs. 4 and 11). As a result, it is found that the correlation between a and l_{gd} in determining BV_{GD}, μ^*, and V^*_P is not very strong.

Moreover, the relation between a, l_{gd}, and l_{sg} (source–gate distance) in determining BV_{GD} is investigated. Although BV_{GD} depends on l_{sg} if a is large compared to the gate depletion layer width, it is almost independent of l_{sg} if a is small. This is because, the drain and source regions are completely divided by the gate depletion if a is small. The channel thickness of triode-like and mixed characteristic FET's are relatively small. Consequently, it is found that BV_{GD} is determined from the gate–drain distance l_{gd} only and the contribution from the other device design parameters to BV_{GD} is negligibly small. Since the source electrode is symmetric to the drain with respect to the gate, BV_{SG} (source–gate breakdown voltage) can be easily obtained by changing l_{gd} for l_{sg}. There is no need to calculate BV_{SG} independently.

IV. Optimum Design of Triode-Like JFET's

A. Optimum Channel Thickness

Since the differential output resistance r_{out} of triode-like I–V characteristics is small, a low-resistance-load direct drive is possible. As seen in Fig. 3, the channel thickness a or gate-spacing $d(= 2a)$ of triode-like JFET's must be small. This is because if a is large, the channel is not pinched-off under the zero biased condition, that is, the equivalent vacuum state necessary for triode-like operation of JFET's cannot be realized. Thus the following relation is obtained as a necessary condition for triode-like operation

$$a < a_0 \tag{10}$$

where a_0 is the zero bias depletion layer width and is given by

$$a_0 = f(V_{bi}) \equiv \sqrt{\frac{2\epsilon}{qN_D} V_{bi}} \tag{11}$$

with V_{bi} being the built-in potential.

If a becomes small enough, the on-resistance R_{on} increases steeply. A large R_{on} increases the dead region,[2] that is, the ac output power efficiency, $\eta \equiv$ (ac output power)/(dc input power), decreases. Thus, an optimum a exists for circuit applications. The following figure-of-merit G is introduced to evaluate device performance,

$$G = g_m{}^*\eta^* \tag{12}$$

where $g_m{}^*$ is the effective transconductance for a given load line. The η^* is a factor proportional to the ac power efficiency η and defined as follows (see Fig. 12):

$$\eta^* = \frac{V_{BB} - V_{DSS}}{V_{BB}}. \tag{13}$$

Assuming $R_L = 8\ \Omega$ and the matching condition ($r_{out} = R_L$), the figure-of-merit G as a function of the reduced channel thickness a/a_0 is shown in Fig. 13. It is found that G peaks in the region $a/a_0 = 1 \sim 1.5$. When a/a_0 is greater than unity but not very large, FET's do not behave with

[2] The dead region is defined by the area of the left-hand side of the I–V curve at $V_G = 0$ V (see Fig. 12). The device is inactive.

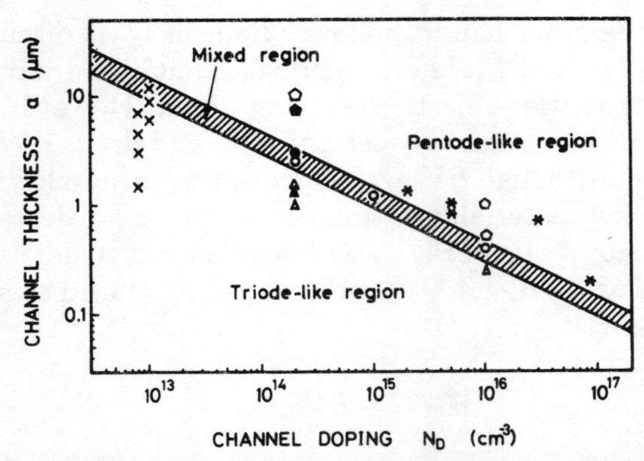

Fig. 14. Classification of JFET operation on the $N_D - a$ plane. Present analysis is denoted by the triangles, circles and pentagons which stand for the triode-like, mixed, and pentode-like characteristics, respectively. The filled symbols (▲, ● and ⬟) correspond to the parameters commonly used in the theory and experiment shown in Fig. 3. The crosses × and asterisks ∗ denote the experimental data for triode-like JFET's reported by Nishizawa *et al.* [6] and normally employed JFET's with saturating *I–V* characteristics, respectively.

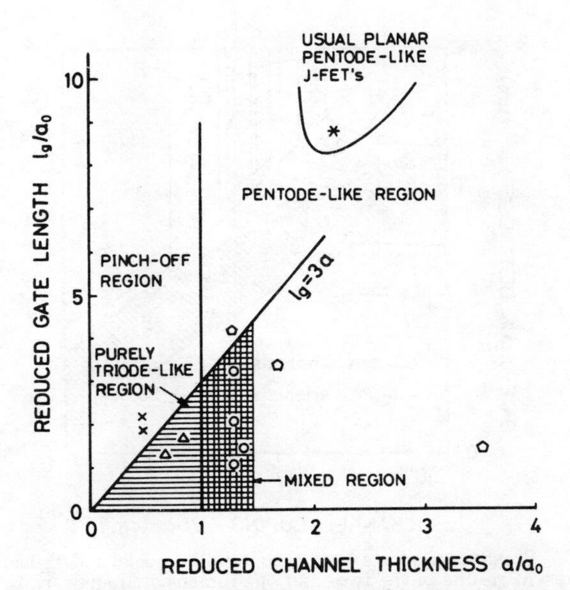

Fig. 15. Gate length and JFET operation. Triangles, circles, and pentagons represent, respectively, the parameters of triode-like, mixed, and pentode-like characteristic FET's analyzed in the present work. The drain current of FET's denoted by crosses × does not reach practical levels before the breakdown occurs. The ratio l_g/a of usual planar pentode-like JFET's is large enough ($\gtrsim 5$).

purely triode-like characteristics shown in Fig. 3(a) but with mixed ones as shown in Fig. 3(b). The mixed characteristics are considered more desirable in comparison to purely triode-like ones which are realized under condition (10). This is because the dead region is reduced and the merits of the triode-like characteristics, i.e., low output resistance and low distortion, are maintained. Since the channel must be pinched-off under a small reverse gate bias (V_{gss}), the mixed characteristics are obtained by

$$f(V_{bi}) < a < f(V_{bi} + V_{gss}). \tag{14}$$

Several N_D and a sets analyzed theoretically and experimentally are shown on the $N_D - a$ plane in Fig. 14. The hatched region stands for condition (14), where V_{gss} is set equal to V_{bi}. This region coincides with the region in which G peaks. The types of *I–V* characteristics analyzed in this paper are plotted on the plane by the following symbols; triangles, circles and pentagons which denote the triode-like, mixed and pentode-like characteristics, respectively. The filled symbols (▲, ●, and ⬟) stand for the parameters used in the theory and experiment shown in Fig. 3. The classification of *I–V* characteristics given by (10) and (14) with $V_{gss} = V_{bi}$ agrees quite well with the experimental and numerical analyses as seen in Figs. 3 and 14. It is also shown for a wide range of N_D values that the mixed characteristics attain excellence in low on-resistance conditions, i.e., the high ac power efficiency, compared to the purely triode-like characteristics [16]. For reference, experimental data of triode-like JFET's reported by Nishizawa *et al.* [6] and of typical planar pentode-like JFET's are added. As a results, it is concluded that the optimum channel thickness of triode-like JFET's is represented by the hatched region in Fig. 14.

B. Restriction of Gate Length

As seen in Fig. 6, *I–V* characteristics are drastically changed when the gate length is varied. FET's with longer l_g do not behave as triodes. This is because the drain field is screened strongly by the gate depletion and cannot modulate the source field through the depletion layer if the gate length is long enough. The gate–drain breakdown limits the practical device operation at high drain voltages. For example, the maximum field of FET's shown in Fig. 6 reaches about 500 kV/cm at $V_D = 40$ V.

Gate lengths and channel thicknesses of the analyzed devices are shown in Fig. 15 with symbols representing their modes of operation. From this figure and Figs. 6 and 7, the upper limit of the gate length is deduced to be about three times the channel thickness. Therefore, the following relation is obtained as one of the conditions on device parameters:

$$l_g < 3a. \tag{15}$$

If condition (15) is not satisfied and the channel thickness is thin, FET's are completely pinched-off by the gate depletion and do not operate as triodes.

C. Optimum Gate-Drain Distance l_{gd}

The strong electric field in the region between the gate and drain causes two important effects on the FET operation; the injection of mobile carriers into the depleted region and the multiplication of electron–hole pairs. The former is necessary for triode-like operation and the latter limits the device operation. In order to clarify the actual operation area where the breakdown is avoided, the breakdown analysis described in Section II is carried out. Calculated results are shown on the $N_D - l_{gd}$ plane in Fig. 16 by several contour lines (solid lines), where the optimum channel thickness shown in Fig. 14 is assumed for each N_D value. When N_D is given, there exist the maximum breakdown voltage and maximum depletion layer width $W_{\max}$. Thus, the contour lines of BV_{GD} is bended to the

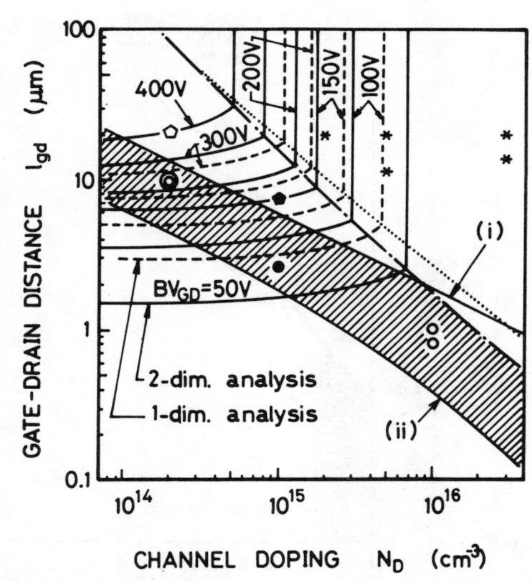

Fig. 16. Breakdown voltage (several contour lines; solid and dashed lines represent results of the two- and one-dimensional analysis, respectively) and optimum gate-drain distance l_{gd} (the hatched area) for triode-like JFET's. The present analysis is introduced by the pentagons and circles which denote, respectively, the pentode-like and mixed characteristics (coaxial circle ⊙ represents the parameters used in Fig. 3(b)). The filled symbols (● and ⬟) stand for the parameters used in Fig. 9. The asterisks * denote the parameters for normal planar JFET's with saturating I–V characteristics.

vertical direction crossing the borderline given by the maximum depletion layer width (chain line). Since the depleted region must reach the drain without the breakdown, the gate–drain distance l_{gd} is restricted to this maximum depletion layer width. For reference, the results of one-dimensional analysis are shown by the dashed lines (BV_{GD}) and dotted line ($W_{\max}$). $W_{\max}$ given by the two-dimensional approach is smaller than that by one-dimensional approach.

The above breakdown limit is one upper limit of l_{gd}. The other upper limit is a length determined from a depletion layer width under an actual operation. If an FET operates as a triode under a supplied voltage V_{BB}, the depleted channel has to be formed before V_D reaches V_{BB}. If not so, triode-like, i.e., enhancement type operation cannot be achieved. For example, two values of l_{gd} used in Fig. 9 are plotted in Fig. 16 by the symbols ● and ⬟. Assuming V_{BB} = 40 V, the depleted region in the device with longer l_{gd} reaches the drain at $V_D \approx V_{BB}$. In this case, the device does not behave as a triode. Comparing to the operation of the FET with shorter l_{gd},[3] it is inferred that the gate–drain distance l_{gd} should be smaller than the width of the depletion layer expanded under the bias condition $V_{BB}/2$. Therefore, the following inequality is obtained:

$$l_{gd} \le f\left(V_{bi} + \frac{V_{BB}}{2}\right). \tag{16}$$

Solid line (i) in Fig. 16 stands for the upper limit given by (16) where V_{BB} = 40 V is assumed. Of course, higher V_{BB} can be used if the channel doping is low, and the boundary given by line (i) is shifted upward.

[3] Quantitatively speaking, the depleted region of the device reaches the drain at 4 ~ 5 V.

On the other hand, the lower limit of l_{gd} is discussed from BV_{GD} and g_m. Short l_{gd} brings about high drain field and the carrier injection is accelerated. However, the breakdown voltage between the gate and drain BV_{GD} is lowered. Although BV_{GD} can increase by using a low concentration material, the transconductance g_m decreases compared to that of FET's with high doping channels. So, the following $BV_{GD} - g_m$ product is introduced as a figure-of-merit:

$$BVG = BV_{GD}g_m. \tag{17}$$

Assuming BVG = 100 V℧ (with unit channel width 1 m) from practical use, the criterion is depicted by solid line (ii) on the $N_D - l_{gd}$ plane. As a whole, it is found that the optimum gate-drain distance is represented by the hatched region in Fig. 16. l_{gd} values used in the present analysis are added on the plane by the pentagon and circles which stand for, respectively, pentode-like and mixed characteristics. The coaxial circle ⊙ represents the parameters used in the theory and experiment shown in Fig. 3(b). Experimental data of normal JFET's with saturating I–V characteristics are denoted by asterisks. These examples illustrate successfully the present characterization.

V. Negative Feedback Effect and Triode-Like Operation

The channel resistance r_{sg} between the source and gate has a negative feedback effect due to the self-bias on FET operation. The structure with the remarkably low-resistance source region [6] is proposed to reduce the resistance r_{sg} and obtain triode-like operation. In order to study the effect, the equivalent circuit analysis (insert in Fig. 17) is carried out. This simulation is a test of the parasitic part while the channel resistance consists of the parasitic and intrinsic channel resistance. Calculated I–V characteristics are shown in Fig. 17 by the solid (R_s = 1Ω) and dashed lines (R_s = 0). By the addition of the resistance R_s, i.e., an increase in the channel resistance r_{sg}, the steep increase in triode-like current (dashed lines) is weakened and the differential output resistance increases. It is found that an addition of resistor to the source causes I–V curves to approach the saturating characteristics.

Reduction of r_{sg} does not necesarily result in triode-like operation. Fig. 18 shows I–V curves of two FET's in which only the source-gate distance l_{sg} is different from each other. The channel thickness of the FET's analyzed here is 0.5 μm and the operational mode belongs to the pentode-like characteristics (see Fig. 14). Calculated results show that both drain currents are saturated in large V_D region, even if $l_{sg} \approx 0$.

Larger l_{sg} or r_{sg} causes the transition from triode- to pentode-like operations, and smaller l_{sg} or r_{sg} is necessary to realize the triode-like operation. However, the reduction of l_{sg} or r_{sg} cannot lead the triode-like operation if the intrinsic FET has a characteristic of pentodes. As discussed previously, the appropriate channel thickness (Fig. 14) and

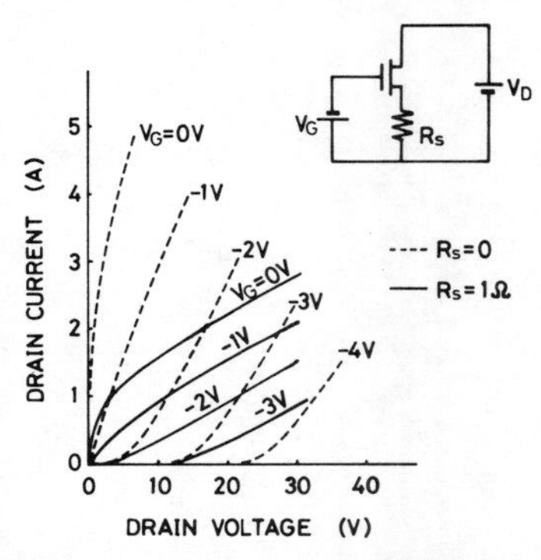

Fig. 17. Transition from the triode- to pentode-like characteristics by inserting resistor R_s.

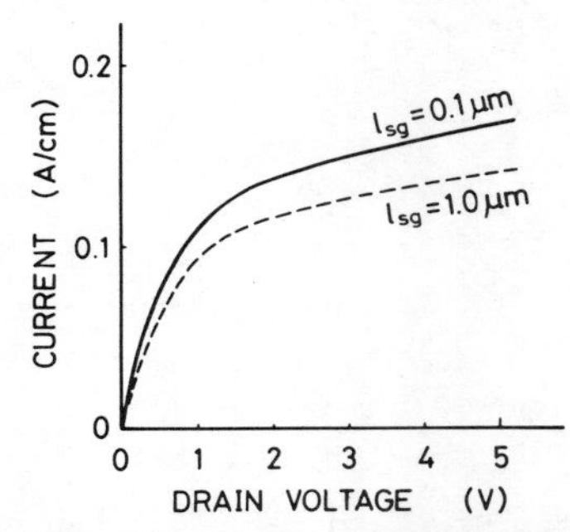

Fig. 18. Comparison of two-FET's which have different source–gate distances l_{sg}. Common parameters used here are $N_D = 1 \times 10^{16}$ cm^{-3}, $l_g = 1$ μm and $\mu = 1100$ cm^2/V · s.

gate-drain distance (Fig. 16), and short gate (condition (15)) are the most important conditions for the design of triode-like JFET's.

VI. Conclusion

The triode-like operation of JFET's is analyzed in comparison with the pentode-like operation and an optimum design of triode-like JFET's is discussed. The operation can be physically featured by the enhancement type behavior of the carrier motion, that is, the triode-like current is caused by the carrier injection into the depleted region. Such an operation is possible in short gate devices. This is because if the gate length is long, the drain field which causes the carrier injection cannot penetrate into the depleted channel under the gate. The physical picture revealed through the analysis provides a qualitative design guide line.

The device characteristics, for example, the current–voltage and pinch-off ones, are analyzed for a wide range of device design parameters. Considering low-resistance-load direct drive circuits, it is found that the mixed characteristics are superior in the ac power efficiency and transconductance product to the purely triode-like characteristics. Moreover, on the basis of the electric field distribution obtained from the two-dimensional numerical analysis, the breakdown phenomena are taken into account by means of the ionization integral and the breakdown voltage is obtained with respect to device geometry. As a whole, quantitative design criteria of triode-like JFET's are made clear.

These numerical analyses are compared to experimental data. Reasonable agreement is found between the calculations and experiments without any adjustable parameters, and validity of the present numerical approach is supported. Thus, the criteria are considered valuable as a design aid.

Acknowledgment

The authors wish to thank T. Toyabe, Dr. S. Asai and Dr. M. Nagata for their valuable discussions with regard to this work. We also express gratitude to M. Ohta, F. Kumagaya, and H. Ito for their experimental support work.

References

[1] D. P. Lecrosnier and G. P. Pelous, "Ion-implanted FET for power application," *IEEE Trans. Electron Devices,* vol. ED-21, pp. 113–118, Jan. 1974.
[2] O. Ozawa and K. Aoki, "A multi-channel FET with a new diffusion type structure," in *Proc. 7th Conf. Solid State Devices* (Tokyo, Japan, 1975), *Japan. J. Appl. Phys.,* vol. 15, Suppl. 15-1, pp. 171–177, 1976.
[3] Y. Watanabe and J. Nishizawa, Japanese patent 205 068, published No. 28-6077, application Date Dec. 1950.
[4] W. Shockley, "Transistor electronics: Imperfections, unipolar and analog transistors," *Proc. IRE,* vol. 40, pp. 1289–1313, 1952
[5] "Audio F.E.T. power transistors," *Wireless World,* vol. 80, no. 1463, pp. 223–224, 1974.
[6] J. Nishizawa, T. Terasaki, and J. Shibata, "Field-effect transistor versus analog transistor (Static induction transistor)," *IEEE Trans. Electron Devices,* vol. ED-22, pp. 185–197, Apr. 1975.
[7] G. F. Newmark and E. S. Rittner, "Transition from pentode- to triode-like characteristics in field effect transistors," *Solid-State Electron.,* vol. 10, pp. 299–304, 1967.
[8] J. A. Geurst, "Theory of insulated-gate field-effect transistors near and beyond pinch-off," *Solid-State Electron.,* vol. 9, pp. 129–142, 1966.
[9] K. Yamaguchi, T. Toyabe, and H. Kodera, "Two-dimensional analysis of triode-like operation of junction gate FET's," *IEEE Trans. Electron Devices,* vol. ED-22, pp. 1047–1049, Nov. 1975.
[10] K. Yamaguchi, T. Toyabe, and H. Kodera, "Two-dimensional analysis of vertical junction gate FET's," in *Proc. 7th Conf. Solid State Devices* (Tokyo, Japan, 1975), *Japan. J. Appl. Phys.,* vol. 15, suppl. 15-1, pp. 163–168, 1976.
[11] R. Zuleeg, "Multi-channel field-effect transistor theory and experiment," *Solid-State Electron.,* vol. 10, pp. 559–576, 1967.
[12] K. Yamaguchi and H. Kodera, "Drain conductance of junction gate FET's in the hot electron range," *IEEE Trans. Electron Devices,* vol. ED-23, pp. 545–553, June 1976.
[13] F. N. Trofimenkoff, "Field-dependent mobility analysis of the field-effect transistor," *Proc. IEEE,* vol. 53, pp. 1765–1766, 1965.
[14] W. C. Niehause, T. E. Seidel, and D. E. Iglesias, "Double-drift IMPATT diodes near 100 GHz," *IEEE Trans. Electron Devices,* vol. ED-20, pp. 765–771, Sept. 1973.
[15] S. M. Sze, *Physics of Semiconductor Devices.* New York: Wiley, 1969.
[16] J. Nishizawa *et al.,* Japanese Patent published No. 46-57768, in which condition (10) has been proposed.

Characteristics of High-Power and High-Breakdown-Voltage Static Induction Transistor with the High Maximum Frequency of Oscillation

MICHIO KOTANI, YUKIO HIGAKI, MARI KATO, AND YOSHINORI YUKIMOTO

***Abstract*—The design and fabrication of high-power and high-breakdown-voltage static induction transistor (SIT) with a high maximum frequency of oscillation are described and then the experimental characteristics are presented.**

A field plate is used to make the breakdown voltage high, and a fine stripe structure is adopted to make the maximum frequency of oscillation high.

As a result, the gate-drain breakdown voltage of 300 V, the gate-source breakdown voltage of 70 V, and the maximum frequency of oscillation of 700 MHz are obtained.

The maximum output power of 216 W with 7.5-dB gain and 55-percent drain efficiency is obtained at 100 MHz without a thermal runaway from an amplifying SIT with four pellets mounted in a single package.

Manuscript received June 3, 1980; revised July 15, 1981. This work was supported by the National Research Development of Japan.

The authors are with the Semiconductor Development Division, LSI Research Laboratories, Mitsubishi Electric Corporation, Itami, Hyogo 664, Japan.

I. Introduction

A HIGH-POWER and high-breakdown-voltage transistor at HF to VHF band with a high maximum frequency of oscillation has been requested for application in RF heating, high-power broadcasting, and plasma-generating equipment.

The static-induction transistor (SIT) which has been developed as a new high output power and high-frequency transistor [1]-[8] fits these requirements.

The SIT has the advantage of ease in making the breakdown voltage high and then obtaining high output power without a substantial lowering of the maximum frequency of oscillation. Considering the operating principle of the SIT, [2], [3], the drain current is composed of majority carriers and the carriers drift from the source to the drain at saturation velocity owing to the strong electric field in the depleted epitaxial layer the thickness of which determines the gate-drain breakdown voltage. Therefore, the effect of transit time can be negligible

Reprinted from *IEEE Trans. Electron Devices*, vol. ED-29, pp. 194-198, Feb. 1982.

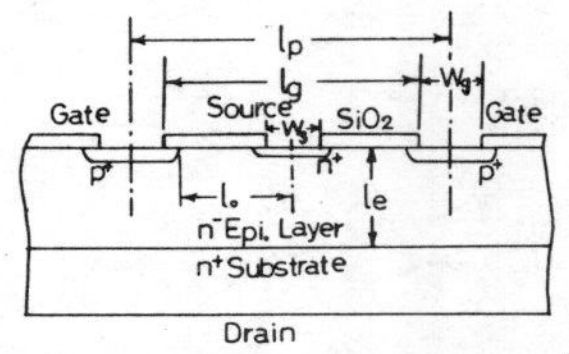

Fig. 1. Cross-sectional view of SIT. $n_d^- = 1 \times 10^{14}/cm^3$; $l_e = 20$ μm; $l_p = 15$ μm; $l_g = 11$ μm.

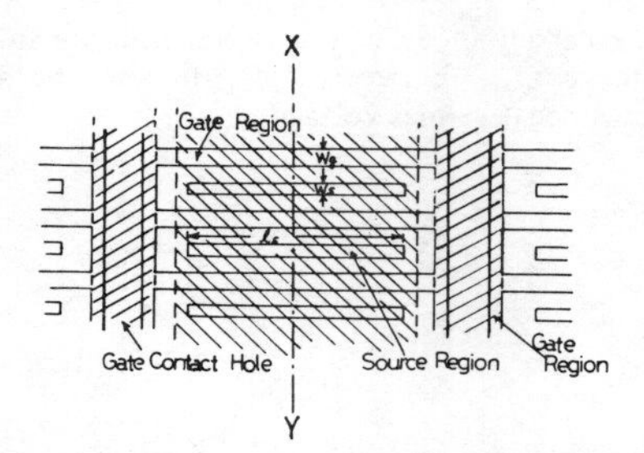

Fig. 2. Top view of SIT. $l_s = 50$ μm; $W_g = 4$ μm; $W_s = 3$ μm.

at the operating frequency of the HF to VHF band even though the depleted epitaxial layer is made thick to obtain the high breakdown voltage. The SIT does not show a thermal instability because its drain current has a negative temperature coefficient [3].

Consequently, it is understood that the SIT is suitable for a high-power and high-breakdown-voltage transistor with a high maximum frequency of oscillation.

In this paper, the design and fabrication of a high-power and high-breakdown-voltage SIT with a high maximum frequency of oscillation are described and also the experimental results of oscillation and amplification at the HF to VHF band are presented.

II. Design and Fabrication of SIT

Figs. 1 and 2 show the cross-sectional and top views of a high-power and high-breakdown-voltage SIT, respectively. Here, Fig. 1 is indicated along the X–Y axis in Fig. 2.

In the design of a high-power and high-breakdown-voltage SIT, the doping density and the thickness of the n^- epitaxial layer are given in order to determine the gate–drain breakdown voltage. Then the doping density and the thickness of the n^- epitaxial layer are selected as values of $n_d^- = 1 \times 10^{14}/cm^3$ and $l_e = 20$ μm, respectively. In this case, the gate–drain breakdown voltage BV_{gdo} is calculated as a value of 400 V for the abrupt plane junction, following the equation

$$BV_{gdo} = \frac{l_e}{W}\left(2 - \frac{l_e}{W}\right) \cdot V_B \tag{1}$$

where $V_B = 1500$ V and $W = 140$ μm are the avalanche breakdown voltage and the depletion layer thickness at V_B for the doping density $n_d^- = 1 \times 10^{14}/cm^3$ [9], [10], and $l_e = 20$ μm is the n^- epitaxial layer thickness of SIT, respectively.

In Fig. 1, in the fabrication, the n^+ source region, surrounded by the p^+ gate region, is formed by $POCl_3$ diffusion through the window of $W_s = 3$ μm and diffused to a depth of about 0.6 μm on the n^- epitaxial layer. The p^+ gate region is made by diffusion using boron nitride through the meshed windows of $W_g = 4$ μm and of $W_d = 16$ μm and diffused to a depth of about 2.0 μm. The depth of the p^+ gate region is made greater than that of the n^+ source region in order to make the depletion region between the gate and the source. The gate pitch l_p and the length l_g between the gate and the gate edges are 15 and 11 μm, respectively. The length l_0 between the edge of the p^+ gate and the center of the n^+ source region is 3.8 μm. For this length, the ideal negative bias voltage V_p necessaiy to deplete the n^- region between the gate and the source is calculated as a value of 1.57 V, following the equation

$$l_0 = \sqrt{\frac{2 \cdot \epsilon (V_p + V_{bi})}{q \cdot n_d^-}} \tag{2}$$

where $\epsilon = 1.06 \times 10^{-12}$ F/cm is the dielectric constant of silicon, q is an electronic charge, and V_{bi} is a built-in voltage [10]. Here, it is assumed that (2) for the abrupt plane junction can be adopted since the depth of the p^+ gate region is greater than that of the n^+ source region. Over the designed voltage V_p of 1.57 V, the n^- region between the gate and the source is fully depleted. Similarly, the ideal voltage necessary to deplete the n^- region between the gate and the drain can be calculated as a value of 48.6 V.

It is understood from Fig. 2, that the source region, the length l_s of which is 50 μm, is surrounded by the gate region. The meshed gate structure is better in reducing the feedback capacitance C_{gdo} between the gate and the drain than the interdigital stripe-type structure the metal electrodes of which are arranged in parallel along the length of the source fingers [11].

The metallization of the gate and the source which correspond to the shaded regions in Fig. 2 are done by evaporated Al and chemical vapor deposited poly-Si and they are perpendicular to the length of the source fingers as shown in Fig. 2. The source metal electrode is insulated by SiO_2 from the narrower gate region.

The gate–drain breakdown voltage of the SIT lowers from the ideal value of 400 V calculated from (1) because the expitaxial layer length l_e between the gate and the drain becomes a little shorter by the p^+ gate diffusion depth, and the electric field crowds and becomes strong at the curvature edge of the p^+ gate region near the surface. The first reason is not so important since the epitaxial layer thickness can be controlled in the fabrication process, but the second reason is instrumental in lowering of the gate–drain breakdown voltage [12]. Therefore, in order to make the breakdown voltage high, the electric field at the curvature edge of the p^+ gate region must be lowered.

In the fabrication, for this purpose, a field plate with a length W_{fp} of 60 μm is located outside the p^+ gate region near the scribe line [12], as shown in Fig. 3. Owing to this field plate, the strength of the electric field at the curvature edge of the p^+ gate region is weakened and also the surface charge is neutralized. The field plate in our experiments is connected to the source to make the feedback capacitance C_{gdo} small.

In addition, PSG (Phosphorus Silicate Glass) film which has an effect of gettering the movable alkali ions in the SiO_2

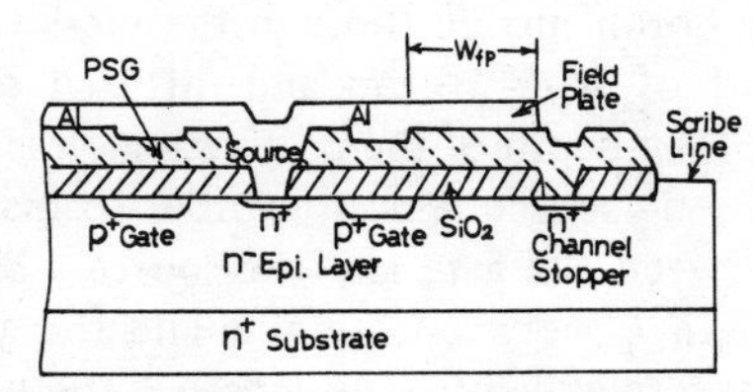

Fig. 3. Field-plate structure of SIT. $W_{fp} = 60\ \mu$m.

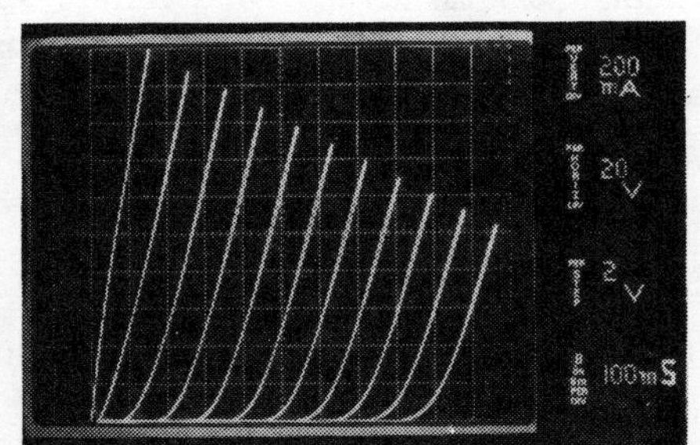

Fig. 4. Static characteristics of one-pellet SIT. $I_{dss} = 1.0$ A ($V_g = 0$ V, $V_d = 10$ V); $\mu = 9$; $g_m = 290$ mS.

TABLE I
DIMENSIONS OF THE ONE-PELLET SIT

One pitch of the SIT	lp	15 μm
Length between gate and gate	lg	11 μm
Width of gate diffusion	Wg	4 μm
Width of source diffusion	Ws	3 μm
Width of field plate	Wfp	60 μm
Epitaxial layer thickness	le	20 μm
Length of source finger	ls	50 μm
Total source length		13.92cm
Size of the 1-pellet SIT		1.8x5.5mm²

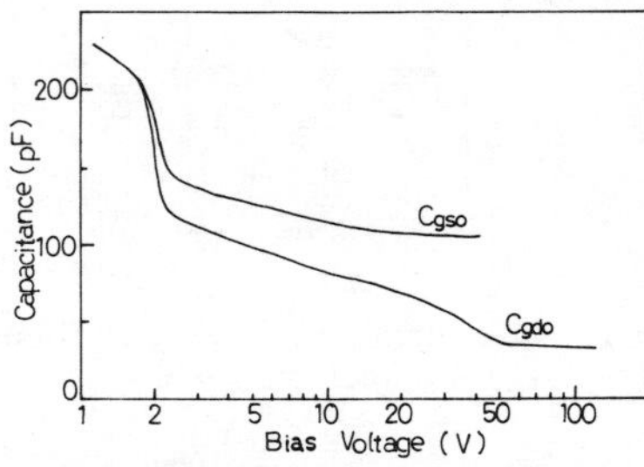

Fig. 5. Feedback capacitance C_{gdo} between the gate and the drain and input capacitance C_{gso} between the gate and the source of one-pellet SIT versus negative-bias voltage.

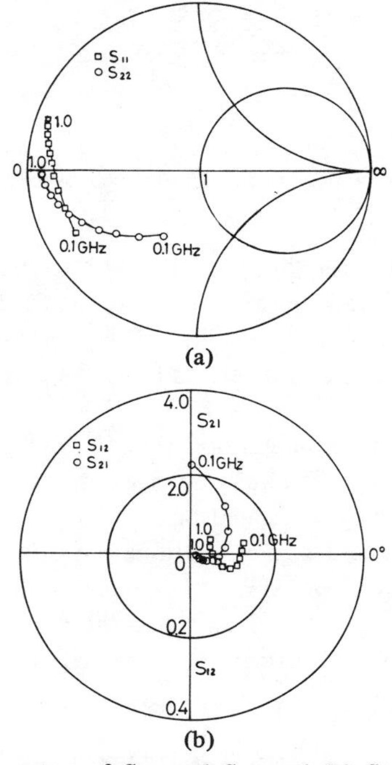

Fig. 6. (a) S-parameters of S_{11} and S_{22} and (b) S-parameters of S_{12} and S_{21} for the one-pellet SIT. $V_d = 100$ V; $I_d = 400$ mA; $V_g = -6$ V.

is deposited on the SiO_2 in order to avoid the drift of the breakdown voltage. Furthermore, the n^+ channel stopper is adopted in order to reduce the leaky current between the gate and the drain which may be induced from the recombination centers near the scribe line, as shown in Fig. 3.

Table I shows a summary of the dimensions of one-pellet SIT described up to now.

III. Static and Small-Signal Characteristics

Fig. 4 shows the drain current versus the drain voltage as a function of negative gate bias voltage. The triode-like characteristics are observed. It can be seen from Fig. 4 that the voltage amplification factor μ is about 9, the mutual conductance g_m is about 290 mS, and the drain current I_{dss} biased at $V_g = 0$ V and $V_d = 10$ V is nearly 1.0 A. The SIT, shown in Fig. 4, has a gate-drain breakdown voltage BV_{gdo} of 300 V and a gate-source breakdown voltage BV_{gso} of 70 V, both measured at a current of 0.1 mA. The experimental breakdown voltage BV_{gdo} of 300 V is lower than the ideal breakdown voltage of 400 V by 100 V. As mentioned before, in Section II, the reasons for this difference are due to the facts that the epitaxial layer becomes shorted by the depth of the p^+ diffusion and that the electric field becomes strong near the surface.

Fig. 5 shows the input capacitance C_{gso} versus the gate-source negative bias voltage and the feedback capacitance C_{gdo} versus the gate-drain negative bias voltage, respectively. In measurements of C_{gso} and C_{gdo}, the third terminal is opened. C_{gso} and C_{gdo} decrease rapidly at about -2 V. C_{gso} becomes nearly constant over the range of several volts where the n^- gate-source region is fully depleted. In Fig. 5, the voltage of -2 V, where the capacitance C_{gso} and C_{gdo} decrease rapidly, corresponds to the experimental voltage to deplete the n^- gate-source region and nearly equals the foregoing designed value of -1.57 V in (2). On the other hand, C_{gdo} decreases gradually and becomes constant over the range of 50 V where the n^- gate-drain region is fully depleted. This value is nearly equal to the calculated value of 48.6 V in (2).

Fig. 6(a) and (b) shows the S-parameters of S_{11}, S_{12}, S_{21}, and S_{22}, respectively. As a matter of course, the input and output impedances show the capacitive high impedance at the HF to VHF band because the p^+ gate is negatively biased both to the n^+ source and to the n^+ drain.

Fig. 7 plots the small-signal power gains and the stability factor as a function of operating frequency, calculated from the S-parameters in Fig. 6 [13]. It is known that the maxi-

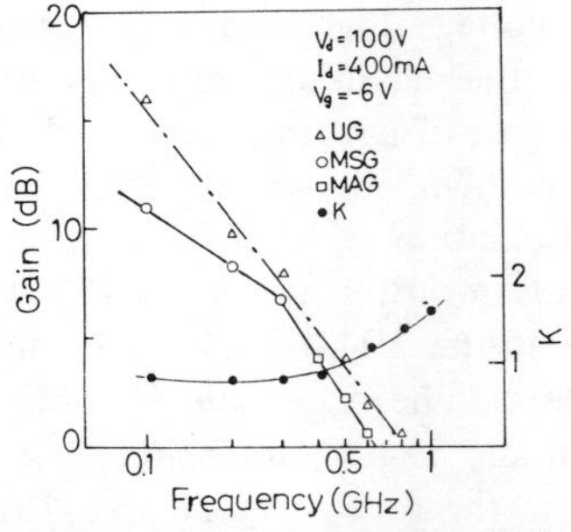

Fig. 7. Gain and stability factor of one-pellet SIT versus the operating frequency. $V_d = 100$ V; $I_d = 400$ mA; $V_g = -6$ V.

TABLE II
CHARACTERISTIC VALUES OF THE ONE-PELLET SIT

Breakdown voltage between gate and drain	BV_{gdo}	300 V
Breakdown voltage between gate and source	BV_{gso}	70 V
Capacitance between gate and drain	C_{gdo}	35 pF
Capacitance between gate and source	C_{gso}	110 pF
Mutual conductance	g_m	290 mS
Factor of voltage amplification	μ	9
Saturation current (Vg = 0V, Vd = 10V)	Idss	1.0 A
The maximum frequency of oscillation	f_{max}	700 MHz

mum frequency of oscillation f_{max} is about 700 MHz and that the small-signal gain of the SIT at 100 MHz is more than 12 dB. The maximum frequency of oscillation $f_{max} \simeq 700$ MHz is very high in spite of the high breakdown voltage of $BV_{gdo} \simeq 300$ V and $BV_{gso} \simeq 70$ V and is high enough to operate the SIT at the HF to VHF band.

Table II shows the summary of the one-pellet SIT characteristics described up to now.

IV. MULTIPELLET OPERATION OF SIT'S

The multipellet operation of SIT's such as two-pellet SIT's and four-pellet SIT's mounted in a single package are carried out in order to obtain high output power at the HF to VHF band. The pellets for the experiment were picked up from a lot of SIT's which showed reasonable one-pellet performance.

Fig. 8(a) and (b) shows the flat-type packages on which two pellets and four pellets of the SIT are mounted. The drain current I_{dss} at $V_g = 0$ V and $V_d = 10$ V and the mutual conductance g_m of the two-pellet and four-pellet devices are twice and four times those of the one-pellet device, respectively. The maximum frequencies of oscillation of the two-pellet and four-pellet devices are 600 and 450 MHz, respectively, by measuring the gain characteristics similarly as in Fig. 7. It is found that f_{max} does not decrease substantially as compared to the value of the one-pellet SIT. The gradual decrease of f_{max} with the increase of number of pellets is attributed to the increase of the capacitance of C_{gso} and C_{gdo}, because the input and output package capacitances are negligible in comparison with those of the SIT.

Fig. 9 shows the characteristics of a 30-MHz-band Clapp oscillator with an autobiased circuit between the gate and the source using the two-pellet device of Fig. 8(a). The autobiased circuit is constructed with resistance and capacitance in parallel. In Fig. 9, the threshold voltage of oscillation is about 8 V where the drain current decreases rapidly. The maximum oscillation output power of 150 W with 68-percent drain efficiency is obtained to the 50-Ω load directly without a thermal runaway. It is considered that the thermally stable oscillation of high output power is a merit for application in RF heating and plasma-generating equipment, in comparison with the case of the multistage power amplifier, in cost and circuit simplicity.

Fig. 10 shows the characteristics of a 30-MHz-band amplifier, using the two-pellet device. The input and output circuits are matched by inductance and capacitance circuit to the 50-Ω input source and to the 50-Ω load, respectively. The maximum output power of 200 W with 12-dB gain and 70-percent drain efficiency is obtained at 125-V drain bias voltage without thermal instability.

Fig. 11 also shows the amplifier characteristics both in case of the two-pellet operation and in case of the four-pellet operation at a 100-MHz band. The input and output circuits are constructed with the 50-Ω system same as that in Fig. 10. In the two-pellet operation, the maximum output power of 115 W with 10-dB gain and 65-percent drain efficiency is ob-

(a)

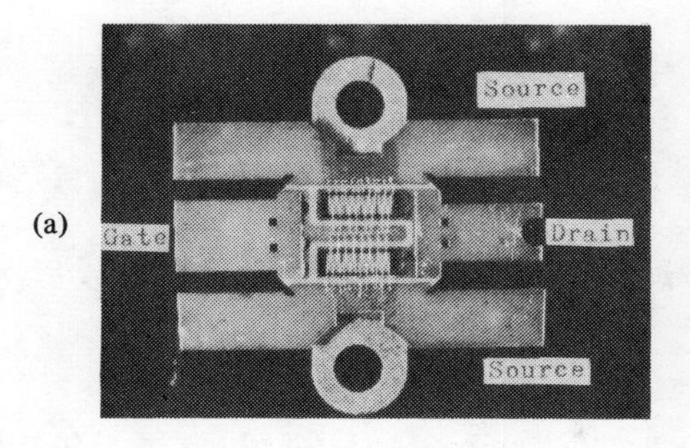

(b)

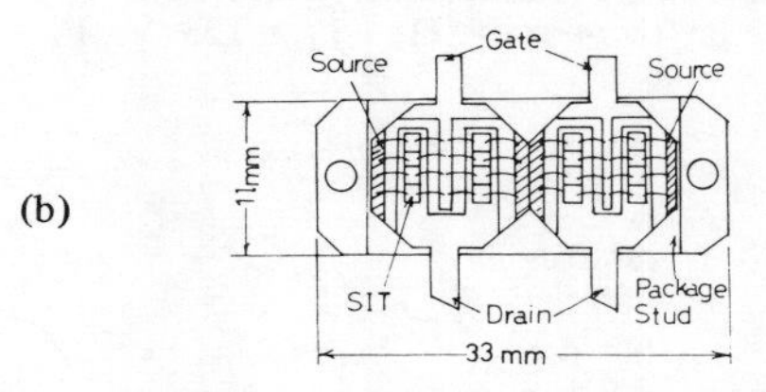

Fig. 8. (a) Photograph of a two-pellet mounted package. (b) Diagram of a four-pellet mounted package.

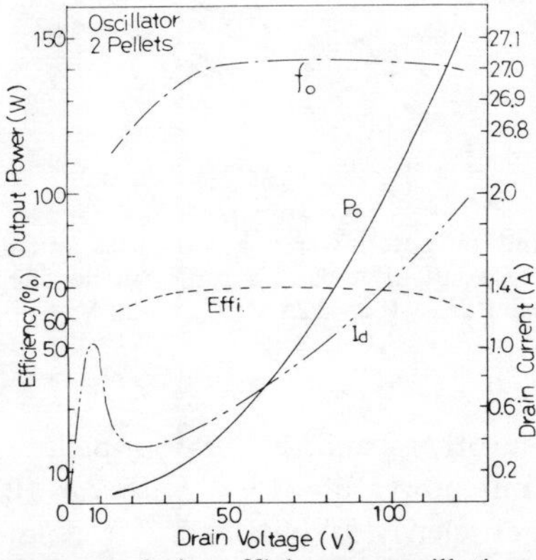

Fig. 9. Output power, drain efficiency, oscillation frequency, and drain current versus drain voltage of the Clapp oscillator with an autobiased circuit, using the two-pellet SIT.

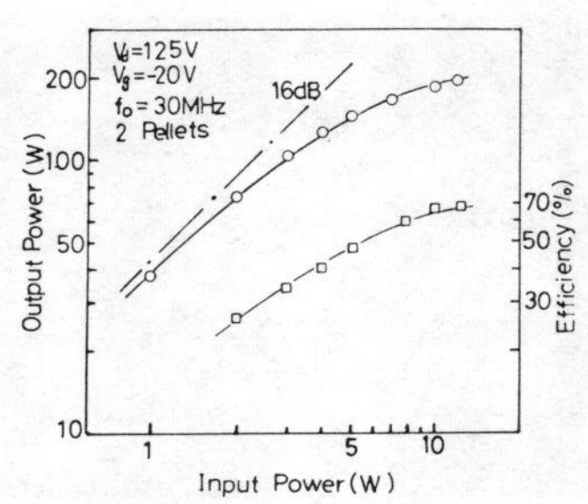

Fig. 10. Input and output power characteristics of a 30-MHz-band amplifier utilizing the two-pellet SIT. $V_d = 125$ V; $V_g = -20$ V; $f_o = 30$ MHz.

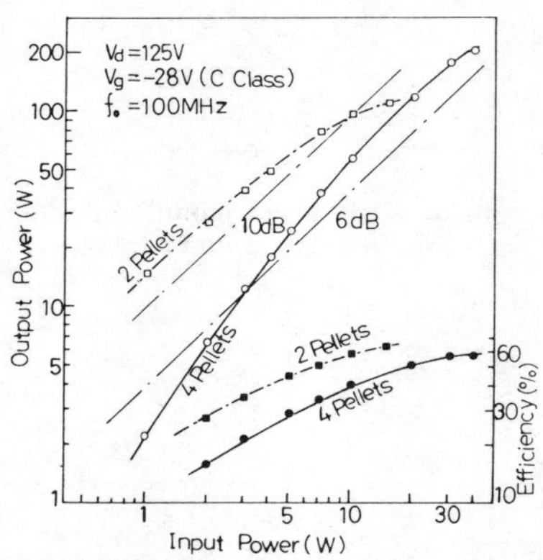

Fig. 11. Input and output power characteristics and drain efficiency characteristics of a 100-MHz-band amplifier using the two-pellet SIT and the four-pellet SIT. $V_d = 125$ V; $V_g = -28$ V; $f_o = 100$ MHz.

tained. On the other hand, in the four-pellet operation, the maximum output power of 215 W with 7.5-dB gain and 55-percent drain efficiency is obtained. It is known that the maximum output power of the four-pellet operation is nearly twice that of the two-pellet operation, although both the drain efficiency and the gain decrease. The decrease in efficiency and gain in the four-pellet operation is caused by the fact that the maximum frequency of oscillation of the four-pellet device is lower than that of the two-pellet device. Accordingly, from those experimental results, it is understood that the SIT is suitable for simple multipellet parallel running in a single package to obtain high output power at the HF to VHF band without thermal runaway.

V. Conclusion

A high-power and high-breakdown-voltage SIT with a high maximum frequency of oscillation was designed and fabricated, and the oscillation and amplifier characteristics were described experimentally. A field-plate structure was employed to make the breakdown voltage high and a fine-stripe structure was adopted to make the maximum frequency of oscillation high. As a result, the gate-drain breakdown voltage of 300 V, the gate-source breakdown voltage of 70 V, and the maximum frequency of oscillation of 700 MHz were obtained. The maximum oscillation output power of 150 W with 68-percent drain efficiency at the 30-MHz band was obtained without a thermal instability. The maximum output power of 216 W with 7.5-dB gain and 55-percent drain efficiency was obtained at 100 MHz from the four-pellet device. From these results, it was clarified that the SIT was suitable for a high output power transistor which has a high breakdown voltage and high maximum frequency of oscillation.

Acknowledgment

The authors wish to thank Prof. J. Nishizawa of Tohoku University and Dr. K. Shirahata, manager of the Semiconductor Research Laboratory for their useful discussions and suggestions.

References

[1] Y. Watanabe and J. Nishizawa, Japanese Patent 205 068 (published) No. 28-6077, application date, Dec. 1950.
[2] J. Nishizawa, T. Terasaki, and J. Shibata, "Field effect transistor versus analogue transistor (static induction transistor)," *IEEE Trans. Electron Devices*, vol. ED-22, pp. 185-197, Apr. 1975.
[3] J. Nishizawa and K. Watanabe, "High frequency high power static induction transistor," *IEEE Trans. Electron Devices*, vol. ED-25, pp. 314-322, Mar. 1978.
[4] S. Teszner, "Gridistor development for the microwave power region," *IEEE Trans. Electron Devices*, vol. ED-19, pp. 355-364, Mar. 1972.
[5] Y. Mochida, J. Nishizawa, T. Ohmi, and R. K. Gupta, "Characteristics of static induction transistor–Effect of series resistance," *IEEE Trans. Electron Devices*, vol. ED-25, pp. 761-767, July 1978.
[6] Y. Kajiwara, M. Aiga, Y. Higaki, Y. Yukimoto, and K. Shirahata, "A 100 W static induction transistor operating at 1 GHz," in *Proc. 11th Conf. on Solid State Devices* (Tokyo, Japan), pp. 305-308, 1979.
[7] T. Shino, H. Kamo, S. Oka, and K. Aoki, "New SIT structure exceeds 10 W at 2 GHz," *Micro.*, vol. 19, no. 2, pp. 48-53, Feb. 1980.
[8] B. J. Baliga, "A power junction gate field effect transistor structure with high blocking gain," *IEEE Trans. Electron Devices*, vol. ED-27, pp. 368-373, Feb. 1980.
[9] A. B. Phillips, *Transistor Engineering.* New York: McGraw Hill, 1962, p. 209.
[10] S. M. Sze, *Physics of Semiconductor Devices.* New York: Wiley, 1969.
[11] Unpublished data from private communication.
[12] A. S. Grove, *Physics and Technology of Semiconductor Devices.* New York: Wiley, 1967, p. 311.
[13] "S-parameters circuits analysis and design," Hewlett Packard application note, no. 95, Sept. 1968.

The Asymmetrical Field-Controlled Thyristor

B. JAYANT BALIGA, SENIOR MEMBER, IEEE

Abstract—**A new device structure has been developed for field-controlled thyristors. In this structure, the uniformly doped n base of the conventional device has been replaced with a very lightly doped region near the gate and a more heavily doped region at the anode. This change in the base doping profile results in a significant improvement in the tradeoff between the forward-blocking voltage capability and the on-state forward-voltage drop. In addition, the high-resistivity region around the gate area allows the device to pinch off anode current flow at zero gate bias due to the built-in potential of the gate junction. The devices can, however, be triggered to the on-state by applying a small forward gate voltage and exhibit a forward voltage drop in the on-state which is much lower than that of conventional devices. The high resistivity of the channel area between gates also results in these devices having dc blocking gains in excess of 60, which is the highest value achieved in devices of this type. Further, because these devices have been fabricated using conventional planar processing techniques, this structure is suitable for high-volume production with high processing yields.**

INTRODUCTION

IN RECENT YEARS, rapid developments in the area of power field-effect devices have been reported in the literature. Although most of the attention has been focused on the power MOS gate field-effect transistor, another promising device being developed for power switching applications is the field-controlled thyristor. This device has also been called the gridistor and the static-induction transistor. The basic structure of a field-controlled thyristor consists of a p^+-n-n^+ rectifier with the n base containing heavily doped p^+-gate regions. The device is operated by maintaining the n^+ cathode at ground potential and by applying bias voltages on the anode and gate terminals. In the absence of any applied potential on the gate regions, the device behaves like a conventional rectifier. As a result, the device is capable of conducting large currents with a low forward-voltage drop when a positive bias is applied to the anode terminal, and can block current flow when a negative bias is applied to the anode terminal. In addition, the gate regions of the device are designed to pinch off the current flow between anode and cathode when a negative bias is applied between the gate and the cathode. The presence of the gate regions, therefore, provides the device with a forward-blocking capability similar to that of a conventional thyristor. In order to obtain devices with good operating characteristics it is essential that large anode voltages can be blocked with the use of small gate bias voltages. Consequently, the ratio of the anode voltage, at the onset of anode current flow, to the applied gate control voltage is an important parameter for these devices and is called the blocking gain of the device. This blocking gain is a strong function of the geometrical shape of the channel formed between the gate regions and also depends upon the resistivity of the n base. In addition to providing the device with the forward-blocking capability, the gate regions can also be used to turn off the anode current when the device is in its on-

Manuscript received November 7, 1979; revised March 3, 1980.
The author is with General Electric Company, Corporate Research and Development Center, Schenectady, NY 12301.

Reprinted from *IEEE Trans. Electron Devices*, vol. ED-27, pp. 1262–1268, July 1980.

state. If a reverse bias is applied to the gate when the device is conducting anode current, the gate depletion layer extends into the channel and pinches off the anode current. This gate voltage-induced turn-off feature provides the device with a high-speed switching capability and extends the application of these devices to higher frequencies than achievable with conventional thyristors.

Two basic types of field-controlled thyristor gate structures have been explored in the past. These are the buried-grid gate structure and the surface-grid gate structure. The buried-grid devices have been fabricated by the masked diffusion of p^+ grid fingers (or a mesh) into an n-type substrate, followed by the growth of an n-type epitaxial layer over the diffused area by using either vapor-phase epitaxy or liquid-phase epitaxy [1]-[3]. The buried-grid structure has also been achieved by ion implantation of boron into an n-type substrate [4]. These devices have been found to exhibit moderate blocking gains (20-40), but have a limited gate turn-off capability due to the inherently high resistance of even heavily doped buried-grid fingers which causes debiasing of the grid fingers during turn-off. This has motivated the development of surface-gate devices with very low gate resistances achieved by contacting the gate fingers along their entire length with patterned metallization. The first such surface-gate devices were fabricated by planar diffusion of boron into 60-$\Omega \cdot$ cm n-type substrates [5]. In spite of a diffusion depth of 10 μm, the cylindrical shape of the planar diffused gate regions resulted in an open-channel structure which had a low blocking gain (5-7). In order to increase the blocking gain, an improved surface-gate structure was recently developed with steep vertical-gate walls by using orientation-dependent etching and selective vapor-phase epitaxial growth in the etched grooves [6], [7]. With this device structure, blocking gains as high as 45 were acheived at groove depths of about 40 μm. These devices also exhibited very fast gate turn-off speeds. However, the fabrication process for these devices was complex and resulted in a low device yield.

This paper describes a new structure for power field-controlled thyristors with which significant improvements in blocking gain, turn-off speed, and on-state forward voltage drop can be achieved. This device structure, however, has a limited reverse-blocking capability and has, therefore, been called the asymmetrical field-controlled thyristor. This reduced reverse-blocking capability is not a serious disadvantage in many inverter applications where antiparalled diodes are used. Although both surface- and buried-gate devices can be fabricated with the new device structure, it is of particular importance to the surface-gate devices because very high blocking gains can be achieved with this structure in spite of the use of conventional planar processing for device fabrication. This greatly simplifies the device processing and results in a higher yield during device fabrication in comparison with the earlier surface-gate devices.

Device Structure

The new field-controlled thyristor-device structure is contrasted with the previously developed (conventional) device structure [5] in Fig. 1. This figure also shows the doping profile of the devices and illustrates the electric-field distribution in the devices during both forward and reverse blocking. It

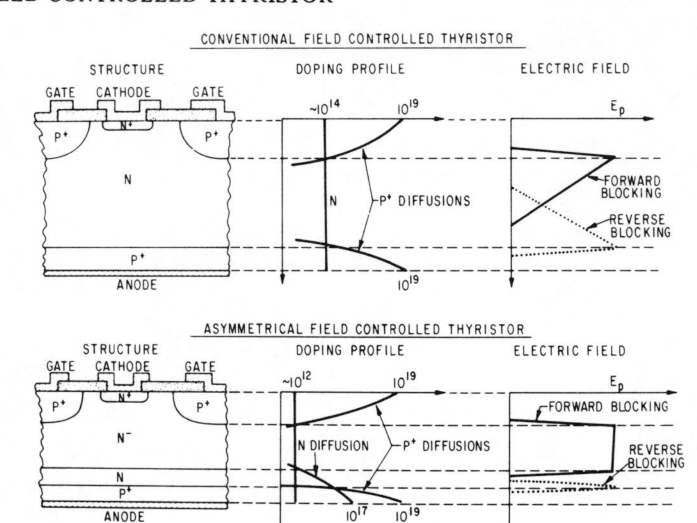

Fig. 1. Cross sections of the asymmetrical field-controlled thyristor and the conventional field-controlled thyristor illustrating the differences in base doping profiles and electric-field distributions in the forward- and reverse-blocking modes of operation.

can be seen that the new device structure is similar to that of a conventional device except for the replacement of the uniformly doped n base of the previous device structure with a two-layer n base consisting of a very lightly doped region near the gate/cathode surface and a more heavily doped region at the anode surface. This modification of the doping profile has a very strong influence upon the electric-field distribution in both the forward- and reverse-blocking stages of device operation. This change in the field distribution results in significant improvement in forward-blocking capability as described later. The change in doping profile also results in a significant improvement in the forward voltage drop.

Due to the parasitic p-n-p transistor formed between the gate and the anode, the breakdown voltage of the field-controlled thyristor is limited by open-base transistor breakdown. In the design of the conventional device structure, the optimum breakdown voltage with the lowest forward voltage drop is achieved by the use of an n-base width of about one diffusion length longer than the maximum depletion width at avalanche breakdown of the base material. In the new device structure, the transistor punch through breakdown is prevented by using the more heavily doped region in the n base at the anode side of the device. This allows using a very low doping level for the rest of the n base without encountering the open-base transistor breakdown effect. When the gate junction is reverse biased during the forward blocking mode of device operation, the gate depletion layer contains a low-ionized impurity density which results in maintaining a high electric field throughout the lightly doped n base as shown in Fig. 1. The electric field then decreases very sharply in the more heavily doped region, thus preventing punch through breakdown. In contrast, the uniform doping of n base in the conventional device structure produces a steadily decreasing electric field with its peak at the gate junction. Since the voltage supported between the gate and the anode terminals is given by the area under the electric-field profile, it is evident from Fig. 1 that, under forward-blocking conditions, the new device structure will have a substantially higher breakdown voltage for the same total wafer thickness and junction diffusion depths. There will, however, be no substantial difference in the forward voltage drop during current conduction since the wafer thickness and the junction depths of anode and cathode regions are unchanged. Thus the new n base doping profile results in a higher forward-blocking capability while maintaining the same forward voltage drop during current condition. A quantitative estimate of this improvement in breakdown voltage can be obtained by using a one-dimensional analysis with abrupt junctions. If a 1000-V conventional device design is considered, the n-base resistivity would be 35 $\Omega \cdot$ cm. At avalanche breakdown, this material has a depletion width of 90 μm. Since a diffusion length of at least 40 μm is required to maintain a low forward voltage drop, the total n-base width of the conventional device would be 130 μm. In the new device structure, the thickness of the more heavily doped n-base region at the anode is about 30 μm to prevent gate depletion-layer punchthrough to the anode. Consequently, for the same wafer thickness, the lightly doped layer of the new device structure would have a thickness of 100 μm. The low doping level of this region results in a uniform electric field as shown in Fig. 1, which has a value of about 2×10^5 V/cm at breakdown [8]. As a result, the forward-blocking capability of the new device structure would be extended to about 2000 V. Thus a two-fold[1] increase in the blocking voltage can be expected while maintaining the same forward voltage drop.

In addition, the use of a lightly doped n-base layer in the gate area of the devices results in a very large increase in the blocking gain in these devices. When the doping level of the n base decreases, the depletion layers from the gate regions extend into the channel area at lower gate voltages. As a result, the anode–cathode current is pinched off at a lower gate voltage, thus increasing the blocking gain. Moreover, if the channel width is small and the n base is very lightly doped, the built-in potential of the junction can be sufficient to cause gate depletion-layer punchthrough under the channel. Under these conditions, the anode–cathode current flow will be terminated with no bias applied to the gate. However, the potential barrier created in the channel at zero gate bias is small and can only prevent anode–cathode current flow at anode voltages below 100 V. Further, due to this depletion-layer punchthrough at low gate voltages, high blocking gains can be expected in the asymmetrical field-controlled thyristors even for the open-channel gate structure achieved by planar diffusion. This eliminates the use of the more complex epitaxial-refill technology which was required to achieve high-blocking gains and, consequently, an expected improvement in the device processing yield.

Another significant advantage of the new structure is the expected improvement in gate turn-off capability. In the conventional device structure, gate turn-off is achieved at gate voltages which create a depletion layer that pinches off the anode–cathode current, but with the depletion layer extending only a short distance from the gate junction. As a result, the gate depletion layer only sweeps out a small fraction of the charge which is injected into the n base during forward conduction. The remaining charge in the n base must then decay by recombination. This causes a long decay tail in the anode-cathode current waveform and decreases the speed of the device [5]. In contrast, the very low doping level of the upper n base layer in the asymmetrical field-controlled thyristor results in the gate depletion layer extending through the entire upper n base layer even at low gate voltages. As a result, most of the charge injected into the n base during forward conduction will be swept out when the gate is reverse biased. This leaves only a small fraction of the charge to decay by recombination, and thus improves the switching speed of the devices.

[1]In practice, slightly less than two-fold increase may be expected due to the decrease in the critical electric field for breakdown with decreasing doping, and the reduced anode injection efficiency due to the presence of the more heavily doped n-base region near the anode.

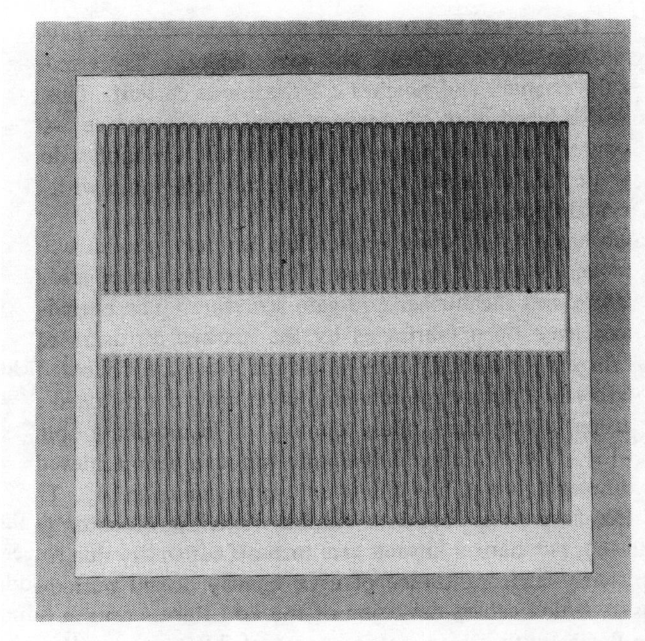

Fig. 2. Photomicrograph of the surface of the asymmetrical field-controlled thyristor. The overall chip dimensions are 120 mils by 120 mils. The device contains 92 cathode fingers interdigitated with the gate area.

Device Fabrication

In order to develop the asymmetrical field-controlled thyristor, a 2000-$\Omega \cdot$ cm phosphorous-doped ingot was used as starting material. This ingot was sliced and the wafers polished to thicknesses ranging from 5 to 10 mils. A phosphorous diffusion was then performed on the back (anode) side of the wafers to obtain an n-type diffused layer with a surface concentration of 2×10^{17} cm^{-3} and a depth of 50 μm. This layer was designed to prevent the punchthrough breakdown discussed in the earlier section. The oxide on the front surface of the wafers was then patterned to perform a boron diffusion to a depth of 10 μm in the gate area with a separation of 40 μm between the diffusion windows. This diffusion was simultaneously performed on the backs of the wafers to obtain the anode region. This was followed by opening windows between the gate areas, and a phosphorous diffusion was performed in these windows to fabricate the cathode regions. Contact windows were then opened over both the cathode and gate areas. Aluminum was evaporated on the front surface, patterned with photolithography, and sintered at 400°C to obtain a good contact to both the gate and the cathode areas. A photomicrograph of the completed device is shown in Fig. 2. The gate metallization of the device extends around the perimeter

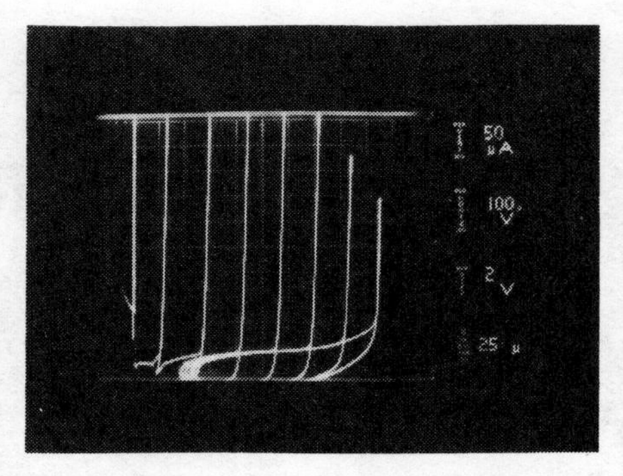

Fig. 3. Photograph of the characteristics of an asymmetrical field-controlled thyristor taken on a curve tracer under forward-blocking conditions. The gate voltage has been increased in seven steps (2 V per step).

of the device, and the cathode contact pad is located in the middle of the chip. The device consists of 92 cathode fingers which are interdigitated with the gate fingers. Each cathode has a width of 10 μm and a length of 850 μm, thus providing the device with a total cathode area of 7.8 $\times$ 10^{-3} cm^2. The overall dimensions of the chip are 120 mils by 120 mils. After device fabrication, the devices were mounted in TO-3 packages for evaluation of their characteristics.

Device Characteristics

The typical blocking characteristics of a device fabricated from a wafer with a thickness of 10 mils is shown in Fig. 3. It can be seen that the devices exhibit the expected triode-type characteristics. For this wafer thickness, the devices can block an anode voltage of over 900 V with an applied gate bias of only 14 V. Thus the dc blocking gain is more than 65. This is the highest dc blocking gain as yet reported for surface-gate devices. Although the ac blocking gain of the deep-groove field-controlled thyristors was found to be over 200, those devices have a dc blocking gain of less than 50 because the onset of blocking (depletion-layer punchthrough in the channel) does not occur until a gate bias of more than 10 V is applied. Thus the asymmetrical FCT devices fabricated in this study have superior blocking characteristics.

It has been found that the blocking gain of the asymmetrical field-controlled thyristors fabricated in this study is a function of the thickness of the n base in spite of keeping the gate junction depth unchanged. The detailed blocking characteristics of devices fabricated from wafers with thicknesses of 5, 6, 7, 8, and 10 mils are shown in Fig. 4. It can be seen that the maximum anode-blocking voltage increases with n-base width. This increase in the maximum anode-blocking voltage is to be expected, as discussed in the previous section, because the high electric field is maintained over a wider n base as the wafer thickness increases. In addition, the blocking gain of these devices is also found to be dependent upon the n-base width. The dc blocking gain measured at a gate bias of -10 V is plotted in Fig. 5. It can be seen that the dc blocking gain is a strong function of the n-base width and increases by more than a factor of 3 in going from a base width of 65 μm to a base width of 190 μm. This dependence of the blocking gain on the n-base width arises from the very low doping level in the n base. This low doping concentration in the n base results in the gate depletion layer extending across the entire n base

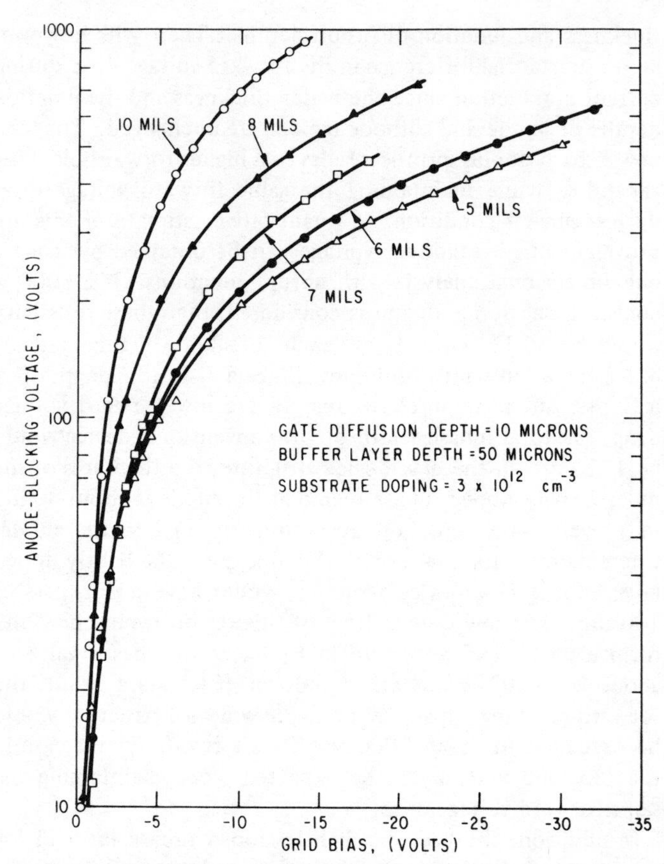

Fig. 4. Forward-blocking characteristics of asymmetrical field-controlled thyristors fabricated from wafers with thicknesses of 5-10 mils. Note the substantial increase in the anode-blocking voltage capability with increasing wafer thickness. These blocking voltages were measured on a curve tracer at an anode leakage current of 100 μA.

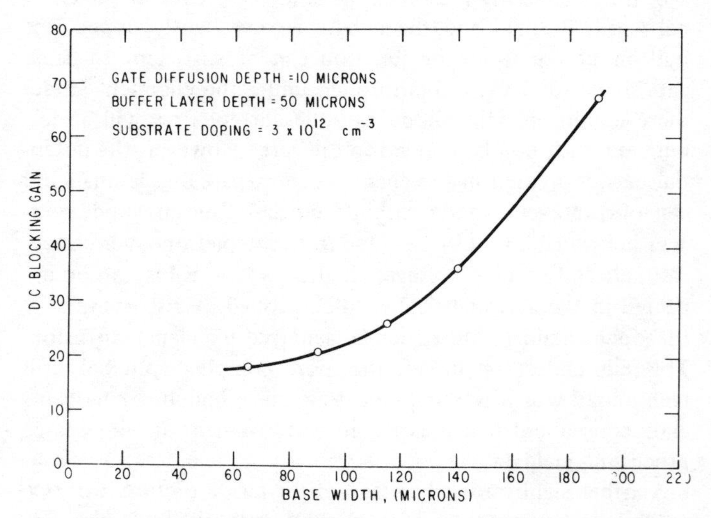

Fig. 5. Dependence of the dc blocking gain of the asymmetrical field-controlled thyristor upon the width of the lightly doped region in the n base.

of these devices for all the wafer thicknesses under forward-blocking conditions. The pinch off region in the channel, therefore, extends all the way from the cathode to the n-type diffused buffer layer on the anode side of the device. Consequently, the width of the pinch off region between anode and cathode, which has been treated in the past as the channel length, increases with increase in n-base width. Previous analy-

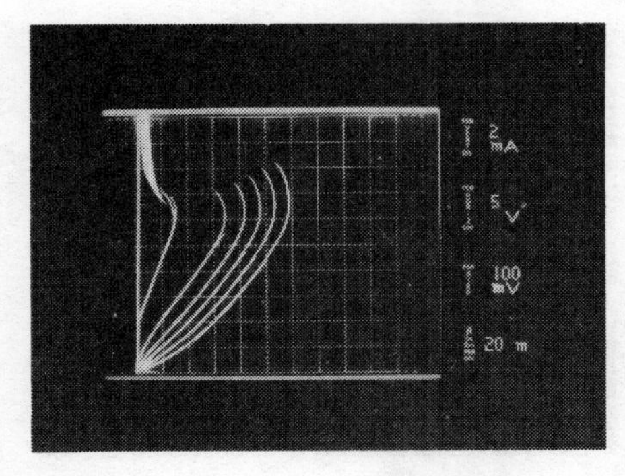

Fig. 6. Observation of regenerative switching in the asymmetrical field-controlled thyristor. The trace with the highest anode breakover (threshold) voltage is observed at zero gate bias. The other traces are obtained by applying increasing forward-gate bias voltages in steps of 0.1 V. The device is fully turned on and exhibits rectifier-diode characteristics when the applied gate bias exceeds 0.6 V. This device was fabricated from a 5-mil thick wafer.

ses have demonstrated that the blocking gain will increase with channel length [9], [10]. The observed increase in the blocking gain of the asymmetrical FCT devices with increase in wafer thickness is, therefore, consistent with these analyses of device operation.

As noted in the earlier section, these devices are designed to allow the gate depletion layer to punch through in the channel region without any applied gate bias. This results in the ability of these devices to block anode current even without an applied reverse gate bias. However, the potential barrier created in the channel by the built-in junction potential of the gate is only sufficient to block current flow at low anode voltages. It has been found that as the anode voltage is increased, the device switches from the blocking state to the on-state. This can be seen in Fig. 6 which shows the characteristics of the device at both zero gate bias and at various positive gate voltages. As the positive gate voltage is increased, the channel-potential barrier decreases and this, in turn, reduces the anode threshold voltage for switching to the on-state. A plot of the variation of the anode threshold voltage with applied forward gate bias for two typical devices is shown in Fig. 7. It can be observed that the threshold voltage decreases slowly with increasing gate bias up to a gate voltage of about 0.4 V and then decreases much more rapidly. The initial slow decrease in the threshold voltage at positive gate voltages of less than 0.4 V can be ascribed to the decrease in channel-potential barrier height as the gate voltage is increased. This decrease in barrier height is sensitive to the resistivity of the channel region because it is dependent upon the rate of variation of the gate depletion layer with gate voltage. The observed difference in the characteristics of the 5- and 8-mil wafers, is, therefore, believed to arise from differences in the resistivity of the starting material. When the gate bias exceeds 0.4 V, injection of the holes begins to occur at the gate. This causes the much more rapid decrease in the anode threshold voltage because of the onset of thyristor-type device turn-on. Once thyristor-type turn-on occurs, the threshold voltage is seen to be independent of the wafer resistivity and thickness.

When the device is operating in the on-state, the device characteristics are similar to those of a rectifier diode as shown in Fig. 8. At low anode currents, the device current increases exponentially with the anode voltage and the devices exhibit a

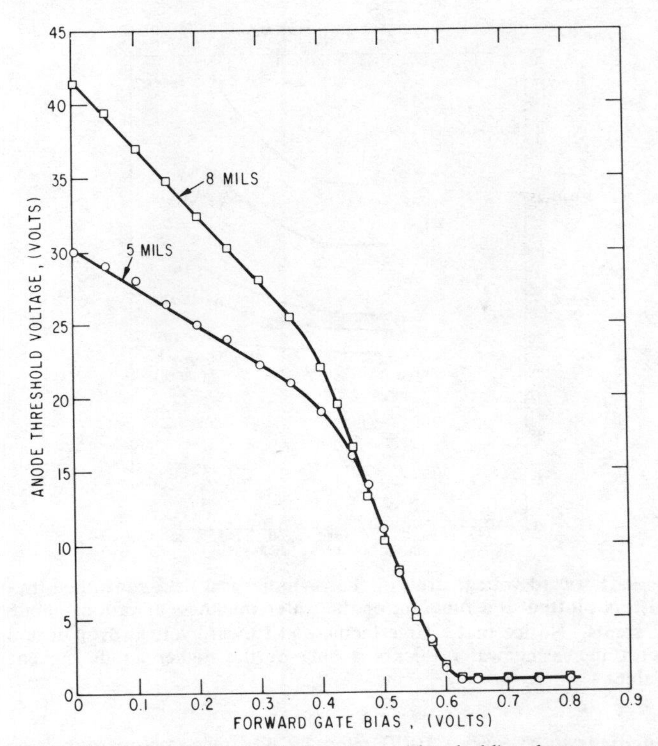

Fig. 7. Variation of the anode breakover (threshold) voltage upon applied forward gate bias for devices fabricated from wafers with a thickness of 5 and 8 mils. Note the change in the rate of variation of the threshold voltage with gate bias above and below 0.4 V.

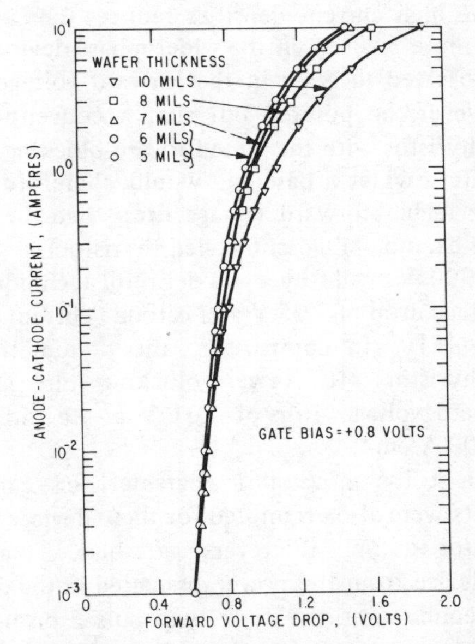

Fig. 8. Forward-conduction characteristics of asymmetrical field-controlled thyristors fabricated from wafers with the thicknesses ranging from 5 to 10 mils. All the characteristics have been measured at a forward gate bias of 0.8 V to ensure complete device turn-on.

low forward voltage drop due to conductivity modulation of the n base by the injected holes and electrons from the anode and cathode regions, respectively. This conductivity modulation decreases as the n base width increases and results in an increase in the forward voltage drop with increasing basewidth as shown in Fig. 9. It can be seen that the effect of increasing the basewidth on the forward voltage drop is small at low current densities, but becomes quite severe at the higher current densities. At a device current of 10 A, the cathode current

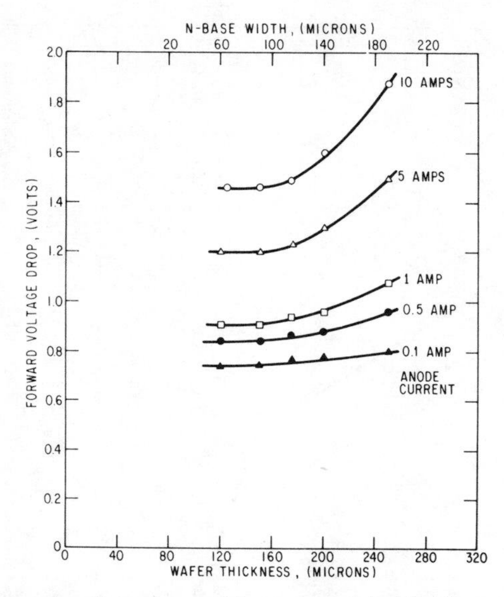

Fig. 9. Forward voltage drop of the asymmetrical field-controlled thyristors plotted as a function of the wafer thickness at various anode currents. Notice that a large increase in forward voltage drop occurs with increasing wafer thickness only at the higher anode current values.

density is in excess of 1000 A/cm^2. At these current densities the effects of carrier–carrier scattering and Auger recombination play a significant role in decreasing the diffusion length of the carriers in the n base. This shortening of the diffusion length at the high current densities reduces the conductivity modulation more severely in the wider n-base devices and produces the observed increase in the forward voltage drop. It should, however, be pointed out that a conventional field-controlled thyristor with the same forward-blocking capability would require a wider n base and would, therefore, exhibit a considerably higher forward voltage drop than the preceding devices. For example, field-controlled thyristors with blocking voltages of 900 V made by epitaxial refill techniques have a forward voltage drop of 1.25 V at a cathode current density of 100 A/cm^2 [6]. In comparison, the asymmetrical field-controlled thyristors with forward blocking voltages of 900 V have a forward voltage drop of 1.01 V at a cathode current density of 100 A/cm^2.

In addition to the above static characteristics, gate turn-off measurements were also attempted for these devices by using a pulse generator to apply the reverse gate bias. To avoid heating up the device from the power dissipated in the device during current conduction, the device was pulsed on using a second pulse generator. A typical set of waveforms is shown in Fig. 10. It is observed that the device can be turned off by the applied gate bias. However, a large gate current flow occurs during turn-off which results in a severe distortion of the gate pulse shape. This large current flow is due to the removal of all the stored charge in the lightly doped n base, as well as the depletion-layer charge created by the reverse gate bias, once the blocking condition is achieved. As discussed in the section which describes the device structure, the gate depletion region extends through the entire lightly doped n base and is, consequently, capable of removing most of the stored charge. In the measurements performed here, the 50-Ω internal impedance of the pulser prevented the rapid removal of this charge and caused the severe distortion of the gate pulse waveform.

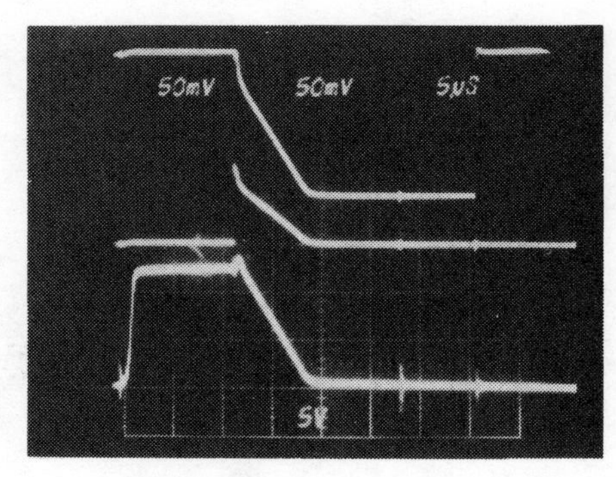

Fig. 10. Gate turn-off waveforms for a typical asymmetrical field-controlled thyristor. The traces from top to bottom in the photograph are in the order of gate voltage at 5 V/cm, gate current at 0.2 A/cm, and anode current at 0.2 A/cm. Note the gradual rise in the gate voltage during gate current flow due to high internal impedance of the gate drive pulser.

In this experimental setup, it was found that the gate turn-off time was independent of the gate pulse height after turn-off because the gate current waveform remained unchanged as the gate bias was increased. Further, it remained the same for the devices with different n-base thickness. Consequently, although the observed turn-off time is about 6 μs, shorter turn-off times can be expected by using a gate pulser with lower internal impedance.

Conclusions

This paper has described a new structure for the field-controlled thyristor. This structure is similar to that of the conventional field-controlled thyristor made by planar diffusion techniques, except for a modified doping profile in the n base of the device. This change in doping profile has been demonstrated to improve the electric-field distribution in the device during the forward-blocking operation, and to result in significant improvements in the blocking gain. The new structure is also designed to improve the tradeoff between the forward-blocking capability and the forward voltage drop in the on-state. This structure should also improve the switching speed of these devices and is capable of forced gate turn-off of the anode current.

Experimental devices made from wafers of various thicknesses cut from an ingot with a resisivity of 2000 Ω · cm show high forward-blocking capability with the highest dc blocking gain as yet reported for devices of this type. In addition, these devices have much lower forward voltage drops during current conduction as compared to the conventional devices. Since the devices have been fabricated by conventional planar diffusion techniques, the processing is simple and produces devices with a high yield. These features of the device can be expected to accelerate their application to power switching circuits.

References

[1] J. I. Nishizawa, T. Terasaki, and J. Shibata, "Field-effect transistor versus analog transistor (static induction transistor)," *IEEE Trans. Electron Devices*, vol. ED-22, pp. 185–197, Apr. 1975.

[2] R. Baradon and P. Laurenceau, "Power bipolar gridistor," *Electron. Lett.*, vol. 12, pp. 486–487, 1976.

[3] B. J. Baliga, "Power field controlled thyristors fabricated using silicon liquid phase epitaxy," presented at the Device Res. Conf., June 1979, Paper WP-B7.

[4] D. P. Lecrosnier and G. P. Pelous, "Ion implanted FET for power applications," *IEEE Trans. Electron Devices*, vol. ED-21, pp.

113–118, Jan. 1974.
[5] D. E. Houston *et al.*, "A field terminated diode," *IEEE Trans. Electron Devices*, vol. ED-23, pp. 905–911, Nov. 1976.
[6] B. W. Wessels and B. J. Baliga, "Vertical channel field-controlled thyristors with high gain and fast switching speeds," *IEEE Trans. Electron Devices*, vol. ED-25, pp. 1261–1265, Oct. 1978.
[7] B. J. Baliga, "Grid depth dependence of the characteristics of vertical channel field controlled thyristors," *Solid-State Electron.*, vol. 22, pp. 237–239, 1979.
[8] S. M. Sze and G. Gibbons, "Avalanche breakdown voltages of abrupt and linearly graded p-n junctions in Ge, Si, GaAs and GaP," *Appl. Phys. Lett.*, vol. 8, pp. 111–113, 1976.
[9] M. S. Adler, "Factors determining the forward voltage drop in the field-terminated diode," *IEEE Trans. Electron Devices*, vol. ED-25, p. 529, May. 1978.
[10] K. Yamaguchi and H. Kodera, "Optimum design of triode-like JFET's by two-dimensional computer simulation," *IEEE Trans. Electron Devices*, vol. ED-24, pp. 1061–69, Aug. 1977.

Electron Irradiation of Field-Controlled Thyristors

B. JAYANT BALIGA, SENIOR MEMBER, IEEE

Abstract—The influence of 3-MeV electron irradiation upon the characteristics of asymmetrical field-controlled thyristors has been examined for fluences of up to 16 Mrad. In addition to the lifetime reduction due to the radiation damage, carrier removal effects have also been observed in the very lightly doped n-base region of these devices. The leakage current, even after radiation at the highest fluence, is not significantly increased and the blocking characteristics of these devices are not degraded. In fact, a small improvement in the blocking gain has been observed at low gate voltages.

The electron irradiation has been found to increase the forward voltage drop during current conduction and to reduce the forced gate turn-off time. Gate turn-off times of less than 500 ns have been achieved by irradiation with a fluence of 16 Mrad. However, this is accompanied by a large increase in the forward voltage drop. Tradeoff curves between the forward voltage drop and the gate turn-off time have been obtained. From these curves, it has been determined that gate turn-off times of 1 μs can be obtained without a significant increase in the forward voltage drop for devices capable of blocking up to 600 V.

Introduction

FIELD-CONTROLLED THYRISTORS (FCT's) represent a recently developed class of semiconductor devices which can be used for power switching applications. These devices have also been called field terminated diodes, static induction thyristors, and bipolar gridistors [1]-[4]. All of these devices consist of a p-i-n rectifier structure into which a gate region has been added in order to control the flow of the anode current during the application of a forward bias on the anode. Within this class of devices, two types of gate structures have been developed, namely, the buried grid structure [3]-[5] and the surface grid structure [1], [2], [6]. Due to the large inherent resistance of the buried grids, devices with this type of structure have inferior forced gate turn-off characteristics when compared with devices made with surface gate structures. Until recently, the buried gate devices have had the advantage of exhibiting higher blocking gains than the surface gate devices. However, the asymmetrical FCT reported in 1980 [6] has a surface gate structure and also exhibits very high blocking gains. A detailed comparison of these device structures can be found in [7].

Although the asymmetrical FCT have been found to exhibit favorable forward-blocking characteristics as well as low forward voltage drops during current conduction at high current densities, the gate turn-off time in these devices has been found to be longer than that observed in the case of devices made using the epitaxial refill process [2], [8]. In this paper, the results of the influence of high energy electron irradiation to improve the gate turn-off characteristics of these devices are reported. The high energy electron irradiation of silicon power devices has been previously used to improve the switching characteristics of rectifiers and conventional thyristors [9], [11]. Electron irradiation of silicon has been found to create many defects in the lattice which include complexes between the impurities in the silicon and the lattice vacancies created by the high energy particle bombardment [12]-[15]. In the case of n-type silicon, it has been determined that one of the divacancy induced deep-levels controls the recombination lifetime without any significant influence of impurity-vacancy complexes [15]. Thus the electron irradiation technique can be generally used to control lifetime in n-base silicon power devices.

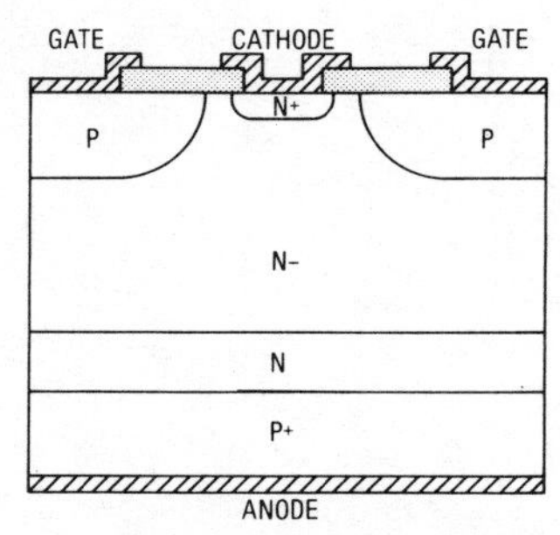

Fig. 1. Cross section of the asymmetrical field-controlled thyristor structure.

Lifetime control in power switching devices has also been done by using gold and platinum diffusion [10], [16]. Although these techniques, as well as electron irradiation, have been found to increase the switching speed, the forward voltage drop and the reverse blocking leakage current of the junctions of the devices also increases. Theoretical analysis, as well as experimental data, indicate that the best tradeoff between the increase in the forward voltage drop and the increase in the switching speed is obtained by using gold diffusion [10], [17]. However, the dominant recombination level for gold lies close to midgap in silicon. This results in a larger increase in the device leakage current as compared with electron irradiation. Electron irradiation also offers many other advantages, such as processing simplicity, improved lifetime control, and more uniform lifetime distribution, when compared with gold or platinum diffusion. For these reasons, in this study the enhancement in the switching performance of the FCT's was achieved by using 3-MeV electron irradiation.

Device Structure and Fabrication

The asymmetrical FCT used in this study had the surface gate structure shown in Fig. 1. These devices were fabricated by starting with 2000 $\Omega \cdot$cm, (111) oriented, n-type, phos-

Manuscript received May 26, 1981.

The author is with General Electric Company, Corporate Research and Development Center, Schenectady, NY 12345.

Reprinted from *IEEE Trans. Electron Devices*, vol. ED-29, pp. 805-811, May 1982.

phorus doped, silicon wafers. The silicon wafer thickness was varied from 6 to 10 mil in order to examine the influence of the n-base width upon the device characteristics. All the wafers were simultaneously processed to obtain identical diffusion profiles. The n-type buffer layer on the bottom of the wafer (anode side) was first obtained by using a phosphorus diffusion with a surface concentration of 2×10^{17} cm^{-3} and a depth of 50 μm. This layer was designed to prevent punchthrough breakdown as discussed in [6]. The oxide on the front surface of the wafers was then patterned and a boron diffusion with a depth of 10 μm was performed to form the gate regions. This diffusion was simultaneously done on the back of the wafers to form the anode region. The cathode n^+ regions were then made by the diffusion of phosphorus through windows opened in the oxide between the gate regions. After opening contact windows, aluminum was evaporated and patterned to define the cathode and gate metallization. The device consisted of 92 cathode fingers interdigitated with gate fingers. The overall dimension of the chip was 120 mil × 120 mil. For a more detailed description of the fabrication of the chip and its surface topography, [6] should be consulted. After device fabrication, the chips were mounted on TO-8 headers and 10-mil gold leads attached to the gate and cathode bonding pads.

Pre-Radiation Device Characteristics

The characteristics of the unradiated FCT devices have been discussed in detail in [6]. In that paper, the effect of the device thickness upon the forward blocking and forward conduction characteristics have been discussed in detail. It was shown that, in addition to the increase in the forward blocking voltage capability, the blocking gain increases with device thickness. It was also found that although the devices with thicknesses of 5 and 6 mil had the same forward voltage drop at up to 10 A, devices fabricated from thicker wafers exhibited an increase in the forward voltage drop with increasing wafer thickness. In the case of the gate turn-off tests, it was also reported that the turn-off speed was limited by the impedence of the gate drive circuit. Typical gate turn-off times ranged from 10 to 15 μs. The electron irradiation was expected to significantly improve this gate turn-off speed.

Electron Irradiation

Electron paramagnetic resonance and lifetime measurements [12], [13] performed using n-base silicon diodes have demonstrated that the electron radiation damage coefficient (K) for silicon increases rapidly when the electron-beam energy is raised from 1 to 2 MeV but then becomes a weak function of the beam energy. All the electron irradiation in this study was, therefore, performed at an electron-beam energy of 3 MeV and the lifetime after irradiation was controlled by varying the fluence received by each device.

The electron irradiation was performed by using devices made from wafers with thicknesses of 6, 8, and 10 mil. In each case, six devices which exhibited identical forward voltage drops and blocking gains before irradiation were selected. Electron irradiation was then performed on five of these devices to a fluence of 1, 2, 4, 8, and 16 Mrad, and one device of each thickness was left unirradiated. All of the device characteristics were then remeasured to quantify the influence of the electron irradiation upon the FCT characteristics.

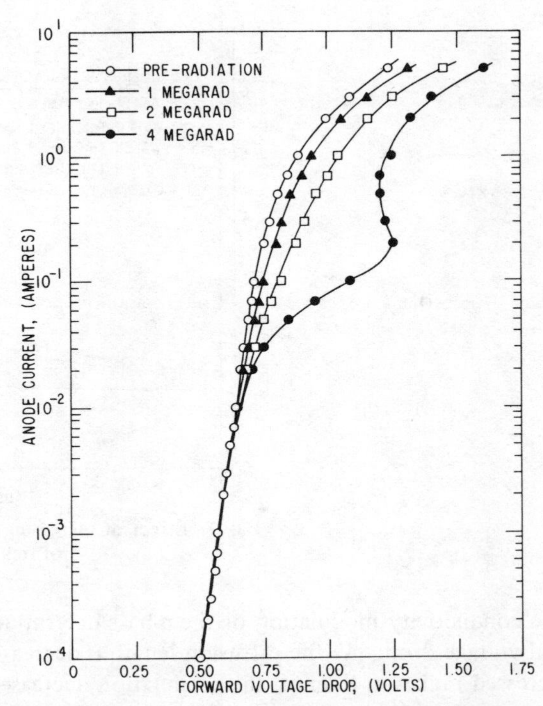

Fig. 2. Forward conduction characteristics of field-controlled thyristors fabricated from wafers with 6-mil thickness. The gate had a forward bias of 0.8 V during these measurements.

Results and Discussion

A. *Forward Conduction Characteristics*

It has been previously demonstrated that when the FCT is operating in the forward conducting mode, its characteristics are similar to those of a p-i-n rectifier [2], [6]. The lifetime in the n-base at high injection levels is, therefore, expected to significantly affect the forward characteristics. Theoretical analysis of the forward characteristics of p-i-n rectifiers has shown that the contribution to the forward voltage drop from the n-base region will increase rapidly when the diffusion length becomes less than half the width of the n-base region [18]. An increase in the forward voltage drop of the FCT devices with increasing radiation fluence is, therefore, to be expected. Further, this effect is expected to be more pronounced in the case of devices with larger base widths.

The forward conduction characteristics of the FCT's fabricated from wafers with a thickness of 6 mil are shown in Fig. 2 for the case of devices irradiated with fluences of up to 4 Mrads. Although the electron irradiation has a negligible influence on the forward characteristics at anode currents of less than 10 mA, it has a strong influence on the forward voltage drop at high current densities. This increase in the forward voltage drop at higher currents with increasing electron fluence is due to the increase in the ratio of the n-base width to the diffusion length. For these FCT, the average current density exceeds 100 A/cm^2 at a forward current of 5 A. At these current densities, high level injection previals in the n-base

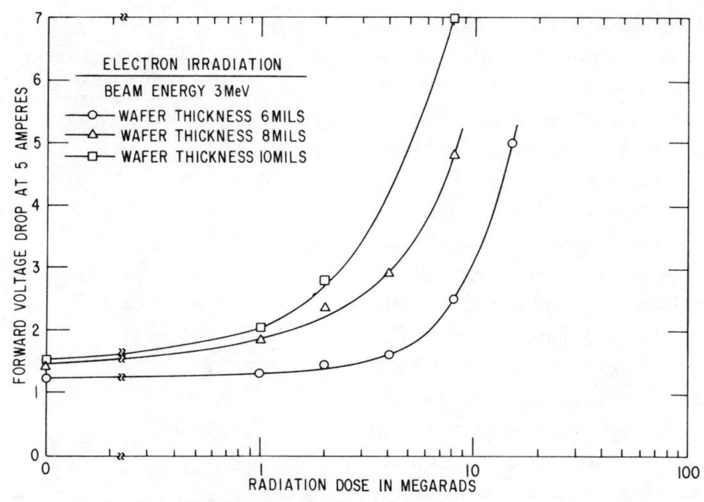

Fig. 3. Effect of increasing radiation dose on the forward voltage drop of field-controlled thyristors.

and the conductivity modulation of the n-base determines the forward voltage drop. As the diffusion length is decreased by the increased radiation fluence, the modulation decreases and causes an increase in the forward voltage drop. The observed breakover in the forward characteristic of the device radiated to a fluence of 4 Mrads is caused by an increase in the silicon resistivity due to carrier removal effects. This is discussed in detail in the next section.

The increase in the forward voltage drop of the FCT with increasing radiation dose can be seen more clearly in Fig. 3. In this figure, the forward voltage drop, measured at an anode current of 5 A, has been plotted as a function of the radiation dose for devices with thicknesses of 6, 8, and 10 mil. For the 6-mil devices, the forward voltage drop shows a sharp increase only when the dose exceeds 8 Mrads. In contrast to this, the devices made from wafers with a thickness of 8 and 10 mil show a rapid increase in the forward drop when the dose exceeds 1 Mrad. This behavior is consistent with p-i-n diode theory since devices with larger base widths will have a larger ratio of the base width to the diffusion length for any given radiation dose. To avoid excessive temperature rise in these devices due to power dissipation during current conduction, the power dissipation must be maintained below 15 W. Based upon this criterion, it can be concluded that the maximum radiation dose must be below 10 Mrad for the 6-mil devices, 4 Mrad for the 8-mil devices, and 2 Mrad for the 10-mil devices. This has an impact on the maximum turn-off speed that can be achieved in these devices as will be discussed later in this paper.

B. *Carrier Removal Effects*

As discussed earlier in the paper, electron irradiation introduces deep levels in the silicon bandgap. In addition to acting as recombination centers, these deep levels also compensate the shallow donors in the n-base. This compensation due to the presence of acceptor-type deep levels leads to the carrier removal effects reported in the literature [19]. As expected, the carrier removal effects are more pronounced for silicon with a higher initial resistivity. Hall-effect measurements on high purity n-type silicon [20] indicate that the Fermi level moves from above midgap towards the divacancy induced deep level at (E_v + 0.39 eV) with increasing radiation fluence. Thus as the fluence increases, the high resistivity n-type silicon will convert to p-type silicon.

The asymmetrical FCT used in this study were fabricated by starting with 2000 $\Omega \cdot$ cm, n-type, float zone silicon. The conversion of the high resistivity, n-base region of these devices to p-type due to the electron irradiation was consequently of significant concern since it would have a strong influence on the device characteristics. The reverse blocking characteristics of the gate junction were, therefore, examined as a function of the fluence to monitor the carrier removal effects. This was done by examining the breakdown and leakage characteristics between the gate and the cathode, as well as the gate and the anode. In addition, the reverse bias capacitance-voltage characteristic of the gate junction was measured in order to determine if any change in the background doping level was taking place.

Measurements of the reverse bias gate characteristics of all the devices as a function of the fluence indicated that the n-base region of the FCT devices was not converted to p-type even at the highest radiation fluence of 16 Mrad used in this study. The capacitance-voltage characteristics, however, did reveal significant carrier removal effects at the higher fluences. The reverse biased gate-to-anode C–V curves of devices made from 8-mil thick wafers are shown in Fig. 4. It can be seen that very little change in the background doping was observed at fluences of up to 4 Mrad. At the fluence of 8 and 16 Mrad, a decrease in the zero bias capacitance was observed and the capacitance was observed to decrease at a smaller rate with increasing gate–anode potential. An accurate measurement of the change in the background doping cannot be obtained from these data due to the graded doping profile of the gate diffusion which causes the gate depletion layer to extend on both sides of the metallurgical junction. However, it can be estimated that the doping concentration in the n-base has been decreased by about 35 percent at a fluence of 8 Mrad and by about 80 percent at a fluence of 16 Mrad. These carrier re-

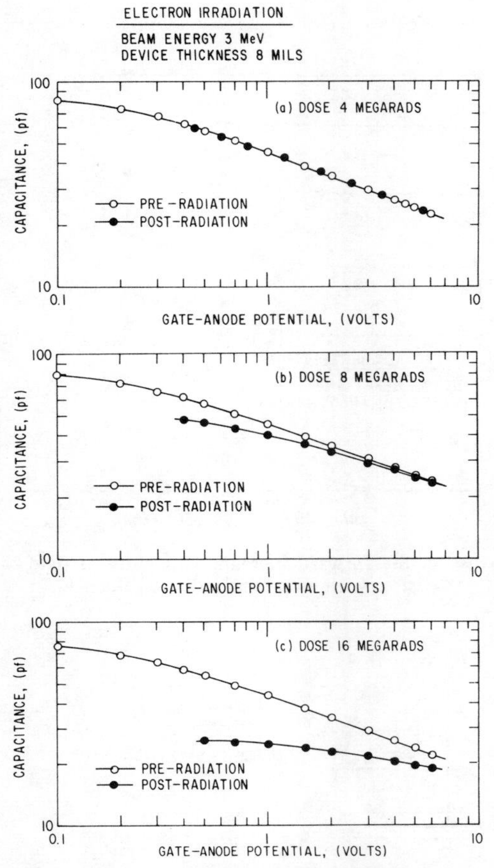

Fig. 4. Gate-to-anode capacitance measured as a function of the reverse bias voltage for devices made from 8-mil thick wafers.

moval effects increase the gate depletion-layer width for any given gate bias voltage. This increase in the gate depletion width lowers the channel pinchoff voltage in these devices. It also results in an increase in the channel potential-barrier height especially at the smaller gate-to-cathode voltages. The influence of this will be discussed in the section on blocking characteristics.

C. Junction Leakage Current

Earlier studies of the effect of electron irradiation upon the reverse bias leakage current of rectifiers [10] have shown that the introduction of deep levels due to electron irradiation produces an increase in the generation current in the junction depletion layer. Although this increase is not as severe as in the case of gold doping, it is an important parameter which can limit the maximum radiation fluence. The reverse bias leakage current between the gate and the cathode, as well as between the gate and the anode was consequently measured as a function of the fluence.

In the FCT it can be seen from Fig. 1 that a p-i-n diode exists between the gate and the cathode. The leakage current at room temperature is determined by the space-charge generation of carriers in the depletion layer of this p-i-n diode. It has been previously shown that the diffusion current component of the leakage current is negligible at room temperature and becomes of importance only when the ambient temperature exceeds 150°C [20]. The leakage current of an FCT

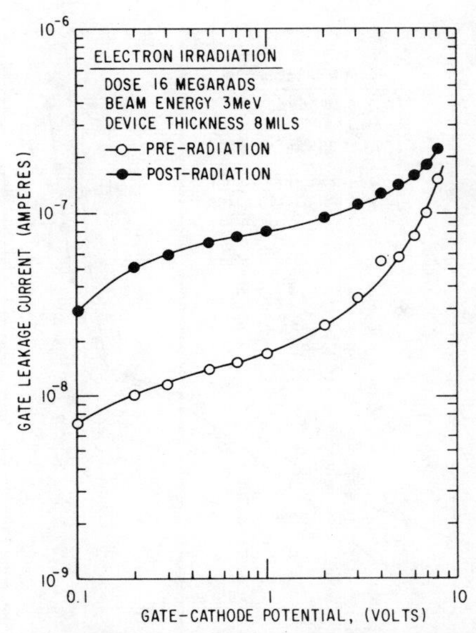

Fig. 5. Gate-to-cathode leakage current of a field-controlled thyristor before and after electron irradiation to a fluence of 16 Mrad.

device with an 8-mil thickness is shown as a function of the reverse bias voltage applied to the gate before and after irradiation to a fluence of 16 Mrad is shown in Fig. 5. An increase in the leakage current by about a factor of five is observed at the highest fluence used in this study. However, even after irradiation up to a fluence of 16 Mrad, the gate-to-cathode leakage current is well below 1 μA. Consequently, the observed increase in the leakage current does not impose any limitation upon the radiation fluence within the range used in this study.

In the case of the gate-to-anode leakage current, it can be seen from Fig. 1 that the FCT structure contains a parasitic p-n-p transistor. The space-charge generation current of the reverse biased gate junction is amplified by the current gain of this open-base bipolar transistor. As the reverse bias increases, the width of the undepleted base decreases and the base transport factor increases. This effect is significant for the asymmetrical FCT devices used in this study due to the very low doping in the n-base near the gate which results in the complete depletion of the lightly doped portion of the n-base when the reverse bias exceeds 200 V. The variation of the gate-to-anode leakage current of a typical FCT device as a function of the reverse bias potential is shown in Fig. 6 both before and after irradiation. A sharp increase in the leakage current is observed in both cases when the voltage exceeds 200 V. It can be seen that the electron irradiation causes an increase in the leakage current at the low anode voltages. This increase in the leakage current arises from the increase in the space-charge generation current due to the introduction of the deep levels by the electron irradiation. At anode voltages in excess of 200 V, the leakage current after irradiation is found to be less than that prior to irradiation. This decrease in the leakage current is caused by the decrease in the minority-carrier diffusion length by the irradiation which leads to a reduction in the gain of the open-base bipolar transistor. Thus the gate-to-anode leakage current can be either increased or reduced by the radiation depending upon the reverse bias voltage and the fluence.

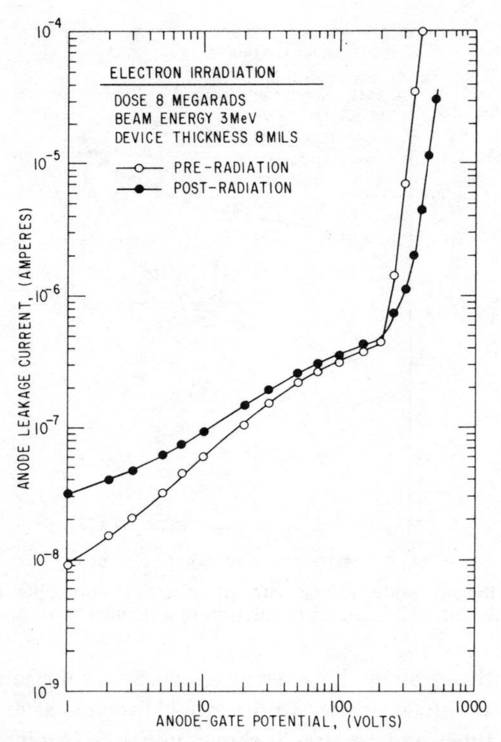

Fig. 6. Gate-to-anode leakage current of a field-controlled thyristor before and after electron irradiation to a fluence of 8 Mrad.

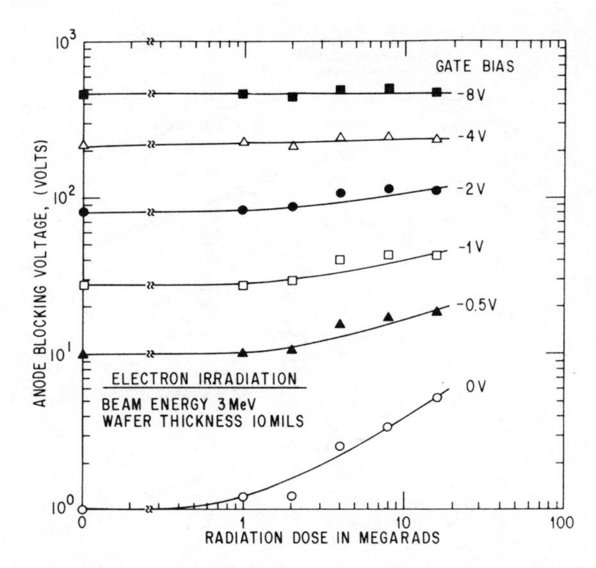

Fig. 7. Increase in the forward blocking capability of field-controlled thyristors with increasing radiation fluence.

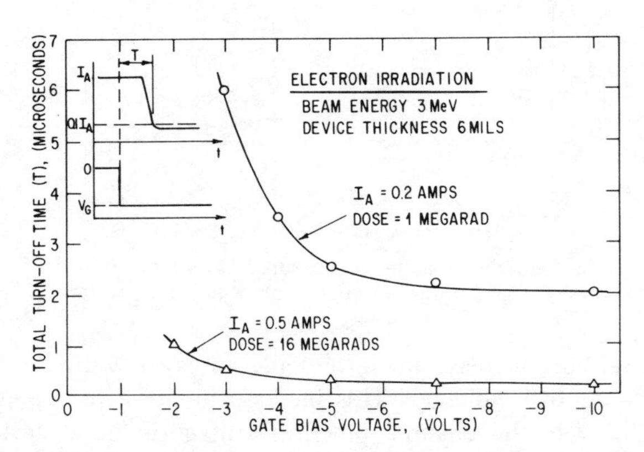

Fig. 8. Dependence of the gate turn-off time of field-controlled thyristors upon the gate reverse bias voltage after irradiation to fluences of 1 and 16 Mrad.

D. Forward Blocking Characteristics

Having established the integrity of the gate junction after the radiation, the forward blocking characteristics of the FCT were measured by reverse biasing the gate-to-cathode junction and measuring the anode voltage at which the anode current increases to 100 μA. As discussed in an earlier section, the carrier removal due to the radiation reduces the gate voltage required to establish a potential barrier in the channel. As a result of this, the blocking gain will be increased by the radiation. This has been found to be prominent at the low gate voltages. From Fig. 7, it can be seen that, for a given gate bias voltage, the anode current can be blocked at larger anode voltages for larger radiation fluences. However, at high anode voltages the change in the blocking voltage capability is negligible. It can be observed that the anode blocking voltage increases by a factor of 5 after electron irradiation to a fluence of 16 Mrad for the zero gate bias case. However, at larger gate voltages, there is a negligible increase in the anode blocking voltage capability. These data demonstrate that electron irradiation of FCT tends to enhance their forward blocking capability.

E. Gate Turn-Off Characteristics

As discussed in the introduction, the asymmetrical FCT's were found to exhibit gate turn-off times of about 15 μs [6]. The primary intention for performing the electron irradiation was to reduce the turn-off time. Improvement in the turn-off time of other power devices has been successfully achieved by using electron irradiation [9]-[11]. The introduction of recombination centers by the electron irradiation was also expected to improve the turn-off speed of the FCT.

The gate turn-off time of FCT's irradiated to a fluence of 1 and 16 Mrad is shown in Fig. 8. Note that this turn-off time includes the storage and fall times. In the case of the device irradiated to a fluence of 16 Mrad, gate turn-off times of less than 250 ns were observed for an anode current of 0.5 A. In comparison, after irradiation to a fluence of 1 Mrad, the gate turn-off time was found to be about 2 μs even at a reduced anode current of 0.2 A.

The improvement in the gate turn-off speed with increasing radiation fluence, for an anode current of 1 A, can be more clearly observed from Fig. 9. In this figure, both the total turn-off time and the fall time have been plotted as a function of the radiation fluence. It can be seen that both the fall time and total turn-off time decrease with increasing radiation fluence. Although the storage time has not been plotted in Fig. 9, from the data it can be deduced that the storage time is also reduced by the electron irradiation. From these data it can, therefore, be concluded that high-speed forced gate turn-off can be achieved by using electron irradiation to reduce the lifetime.

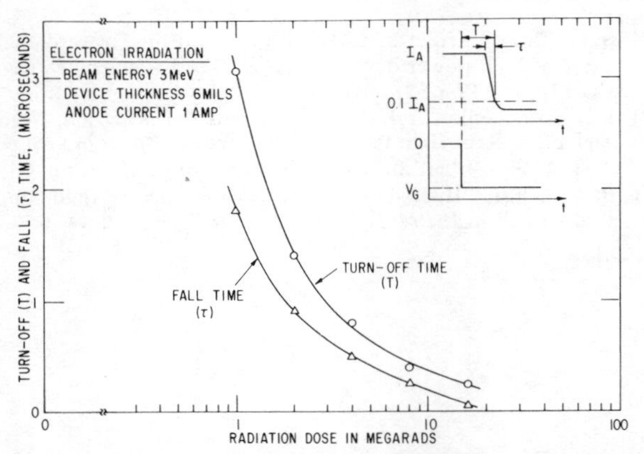

Fig. 9. Improvement in the gate turn-off speed of field-controlled thyristors with increasing radiation fluence.

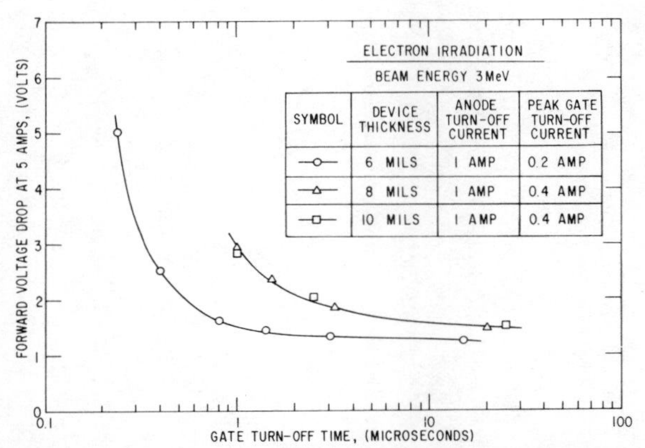

Fig. 10. Tradeoff curves of forward voltage drop versus gate turn-off time for field-controlled thyristors fabricated from wafers with thicknesses of 6, 8, and 10 mil.

F. Forward Drop Turn-Off Time Tradeoff Considerations

The data discussed thus far show that although the gate turn-off time decreases with increasing radiation fluence, this is accompanied by an increase in the forward voltage drop. In choosing the electron irradiation fluence it is, consequently, necessary to tradeoff these device parameters. In order to simplify the tradeoff process, the forward voltage drop measured at a forward current of 5 A has been plotted as a function of the gate turn-off time in Fig. 10. From this plot it can be concluded that, for the devices fabricated from wafers of 8 and 10 mil in thickness, gate turn-off times of between 2 and 3 μs can be achieved without a large increase in the forward voltage drop. In comparison, for devices fabricated from wafers with a thickness of 6 mil, gate turn-off times of less than 1 μs can be achieved without a large increase in the forward voltage drop. It is also worth noting that these devices require a lower peak gate current during turn-off. This simplifies the design of the gate turn-off circuitry and reduces the gate drive power requirements.

Conclusions

The influence of 3-MeV electron irradiation upon the characteristics of FCT has been examined. Although carrier removal, due to the introduction of compensating deep levels by the radiation damage, has been observed it does not produce any adverse effect on the device characteristics. In fact, a slight improvement in the anode blocking capability is observed especially at low gate bias voltages. The deep levels introduced by the electron irradiation do cause the expected increase in the leakage current due to additional space-charge generation. However, this increase is small and does not significantly degrade the blocking characteristics of the devices.

The strongest effect of the electron irradiation has been found to be upon the forward voltage drop and the gate turn-off time. The forward voltage drop increases with increasing radiation fluence while the turn-off time decreases with increasing radiation fluence. Tradeoff curves between these important device parameters have been obtained. These curves show that turn-off times of less than 1 μs can be achieved without a significant increase in the forward voltage drop for devices fabricated from wafers with a thickness of 6 mil. These devices are capable of a forward blocking capability of over 600 V. In the case of higher voltage (up to 1000 V) devices fabricated from wafers with thicknesses of 8 and 10 mil, the gate turn-off time is limited to between 2 and 3 μs unless a large increase in the forward voltage drop can be tolerated.

Acknowledgment

The author wishes to acknowledge the assistance received from Dr. R. O. Carlson and Dr. P. Campbell in performing the electron irradiation, and from Ms. N. Waldron in the preparation of the manuscript.

References

[1] D. E. Houston *et al.*, "A field terminated diode," *IEEE Trans. Electron Devices*, vol. ED-23, pp. 905–911, 1976.

[2] B. W. Wessels and B. J. Baliga, "Vertical channel field-controlled thyristors with high gain and fast switching speeds," *IEEE Trans. Electron Devices*, vol. ED-25, pp. 1261–1265, 1978.

[3] J. I. Nishizawa, T. Tersaki, and J. Shibata, "Field effect transistor versus analog transistor (static induction transistor)," *IEEE Trans. Electron Devices*, vol. ED-22, pp. 185–197, 1975.

[4] R. Baradon and P. Laurenceau, "Power bipolar gridistor," *Electron Lett.*, vol. 12, pp. 486–487, 1976.

[5] B. J. Baliga, "Buried grid field-controlled thyristors fabricated using silicon liquid phase epitaxy," *IEEE Trans. Electron Devices*, vol. ED-27, pp. 2141–2145, 1980.

[6] —, "The asymmetrical field-controlled thyristor," *IEEE Trans. Electron Devices*, vol. ED-27, pp. 1262–1268, 1980.

[7] —, *Silicon Power Field Controlled Devices and Integrated Circuits*, D. Kahng, Ed. New York: Academic Press, (Applied Solid-State Science Series, Supplement 2B), 1981.

[8] —, "Grid depth dependence of the characteristics of vertical channel field controlled thyristors," *Solid-State Electron.*, vol. 22, pp. 237–239, 1979.

[9] K. S. Tarneja and J. E. Johnson, "Tailoring the recovered charge in power diodes using 2 MeV electron irradiation," in *Electrochem. Soc. Meet.*, paper 261RNP, 1975.

[10] B. J. Baliga and E. Sun, "Comparison of gold, platinum, and electron irradiation for controlling lifetime in power rectifiers," *IEEE Trans. Electron Devices*, vol. ED-24, pp. 685–688, 1977.

[11] R. O. Carlson, Y. S. Sun, and H. B. Assalit, "Lifetime control in silicon power devices by electron or gamma irradiation," *IEEE Trans. Electron Devices*, vol. ED-24, pp. 1103–1108, 1977.

[12] A. O. Evwaraye and E. Sun, "Electron irradiation induced divacancy in lightly doped silicon," *J. Appl. Phys.*, vol. 47, pp. 3776–3780, 1976.

[13] G. D. Watkins and J. W. Corbelt, "Defects in irradiated silicon," *Phys. Rev.*, vol. 134, pp. A1359–A1377, 1964.

[14] A. O. Evwaraye and B. J. Baliga, "Use of electron irradiation for lifetime control in p-base silicon devices," in *151st Electrochem. Soc. Meet.*, paper 65, pp. 187–189, 1977.

[15] A. O. Evwaraye and B. J. Baliga, "The dominant recombination centers in electron irradiated semiconductor devices," *J. Electrochem. Soc.*, vol. 124, pp. 913–916, 1977.

[16] M. D. Miller, "Limitations of the use of the platinum in power devices," in *IEDM Tech. Dig.*, abstract 20.3, pp. 491–494, 1976.

[17] B. J. Baliga and S. Krishna, "Optimization of recombination levels and their capture cross-sections in power rectifiers and thyristors," *Solid-State Electron.*, vol. 20, pp. 225–232, 1977.

[18] S. C. Choo, "Effect of carrier lifetime on the forward characteristics of high power devices," *IEEE Trans. Electron Devices*, vol. ED-17, pp. 647–652, 1970.

[19] I. D. Konozenko, A. K. Semenyuk, and V. I. Khivrich, "Radiation Defects in Si of High Purity," in *Radiation Effects in Semiconductors*, J. W. Corbelt and G. D. Watkins, Eds., 1971, pp. 249–255.

[20] B. J. Baliga, "High temperature performance of field-controlled thyristors," in *IEDM Tech. Dig.*, paper 25.5, pp. 654–657, 1980.

HIGH GAIN POWER SWITCHING USING FIELD CONTROLLED THYRISTORS

B. JAYANT BALIGA
General Electric Company, Corporate Research and Development Center, Schenectady, NY 12301, U.S.A.

(*Received* 26 *May* 1981; *in revised form* 6 *July* 1981)

Abstract—A technique for high gain power switching using field controlled thyristors is described. This technique uses a MOSFET connected in series with the FCT to control the current flow. The circuit exhibits normally-off behavior and is capable of operation at high voltages. The current through the FCT can be turned on and off by the application of a low voltage gate signal to the MOSFET. Turn-on and turn-off times of less than 1 μs have been observed at a current gain of over 30. The new gating technique offers the advantage of the large operating current density of the FCT even at high breakdown voltages and the high input impedance of the MOS gate used to trigger the device during power switching.

INTRODUCTION

Field Controlled Thyristors are recently developed power semiconductor devices which can be used for power switching applications. Like conventional thyristors, these devices can block current flow for both polarities of applied anode voltage and can also conduct forward current at a high current density with a low forward voltage drop. These devices have also been shown to exhibit gate turn-off capability with turn-off times of less than one microsecond.

The basic field controlled thyristor (FCT) structure is shown in Fig. 1. The device consists of a *p–i–n* rectifier structure in which a grid region is incorporated in order to control the forward current flow. In the absence of a gate bias, the device operates like a *p–i–n* rectifier between the anode and cathode terminals. Consequently, with negative voltages applied to the anode, the device blocks current flow across the reverse biased anode-base junction ($J1$) producing the reverse blocking characteristics. With positive voltages applied to the anode, the *p–n* junction $J1$ becomes forward biased and the device conducts current at a high current density due to the modulation of the conductivity of the *n*-base by the carriers injected from the anode and the cathode regions. In order to control the power delivered to a load it is also desireable to be able to block current flow for positive applied anode voltages. To obtain these forward blocking characteristics in this device, it is necessary to apply a negative bias on the gate. This reverse biases the gate junction ($J2$) and causes its depletion layer to extend under the cathode. When the depletion layers of adjacent gate regions punch-through under the cathode, a potential barrier is formed between the anode and the cathode.[1] This potential barrier prevents the injection of electrons from the cathode to the anode and, thus, allows the device to block current flow. However, as the anode voltage increases, the potential barrier height decreases. This causes anode current flow to commence when the anode voltage is increased beyond a certain value for each applied gate bias voltage. The ratio of this anode voltage to the applied gate bias is defined as the blocking gain of the device. Thus, the field controlled thyristor has a normally-on characteristic and requires the application of a gate voltage to maintain it in the off-state.

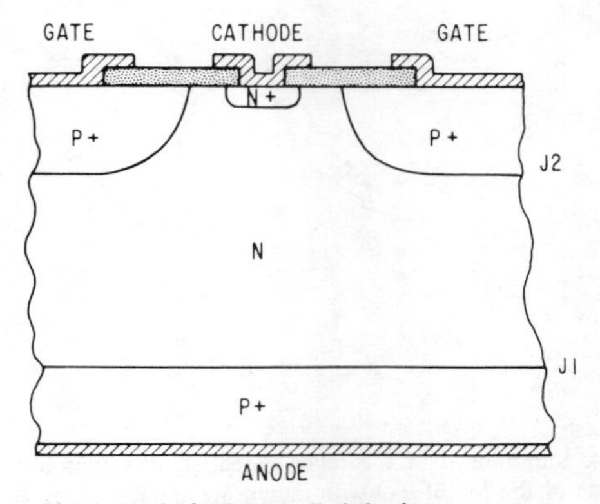

Fig. 1. Basic field controlled thyristor structure.

These devices can also be switched rapidly from the conducting mode to the forward blocking mode (forced gate turn-off) by the application of a negative gate voltage while the anode current is flowing. During gate turn-off, sufficient gate current must be supplied by the gate drive circuit so as to remove the minority carrier stored charge in the *n*-base and to allow the gate depletion layer to extend under the cathode to pinch off the anode current flow. It has been found that turn-off times of less than one-microsecond can be achieved when the peak gate turn-off current is comparable to the anode current.[2]

The conventional circuit[3] that has been used to operate field controlled thyristors is shown in Fig. 2. In this circuit the current supplied to the load by the power supply (V_L) can be controlled by gating the field controlled thyristor using switch $S1$. When switch $S1$ is open, the FCT is in its on-state and current is supplied to the load. When switch $S1$ is closed, the gate voltage (V_G)

Reprinted with permission from ***Solid-State Electron.***, vol. 25, no. 5, pp. 345–351 and 353, May 1982.

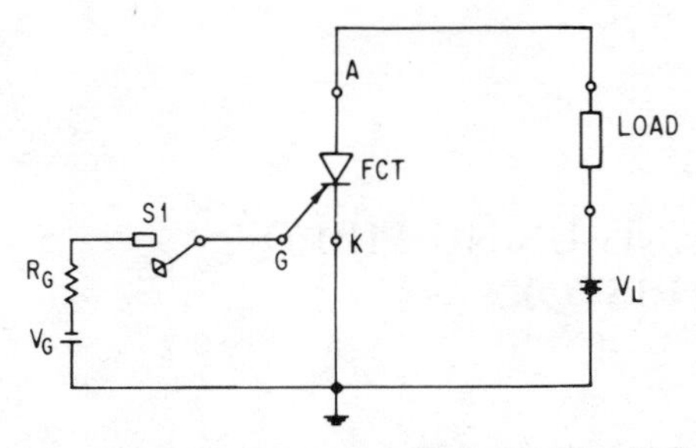

Fig. 2. Conventional circuit used for power switching applications using a field controlled thyristor.

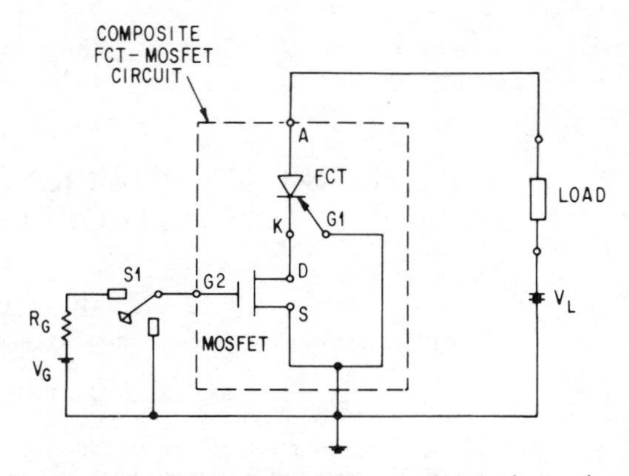

Fig. 3. New gating circuit for field controlled thyristor using a MOSFET.

is used to maintain the FCT in its forward blocking mode. The turn-off speed of the FCT is controlled by the peak gate current during turn-off. This current can be controlled by the gate resistance (R_G).

Several problems have been encountered with the use of this gating technique. Firstlv, the devices are normally-on in the absence of a gate bias voltage and the circuit cannot ensure fail-safe start up and operation. Secondly, the devices require a substantial gate voltage (V_G) in order to operate them at large forward blocking voltages. This problem has been partially overcome by improvements in the device structure which have allowed the development of devices with high blocking gains. For example, the devices reported in 1976[3] required gate bias voltages of over 150 V in order to block anode current flow at an anode voltage of 600 V. In contrast, the more recently developed device structures reported in 1979[4] and 1980[5] are capable of blocking current flow at anode voltages as high as 1000 V with gate bias voltages of less than 30 V. In spite of these improvements in the blocking gain, it is worth noting that some gate bias voltage is necessary to maintain these devices in their forward blocking mode. Thirdly, these devices require substantial gate drive currents to switch them from the on-state to the blocking state. Thus, although submicrosecond turn-off times have been observed[2, 4], gate turn-off current gains of less than 5 are typically necessary to achieve these high turn-off speeds. These drawbacks of the field controlled thyristors have been primarily responsible for its limited application to power switching applications.

NEW GATING TECHNIQUE

The problems discussed above in the conventional gating technique for field controlled thyristors can be solved by using the new gating technique discussed here. The new gating circuit is shown in Fig. 3. In this circuit, an n-channel, normally-off, MOS gated field effect transistor is used to control the current flow through the field controlled thyristor. To accomplish this, the drain (D) of the MOSFET is connected to the cathode (K) of the FCT and the source (S) of the MOSFET is connected to the gate ($G1$) of the FCT. The current conduction through the load is then controlled by the applied bias (V_G) to the gate ($G2$) of the MOSFET.

If the load voltage (V_L) is applied with the switch $S1$ connecting the gate ($G2$) of the MOSFET to the ground potential, the MOSFET remains in its off-state and does not allow current flow between its drain and source terminals unless the MOSFET breakdown voltage is exceeded. At the same time, the application of the positive voltage (V_L) to the anode of the FCT reverse biases its gate junction (junction $J2$ in Fig. 1). The depletion layer of this junction then spreads under the cathode of the FCT and establishes a potential barrier between the anode and the cathode. This process shields the cathode potential from the anode potential. This can be clearly observed in Fig. 4 which shows a plot of the measured FCT gate-to-cathode potential (also the drain-to-source potential of the MOSFET) as a function of the applied anode voltage. These measurements were obtained by using an asymmetrical field controlled thyristor described in Ref. [5]. It can be observed that the drain voltage of the MOSFET remains at less than 50 V for anode voltages of over 600 V. This feature is extremely important for the operation of the circuit because it allows the use of a low breakdown voltage MOSFET device for controlling the load current. As will be discussed below, in the circuit shown in Fig. 3, the MOSFET is required to carry the load current during the conducting phase. Consequently, the MOSFET used in this circuit must

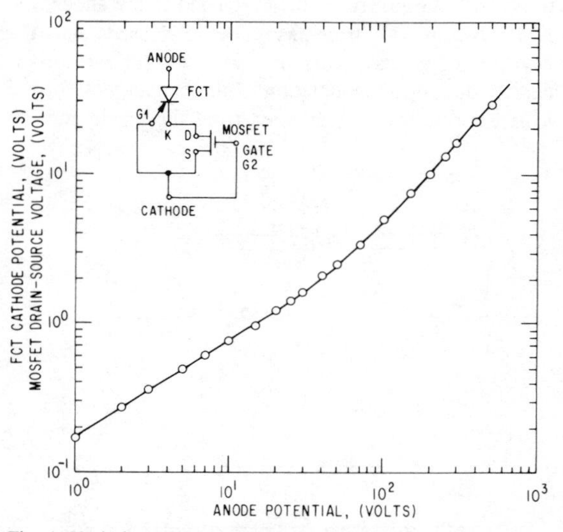

Fig. 4. Variation of FCT cathode potential (also drain-to-source voltage of the MOSFET) with increasing anode voltage applied to the composite circuit.

have a low on-resistance in order to maintain a small forward voltage drop in the device and to minimize the power dissipation. It has been shown that the on-resistance per unit area for these unipolar devices will increase as a 2.5 power of the breakdown voltage[6]. Thus, the ability to use a low breakdown voltage MOSFET in the circuit due to the shielding of the cathode potential is extremely important for maintaining a low device forward drop and a low power dissipation. With the gate ($G2$) connected to the ground potential, the new gating circuit then maintains the FCT in a current blocking condition. This gating circuit, thus operates the FCT device in a normally-off mode without the need for a gate bias voltage and thus assures fail-safe start up and operation.

In order to deliver current to the load, the switch $S1$ is used to connect the gate ($G2$) to the positive gate power supply (V_G). This positive gate bias (V_G) must exceed the threshold voltage of the MOSFET in order to switch it to the conducting state. When the MOSFET is in its conducting state, the load current can now flow through the FCT from the anode to the cathode and then via the drain of the MOSFET to the ground terminal. Thus, as mentioned earlier, in this circuit the full load current flows through the MOSFET. As a result, the on-resistance of this device must be small to minimize the forward voltage drop and the power dissipation. Since the forward voltage drop of the FCT is typically about 1.5 V it is desireable to keep the forward voltage drop of the MOSFET below 0.5 V. This can be readily accomplished because the MOSFET used in this gating circuit can have a low breakdown voltage.

The gating circuit shown in Fig. 3 also has the advantage of achieving forced gate turn-off of the anode current with a very high gate turn-off current gain. In order to turn off the load current, the switch $S1$ is used to connect the gate $G2$ of the MOSFET to the ground potential. When this is done, the MOSFET switches from its conducting state to its blocking state. In order to accomplish this, a displacement current must flow to discharge the input gate capacitance of the MOSFET. Once the MOSFET turns-off, its drain potential rises. This produces a reverse bias on the gate junction ($J2$) of the FCT. The minority carriers stored in the n-base of the FCT are then removed via the gate until the FCT device supports the load voltage (V_L). The current flowing out of the gate ($G1$) is equal to the anode current of the FCT until the gate junction ($J2$) begins to support the anode voltage (V_L). This process is similar to the reverse recovery process in a p–i–n rectifier[7]. Thus, the removal of the stored charge in the FCT occurs under conditions similar to unity current gain turn-off for the conventional circuit shown in Fig. 2. As discussed in the introduction, this has been observed to ensure very rapid turn-off in the FCT. However, in the new gating circuit shown in Fig. 3, the gate drive current required during turn-off is determined by the displacement current required for discharging the gate capacitance of the MOSFET. Since this gate current can be made small, the new gate circuit offers very high gate turn-off current gains at very fast turn-off speeds.

EXPERIMENTAL RESULTS

In order to obtain experimental verification of the operating characteristics of the new gating circuit, 500 V asymmetrical field controlled thyristors were used to switch current delivered to a resistive load. With the gate shorted to the cathode terminal, these field controlled thyristors exhibit a forward voltage drop of less than 2 V at an anode current of 1 A. The forward characteristic of a typical FCT obtained on a curve tracer is shown in Fig. 5. The breakover in the characteristics of these devices at low anode currents is due to the potential barrier created in the channel by the gate junction diffusion potential as discussed in Ref. [5].

The gating of the FCT was done by using a 90 V MOSFET connected to the FCT terminals as shown in Fig. 3. The characteristics of this n-channel MOSFET are shown in Fig. 6. This device has an on-resistance of less than 0.2 Ω. The threshold voltage for turn-on of the device was about 3 V. However, a gate bias voltage of more than 6 V was necessary to ensure a forward voltage drop of less than 0.2 V at a drain current of 1 A.

The FCT and MOSFET devices were connected together as shown in Fig. 3 within the dashed lines and the characteristics of the composite circuit were first examined on a curve tracer. Figure 7 shows the characteristics obtained by using the same bias voltage to drive the MOSFET as used in Fig. 6 to display the MOSFET characteristics. As expected, the composite circuit exhibits characteristics obtained by the addition of the forward voltage drops of the FCT and MOSFET at any given forward current through these devices connected in series.

The power switching capability of the composite circuit was then examined by using a resistive load. To avoid excessive temperature rise in the devices, a pulser was used to derive the load current with a duty cycle of less than 10%. The MOSFET gate drive was also obtained by using a pulser to supply the gate voltage. The gate pulser had an internal impedance of 50 Ω which controlled the peak input gate current to the MOSFET. Typical device current and voltage waveforms are shown in Fig. 8. In this figure, the anode voltage is applied at time t_1 with the gate at zero bias. No current flow is observed through the load due to the normally-off behavior of the composite circuit. At time t_2, a gate voltage is applied to the MOSFET. This results in a pulse of gate current with a peak amplitude of 0.06 A and a duration of about 1 μs. This gate current is used to charge the input capacitance of the MOSFET. As the MOSFET turns on, the anode current rises to 1 A and the anode voltage falls to less than 2 V (the forward voltage drop of the composite circuit). The total turn-on time, from the application of the gate voltage to the rise of the anode current to 90% of the steady state value, is less than 1 μs. Faster turn-on can be achieved by using larger gate input drive currents to charge the MOSFET input capacitance at a faster rate. However, this will reduce the current gain of the composite circuit. The turn-on time has also been found to be dependent upon the anode current as shown in Fig. 9. It can be seen that the increase in the turn-on time with increasing anode

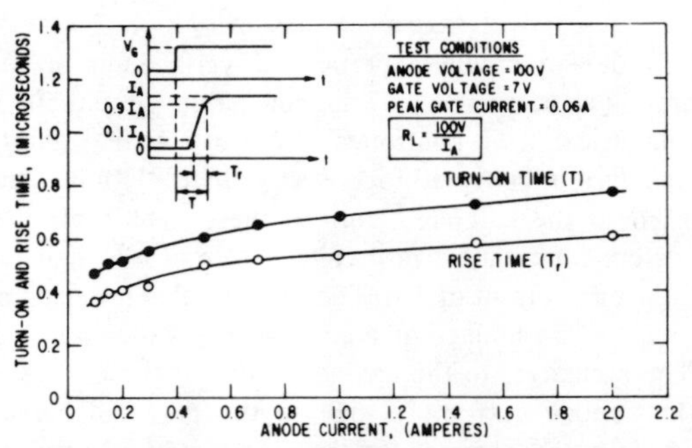

Fig. 9. Dependence of the turn-on and rise times of the composite FCT-MOSFET circuit upon the anode current. In this test, the anode supply voltage has been kept constant at 100 V and the anode current varied by changing the series load resistance. The peak gate current does not vary with anode current or voltage.

current arises from an increase in the rise time (T_r). The delay time is determined by the time taken for the MOSFET gate voltage to reach its threshold voltage.

After the anode current has reached its steady state value of 1 A, the gate voltage is turned off at time t_3 in Fig. 8. In the test circuit, this is done by bringing the gate pulser voltage to zero, and results in a reverse gate current flow limited by the 50 Ω pulser impedance. This gate current discharges the MOSFET gate capacitance. As soon as the MOSFET turns-off, the anode current begins to fall due to removal of the minority carriers from the gate of the FCT as described earlier. The anode current turn-off transient can be seen more clearly in Fig. 10. The total turn-off time for the anode current, from the end of the application of the positive gate pulse to the decrease of the anode current to 10% of its steady state value, is less than 1 μs. The dependence of the turn-off and fall times upon the anode current is shown in Fig. 11. It can be seen that the turn-off time increases with decreasing anode current. This increase in turn-off time occurs due to an increase in the fall time with decreasing anode current. Under the test conditions used to obtain the data shown in Fig. 11, the decrease in the anode current was achieved by increasing the load resistance (R_L). This increases the *RC* charging time constant of anode capacitance of the FCT which causes the observed increase in the turn-off time at the lower anode currents.

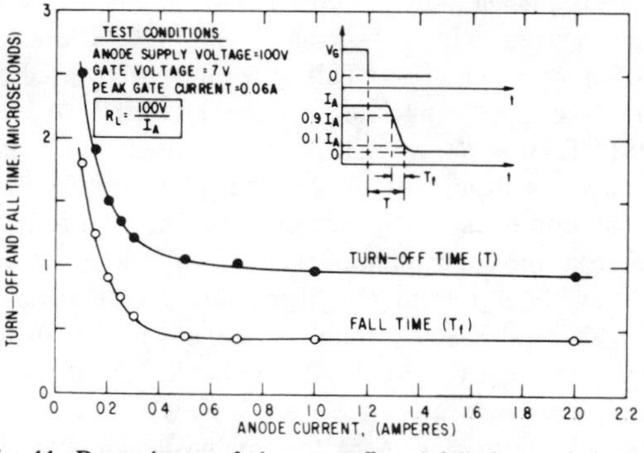

Fig. 11. Dependence of the turn-off and fall times of the composite FCT-MOSFET circuit upon the anode current. In this test, the anode supply voltage has been kept constant at 100 V and the anode current varied by changing the series load resistance.

In all the tests performed using a resistive load it was found that the peak gate current was only a function of the gate voltage and not dependent upon the anode current or voltage. The tests performed here were limited to a maximum anode current of 2 A due to the high impedance of the pulser (50 Ω). At this current, the composite circuit exhibits turn-on and turn-off times of less than one microsecond at a current gain (anode current to peak gate current ratio) of over 30. This current gain, especially for gate turn-off conditions, is much superior to that achievable by using the conventional circuit for gating the FCT. The current gain is also superior to that achievable using bipolar transistors or GTOs. It is worth noting that since the gate current is independent of the anode current and voltage, and since the FCT and MOSFET used here in the test circuit can handle up to 10 A of anode current[5], gate turn-off current gains of over 100 are anticipated. This feature will greatly simplify the gate circuitry that is required to drive the composite circuit shown in Fig. 3 in power switching applications. Results of tests currently being performed under inductive load conditions will be reported in the future.

CONCLUSIONS

A new gating technique has been introduced for power switching applications using field controlled thyristors. This circuit utilizes a low breakdown voltage, high current MOSFET to control current flow through a high breakdown voltage, high current FCT. In the circuit, the drain of the MOSFET is connected to the cathode of the FCT, and the source of the MOSFET is connected to the gate of the FCT. The circuit exhibits normally-off characteristics and can be used at high operating voltages because the FCT can be designed for high breakdown voltage capability with a low forward voltage drop due to conductivity modulation of its base region during forward conduction. The current through the FCT can be turned on and off by the application of a low gate bias voltage (typically less than 10 V) at relatively small gate drive currents (typically less than 0.1 A). As a result, the circuit is capable of switching load current with turn-on and turn-off times of less than one microsecond at a current gain of greater than 100. Further, since both the FCT and the MOSFET have been demonstrated to be capable of operating at elevated temperatures[8], the gating circuit described here should be useful for applications with high ambient temperatures.

This gating technique allows the operation of the FCT in a normally-off mode for the first time. This makes it a viable candidate for the lower frequency power switching applications, such as motor drives, where bipolar transistors, Gate turn-off thyristors (GTO), and MOSFETs are being considered. The new gating method for the FCT offers advantages with respect to all of these devices. In comparison with the high voltage MOSFET, the FCT operates at ten times the current density. Consequently, much smaller silicon chips are needed for any given application with attendant savings in cost. With the gating technique described in this paper all the advantages of the high impedance MOS gate that are

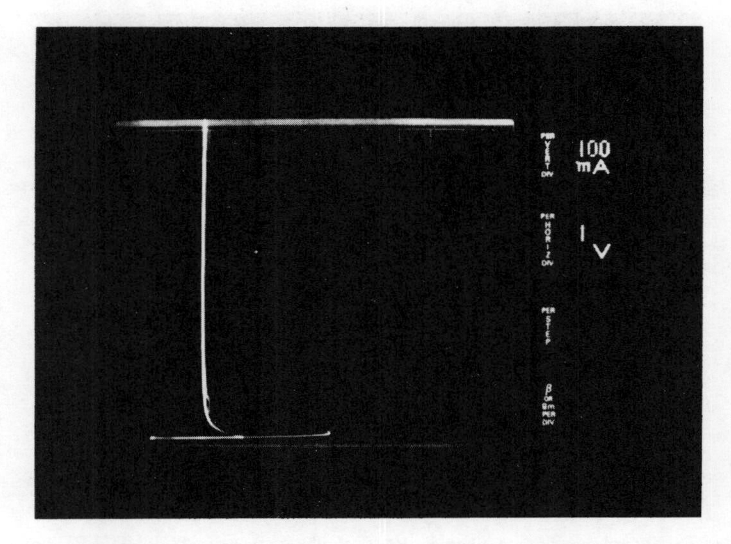

Fig. 5. Forward conduction characteristics of an asymmetrical FCT device with gate shorted to the cathode.

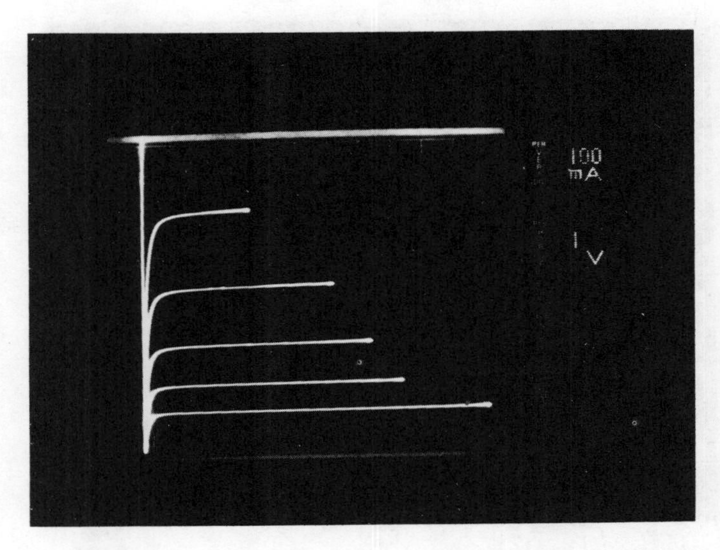

Fig. 6. Forward characteristics of a 90 V MOSFET used for gating the FCT.

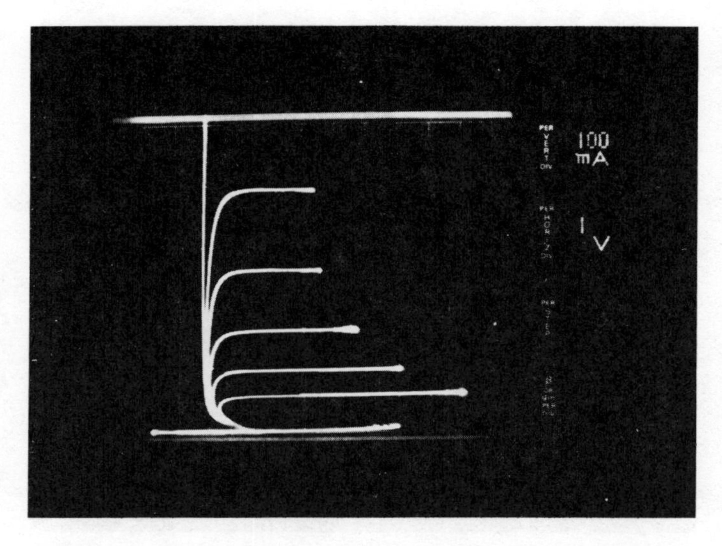

Fig. 7. Forward conduction characteristics of the composite FCT-MOSFET circuit.

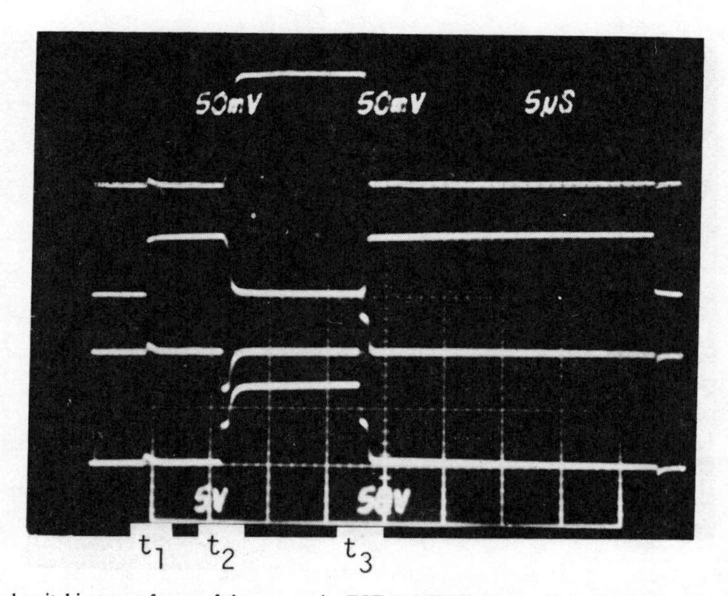

Fig. 8. Typical switching waveforms of the composite FCT-MOSFET circuit obtained by using a resistive load. The traces from top to bottom in the photograph are in the order of anode current at 0.5 A per division, anode voltage at 50 V per division, gate current at 0.1 A per division, and gate voltage a 5 V per division.

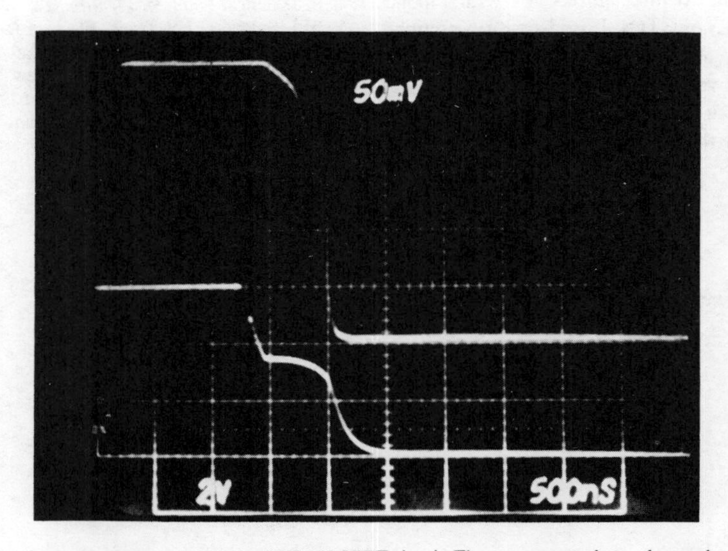

Fig. 10. Gate turn-off transients of composite FCT-MOSFET circuit. The upper trace shows the anode current at 0.2A per division and the lower trace shows the gate voltage at 2 V per division.

ascribed to the MOSFETs are also retained because the FCT gating is performed via a low breakdown voltage MOSFET. In comparison with the bipolar transistor, the new FCT gating techniques offers the above advantages of a high impedance MOS gate while also offering a three times higher operating current density. This will also simplify the circuit and decrease chip cost. Finally, in comparison with GTOs, the new FCT gating technique offers much higher (10 to 100 times) gate turn-off current gains together with higher switching speeds. For these applications requiring switching speeds of below 10 to 20 KHz (compatible with the one microsecond turn-off times), it can be concluded that the new gating technique will offer significant performance and cost advantages when compared with existing approaches with bipolar transistors, MOSFETS or GTOs. It is also worth pointing out that the advantages of using FCTs with the new gating technique are even greater for circuits operating at higher voltages because the current density in these devices decreases as the square root of the breakdown voltage, while the current density of both the bipolar transistor and the MOSFET decreases more precipitously as the 2.5 power of the breakdown voltage.

Acknowledgements—Discussions with Dr. M. S. Adler were helpful during the analysis of the circuit performance. The author also wishes to thank Ms. N. Waldron for her assistance with manuscript preparation.

REFERENCES

1. B. J. Baliga, *Int. Electron Dev. Meet.* Paper 4.2, pp. 76–78 (1979).
2. B. W. Wessels and B. J. Baliga, *IEEE-Trans. Electron Dev.* ED-25, 1261 (1978).
3. D. E. Houston, S. Krishna, D. E. Piccone, R. J. Finke and Y. S. Sun, *IEEE-Trans. Electron Dev.* **ED-23**, 905 (1976).
4. B. J. Baliga, *Solid-St. Electron.* **22**, 237 (1979).
5. B. J. Baliga, *IEEE-Trans. Electron Dev.* **ED-27**, 1262 (1980).
6. B. J. Baliga, *Appl. Solid St. Sci. Se.*, Supplement 2B (Edited by D. Kahng). Academic Press, New York (1981).
7. H. Benda and E. Spenke, *Proc. IEEE* **55**, 1331 (1967).
8. B. J. Baliga, *Int. Electron Dev.* Meet. Paper 25.5, pp. 654–657 (1980).

NORMALLY-OFF TYPE HIGH SPEED SI-THYRISTOR

Y. Nakamura, H. Tadano, S. Sugiyama, I. Igarashi
T. Ohmi* and J. Nishizawa*

TOYOTA Central Research and Development Labs., Inc.
Nagakute-cho, Aichi-ken, 480-11, Japan
*Research Institute of Electrical Communication, Tohoku Univ.
Sendai, 980, Japan

ABSTRACT

The design, fabrication, and characterization of a normally-off type static induction thyristor (SIThy) with a surface gate structure were made with an intention to use it in high speed switching at high currents. The device chip had a size of 7 × 10 mm^2 and included 9,000 channels, where each channel stripe had a width of 1.5 μm and a length of 250 μm. The double LOCOS technique was used in the fabrication process of the device. By mounting the chips in specially designed packages, a very low forward voltage drop of 1.2 V at 100 A and a high switching speed of 300 ~ 600 nsec in turn-on and turn-off times were obtained. A bipolar mode static induction transistor (BSIT) was also made by using the same photomasks and the same fabrication process as SIThy. It also exhibited a low forward voltage drop of 0.7 V at 100 A and a high switching speed of 100 ~ 200 nsec in turn-on and turn-off times.

INTRODUCTION

High power semiconductor devices have been recently used for high efficiency control of various power control equipments such as electric vehicles, robots, DC to AC converters for solar cells, and so on. More excellent switching performance is required of the semiconductor devices, in order to make the control frequency higher than the audio region and to reduce the switching loss. In these respects, static induction thyristors (SIThy) (1-4) have been particularly noted for their high power, high speed and low loss (5-7). These excellent features are due to their high dI/dt and dV/dt capabilities, negative temperature coefficient of anode current in high current region, very low forward voltage drop, and high speed operation. The quite same device is called field controlled thyristor (8,9) or bipolar gridistor (10).

The present study deals with the design, fabrication and characteristics of a normally-off SIThy with a surface gate structure. A normally-off SIThy has such advantages that it can replace a bipolar transistor without modification of a circuit, and it can make high speed and low loss switching operation with a small control power as well as a normally-on SIThy (7). The gate resistance of SIThy must be designed as small as possible in order to absorb higher instantaneous inverse gate current for achieving higher turn-off speed. Therefore, we have adopted the surface gate structure which can reduce the gate resistance. New packages were designed for an improvement of high speed switching.

Bipolar mode SIT's (BSIT) (11-14) were also made by the same photomasks and the same fabrication process as SIThy. A lower forward voltage drop and a higher switching speed than SIThy was obtained for BSIT. However, an application of BSIT is limited to drain voltage range lower than 600 V.

DESIGN AND FABRICATION

<u>Design</u> Fig. 1 (a) shows a schematic presentation of the cross-sectional view of a channel structure of the surface gate SIThy; 2a is the width of a channel, l_{GK} is the distance of two front ends of the cathode and the gate diffusion, l_{KA} is the distance between the cathode and the edge of the depletion layer in the gate to the anode region and N_D is the impurity concentration in the channel and the high resistivity region. These parameters are the most important factors to obtain the normally-off characteristic and dominate the performance of SIThy. Under the condition that the impurity concentration N_D is low enough to sufficiently deplete the channel and the high resistivity region, the voltage blocking gain μ is experimentally given by (15)

$$\mu = \frac{l_{GK} \cdot l_{KA}}{a^2} \quad . \qquad (1)$$

To obtain the normally-off characteristic, the voltage blocking gain μ must be selected to be high far from 100. The values described in Fig. 1 (a) were determined considering the above mentioned conditions and the gate to anode breakdown voltage.

The schematic surface view of a channel structure is shown in Fig. 1 (b), where l_{ch} and w_G are the length of a channel stripe and the width of a gate, respectively. The values were determined from the following consideration; First, for l_{ch}, the current density J of a channel has been selected to be 800 A/cm^2 in order to lower the forward voltage drop to about 1.2 V and obtain an appropriate proportion of the channel area in a chip. Secondly, for w_G, the gate resistance is made as small as possible so that the voltage drop caused by the gate resistance may not exceed 0.1 V at the center of a gate stripe when the current of $a \cdot l_{ch} \cdot J$, corresponding to the amount of the hole

Reprinted from *IEEE Int. Electron Devices Meet.*, 1982, pp. 480-483.

current, is drawn in turn-off process. The SIThy having a chip size 7 × 10 mm^2 and comprising 9,000 channels was designed, by taking account of the current uniformity and the resistance of the electrode metal. This chip can be used for 25 A switching. Thus, this design of the multi-channel structure and the reduced gate resistance has been adopted to shorten a turn-off time and obtain higher dI/dt and dV/dt capabilities. In applications where a slight increase of forward voltage drop is allowed, the SIThy chip can be available for the use of higher currents exceeding 100 A.

Fabrication In order to realize the cross-sectional structure as illustrated in Fig. 1 (a), the double LOCOS technique was applied in the fabrication. The process flow chart is shown in Fig. 2. Firstly, for formation of p^+-anode, boron was diffused into the backside of the wafer having a thickness of 450 μm and an impurity concentration of about 1×10^{14} cm^{-3}. Secondly, the first LOCOS process was made to form an SiO_2 layer of about 1 μm on the wafer surface. Then, the SiO_2 layer was perfectly removed to form the recessed parts for gate. Thirdly, for formation of p^+-gates, boron was deposited on the wafer surface, followed by simultaneous application of the second LOCOS process and diffusion of boron onto the recessed parts. As a consequence of the double LOCOS technique, the designed width of channels was obtained. The SEM photograph of the cross-section in the neighborhood of the surface is shown in Fig. 3 (a), and the microscopic photograph of the gates and channels enhanced by angle lapping and stain etching is shown in Fig. 3 (b). Next, phosphorus-doped poly-Si was deposited on the wafer by CVD, and the phosphorus was made to diffuse to form the cathodes. Lastly, poly-Si was etched by R.I.E., followed by Al deposition, to form the electrodes.

Fabrication of BSIT can start at the first LOCOS by using an n^--epitaxial wafer grown on an n^+-substrate. The thickness of the epitaxial layer was 45 μm. The impurity concentration was $5 \times 10^{13} cm^{-3}$. Thus, BSIT could be fabricated by using the same photomasks and the same process as SIThy by replacing the p^+-anode by the n^+-drain.

MEASUREMENTS AND RESULTS

Blocking voltage, BV_{AKS} of SIThy and BV_{DSS} of BSIT, were both about 500 V. In order to futher increase the blocking voltages, it may be very effective to form a p^--layer in front of the cathode. This may also improve the current gain.

The I-V and switching characteristics of the devices were measured by mounting the fabricated chips in two kinds of specially designed packages for 1 chip and for 4 chips. The package for 4 chips was made applicable for high current use because a parallel connection of several chips can be employed without ballasting resistor owing to the negative temperature coefficient of the anode current. Fig. 4 shows pakages which have very wide electrodes and are constructed so as to connect many short wires between the bonding pads and the electrodes. As a result, remarkable improvement of the switching characteristics can be achieved due to very small inductance which exhibit significant role for high current and high speed switching.

The I_A-V_{AK} characteristic of the 4-chips SIThy is shown in Fig. 5. The forward voltage drop was 1.2 V at 100 A and 1.6 V at 200 A, where the current density was about 800 A/cm^2 and 1,600 A/cm^2, respectively. Fig. 6 shows the I_D-V_{DS} characteristic of the 4-chips BSIT, where the forward voltage drop was 0.7 V at 100 A. The anode current density J_A v.s. V_{AK} characteristic measured at various temperatures by use of the SIThy including 30 channels is shown in Fig. 7. It shows a negative temperature coefficient of anode current at current densities higher than 20 A/cm^2.

Next, in the measurement circuit shown in Fig. 8 which was specially devised, as well as the package, so as to make high speed switching at high currents. In high speed switching at high currents, it is regard to make uniform current flow through the wide area in the circuit. If the current flow is concentrated in a narrow path, the inductance of the path will delay the transient time and a fine waveform may not be expected. From the specially designed package and the measurement circuit, the excellent results were obtained as summarized in Tab. 1. The values of the table were obtained from the anode or the drain current waveforms.

Table 1

Device		t_{on}(μsec)	t_{off}(μsec)	t_{stg}(μsec)	I_A, I_{DS}(A)	Note
SIThy	1 chip	0.2	0.3	0.3	25	
		0.2	0.4	0.7	100	Fig.9
	4 chips	0.6	0.6	1.5	200	
BSIT	1 chip	0.1	0.1	0.4	50	
	4 chips	0.2	0.2	0.4	200	Fig.10

V_{GK} or V_{GS} = −5(V)

Fig. 8 shows the switching waveform of I_A and V_{AK} of a 1-chip SIThy. Fig. 9 shows the switching waveform of I_D and V_{DS} at the turn-off of the 4-chips BSIT. These observed turn-on and turn-off speeds can be said much faster than those of conventional GTO such as 18 μsec (16). The observed switching times are limited by the presence of neutral region as long as about 350 μm in the high resistivity region, resulting from a smaller depletion layer width at the applied anode voltage in the measurement. A higher switching speed will be achieved by many ways, as an example, by lowering the impurity concentration in that region, where the depletion layer tends to spread just near to the p^+-anode region. This will reduce the forward voltage drop to about 1 V (17). The storage time depends on the negative gate bias voltage-V_G. In order to shorten the storage time by increasing the gate bias voltage, gate to cathode breakdown voltage BV_{GKO} must be made higher.

CONCLUSION

A very low forward voltage drop and a very high switching speed have been realized with a normally-off type SIThy fabricated by use of the double LOCOS technique. This type SIThy is very promising in application to power control equipments, owing to low drive power, low loss and the possibility of making simpler circuit. The SIThy with such low power dissipation provides an extremely high power efficiency and will be a most suitable device for application at higer currents and higher blocking voltages. Although BSIT fabricated in this study has indicated lower forward voltage drop and higher switching speed than SIThy, its use will be limited to the appropriate range of blocking voltage. These devices, SIThy and BSIT, will significantly improve the efficiency of high power control equipments.

REFERENCES

(1) J. Nishizawa, T. Terasaki and J. Shibata, "Field-Effect Transistor Versus Analog Transistor (Static Induction Transistor)," IEEE Tran. Electron Devices, Vol.ED-22, P.185, Apr. 1975.

(2) J. Nishizawa and K. Nakamura, "Characteristics of new thyristors," in Proc. of 1976 Int. Conf. Solid State Devices, Japan J. Appl. Phys. Suppl., No.16-1, p.541

(3) Y. Kajuwara, Y. Watakabe, M. Bessho, Y. Yukimoto and K. Shirahata, " A SHIGH SPEED, HIGH VOLTAGE STATIC INDUCTION THYRISTOR," Tech. Dig. 1977 IEDM, p.38, 1977.

(4) J. Nishizawa, K. Nakamura, " STATIC INDUCTION THYRISTOR," Rev. de Physiquee Appliquee, T.13, p.725, 1978.

(5) T. Terasawa, K. Miyata, S. Murakami, T. Nagano and M. Okamura, "HIGH POWER STATIC INDUCTION THYRISTOR," Tech. Dig. 1979 IEDM, p.250, 1979.

(6) T. Ohmi, "POWER STATIC INDUCTION TRANSISTOR TECHNOLOGY," Tech. Dig. 1971 IEDM, p.84, 1979.

(7) J. Nishizawa and Y. Otsubo, "EFFECTS OF GATE STRUCTURE ON STATIC INDUCTION THYRISTOR," Tech. Dig. 1980 IEDM, p.658, 1980.

(8) D. E. Houston, S. Krishna, D. Piccone, R, J, Finke and E. Sun, " A Field Terminated Diode," IEEE Trans. Electron Devices, Vol.ED-23,

(9) B. W. Wessels and B. J. Baliga, "Vertical Channel Field - Controlled Thyristors with High Gain and Fast Switching Speeds," ibid., Vol.ED-25, p.1261, 1978.

(10) R. Barandon and P. Lawrencean, "Power bipolar gridistor," Electron. Lett., Vol.12, p.486, 1976.

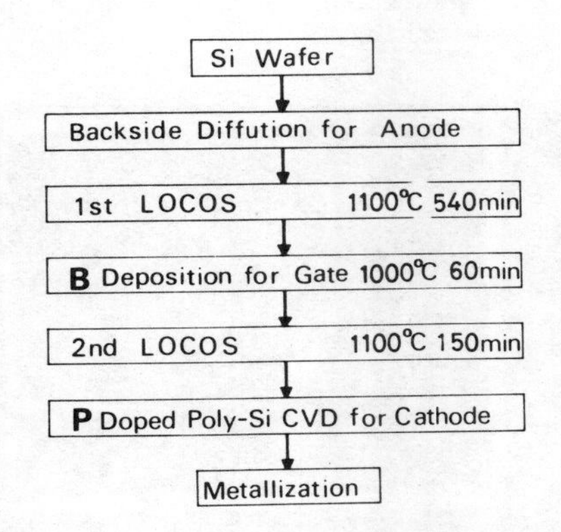

Fig. 2 Process flow chart.

(11) J. Nishizawa, T. Ohmi, Y. Mochida, T. Matsuyama and S. Iida, "BIPOLAR MODE STATIC INDUCTION TRANSISTOR (BSIT) -HIGH SPEED SWITCHING DEVICE," Tech. Dig. 1978 IEDM, p.676, 1978.

(12) J. Nishizawa, T. Ohmi, Y. Mochida and T. Matsuyama, "Bipolar Mode Static Induction Transistor," Proc. 1979 Int. Conf. Solid State Devices: Japan J. Appl. Phys. Supplement, Vol.19-1, p.289, 1980.

(13) T. Ohmi, "Punching Through Device and Its Integration - Static Induction Transistor," IEEE Trans. Electron Devices, Vol.ED-27, p.536, 1980.

(14) J. Nishizawa, T. Ohmi and H. L. Chen, "Analysis of Static Characteristics of a Bipolar-Mode SIT (BSIT)," ibid., Vol.ED-29, Aug. 1982.

(15) J. Nishizawa, T. Ohmi and Y. Otsubo, Japanese Patent Application, 55-163382.

(16) M. Azuma, K Takigami and M. Kurata, "2500 V 600 A Gate Turn-off Thyristor (GTO)," Tech. Dig. 1979 IEDM, p.246, 1979.

(17) J. Nishizawa, T. Ohmi, M. S. Hsieh and K. Motoya, "ANALYSIS OF CHARACTERISTIC OF STATIC INDUCTION THYRISTOR," Papers on Tech. Group of Electron Devices of IEEE Japan, ED-81-84, Nov. 1981.

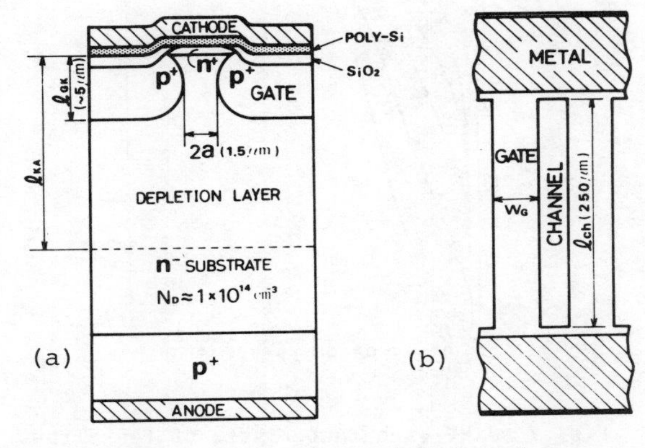

Fig. 1 Schematic views of a channel; (a) cross-section and (b) surface.

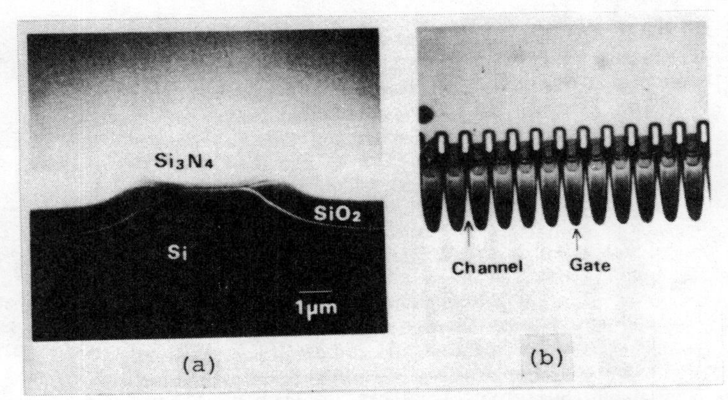

Fig. 3 (a) SEM photograph of the cross-section, (b) Microscopic photograph of gates and channels.

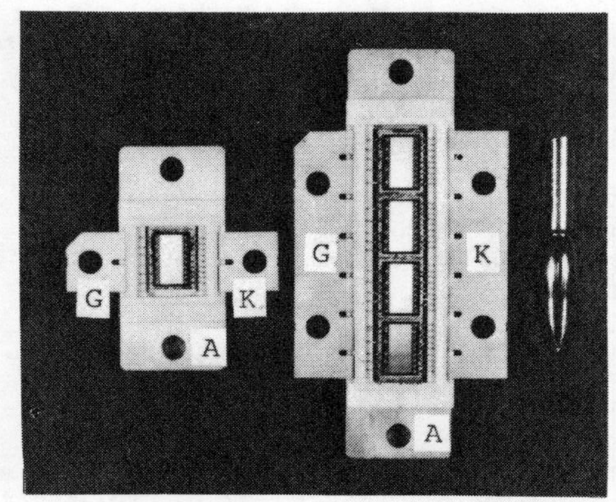

Fig. 4 Specially designed packages.

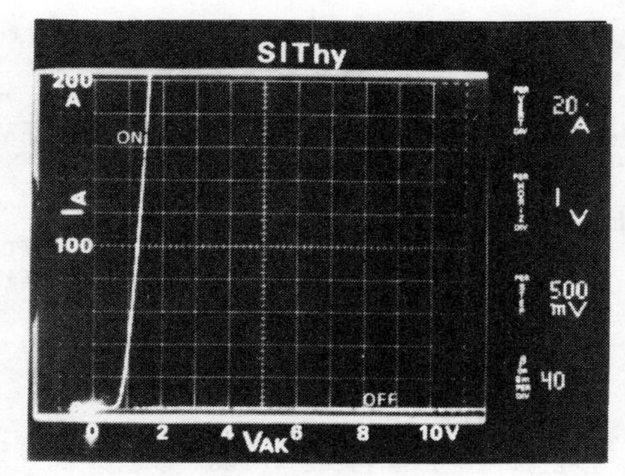

Fig. 5 I_A-V_{AK} characteristic of the 4-chips SIThy.

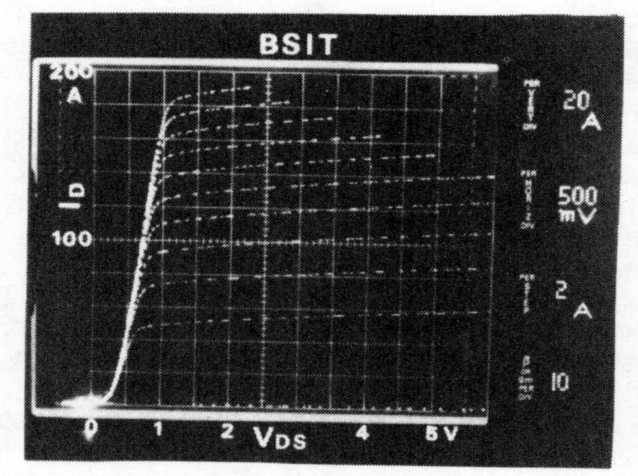

Fig. 6 I_D-V_{DS} characteristic of the 4-chips BSIT.

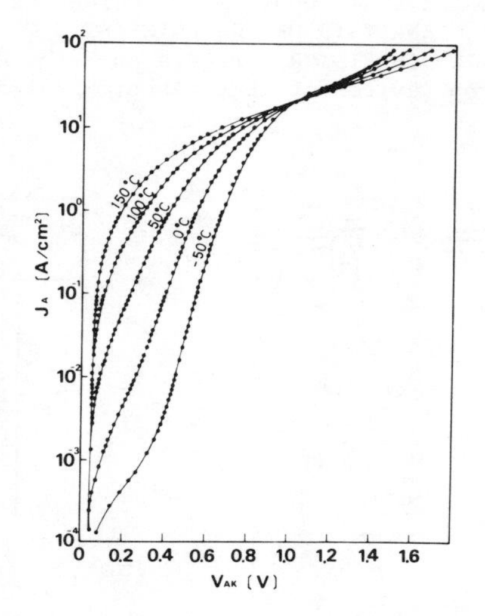

Fig. 7 J_A-V_{AK} characteristic of the SIThy.

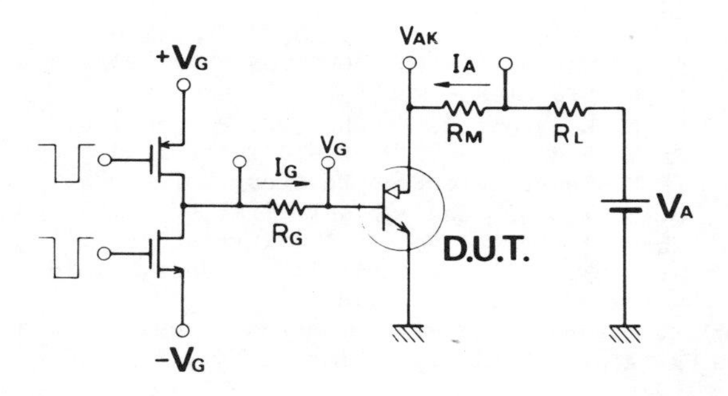

Fig. 8 Measurement curcuit of high speed switching.

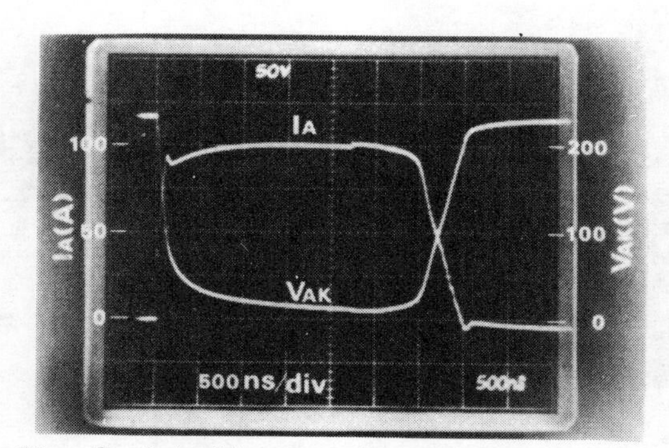

Fig. 9 Switching waveform of 1-chip SIThy.

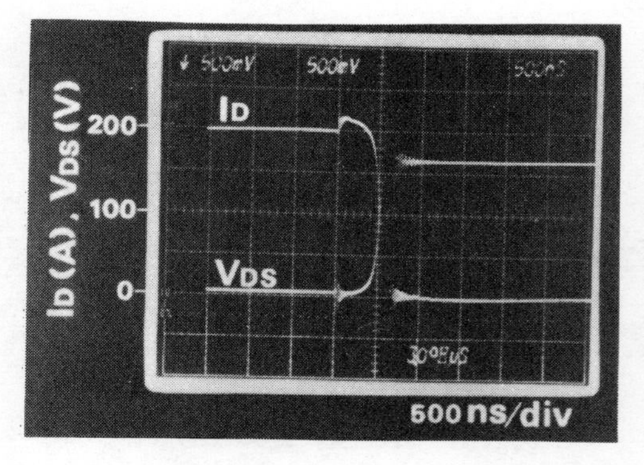

Fig. 10 Switching waveform of 4-chips BSIT.

A New FET–Bipolar Combinational Power Semiconductor Switch

D.Y. CHEN
S. CHANDRASEKARAN
S.A. CHIN
Virginia Polytechnic Institute and State University

A novel FET–BJT combinational transistor configuration is proposed and demonstrated using discrete devices. This new transistor features fast switching, very simple drive requirement, elimination of reverse bias second breakdown, and good utilization of semiconductor chip area. Initial results indicate that power hybrid construction of the device is essential to enhance the current rating of the device.

I. INTRODUCTION

The progress made in power transistor technology, for both field-effect transistors (FET) and bipolar junction transistors (BJT) has been truly remarkable in recent years. Devices are fabricated to handle higher voltage and more current, to switch faster and more ruggedly, and to be cost effective. In general, FET has advantages over BJT in driving simplicity, switching speed, and second breakdown ruggedness. However, the chip area current density requirement is satisfied by the bipolar transistor when compared on the basis of voltage rating and conduction voltage drop. In other words, it requires significantly more chip area to fabricate an FET than to fabricate a BJT to achieve comparable ratings of voltage, current, and conduction voltage drop. This is the fundamental reason why high power FET is not cost competitive with BJT. Recently, several FET–BJT combination transistors have been proposed and reported in the literature [1–6]. In most of these combination transistors, the FET is used to simplify the drive requirement or to increase the switching efficiency, and the BJT is used for current-carrying capability. While none of these combinational transistors would solve all the problems, each does find suitable applications.

In this paper, a new FET–BJT combination transistor is proposed and the test results are presented [7]. In this new transistor, two FETs and one BJT are combined to form a three-terminal device. Since the BJT is gated on and off by the two FETs, this new device is referred to as an FET-gated bipolar transistor (FGT).

Manuscript received October 3, 1982, revised October 4, 1983.

This work was supported by NASA Lewis Research Center under Grant NAG3-340.

Authors' address: Department of Electrical Engineering, Virginia Polytechnic Institute and State University, Blacksburg, VA 24061.

II. FET-GATED BIPOLAR POWER TRANSISTORS

A. Transistor Configuration

Fig. 1 shows the circuit diagram of the FET-gated bipolar transistor in which Q_1 is the main bipolar power transistor, Q_2 and Q_3 are power FETs, and D is a Zener diode. Q_2 is in a Darlington configuration with Q_1, and Q_3 is in an emitter-open configuration with Q_1. When the gate voltage is high, both Q_2 and Q_3 conduct; Q_1 therefore conducts and FGT conducts. When the gate voltage is low, both Q_2 and Q_3 cut off and Q_1 is cut off by emitter-open and the FGT is rapidly cut off [1–3]. A Zener diode D is used for two purposes: to provide a reverse current path for the base of Q_1 at emitter-open turn-off, which clamps the base terminal voltage at the Zener clamping voltage; and to provide a forward current path for discharging Q_2 gate capacitance at turn-off.

Q_2 is voltage rated as BV_{CBO} of Q_1, but has a current rating of approximately $1/\beta_{Q1}$ of I_C. Q_3 carries about the same current as Q_1 but is a low voltage device, typically below 50 V. The Zener diode breakdown voltage must be higher than $V_{BE}(Q_1) + V_{DS}(Q_3)$ such that the base current of Q_1 is not diverted through the Zener diode during the conduction period.

Reprinted from *IEEE Trans. Aerosp. Electron. Syst.*, vol. AES-20, pp. 104–110, Mar. 1984.

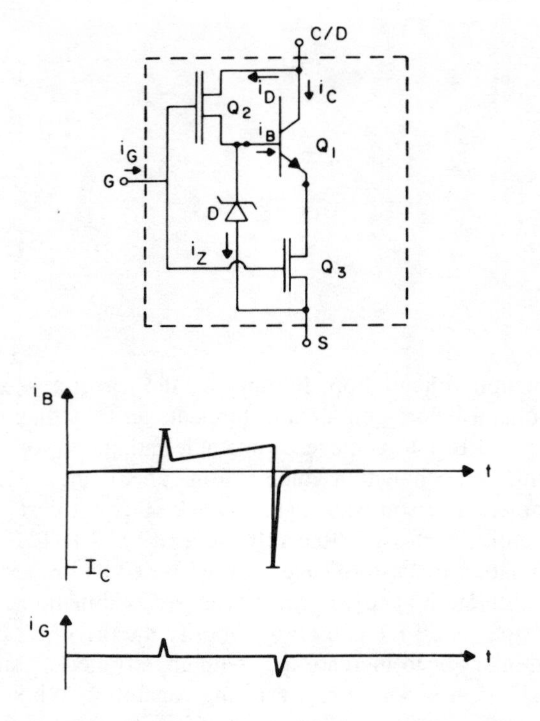

Fig. 1. FET-gated bipolar transistor and associated waveforms.

B. Advantages of the FET-Gated Bipolar Power Transistor

1) *The transistor is free from reverse bias second breakdown.* A reverse bias second breakdown (RBSB) of a bipolar transistor is due to emitter current crowding in the case of turning off an inductive load. Since Q_1 is turned off by emitter-open, RBSB can be avoided [1, 3].

2) *Drive requirement is simple when compared with other types of transistors.* As compared with a bipolar transistor of the same rating, an FGT requires essentially zero steady state drive current. The only appreciable drive current required is the surge current for charging and discharging the gate capacitance of Q_2 and Q_3 at the instant of turn-on and turn-off. As compared with a power FET of the same rating, the chip areas of both Q_2 and Q_3 are much smaller, and so are the gate capacitances. Therefore, the gate drive surge current requirement for an FGT is much less. An estimate of the relative chip sizes of Q_2 and Q_3 are given in a later section.

3) *Switching speed is very fast.* The switching speed of an FGT is limited primarily by the speed of Q_1. However, Q_1 is inherently provided with a nearly ideal base drive current waveform, as shown in Fig. 1, so Q_1 can be switched very rapidly. As is explained in a later section, base current "kick" is inherently built in for both turn-on and turn-off, which speeds up the switching of the FGT.

4) *Snubber circuit may not be necessary.* In general, a snubber circuit is used to avoid reverse bias second breakdown or to reduce dv/dt at turn-off. A snubber circuit may not be necessary, or, if a snubber circuit is used, the amount of snubbering can be reduced for this device because of two reasons: the device is free from RBSB; the dv/dt effect is much reduced due to small gate Miller capacitance of Q_2.

5) *It requires only slightly larger chip area than a comparable bipolar Darlington transistor but much less chip area than a comparable FET.* The chip areas of a transistor, either bipolar or FET, increase rather rapidly with voltage rating if other transistor parameters are kept constant [5, 6]. Since Q_1 is free from reverse bias second breakdown and can be operated up to BV_{CBO} instead of $BV_{CEO(SUS)}$, the voltage rating of the same chip is significantly enhanced. Therefore, for the same voltage rating, the Q_1 chip area can be significantly reduced as compared with a comparable bipolar Darlington transistor with conventional reverse bias turn-off. An estimate of the relative sizes of total chip areas is given in the appendix.

C. Experimental Switching Characteristics

The switching characteristics reported in this paper were obtained from a typical inductive load switching circuit. A snubber circuit may or may not be used, depending on the test. Three sets of discrete transistors were used to form the FGT. A low current setting is rated at 2–3 A, a medium current setting is in the range of 8 A, and the high current one is about 25 A.

Fig. 2 shows a typical base current waveform for the bipolar transistor Q_1 and the waveform of the sum of

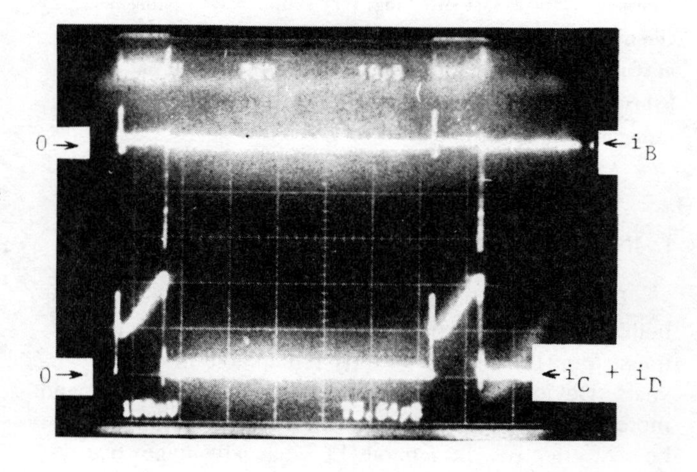

Fig. 2. Operating current waveforms for FGT for continuous-mode operation (with snubber). Parameters—i_B : 2A/div, $i_c + i_D$: 2A/div, time: 10 μs/div; Q_1: General seminconductor 2N6655 (450 V, 10 A), Q_2: IRF 730, Q_3: IRF 530, D: TCG 138A (5.1 V, 1 W).

collector and drain currents of an FGT. The turn-on C/D current spike is caused primarily by the reverse recovery of the flyback diode in the test circuit. As can be seen from the figure, at the instant of turn-on, a large base current kick is provided to rapidly turn on Q_1. At the instant of turn-off, a large reverse base current, equal to the collector current, is provided to shut off Q_1 very rapidly.

During the current conduction period, the base current is approximately proportional to the collector current, which prevents oversaturation of Q_1. From the viewpoint of driving a bipolar transistor, the base current waveform shown in Fig. 2 is close to ideal. Such a waveform is difficult to achieve in a conventional transistor because of the limitation of the base drive power supply, the loop impedance of the base circuit. In an FGT, however, this driving waveform is an inherent characteristic of the transistor. As will be explained in the following subsections on turn-on and turn-off, the base current of Q_1 is not derived from any base drive power supply as in the case of a conventional transistor, but rather from the load circuit. And this is accomplished with a simple gate drive circuit which requires essentially zero on-state current and relatively small surge current at turn-on and turn-off.

(1) *Turn-On Characteristics.* As mentioned earlier, it is important to provide a base current spike at the turn-on of a bipolar transistor to minimize the turn-on loss if the circuit is operated at the continuous inductor current mode. In a conventional transistor, this base current spike must be derived from the base drive power supply or from a speed-up capacitor. However, this current kick is quite often limited by the loop impedance of the base circuit, including the lead inductance, the device base-emitter impedance, and the power supply capability. In an FGT, the base current kick for Q_1 is derived from the load inductance through FET Q_2. In the case of a continuous mode of operation in which an FGT turns on when the inductive load current is nonzero, the inductive current is shared by Q_1 and Q_2. Initially at turn-on, before Q_1 is completely conducting, a good portion of the inductive current flows through Q_2 to provide a base current kick to rapidly turn on Q_1. After Q_1 is completely turned on, most of the load current flows through Q_1 and approximately $1/\beta Q_1$ of I_C flows through Q_2. The turn-on current gain (i.e., I_C/I_B at turn-on) of Q_1 depends on the ratio of the terminal impedance of Q_1 and Q_2 at the instant of turn-on. Figs. 3 and 4 show turn-on

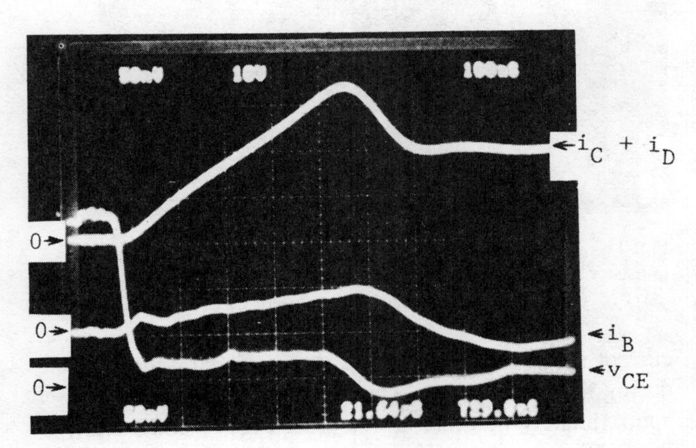

Fig. 3. Turn-on waveforms (with snubber). Parameters: $i_C + i_D$: 1 A/div, i_B: 1 A/div, V_{CE}: 10 V/div, time: 100 ns/div. Parts used: Same as in Fig. 2.

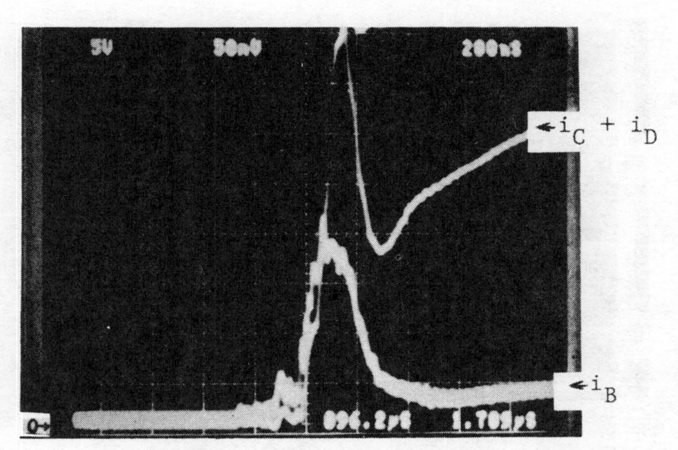

Fig. 4. Turn-on current waveforms using large current device as main transistor (with snubber). Parameters: $i_C + i_D$: 2 A/div, i_B: 2 A/div, time: 200 ns/div; Q_1: Westinghouse D-60-T (450 V, 40 A), Q_2: IRF 730, Q_3: IRF 150, D: Motorola HEPZ3516 (15 V, 10 W).

waveforms on an expanded time scale for two different transistors. As can be seen from Fig. 3, V_{CE} drops rapidly at turn-on. In the case of a discontinuous mode of operation, the inductive load current is zero at the instant of turn-on and the base current kick is also zero. Q_1 is therefore driven with base current kick only if needed, which minimizes the turn-on loss under variable load conditions. Fig. 5 shows the waveforms for a discontinuous mode of operation.

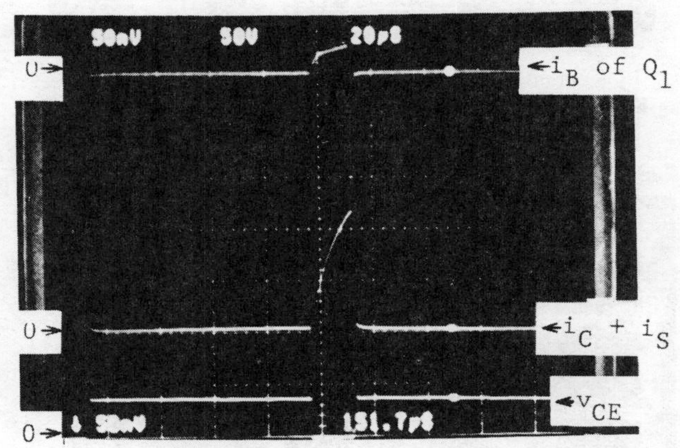

Fig. 5. Operating waveforms for FGT for discontinuous-mode of operation (with snubber). Parameters: i_B of Q_1: 1 A/div, $i_C + i_D$: 1 A/div, V_{CE}: 50 V/div, time: 20 μs/div; Q_1: Toshiba T-1115 (1500 V, 3 A), Q_2: IRF 730, Q_3: IRF 530, D: TCG 138 A (5.1 V, 1 W).

(2) *Turn-Off Characteristics.* Turn-off characteristics of an FGT depend primarily on the characteristics of Q_1. Several types of transistors were used as Q_1 in the test. Figs. 6 and 7 show the turn-off waveforms of an FGT using a high voltage slow Q_1. As can be seen from Fig. 6, the collector/drain current turns off approximately 1 μs after the emitter current falls. The *C/D* current fall time is about 100 ns, which is very fast for the transistor used. During the storage time, that is, the time duration between the emitter current turning off

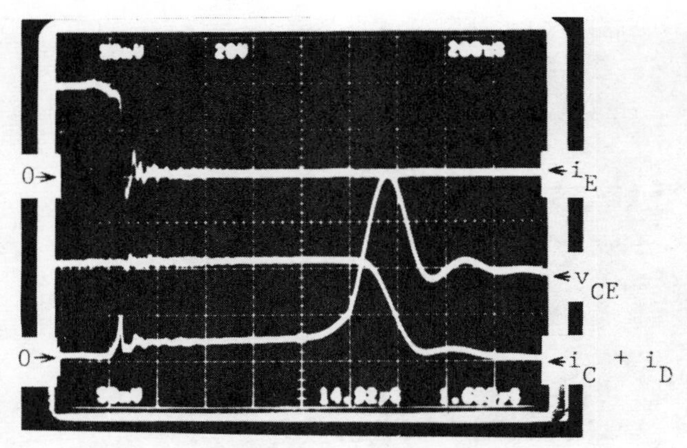

Fig. 6. Turn-off waveforms using high voltage device as main transistor (without snubber). Parameters: i_E: 1 A/div, V_{CE}: 20 V/div, $i_C + i_D$: 1 A/div, time: 200 ns/div. Parts used: Same as in Fig. 5.

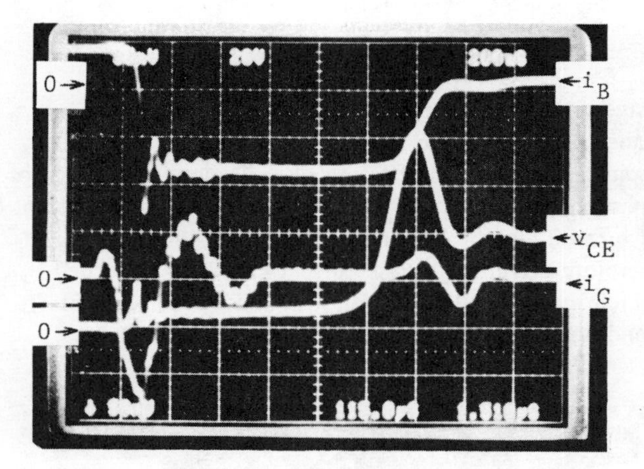

Fig. 7. Turn-off waveforms using high voltage device as main transistor (without snubber). Parameters: i_B: 1 A/div, V_{CE}: 20 V/div, i_G: 200 mA/div, time: 200 ns/div. Parts used: Same as in Fig. 5.

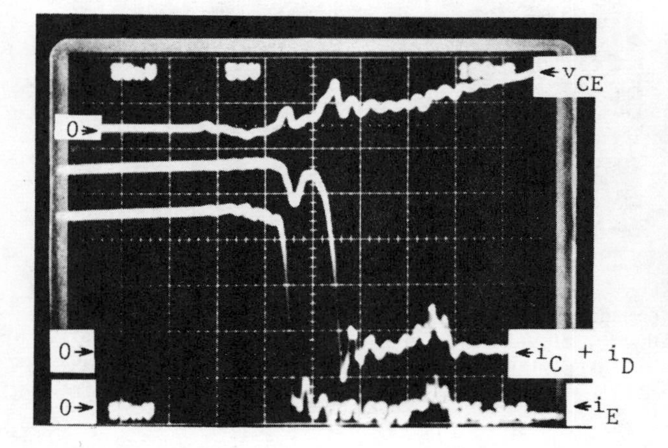

Fig. 8. Turn-off waveforms using fast device as main transistor (with snubber). Parameters: V_{CE}: 50 V/div, $i_C + i_D$: 1 A/div, i_E: 1 A/div, time: 100 ns/div. Parts used: Same as in Fig. 2.

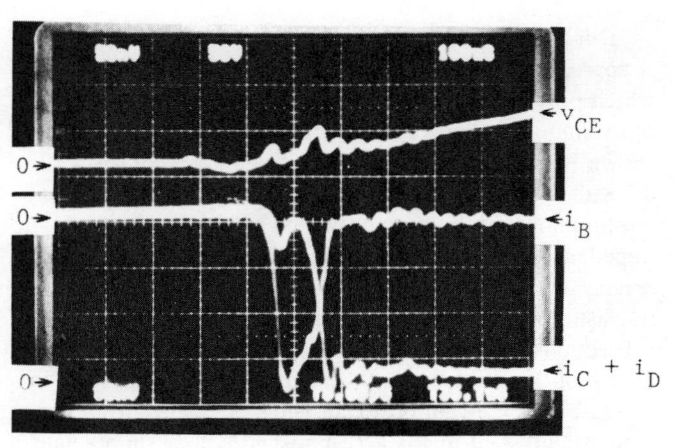

Fig. 9. Turn-off waveforms using fast device as main transistor (with snubber). Parameters: V_{CE}: 50 V/div, i_B: 1 A/div, $i_C + i_D$: 1 A/div, time: 100 ns/div. Parts used: Same as in Fig. 2.

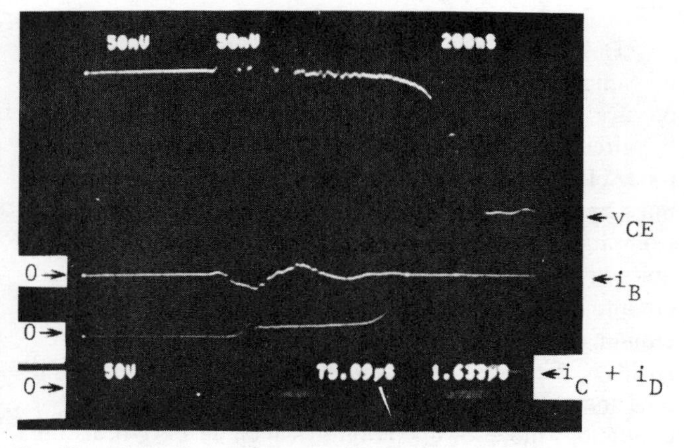

Fig. 10. Turn-off waveforms using high voltage, medium current device as main transistor (without snubber). Parameters: V_{CE}: 50 V/div, i_G: 1 A/div, $i_D + i_C$: 1 A/div, time: 200 ns/div; Q_1: Motorola MJ12004 (1500 V, 4 A), Q_2: IRF 730, Q_3: IRF 530, D: TCG 138 A (5.1 V, 1 W).

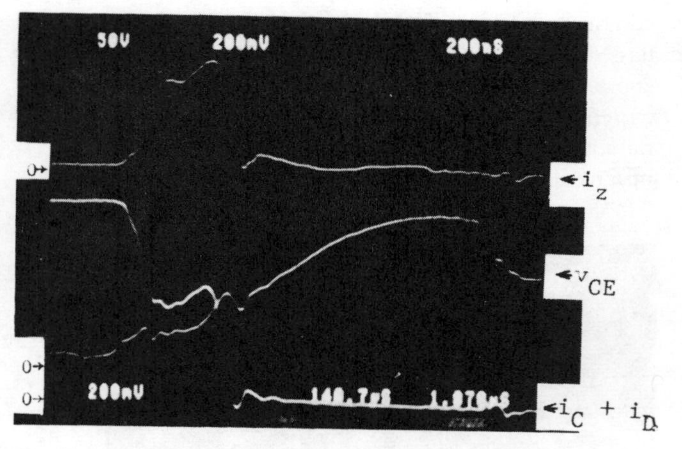

Fig. 11. Turn-off waveforms using high current device as main transistor (with snubber). Parameters: i_Z: 6 A/div, V_{CE}: 50 V/div, $i_C + i_D$: 6 A/div, time: 200 ns/div. Parts used: Same as in Fig. 4.

and the C/D current falling off, V_{CE} is clamped at the Zener voltage, about 8 V. V_{CE} starts to rise a significantly long time after the emitter current is cut off, which eliminates the possibility of reverse bias second breakdown. Fig. 7 shows the base current and the gate current waveforms. As can be seen, the reverse base current is slightly less than the drain/collector current during the storage time. The reverse gate current is approximately 600 mA in this case. Figs. 8 and 9 show the FGT turn-off waveforms using a fast transistor as Q_1. Because of the fast switching speed, a snubber circuit was used in this case to minimize the effect of the lead

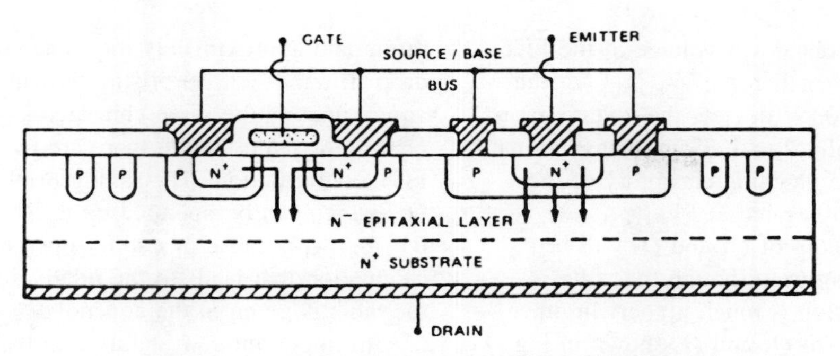

Fig. 12. Device structure for integrating FET and bipolar transistor.

inductance in the FGT assembly. As can be seen from Fig. 8, the storage time is about 100 ns and the fall time is about 10 ns, switching approximately at 4 A. Fig. 9 shows the waveform for the base current of Q_1. Fig. 10 shows the waveforms for the case of using a high voltage medium current slow transistor as Q_1. Turning off a current level of 8 A without snubber, V_{CE} rises to a 250 V peak, the storage time is about 800 ns, and the fall time is about 200 ns.

Fig. 11 shows the waveforms for the case of using a large current device as Q_1. The turn-off was conducted at a *C/D* current of approximately 25 A. A snubber circuit was used in this test. At the first glance, the *C/D* current waveform seems to exhibit a two-step turn-off. However, the first drop of the *C/D* current waveform corresponds to the first step rise of V_{CE}. The snubber circuit takes the current away from the inductive load current, which results in a step change of the *C/D* current.

The second drop of *C/D* current is the real fall time of the transistor. As mentioned earlier, the first step rise of V_{CE} is caused by the Zener clamping voltage. A 15 V Zener diode was used in this test not because of necessity but because of availability. As can also be seen from this figure, the Zener diode conducts during the storage time.

(*a*) *Storage time*. The storage time of an FGT depends not only on the characteristics of Q_1 but also on the choice of Q_2. Because of the Darlington configuration, Q_1 is never in deep classical saturation. The degree of saturation of Q_1 depends on the V_{DS} of Q_2. A choice of small area Q_2 decreases the degree of saturation of Q_1 and therefore decreases the storage time. But the trade-off is conduction voltage drop. Because of the built-in emitter-open turn-off, the storage time of an FGT is significantly reduced as compared with the case when Q_1 is under conventional turn-off.

(*b*) *Fall time*. As reported in [1], the fall time of a bipolar transistor can be significantly reduced by using emitter-open turn-off. Therefore, the fall time of an FGT is as short as Q_1 fall time can be. It was reported that the collector current tailing phenomenon may occur in a bipolar transistor turn-off if the transistor is in deep saturation before turn off and if the reverse base current is too large [8]. This phenomenon is device structure limited to begin with and may be minimized by more deliberated device design. But whatever device design Q_1 may have, the current tailing phenomenon would not occur because Q_1 is never in deep saturation.

(*c*) *Power loss in Zener diode D*. As shown in Fig. 11, Zener diode *D* conducts during the storage time. The power loss incurred can be calculated by using $P_{L.Z} = V_Z \times I_C \times t_S \times f$, where V_Z is the Zener clamping voltage, I_C is the collector current of Q_1 at turn-off, t_S is the storage time, and f is the switching frequency. Therefore, the turn-off power loss of an FGT occurs at the Zener diode during the storage time and occurs mainly at Q_1 during the fall time. V_Z should be kept low to minimize such loss.

III. FABRICATION OF FGT TRANSISTORS

Transistors Q_1 and Q_2 can be separate chips or an integrated chip, and Q_3 is a separate chip. A device structure for integrating Q_1 and Q_2 has been reported in [6]. Fig. 12 shows such a device structure in which *P* type diffusion forms both the body of the FET and the base of the bipolar transistor. The N^+ diffusion forms both the source of the FET and the emitter of the bipolar, and the N^+ substrate forms the collector and the drain. At the perimeter of *P* diffusion are field-limiting rings for enhancing device voltage capability. Fig. 13 shows the collector and drain characteristics for a bipolar transistor, and an FET having the same *n*-region doping and thickness.

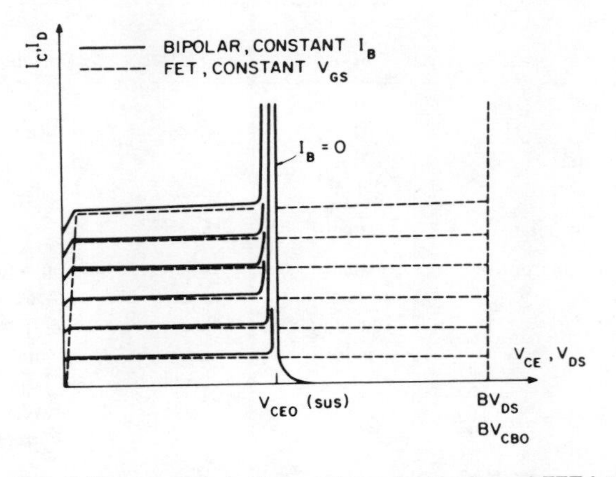

Fig. 13. Collector and drain characteristics for bipolar and FET having same *n*-region doping and thickness.

In this structure, the breakdown voltage of the FET portion BV_{DS} is much higher than the $V_{CEO(SUS)}$ of the bipolar transistor. For a 400 V device, the voltage ratio (BV_{DS} to BV_{CEO}) is typically 2 to 1. Using conventional turn-off of Q_1, the voltage blocking capability of the bipolar transistor is normally rated at $BV_{CEO(SUS)}$. Therefore, the integrated chip of Q_1 and Q_2 will be limited to $BV_{CEO(SUS)}$ rating even though the voltage capability of the FET portion is much higher. In other words, the integrated chip of Q_1 and Q_2 shown in Fig. 11 is not fully utilized if a conventional turn-off method is used.

Using the FGT configuration, however, the bipolar transistor blocking capability can be enhanced to be BV_{CBO}. Therefore, the voltage blocking capability of the bipolar transistor coincides with an FET breakdown voltage BV_{DS}. The integrated chip of Q_1 and Q_2 can then be utilized to its full voltage capability of BV_{DS} in an FGT.

It was observed that the lead length of the interconnecting wires plays an important role in the operation of the device. For low current applications, discrete transistors and Zener diode can be used to form an FGT without problems. For high current applications, power hybrid construction must be used for the device to function properly.

IV. COMPARISON WITH OTHER TRANSISTOR TYPES

Table I summarizes the results of a comparison between the new FGT with the power FET and the conventional Darlington transistors. As can be seen from the table, the total semiconductor chip area requirement is much against an FET. An FGT requires essentially the same chip area as the conventional Darlington transistor. This comparison is based upon 400 V device voltage rating and approximately the same conduction voltage drop. It may seem surprising that an FGT requires approximately the same chip area as a comparable Darlington even though there are two FET chips in the FGT configuration. The saving of chip area comes from Q_1, which can be operated up to BV_{CBO} rather than $BV_{CEO(SUS)}$ because of emitter-open turn-off. A detailed calculation that leads to the relative chip sizes shown in the table is given in the appendix.

An important consideration in the comparison between an FET and an FGT is the gate capacitances which affect not only the gate driving requirement, gate circuit power losses at high frequency operation, but also device turn-off dv/dt capability. At device turn-off, the displacement current due to "Miller capacitance" may pull the gate voltage high enough to false trigger the device unless the gate circuit impedance is kept very low. This is not a trivial problem in real practice, especially in a circuit such as a bridge inverter in which multiswitches are connected in a "totem-pole" fashion. In an FGT, the gate capacitances are significantly less than the counterparts of a comparable FET because the physical chip size of both Q_2 and Q_3 are much smaller than that of a comparable FET. Both Q_2 and Q_3 affect the size of the overall gate capacitance, but only Q_2 is responsible for the "Miller capacitance" of an FGT.

TABLE I
Comparison of FGT, FET, and Conventional Darlington Transistors

		FGT	FET	Darlington
Relative	*Q1*	0.4	2.5	1
Active	*Q2*	0.5	—	0.2
Chip areas	*Q3*	0.4	—	—
Total		1.3	2.5	1.2
Gate capacitance		small	large	
Drive requirement		small gate current pulse	relatively large gate current pulse	on-state forward current and large reverse current
Relative switching speed		medium	high	low

V. CONCLUSIONS

The FET-gated transistor concept proposed and demonstrated in this paper provides the designers with a practical means of combining the merits of a bipolar transistor and an FET into a single three-terminal device. The new device features simple drive requirement, fast switching, reverse second breakdown ruggedness, and good utilization of semiconductor chip area. Since the driving requirement is considerably reduced, even compared with an FET of comparable ratings, this device opens up the possibility of direct control of a high power transistor with low level logic signals.

The switching performance of the new transistor is very much affected by the lead length of the interconnecting wires among the chips. For low current applications, discrete devices can be used to form the configuration to achieve the performance. For high current applications, it is essential to minimize the lead length for the device to function properly. Power hybrid construction should provide a means of packaging the new device.

APPENDIX

Generally speaking, for device voltage rating above 250 V, the chip active area is proportional to 2.5 the power of the voltage rating to achieve the same conduction voltage drop [6]. For lower voltage rating, the chip area increases less rapidly with increasing voltage rating, because the conduction voltage drop due to device bulk resistance is less predominant for low voltage devices. This is true for both FETs and bipolar transistors. However, because of conductivity modulation in the conduction mechanism of a bipolar transistor, it normally requires much less chip size to fabricate a BJT than to fabricate a comparable FET. For a 400 V device the area ratio is typically between 2 and 3 in favor of BJT.

From the general information described above, a comparative estimate of the chip sizes will be given for an FGT, an FET, and a conventional Darlington for devices rated above 400 V. Assuming the chip area of the output transistor of the conventional Darlington is 1 unit, the driver transistor is 0.2 unit and the total chip area is 1.2 unit. For an FET of comparable rating, the area is approximately 2.5 unit. For the new FGT, Q_1 requires only 0.2 unit (1 unit $\times$ $(1/2)^{2.5} \approx 0.18$) because Q_1 can be operated at BV_{BCO} which is about 2 times BV_{CEO} of the output transistor in a conventional Darlington transistor. Q_2 is 2.5 times the Darlington drive transistor and therefore requires an 0.5 unit. Q_3 is a low voltage FET rated below 50 V and is estimated to be 0.2 unit [6]. To minimize the conduction drop, the area of both Q_1 and Q_3 is increased by a factor of 2. This leads to the chip areas shown in Table II.

REFERENCES

[1] Chen, D., and Jackson, B. (1981)
Turn-off characteristics of power transistors using emitter-open turn-off.
IEEE Transactions on Aerospace and Electronic Systems, AES-17, 3 (May 1981), 386–391.

[2] Skanadore, R. (1980)
A new bipolar high frequency power switching technology eliminates load-line shaping.
In *Proceedings of PowerCon 7*, 1980.

[3] Chen, D., and Walden, J.P. (1981)
Application of emitter-open turn-off technique for high voltage inverter applications.
Presented at the IEEE Power Electronics Specialists Conference, June 1981.

[4] Blanchard, R., Baker, R. and White, K. (1981)
A new high-power MOS transistor for very high current, high voltage switching applicatins.
In *Proceedings of PowerCon 8*, April 1981.

[5] Adler, M.S. (1982)
A comparison between bimos device types.
Presented at the IEEE Power Electronics Specialists Conference, June 1982.

[6] Zommer, N. (1981)
The monolithic HV BIPMOS.
Presented at the IEEE International Electron Device and Material Conference, Dec. 1981.

[7] Chen, D. (1981)
A bipolar-FET hybrid power semiconductor switch.
Invention Disclosure, Virginia Polytechnic Institute and State University, Blacksburg, Sept. 1981.

[8] Hetterscheid, W.T. (1974)
Turn-on and turn-off behavior of high voltage switching transistor. Mullard Technical Communications, Application Note 124, 1974.

THE MONOLITHIC HV BIPMOS

NATHAN ZOMMER

INTERSIL INC.
CUPERTINO, CALIF. 95014

ABSTRACT

An Integrated switch concept was developed which combines the advantages of the BJT and the MOSFET as power switches in one chip...the BIPMOS. Two types will be described, the "Uncommitted" and the "Shunt". Device layouts, mode of operation, fabrication, and application are described. The devices are fabricated in a self alligned MOS process, where the BJT is fully merged within the MOSFET. With the "Uncommitted" version a BIPMOS Darlington or a "Zenered" MOSFET are built. The "Shunt" BIPMOS is a 450v device with a current capability of 260A/Cm^2. It offers efficiency merits over a BJT or a MOSFET when used in typical switch mode power supplies. A SMPS diagram using the "Shunt" BIPMOS is described with an efficiency analysis.

I. INTRODUCTION

Several articles have been already published which describe the advantages of a power MOS device relative to a BJT in power conversion circuits. The BJT has a higher transconductance and a lower effective RON per unit chip area when compared to a MOSFET. However the BJT is "slower" needs higher driving energy and its SOA is limited due to reverse 2nd breakdown (RSB). In inductive load switching, large safety margins for device VBD are used when using a proper BJT. The power MOSFET on the other hand is faster, requires less drive energy and is less susceptible to RSB, but its transductance is lower and RON much higher, [1]-[3]. (A secondary result of the work described in this paper is a proper quantitative comparison between a power BJT and a power MOSFET in terms of RON and switching time. Both device were fabricated simultaneously with the same process on the same Si chip ans were designed with the same active area). It has become apparent that the BJT is better suited for low frequency high current applications and the MOSFET for high frequency, and presently, for low to medium current application. The ultimate goal of the technologist is to develop an ideal switch that would merge the speed and the low energy drive of the MOSFET with the low voltage drop of the BJT and would have no RSB limitation. Short of creating the ideal switch, this paper describes a monolithic device which enables the system level, the designer to utilize a quasi-ideal switch. A hybrid version of the concept was described earlier [4].

II. BIPMOS PROCESS & DESIGN

The BIPMOS is fabricated with the conventional vertical double diffused MOS process. The conventional power MOS device contains in itself a vertical BJT. For this discussion, we will refer to n-channel FETs and npn BJT's. The FET's n+ source p-body and n^-/n^+ drain constitute the BJT's emitter, base and collector. The usual practice in power MOSFET design is to quench the activity of this parasitic BJT. This is done by shorting, on the surface, the n+ source (emitter), with the p-body (base), thus in effect turning the BJT into a parasitic diode. The modification to the process in order to turn it into a BIPMOS process, is not shorting the n+ source (emitter) to the p-body (base), in the same contact opening. The former MOSFET structure, consisted on the surface of the gate metal bus and the source metal bus (Al). The drain isaccessed via the back of the die. In the non-shotred structure the n+ region and the p-body region are contacted in separate locations on the chip. Thus enabling with the proper metal mask to access the n+ and the p separate from each other. The n+ region can be further partitioned to separate physically the BJT from the MOSFET, thus creating these two devices side by side to each other. This was done in the first BIMOS design--the "Uncommitted" BIPMOS switch, fig.1 the device was equally partitioned between the BJT and the MOSFET, with four bonding pads; base and emitter of the BJT and source and gate of the MOSFET, the drain and collector were merged. With the four top bonding pads that are provided, the user can interconnect them in any combination to create a BIMOS"Darlington", a"Zener" protected MOSFET or a shunt BIMOS switch. Further, it lays ground for"custom" type BIPMOS devices in which based on the application the proper interconnections can be made by metal mask options. 2nd generation BIPMOS devices were designed in which the BJT and the MOSFET were fully merged in the same active area. This approach was used for the BIPMOS "Shunt", fig. 2, where the source and the emitter are combined. More will be said about the novel "shunt" device.

Reprinted from *IEEE Int. Electron Devices Meet.*, 1981, pp. 263-266.

III CHARACTERISTICS AND APPLICATION

As mentioned earlier, the 1st generation BIPMOS device was a mere side by side structure of a BJT and a MOSFET, occupying same active area of $11.3mm^2$. With the four probing contacts on the chip, the BJT and/or the MOSFET can be turned on separately to study their relative conductance. Typical gains of the BJT were 40 at 71 A/Cm^2, with effective RON of 4 $mohm\text{-}cm^2$. On the other hand, the FET had transductance of 9.5 mho/Cm^2 at 26 A/Cm^2, and RON of 0.14 $ohm\text{-}cm^2$. Notice that the data is for 450v breakdown devices. The data clearly indicated the superiority of the BJT due to its conductivity modulation in the epi region. An inverter circuit was built with 100 ohm load, and the BIPMOS as the active switch to evaluate switching time and frequency response. Typical measured ton and toff were within the 100 nsec range. The inverter responded perfectly to switching frequencies up to 20 MHz, fed by a signal generator to its gate and base inputs. A BIPMOS Darlington was also tested where the FET's source was connected to the base of the BJT as an input. The FET is the 1st amplifying stage converting input voltage to input current for the BJT. This Darlington had an effective transductance of 60 mho/Cm^2 at 220 A/Cm^2. (1 volt step input to the gate created 60 A/Cm^2 step in output current). Switching characteristics of the Darlington are limited by the BJT output [4]. The uncommitted BIPMOS switch can also be connected as a zener protected MOSFET, where the emitter is tied to the gate of the FET, which is a standard practice. However, the more interesting switch concept in terms of applications is the parallel combination of the BJT and the MOSFET -- the BIPMOS "Shunt" [4], It is a four terminal device with separate base and gate leads, thus there is full flexibility in turning on/off either the BJT or the FET. The device designed was at 450v, its conductance characteristics are shown in fig. 3. The BJT and FET share the same active area, thus a full comparison between the two technologies in terms of conductance and speed was given. The BJT exhibited six times the current handlin capability of the MOSFET at a minimal current gain of 10. Similarly, the MOSFET switched at six times the speed of the BJT (15 nsec vs 100 nsec). The uniqueness of the BIPMOS "Shunt" is that in the system level with the proper drive, it performs much like a quasi ideal switch. The problem of RSB in inductive load switching is eliminated, switching and conduction losses are reduced too. To further clarify it, we chose a DC converter circuit that is typical to off line switch mode power supplies (SMPS), fig. 4. The input voltage to the converter is rectified line voltage up to 380v. S_1 is the BIPMOS shunt switch. The smart controller is a PWM controller that is capable of providing the necessary timing signals to drive the BIPMOS. The FET of S_1 is turned on first (driver 1) which greatly reduces the voltage drop across the BJT. This voltage is determined by the RON of the FET The BJT then may be turned on (driver 2) to take advantage of its higher conductance and lower voltage drop. On turn on, a short delay is sufficient between the FET's turn on and the BJT's, this can be naturally provided by the BJT's inherent delay. On turn off, the base drive is just removed and after a delay sufficient to let the BJT turn off (less the 1 us) the MOSFET is switched off. Thus the switching across the high voltage is done by the MOSFET, the heavy current conduction is done by the BJT for the proper durations. The timing diagram for such a BIPMOS converter is shown in fig.5, for 100 khz application and 40% duty cycle. Ta and the Tc are the short times the MOSFET is only on, and Tb is when both the FET and BJT supply the current to the load. An efficiency analysis of the BIPMOS converter versus a MOSFET and a BJT converter was done to quantify the superiority of the BIPMOS. Note also, that due to the switching of the BJT across a much reduced voltage, while the FET is on, eliminates its RSB failure mode. The efficiency calculations are based on the following criteria:

1) A one ampere current is switched across a 380v line with 50% duty cycle.
2) The BJT has tr=tf=ts=.3u sec and Vce (sat)=.6v
3) The FET is a four ohm device with tr=tf=15nsec
4) The BIPMOS shunt has a 15 ohm MOSFET with tr=tf=15nsec, and the BJT is conservatively rated as the BJT at 2). The switching delay between the BJT and the MOSFET is one usec.

The switching losses are approximated by:

$Ps=Vi \times I \times (tr+tf) \times f/2$ where 'f' is the frequency of the converter.

Conduction losses are given by:

$Pc=I \times Von \cdot D$; for the BJT and MOSFET alone Pc is not dependant on frequency. However, for the BIPMOS Pc is frequency dependant since the conduction period of the MOSFET half is an increasing share of the duty cycle as frequency increases. The overall losses of each switch as a function of frequency is given in fig. 6 (H stands for the hybrid BIPMOS & M for the Monolithic device). The losses of the BJT which are at low 'f', conduction losses mainly, increase rapidly and are dominated by its switching losses. The MOSFET on the other hand, maintains its high losses fairly constant up to 150 KHz, which after the switching losses add on with increasing frequency. The BIPMOS is similar to the BJT at low frequencies, but its switching losses do not rise as rapidly as the BJT's. At high frequencies, the BJT of the BIPMOS starts to limit the efficiency of the switch. It should ne noted, that the efficiency comparison presented was for a BIPMOS device for which the MOSFET was significantly down-rated in terms of RON. In reality for the Monolithic BIPMOS switch of 80x91 mils square (for 1A and 380v circuit condition) BJT Vce sat was 0.4v and RON of MOSFET was 4 ohms. The efficiency of such a device is added to fig.6. which indicates its applicability to the 300 KHz range.

IV SUMMARY

The Monolithic BIPMOS power device concept was described to combine the best performance features of the BJT and MOSFET in one monolithic chip. It was also shown that with a metal mask option, an uncommitted BIPMOS can be custom made for specific functions. The BIPMOS shunt for power control circuits offers significant efficiency advantages when compared to BJT or MOSFET alone topologies.

V ACKNOWLEDGEMENT

The author would like to extend his thanks to Intersil's application Engineering Group, device processing group, and to Jack Gifford for continuous support. Special thanks to Michelle Connolly for editing the manuscript.

REFERENCES

1. P.L. Hower and K.S. Tarneja, "The influence of circuit and device parameters on the switching performance of power transistors", presented at Power Conversion '79, in Munich, West Germany. Pc-79-6.7 p.p. 1-19

2. P.L. Hower, "A comparison of BiPolar and FET as power switches", Power Conversion International, Jan 1981 p.p. 45-53

3. B.R. Pelly and Z. Zansky, "High voltage power MOSFETS meet off line converter needs", EDN. Nov. 1979 p.p. 207-212

4. J. Meador and N. Zommer, "Using BiPolar-MOSFET combinations to optimize the switching transistor function", Power Con. #7, 1981, Dallas Texas.

FIG. 1
BIMOS SWITCH CONCEPT

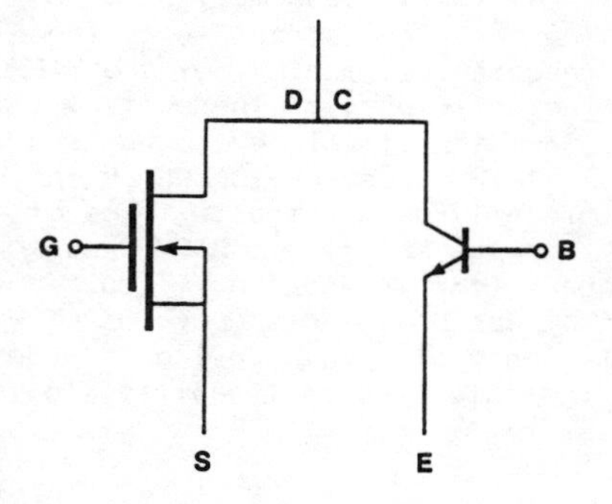

FIG. 2
BIMOS SHUNT

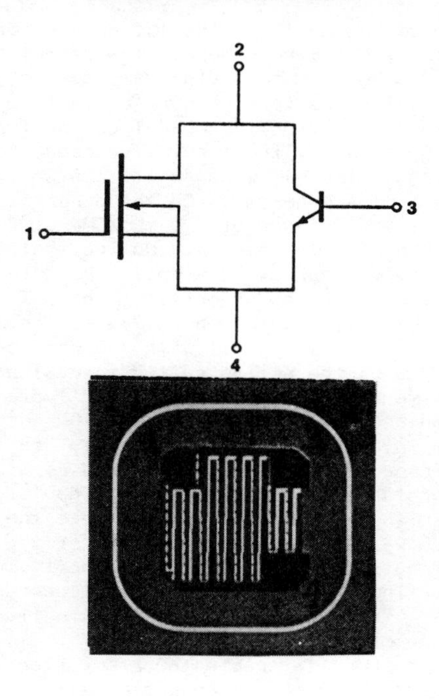

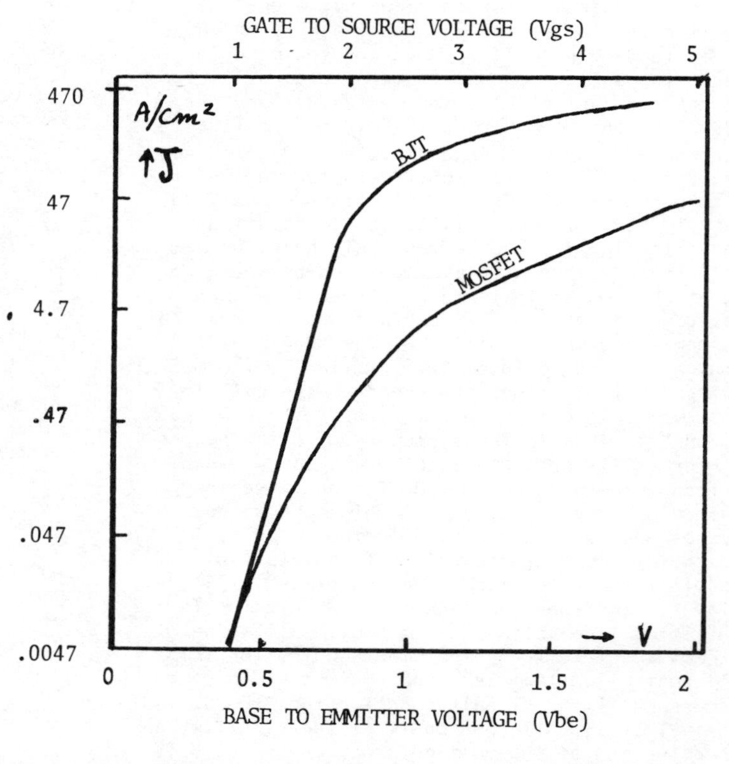

FIG. 3 Conductance comparison between MOSFET and BJT fabricated insame process as part of the BIPMOS.

FIG. 4

BIMOS CONVERTER

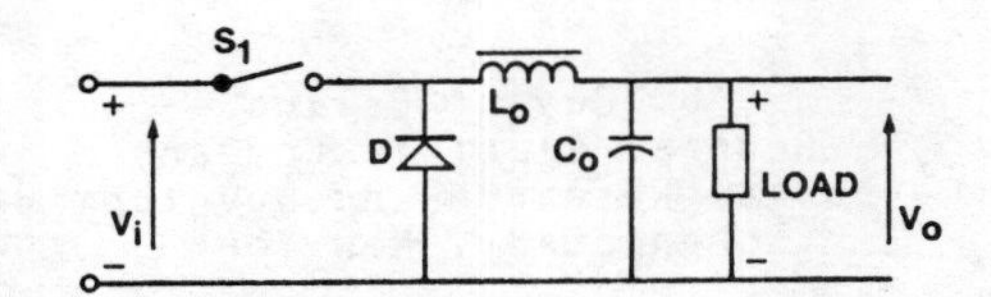

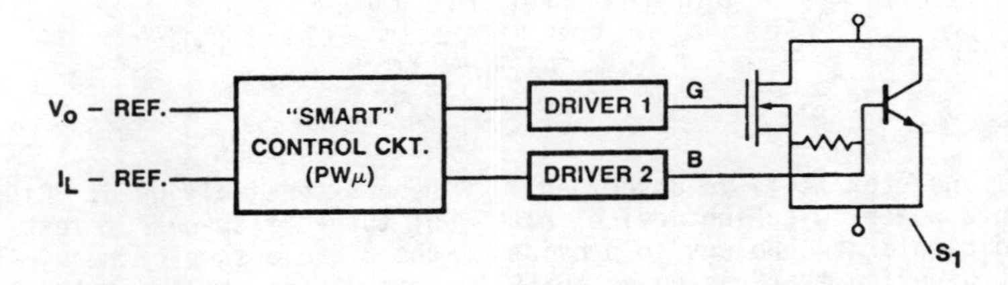

FIG. 5

TIMING DIAGRAM AND CURRENT FOR BIMOS CONVERTER

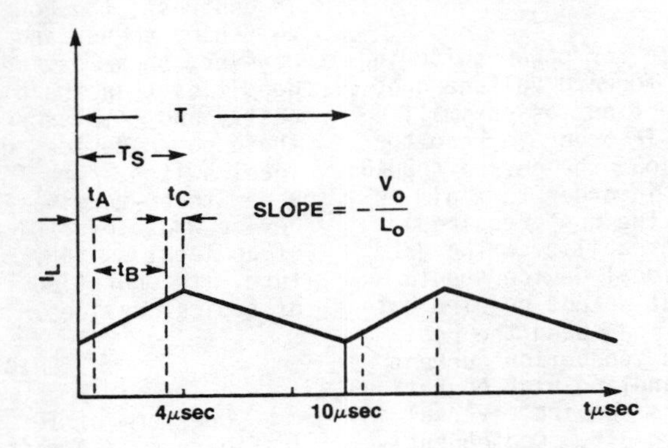

FIG. 6

BIMOS SHUNT EFFICIENCY

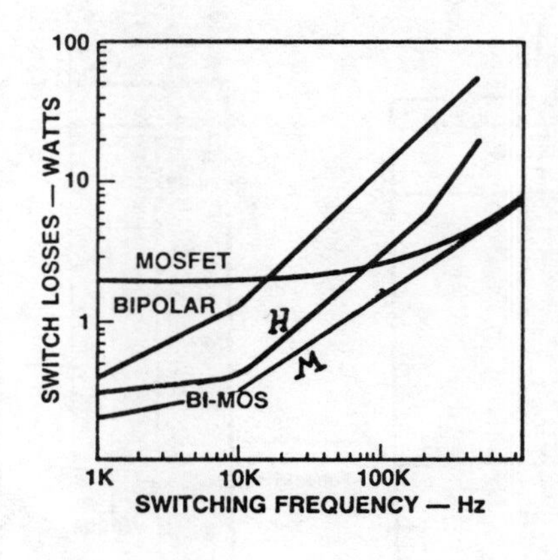

(1) A one amp current is switched across a 380 volt line with a 50% duty cycle.

(2) The BJT has $t_r = t_f = t_s = .3\mu s$ and $V_{CE(SAT)} = .6V$.

(3) The MOSFET is a four ohm device with $t_r = t_f = 15ns$.

(4) The BI-MOS uses a 15 ohm MOSFET with $t_r = t_f = 15ns$ and the BJT of (2), with the MOSFET switched off $1\mu s$ after the BJT.

THE INSULATED GATE TRANSISTOR (IGT) - A NEW POWER SWITCHING DEVICE

By

B. Jayant Baliga
General Electric Company
Corporate Research and Development
Schenectady, New York

Mike Chang, Peter Shafer, and Marvin W. Smith
General Electric Company
Semiconductor Products Department
Auburn, New York

The Insulated Gate Transistor (IGT) is described in this paper. It is a new power switching device that combines Power MOSFET and bipolar technology to provide the circuit designer with a device that has Power MOSFET input characteristics and bipolar output characteristics. Device ratings, characteristics and applications are discussed. Practical operating frequencies, voltage, and current limits, as well as interface requirements with other devices are defined.

Introduction

The ideal semiconductor device for power switching applications should exhibit a low forward voltage drop in order to keep the forward conduction losses small and a high turn-on/turn-off speed in order to keep the switching losses small. In addition, the device should operate at a high-current density in order to minimize the size of the chip (and, hence, the cost required) for any desired current handling capability. The gate drive power requirements for the ideal device should also be low. For applications that do not require gate turn-off capability, the thyristor has been the best choice because of its high forward conduction current density, even when designed for handling high operating voltages. However, many thyristors require reversal in the anode potential to achieve turn-off. Consequently, for applications where the load current is controlled by the input signal, the power bipolar transistor has been extensively used. Bipolar transistors are capable of turn-off speeds of less than 1 microsecond, but have the disadvantage of requiring high base drive current both during the on-state and during turn-off. Other competitive devices that have been developed for gate turn-off applications are the gate turn-off thyristor (GTO) and the Power MOSFET. The GTO has the advantage of operating at higher current densities than the bipolar transistor but cannot operate at high switching speeds and requires very high gate turn-off currents. In contrast, the Power MOSFET is capable of very high switching speeds and requires low gate drive currents. However, these devices operate at much lower current densities than the bipolar transistor and have more costly and complex processing requirements. Thus, these power devices do not meet the requirements of an ideal switch. The Insulated Gate Transistor (IGT) is a new three-terminal gate turn-off power semiconductor device whose electrical characteristics approach those of an ideal switch. This paper discusses the IGT structure, its operation, and presents the characteristics of typical devices.

IGT Structure

One form of IGT structure is shown in cross-section in Figure 1. In this structure, current flow is prevented when negative biases are applied to the collector with respect to the emitter because the lower (collector)

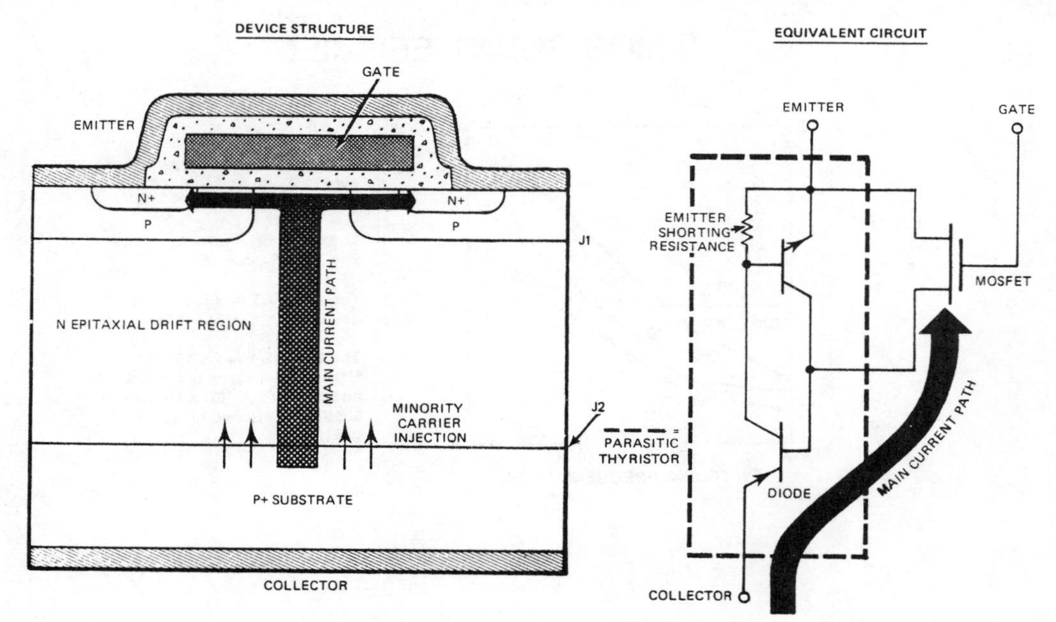

FIG. 1 Insulated Gate Transistor (Power MOS IGT)

Reprinted from *IEEE Ind. Appl. Soc. Meet.*, 1983, pp. 794-803.

junction, (J2), becomes reverse biased. This provides the device with its reverse blocking capability (see Figure 2). When positive voltages are applied to the collector with the gate biased at the emitter potential, the upper junction, (J1), becomes reverse biased and the device operates in its forward blocking mode. However, if a positive gate-to-emitter voltage is applied

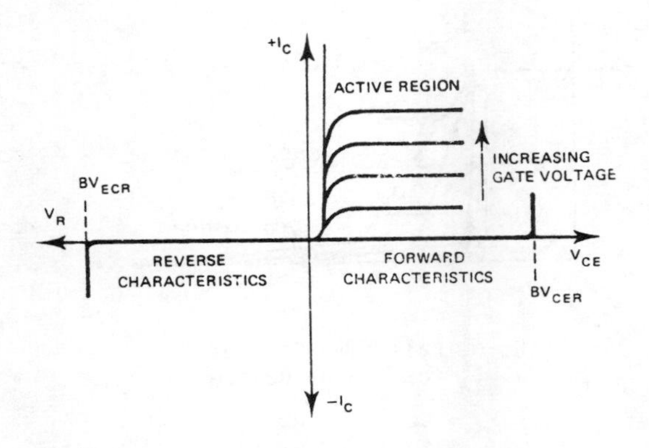

FIG. 2 Blocking Characteristics.

of sufficient magnitude to invert the surface of the p-base region under the gate, the device switches to its forward conducting state because current can now flow from the emitter N+ region into the n-base region via the inversion layer. In this forward conducting state, the junction J2 becomes forward biased and the collector P+ region injects holes into the n-base region. In this regime of operation, the device operates at high current densities like a forward biased p-i-n diode.

The significantly improved current conduction of the IGT is illustrated in Figure 3 where its forward conduction current density is compared to that of the bipolar transistor and Power MOSFET. Due to its rectifier-like forward conduction characteristics, the current density of the IGT exceeds that of the bipolar transistor and Power MOSFET for forward voltage drops above 1 Volt. At a forward voltage drop of 2 Volts, the IGT current density is observed to be 5 times that of the bipolar transistor and 20 times that of the Power MOSFET.

The IGT is also designed to obtain the high input impedance gate controlled characteristics of Power MOSFET's. This allows controlled turn-on, as well as gate controlled turn-off. Devices with intrinsic gate turn-off time capabilities ranging from 0.2 to 20 microseconds have been constructed. It should be noted that decreasing the turn-off time of the IGT results in a sacrifice in forward current capability per unit area of silicon (current density). This is indicated in Figure 3 where curves for IGT's with gate turn-off times of 15, 1 and 0.25 microseconds are compared to that of the bipolar transistor and Power MOSFET. These characteristics point out two notable advantages of the IGT. First, in these devices, we have a unique ability to tailor its characteristics which allows trading-off conduction losses with switching losses in power circuits. Second, even after reducing the gate turn-off time of the IGT devices to 0.25 microseconds, the current density is still substantially higher than that of the Power MOSFET. As an example, at a current density of 100 Amps/cm^2, IGT (C) exhibits a forward drop of 5 Volts compared to 35 Volts for the equivalent (same die size and blocking voltage) MOSFET.

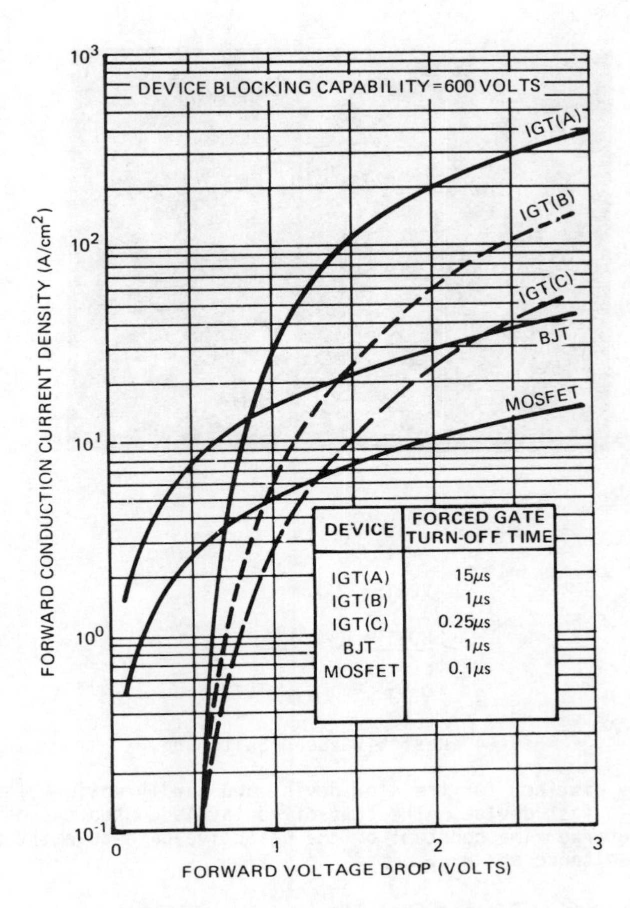

DEVICE	FORCED GATE TURN-OFF TIME
IGT(A)	15μs
IGT(B)	1μs
IGT(C)	0.25μs
BJT	1μs
MOSFET	0.1μs

FIG. 3 Forward Voltage Drop vs. Current Density.

It is worth pointing out that the IGT structure shown in Figure 1 contains a parasitic p-n-p-n thyristor structure between the collector and the emitter terminals as illustrated in the equivalent circuit. If this thyristor latches on, the current can no longer be controlled by the MOS gate. Consequently, it is important to design the device with a low emitter shorting resistance to suppress this thyristor action. This has been achieved in the IGT devices using high resolution Power MOS technology to provide margin over the gate controlled current density.

IGT Operation

Switching Properties

The IGT is designed such that the turn-on and turn-off times of the device can be controlled by the gate-to-emitter source impedance. Its equivalent input capacitance is lower than a Power MOSFET with a comparable current and voltage rating. The device is turned on by applying a positive voltage between the gate and emitter terminals. When V_{GE} is greater than $V_{GE(th)}$, collector current flows. In switching applications where $V_{GE} >> V_{GE(th)}$, the device saturates.

The IGT is similar to a Power MOSFET during turn-on and does not exhibit a significant storage time during turn-off. However, during turn-off, it exhibits a fall time that consists of two distinct time intervals - designated hereafter as t_{f1} and t_{f2}. Typical switching waveforms for a resistive load are shown in Figure 4 using two types of IGT's. The two time intervals are

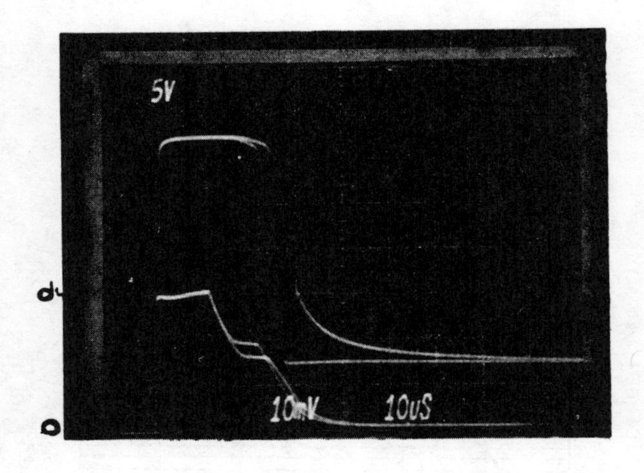

TOP (I_C)

VERT = 2A/cm

BOTTOM (V_{GE})

VERT = 5V/cm

FIG. 4 Resistive Load Switching.

very distinct for the slow device and hardly noticeable for a fast device. The turn-off delay is caused by the discharge time constant of the effective gate-to-emitter capacitance and R_{GE}.

Controlling Current Fall Time

The current fall time of the IGT can be controlled by use of external circuitry. t_{f1} is directly controlled by the value of R_{GE} (Figure 5). This dependence is shown in Figure 6. t_{f2} is not controllable and is an inherent characteristic of the type of IGT that is selected (that is, - slow, medium or fast switching types).

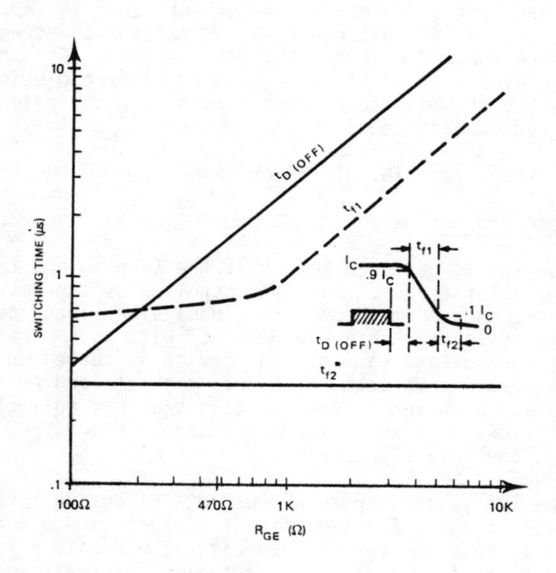

FIG. 5 Typical Turn-off Time vs. R_{GE}.

Since t_{f1} and t_{f2} contribute significantly to switching losses, the control feature of t_{f1} by a resistor (R_{GE}) is a definite advantage. For example, in the case of an inductive load, the fall time can be controlled to the extent that snubberless operation is possible, since E_L = -L dI/dt. Figures 6 and 7 are idealized representations of the two phases of the device turn-off. That is, a slow device can be used for d.c. and low fre-

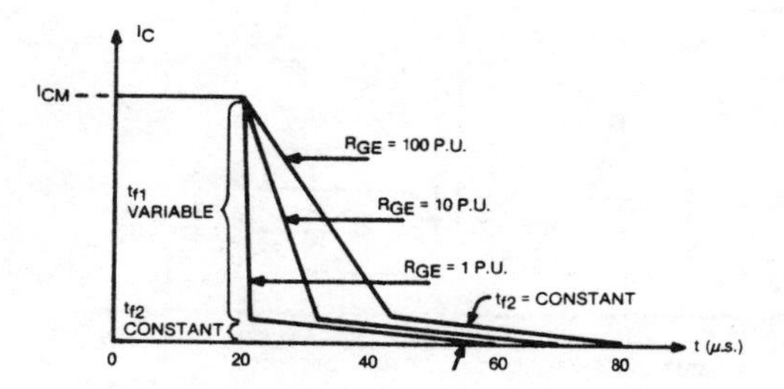

FIG. 6 Fall Time Control for a Slow Device.

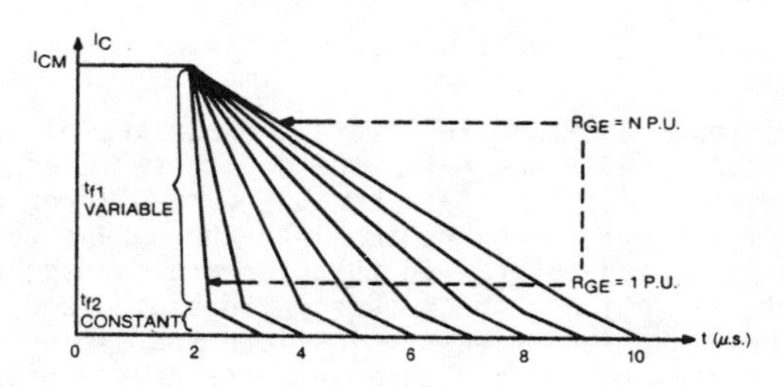

FIG. 7 Fall Time Control for a Fast Device.

quency applications with minimal gate turn-off current or a fast device can be used with a linear turn-off characteristic [R_{GE} = N per unit (p.u.), and N > 1 Figure 7]. For higher frequency operation, a fast device with R_{GE} = 1 p.u. will minimize switching losses due to t_{f1} and t_{f2}.

Conduction and Switching Losses

In switching applications, power losses in the IGT will consist of: (1) drive losses, (2) conduction losses, (3) off-state losses, and (4) switching losses. In dc or low frequency applications, where the total switching times are much less than the period, switching losses are generally negligible. Also, if the ambient temperature extremes are limited, the off-state losses are generally insignificant.

Using the waveforms in Figure 8 as a reference, switching losses can be determined as a function of time, current, and voltage. Note that this is a general procedure and that individual applications may differ. However, the procedure for calculating losses would be similar for any given set of waveforms. Figure 8(a) and (b) show typical resistive and clamped inductive load switching waveforms. For purely inductive loads, turn-n losses are very small because the transistor turns on at essentially zero current each cycle. Therefore, for a purely inductive load, switching losses will be determined by the turn-off losses.

Note the deviation from normal rise and fall time definitions (i.e., instead of 10% to 90% and 90% to 10%, 0 to 100%, and 100% to 0 is used to determine power dissipation).

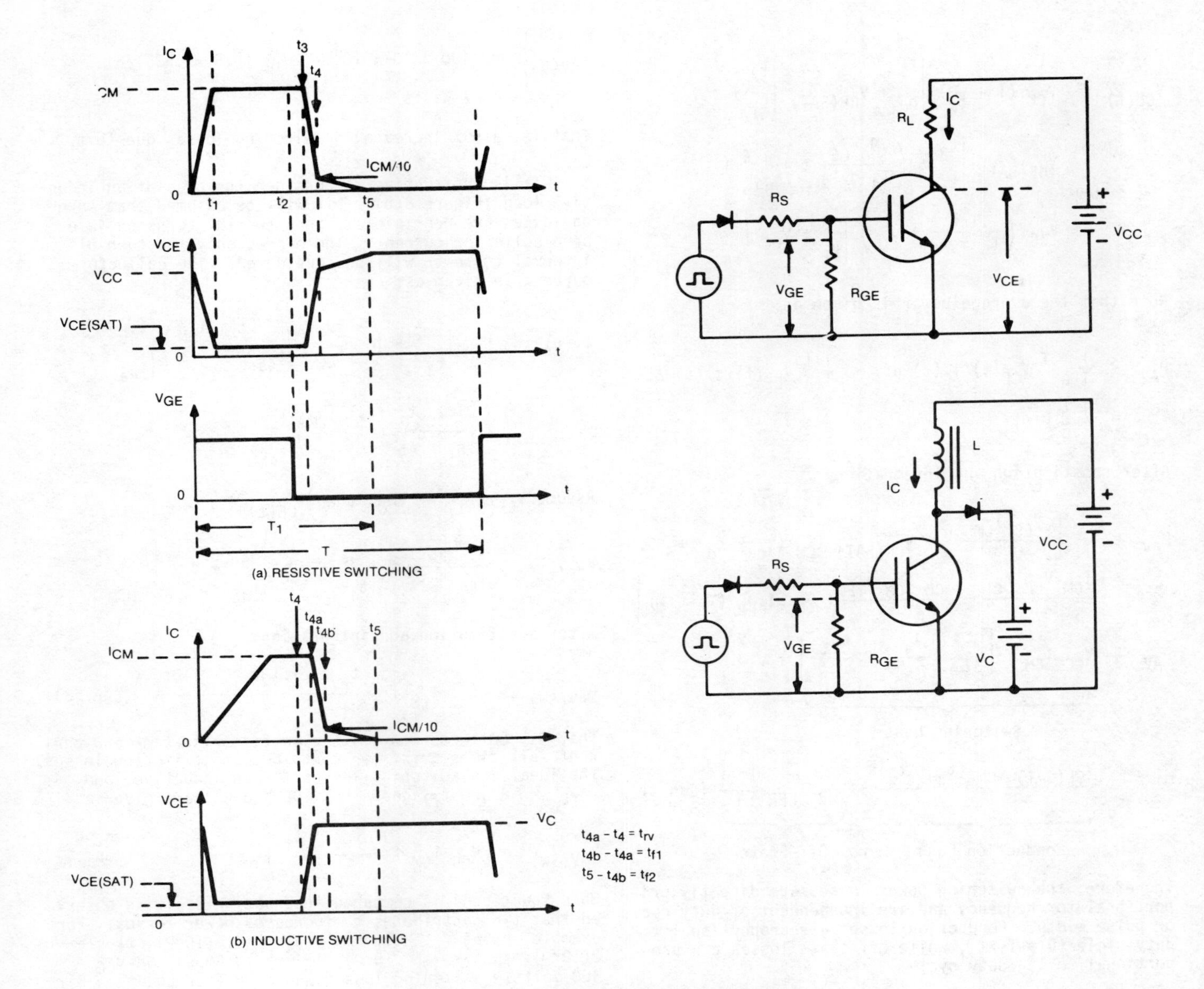

FIG. 8 Idealized Switching Waveforms.

$0 - t_1 = t_r$ = rise time, $t_2 - t_1 = t_c$ = conduction time;

$t_3 - t_2 = t_d$ = turn-off delay,

$t_4 - t_3 = t_{f1}$ = fall time one;

$t_5 - t_4 = t_{f2}$ = fall time two;

$T = 1/f$ = period, I_{CM} = maximum collector current;

$V_{CE(SAT)}$ = collector-to-emitter saturation volts;

V_{GE} = gate-to-emitter volts, V_{CC} = collector-to-emitter power supply voltage.

Switching Mode Application. When gate voltage is applied to an IGT operating in a saturated switching mode, the collector-to-emitter voltage decreases and goes into hard saturation if sufficient gate voltage is available. The sequence of events for a resistive load is shown in Figure 8(a). Notice the conduction region defined by $V_{CE(SAT)}$ and $I_{C(MAX)}$ during time interval $t_3 - t_1$. The switching losses can be summarized and are contained in intervals 0 to t_1 and $t_3 - t_5$ [see Figure 8(a)].

Total Power Dissipation. Instantaneous collector power is defined as: $P_p = [V_{CE}(t)]\,[I_{CE}(t)]$. The equation for collector current during the period T [assume $I_C >> I_{CER}$] is:

$$i_C(t) = \frac{I_{CM}t}{t_r}\Bigg|_{o}^{t_1} + I_{CM}\Bigg|_{t_1}^{t_2} + I_{CM}\Bigg|_{t_2}^{t_3} + \ldots$$

$$I_{CM}\left(\frac{-.9\,t}{t_4 - t_3} + 1\right)\Bigg|_{t_3}^{t_4} + \ldots$$

$$\frac{I_{CM}}{10}\left(\frac{-t}{t_5 - t_4} + 1\right)\Bigg|_{t_4}^{t_5} + I_{CER}\Bigg|_{t_5}^{T}$$

While the corresponding $V_{CE}(t)$ is (assume $V_{CE(SAT)} << V_{CC}$):

$$V_{CE(t)} \cong V_{CC}\left(1 - \frac{t}{t_r}\right)\Big|_{o}^{t_1} + V_{CE(SAT)}\Big|_{t_1}^{t_2} + \ldots$$

$$V_{CE(SAT)}\Big|_{t_2}^{t_3} + \left(\frac{.9\,V_{CC}}{t_4 - t_3}\,t\right)\Big|_{t_3}^{t_4} + \ldots$$

$$V_{CC}\left(\frac{.1\,t}{t_5 - t_4} + .9\right)\Big|_{t_4}^{t_5} + V_{CC}\Big|_{t_5}^{T}$$

Note that the average power is given as:

$$P_{AV} = \frac{1}{T}\int_o^T V_{CE}(t)\,i_C(t)\,dt = \frac{1}{T}\sum_o^T V_{CE}(t)\,i_C(t)\,\Delta t$$

$$= \frac{1}{T}\sum_o^T P\,\Delta t$$

After substitution and integration,

$$P_{AV} = \frac{1}{T}\left[\frac{V_{CC}\,I_{CM}\,t_r}{6} + V_{CE(SAT)}\,I_{CM}\,(t_c + t_d) + \ldots \frac{I_{CM}\,V_{CC}\,t_{f1}}{5.55} + \frac{V_{CC}\,I_{CM}\,t_{f2}}{21.5} + V_{CC}\,I_{CER}\,(T - T_1)\right]$$

$$P_{AV} = \underbrace{V_{CC}\,I_{CM}\,f\left[\frac{t_r}{6} + \frac{t_{f1}}{5.55} + \frac{t_{f2}}{21.5}\right]}_{\text{Switching Loss}} + \ldots$$

$$\underbrace{\frac{V_{CE(SAT)}\,I_{CM}\,(t_c + t_d)}{T}}_{\text{Conduction Loss}} + \underbrace{V_{CC}\,I_{CER}\,\frac{T-T_1}{T}}_{\text{Off-State}} \quad \text{(Eqn. 1)}$$

Therefore, the switching power losses are directly proportional to frequency and are independent of duty cycle or pulse width. Conduction losses are proportional to duty cycle ($D = T_1/T$), while off-state losses are proportional to 1 - duty cycle.

In order to assess the significance of Equation 1, idealized switching losses for a resistive load are given in Equation 2.

$$P_{AV} = V_{CC}\,I_{CM}\,f\left(\frac{t_r + t_f}{6}\right) \quad \text{(Eqn. 2)}$$

Where t_r, t_f = current rise and fall times from 0 to 100% and 100% to zero, respectively.

By inspection of Equation 1, it is observed that t_{f2} adds another component to the switching losses and will be a limiting parameter for high speed switching applications. For example, if the switching frequency is 10 kHz, $t_r = t_{f1} = .5(10^{-6})$, $t_{f2} = 2(10^{-6})$, I_C = 10 Amps, and V_{CC} = 400 Volts. Switching losses would be (per Equation 1):

$$P_{AV(SW)} = 400 \times 10 \times 10^4 \times 10^{-6}\left[\frac{.5}{6} + \frac{.5}{5.55} + \frac{2}{21.5}\right]$$

$$= 10.72\ \text{W}$$

Now, per equation 2 (let t_f = .555 μs to account for 100% fall time):

$$P_{AV(SW)} = 400 \times 10 \times 10^4 \times 10^{-6}\,[(.5 + .555)/6]$$

$$\cong 7\ \text{Watts}$$

That is, a 53% increase in switching losses due to t_{f2}.

It is instructive to consider the case of an inductive load [Figure 8(b)]. It will be assumed that turn-on losses are negligible, since the IGT turns on into zero collector current. Therefore, only the turn-off interval t_3 to t_5 will be considered. The collector current is given as:

$$i_C(t) = I_{CM}\Big|_{t_4}^{t_{4a}} + \left(-\frac{.9\,I_{CM}}{t_{4b} - t_{4a}}\,t + I_{CM}\right)\Big|_{t_{4a}}^{t_{4b}} + \ldots$$

$$\left(-\frac{.1\,I_{CM}t}{t_5 - t_{4b}} + .1\,I_{CM}\right)\Big|_{t_{4b}}^{t_5}$$

Assume $V_{CE(SAT)} >> V_{CC}$, then $V_{CE(t)}$ is:

$$V_{CE(t)} = \frac{V_{CC}}{t_{4a} - t_4}\,t\,\Big|_{t_4}^{t_{4a}} + V_{CC}\Big|_{t_{4a}}^{t_{4b}} + V_{CC}\Big|_{t_{4b}}^{t_5}$$

After substitution and integration:

$$P_{AV(SW)} = V_{CC}\,I_{CM}\,f\left(\frac{t_{rv} + 1.1\,t_{f1} + .1\,t_{f2}}{2}\right) \quad \text{(Eqn. 3)}$$

That is, collector-to-emitter voltage rise time and current fall times $t_{f1} + t_{f2}$ contribute to switching losses. The idealized switching losses for an inductive load (t_{f2} = 0 and t_{f1} goes from 100% to 0%) are given as:

$$P_{AV(SW)} = V_{CC}\,I_{CM}\,f\left(\frac{t_{rv} + t_f}{2}\right) \quad \text{(Eqn. 4)}$$

When Equation 4 is compared to Equation 3, it is observed that t_{f2} contributes to increased power losses. For example, using f = 10 kHz, $t_{f1} = .5(10^{-6})$, $t_{f2} = 2(10^{-6})$, $t_{rv} = .5\,(10^{-6})$, I_C = 10 Amps, and V_{CC} = 400 Volts, switching losses would be (per Equation 3):

$$P_{AV(SW)} = 20\,(.5 + .55 + .2) = 25\ \text{Watts}$$

per Equation 4,

$$P_{AV(SW)} = 20\,(.5 + .55) = 21\ \text{Watts}$$

That is, the contribution of power loss due to t_{f2} is not as much as one would intuitively believe at first glance - a 19% increase when compared to devices with a linear current fall time from 100% to zero.

Gate Drive

The input characteristics of the IGT are similar to a Power MOSFET. That is, it has a gate-to-emitter threshold voltage and a capacitive input impedance. In order to turn the device "on", the input capacitance must be charged up to a value greater than $V_{GE(th)}$ before collector current can begin to flow. The collector-to-emitter saturation voltage decreases with an increase in magnitude of V_{GE}. That is, for lowest values of "on" state voltage, V_{GE} should be much greater than $V_{GE(th)}$. Typical output characteristics are shown in Figure 9.

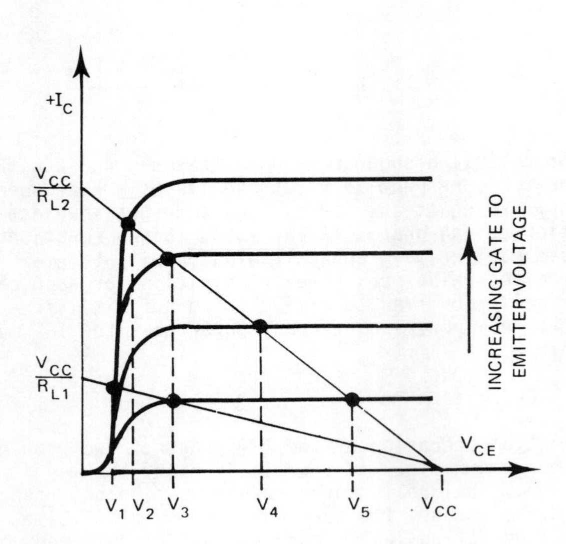

FIG. 9 Output Characteristics.

In order to turn the IGT off, a resistor between gate and emitter is all that is required. This resistor provides a path for the gate-to-emitter input capacitance to discharge. It must be emphasized that R_{GE} has a lower limit that cannot be exceeded. The IGT has a maximum controllable collector current that is dependent on the gate-to-emitter dV/dt. That is, the higher the gate-to-emitter turn-off dV/dt, the lower the controllable collector current. A typical device will have a minimum value of R_{GE}, which will vary from slow device types to fast device types.

The controllable current will also be dependent on the nature of the load. That is, the controllable current will be specified with an inductive and a resistive load at V_{CE} maximum.

Some typical drive circuits are shown in Figures 10(a) thru 10(d). There can be many variations in the drive circuit. It is safe to conclude that most circuits common to Power MOSFETs can be used to drive the IGT with minor modifications. In most cases, a negative turn-off voltage is not required between gate and emitter.

In Figure 10(a), a symmetrical drive is utilized to drive a lamp load. R_{GE} would be quite large and would minimize drive current. Required turn-on and turn-off times would be slow, tens of microseconds.

In Figure 10(b), a high speed asymmetrical drive is utilized. Isolation between primary and secondary is realized by the transformer T1. A diode is used in conjunction with R_{S1} to provide fast turn-on time without effecting turn-off time. Therefore, R_{S1} and R_{GE} would determine turn-on and turn-off times, respectively.

In Figure 10(c), an asymmetrical drive is utilized for a relay or solenoid driver. R_S determines turn "on" time, while R_{GE} determines turn-off time. CR1 is normally required for this type of application in order to prevent excessive collector-to-emitter voltage due to effects of L dI/dt during turn-off of Q_1. Since t_{f1} can be controlled, it is possible to eliminate CR1 in some applications or, in many cases, the need for critical layouts is minimized when CR1 must be used.

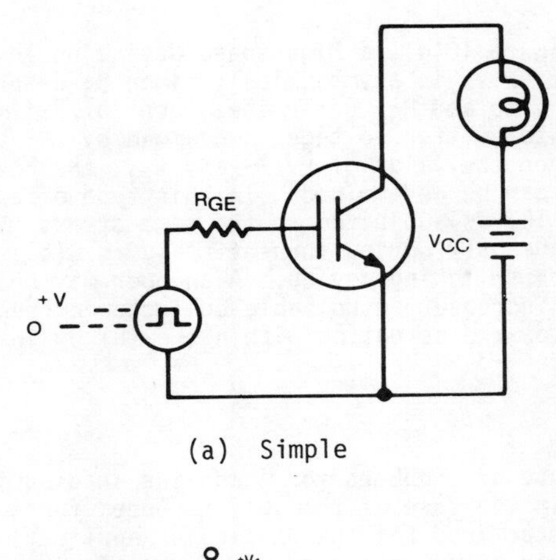

(a) Simple

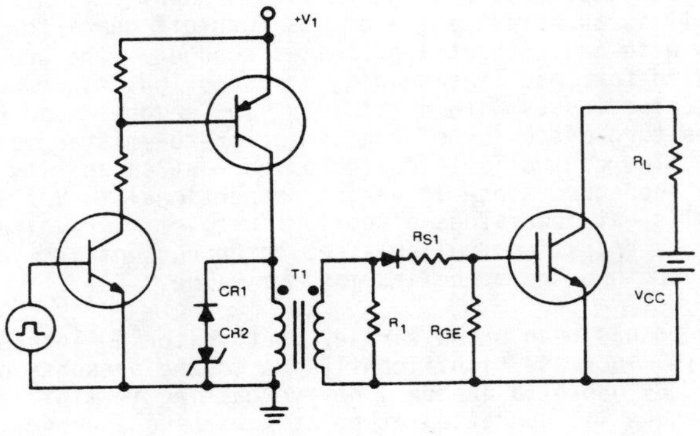

(b) Isolation

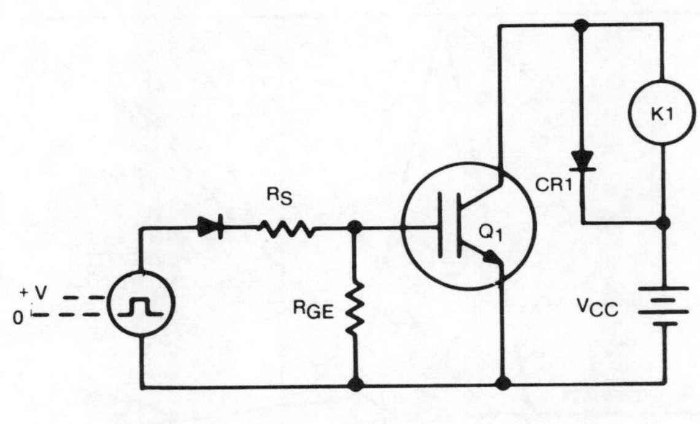

(c) Asymmetrical

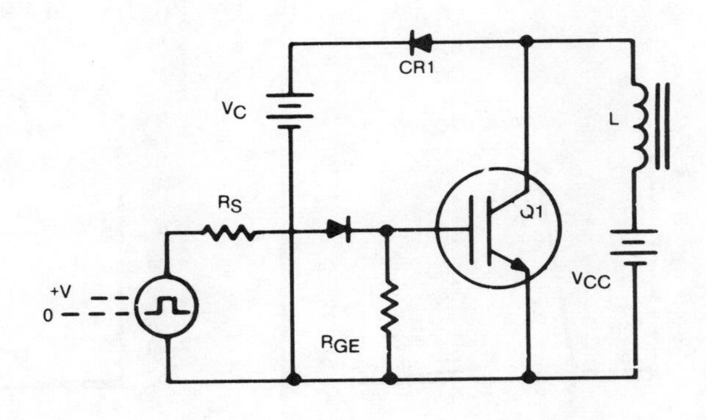

(d) Clamped Inductive

FIG. 10 IGT Drive Circuits.

In Figure 10(d), a high speed device would be used. The drive is asymmetrical, since R_S determines turn "on" time and R_{GE} determines turn "off" time. The collector-to-emitter voltage is clamped by CR1 and V_C. Depending on the value of L, I_C and t_{f1}, the need for the clamp can be determined. In this type of application, the IGT is subjected to the same stress [V_{CE} and I_C simultaneously during turn-off [Figure 8(b)] as any other power switching device. A snubber may be required to increase controllable collector current or to decrease power dissipation within the IGT by load line shaping.

Snubbers

The use of snubbers for load line shaping is well-known.[1] In the case of the IGT, the need for snubbers must be determined for the particular application. Figure 11 is an illustration of IGT turn-off characteristics with and without a polarized snubber. The snubber is used for load line shaping and functions to reduce switching losses within the IGT. The snubber also reduces turn-off dV/dt of the collector-to-emitter voltage. The controllable collector current capability is also increased since it varies proportional to V_{CE}. That is, at low values of collector-to-emitter voltage ($V_{CE} \ll V_{CER}$), controllable collector current is much greater than the specified maximum value.

It has been shown earlier that switching losses in the IGT increase significantly due to the presence of t_{f2}. By use of a snubber, device heating is minimized. In Figure 11, operation at point A with no snubber, t_{rv1}, t_{f1} and t_{f2}, the IGT has switching losses equal to:

$$P_{AV(SW)} = V_{CC} I_{CM} f \frac{t_{rv1} + 1.1 t_{f1} + .1 t_{f2}}{2}$$

When the polarized snubber is used (cases B and C with t_{rv2} and t_{rv3}, respectively), these losses are reduced. One can write equations for each case to get an exact expression or use graphical analysis. Since equations were used earlier, a graphical solution will be employed to determine peak power dissipation for each case. It is readily observed that snubbering definitely reduces the peak power and average power that the IGT must dissipate.

Parallel Operation

Parallel operation of the IGT presents two problems --static and dynamic. In the static case, the magnitude of the individual collector currents must be balanced. In the dynamic case, the turn-on time, turn-off time, and magnitude of collector current must be balanced. In order to assess the performance of the IGT, the "on" state voltage as a function of current, turn-on time, and turn-off times must be evaluated over the operating temperature range of the device. Figure 13 shows typical "on" state voltage as a function of current and temperature with constant V_{GE}. It is of interest to note that a typical IGT rated at 10 Amps has a negative temperature coefficient at collector currents less than 7 Amps. ΔV_{CE} varies from -.5 mV/°C at I_C = .5 Amps to a zero temperature coefficient at approximately 7 Amps.

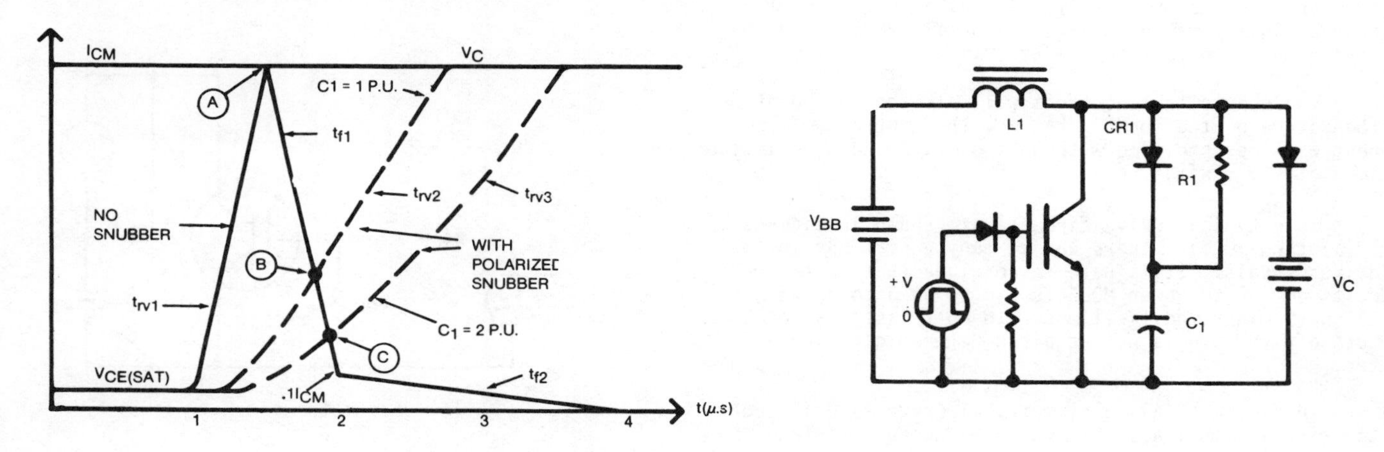

FIG. 11 IGT Turn-off with/without Polarized Snubber.

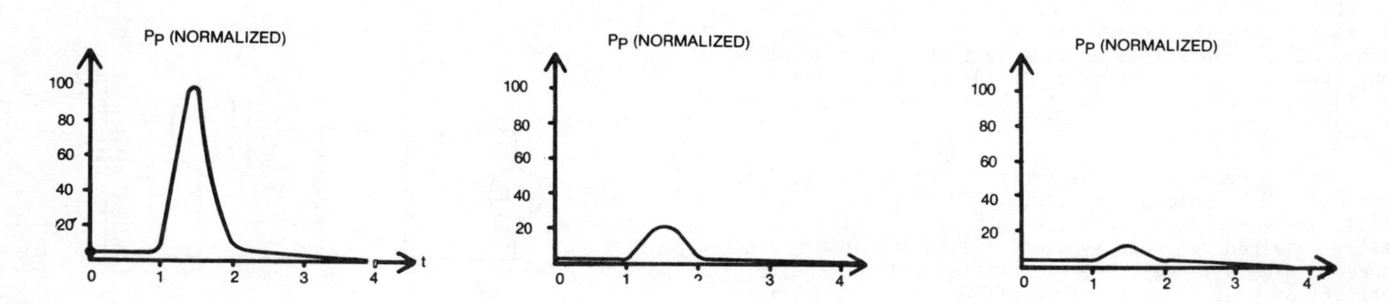

FIG. 12 IGT Peak Power with and without Snubbers.

[1]GE Transistor-Diode Manual, 1st Edition, M. Smith, editor, 1982, Chapter 8.

For currents greater than 7 Amps, the temperature coefficient is positive. At an I_C of 9 Amps, the temperature coefficient is approximately +.75 mV/°C. Therefore, the IGT behaves similar to a Power MOSFET and a bipolar. If the IGT is compared to a Power MOSFET with equivalent current and voltage ratings, it is clearly superior at elevated junction temperatures - since the on-state voltage of the Power MOSFET increases as much as 2.5 times from T_J = 25°C to T_J = 150°C.

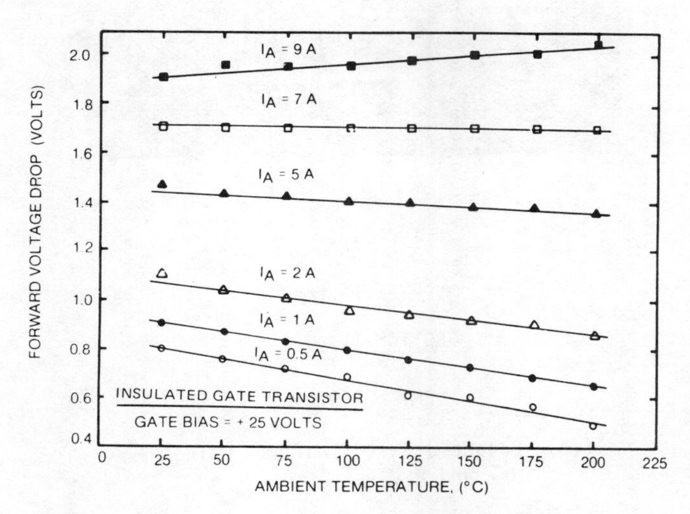

FIG. 13 V_{CE} vs. I_C and T_J for a 10 Amp Device.

Two IGT's can be connected in parallel. In order to achieve the best current sharing, the devices should be matched for gate-to-emitter threshold voltage $[V_{GE(th)}]$ and transconductance $\left(g_{fs} = \frac{I_C}{V_{GE}} \right)$. In addition, the physical layout must be such that there is geometric balance in the gate-to-emitter and collector-to-emitter areas. This is illustrated in Figure 14. In low frequency applications, the dynamic unbalance due to differential turn-on and turn-off times is insignificant and only the degree of static current unbalance between Q_1 and Q_2 is of concern. However, as the operating frequency increases, package inductance and series resistance [R and L in Figure 14(b)] in each lead of the device can cause dynamic unbalance. R_{S1}, R_{S2}, R_{GE1} and R_{GE2} are recommended to prevent oscillation and to tailor turn-on and turn-off times of each device. All connecting leads to the devices must be as short and symmetrical as possible.

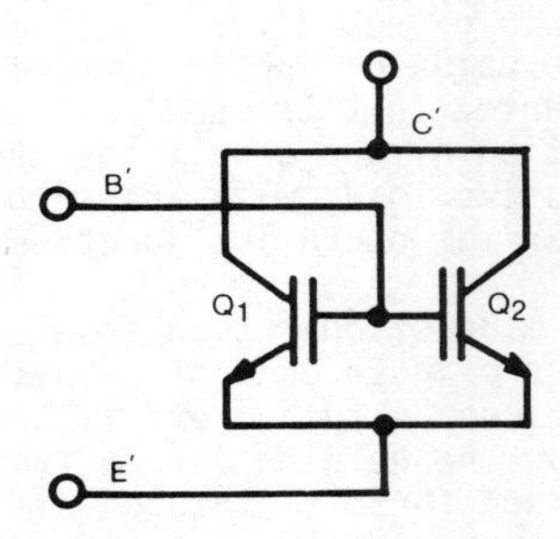

14(a) Low Frequency Equivalent Circuit.

For high frequency switching, the magnitude of the collector current is not a problem. However, differential switching times could cause a problem. The problem is illustrated in Figure 15.

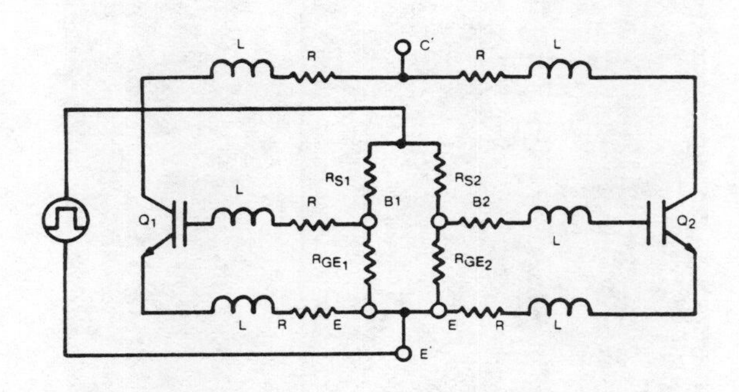

14(b) High Frequency Equivalent Circuit.

FIG. 14 Parallel Connection.

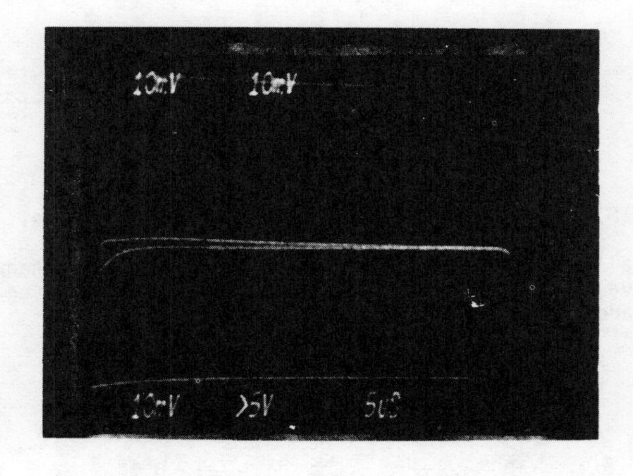

TOP (I_T)

VERT = 2A/cm

MIDDLE ($I_1 + I_2$)

VERT = 2A/cm

BOTTOM

V_{GE} = 20V/cm

15(a) Switching Times Equal.

In Figure 16(b), an extreme unbalance exists ($R_{S1} \neq R_{S2}$ and $R_{GE1} \neq R_{GE2}$) due to Q_1 turning on before Q_1 and Q_1 turning off before Q_2. This problem can be minimized by matching transconductance and tailoring the drive circuitry to provide equal turn-on and turn-off times. That is, adjust R_{S1} and/or R_{S2} [Figure 14(b)] for equal turn-on time and R_{GE1} and/or R_{GE2} for turn-off time. The results of balancing switching times by use of gate resistors using two units selected at random are shown in Figure 16(b).

TOP (I_C)

VERT = 2A/cm

BOTTOM (V_{GE})

VERT = 5V/cm

15(b) Unequal Switching Times.

FIG. 15 Parallel Operation in the Switching Mode.

If a further degree of balance is required other than drive compensation, bucking inductors may be employed in addition to matching transconductance.

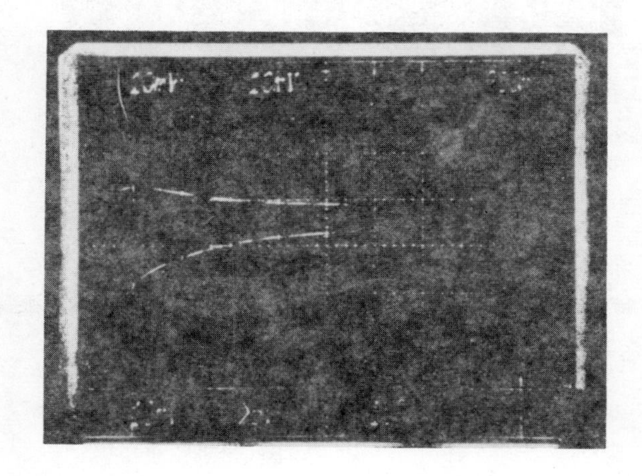

TOP (I_{TOTAL})

VERT = 2A/cm

BOTTOM ($I_{C1} + I_{C2}$)

VERT = 2A/cm

16(a) Switching Time Unbalance.

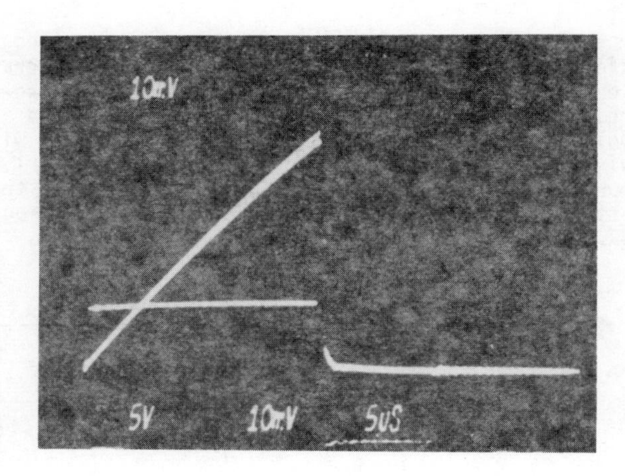

TOP (I_C)

VERT = 2A/cm

BOTTOM (V_{GE})

VERT = 10V/cm

16(b) Balanced Switching Time Using R_{GE} and R_{S1}.

FIG. 16 Parallel Operation.

Switching Times as a Function of Temperature

The IGT has a positive temperature coefficient associated with its fall time. It is approximately .27%/^{o}C. In critical applications where increased fall time will result in excessive power dissipation, the device turn-off time can be controlled by choosing the temperature coefficient of R_{GE} to compensate for device characteristics. If R_{GE} is a thermistor, it should be sized to minimize the effects of self-heating due to the application of constant power. If the thermistor is properly chosen, minimal change in the fall time will result - a definite advantage. The rise time of the IGT is relatively constant over temperature and is similar to a Power MOSFET. Therefore, no temperature compensation is required for turn-on time.

Protection

The IGT can latch "on" if the maximum controllable collector current is exceeded for a specified minimum value of R_{GE}. The value of R_{GE} is specified as a minimum value to guarantee that the device cannot latch for any rated combination of use conditions which include a resistive or inductive load, with and without snubbers and permissible maximum operating junction temperatures.

The latching mode of operation should be generally avoided, since loss of gate control results. If the device latches, it can be turned off if the collector current falls below the holding current for the device, or if the devices are force commutated similar to an SCR. The device can be easily controlled to prevent latching. Since the magnitude of collector current is controlled by V_{GE}, a limit on $V_{GE(max)}$ will not allow excessive collector current to flow. However, current source operation results if the load is shorted, thereby forcing the device to dissipate excessive power until it can be turned off. This is entirely possible in inverter circuits. The FBSOA characteristics are such that the device can survive full voltage and current for 5 μs. During the maximum intrapulse period, the device can be safely turned off to insure device

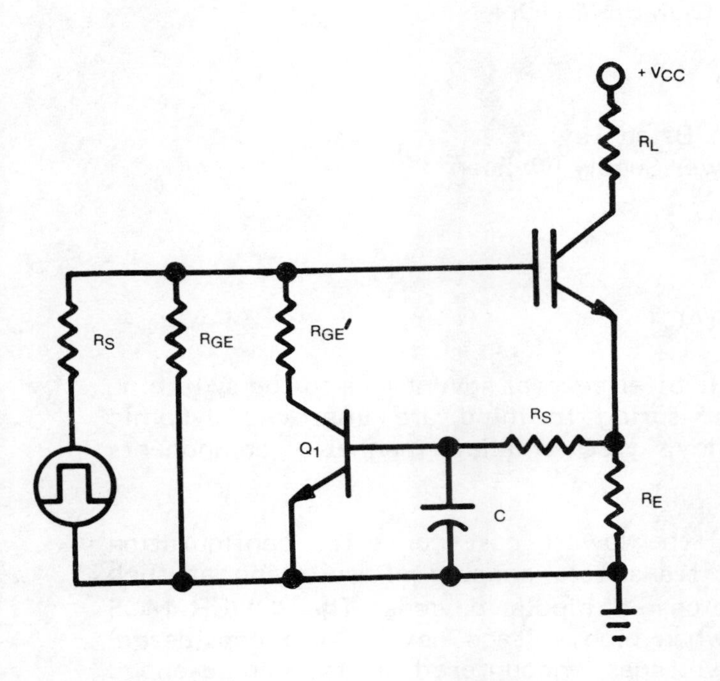

17(a) Current Viewing (Low Frequency)

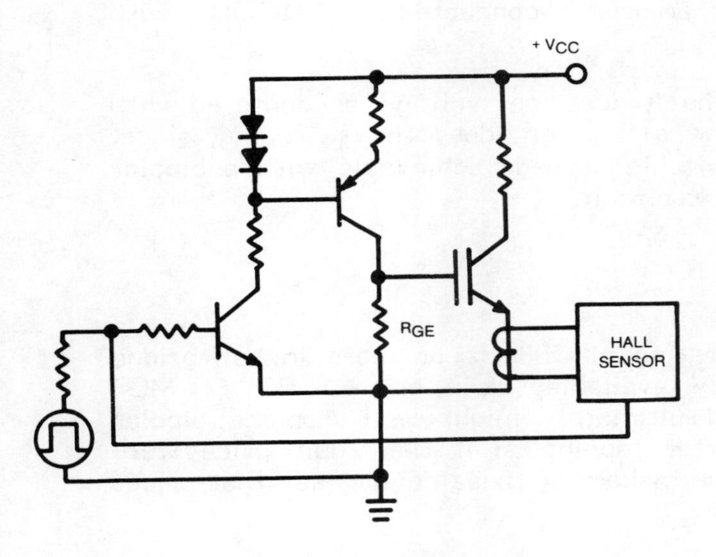

17(b) Hall Effect

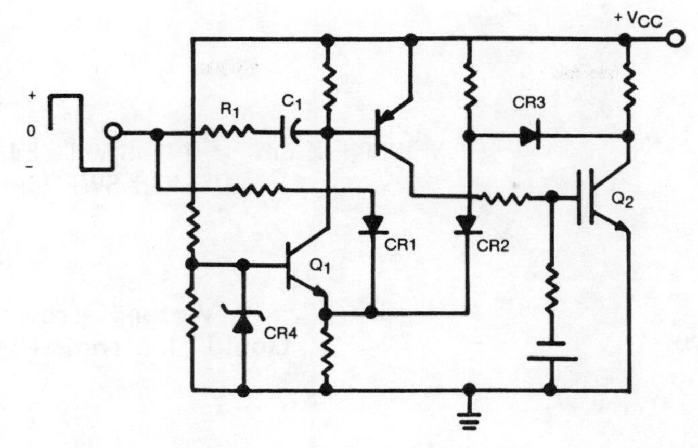

17(c) Intrapulse Sensing

FIG. 17 Typical Current Limiting Circuits.

survival. Current sensing can be implemented by use of a current viewing resistor in the emitter circuit. Hall Effect devices may also be used to eliminate excessive power dissipation in the resistor. Some typical current limiting circuits are shown in Figure 17. All circuits turn off the IGT when excess collector current exists. The circuit of Figure 17(a) functions to remove drive from the IGT when I_C R_E is greater than V_{BE} of Q_1. The disadvantage of this scheme is power dissipation in R_E. The Hall sensor in Figure 17(b) is free of excess power dissipation but is costly. The circuit of Figure 17(c) senses excessive current by turning off Q_1 if $V_{CE(SAT)}$ of Q_2 rises due to excessive current or collector-to-emitter voltage during the intrapulse period. If during the negative portion of the input voltage (circuit requires negative pulse for turn-on) a fault develops, the V_{CE} of Q_2 rises. This increase in V_{CE} reverse biases Q_1 and provides instantaneous shutdown of Q_2.

Conclusion

The Insulated Gate Transistor (IGT) is a new power switching device that features simple control circuitry, low on-state losses, fast (or slow) switching speeds and has FBSOA and RBSOA characteristics similar to a Power MOSFET. Devices have been constructed which are rated at 25 Amps and 600 Volts. Typical applications for the device would include relay drivers, motor speed control, switching power supplies, lamp drivers and fluorescent lighting. Samples of the device are available now.

A 300KHZ OFF-LINE SWITCHING SUPPLY USING A UNIQUE BI-MOS SWITCH COMBINATION

by

Victor Farrow & Brian Taylor
Gould Electronic Power Supply Division

ABSTRACT

Switching at frequencies in excess of 100kHz offer several advantages to the switching power supply. Two such advantages that spring to mind are improved dynamic response and higher volumetric efficiency, viz., smaller magnetic components and output filter capacitors.

For low output powers (up to 300 Watts) the lowest cost converter configuration should be the single-ended converter, but transistors capable of switching at such frequencies tend to be very expensive in the case of bipolar devices. The POWER-MOS (also very expensive - at the moment - when high voltage devices are considered) is currently incapable of handling the voltages encountered in the single-ended converter during 220V off-line operation.

The converter described here, in fact, fulfils all the necessary requirements. It utilises a low-cost high voltage bipolar transistor connected in CASCODE with a low-voltage POWER-MOS transistor.

The bipolar device in effect 'holds-off' the 'twice line' voltage encountered with this type of circuit, while the MOS device effectively does the switching, since, as can be shown, the highest possible switching speeds achievable with a bipolar transistor are realised by 'opening' the emitter circuit.

INTRODUCTION

Planar bipolar transistors capable of operating at 300kHz or more in half-bridge or full-bridge configurations are currently available, at a price. POWER-MOS devices are also becoming available, and ultimately might well displace bipolar transistors currently used in these converter topologies; if their unit price were to be considerably reduced. To be fair, the makers of these devices say that prices will fall.

For use in the single-ended converter (and in this context both forward and flyback arrangements should be included) it is almost always the MESA transistor that inherently becomes the first-choice candidate, due to the device's inherent ability to handle 'twice-rail' voltages economically. Unfortunately the MESA transistor has one serious drawback that would normally eliminate its inclusion in a high-frequency switching supply. This drawback is namely its reluctance to operate

Reprinted with permission from *Proc. of the 2nd Annual Int. Powerconversion Conf.*, 1980, pp. 3A6.1–3A6.10.

at high frequencies efficiently, i.e., saturated switching. Operated in a non-saturated mode with conventional drive circuits the MESA transistor can be coaxed to operate inefficiently.

Operating the MESA transistor (and this applies with equal validity to any high-voltage bipolar transistor, regardless of its geometry) in COMMON-BASE as against the more conventional COMMON-EMITTER connection, will allow the device to switch cleanly and efficiently at frequencies approaching 1MHz and beyond if so desired. COMMON-BASE operation is achieved by connecting the high-voltage bipolar in a CASCODE arrangement with a low-voltage high speed switching transistor. The choice of the low-voltage switch will of necessity be totally arbitrary and can be either bipolar or FET (it is immaterial whether the FET is a junction or a MOS technology device). The relative merits of the low-voltage switch can be simplified to a straight choice of low rise and fall-times and low VCE(sat) on the one hand for the bipolar, to slower rise and fall times, but total lack of storage time on the other hand for the FET. It was because of the lack of storage time that the MOS device was chosen for the converter described in this paper. It is for this reason that the switch combination is referred to as the BI-MOS switch.

The penalty for this high frequency capability is a slight increase in circuit complexity with the added constraint that the low-voltage transistor must also be capable of switching the full 'load' current.

Conversely there is no penalty where efficiency is concerned since both transistors are being switched into and out of 'saturation'. Of greater significance is the almost total freedom from Reverse bias Second Breakdown in the high-voltage transistor. The lack of Reverse bias Second Breakdown will be explained later, but also of significance is that breakdown voltage parameters for the selection of the high-voltage transistor are primarily centred around BV_{CBO}. There is also the added benefit that low-voltage high-speed switches are readily available and the cost of these devices is such that this switch topology can be realised economically and can be utilized in any converter configuration currently in use.

This paper shows how the BI-MOS SWITCH has been utilized in a 50 to 100W single-ended off-line converter. The choice of bipolar transistor, converter configuration and output power was deliberate and not imposed. The high-voltage transistor was chosen for its 'low switching speed' which occurs at low collector currents, viz., 0.5 Ampere to 1 Ampere. This of necessity defined the output power of the converter. The typical storage time of the transistor for this range of collector currents is of the order of 5 micro-seconds, while t rise and t fall are typically 500 nano-seconds. These times reduce considerably at higher collector currents but certainly do not fall below 2 micro-seconds total switching time. With pulse width modulation imposed by line and load regulation requirements the normally 'safe' maximum frequency of operation would be in the region of 100kHz for a single-ended configuration and slightly higher for push-pull.

The very significant improvement achieved with this transistor when implemented in a cascode arrangement can therefore be readily appreciated. Conversely the implication that limited power handling is inherent with this switch circuit is in fact unfounded. Experimental investigation with 15 Ampere transistors - switching their rated ICsat - has been carried out successfully and this in turn implies that the BI-MOS SWITCH can in fact be used in converters of any power rating where transistors would be considered suitable.

The high-frequency capability of the BI-MOS SWITCH circuit allows its inclusion for consideration in lower frequency switching applications where its improved performance and lack of Reverse bias SOAR limitations may be regarded as desirable if not essential.

OPERATION OF THE BI-MOS SWITCH

To understand the operation of the BI-MOS switch it is necessary to examine the two operational states as would occur with the simplified drive circuit of Fig.1. (The BI-MOS SWITCH is quite capable of operating in the converter with this arrangement but performance and efficiency is marginally impaired with the BUX48 as the bipolar half of the switch.) The two operational states of major concern in this instance are dynamic, i.e., the OFF-ON transition and vice-versa. The static states are of no real significance.

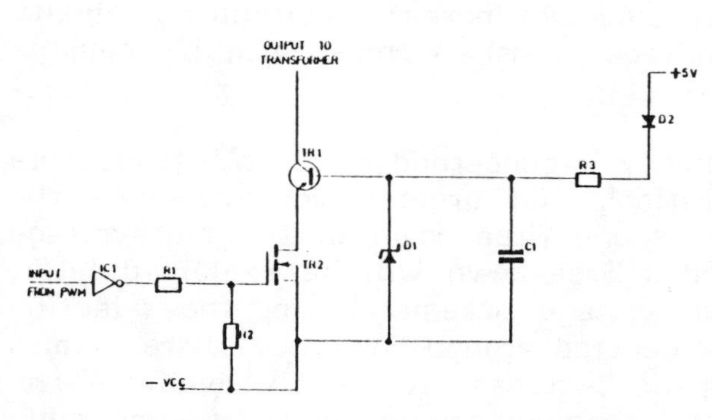

SIMPLIFIED DRIVE CIRCUIT FOR BI-MOS SWITCH. FIG.1.

(a) OFF-ON transition:

Assume quiescent conditions are prevailing after the preceding ON-OFF transition which will be considered in the next section. (All voltages and their polarities are referred to $-V_{CC}$.)

The base of TR1 is at a voltage determined by the leakage current of D1 and corresponding IR drop across R3, but will nonetheless have settled to about +14V. The gate of TR2 will be well below its threshold voltage as a result of the 'low' output of the TTL gate IC1.

A negative going pulse from the Pulse Width Modulator will force the output of IC1 into its 'high' state, and rapidly start the charge of the Gate-Source capacitance C_{GS} of TR2. (This necessitates the use of a buffered TTL line driver.) Upon exceeding V_{GS} threshold of TR2 the MOS transistor turns on pulling the emitter of TR1 down to about +1.5V. This action causes a large current to flow from C1 into the base of TR1 driving it very rapidly into conduction (t rise is approximately 50 nano-seconds - see photograph 1) and into its fully bottomed state. Base current of TR1 is maintained via D2 and R3 and is adjusted to maintain TR1 in saturation. This steady state will be maintained as long as the input to IC1 is held low.

If required, proportional drive to the base of TR1 could be employed by the use of a current transformer, with the secondary connected across C1 and a single turn primary in the collector circuit. It should be borne in mind that recovery of the energy in the core of the current transformer will take place during the off-time of the BI-MOS SWITCH and due allowance must be made for this energy recovery process.

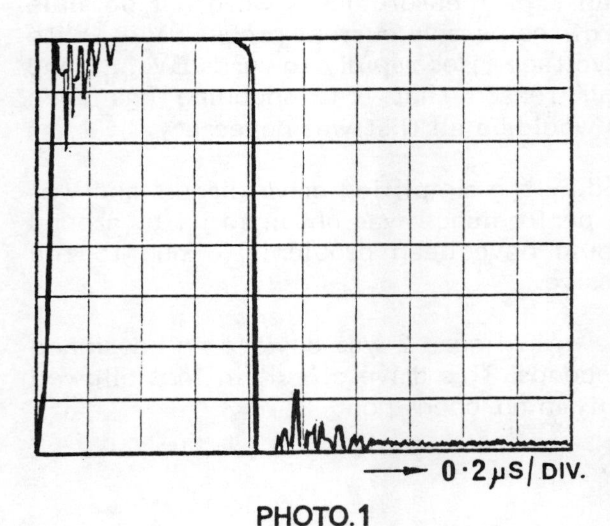

PHOTO.1

(b) ON-OFF transition:

At the cessation of the negative drive pulse to the input of IC1 the output of this TTL gate goes low and charge removal from C_{GS} and C_{DG} ot TR2 commences. (It was found in practice that speed up to the gate of TR2 was unnecessary.) When V_G falls below the off threshold of the device $I_{DS(ON)}$ goes into its fall time domain. This in turn commences the removal of forward base current of TR1 with a di/dt determined by the rate of fall of $I_{DS(ON)}$.

The combination of di/dt and any stray inductance in the emitter circuit of TR1 causes its VCE to change, and this in turn causes the collector to emitter capacitance of TR1 to commence passing an ever increasing current to start the 'sweeping out' of the charge in the base-emitter region of TR1.

Upon the cessation of $I_{DS(ON)}$ the full collector current of TR1 is now passed by the collector to emitter capacitance to complete the base-emitter charge removal (Note that this action must not be confused with 'current crowding' or 'current pinching' away from the periphery of the emitter, 1.3, normally associated with conventionally driven high-voltage bipolar transistors) causing the base of TR1 to start rising towards the clamp voltage imposed by D1. This sweeping out of charge is the storage time of TR1 and approximates to a maximum of 400 nano-seconds at a collector current of 0.5 Ampere and is proportional to the magnitude and duration of forward base and collector currents. (The figure of 400 nano-seconds should be compared with the 5 micro-seconds mentioned in the Introduction.) It should be pointed out at this point that if TR2 had been a bipolar transistor, reverse biasing of its base-emitter junction would be necessary to prevent 'Miller turn-on'.

At the end of the storage time the collector voltage of TR1 starts to rise at a rate dependent upon external circuit inductances in association with C_{OB} of TR1 and any stray capacitances. Since no emitter current is flowing, no Reverse bias Second Breakdown phenomenon can be said to exist.

As the collector voltage of TR1 continues to rise the collector-base region of TR1 goes into its reverse recovery mode, since it can be likened to a diode that had been forward biased during the transistors saturated 'on time' state, and the corresponding reverse recovery current flows into C1 and D1 causing the base of TR1 to rise finally to the clamp voltage of D1. (D1 is necessary to prevent the maximum rated Drain-Source voltage of TR2 being exceeded.)

Upon completion of collector-base reverse recovery (once again this recovery time is dependent on forward base current and conduction time) collector current ceases abruptly - t fall is about 40 nano-seconds, see photograph 1 again. With some Planar transistors that were used experimentally, fall time measurements were not possible due to the 60 Megahertz bandwidth limitation of the passive current probes used. With the cessation of collector current, collector voltage rises rapidly towards BV_{CBO} and possible avalanche conditions. It is for this reason that R-C snubbing has been employed. In a bridge converter clamp diodes would be all that was necessary.

The operation as described above was provided by the simplified drive circuit and was felt to be unacceptable, although reasonable performance was obtained up to around 300 kilohertz, but closed-loop regulation would have been problematic due to the storage time which was considered to be excessive.

It was for this reason that the drive circuit shown in figure 2 was devised and designed for a maximum total 'on' time of 900 nano-seconds. This drive circuit in fact allowed the converter to operate perfectly satisfactorily in an 'open-loop'

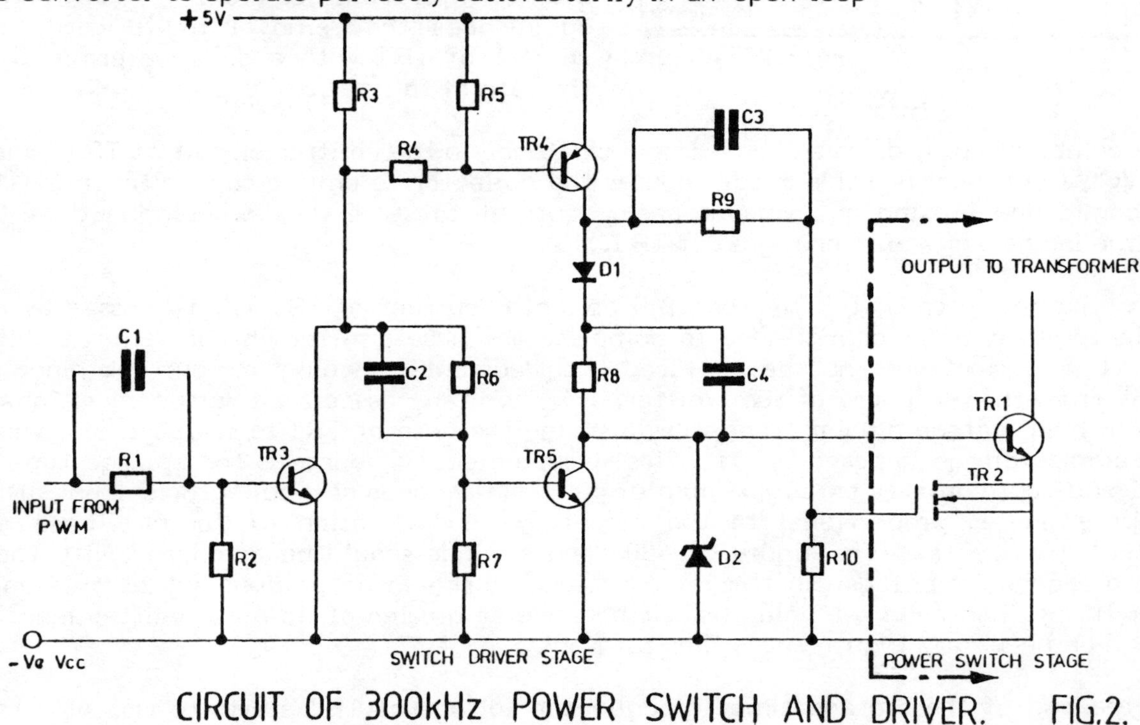

CIRCUIT OF 300kHz POWER SWITCH AND DRIVER. FIG.2.

state at a frequency of 500 kilohertz. It is interesting to note that at 500 kilohertz operation that the overall efficiency improved. (See section OVERALL PERFORMANCE.)

Returning to the pulse-drive circuit for the BUX48 shown in Figure 2, one finds that a positive going drive pulse from the pulse width modulator turns on TR3, but not into saturation; which in turn rapidly turns off TR5 and only slightly more leisurely switches on TR4. This action occurs within the first 100 nano-seconds - See Vbase and Ibase waveforms on photograph 2 - and results in TR1 being driven into its fully bottomed state.

For the maximum on time of 900 nano-seconds, after a time interval set by C2 and resistors R3, R6 and R7 in parallel with the input resistance of TR5 - and approximating to roughly 500 nano-seconds - TR5 is turned on in a progressive manner thus starving the base of TR1 of forward drive current, by linearly ramping down this base drive. This progressive base starvation to zero base current takes approximately a further 200 nano-seconds. From this point the base drive reverses and over the next 200 nano-seconds reverse base drive is linearly increased to the point in time when TR4 suddenly switches off, causing TR2 to abruptly turn off. The interval from the commencement of TR1 reverse base current to the turn off of TR2 may be regarded as the storage time of TR1 (which in turn is preceded by a period when the transistor was progressively taken out of saturation), and since this storage time lengthens with increasing conduction time, it may be assumed that the progressive turn on of TR5 crudely compensates for TR1 storage time, i.e., TR5's early turn-on only occurs for 'on times' of the BI-MOS switch that are greater than 500 nanoseconds. For shorter times than the 500 nano-seconds the full ramping down process described above is curtailed.

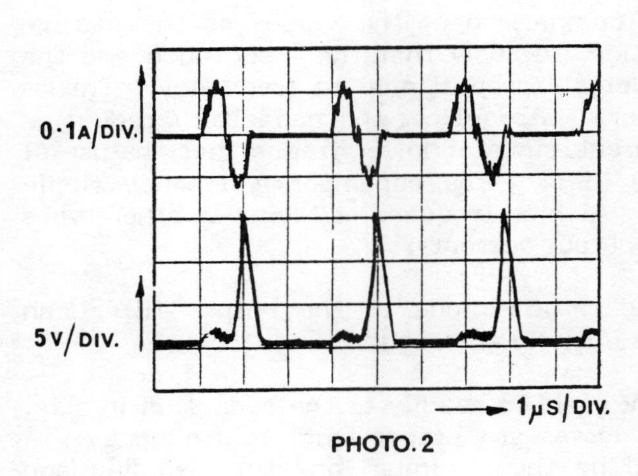

PHOTO. 2

At the termination of its storage time TR1's collector-base junction enters its reverse recovery mode - as described previously - leading to the rapid increase in voltage at the base of TR1. Examination of the voltage waveform for the base of TR1 in photograph 2 shows a peak value of +17V for an operational rail of +5V for the driver. The +17V is the clamp voltage of D2. This rapid increase in base voltage of TR1 is blocked by D1, thus preventing the ve application of gate drive to TR2.

TABLE 1.

ON TIME	STORAGE TIME	
	FIG. I. DRIVE CIRCUIT	FIG. 2. DRIVE CIRCUIT
0·3μs	NEGLIGIBLE	NEGLIGIBLE
0·5μs	NEGLIGIBLE	100 ns
0·7μs	100 ns	200 ns
0·9μs	200 ns	400 ns

The action of this pulse drive circuit proves a trade-off of slightly increased dissipation in TR1 during the time it is conducting when out of saturation for the benefit of greater control in turn-off of the BI-MOS SWITCH for shorter conduction intervals. The improved control can best be summarised in Table 1.

THE DESIRABILITY OF HIGH FREQUENCY SWITCHING

High frequency converters have been covered in considerable depth by numerous papers and articles in the past. Nevertheless it is felt desirable to touch upon a few pertinent points and in some cases dispel a few fears fostered by the 'mystique' surrounding high frequency converters.

Reiterating, the major attractions of high frequency switching are the use of smaller magnetic and capacitive components and thus realise a significant improvement in dynamic response.

The reduction in size of these components can also be realised for the EMI/RFI filter for conducted interference, and with ever more stringent requirements coming into effect, high frequency switching becomes even more desirable.

With respect to output filtering, a simple comparison of the values of the passive components will serve, by way of illustration, to show that the inductance and the capacitance of the output L-C filter is inversely proportional to frequency. This is found to be true in the case of the inductor. In practice, the capacitor's reduction with increasing frequency is even more apparent, since at lower frequencies the use of electrolytic capacitors is the norm where ESR of the capacitor and hence ripple current capability are the dominant factors. At high frequencies Mylar or other types of film capacitors would be the rule for low output currents.

The consequential improvement in dynamic response due to the output filter can therefore be greater proportionally than was initially envisaged.

The transformer, on the other hand, does not yield so readily to the reduction in size. This anomalous situation arises since core losses are proportional to frequency. A further constraint on transformer size will be the minimum one turn winding and available window area. On the other hand conventional ferrite cores and conventional winding techniques prove equally successful at high frequencies as they do at frequencies currently in vogue. The only precautions that should be observed for all windings is minimising the problems caused by skin effect and proximity. Exotic transformer construction techniques can sometimes create more problems than they solve.

When conversion frequencies approaching 1 Megahertz are considered, very occasionally is any reference made to the performance of the rectifiers associated with the transformer secondary. In converters operating at a few tens of kilohertz the reverse recovery time of fast recovery devices are found to be perfectly adequate. At frequencies of hundreds of kilohertz the reverse recovery time becomes a significant percentage of the available conduction time, and this could well be an embarrassment.

THE CONVERTER

Referring to Figure 3, it immediately becomes apparent that with the exception of the POWERSWITCH, the converter design and configuration is quite conventional. The transformer design and construction posed fewer problems than was originally envisaged and, as stated previously, should pose few serious problems for other configurations.

The transformer in fact uses a normal high-permeability RM10 ferrite core and only two precautions were deemed to be necessary:-

(a) The primary and secondary windings were interleaved to minimise leakage inductance, which when measured was found to be 12 micro-henries referred to the primary. In fact this seemingly high figure for leakage inductance was less of an embarrassment than was originally envisaged since it did limit primary current at the initiation of the 'Power stroke' when there would normally be a virtual short on the secondary due to the reverse recovery time of the flywheel diode D6.

(b) The primary and secondary windings were small-diameter multi-strand conductors to minimise 'skin effect'.

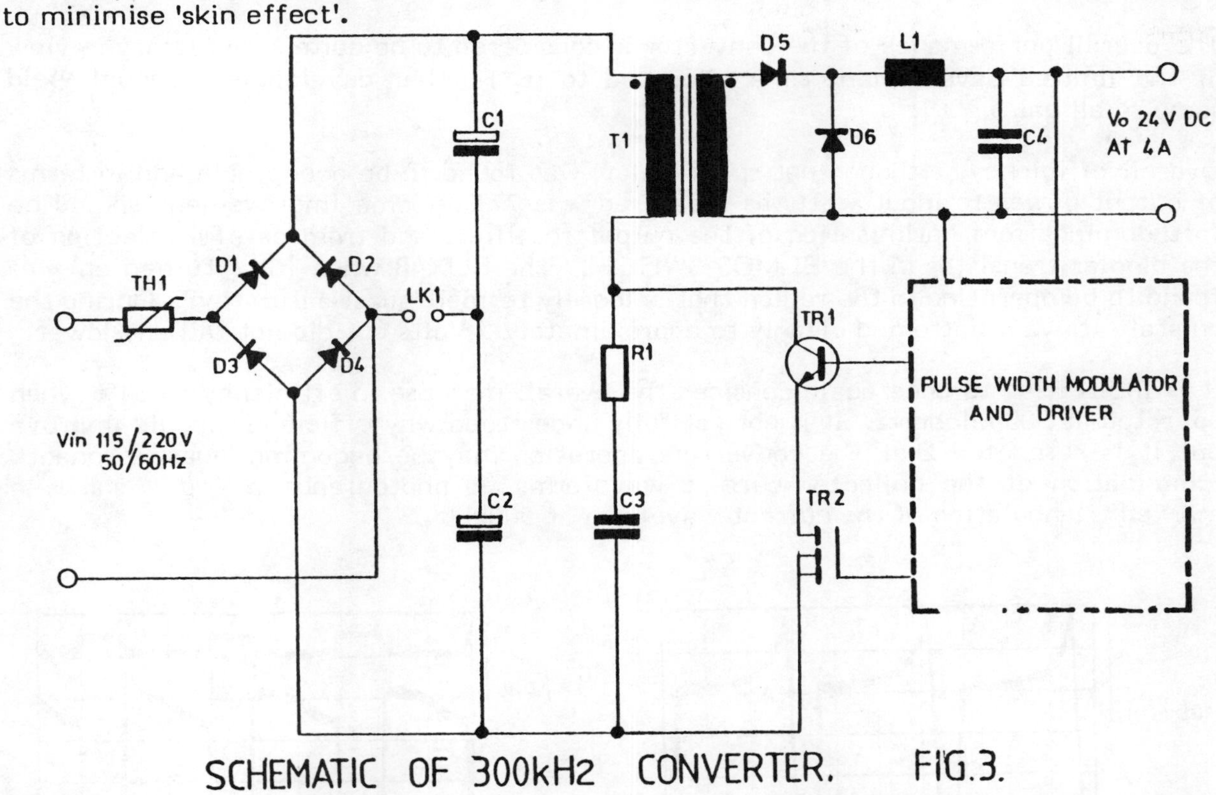

SCHEMATIC OF 300kHz CONVERTER. FIG.3.

As stated previously the freedom from reverse-bias second breakdown precluded the necessity of turn-off load line shaping but the R-C 'snubber' R1 and C3 was incorporated to keep the collector voltage of the transistor within its V_{CBO} rating.

The choice of secondary rectifiers provided the most serious challenge, as initially supposed, and the final choice of the 'epitaxial' devices that were used was settled on since these were the only devices conveniently available. This explains their being installed in a circuit requiring lower rated current devices. Faster recovery rectifiers would almost certainly have contributed to an improvement in performance.

It is interesting to point out that the use of MOS transistors as synchronous rectifiers was considered and investigation in this direction is planned for the future. The use of Ion implant devices is also worth considering due to faster recovery times over epitaxial rectifiers that are claimed for the Ion implant rectifiers. Unfortunately the cost of these devices tends to be fairly high.

For 5 Volt outputs Schottky barrier rectifiers should prove perfectly satisfactory.

No provision was made for overload protection for the converter but all options are open including the use of a fast cycle by cycle primary current limit to capitalise on the BI-MOS SWITCH's capability to turn off very rapidly and thus provide protection to the switch for virtually any eventuality.

OVERALL PERFORMANCE

The overall performance of the converter is considered to be quite satisfactory in view of the limited development time allocated to it. Further development should yield some small gains.

Overall efficiency, although not spectacular, was found to be acceptable and in terms of output power to input watts as measured was 77%. Some improvement should be forthcoming from optimisation of the output rectifiers and from careful selection of the bipolar transistor in the BI-MOS SWITCH. The BUX48 when fully 'turned on' was found to be operating in the region that is loosely termed 'quasi-saturation'. During the on state its VCE bottomed cleanly to approximately 3 Volts but did not fall any lower.

It is interesting to once again consider the overall increase in efficiency to 81% when operated at 500 kilohertz. It is not yet fully understood why efficiency should improve but it is suspected that the converters operation may be becoming 'quasi-resonant'. Examination of the collector current waveforms of photographs 3 and 4 shows a marked triangulation of the current waveform at 500kHz.

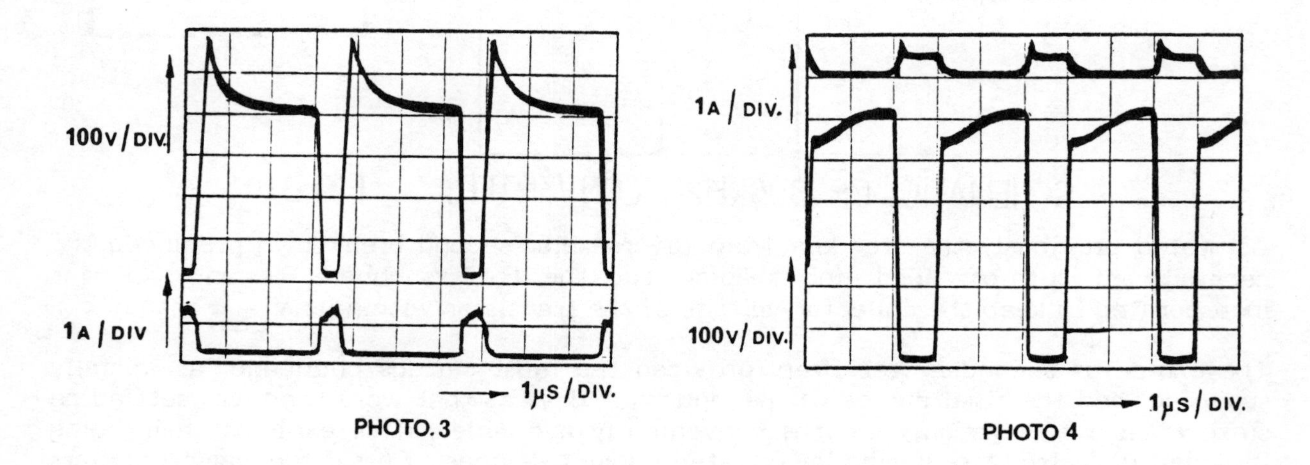

PHOTO. 3

PHOTO 4

A MOS transistor with lower rise and fall times would give a considerable reduction in switching losses or, alternatively, a very fast bipolar could be used with the consequential trade-off of storage time. But these improvements to the low voltage half of the SWITCH would only give marginal benefits.

Optimisation of the 'snubber' should also provide some benefit but this would be more difficult to achieve without incurring the penalty of additional cost.

Performance in terms of regulation was perfectly acceptable. The target specification was easily complied with but this was in some small measure due to its relatively conservative rating, and could most certainly be improved.

The transient response time of 36 microseconds puts the converter beyond the capability of 'run of the mill' commercially available switchers and begins to encroach upon the domain of 'linears'.

CONCLUSION

This paper has set out initially to show that no exotic or costly hardware is required in the design and construction of a high-frequency converter. It is true that this has in a small way been made possible by the Cascode connection of the BI-MOS SWITCH. It is equally true to say that no new technology is involved and that everything applicable to this converter is equally applicable to lower frequency converters.

REFERENCES

1. 'High Voltage High Performance Power Switching Transistors'
   By Lloyd H. Dixon, Jnr. Application Note U-75, Unitrode Corporation.

2. Soft Ferrites - Properties and Applications.
   E.C. Snelling. London ILIFFE Books Ltd.

3. A New Power Transistor Structure for Improved Switching Performances
   By K. Owyang, P. Shafer. General Electric Company, Discrete Semiconductor Device Center, New York.

4. Design of High Efficiency Off-Line Converters above 100kHz.
   By Rudolf Severns and David Sommers. Intersil Inc.

NOTE:
The CASCODE connection of the BI-MOS SWITCH is the subject of patent applications by Gould Electronic Power Supply Division.

Appendix
Designing Device Blocking Voltage Capability

One of the most important transistor characteristics is its blocking voltage capability. As discussed in the Introduction, the blocking voltage capability is firstly determined by proper selection of the resistivity (or doping level) and thickness of the n-base region. This Appendix provides data that can be used by the device designer to perform this task. The first graph (Fig. A) gives the variations in the breakdown voltage and the depletion width with the doping level of the n-base region. This graph can be used to design the n-base region of the power bipolar transistor, the power MOSFET and the power JFET. Since the epitaxial regions used during device fabrication are often specified by their resistivity, the relationship between the resistivity and the doping level are provided in Fig. B.

The design of the n-base region of the GTO, the power FCT and the power MOS-IGT must be performed differently by allowing for the open base PNP transistor that exists in these device structures. To perform this design, Fig. C illustrates the effect of depletion layer punch-through upon the blocking voltage capability. Here the blocking voltage is limited by avalanche breakdown at the higher doping levels and by punch-through of the depletion layer at the lower doping levels. This punch-through breakdown voltage is determined by the n-base width (W_n). To achieve the desired forward and reverse blocking capability, it is necessary to optimize the doping level and width of the n-base region. As an example in Fig. C, a blocking voltage capability of 1500 V can be achieved by using an n-base width of 200 microns and a doping level of 7×10^{13} per cm^3. However, it must be pointed out that the breakdown voltage of power transistors is determined by how the device structure is terminated at the edges of the chip. In order to prevent breakdown from occurring at the edges of transistors at low voltages, many device termination techniques have been developed. These techniques have been recently reviewed in a paper by Baliga that is provided here for reference.

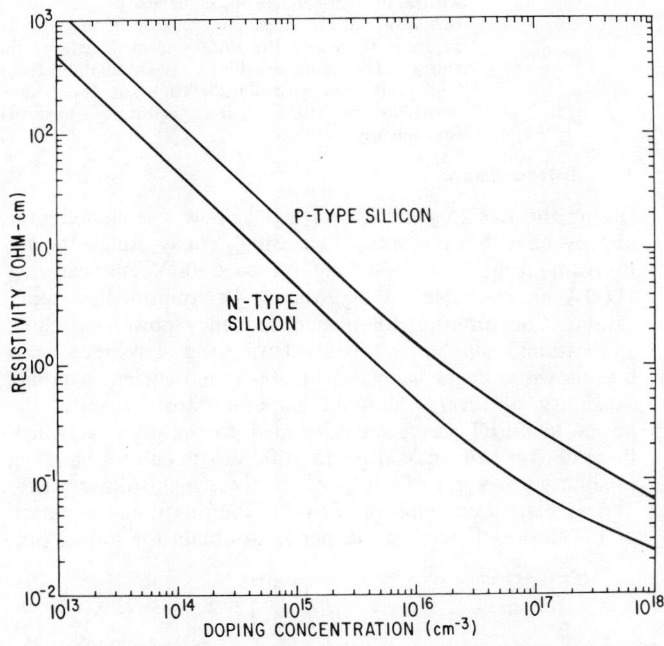

Fig. B. Dependence of resistivity of p and n type silicon upon the background doping level.

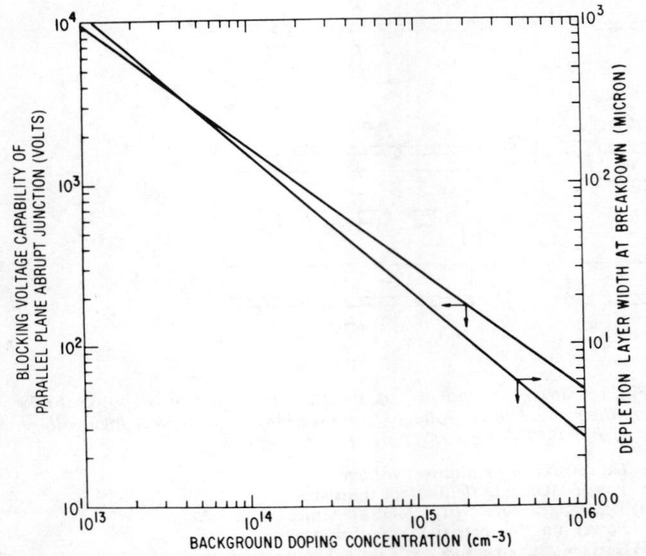

Fig. A. Variation of breakdown voltage and depletion layer width at breakdown with background doping level of abrupt junction diodes.

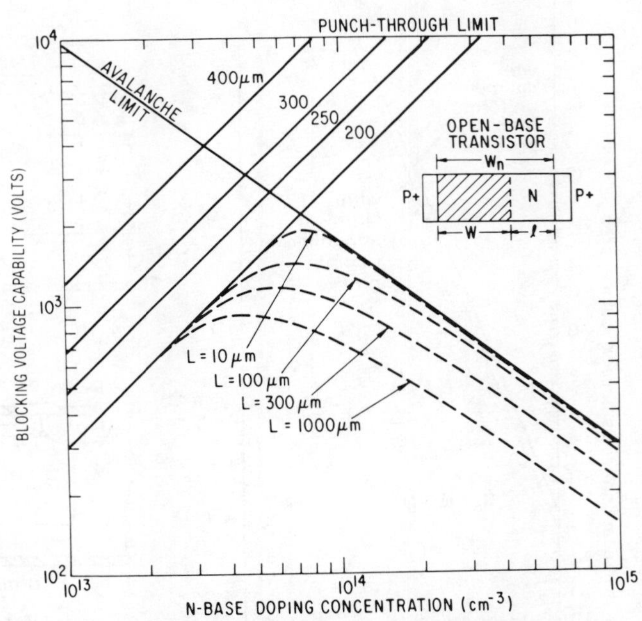

Fig. C. Blocking voltage capability of an open base transistor as a function of the base doping level and base width.

High-voltage device termination techniques
A comparative review

B. Jayant Baliga, B. Tech., M.S., Ph.D.

Indexing terms: *Transistors, Thyristors, Metal-oxide-semiconductor structures, Power electronics*

Abstract: High-voltage power device performance is often limited by the ability to approach nearly ideal behaviour at the edges of the chip. Consequently, a large number of termination techniques have been explored to reduce the surface electric field at the edges of devices, and so to maximise the breakdown voltage. The paper provides a review of these techniques. A comparison between the various approaches is then performed with consideration for device type (thyristors, field-effect transistors, transistors etc.) and device die size. This comparison is intended to serve as a guide to choosing the device termination appropriate for each application.

1 Introduction

During the last 25 years, the ratings of power semiconductor devices have been steadily increasing. Today, single power thyristors capable of blocking up to 6500 V and carrying 1000 A are available for high-voltage DC transmission applications. Concurrently, for higher frequency power switching applications, bipolar transistors have been developed with breakdown voltages in excess of 500 V and current handling capability of several hundred amperes. Most recently, the power MOSFET has been developed to the point at which these devices can work at up to 1000 V with current handling capability in excess of 10 A. All of these high-voltage power devices share a common problem in requiring the termination of the devices in such a manner as to obtain the lowest possible peak electric field at the edges. Unless this can be achieved with economical utilsation of the chip area, the device ratings will be substantially worse than in the case of an equal area device fabricated with ideal termination. This problem of effective edge termination in high-voltage power devices is particularly severe in the case of unipolar power devices, such as the power MOSFET and the power junction gate field-effect transistor. In these devices, the on resistances, which limit their maximum current handling capability, are determined by the resistivity and thickness of the drift region. Typical cross-sections of these devices are shown in Figs. 1*b* and *c* with the drift region indicated. As the breakdown voltage is increased, the depletion width W_D must be increased and the doping level N_D in the drift region must be reduced to prevent avalanche breakdown. In the ideal case, the on resistance will then increase as the 2.5 power of the breakdown voltage, as shown in Fig. 1*a* [1]. This highlights the extreme importance of achieving near-ideal breakdown in these unipolar devices. If the device termination is poorly designed, breakdown voltages of less than 50% of the ideal case can easily occur. This would cause an undesirable increase in the device on resistance by

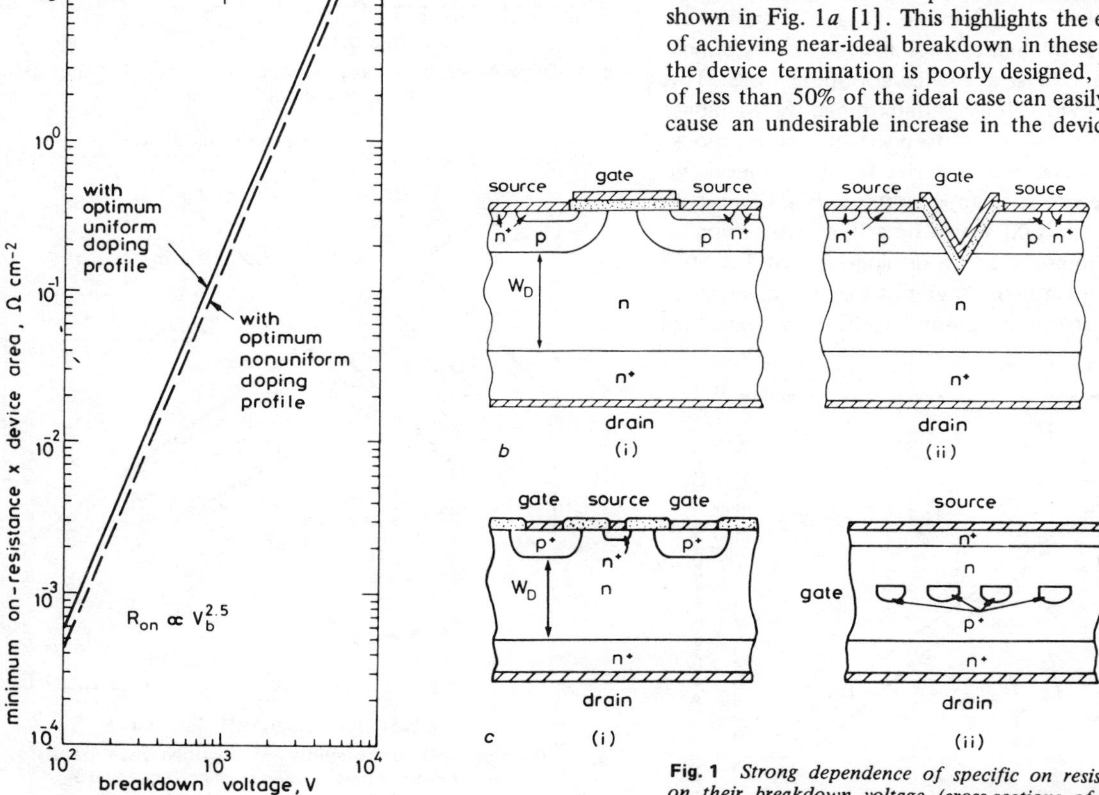

Fig. 1 *Strong dependence of specific on resistance of power FETs on their breakdown voltage (cross-sections of a typical high-voltage power MOSFET and a JFET are given for reference)*

a On-resistance/breakdown-voltage
b Power MOS gate field-effect transistor
(i) DMOS structure (ii) VMOS structure
c Power junction gate field-effect transistor
(i) Surface gate structure (ii) Buried gate structure

Paper 2135 I, first received 28th May and in revised form 10th August 1982. Invited paper

Dr. Baliga is with the General Electric Company Corporate Research & Development Center, PO Box 43, Schenectady NY 12301, USA

more than a factor of five. Thus, achieving close to ideal breakdown in unipolar devices is particularly critical to achieving good performance. Although, it may seem that achieving close to ideal breakdown in bipolar devices, such as power thyristors, is less critical because their current ratings scale inversely as the square root of the breakdown voltage, this can still be a very important factor in the high-voltage devices developed for high-voltage DC transmission networks. In these applications, hundreds of devices must be connected in series and parallel to form the power switching network required to handle the transmission line potential. Consequently, small increases in the breakdown voltage ratings of these devices can result in a substantial reduction in the number of devices required in the power switching network. This not only reduces the cost of the devices but allows considerable savings in the protective networks and gate drive circuitry required for each device.

The appreciation by the semiconductor power device industry of the importance of achieving close to ideal breakdown in power devices is evident because of the large number of device termination techniques that have been explored [2–18]. This paper will review these termination techniques with particular emphasis on their applicability to specific device types. A comparison of the effectiveness of each of these methods is then performed at the end of the paper. This review is intended to serve as a guide to the selection of the device termination technology which will be optimally suitable for each application.

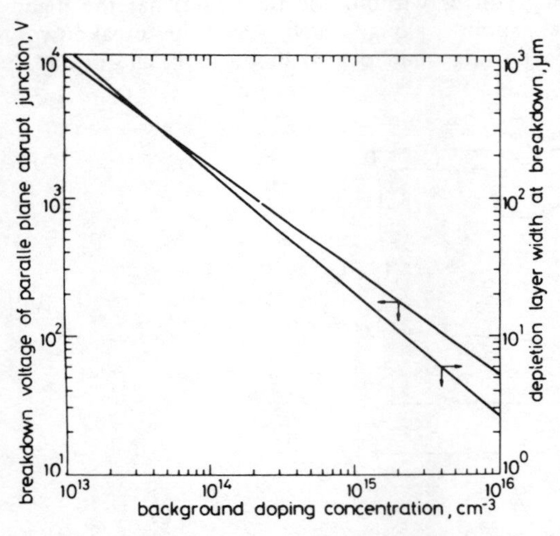

Fig. 2A *Breakdown voltage and depletion layer width at breakdown for abrupt parallel plane junction as function of background doping level*

These curves are used as the normalisation parameters to simplify comparison between the various device termination methods

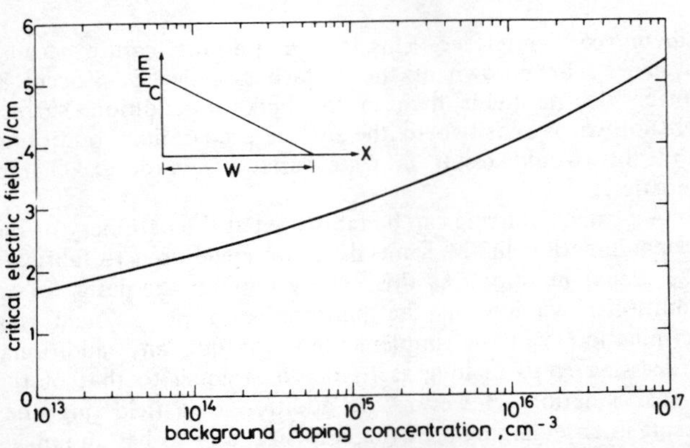

Fig. 2B *Peak electric field at breakdown for the abrupt parallel plane junction as function of background doping level*

This field is also used as a normalisation parameter to simplify comparison between the various device termination methods

2 Abrupt parallel plane junction

The abrupt parallel plane junction is an ideal case where one side of the junction is very highly doped, with the other side of junction having a uniform doping level. This junction is also assumed to extend infinitely on both edges so as to eliminate any edge termination considerations. It can, therefore, be treated as the ideal case for which the highest breakdown voltage will be observed, and has been extremely useful for obtaining a normalised representation of the breakdown voltages and electric fields for the various edge termination techniques for purposes of comparison.

Based on the above idealisation, Poisson's equation for the abrupt parallel plane junction can be written as

$$\frac{d^2V}{dx^2} = -\left(\frac{dE}{dx}\right) = -\left(\frac{qN_B}{\epsilon\epsilon_o}\right) \tag{1}$$

where V is the applied voltage, E is the electric field, q is the electronic charge, N_B is the background doping $(N_D - N_A)$ and $\epsilon\epsilon_o$ is the dielectric constant. For the uniformily doped case, the electric field profile is as shown in the inset of Fig. 2B:

$$E(x) = \frac{qN_B}{\epsilon\epsilon_o}(x - W) \tag{2}$$

and

$$V(x) = \frac{qN_B}{\epsilon\epsilon_o}(2Wx - x^2) \tag{3}$$

where W is the depletion layer width. To obtain the breakdown voltage, the integral of the ionisation coefficient α should be equal to unity. Using Fulops approximation

$$\alpha = 1.8 \times 10^{-35} E^7 \tag{4}$$

it can be shown that the breakdown voltage is given by

$$BV_{APP} = 5.34 \times 10^{13} N_B^{-3/4} \tag{5}$$

$$E_{CPP} = 4010\, N_B^{1/8} \tag{6}$$

and

$$W_c = 2.67 \times 10^{10} N_B^{-7/8} \tag{7}$$

where E_{CPP} is the peak electric field at breakdown and W_{CPP} is the depletion layer width at breakdown. The breakdown voltage, the peak electric field and the depletion width for the ideal abrupt parallel plane junction are plotted in Fig. 2 as a function of the background doping level. These values are useful for the design of device terminations, and will be used as normalisation parameters to evaluate the effectiveness of each technique.

3 Planar junctions

One of the most commonly used techniques for terminating a p–n junction is to perform the diffusion through a window in a masking layer, as indicated in the inset of Fig. 3. These junctions are known as planar junctions, and can be bounded by either cylindrical surfaces at the edges of the window or by spherical surfaces at the sharp corners of the window. It has been demonstrated that the electric field at these cylindrical and spherical surfaces is substantially greater than in the parallel plane case [2, 3]. Consequently the breakdown voltage of these junctions is lower than for the ideal parallel plane case. A normalised analytical solution to the breakdown voltage of the planar junction has been derived [4]:

$$\frac{BV_c}{BV_{APP}} = \frac{1}{2}\left\{\left(\frac{r_j}{W_c}\right)^2 + 2\left(\frac{r_j}{W_c}\right)^{6/7}\right\} \ln\left\{1 + 2\left(\frac{W_c}{r_j}\right)^{8/7}\right\} - \left(\frac{r_j}{W_c}\right)^{6/7} \tag{8}$$

$$\frac{BV_s}{BV_{APP}} = \left(\frac{r_j}{W_c}\right)^2 + 2.14\left(\frac{r_j}{W_c}\right)^{6/7} - \left\{\left(\frac{r_j}{W_c}\right)^3 + 3\left(\frac{r_j}{W_c}\right)^{13/7}\right\}^{2/3} \tag{9}$$

where r_j is the junction depth, and BV_c and BV_s are the breakdown voltages of the cylindrical and spherical junctions, respectively. These normalised breakdown voltages are plotted in Fig. 3 as a function of the normalised radius of curvature (r_j/W_c). It can be seen that the spherical junction termination is significantly inferior to the cylindrical junction termination. Consequently, it is important when designing the diffusion windows of high-voltage devices to avoid the formation of any sharp corners which could lead to spherical junction surfaces.

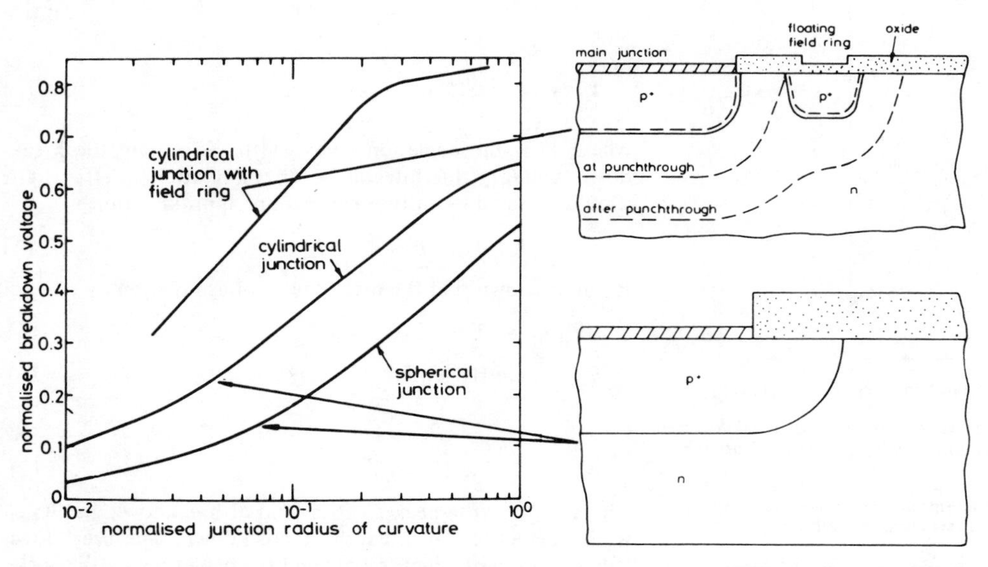

Fig. 3 *Normalised breakdown voltage of spherical, cylindrical and planar diffused junctions with single optimally located floating field ring as function of the normalised radius of curvature*

For high-voltage devices with breakdown voltages in excess of a few hundred volts, the depletion width at breakdown exceeds $10\,\mu$m, as can be read from Fig. 2A. Using conventional planar diffusion technology, with boron as the dopant, it is difficult to achieve diffusion depths that approach such large values. As the radius of curvature r_j is approximately equal to the diffusion depth, planar junctions fabricated using conventional technology will have normalised radii of curvature of less than unity. From Fig. 3 it is clear that such junctions will then be limited to breakdown voltages of less than 60% of the ideal case. This low breakdown voltage is unacceptable for most power devices. To overcome this limitation the planar field ring termination has been developed [5, 6].

4 Planar field ring termination

The addition of a field ring to the planar junction structure, as shown in the inset of Fig. 3, has been found to significantly raise the breakdown voltage. This increase in the breakdown voltage stems from the increased spreading of the depletion layer at the periphery of the device, and this alleviates the electric field crowding at the planar junction. For the field ring to be fully effective it must be optimally located so that it shares the potential applied to the planar junction. If the field ring is placed too far from the planar junction very little of the potential is transferred to it, and the breakdown occurs due to field crowding at the planar junction. If the field ring is placed too close to the planar junction most of the potential is transfered to it, and breakdown occurs due to field crowding at the field ring. With optimal field ring spacing breakdown occurs simultaneously at the planar junction and the field ring. The increase in breakdown voltage achieved by the addition of an optimally spaced field ring is shown in Fig. 3 [6]. It should be noted that, at the smaller radii of curvature, the addition of a field ring can result in doubling the breakdown voltage, Under typical design conditions, breakdown voltages of up to 80% of the ideal value can be achieved with the field ring termination technique.

It is worth pointing out that the planar junction termination (with or without the field ring) has the highest electric field point, along which avalanche breakdown is expected to be favoured, located in the bulk and not at the silicon/oxide interface. This is an important feature because avalanche breakdown at the surface is believed to occur at lower electric fields than in the bulk. In addition, surface breakdown is sensitive to the surface preparation conditions, and this would result in wide variations in device characteristics.

A planar field ring can be fabricated at the periphery of the planar junction in the same diffusion cycle used to fabricate the planar junction. As this merely requires the design of an additional window in the photomask, a planar field ring termination can be implemented without any additional processing steps as long as its depth is equal to that of the planar junction. However, the addition of a field ring does result in an enlargment in the chip size, and so has an impact on the cost of the device. The improvement in the breakdown voltage resulting from the addition of an optimally placed field ring is, however, usually well worth the additional area on the chip. This is particularly true for state-of-the-art power

MOSFET and JFET devices whose on resistances increase as the 2.5 power of the breakdown voltage. In these devices, increasing the breakdown voltage from 50% to 80% of the ideal value by the addition of a field ring will result in a significant reduction in chip size, in spite of the added area consumed by the field ring.

5 Planar junction with field plate

This device termination technique is sometimes used instead of the field ring approach, and often in conjunction with it. It has been experimentally and theoretically demonstrated that controlling the surface potential at the edges of the planar junction allows an increase in its breakdown voltage [7].

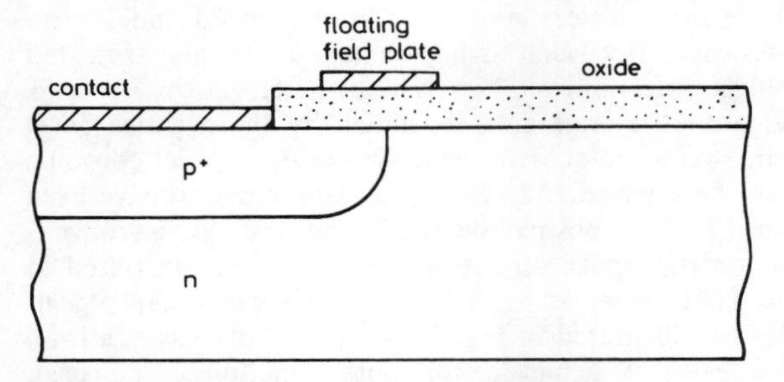

Fig. 4 *Planar junction termination with field plate*

Although this can be achieved by using a separate power supply to control the field plate potential, this is not usually practical. Instead, the field plate is either left floating, as illustrated in Fig. 4, or is formed as an extension of the planar junction metallisation [8, 9]. Some studies on the use of resistive field plates instead of the metallisaton have also been reported [10]. These studies have demonstrated that the addition of a field plate to a planar junction termination can result in achieving between 50 and 60% of the ideal breakdown voltage. Thus, although this device termination technique is compatible with planar device processing, the breakdown voltages achieved with it are inferior to those achievable by the addition of a field ring.

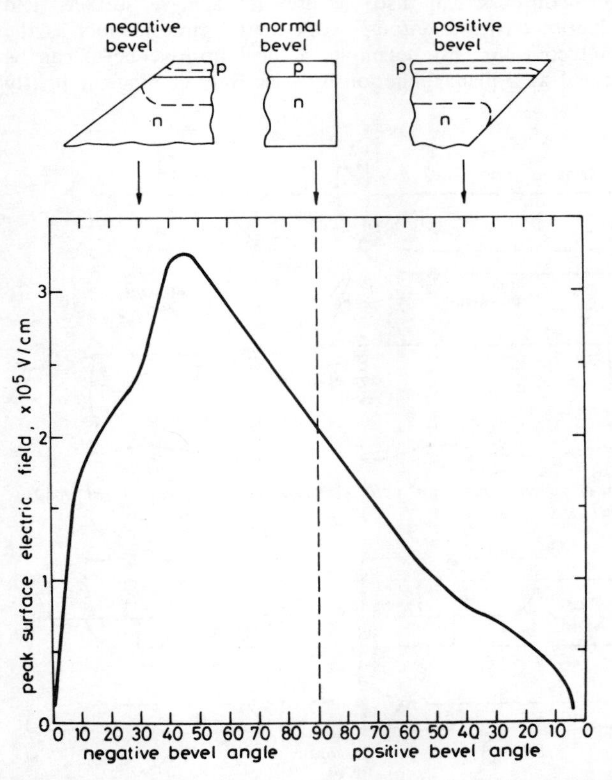

Fig. 5 *Dependence of peak surface electric field on bevel angle for positive and negative bevel contours*

6 Bevelled surfaces

During reverse biasing of a p–n junction, the charges on both sides of the junction must balance each other to maintain charge neutrality. The depletion layer widths on the two sides of the junction must consequently adjust themselves to achieve this charge balance. In the bulk, the space-charge distribution is determined solely by the ionised impurity concentrations and profiles on the two sides of the junction. However, at the edges, the shape of the junction edge plays a significant role in determining the space-charge-layer widths on the two sides of the junction. By appropriately contouring the edges of the device, it is possible to achieve a reduction in the surface electric field, and so to enhance the breakdown voltage.

The effect of the surface bevel angle on the highest surface electric field is shown in Fig. 5 [11]. If the area of the semiconductor decreases when proceeding from the heavily doped side of the junction to the lightly doped side, the junction is considered to have a positive bevel. In these cases, more material is removed from the lightly doped side than from the heavily doped side. Consequently, to maintain charge balance, the depletion layer spreading on the lightly doped side exceeds the depletion layer reduction on the heavily doped side. This process results in a net reduction in the peak surface electric field, as can be observed in the plot of the calculated peak electric field as a function of the positive bevel angle. As typical bevel angles of 45° can be achieved in practice, the surface electric field can be reduced to less than half that in the bulk. This reduction in the surface electric field is usually sufficient to ensure that surface breakdown does not occur before bulk breakdown when state-of-the-art passivation techniques are used. The positive bevelling technique, therefore, allows achievement of the ideal parallel plane breakdown voltages.

The fabrication of a positively bevelled surface contour is

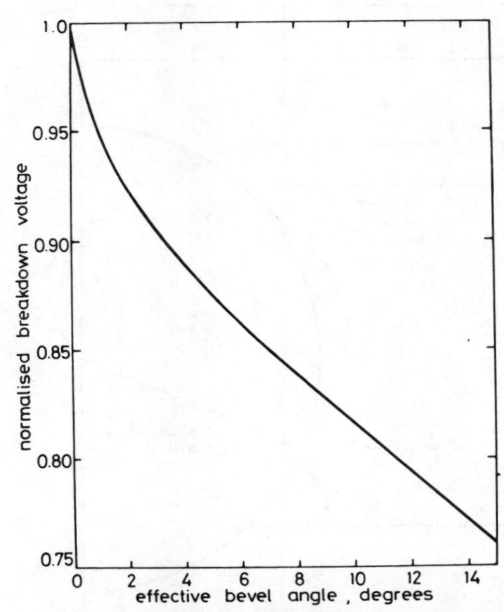

Fig. 6 *Normalised breakdown voltage of negative bevel surface contours as function of effective bevel angle*

The effective bevel angle (θ_{eff}) is related to the actual bevel θ by $\theta_{eff} = 0.04 \times \theta (W/d)^2$, where W is the depletion width on the lightly doped side and d is the depletion width on the diffused side, under breakdown conditions for the parallel plane case

usually achieved by either lapping the wafer at the appropriate angle to its surface, by using grit blasting, or by sawing the edges of the chip at the appropriate angle. In all these cases, the physical damage at the bevelled surface must be chemically removed prior to passivation. In practice, the positive bevelling technique is used only for large-area devices where the chip dimensions usually exceed 25 mm; typical examples of devices are high-current rectifiers and thyristors.

Another contour used to reduce the surface electric field is the negative bevel, also shown in Fig. 5. In this case more material is removed from the heavily doped side of the junction than from the lightly doped side. Consequently, the depletion layer on the lightly doped side shrinks, while that on the heavily doped side expands. For negative bevel angles greater than 10°, and for abrupt junction devices, this contour results in an increase in the surface electric field. However, at very shallow negative bevel angles and for graded junctions, the surface electric field is reduced to below that in the bulk. This occurs because, at these shallow angles, the depletion layer edge on the lightly doped side is pinned at the junction, while the depletion width on the heavily doped side at the bevel surface expands, eventually leading to a surface field reduction. It is important to note that to achieve effective depletion layer spreading on the heavily doped side of the junction it should have a graded diffusion profile.

Modelling studies [12] have indicated that, although the peak surface electric field is reduced with a negative bevel angle, the peak electric field inside the semiconductor near the negatively bevelled surface exceeds that in the parallel plane case [12]. The breakdown voltage of the negatively bevelled junction is consequently always lower than for the parallel plane case. The dependence of the breakdown voltage on the negative bevel angle is shown in Fig. 6 in a normalised form [13]. The effective bevel angle is defined in Fig. 6, where θ is the actual bevel angle, and d and W are the depletion layer widths on the heavily and lightly doped sides of the junction, respectively.

The negatively bevelled contour is usually fabricated by using shallow angle lapping at the edges of the device. Typical bevel angles range from 0.5 to 5°. Using these techniques, breakdown voltages of 90% of the parallel plane case are typically achieved. However, at these shallow bevel angles, considerable area is consumed at the edges of the chip. This restricts the application of the negative bevel contour to very large area devices containing both forward and reverse blocking capability, such as high-voltage thyristors fabricated from single wafers of over 25 mm in diameter.

Owing to the large area consumed by the negative bevel contour, two techniques for achieving positive bevel contours on both the forward and reverse blocking junction have been developed (double positive bevel). In the first case a groove is formed on the upper surface of the device, as illustrated in Fig. 7*a* [14], whereas in the second case the wafer edge is rounded, as illustrated in Fig. 7*b* [15]. In both cases, a local positive bevel is achieved for both junctions. The peak surface electric field is then reduced to less than 80% of that in the bulk, and nearly ideal breakdown voltages can be achieved. However, as this peak electric field is higher than in the case of the single positive bevel angle, these double positive bevel contours are more difficult to passivate. Furthermore, owing to the complexity of the processes used for the fabrication of these contours, they are only used for the fabrication of very-large-area high-voltage thyristors fabricated from wafers over 25 mm in diameter.

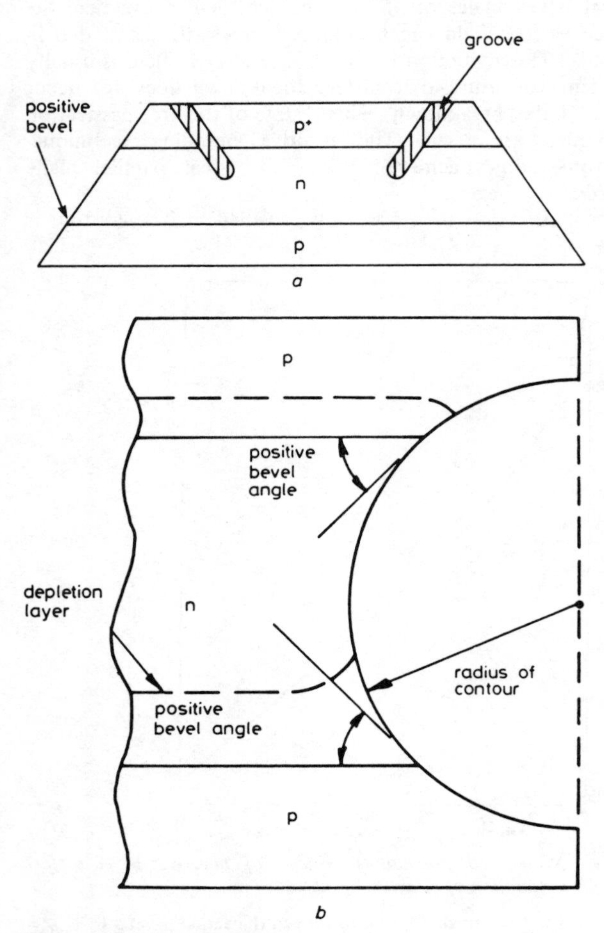

Fig. 7 *Double positive bevel structures*
a Angled groove *b* Wafer edge contouring

7 Etch contours

Etch contours can also be used to achieve surface field reduction. These contours work in the same manner as the bevel contours. For example, a local positive bevel can be created at a planar junction by selectively etching the lightly

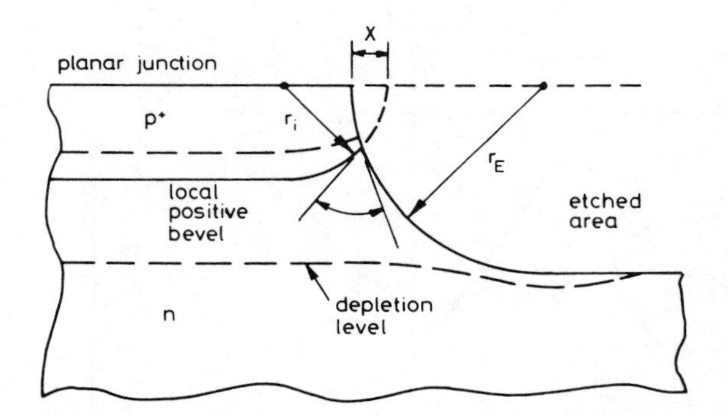

Fig. 8 *Planar junction with etch contour to create local positive bevel*

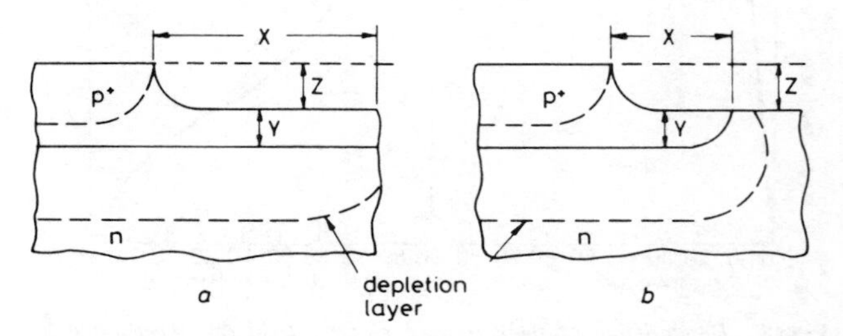

Fig. 9 *Parallel plane and planar junctions with surface etch contours*

doped side, as shown in Fig. 8. The selective removal of material from the lightly doped side of the junction increases the spreading of the depletion layer, as in the case of the positive bevel contours, and so leads to a surface field reduction. The edge of the etched region does not have to intersect the planar junction. Numerical analysis [16] indicates that when the etched region is far from the junction edge, the peak electric field occurs at the upper surface as in the case of the planar junction. As the etched region is brought closer to the junction edge, the surface electric field decreases to a minimum of about 60% of the bulk, and then again increases when the etched region intersects the junction. Using this etch contour, up to 90% of the ideal breakdown voltage can be achieved. As the surface electric field is also reduced to lower values than in the planar junction case, even with an optimally placed field ring, this method is expected to exhibit greater stability. This type of contour is also convenient for small-area devices as all the devices on a wafer can be processed simultaneously.

The complementary negative etch contour for planar junctions is illustrated in Fig. 9. In these cases, the etch removes more material from the heavily doped side of the junction. Numerical modelling [17] of these structures indicates that breakdown voltages of up to 80% of the ideal case can be achieved with a surface field of 60% of the bulk. However, the surface field reduction and the breakdown voltage achieved with the negative etch contour are very sensitive to the etch depth which determines the amount of material removed from the heavily doped side of the junction. The negative etch contour is consequently not as favourable as the positive etch contour.

8 Junction termination extension

In the bevel and etch contours, charge is selectively removed from either the heavily or lightly doped side of the junction at the edges. Another approach, called junction termination extension, utilises the addition of charge to the heavily doped side of the junction [18]. This can be done with precision by ion implantation, as illustrated in Fig. 10 for both a planar and a parallel plane junction with an etched area. It has been theoretically and experimentally verified that up to 95% of the breakdown voltage can be achieved when the active doping concentration in the ion-implanted region is within 60–80% of charge obtained by taking the product of the dielectric constant of silicon and the peak electric field at breakdown in the parallel plane case. However, it should be noted that high leakage currents have been observed with the junction termination extension technique. It has also been found that the surface passivation plays a major role in controlling the effectiveness of this technique. This may lead to problems in achieving good reproducibility.

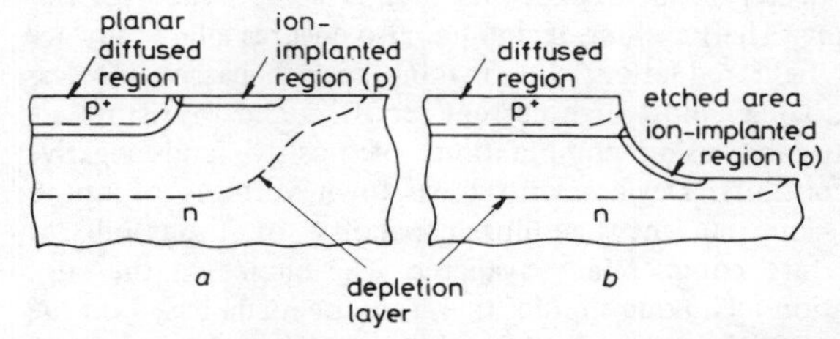

Fig. 10 *Planar junctions with junction termination extension obtained using ion implantation*

9 Summary

The large variety of high-voltage device termination techniques that have been developed have been reviewed. These techniques are compared in Table 1, on the basis of the typical breakdown voltage achievable using each technique and the surface electric field reduction. It can be seen that the positive bevel contour offers the best results. However, the selection of a specific technique for any particular application is dependent on several other important factors. First, it depends on the size of the device. In the case of small devices, bevelling techniques are not practical. The most commonly used field termination method for small dies is to use planar junctions with field rings. This allows achieving up to 80% of the ideal breakdown voltage of the parallel plate junction. This is adequate for most bipolar devices, such as the bipolar transistor or the thyristor. In the case of power MOSFET and JFET devices, where the on resistance is more strongly influenced by the breakdown voltage, as indicated in Fig. 1,

Table 1: High-voltage device termination techniques

Technique	Typical breakdown voltage as percentage of parallel plane case	Peak surface electric field as percentage of bulk	Typical device size of application	Typical device types	Remarks
	%	%			
Planar junction	50	80	small (< 100 mils)	BJT, MOSFET	not used often for high voltage devices
Planar junction with field ring	80	80	medium (up to 1 in)	BJT, MOSFET, SCR	well suited to large number of devices per wafer
Planar junction with field plate	60	80	medium (up to 1 in)	BJT, MOSFET	usually used in conjunction with field plate
Positive bevel	100	50	large (> 1 in)	rectifier, SCR	well suited to single device per wafer
Negative bevel	90	60	large (> 1 in)	SCR	well suited to single device per wafer only
Double positive bevel	100	80	large (> 1 in)	SCR	well suited to single device per wafer only
Positive etch contour	90	60	all	BJT, MOSFET, SCR	well suited to large number of devices per wafer
Negative etch contour	80	60	all	BJT, MOSFET, SCR	well suited to large number of devices per wafer
Junction termination extension	95	80	all	BJT, MOSFET, SCR	well suited to both single devices and large number of devices per wafer; high leakage current; passivation sensitive

the junction termination extension offers the best promise, albeit at higher leakage currents. Alternatively, the positive etch contour can be used to achieve close to 90% of the ideal breakdown voltage. Secondly, the planar junction with field ring and planar junction with junction termination extension can be used to simultaneously process a large number of small devices on each wafer at the same time. In contrast, the bevelling methods require handling individual devices, and are thus not cost effective for small devices. However, for large-area devices fabricated, for instance, from individual silicon wafers, the bevelling techniques are the most promising owing to the nearly ideal breakdown voltages achievable with this technique. The bevelling technique also ensures a large surface electric field reduction, thus making surface passivation less critical. Today most high current rectifiers and thyristors are fabricated by using combinations of positive and negative bevel contours. Devices with breakdown voltages of up to 6500 V and current handling capability of thousands of amperes are commercially available. The choice of the edge termination technique should, therefore, be made based on the die size, on the sensitivity of the other electrical characteristics of the device, on the breakdown voltage, and on the ease of surface passivation of the edges.

10 Acknowledgment

The author wishes to thank N. Waldron for assistance with manuscript preparation.

11 References

1 BALIGA, B.J.: 'Silicon power field controlled devices and integrated circuits' *in* KHANG. D. (Ed.): 'Silicon integrated circuits. Applied Solid-State Science Series, Supplement 2B' (Academic Press, 1981)

2 SZE, S.M., and GIBBONS, G.: 'Avalanche breakdown voltages of abrupt and linearly graded *p–n* junctions in Ge, Si, GaAs and GaP', *Applied Phys. Lett.*, 1966, **8**, pp. 111–113

3 TEMPLE, V.A.K., and ADLER, M.S.: 'Calculation of the diffusion curvature related avalanche breakdown in high voltage planar *p–n* junctions', *IEEE Trans.*, 1975, **ED-22**, pp. 910–916

4 BALIGA, B.J., and GANDHI, S.K.: 'Analytical solutions for the breakdown voltage of abrupt cylidrical and spherical junctions', *Solid-State Electron.*, 1976, **19**, pp. 739–744

5 KAO, Y.C., and WOLLEY, E.D.: 'High voltage planar *p–n* junctions', *Proc. IEEE*, 1967, **55**, pp. 1409–1414

6 ADLER. M.S., TEMPLE, V.A.K., FERRO, A.P., and RUSTAY, R.C.: 'Theory and breakdown voltage for planar devices with a single field limiting ring', *IEEE Trans.*, 1977, **ED-24**, pp. 107–113

7 GROVE, A.S., LEISTIKO, O., and HOOPER, W.W.: 'Effect of surface fields on the breakdown voltage of planar silicon *p–n* junctions', *ibid.*, 1967, **ED-24**, pp. 157–162

8 CONTI, F. and CONTI, M.: 'Surface breakdown in silicon planar diodes equipped with field plate', *Solid-State Electron.* 1972, **15**, pp. 93–105

9 TEMPLE, V.A.K., and ADLER, M.S.: 'Enhancement of breakdown voltage using floating metal field plates', *Int. J. Electronics*, 1976, **40**, pp. 293–303

10 CLARK, L.E., and ZOROGLU, D.S.: 'Enhancement of breakdown properties of overlay annular diodes by field shaping resistive films', *Solid-State Electron.*, 1972, **15**, pp. 653–657

11 DAVIES, R.L., and GENTRY, F.E.: 'Control of electric field at the surface of P-N junctions', *IEEE Trans.*, 1964, **ED-11**, pp. 313–323

12 CORNU, J.: 'Field distribution near the surface of beveled P-N junctions in high voltage devices', *ibid.*, 1973, **ED-20**, pp. 347–352

13 ADLER, M.S., and TEMPLE, V.A.K.: 'A general method for predicting the avalanche breakdown voltage of negative bevelled devices', *ibid.*, 1976, **ED-23**, pp. 956–960

14 OTSUKA, M.: 'A new edge contour for Si high voltage thyristors'. *IEE Conf. Publ. 53*, 1969, pp. 32–38

15 CORNU, J., SCHWEITZER, S., and KUHN, O.: 'Double positive beveling: A better edge contour for high voltage devices', *IEEE Trans.*, 1974, **ED-21**, pp. 181–184

16 TEMPLE, V.A.K., BALIGA, B.J., and ADLER, M.S.: 'The planar junction etch for high voltage and low surface fields in planar devices', *ibid.*, 1977, **ED-24**, pp. 1304–1310

17 TEMPLE, V.A.K., and ADLER, M.S.: 'The theory and application of a simple etch contour for near ideal breakdown voltage in plane and planar P-N junctions', *ibid.*, 1976, **ED-23**, pp. 950–955

Bibliography

Bipolar Power Transistors

A. Design Considerations

[1] H. C. DeGraff, J. W. Slotboom, and A. Schmitz, "The emitter efficiency of bipolar transistors," *Solid State Electron.*, vol. 20, pp. 515–1259, 1977.

[2] P. L. Hower, "Application of a charge control model to high voltage power transistors," *IEEE Trans. Electron Devices*, vol. ED-23, pp. 863–869, Aug. 1976.

[3] G. Bosch, "Anomalous current distributions in power transistors," *Solid State Electron.*, vol. 20, pp. 635–640, 1977.

[4] L. J. Turgeon and D. H. Navon, "Two dimensional nonisothermal carrier flow in a transistor structure under reactive circuit conditions," *IEEE Trans. Electron Devices*, vol. ED-25, pp. 837–843, July 1978.

[5] R. A. Sunshine, "Multidimensional current flow in silicon power transistors operating in the saturation mode," in *Proc. IEEE Power Electron. Specialists Conf.*, June 1974, pp. 154–161.

B. Switching Dynamics

[6] C. Hu and M. J. Model, "A model of power transistor turn-off dynamics," in *Proc. IEEE Power Electron. Specialists Conf.*, June 1980, pp. 91–96.

[7] P. L. Hower, "A model for turn-off in bipolar transistors," in *Proc. IEEE Int. Electron Devices Meet.*, Dec. 1980, p. 289.

[8] P. L. Hower, J. B. Brewster, and M. Morozowich, "A new method of characterizing the switching performance of power transistors," in *Proc. IEEE Indust. Appl. Soc.*, Oct. 1978, pp. 1044–1049.

C. Device Characteristics

[9] T. Matsushita, T. Aoki, T. Ohtsu, H. Yamoto, H. Hayashi, M. Okayama, and Y. Kawana, "High reliability, high voltage, transistors by use of the SIPOS process," *IEEE Trans. Electron Devices*, vol. ED-23, pp. 826–830, Aug. 1976.

[10] T. Suzuki, M. Ura, and T. Ogawa, "Application of high resistivity silicon epitaxial technique to high voltage power transistors," *IEEE Trans. Electron Devices*, vol. ED-23, pp. 982–983, Aug. 1976.

[11] K. S. Tarneja and P. L. Hower, "A new high power switching transistor," in *Proc. IEEE Indust. Appl. Soc.*, Oct. 1981, pp. 754–756.

[12] R. Eaton, S. W. Kessler, and R. E. Reed, "Development of a transcalent silicon power switching transistor," in *Proc. IEEE Indust. Appl. Soc.*, Oct. 1977, pp. 1032–1041.

D. Darlington Structures

[13] K. Owyang, "Recent developments in high power switching Darlington transistors," in *Proc. IEEE Indust. Appl. Soc.*, Oct. 1980, pp. 715–719.

[14] S. Saeki, "Structures and characteristics of 400 A–300 V monolithic high power transistors," in *Proc. IEEE Power Electron. Specialists Conf.*, June 1975, pp. 274–281.

[15] S. Krishna and A. J. Yerman, "The development of a power Darlington transistor," in *Proc. IEEE Power Electron. Specialists Conf.*, June 1979, pp. 55–61.

Gate Turnoff Thyristor/Latching Transistor

A. Design Considerations

[1] M. Naito, T. Nagano, H. Fukui, and Y. Terasawa, "One-dimensional analysis of turn-off phenomena for gate turn-off thyristor," *IEEE Trans. Electron Devices*, vol. ED-26, pp. 226–231, Mar. 1979.

[2] Y. Shimizu, M. Naito, M. Odamura, and Y. Terasawa, "Numerical analysis of turn-off characteristics for a gate turn-off thyristor with a shorted anode emitter," *IEEE Trans. Electron Devices*, vol. ED-28, pp. 1043–1047, Sept. 1981.

B. Device Characteristics

[3] T. Nagano, M. Okamura, and T. Ogawa, "A high power, low forward drop gate turn-off thyristor," in *Proc. IEEE Indust. Appl. Soc.*, Oct. 1978, pp. 1003–1006.

[4] O. Aina, P. O. Shafter, and E. D. Wolley, "Characteristics of a 25 amp, 800 volt latching transistor (GTO)," in *Proc. IEEE Indust. Appl. Soc.*, Oct. 1978, pp. 1056–1062.

[5] M. Azuma, M. Kurata, and K. Takigaini, "2500 V–600 A gate turn-off thyristor (GTO)," *IEEE Trans. Electron Devices*, vol. ED-28, pp. 270–274, Mar. 1981.

[6] T. Nagano, T. Yatsuo, and M. Okamura, "Characteristics of a 3000 V, 1000 A gate turn-off thyristor," in *Proc. IEEE Indust. Appl. Soc.*, Oct. 1981, pp. 750–753.

[7] A. Tada and H. Hagino, "A high voltage, high power, fast switching gate turn-off thyristor," in *Proc. IEEE ISPC Conf.*, May 1982, pp. 66–73.

C. Device-Circuit Interactions

[8] R. E. Locker, "Use of latching transistors for power control and conversion," in *Proc. IEEE Power Electron. Specialists Conf.*, June 1978, pp. 202–209.

[9] K. Kishi, M. Kurata, and K. Imai, "High power gate turn-off thyristors (GTOs) and GTO-VVF inverter," in *Proc. IEEE Power Electron. Specialists Conf.*, June 1977, pp. 268–274.

[10] T. Jinzenji, T. Kanzaki, F. Moriza, and M. Azuma, "Three phase static power supplies for air-conditioned electric coaches using high power GTOs," in *Proc. IEEE Indust. Appl. Soc.*, Oct. 1979, pp. 1088–1098.

Power MOSFET's

A. Design Considerations

[1] C.A.T. Salama and J. G. Oakes, "Nonplanar power field effect transistors," *IEEE Trans. Electron Devices*, vol. ED-25, pp. 1222–1228, Oct. 1978.

[2] K. P. Lisiak and J. Berger, "Optimization of nonplanar power MOS transistors," *IEEE Trans. Electron Devices*, vol. ED-25, pp. 1229–1234, Oct. 1978.

[3] C. Hu, "Optimum doping profile for minimum on-resistance and high breakdown voltage," *IEEE Trans. Electron Devices*, vol. ED-26, pp. 243–244, Mar. 1979.

B. Device Characteristics

[4] Y. Monta, H. Takahashi, H. Matyoshi, and M. Fukuta, "Si UHF MOS high power FET," *IEEE Trans. Electron Devices*, vol. ED-21, pp. 733–734, Nov. 1974.

[5] I. Yoshida, M. Kubo, and S. Ochi, "A high power MOSFET with a vertical drain electrode and a meshed gate structure," *IEEE J. Solid State Circuits*, vol. SC-11, pp. 472–477, Aug. 1976.

[6] H. Ikeda, K. Ashikawa, and K. Urita, "Power MOSFETs for medium-wave and short wave transmitters," *IEEE Trans. Electron Devices*, vol. ED-27, pp. 330–334, Feb. 1980.

[7] V.A.K. Temple, R. P. Love, and P. V. Gray, "A 600 volt MOSFET designed for low on-resistance," *IEEE Trans. Electron Devices*, vol. ED-27, pp. 343–349, Feb. 1980.

[8] J. P. Stengle, H. Strack, and J. Tihanji, "Power MOS transistors for 1000 V blocking voltage," in *Proc. IEEE Int. Electron Devices Meet.*, Dec. 1981, p. 422.

C. Device-Circuit Interaction

[9] P. Freundel, "Power MOSFETs or bipolar power transistors for converter circuits," in *Proc. IEEE ISPC Conf.*, May 1982, pp. 38–44.

[10] T. M. Undeland, "Application of switch mode controlled vertical MOSFET for power conversion in dc and ac motor drives," in *Proc. IEEE Indust. Appl. Soc.*, Oct. 1981, pp. 676–681.

Power Junction FET's/Static Induction Transistors (SIT's)

[1] J. I. Nishizawa, T. Terasaki, and J. Shibata, "Field effect transistor versus analog transistor (static induction transistor)," *IEEE Trans. Electron Devices*, vol. ED-22, pp. 185–197, Apr. 1975.

[2] Y. Mochida, J. I. Nishizawa, T. Ohmi, and R. K. Gupta, "Characteristics of static induction transistors: Effects of series resistance," *IEEE Trans. Electron Devices*, vol. ED-25, pp. 761–767, July 1978.

[3] K. Sakai, Y. Komatsu and H. Kobayaski, "Complementary power FETs with vertical structures," in *Proc. IEEE Power Electron. Specialists Conf.*, June 1974, pp. 214–221.

[4] O. Ozawa, H. Iwasaki, and K. Muramoto, "A vertical channel JFET fabricated using silicon planar technology," *IEEE J. Solid State Circuits*, vol. SC-11, pp. 511–517, Aug. 1976.

[5] O. Ozawa and H. Iwasaki, "A vertical FET with self aligned ion implanted source and gate regions," *IEEE Trans. Electron Devices,* vol. ED-25, pp. 56–57, Jan. 1978.

[6] J. I. Nishizawa and K. Yamamoto, "High frequency, high power static induction transistor," *IEEE Trans. Electron Devices*, vol. ED-25, pp. 314–322, Mar. 1978.

[7] B. J. Baliga, "A power junction gate FET structure with high blocking gain," *IEEE Trans. Electron Devices*, vol. ED-27, pp. 368–373, Feb. 1980.

Power Field Controlled Thyristors

[1] D. E. Houston, S. Krishna, D. E. Piccone, R. J. Finke, and Y. S. Sun, "A field terminated diode," *IEEE Trans. Electron Devices*, vol. ED-23, pp. 905–911, Aug. 1976.

[2] J. I. Nichizawa and K. Nakamura, "Characteristics of new thyristors," in *Proc. Int. Solid State Devices Conf.*, pp. 193–194, Sept. 1976.

[3] B. W. Wessels and B. J. Baliga, "Vertical channel field controlled thyristors with high gain and fast switching speeds," *IEEE Trans. Electron Devices*, vol. ED-25, pp. 1261–1265, Oct. 1978.

[4] B. J. Baliga, "Grid depth dependence of the characteristics of vertical channel field controlled thyristors," *Solid State Electron.*, vol. 22, pp. 237–239, 1979.

[5] J. I. Nishizawa and Y. Ohtsubo, "Effect of gate structure on static induction thyristor," in *Proc. IEEE Int. Electron Devices Meeting*, Dec. 1980, pp. 658–660.

[6] B. J. Baliga, "Buried-grid field controlled thyristors fabricated using silicon liquid phase epitaxy," *IEEE Trans. Electron Devices*, vol. ED-27, pp. 2141–2145, Nov. 1980.

[7] R. Baradon and P. Laurenceau, "Power bipolar gridistor," *Electron. Lett.*, vol. 12, pp. 486–487, 1976.

[8] B. J. Baliga, "Temperature dependence of field controlled thyristor characteristics," *IEEE Trans. Electron Devices*, vol. ED-28, pp. 257–264, Mar. 1981.

[9] ——, "The *di/dt* capabiility of field controlled thyristors," *Solid State Electron.*, vol. 25, pp. 583–588, July 1982.

[10] ——, "The breakover phenomena in field controlled thyristors," *IEEE Trans. Electron Devices*, vol. ED-29, pp. 1579–1587, Oct. 1982.

[11] ——, "The *dV/dt* capability of field controlled thyristors," *IEEE Trans. Electron Devices*, vol. ED-30, pp. 612–616, June 1983.

Emerging Power Device Technologies

[1] B. J. Baliga, "Enhancement and depletion mode vertical channel MOS gated thyristors," *Electron. Lett.*, vol. 15, p. 645, Sept. 1979.

[2] J. Tihanji, "Functional integrations of power MOS and bipolar devices," in *Proc. IEEE Int. Electron Devices Meet.*, Dec. 1980, p. 75.

[3] L. Leipold, W. Baumgartner, W. Ladenhauf, and J. P. Stengl, "A FET controlled thyristor in SIPMOS technology," in *Proc. IEEE Int. Electron Devices Meet.*, Dec. 1980, p. 79.

[4] J. D. Plummer and B. W. Scharf, "Insulated gate planar thryristors," *IEEE Trans. Electron Devices*, vol. ED-27, pp. 380–394, Feb. 1980.

[5] B. J. Baliga, "Bipolar operation of power junction FET's," *Electron. Lett.*, vol. 16, pp. 300–301, Apr. 1980.

[6] T. Tamama, M. Sakaue, and Y. Mizushima, "Bipolar mode transistors on a voltage controlled scheme," *IEEE Trans. Electron Devices*, vol. ED-28, pp. 777–783, July 1981.

[7] J. Nishizawa, T. Ohmi, Y. Mochida, T. Matsuyama, and S. Iida, "Bipolar mode static induction transistor," in *Proc. IEEE Int. Electron Devices Meet.*, Dec. 1978, pp. 676–679.

[8] B. J. Baliga, "The MAJIC-FET: A high speed power switch with low on-resistance," *IEEE Electron Device Lett.*, vol. EDL-3, pp. 189–191, July 1982.

[9] P. S. Hsu, "A new, high speed, unique switch," in *Proc. IEEE Power Electron. Specialists Conf.*, June 1982, pp. 378–382.

[10] H. Kondo and Y. Yukimoto, "A new bipolar transistor—GAT," *IEEE Trans. Electron Devices*, vol. ED-27, pp. 373–379, Feb. 1980.

[11] B. J. Baliga, M. S. Adler, P. V. Gray, and R. P. Love, "The insulated gate rectifier," presented at IEEE Int. Electron Devices *Meet.*, Dec. 1982.

[12] B. J. Baliga, "Semiconductors for hig voltage vertical channel FETs," *J. Appl. Physics*, vol. 53, pp. 1759–1764, Mar. 1982.

[13] P. M. Campbell, R. S. Ehle, P. V. Gray, and B. J. Baliga, "150 V vertical channel GaAs FET," presented at IEEE Int. Electron Devices Meet., Dec. 1982.

Device Safe-Operating Area

[1] S. Rubin and D. L. Blackburn, "A test unit for nondestructive determination of forward-biased safe-operating-area circuits for power transistors," in *Proc. IEEE Indust. Appl. Soc.*, Oct. 1977, pp. 666–673.

[2] P. L. Hower, "Collector charge dynamics and second breakdown energy of power transistors," in *Proc. IEEE Power Electron. Specialists Conf.*, June 1974, pp. 149–153.

[3] B. A. Beatty, S. Krishna, and M. S. Adler, "Second breakdown in power transistors due to avalanche injection," *IEEE Trans. Electron Devices*, vol. ED-23, pp. 851–857, Aug. 1976.

[4] S. Leung and T. C. New, "Improved SOA in low cost power transistors," in *Proc. IEEE Power Electron. Specialists Conf.*, June 1978, pp. 195–201.

[5] T. Asakawa and N. Tsubonchi, "Second breakdown in MOS transistors," *IEEE Trans. Electron Devices*, vol. ED-13, pp. 811–812, Nov. 1966.

[6] S. Krishna, "Second breakdown in high voltage MOS transistors," *Solid State Electron.*, vol. 20, pp. 875–878, 1977.

[7] I. Yoshida, T. Okabe, M. Katsueda, S. Ochi, and M. Nagata, "Thermal stability and secondary breakdown in planar power MOSFETs," *IEEE Trans. Electron Devices*, vol. ED-27, pp. 395–398, Feb. 1980.

[8] T. Nagano, H. Fujui, T. Yatsuo, and M. Okamura, "A snubberless GTO," in *Proc. IEEE Power Electron. Specialists Conf.*, June 1982, pp. 383–387.

Thermal Considerations

[1] W. E. Newell, "Transient thermal analysis of solid state power devices," in *Proc. IEEE Power Electron. Specialists Conf.*, June 1975, pp. 234–251.

[2] D. L. Blackburn and F. F. Oettinger, "Transient thermal response measurements of power transistors," in *Proc. IEEE Power Electron. Specialists Conf.*, June 1974, pp. 140–148.

[3] S. Rubin, "Thermal resistance measurements on monolithic and hybrid Darlington power transistor," in *Proc. IEEE Power Electron. Specialists Conf.*, June 1975, pp. 252–261.

[4] F. F. Oettinger, D. L. Blackburn, and S. Rubin, "Thermal characterization of power transistors," *IEEE Trans. Electron Devices*, vol. ED-23, pp. 831–838, Aug. 1976.

[5] M. S. Adler and H. H. Glascock, "Investigation of the surge characteristics of power rectifiers and thyristors in large-area press packages," *IEEE Trans. Electron Devices*, vol. ED-26, pp. 1085–1091, July 1979.

[6] B. B. Adams, S. W. Kessler, and R. E. Reed, "Unique liquid cooled solid state power device development," in *Proc. IEEE Indust. Appl. Soc.*, Oct. 1979, pp. 1071–1087.

[7] N. Zommer, "Designing the power handling capabilities of MOS power devices," *IEEE Trans. Electron Devices*, vol. ED-27, pp. 1290–1296, July 1980.

[8] D. L. Blackburn and D. W. Berning, "Power MOSFET temperature measurements," in *Proc. IEEE Power. Electron. Specialists Conf.*, June 1982, pp. 400–407.

Device Reliability

[1] R. F. Chick and W. H. Karstaedt, "Reliability in the application of high power semiconductors," in *Proc. IEEE Indust. Appl. Soc.*, Oct. 1978, pp. 1050–1055.

[2] W. R. Comstock and R. E. Locher, "High current diode and SCR reliability considerations," in *Proc. IEEE Power Electron. Specialists Conf.*, June 1975, pp. 224–233.

[3] J. B. Brewster and G. F. Sherbondy, "Complete characterization studies provide verification of RBDT (RSR) reliability," *IEEE Trans. Electron Devices*, vol. ED-26, pp. 1462–1468, Oct. 1979.

[4] N. D. Zommer, D. L. Feucht, and R. W. Heckel, "Reliability and thermal impedance studies in soft soldered power transistors," *IEEE Trans. Electron Devices*, vol. ED-23, pp. 843–850, Aug. 1976.

Device Gating Techniques

[1] J. M. Joyce and J. E. Kress, "Power transistor switching with a controlled regenerative current mode transformer," in *Proc. IEEE Power Electron. Specialists Conf.*, June 1977, p. 148.

[2] B. Jackson and D. Y. Chen, "Effects of emitter open switching on the turn-off characteristics of high voltage power transistors," in *Proc. IEEE Power Electron. Specialists Conf.*, June 1980, pp. 147–154.

[3] D. Y. Chen and J. P. Walden, "Application of transistor emitter open turn-off scheme to high voltage power inverters," in *Proc. IEEE Power Electron. Specialists Conf.*, June 1981, pp. 252–257.

[4] T. H. Sloane, H. A. Owen, and T. G. Wilson, "Switching transients in high frequency high power convertors using power MOSFETs," in *Proc. IEEE Power Electron. Specialists Conf.*, June 1979, p. 244.

Device CAD Circuit Modeling

[1] B. W. Williams, "Determination of power semiconductor parameter values from structure data," *Solid State Electron.*, vol. 25, pp. 395–410, May 1982.

[2] P. M. Wilson, R. T. George, H. A. Owen, and T. G. Wilson, "A DC model for power transistors suitable for CAD and analysis," in *Proc. IEEE Power Electron. Specialists Conf.*, June 1979, pp. 428–436.

[3] H. A. Nienhaus, J. C. Bowers, and P. C. Herren, "A high power MOSFET computer model," in *Proc. IEEE Power Electron. Specialists Conf.*, June 1980, pp. 97–103.

Author Index

A

Amano, H, 184

B

Baliga, B. J., 257, 292, 317, 324, 331, 354, 375
Becke, H. W., 135
Berning, D. W., 55
Blackburn, D. L., 42, 55

C

Chandrasekaran, S., 343
Chang, M., 354
Chen, D. Y., 60, 343
Chi, M.-H., 236
Chin, S. A., 343
Clemente, S., 243, 273

D

Der, C. F., 285

E

Einthoven, W. G., 33

F

Farrow, V., 364
Ferraro, A., 65, 103
Frank, W. E., 285
Fukui, H., 140, 184

H

Hancock, D. J., 77
Higaki, Y., 312
Hower, P. L., 19, 42
Hu, C., 225, 236

I

Ichikawa, K., 150, 161
Igarashi, I., 339
Isidori, A., 243

J

Jackson, B., 60

K

Kato, M., 312
Kodera, H., 303
Kotani, M., 312

M

Matsuzaki, K., 161
Mattern, K. E., 190
McGrath, E. J., 28
Misra, R. P., 135
Miya, H., 184

N

Nagano, T., 124, 140
Nakamura, Y., 339
Navon, D. H., 28
Nishizawa, J., 339

O

Oettinger, F. F., 42
Ogawa, T., 124
Ohmi, T., 339
Okamura, M., 124, 140

P

Paice, D. A., 190
Peak, S. C., 96
Pelly, B. R., 199, 243, 273
Plummer, J. D., 213
Plunkett, A. B., 96

R

Robson, R. R., 77
Rubin, S., 42

S

Sakurada, S., 140
Seki, N., 150, 161
Severns, R., 266
Shafer, P., 354
Smith, M. W., 354
Steigerwald, R. L., 89, 103, 117, 174
Sugiyama, S., 339
Sun, S. C., 213

T

Tadano, H., 339
Taylor, B., 364
Tsuruta, Y., 150, 161
Turnbull, F. G., 103

W

Walden, J. P., 257
Wheatley, C. F., Jr., 33

Y

Yamaguchi, K., 303
Yatsuo, T., 140
Yukimoto, Y., 312

Z

Zommer, N., 350

Subject Index

A

Anode shorting, 123
  in GTO, 135, 140
Antiparallel diodes, 1
Armature choppers, 89
Auger recombination
  in power transistors, 28

B

Base-drive circuits, 77
Battery chargers, 19
  pulse, 117
Bibliographies
  general, 383
BI-MOS switch circuits, 364
BIPMOS
  power switches, 350
Bipolar junction transistors, 291
Bipolar transistors, 343
  parasitics, 65
  switches, 266
BJF-FET combination, 291
BJT
  see Bipolar junction transistors
Blocking voltage capability
  of transistors, 375
Bridge inverters, 1, 19

C

CAD
  see Computer-aided design
Case Temperature
  effect on reverse-bias second breakdown, 55
Cathode wiring
  of GTO, 184
Chip area allocation, 33
Choppers
  dc, 19, 89, 273
  for motor speed control, 273
  two-quadrant, 89
Circuit breakers, 124
Circuit modeling
  bibliography, 383
Circuit topology, 1
Common source inductance
  of MOSFET, 243
Computer-aided design
  bibliography, 383
Computer simulation
  of triodelike JFET, 303
Converters
  ac/dc, 190
  dc-to-dc, 1
  power, 124, 364
  resonant, 77
  transistor, 65
Current crowding
  in transistors, 21
Current gain
  limiting factors, 28
  measurement, 28

D

Darlington transistors, 19, 190
  design, 33
  models, 33
DC motors
  speed control, 273
Diode reverse recovery, 257
Diodes
  antiparallel, 1
Discrete transistors, 33
DMOS, 197, 236
dv/dt effects, 1
  in bipolar junction transistors, 266
  in MOSFET, 266

E

Electric railroads
  use of thyristor inverters, 161
Electric vehicles
  battery charging, 117
  two-quadrant transistor chopper, 89
Electron irradiation
  of field-controlled thyristors, 324
  of silicon power MOSFET, 257
Emitter balancing, 42, 123
Emitter-open turnoff, 60

F

FCT
  see Field-controlled thyristors
FET
  see Field effect transistors
FET-BJT combination transistors
  fast switching, 343
FET-gated bipolar power transistors
  switching, 343
Field-controlled thyristors, 1, 291
  asymmetrical, 317
  design, 375
  electron irradiation, 324
  fabrication, 317
  power switching, 331
Field effect transistors, 197
  junction gate, 291, 292, 303
  terminations, 376

G

Gate coupling
  of GTO, 184
Gate decoupling
  of GTO, 184
Gate turnoff thyristors, 1, 123, 375
  application to ac motor drives, 190
  bibliography, 383
  current status, 124
  high-current, 140
  high-power, 124, 140, 161, 174
  high-speed, 135
  high-voltage, 135, 140
  paralleling, 184
  ultra high voltage, 140
Gating circuits
  for high-power thyristors, 150, 161

Gating techniques
  bibliography, 383
GTS's
  see Gate turnoff thyristors

H

Heating
  induction, 285
HEXFET, 199
  power, 273
High-current transistors
  for dc–dc chopper, 89
  power MOSFET, 225
High-gain power switching
  using field-controlled rectifiers, 331
High-power GTO
  applications, 161, 174
  design, 124, 140
  gating circuits, 150
High-power transistors
  static induction, 312
High-voltage transistors
  in inverter circuits, 21
  power JFET, 291, 292
  power MOSFET, 225
  terminations, 376
  turnoff, 60
High-volume production
  of field-controlled thyristors, 317
Hot spots
  formation, 42

I

IGT
  see Insulated gate transistors
Induction heating, 197
  use of RF generators, 285
Induction motors
  drives, 96, 161
Insulated gate transistors, 1, 291
  switching, 354
Intermediate power rating, 103
Inverters, 1, 19, 123, 161
  half bridge, 1
  high-voltage transistors, 21
  motor drives, 96, 190
Irradiation
  electron, 257, 324

J

Junction gate FET, 1, 291
  optimum design, 303
  triodelike, 303
  with recessed gate, 292
Junction transistors
  bipolar, 266
  junction gate, 291

L

Latching transistors, 1, 123
  bibliography, 383
LDMOS, 197
  on-resistance modeling, 213
Low-loss snubbers
  for transistor converters, 65

M

Markets
  for power MOSFET, 199
Microprocessor interface
  armature chopper, 89
Monolithic circuits
  high-voltage BIPMOS, 350
MOSFET, 197
  dv/dt effects, 266
  for induction heating, 285
  models, 243
  parasitics, 266
  power, 199, 213, 225, 236, 243, 257, 273, 285
  reverse recovery, 257
  switching, 243, 350
Motor control, 1, 96
  choppers, 273
Motor drives, 123
  ac, 1, 19, 190
  dc, 1, 19, 197
  PWM inverter-induction, 96
Multistage Darlington power transistors
  chip area proportioning, 33

N

Negative-bias circuits, 150

O

Off-gating circuits, 150
On-gating circuits, 150
On-state resistance
  MOSFET, 225

P

Paralleling
  of GTO's, 123, 184
Parasitic p-n diodes, 197
Photovoltaic arrays
  power conditioners, 103
Plasma junctions
  terminations, 376
Power bipolar transistors, 21
  avalanche injection, 19
  bibliography, 383
  blocking voltage, 19, 375
  current gain, 19
  Darlington, 1, 33
  diffusion profiles, 19
  emitter injection efficiency, 19
  FET-gates, 343
  maximum forward current, 19
  multistage, 33
  N-base region, 375
  safe operating area, 19, 21
  second breakdown, 19, 42
  1982 market, 1
Power conditioners
  photovoltaic arrays, 103
Power field controlled thyristors, 291, 331
  bibliography, 383
Power junction FET
  bibliography, 383
Power MOSFET, 1
  for induction heating, 285
  HEXFET, 273
  reverse recovery, 257
  switching, 354
Power supplies
  off-line switching, 364
  switching, 1
Power transistors, 1, 124, 197
  current gain, 28
  comparison with GTO, 174

Darlington, 33
fast switching, 55
modeling, 213
MOSFET, 197, 199, 225, 243, 257, 350, 354
performance trade-offs, 21
second breakdown, 42, 55, 236
stable hot spots, 42
turnoff characteristics, 60
water cooled, 117
Protection
of GTO, 190
Pulse transformers, 150
PWM inverter-induction motor drive, 96

R

Recessed-gate structures
JFET, 292
Reliability
bibliography, 383
Residential power rating, 103
Resonant circuits
inverters, 1
10-kW series converter, 77
Reverse-bias second breakdown, 55
Reverse-bias turnoff
of power transistors, 60
Reverse current conduction, 266
Reverse recovery
of silicon power MOSFET, 257
RF generators
for induction heating, 285
solid state, 285

S

Safe operating area, 19, 21
bibliography, 383
MOSFET, 199
Schottky barrier devices, 135
Second breakdown
in power transistors, 42
reverse bias, 55
vertical power MOSFET, 336
Semiconductor switches
power, 343
Schockley–Hall–Read, 28
SHR
see Shockley–Hall–Read
SHR and Auger recombination, 28
SIT
see Static induction transistors
Snubber circuits, 1, 19, 354
classification, 7
high-current, 89
in GTO, 161, 174
low-loss, 65
SOA
see Safe operating area
Space-borne circuits
10-kW series resonant converter, 77
Spurious triggering, 197
minimization, 266
Stable hot spots
in power transistors, 42
Static induction thyristors
high-speed, 339
Static induction transistors, 1
bibliography, 383
bipolar mode, 339
high breakdown voltage, 312
multipellet operation, 312
Storage time
in transistors, 21
Switching circuits
HF, 364
high-speed, 266
IC, 350
insulated gate transistors, 354
power supplies, 1, 364
semiconductor, 343

T

Terminations
selection, 376
Thermal considerations
bibliography, 383
Thermal hysteresis, 42
Thyristors
field-controlled, 317, 324, 331
gate turnoff, 1, 123, 124, 135
power field controlled, 1, 291
static induction, 339
terminations, 376
ultra high voltage, 140
see also Field-controlled thyristors; Gate turnoff thyristors; High-power GTO; Power field-controlled thyristors; Static induction thyristors.
Traction
vehicles, 96
Transistor converters
low-loss snubbers, 65
Transistors
blocking voltage, 364
converters, 65
FFT, 197
high-voltage, 21
latching, 123
MESA, 364
static induction, 312
see also Bipolar junction transistors; Bipolar transistors; Darlington transistors; Discrete transistors; FET-BJT combination transistors; FET-gated bipolar power transistors; HEXFET; High-current transistors; High-power transistors; High-voltage transistors; Insulated gate transistors; Junction gate transistors; Junction transistors; Latching transistors; MOSFET; Multistage Darlington power transistors; Power bipolar transistors; Power junction FET; Power MOSFET; Power transistors; Static induction transistors; Triodelike JFET; Vertical power MOSFET.
Transistor switches
for battery charger, 117
Triodelike JFET
2-dimensional computer simulation, 303
Two-quadrant transistor chopper
for electric vehicle drive, 89

U

Uninterruptible power supplies, 1, 123
Utilities
interfaces, 103

V

Vertical power MOSFET
second breakdown, 236
VMOS, 197
on-resistance modeling, 213

Editor's Biographies

B. Jayant Baliga (S'71–M'74–SM'79–F'83) was born in Madras, India, in 1948. He received the B.Tech. degree in electrical engineering from the Indian Institute of Technology, Madras, in 1969, and the M.S. and Ph.D. degrees from Rensselaer Polytechnic Institute, Troy, NY, in 1971 and 1974, respectively. (His thesis work involved experimental studies of diffusion and epitaxial growth technology for III–V compound semiconductors.)

In 1974 he joined the General Electric Corporate Research and Development Center, Schenectady, NY, and has been working on the physics and technology of silicon and gallium arsenide power devices. His activities have included studies of lifetime and processing upon power device characteristics, and the development of high-power field-controlled devices. He is currently continuing his pioneering studies of MOS-bipolar functional integration for the development of superior power switching devices. In December 1979, he was appointed the technical coordinator of the gallium arsenide power device development program. In May 1981, he was appointed the Acting Manager of the Device Physics Unit. This was followed by his appointment as Manager of the High Voltage Device and Integrated Circuits Unit in April 1982. Since 1974, he has served as an Adjunct Faculty Member of Rensselaer Polytechnic Institute where he has assisted in research on transparent semiconducting films and their application to solar cells. He has written and presented over 150 papers on diffusion and epitaxy for III–V compounds, chemical vapor deposition, lifetime control in power devices, field-controlled devices, and solar cells. He has contributed to several books including an extensive review chapter titled "Silicon Power Field Controlled Devices and Integrated Circuits" published in the *Applied Solid State Science Series* by Academic Press. In April 1982, he taught an IEEE short course based upon this review.

Dr. Baliga is a member of Sigma Xi, the Electrochemical Society, the AIME Electronic Materials Committee, and the IEEE International Electron Devices Subcommittee on Solid State Devices. He is listed in *Who's Who in Engineering, Who's Who in Technology, Who's Who in America, International Who's Who in Engineering, American Men and Women of Science*, and *Outstanding Young Men of America*. He was awarded the Phillips India Award in 1969, the IBM Fellowship in 1972, and the Allen B. Dumont Prize in 1974. In addition, he has been awarded several publication and patent awards at GE. He has served as an officer of the IEEE student branch in 1971, and the IEEE Schenectady Section Electron Devices Chapter in 1975 and 1976. In 1982, he received the IEEE Region I award for his work on power field controlled devices. In 1983 he was the recipient of an IR100 Award, the Dushman Award, and the Coolidge Fellowship Award. He was also elected Fellow of the IEEE in 1983 for his outstanding contributions to power semiconductor devices.

Dan Y. Chen (S'72–M'75–M'79–SM'83) received the B.S. degree in electrical engineering from National Chiao-Tung University, Taiwan, in 1969 and the Ph.D. degree, also in electrical engineering, from Duke University, Durham, NC, in 1975.

From 1975 to 1979 he was employed as a member of the research staff at the General Electric Research and Development Center, Schenectady, NY. Since 1979 he has been on the faculty of the Department of Electrical Engineering, Virginia Polytechnic Institute and State University at Blacksburg, VA, where he is presently an Associate Professor. Since the start of his graduate study in 1970, he has been working in various fields of power electronics. His activities have included work in power semiconductor circuits, circuit-device interactions, device characterization, magnetic devices for power electronic applications, and product applications such as brushless motor robotic drive, electronic ballast, appliance power supply, electric car drive, etc. He has published more than 40 papers on power electronics. Since 1979, he has been a consultant to various private industries and government agencies.

Dr. Chen has been involved in various IEEE activities including paper reviewing, short course teaching, serving as a session and program committee chairman for several national and international conferences. He is currently the chairman of the Power Semiconductor Application Committee of the IEEE Industry Applications Society. He is a co-recipient of 1974 Barry Carlton Honorable Mention Award presented by IEEE Aerospace and Electronics Society.